Modern Physical Metallurgy

Fourth Edition

R. E. Smallman BSc, PhD, DSc, FRS, FIM, CEng
Feeney Professor of Metallurgy and Head of Department of Metallurgy and Materials,
The University of Birmingham.
Formerly Professor of Physical Metallurg
The University of Birmingham

Butterworths
London Boston Singapore Sydney Toronto Wellington

First published 1962
Second edition 1963
Reprinted 1965, 1968
Third edition 1970
Reprinted 1976 (twice), 1980, 1983
Fourth edition 1985
Reprinted (with corrections) 1990

British Library Cataloguing in Publication Data

Smallman, R.E.
Modern physical metallurgy.—4th ed.
1. Physical metallurgy
I. Title
669′.9 TN690

ISBN 0–408–71050–0
ISBN 0–408–71051–9 Pbk

Library of Congress Cataloging in Publication Data

Smallman, R.E.
Modern physical metallurgy.

(Butterworths monographs in metals)
Bibliography: p.
Includes index.
1. Physical metallurgy. I. Title. II. Series:
Butterworths monographs in materials.
TN690.S56 1985 669′.9 84–23884
ISBN 0–408–71050–0
ISBN 0–408–71051–9 (pbk.)

Filmset by Mid-County Press, London SW15
Printed and bound in Great Britain by Courier International Ltd, Tiptree, Essex

Preface to the First Edition

After a period of intense activity and development in nearly all branches of the subject during the past decade, physical metallurgy appears to have again attained a state of comparative quiescence, with the difference that there is now fairly general agreement about the essential principles of the subject. This does not mean that we have a complete understanding of the science of metals, as the accounts on the electronic structure of metals and alloys, and on the work hardening of metals, given in Chapter 4 and Chapter 6 respectively, show. Nevertheless, we have a sufficient understanding to be able, in broad outline, to present a fairly consistent picture of metallic behaviour, and it would appear to be an appropriate moment to attempt to present in a single book the essential features of this picture.

In this book I have tried to give, in the simplest terms possible, the fundamental principles of physical metallurgy, and to show how the application of the principles leads to a clearer understanding of many technologically important metallurgical phenomena. Throughout the text I have adopted a uniform plan of first describing the experimental background of a particular phenomenon, and then attempting to explain the experimental facts in terms of atomic processes occurring in the metal. These processes usually involve the interaction between certain well-established features of the atomic structure of metals and alloys such as solute atoms, vacant lattice sites, dislocations and stacking-faults. In this way it has been possible to treat such problems as work hardening, annealing, creep, fatigue, fracture, yielding and strain-ageing, quenching, precipitation hardening, and radiation damage, and, it is hoped, to give the student some insight into their nature without including a mass of data in the text. The explanations are based substantially upon the accepted general climate of opinion which prevailed at the time of writing, but a personal viewpoint may well have coloured some of the topics discussed. If a theory is still speculative, this fact has been mentioned at the appropriate place in the text.

The book is intended primarily for students who wish to obtain a sound working knowledge of the subject, and I hope it will prove useful to those who are studying for Higher National Certificate and Associateship examinations in the various colleges of technology, and also those reading for a degree in metallurgy at a university.

A collection of questions has been added at the end of the book which should prove useful to the serious student. Questions requiring answers of a purely descriptive nature have not been included in the collection, since this type of question needs to be examined for quality of presentation and content, and, in consequence, is best set and examined by the student's tutor. Good examples of such questions can, of course, be obtained from past papers of the examination for which the student is studying. The questions given in the appendix are, therefore, entirely numerical and it is important that, wherever possible, students should work out the examples completely and obtain definite answers. In the past this aspect of the study of physical metallurgy has often been neglected.

Acknowledgements

I would like to express my thanks to the many friends and colleagues, too numerous to mention individually, who have helped me during the preparation of this book. Particular thanks are, however, due to Professor A.D. McQuillan of the University of Birmingham for his kindness in reading through the manuscript and giving suggestions on many points of presentation.

I am indebted to many associates in other universities and research institutions who have supplied original micrographs for reproduction. Reference to these workers and their work is given in the appropriate place in the text. Acknowledgement is also made to societies and publishers for permission to use certain diagrams whether in their original form or in a modified form.

R.E. Smallman

Preface to the Fourth Edition

It is now almost fifteen years since the last revision of the text and although the basic information contained in the book has stood the test of time remarkably well, there have been some developments in the subject which should be included in any treatise on Modern Physical Metallurgy. These developments have not been particularly dramatic compared with those that took place in the 1950s and 1960s, such as the application of dislocation ideas to alloy design or the advent of transmission electron microscopy. Instead there has been steady progress in the subject across a broad front.

The new edition therefore reflects this general development in our understanding of the subject and contains some fifty new diagrams and approximately thirty per cent new text, in some cases to give a better explanation, in others to include new features and in yet others to produce a more economical presentation. Apart from these additions and alterations the general development of the subject has necessitated a complete re-structuring of the previous material so that there are now fifteen chapters instead of the previous ten and some of the material from one former chapter has been combined with some of the material from a different chapter to produce, in the author's view, a better overall presentation.

Perhaps the most remarkable development has been in the techniques for the assessment of microstructure and its microchemical analysis. These metallographic techniques are basic to the field of physical metallurgy and without them, none of the developments in plasticity, phase transformations, fracture, radiation damage, oxidation and corrosion, and alloy design added to the book, would have been achieved. Chapter 2 is thus largely new and contains text on all the techniques arising from the interaction of electrons with matter, namely x-ray topography, Auger spectroscopy, SEM, TEM, HVEM and STEM, with associated analytic facilities of back-scattering, x-ray analysis (EDX), energy-loss analysis (EELS) and convergent-beam diffraction (CBDPs). Developments such as weak-beam microscopy and direct lattice resolution microscopy are introduced later in a new Chapter 7. Material from the previous Chapter 2 has been included in a new Chapter 3 together with an introduction to ternary equilibrium diagrams, not previously included, and some further developments of solidification. Chapter 4 includes some modifications to the treatment of phase changes and diffusion and in Chapter 6 dislocations in ordered structures has been added. Chapter 8 includes the yielding behaviour of ordered alloys and some new aspects of the structure of grain boundaries. The discussion on point defects in Chapter 9 now includes radiation growth and swelling, radiation-induced segregation and radiation effects in ordered alloys. Work hardening and annealing have been re-cast into a new Chapter 10 which now includes dispersion-hardening and ordered alloys as well as *in situ* observations on recovery and recrystallization.

Phase transformations are now considered in Chapters 11 and 12, with new material on mechanisms of precipitation hardening, vacancies and precipitation, spinodal decomposition, dispersion-strengthening alloys and superalloys in Chapter 11, and HSLA and dual-phase steels in Chapter 12. There are improved treatments of creep and fatigue in Chapter 13 including deformation mechanism maps, structure of persistent slip bands, and fatigue crack formation and propagation. Different aspects of brittle and ductile fracture, fracture toughness and crack propagation are now dealt with in Chapter 14 and are conveniently summarized in fracture mechanism maps. Oxidation and corrosion now treated in Chapter 15 include intergranular voiding, breakaway oxidation, and some further consideration of corrosion failures.

Contents

Chapter 1

The structure of atoms and crystals

1.1 Metallic characteristics

Before beginning the study of the physical principles of metallic behaviour, one must first have a clear idea of what qualities characterize the metallic state. Frequently, a metal is described as having a high lustre, good electrical and thermal conductivity, and as being malleable and ductile. Very many metals do have all these properties, but many non-metallic substances can also exhibit one or more of them. Furthermore, there are large variations in the magnitude of these properties among the metals themselves. One has only to look at the remarkable malleability and ductility of lead at room temperature, and compare it with the hardness and brittleness of tungsten at the same temperature to realize the great diversity of behaviour exhibited by metals.

The most characteristic property of a metal is its high electrical or thermal conductivity. For example, among the best and worst conductors of electricity are copper and lead respectively, yet the resistivity of lead is only twelve times greater than that of copper. A non-metallic material such as diamond has, however, a resistivity which is about one thousand million times greater than copper. The reason for the enormous difference in conductivity between a metal and a non-metal is that electrons, which are negatively charged fundamental particles, can move freely through a metal under the action of a potential difference, whereas in a non-metal this is not the case. Thus the fundamental characteristics of the metallic state must be sought in the electronic structure of a metal, and we must begin the study of physical metallurgy by examining the structure of the atoms out of which a metal is built.

1.2 The atom

The crude picture of the atom conceived by Rutherford, was a positively charged nucleus which carried the greater part of the mass of the atom, with electrons clustering around it. He suggested that the electrons were revolving round the nucleus in circular orbits so that the centrifugal force of the revolving electrons was just equal to the electrostatic attraction between the positively charged nucleus and the negatively charged electrons. In order to avoid the difficulty that revolving electrons should, according to the laws of electrodynamics, continuously give off energy in the form of electromagnetic radiation, Bohr, in 1913, was forced to conclude that, of all the possible orbits, only certain orbits were in fact permissible. These discrete orbits were assumed to have the remarkable property that when an electron was in one of these orbits, no radiation could take place. The set of stable orbits were characterized by the criterion that the angular momenta of the electrons in the orbits were given by the expression $n\boldsymbol{h}/2\pi$, where $\boldsymbol{h}$ is Planck's constant, and n could only have integral values (i.e., $n=1, 2, 3, \ldots$, etc.). In this way Bohr was able to give a satisfactory explanation of the line spectrum of the hydrogen atom, and lay the foundation of modern atomic theory.

In the later developments of the atomic theory, by de Broglie, Schrödinger and Heisenberg, it was realized that the classical laws of particle dynamics could not be applied to fundamental particles. In classical dynamics it is a prerequisite that the position and momentum of a particle is known exactly, but in atomic dynamics, if either the position or the momentum of a fundamental particle is known exactly, then the other quantity cannot be determined. In fact there must exist an uncertainty in our knowledge of the position and momentum of a small particle, and the product of the degree of uncertainty in each quantity is related to the value of Planck's constant [$\boldsymbol{h}=6.6256\times10^{-34}$ J s]. In the macroscopic world this fundamental uncertainty is too small to be measurable, but when treating the motion of electrons revolving round an atomic nucleus, the application of the Uncertainty Principle (enunciated first by Heisenberg) is essential.

The consequences of the Uncertainty Principle is that we can no longer think of an electron as moving in a fixed orbit around the nucleus, but must consider the motion of the electron in terms of a wave function. This function specifies only the probability of finding one electron having a particular energy, in the space surrounding the nucleus. The situation is further complicated by the fact that the electron behaves, not only as if it were revolving round the nucleus, but also as if it were spinning about its own axis. Consequently, instead of specifying the motion of an electron in an atom by a single integer n, as in the case of the Bohr theory, it is now necessary to specify the electron state using four numbers. These numbers, known as quantum numbers, are n, l, m and s, where n is the principal quantum number, l the orbital quantum number, m the inner quantum number and s the spin quantum number. Another basic principle of the modern quantum theory of the atom is the Pauli Exclusion Principle which states that no two electrons in the same atom can have the same numerical values for their set of four quantum numbers.

If we are to understand the way in which the Periodic Table is built up in terms of the electronic structure of the atoms of the various elements, we must now consider the significance of the four quantum numbers and the limitation placed upon the numerical values they can assume. The most important quantum number is the principal quantum number since it is mainly responsible for determining the energy of the electron. The principal quantum number can have integral values beginning with $n=1$, which is the state of lowest energy, and electrons having this value are the most stable, the stability decreasing as n

increases. Electrons having a principal quantum number n can take up integral values of the orbital quantum number between 0 and $(n-1)$. Thus if $n=1$, l can only have the value 0, while for $n=2$, $l=0$ or 1, and for $n=3$, $l=0$, 1 or 2. The orbital quantum number is associated with the angular momentum of the revolving electron, and determines what would be regarded in non-quantum mechanical terms as the shape of the orbit. For a given value of n, the electron having the lowest value of l will have the lowest energy, and the higher the value of l the greater will be the energy.

The two quantum numbers m and s are concerned respectively with the orientation of the electron's orbit round the nucleus, and with the orientation of the direction of spin of the electron. For a given value of l, an electron may have integral values of the inner quantum number m from $+l$ through 0 to $-l$. Thus for $l=2$, m can take on the values $+2$, $+1$, 0, -1 and -2. The energies of electrons having the same values of n and l but different values of m are the same, provided there is no magnetic field present. When a magnetic field is applied, the energies of electrons having different m values will be altered slightly, as is shown by the splitting of spectral lines in the Zeeman effect. The spin quantum number s may, for an electron having the same values of n, l and m, take one of two values $+\frac{1}{2}$ or $-\frac{1}{2}$. The fact that these are non-integral values need not concern us for the present purpose; we need only remember that two electrons in an atom can have the same values for the three quantum numbers n, l and m, and that the two electrons will have their spins oriented in opposite directions. Only in a magnetic field will the energies of the two electrons of opposite spin be different.

1.3 The nomenclature of the electronic states in an atom

Before discussing the way in which the periodic classification of the elements can be built up by considering the electronic structure of the atoms, it is necessary to outline the system of nomenclature which will enable us to describe the states of the electrons in an atom. Since the energy of an electron is determined by the values of the principal and orbital quantum numbers alone, it is only necessary to consider these in our nomenclature. The principal quantum number is simply expressed by giving that number, but the orbital quantum number is denoted by a letter. These letters, which derive from the early days of spectroscopy, are *s*, *p*, *d* and *f* which signify that the orbital quantum numbers l are 0, 1, 2 and 3 respectively*.

When the principal quantum number $n=1$, l must be equal to zero, and an electron in this state would be designated by the symbol 1*s*. Such a state can only have a single value of the inner quantum number $m=0$, but can have values of $+\frac{1}{2}$ or $-\frac{1}{2}$ for the spin quantum number s. It follows, therefore, there are only two electrons in any one atom which can be in a 1*s* state, and these will have the electron spins in opposite directions. Thus when $n=1$, only *s* states can exist and

* The letters, *s*, *p*, *d* and *f* arose from a classification of spectral lines into four groups, termed sharp, principal, diffuse and fundamental in the days before the present quantum theory was developed.

TABLE 1.1 Allocation of states in the first three quantum shells

Shell	*n*	*l*	*m*	*s*	*Number of states*
1st	1	0	0	$+\frac{1}{2}$ $-\frac{1}{2}$	2, 1*s* states
2nd		0	0	$\pm\frac{1}{2}$	2, 2*s* states
	2	1	+1 0 −1	$\pm\frac{1}{2}$ $\pm\frac{1}{2}$ $\pm\frac{1}{2}$	6, 2*p* states
3rd		0	0	$\pm\frac{1}{2}$	2, 3*s* states
		1	+1 0 −1	$\pm\frac{1}{2}$ $\pm\frac{1}{2}$ $\pm\frac{1}{2}$	6, 3*p* states
	3				
		2	+2 +1 0 −1 −2	$\pm\frac{1}{2}$ $\pm\frac{1}{2}$ $\pm\frac{1}{2}$ $\pm\frac{1}{2}$ $\pm\frac{1}{2}$	10, 3*d* states

these can be occupied by only two electrons. Once the two 1*s* states have been filled, the next lowest energy state must have $n=2$. Here l may take up the value 0 or 1, and therefore electrons can be in either a 2*s* or 2*p* state. The energy of an electron in the 2*s* state is lower than in a 2*p* state, and hence the 2*s* states will be filled first. Once more there are only two electrons in the 2*s* state, and indeed this is always true of *s* states irrespective of the value of the principal quantum number. The electrons in the *p* state can have values of $m=+1, 0, -1$, and electrons having each of these values for m can have two values of the spin quantum number, leading, therefore, to the possibility of six electrons being in any one *p* state. This is shown more clearly in *Table 1.1.*

No further electrons can be added to the state for $n=2$ after the two 2*s* and six 2*p* states are filled, and the next electron must go into the state for which $n=3$ which is at a higher energy. Here the possibility arises for l to have the values 0, 1 and 2 and hence, besides *s* and *p* states, *d* states for which $l=2$ can now occur. When $l=2$, m may have the values $+2, +1, 0, -1, -2$ and each may be occupied by two electrons of opposite spin, leading to a total of ten *d* states.

Finally when $n=4$, l will have the possible values from 0 to 4, and when $l=4$ the reader may verify that there are fourteen 4*f* states.

1.4 The Periodic Table (see *Table 1.2*)

The simplest atom is the hydrogen atom which has a single proton as its nucleus, and may, therefore, have only one electron revolving round it in order that the atom shall be electrically neutral. In the free atom, in its lowest energy condition, the electron will be in a 1*s* state. For helium, which has a nucleus made up of two protons and two neutrons, the atomic mass will be four times greater than for hydrogen, but because the nuclear charge is governed solely by the

TABLE 1.2 Periodic Table of the elements

Group I *A*	*B*	*Group II*		*Group III*		*Group IV*		*Group V*		*Group VI*		*Group VII*		*Group VIII*	
H 1														He 2	First Period
Li 3			Be 4		B 5		C 6		N 7		O 8		F 9	Ne 10	Second Period
Na 11			Mg 12		Al 13		Si 14		P 15		S 16		Cl 17	A 18	
K 19		Ca 20		Sc 21		Ti 22		V 23		Cr 24		Mn 25		Fe, Co, Ni 26 27 28	First Long Period
	Cu 29		Zn 30		Ga 31		Ge 32		As 33		Se 34		Br 35	Kr 36	
Rb 37		Sr 38		Yt 39		Zr 40		Nb 41		Mo 42		Tc 43		Ru, Rh, Pd 44 45 46	Second Long Period
	Ag 47		Cd 48		In 49		Sn 50		Sb 51		Te 52		I 53	Xe 54	
Cs 55		Ba 56		La* 57		Hf 72		Ta 73		W 74		Re 75		Os, Ir, Pt 76 77 78	
	Au 79		Hg 80		Tl 81		Pb 82		Bi 83		Po 84		At 85	Rn 86	
Fr 87		Ra 88		Ac 89		Th 90		Pa 91		U 92		†			

* The elements between La (57) and Hf (72) are known as the rare-earths or lanthanides.

† The transuranic elements follow here: e.g. Np (93), Pu (94), Am (95), Cm (96), Bk (97), Cf (98), ...

number of protons, it will have only two electrons in orbit around the nucleus. Both these electrons will have the lowest energy if they are in 1*s* states. The 1*s* states are now filled, and the next atom (lithium) which has a nuclear charge of three, can only accommodate two electrons in the 1*s* states, and the third must enter a 2*s* state having somewhat higher energy. Once the set of states corresponding to a given principal quantum is filled, the electrons in these states are said to form a closed shell, and it is a consequence of quantum mechanics that once a shell is filled, the energy of that shell falls to a very low value so that these electrons are in very stable states. Thus, lithium has two electrons very strongly bonded to the nucleus and one in a 2*s* state which is much less strongly bonded. This electron, frequently referred to as a valency electron, can be removed fairly readily, and hence lithium can form an ion having a single positive charge, and a valency of one. The outer 2*s* electron is, for that reason, relatively free.

Beryllium has a nuclear charge of four, and the electrons will therefore occupy the 1*s* and 2*s* states, leaving the six 2*p* states of higher energy empty. In the next six atoms, having nuclear charges from five to ten, these 2*p* states will be progressively filled, and then at the element having a nuclear charge of ten (neon) all available states having the principal quantum numbers 1 and 2 are filled and now the atom has two closed shells. As in the case of helium the electrons are in low energy states from which they cannot easily be removed. Consequently neon, like helium, does not readily form an ion and hence cannot take part in chemical reactions. Throughout the system of elements, each time an atom has sufficient electrons to fill the closed shells, a non-reactive element is formed, and these are known collectively as the inert gases.

In an analogous manner, the atoms having nuclear charges, or, as they are more often called, atomic numbers, lying between eleven and eighteen, will build up the third shell based on $n=3$ by filling first the 3*s* states and then the 3*p* states. It might be thought that after argon, for which $Z=18$, the next atom of greater atomic number would have an electron in a 3*d* state. This, however, is not the case because it so happens that the energy of an electron in a 4*s* state is lower than in a 3*d* state. In consequence, in potassium for which $Z=19$ the electron of highest energy is in an *s* state, and this element in its chemical behaviour strongly resembles sodium and lithium which also have single electrons in *s* states. Calcium, with $Z=20$, has two electrons in the 4*s* states which are, therefore, now filled and the next, scandium, which might have been expected to have its highest energy electron in a 4*p* state, instead finds that the electron energy is lower if it is placed in the 3*d* state which has been hitherto left empty because its energy was higher than that for a 4*s* state. After scandium, the succeeding elements continue the process of filling the 3*d* states, which is completed at zinc. A complication, however, arises in this process of filling the 3*d* states. In the free atoms it is found that, as the 3*d* states are filled, the electrons first occupy the five states corresponding to the five values of the inner quantum number *m* with electrons all having the same spin quantum number (Hund's rule). When all five states are occupied, their energy drops to a low value, and hence energetically it is advantageous for chromium to use what might have been expected to have been a 4*s* electron to complete the set of five 3*d* states. Hence chromium has only one 4*s* electron and five 3*d* electrons. A similar process takes place in copper, where one of the 4*s* electrons is used to complete the ten 3*d* states, thus filling the third shell and obtaining a considerable reduction in the energy of the electrons in this shell. The elements from scandium to copper, in which the 3*d* states are progressively filled, are known as the transition elements. After copper the next seven elements resume the interrupted process of filling the 4*s* and 4*p* states, and krypton, in which these states are filled, is another inert gas.

The next group of elements, from rubidium to xenon, repeat the previous process by filling the 5*s*, then the 4*d* and finally the 5*p* states. 4*f* states are, however, not filled at this stage because they have a higher energy than the 5*s*, 4*d*, 5*p* and 6*s* states. It is only after lanthanum that it becomes energetically favourable to fill the fourteen 4*f* states and this gives rise to a series of elements known as the rare earths, which lie between lanthanum and hafnium in the Periodic Table of the elements.

After the 4*f* states have been filled, the following elements up to the next inert gas, radon, resume the filling of the 5*d*, and ultimately the 6*p* states. The remaining elements then begin once more with *s* states, namely the 7*s* electrons and continue in an analogous manner to the previous group of metals. However, at this stage there are only six naturally occurring elements remaining, and it has not yet been definitely established if some of these should not be considered to form a group similar to the rare earths.

In this way we can build up a scheme for the presentation of the elements which is illustrated in *Table 1.2*. Here the numbers under the chemical symbols of the elements are their atomic numbers. Horizontal rows in the table are termed periods, and vertical columns are referred to as groups. It will be evident that each period is terminated by an element of the inert gases, in which all electronic states corresponding to a particular value of the principal quantum number have been filled, and that the elements in any one group have the electrons in their outermost shell in the same configuration.

1.5 Chemical behaviour and the metallic bond

The chemical behaviour of the elements can be explained in terms of the great stability which arises when the electron shells are filled. In the inert gases where this condition occurs in the atoms, it is very difficult to remove an electron from the outermost filled shell, and hence produce a positively charged ion, and equally difficult to add an extra electron outside the filled shells, and hence give rise to a negatively charged ion. Thus, the inert gases, since they cannot readily be converted into ions, are unable to enter into chemical combinations.

Those elements having few electrons outside filled shells can readily shed their outer electrons to form positively charged ions (cations). Those which have a large number of outer electrons find it easier to take up a sufficient number of electrons to fill the next shell, and thus form negatively charged ions (anions). For example, lithium, sodium and potassium all have a single outer electron (valency electron) in an *s*-state, which when shed gives rise to singly charged positive ions. These elements are, therefore, univalent. Chlorine, bromine and iodine, on the contrary, all have one electron short of the number required to fill an outer shell,

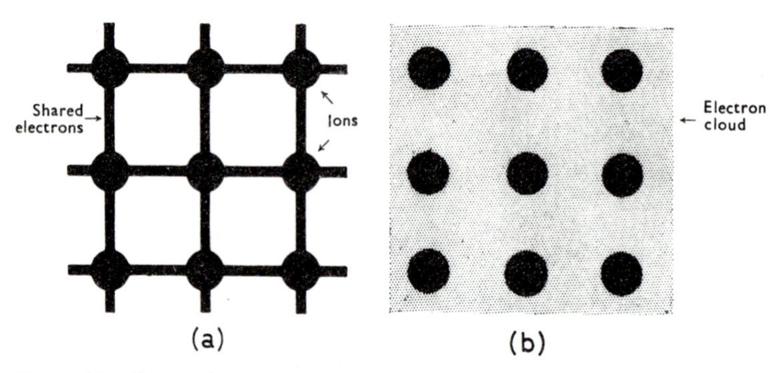

Figure 1.1 Schematic representation of (a) covalent bond and (b) metallic bond

and when they acquire the necessary electron to complete the shell they form highly stable, singly charged negative ions. The chemical affinity between sodium and chlorine can be easily explained in terms of the transfer of the outer electron of sodium to the chlorine atom, when both will now have the former electronic structure characterized by a full shell, and the two ions of opposite charge so produced will be held together by electrostatic attraction. This is the basis of ionic or heteropolar bonding.

The elements on the left of the Periodic Table form positive ions, and have valencies which increase as one proceeds along the periods from one group to the next starting from Group I. Similarly the elements on the right-hand side give rise to negative ions and, in this case, the valencies increase progressively on passing along the periods from right to left. An interesting situation now arises with the elements which have a half-filled outer shell of electrons, such as carbon, which theoretically could form quadrivalent ions having either a positive or negative charge. In fact such elements, although they sometimes enter into ionic bonding with other elements, more frequently make use of another type of bonding known as covalent or homopolar bonding. In this type of bonding neighbouring atoms share their outer valency electrons in such a way that each atom has effectively a full outer shell for part of the time.

In covalent bonding, an atom prefers to share only a single electron with each of its neighbours, as shown in *Figure 1.1(a)* and, therefore, the number of covalent bonds formed by an element is equal to $(8 - N)$ where N is the number of electrons outside the full shell. Another essential feature of covalent bonding is that the shared electrons are in *s*- and *p*-states. Thus carbon, having four outer electrons, can form covalent linkages with four other carbon atoms. In diamond, which is a form of solid carbon, each carbon atom is surrounded by four other atoms symmetrically situated at the corners of a regular tetrahedron. In this way a three-dimensional network of carbon atoms can be built up.

In solids in which the bonding is either ionic or covalent, the electrons are not free to migrate under the action of an applied electromotive force and hence are insulators. As we have seen, the most important characteristic of a metal is its ability to conduct electricity, and therefore the bonding cannot be either ionic or covalent. All elements showing definite metallic characteristics are grouped on the left-hand side of the Periodic Table shown in *Table 1.2*. The atoms of the metallic elements all have a small number of electrons outside a full shell. This is the case for all elements in subgroups I, II and III, for the elements of the three transition series, and for the rare-earth elements.

In a metal the outer electrons are essentially free to move throughout the material, and one must therefore picture a metal as an array of positively charged ions completely permeated by a gas of freely moving electrons (*Figure 1.1*(*b*)). The bond is caused mainly by the attraction of the positive ions to the free electrons. One of the most important consequences of this mode of bonding is that the bonding forces are not spatially directed and the ions will, therefore, group themselves together in whatever geometrical form gives the most economical packing. However, it must be remembered that when two ions approach each other there is a repulsive force brought into play, and this force limits the degree of economical packing, so that in fact the metal ions can be thought of as hard spheres. The problems of crystal structure of metals can, therefore, be considered in terms of the packing of equally sized spheres.

1.6 Arrangement of atoms in metals

Metal ions are very small, and have diameters which are only a few times 10^{-7} mm, i.e. less than a nanometer. A millimetre cube of metal, therefore, contains something like 10^{20} atoms. It has been shown that the ions in solid metal are not randomly arranged, but are packed together in a highly regular manner. In the majority of metals the ions are packed together so that the metal occupies the minimum volume. In all metals, including those in which the ions are not close-packed, the arrangement of ions follows a specific pattern, and the structure of the metal is characterized by a simple unit of pattern, known as a structure cell, which, when it is repeated in a completely regular manner throughout the metal, defines the position of all the ions in the metal crystal.

There are two methods of packing spheres of equal size together so that they occupy the minimum volume. These are the face-centred cubic (f.c.c.) and close-packed hexagonal (c.p.h.) arrangements. The structure cells of the two arrangements are shown in *Figures 1.2*(*a*) and *1.2*(*b*). The other structure cell shown in *Figure 1.2*(*c*) is a method of packing which, although it does not lead to close-packing, is nevertheless adopted by many metals. This method of packing spheres is known as the body-centred cubic (b.c.c.) arrangement.

Apart from the three crystal structures already mentioned, other structures such as orthorhombic (e.g. gallium, uranium), tetragonal (e.g. indium, palladium) and rhombohedral (e.g. arsenic, antimony, bismuth) occur.

To specify completely the structure of a particular metal it is necessary to give not only the type of crystal structure adopted by the metal, but also the dimensions of the structure cell. The number of quantities necessary to define a structure cell clearly depends on the degree of geometric regularity exhibited by the cell. Thus, in the cubic structure cells it is only necessary to give the length of an edge, whereas in a hexagonal cell it is necessary to give the two lengths a and c indicated in *Figure 1.2*(*b*). However, if the structure is ideally close-packed the

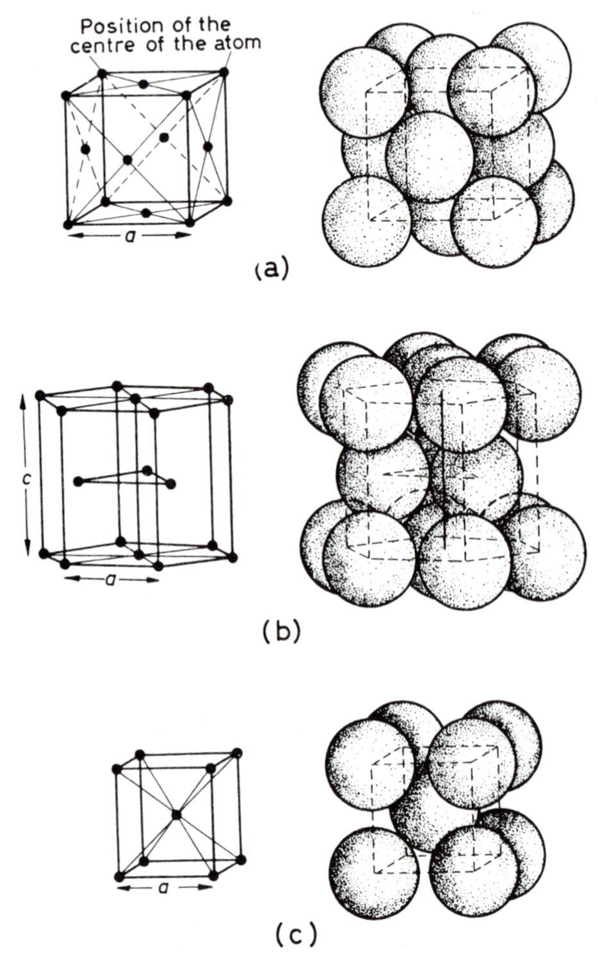

Figure 1.2 Arrangement of atoms in (a) face-centred cubic structure, (b) close-packed hexagonal structure, and (c) body-centred cubic structure

two quantities a and c must be in the ratio $c/a = 1.633$. In metal structures, the ratio c/a, known as the axial ratio, is never exactly 1.633, and the structures are, therefore, never quite ideally close-packed, e.g. for zinc $c/a = 1.86$ and for titanium $c/a = 1.58$. The quantities which give the size of a structure cell are termed the lattice parameters.

A knowledge of the lattice constants permits the atomic radius r of the metal atoms to be calculated on the assumption that they are spherical and that they are in closest possible contact. The reader should verify that in the f.c.c. structure $r = (a\sqrt{2})/4$, and in the b.c.c. structure $r = (a\sqrt{3})/4$, where a is in both cases the lattice parameter. As the quantities a and r are very small, it is customary to measure them in terms of nanometres (10^{-9} m).

An important concept in dealing with crystal structures is the co-ordination number, which is defined as the number of equidistant nearest neighbouring atoms around any atom in the crystal structure. Thus, in the body-centred cubic

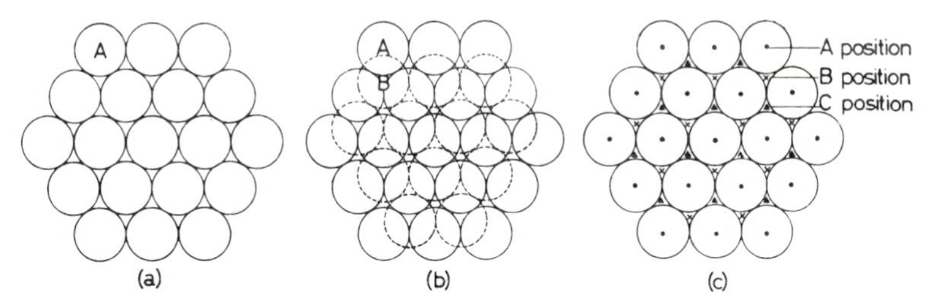

Figure 1.3 (a) Arrangements of atoms in a close-packed plane, (b) positioning of two close-packed planes, and (c) the stacking of successive planes

structure shown in *Figure 1.2(c)*, it is easily seen that the atom at the centre of the cube is surrounded by eight equidistant atoms at its corners, and the co-ordination number is 8. It is perhaps not so readily seen from *Figure 1.2(a)* that the co-ordination number in the face-centred structure is 12. Perhaps the easiest method of visualizing this is to take two f.c.c. cells placed side by side and to consider the atoms surrounding the common face-centring atom. In the close-packed hexagonal structure with the ideal packing ratio for $c/a = 1.633$, the co-ordination number is again 12, as can readily be seen by considering once more two cells stacked one on top of the other, and choosing the centre atom of the common plane. This plane is frequently referred to as the basal plane.

The most densely packed plane in the ideally close-packed hexagonal structure is the basal plane, and has the same atomic arrangement as the closest packed plane in the face-centred cubic structure*. Both the c.p.h. and f.c.c. cubic structure represent two equally good methods of packing spheres; the difference between them arises from the different way in which the close-packed planes are stacked. *Figure 1.3* shows the arrangement of atoms in such a close-packed plane. In laying down a second plane of close-packed atoms, the first atom may be placed in either position B or C, which are entirely equivalent sites. However, once the first atom is placed in one of the two types of site, all other atoms in the second plane must be in similar sites. This is simply because neighbouring sites of the types B and C are too close together for both to be occupied in the same layer. At this stage there is no difference between the c.p.h. and f.c.c. structure, and the difference arises only when the third layer is put in position. In building up the third layer, assuming that sites of type B have been used to construct the second layer, as shown in *Figure 1.3*, sites A or C may be selected. If sites A are chosen, then the atoms in the third layer will be directly above those in the first layer, and the structure will be c.p.h., whereas, if sites C are chosen this will not be the case and the structure will be f.c.c. Thus a c.p.h. structure consists of layers of closely packed atoms stacked in the sequence of ABABAB or, of course equally well, ACACAC. An f.c.c. structure has the stacking sequence ABCABCABC so that the atoms in the fourth layer lie directly above those in the bottom layer.

Table 1.3 gives the crystal structure adopted by the more common metals at room temperature. Some metals, however, adopt more than one crystal structure,

* The closest packed plane of the f.c.c. structure is usually termed a (111) plane as indicated in *Figure 1.9*. The notation used to describe a plane in this way is discussed in the appendix to Chapter 1.

TABLE 1.3 The crystal structures of some common metals at room temperature

Element	*Crystal structure*	*Closest interatomic distance (ångströms or nm × 10)*
Aluminium	f.c.c.	2.862
Beryllium	c.p.h. ($c/a = 1.568$)	2.225
Cadmium	c.p.h. ($c/a = 1.886$)	2.979
Chromium	b.c.c.	2.498
Cobalt	c.p.h. ($c/a = 1.623$)	2.506
Copper	f.c.c.	2.556
Gold	f.c.c.	2.884
Iron	b.c.c.	2.481
Lead	f.c.c.	3.499
Lithium	b.c.c.	3.039
Magnesium	c.p.h. ($c/a = 1.623$)	3.196
Molybdenum	b.c.c.	2.725
Nickel	f.c.c.	2.491
Niobium	b.c.c.	2.859
Platinum	f.c.c.	2.775
Potassium	b.c.c.	4.627
Rhodium	f.c.c.	2.689
Rubidium	b.c.c.	4.88
Silver	f.c.c.	2.888
Sodium	b.c.c.	3.715
Tantalum	b.c.c.	2.860
Thorium	f.c.c.	3.60
Titanium	c.p.h. ($c/a = 1.587$	2.89
Tungsten	b.c.c.	2.739
Uranium	orthorhombic	2.77
Vanadium	b.c.c.	2.632
Zinc	c.p.h. ($c/a = 1.856$)	2.664
Zirconium	c.p.h. ($c/a = 1.592$)	3.17

each of the exhibited crystal forms being stable only over a certain temperature range. The best known example of this phenomenon, called polymorphism, is exhibited by iron which is b.c.c. at temperatures below 910 °C and above 1400 °C, but f.c.c. between 910 °C and 1400 °C. Other common examples include titanium and zirconium which change from c.p.h. to b.c.c. at temperatures of 882 °C and 815 °C respectively, tin which changes from cubic (grey) to tetragonal (white) at 13.2 °C, and the metals uranium and plutonium. Plutonium is particularly complex in that it has six different crystal structures between room temperature and its melting point of 640 °C.

1.7 Electrons in metal crystals

If one imagines atoms being brought together uniformly to form a metallic structure, then, when the distance between neighbouring atoms approaches the interatomic distance in metals, the outer electrons are no longer localized around individual atoms. Once the outer electrons can no longer be considered to be attached to individual atoms but have become free to move throughout the metal then, because of the Pauli Exclusion Principle, these electrons cannot retain the

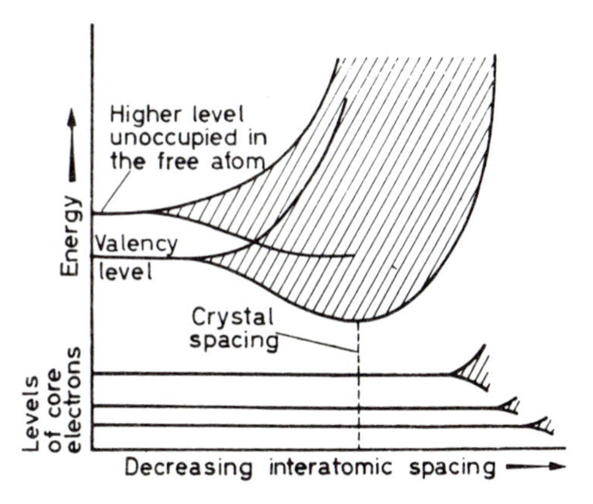

Figure 1.4 Broadening of atomic energy levels in a metal

same set of quantum numbers that they had when they were part of the atoms. As a consequence, the free electrons can no longer have more than two electrons of opposite spin with a particular energy. The energies of the free electrons are distributed over a range which increases as the atoms are brought together to form the metal. If the atoms when brought together are to form a stable metallic structure, it is necessary that the mean energy of the free electrons shall be lower than the energy of the electron level in the free atom from which they are derived. *Figure 1.4* shows the broadening of an atomic electron level as the atoms are brought together, and also the attendant lowering of energy of the electrons. It is the extent of the lowering in mean energy of the outer electrons that governs the stability of a metal. The equilibrium spacing between the atoms in a metal is that for which any further decrease in the atomic spacing would lead to an increase in the repulsive interaction of the positive ions as they are forced into closer contact with each other, which would be greater than the attendant decrease in mean electron energy.

In the metallic structure the free electrons must, therefore, be thought of as occupying a series of discrete energy levels at very close intervals. Each atomic level which splits into a band contains the same number of energy levels as the number N of atoms in the piece of metal. As previously stated only two electrons of opposite spin can occupy any one level, so that a band can contain a maximum of $2N$ electrons. Clearly, in the lowest energy state of the metal all the lower energy levels are occupied.

The energy gap between successive levels is not constant but decreases as the energy of the levels increases. This is usually expressed in terms of the density of electronic states $N(E)$ as a function of the energy E. The quantity $N(E)\,dE$ gives the number of energy levels in a small energy interval dE, and for free electrons is a parabolic function of the energy as shown in *Figure 1.5*.

Because only two electrons can occupy each level, the energy of an electron occupying a low-energy level cannot be increased unless it is given sufficient energy to allow it to jump to an empty level at the top of the band. The energy width of these bands is commonly about 5 or 6 electron volts* and, therefore,

* An electron volt is the kinetic energy an electron acquires in falling freely through a potential difference of 1 volt [$1\,eV = 1.602 \times 10^{-19}$ J; 1 eV per particle $= 23\,050 \times 4.186$ J per mol of particles].

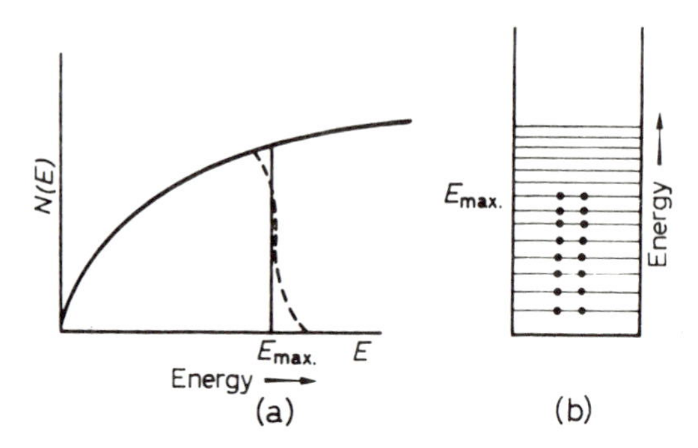

Figure 1.5 (a) Density of energy levels plotted against energy; (b) filling of energy levels by electrons at absolute zero. At ordinary temperatures some of the electrons are thermally excited to higher levels than that corresponding to $E_{max.}$ as shown by the broken curve in (a)

considerable energy would have to be put into the metal to excite a low-lying electron. Such energies do not occur at normal temperatures, and only those electrons with energies close to that of the top of the band (known as the Fermi level and surface) can be excited, and therefore only a small number of the free electrons in a metal can take part in thermal processes. The energy of the Fermi level E_F depends on the number of electrons N per unit volume V, and is given by $(\boldsymbol{h}^2/8m)(3N/\pi V)^{2/3}$.

The electron in a metallic band must be thought of as moving continuously through the structure with an energy depending on which level of the band it occupies. In quantum mechanical terms this motion of the electron can be considered in terms of a wave with a wavelength which is determined by the energy of the electron according to de Broglie's relationship

$$\lambda = \boldsymbol{h}/mv \tag{1.1}$$

where $\boldsymbol{h}$ is Planck's constant and m and v are respectively the mass and velocity of the moving electron. The greater the energy of the electron, the higher will be its momentum mv, and hence the smaller will be the wavelength of the wave function in terms of which its motion can be described. Because the movement of an electron has this wave-like aspect, moving electrons can give rise, like optical waves, to diffraction effects. Moreover, the regular array of atoms on the metallic lattice can behave as a three-dimensional diffraction grating since the atoms are positively charged and interact with moving electrons.

At certain wavelengths, governed by the spacing of the atoms on the metallic lattice, the electrons will experience strong diffraction effects, the results of which are that electrons having energies corresponding to such wavelengths will be unable to move freely through the structure. As a consequence, in the bands of electrons, certain energy levels cannot be occupied and therefore there will be energy gaps in the otherwise effectively continuous energy spectrum within a band.

The interaction of moving electrons with the metal ions distributed on a lattice depends on the wavelength of the electrons and the spacing of the ions in the direction of movement of the electrons. Since the ionic spacing will depend on the direction in the lattice, the wavelength of the electrons suffering diffraction by the ions will depend on their direction. The kinetic energy of a moving electron is a function of the wavelength according to the relationship

$$E = \boldsymbol{h}^2/2m\lambda^2 \tag{1.2}$$

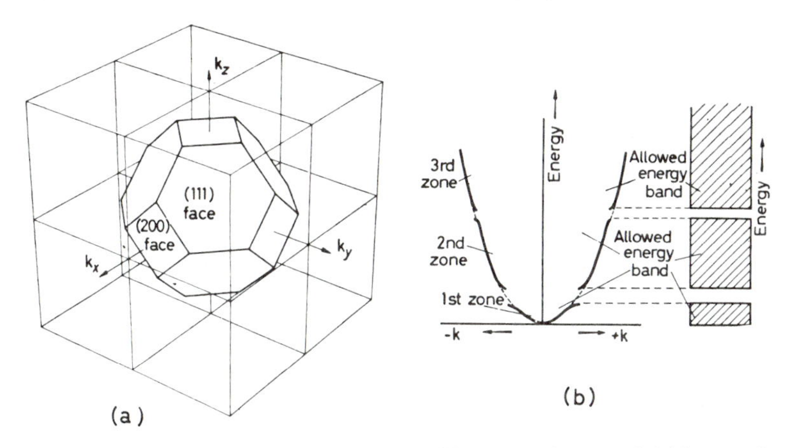

Figure 1.6 Schematic representation of a Brillouin zone in a metal. (a) is reproduced from *Extractive and Physical Metallurgy of Plutonium*, by courtesy of the American Institute of Metallurgical Engineers

and since we are concerned with electron energies, it is more convenient to discuss interaction effects in terms of the reciprocal of the wavelength. This quantity is called the wave number and is denoted by k.

In describing electron–lattice interactions it is usual to make use of a vector diagram in which the direction of the vector is the direction of motion of the moving electron and its magnitude is the wave number of the electron. The vectors representing electrons having energies which, because of diffraction effects, cannot penetrate the lattice, trace out a three-dimensional surface known as a Brillouin zone. *Figure 1.6*(*a*) shows such a zone for a face-centred cubic lattice. It is made up of plane faces which are, in fact, parallel to the most widely spaced planes in the lattice, i.e. in this case the $\{111\}$ and $\{200\}$ planes. This is a general feature of Brillouin zones in all lattices.

For a given direction in the lattice, it is possible to consider the form of the electron energies as a function of wave number. The relationship between the two quantities as given from equation 1.2 is

$$E = \boldsymbol{h}^2 k^2 / 2m \qquad (1.3)$$

which leads to the parabolic relationship shown as a broken line in *Figure 1.6*(*b*). Because of the existence of a Brillouin zone at a certain value of k, depending on the lattice direction, there exists a range of energy values which the electrons cannot assume. This produces a distortion in the form of the E–k curve in the neighbourhood of the critical value of k and leads to the existence of a series of energy gaps, which cannot be occupied by electrons. The E–k curve showing this effect is given as a continuous line in *Figure 1.6*(*b*).

The existence of this distortion in the E–k curve, due to a Brillouin zone, is reflected in the density of states versus energy curve for the free electrons. As previously stated, the density of states–energy curve is parabolic in shape, but it departs from this form at energies for which Brillouin zone interactions occur. The result of such interactions are shown in *Figure 1.7*(*a*) in which the broken line represents the $N(E)$–E curve for free electrons in the absence of zone effects and

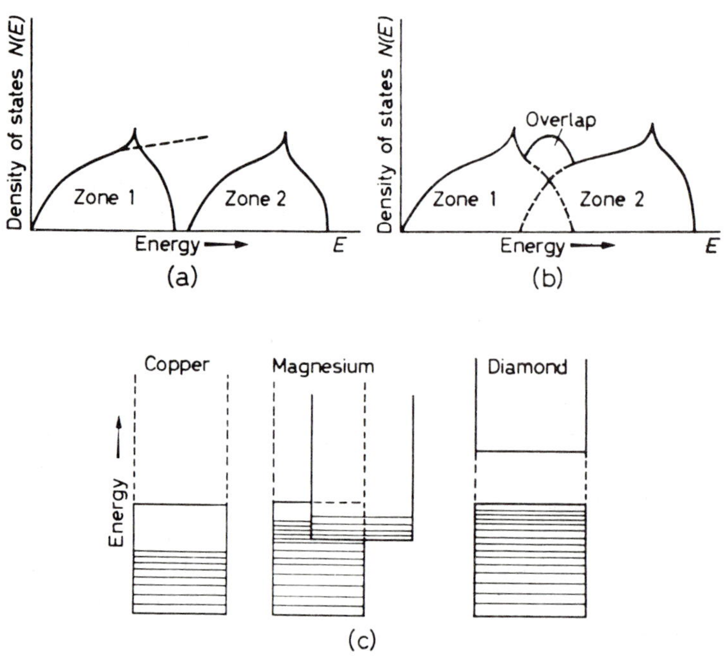

Figure 1.7 Schematic representation of Brillouin zones

the full line is the curve where a zone exists. The total number of electrons needed to fill the zone of electrons delineated by the full line in *Figure 1.7(a)* is $2N$, where N is the total number of atoms in the metal. Thus, a Brillouin zone would be filled if the metal atoms each contributed two electrons to the band. If the metal atoms contribute more than two per atom, the excess electrons must be accommodated in the second or higher zones.

In *Figure 1.7(a)* the two zones are separated by an energy gap, but in real metals this is not necessarily the case, and two zones can overlap in energy in the $N(E)$–E curves so that no such energy gaps appear. This overlap arises from the fact that the energy of the forbidden region varies with direction in the lattice and often the energy level at the top of the first zone has a higher value in one direction than the lowest energy level at the bottom of the next zone in some other direction. The energy gap in the $N(E)$–E curves, which represent the summation of electronic levels in all directions, is then closed (*Figure 1.7(b)*).

1.8 Metals and insulators

For electrical conduction to occur, it is necessary that the electrons at the top of a band should be able to increase their energy when an electric field is applied to materials so that a net flow of electrons in the direction of the applied potential, which manifests itself as an electric current, can take place. If an energy gap between two zones of the type shown in *Figure 1.7(a)* occurs, and if the lower zone is just filled with electrons, then it is impossible for any electrons to increase their energy by jumping into vacant levels under the influence of an applied electric field, unless the field strength is sufficiently great to supply the electrons at the top of the filled band with enough energy to jump the energy gap. Thus metallic conduction is due to the fact that in metals the number of electrons per atom is insufficient to fill the band up to the point where an energy gap occurs. In copper, for example, the 4*s* valency electrons fill only one half of the outer *s*-band. In other metals, e.g. Mg, the valency band overlaps a higher energy band and the electrons near the Fermi level are thus free to move into the empty states of a higher band. When the valency band is completely filled and the next higher band, separated by an energy gap, is completely empty, the material is either an insulator or semiconductor. If the gap is several electron volts wide, such as in diamond where it is 7 eV, extremely high electric fields would be necessary to raise electrons to the higher band and the material is an insulator. If the gap is small enough, such as 1–2 eV as in silicon, then thermal energy may be sufficient to excite some electrons into the higher band and also create vacancies in the valency band, the material is a semiconductor. In general, the lowest energy band which is not completely filled with electrons is called a conduction band, and the band containing the valency electrons the valency band. For a conductor the valency band is also the conduction band. The electronic state of a selection of materials of different valencies is presented in *Figure 1.7*(*c*).

As mentioned at the beginning of the chapter, although all metals are relatively good conductors of electricity, they exhibit among themselves a range of values for their resistivities. There are a number of reasons for this variability. The resistivity of a metal depends on the density of states of the most energetic electrons at the top of the band, and the shape of the $N(E)$—E curve at this point. It also depends on the degree to which the electrons are scattered by the ions of the metal which are thermally vibrating, and by impurity atoms or other defects present in the metal.

1.9 Real crystals and imperfections

In developing our concept of a metal, we have considered the lattice of a metal crystal to be built up from a perfect and regular arrangement of atoms. The zone theory, discussed in the previous section, is based on such an ideal crystal, although it is realized that real crystals are never quite so perfect as we have suggested. The basic structure of a real metal crystal is as regular as we have outlined, but lattice distortions as well as definite imperfections do exist. One source of irregularity can arise because the atoms are not at rest, as we have supposed, but vibrate about their mean lattice positions, with a frequency determined by the interatomic forces and with an amplitude governed by the temperature of the crystal. The specific heat of a metal discussed in Chapters 2 and 3 arises from this effect. A second complication is that the crystal may contain foreign atoms, added either intentionally as alloying elements or unintentionally as impurities which, because of their different atomic size, give rise to local distortions of the solvent lattice. These solute atoms may be dispersed randomly throughout the crystal, as shown in *Figures 1.8*(*a*) and (*b*), when they are said to

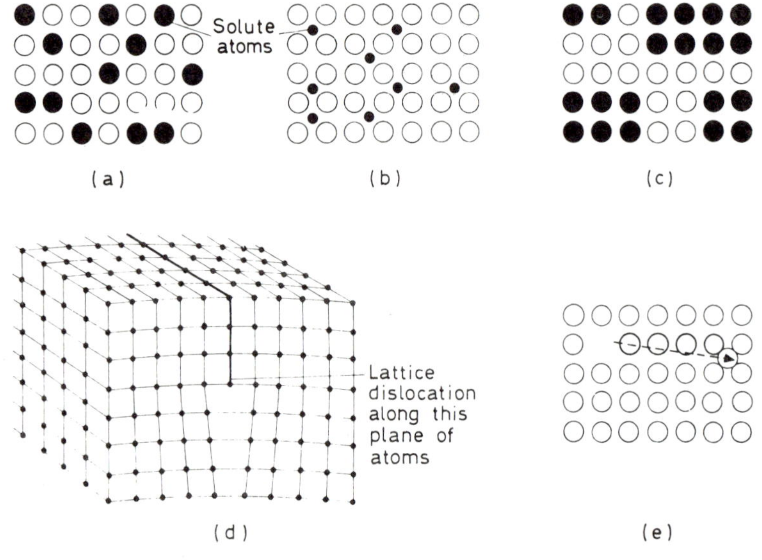

Figure 1.8 Schematic diagrams of (a) substitutional solid solution, (b) interstitial solid solution, (c) phase mixture, (d) dislocation, and (e) vacancy-interstitial pair

be in solid solution, or they may aggregate together to form particles of a second phase (*Figure 1.8*(*c*)).

Apart from foreign atoms, other irregularities, classified generally as lattice imperfections, also occur in metal crystals. These imperfections may take the form of (i) volume defects, such as cracks or voids; (ii) line defects, such as dislocations; or (iii) point defects, such as vacant lattice sites (or simply vacancies) and interstitial atoms (or simply interstitials).

Stacking faults arise because to a first approximation there is little to choose electrostatically between the stacking sequence of the close-packed planes in the f.c.c. metals ABCABC ... and that in the c.p.h. metals ABABAB.... Thus, in a metal like copper or gold, the atoms in a part of one of the close-packed layers may fall into the 'wrong' position relative to the atoms of the layers above and below, so that a mistake in the stacking sequence occurs (e.g. ABCBCABC ...). Such an arrangement will be reasonably stable, but because some work will have to be done to produce it, stacking faults are more frequently found in deformed metals than annealed metals.

Dislocations are also found in real crystals. These imperfections greatly affect the structure-sensitive properties of a crystal, e.g. yield strength, hardness etc., and it is found that the calculated yield strength and breaking strengths of ideal crystals are some 100 to 10 000 times greater than the actual strengths of real crystals. It is because a dislocation line extends over many atomic diameters in the lattice, as shown in *Figure 1.8*(*d*). that it becomes an important centre of weakness. Point defects also affect the mechanical properties, but they have a far greater influence on phenomena such as diffusion, which involve the movement of individual atoms through the crystal. The schematic diagram of the lattice, shown in *Figure 1.8*(*e*), contains both a vacancy, i.e. an unoccupied lattice site which is normally occupied by an atom in a perfect crystal, and an interstitial, i.e. an atom

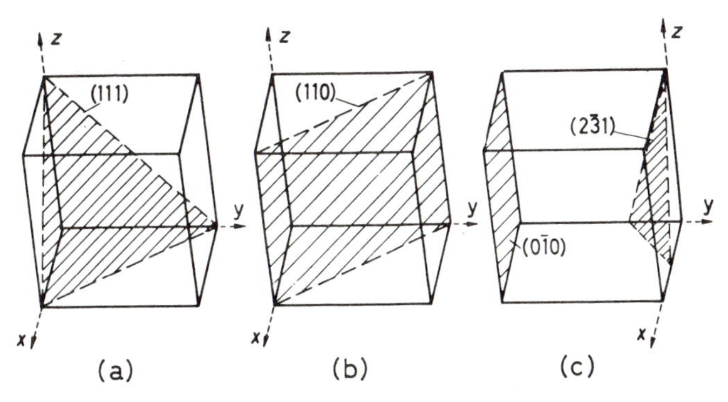

Figure 1.9 Miller indices of planes in cubic crystals. (a) (111), (b) (110), and (c) (2$\bar{3}$1)

that occupies the interstice between atoms in normal lattice sites, and it will be evident that those atoms which surround a vacancy will be able to move more easily than those which do not.

To complete our picture of a metal, it is necessary to point out that a piece of metal in commercial use is not made up from one huge single crystal, but instead contains a large number of small interlocking crystals, i.e. it is polycrystalline. Each crystal or grain, within the mass as a whole, is separated from its neighbours by a three-dimensional surface known as a grain boundary, the shape of which bears no relation to the internal atomic arrangement of the crystal. The orientation of the crystal axes within a grain usually differs from grain to grain, often as much as 30° to 40°, so that the grain boundary may be regarded as a narrow region, about two atoms thick, across which the atoms change from the lattice orientation of the one grain to that of the other.

It is because these lattice defects, i.e. solute atoms, vacancies and dislocations, can be distributed throughout the metal in a variety of ways, that the physics of metals is such a fascinating field of study. A more detailed account of such distributions is given in later chapters.

1.10 The elements of crystallography

Very often it is necessary to refer to specific planes and directions in a crystal lattice. For example, to state that precipitation takes place on those planes parallel to the cube faces, or that a particular metal is softest in a direction parallel to the cube diagonal. To make such a discussion less laborious, a notation is adopted, known as the Miller index system, in which three axes, X, Y and Z are chosen, parallel to the three edges of the crystal cell. To specify a crystal plane it is necessary to determine its intercept on the three axes X, Y and Z, then take the reciprocals of these intercepts. The reciprocal of the intercept will be in the form h/n, k/n, l/n, so that when the integers hkl are enclosed in brackets, they give the Miller indices of the plane (h, k, l).

Figure 1.9 shows some important planes in the cubic system to illustrate the method. The plane shown in *Figure 1.9(a)* makes intercepts of one cell length distance each on the X, Y and Z axes respectively, i.e. 1, 1, 1. Taking reciprocals, these intercept values remain 1, 1, 1, and, since each is already in terms of the lowest whole number possible, we identify the plane as (111). *Figure 1.9(b)* shows a plane which makes intercepts on the X, Y and Z axes at 1, 1 and ∞ respectively. The reciprocal intercepts are 1, 1 and 0, and the plane is called the (110) plane. The final example, *Figure 1.9(c)*, has a plane which makes intercepts of 1/2, $-1/3$ and 1 respectively. The reciprocals of these values are 2, $\bar{3}$, 1, which when enclosed in the appropriate brackets indicates the $(2\bar{3}1)$ plane.

The indices (123), for example, or more generally (hkl), represent not only that plane for which the values of h, k and l have been determined, but the whole family of planes parallel to this particular one. Often it is necessary to specify all planes of a given crystallographic type, e.g. all the cube faces, not merely those parallel to (100) and this is indicated by enclosing the indices obtained in the usual way in different brackets. Thus the class of all cube faces is denoted by $\{100\}$, which includes (100), $(\bar{1}00)$, (010), $(0\bar{1}0)$, (001) and $(00\bar{1})$ respectively.

To define a direction it is necessary to construct a line through the origin parallel to the unknown direction, and then to determine the co-ordinates of a point on this line in terms of lengths of the cell edges. The co-ordinates found in this way are then reduced to whole numbers, and to distinguish these indices from those given to the planes they are enclosed in square brackets. For example, if the co-ordinates are $X=a$, $Y=-2b$, $Z=c/3$, the line is a $[3\bar{6}1]$ direction. For the cubic system, the determination of direction indices is particularly easy, since it is found that a direction defined as above has the same indices as the plane to which it is perpendicular. Thus the X-axis is perpendicular to the (100) plane and is, therefore, the [100] direction, and a direction parallel to the cube diagonal is the [111] direction. To indicate the class of all directions of the same crystallographic type, different brackets are used. Thus, $\langle 100 \rangle$ stands for the all cube edges, and includes the [100], [010], [001], $[\bar{1}00]$, $[0\bar{1}0]$ and $[00\bar{1}]$ directions.

In other crystal systems such as tetragonal or orthorhombic, the Miller index notation is also used, but in hexagonal crystals a modification is necessary. In this system a Miller–Bravais notation is adopted in which four axes are taken, three of them (X, Y and U) are 120° to one another along close-packed directions in the basal plane, while the fourth (Z) is the perpendicular axis. The intercepts of a plane on the X, Y, U and Z axes, are then determined as before, and the Miller–Bravais indices are $(hkil)$. *Figure 1.10* shows some important planes in hexagonal crystals. Let us deduce the plane shown in *Figure 1.10(c)*. It marks off intercepts on the X, Y, U and Z axes equal to 1, 1, $-\frac{1}{2}$ and 1 respectively, and the reciprocals of these intercepts are 1, 1, $\bar{2}$ and 1, which become the Miller–Bravais indices to identify this plane $(11\bar{2}1)$. From this example it will be noted that $(h+k+i)$ is zero, which is a general feature of this system.

For crystallographic directions in the hexagonal system either three or four axes may be used. The direction d_3 described by the three-axis, or Miller, system has indices U, V, W such that

$$d_3 = Ua_1 + Va_2 + Wc$$

Thus the close-packed directions in the basal plane are [100], [110] and [010]. A direction, d_4, described by the four-axis system has indices u, v, t, w such that

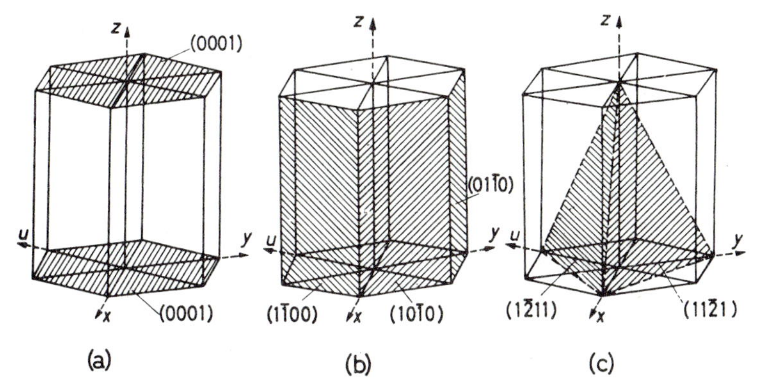

Figure 1.10 Miller–Bravais indices of planes in hexagonal crystals, (a) basal planes $\{1000\}$, (b) prism planes $\{10\bar{1}0\}$, and (c) pyramidal planes $\{11\bar{2}1\}$

$$d_4 = ua_1 + va_2 + ta_3 + wc$$

If the condition that $u+v+t=0$ is imposed, crystallographically similar directions have similar indices, e.g. the close-packed directions become $[2\bar{1}\bar{1}0]$, $[11\bar{2}0]$ and $[\bar{1}2\bar{1}0]$. Directional Miller indices cannot be converted into Miller–Bravais indices simply by inserting the fourth index, t, such that $t = -(u+v)$, but instead the equations

$$U = u - t, \quad V = v - t, \quad W = w$$

or

$$u = \tfrac{1}{3}(2U - V), \quad v = \tfrac{1}{3}(2V - U), \quad t = -(u+v)$$

$$w = W$$

must be used. The Miller–Bravais system of indexing crystallographic planes and directions has the advantage over the three-index system that similar planes and directions have similar indices.

1.11 The stereographic projection

The relationships between planes, directions and angles of a crystal can be represented conveniently on a two-dimensional diagram by the use of projective geometry. The stereographic projection is the most frequently used, particularly in the analysis of markings which appear on polished grains after deformation, i.e. slip lines, twins, cracks, etc., and in the determination of the orientation of single crystals or the preferred orientation of grains in a polycrystalline aggregate.

The crystal is assumed to be located at the centre of a sphere, as shown in *Figure 1.11*(*a*) for a cubic crystal, so that a crystal plane, such as the (111) plane marked, may be represented on the surface of the sphere by the point of intersection, or pole, of its normal P. The angle between the two poles (001) and (111), shown in *Figure 1.11*(*b*), can then be measured in degrees along the arc of the great circle between the poles P and P'. To represent all the planes in a crystal

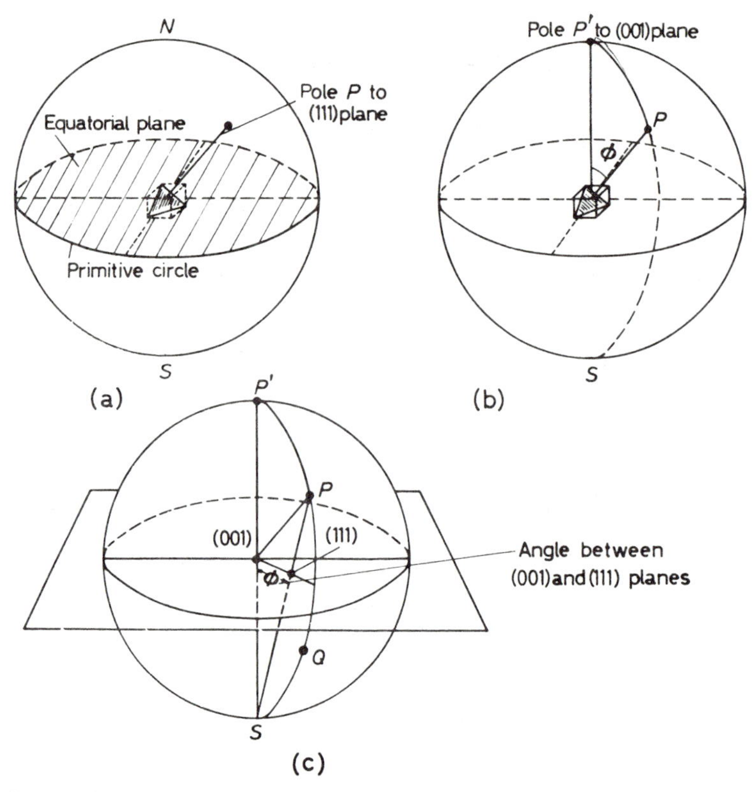

Figure 1.11 Principles of stereographic projection, illustrating (a) the pole P to a (111) plane, (b) the angle between two poles P, P', and (c) stereographic projection of P and P' poles to the (111) and (001) planes respectively

in this way is rather cumbersome, so that in the stereographic projection, the array of poles on the reference sphere, which represents the various planes in the crystal, are projected on to the equatorial plane. The pattern of poles projected on the equatorial or primitive plane then represents the stereographic projection of the crystal. As shown in *Figure 1.11(c)*, those poles in the northern half of the reference sphere are projected on to the equatorial plane by joining the pole P to the south pole S, while those in the southern half of the reference sphere, such as Q, are projected in the same way in the direction of the north pole N. *Figure 1.12(c)* shows the stereographic projection of some simple cubic planes, $\{100\}$, $\{110\}$ and $\{111\}$, from which it can be seen that those crystallographic planes which have poles in the southern half of the reference sphere are represented by circles in the stereogram, while those which have poles in the northern half are represented by dots.

As shown in *Figure 1.11(b)*, the angle between two poles on the reference sphere is the number of degrees separating them on the great circle passing through them. The angle between P and P' can then easily be determined by means of a graduated spherical transparent cap marked out with meridian circles and latitude circles as in geographical work. In the stereographic representation of poles, use is also made of a stereographic net, commonly referred to as a Wulff

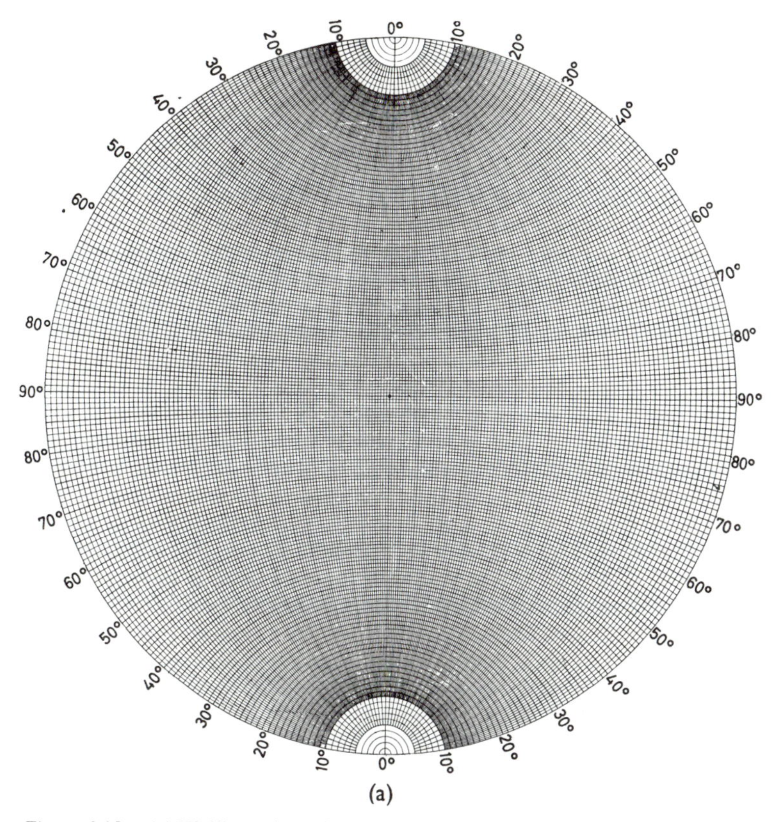

Figure 1.12 (a) Wulff net (from the net prepared in 1888 by the late Admiral C. D. Sigsbee, by courtesy of the Hydrographic Dept, US Navy)

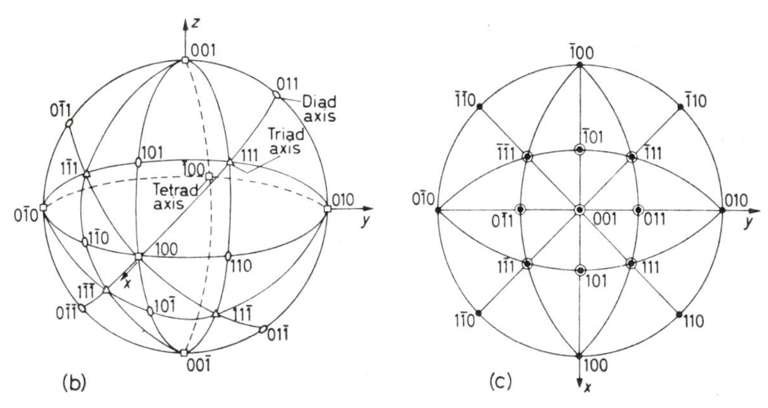

Figure 1.12 Projection of planes in cubic crystals; (b) spherical projection, and (c) stereographic projection

net, to carry out the same operation in the plane of the primitive circle as is carried out by means of the transparent cap on the surface of the sphere. The net, shown in *Figure 1.12*(*a*), is graduated in intervals of 2° and the meridians in the projection extend from top to bottom and the latitude lines from side to side.

Thus, to measure the angular distance between any two poles in the stereogram, the net is rotated about the centre until the two poles lie upon the same meridian, which then corresponds to one of the great circles of the reference sphere. The angle between the two poles is then measured as the difference in latitude along the meridian. The reader may wish to measure the angle between a few planes in this way and check the results by calculating the angle between two planes from the formula

$$\cos\theta = (h_1h_2 + k_1k_2 + l_1l_2)/\sqrt{[(h_1^2 + k_1^2 + l_1^2)(h_2^2 + k_2^2 + l_2^2)]} \qquad (1.4)$$

where (h_1, k_1, l_1) and (h_2, k_2, l_2) are the Miller indices of the two planes being considered.

Some useful crystallographic rules may be summarized:

(i) The zone law: if $hu + kv + lw = 0$, then the plane (hkl) contains the line $[uvw]$. All the different planes containing $[uvw]$ are said to form a zone, of which $[uvw]$ is the zone axis (somewhat like the leaves of an open book relative to the spine). The pole of a plane containing $[uvw]$ must lie at 90° to it. The trace of all such poles is called a zone circle. A zone circle is to a zone as a plane is to a pole. In cubic systems zone circles and plane traces with the same indices lie on top of one another. This is not true of any other crystal system.

(ii) If a zone contains $(h_1k_1l_1)$ and $(h_2k_2l_2)$ it also contains any linear combination of them, e.g., $m(h_1k_1l_1) + n(h_2k_2l_2)$. For example, the zone $[111]$ contains $[1\bar{1}0]$ and $[01\bar{1}]$ and it must therefore contain $[1\bar{1}0] + [01\bar{1}] = [10\bar{1}]$, $[1\bar{1}0] + 2[01\bar{1}] = [11\bar{2}]$, etc. The same is true for different directions in the same plane which are identical for cubic crystals.

(iii) $[u_1v_1w_1] + [u_2v_2w_2]$ lies between $[u_1v_1w_1]$ and $[u_2v_2w_2]$ from the law of vector addition.

(iv) The angle between two directions is given by:

$$\cos\theta = \frac{[u_1v_1w_1].[u_2v_2w_2]}{|u_1v_1w_1|\,|u_2v_2w_2|} = \frac{u_1u_2 + v_1v_2 + w_1w_2}{\sqrt{[(u_1^2 + v_1^2 + w_1^2)(u_2^2 + v_2^2 + w_2^2)]}}$$

which the reader will note is the same as for two planes. This is true only for the cubic system.

When constructing the standard stereogram of any crystal it is advantageous to examine first the symmetry elements of that structure. As an illustration let us consider a cubic crystal, since this has the highest symmetry of any crystal class. A close inspection shows that the cube has thirteen axes, nine planes and one centre of symmetry and that the thirteen axes of symmetry are made up from 3 four-fold or tetrad axes, 4 three-fold or triad axes, and 6 two-fold or diad axes. An n-fold axis of symmetry operates in such a way that after rotation through an angle $2\pi/n$, the crystal comes into an identical or self-coincident position in space. Thus, a tetrad axis passes through the centre of each face of the cube parallel to one of the edges, and a rotation of 90° in either direction about one of these axes turns the cube into a new position which crystallographically is indistinguishable from the old position. Similarly, the cube diagonals form a set of 4 three-fold axes, and each of the lines passing through the centre of opposite edges form a set of 6 two-fold symmetry axes. These symmetry elements are easily seen in the spherical projection of a cubic crystal shown in *Figure 1.12(b)*, where the $\langle 001 \rangle$ tetrad axes are denoted by the symbol □, the $\langle 111 \rangle$ triad axes by the symbol △, and the $\langle 110 \rangle$ diad axes by the symbol (). In the stereographic projection of the crystal, *Figure 1.12(c)*, the planes of symmetry divide the stereogram into 24 equivalent spherical triangles, commonly called unit triangles, which correspond to the 48 (24 on the top and 24 on the bottom) seen in the spherical projection. The two-fold, three-fold and four-fold symmetry about the $\{110\}$, $\{111\}$ and $\{100\}$ poles respectively, is very apparent.

Finally, the construction of the stereogram demonstrates the vector adoption rule which states that the indices of any plane can be found merely by adding simple multiples of other planes which lie in the same zone. For example, it can be seen from *Figure 1.12(b)* that the (011) plane lies between the (001) and (010) planes and clearly $011 = 001 + 010$. Owing to the action of the symmetry elements, it is also evident that there must be a total of 12 $\{011\}$ planes because of the respective three-fold and four-fold symmetry about the $\{111\}$ and $\{100\}$ axes. As a further example, it is clear that the (112) plane lies between the (111) plane and the (001) plane since $112 = 111 + 001$ and that the $\{112\}$ form must contain 24 planes, i.e. it is an icositetrahedron. The plane (123), is an example of the most general crystal plane in the cubic system because its indices h, k and l are all different, lies between the (112) and (011) planes, and the 48 planes of the $\{123\}$ form make up a hexakisoctahedron, i.e. a six-faced octahedron.

Suggestions for further reading

Barrett, C.S. and Massalski, T.B., *Structure of Metals*, 2nd edn, McGraw-Hill, 1980

Cottrell, A.H., *An Introduction to Metallurgy*, Edward Arnold, 1975

Kelly, A. and Groves, G.W., *Crystallography and Crystal Defects*, Longmans, 1970

Phillips, F.C., *Introduction to Crystallography*, Longmans Green, 1949

Chapter 2

The physical examination of metals and alloys

2.1 Introduction

In Metallurgy and Materials Science the responsibility has been assumed of studying the relation between the structure of an alloy and its properties and then using this knowledge to change the structure-dependent properties by modification in structure. In this way materials are provided with the best properties for a specific purpose to meet the increasing requirements of modern technology. The structure and properties of an alloy can be studied in many different ways depending on the nature of the information required; the basic techniques available are described in the following sections.

2.2 Metallography

The light microscope In all branches of physical metalllurgy the use of a microscope is imperative. The simplest of these, the light microscope, consists essentially of three parts: (i) an illuminator, to illuminate the surface of the metal, (ii) an objective, to resolve the structure, and (iii) an eyepiece, to enlarge the image formed by the objective. Microscopic examination of a representative specimen of metal, after polishing and then etching with a suitable chemical reagent, is sufficient to reveal such features as the arrangement and size of grains, the distribution of phases, the results of plastic deformation and the existence of impurities and flaws. Slight chemical attack, or etching of the surface, first reveals the grain boundaries, but further etching produces shades which vary from one grain to the next, owing to the fact that the etching reagent does not attack the metal evenly but along certain crystallographic planes. Facets having the same orientation are then produced in each grain, and, since each grain has a different orientation from that of its neighbours, one grain may reflect light into the objective of the microscope, and consequently appear light, while adjacent grains reflect most of the light in other directions and appear darker (*Figure 2.1*(*a*)).

Figure 8.25 shows the microstructure of a pure metal. No fine-scale atomic details, such as the structure of a grain boundary or the existence of dislocations, are revealed because the resolving power (i.e. the ability of the microscope to

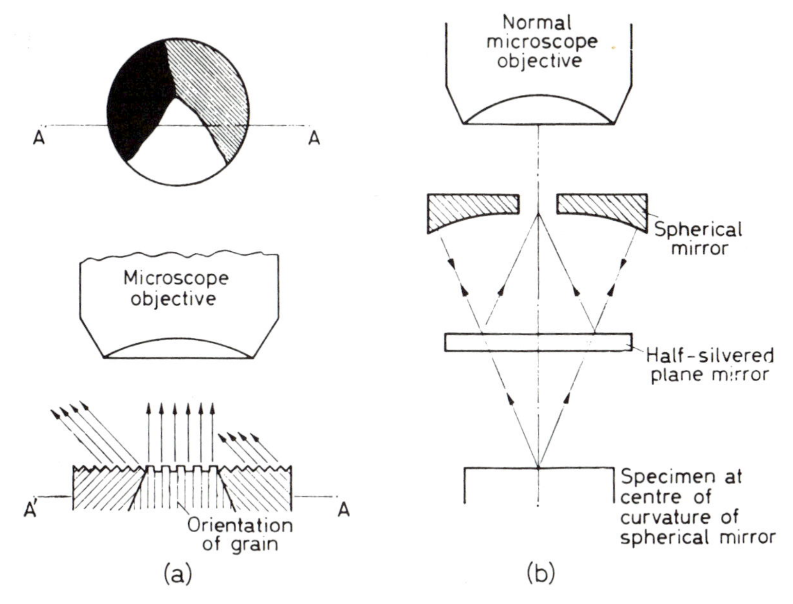

Figure 2.1 Schematic arrangement of microscope system for (a) normal and (b) reflecting objective, microscopy

separate distinctly two close objects) is too poor. The smallest distance δ that may be resolved is controlled by three factors, (i) the wavelength of the illumination λ, (ii) the effective aperture of the objective lens, and (iii) the medium between the lens and the specimen, according to the relation:

$$\delta = 0.5\lambda/\text{Numerical aperture} = \lambda/2\mu \sin \alpha \tag{2.1}$$

It is evident that the numerical aperture, and hence the resolution, is increased by increasing the refractive index of the medium (μ for air = 1, and for cedar oil = 1.5) and the half angle α subtended by the maximum cone of rays entering the objective. However, even in the most favourable case with α approaching 90°, because the wavelength of light is about 500 nm (or 0.5 μm), the resolution cannot exceed 200 nm (or 0.2 μm).

In recent years several techniques have been developed which have greatly increased the usefulness of the simple metallurgical microscope. One of these is the adaptation of a hot-stage for studying a metal specimen at elevated temperatures. The technique is extremely useful for studying structural changes such as recrystallization, grain growth (*see* Chapters 10, 11 and 12) at the temperatures at which they occur. However, since many metals oxidize rapidly in air to form films which would soon obscure the structure, it is usually necessary to surround the specimen with a furnace and contain the complete assembly in a water-cooled chamber which can be evacuated or filled with an inert gas. Observation is then made through an optically worked silica window, and for this reason the working distance of the objective must be sufficient to clear the window. Clearly, a normal 16 mm transmission objective, which has a working distance of about 5 mm, can be used for this application if low magnifications

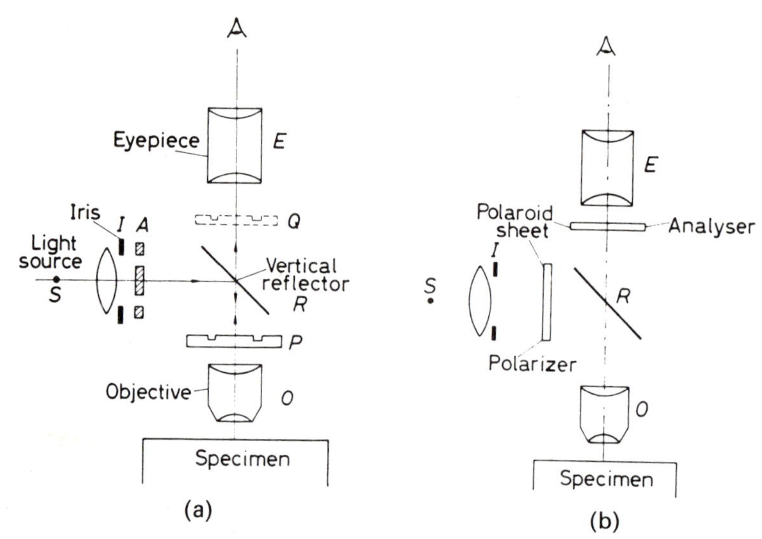

Figure 2.2 Schematic arrangement of microscope system for (a) phase contrast, and (b) polarized light, microscopy

only are required. For high magnification work, however, the working distance of the objective must be improved by employing a reflecting objective such as that designed by Dyson (*Figure 2.1*(*b*)) which enables a 4 mm objective to be used at a distance of about 13 mm from the specimen.

Phase-contrast microscopy is another technique in metallography, and this enables special surface features to be studied even when there is no colour or reflectivity contrast. For example, the light reflected from a small depression in a metallographic specimen will be retarded in phase by a fraction of a light wavelength relative to that reflected from the surrounding matrix and, whereas in ordinary microscopy a phase difference in the light collected by the objective will not contribute to contrast in the final image, in phase-contrast microscopy small differences in phases are transformed into differences in brightness which the eye can detect.

The general uses of the technique include the examination of multi-phased alloys after light etching, the detection of the early stages of precipitation, and the study of cleavage faces, twins and other deformation characteristics. The optimum range of differences in surface level is about 20–50 nm, although under favourable conditions these limits may be extended. A schematic diagram of the basic arrangement for phase contrast in the metallurgical microscope is shown in *Figure 2.2*(*a*). A hollow cone of light produced by an annulus A, is reflected by the specimen and brought to an image in the back focal plane of the objective. A phase plate of suitable size should, strictly, be positioned in this plane but, for the ease of interchangeability of phase plates, the position Q in front of the eyepiece E is often preferred. This phase plate has an annulus, formed either by etching or deposition, such that the light it transmits is either advanced or retarded by a quarter of a wavelength relative to the light transmitted by the rest of the plate and, because the light reflected from a surface feature is also advanced or retarded by approximately $\lambda/4$, the beam is either in phase or approximately $\lambda/2$ or π out of phase with that diffracted by the surface features of the specimen. Consequently, reinforcement or cancellation occurs, and the image intensity at any point depends on the phase difference produced at the corresponding point on the specimen surface, and this in turn depends upon the height of this point relative to the adjacent parts of the surface. When the light passing through the annulus is advanced in phase, positive phase contrast results, and areas of the specimen which are proud of the matrix appear bright and depressions dark; when the phase is retarded negative contrast is produced and 'pits' appear bright and 'hills' dark.

The use of polarized light in the microscopical examination of opaque specimens is also becoming quite common and the basic arrangement for its use is shown in *Figure 2.2(b)*. The only requirements of this technique are that the incident light on the specimen be plane-polarized and that the reflected light be analysed by a polarizing unit in a crossed relation with respect to the polarizer, i.e. the plane of polarization of the analyser is perpendicular to that of the polarizer.

The application of the technique depends upon the fact that plane-polarized light striking the surface of an optically isotropic metal is reflected unchanged if it strikes at normal incidence. If the light is not at normal incidence the reflected beam may still be unchanged but only if the angle of incidence is in, or at right angles to, the plane of polarization, otherwise it will be elliptically polarized. It follows that the unchanged reflected beam will be extinguished by an analyser in the crossed position whereas an elliptically polarized one cannot be fully extinguished by an analyser in any position. When the specimen being examined is optically anisotropic, the light incident normally is reflected with a rotation of the plane of polarization and as elliptically polarized light; the amount of rotation and of elliptical polarization is a property of the metal and of the crystal orientation.

If correctly prepared, as-polished specimens of anisotropic metals will 'respond' to polarized light and a grain contrast effect is observed under crossed polars as a variation of brightness with crystal orientation. Metals which have cubic structure, on the other hand, will appear uniformly dark under crossed polars, unless etched to invoke artificial anisotropy, by producing anisotropic surface films or well-defined pits. An etch pit will reflect the light at oblique incidence and elliptically polarized light will be produced. However, because such a beam cannot be fully extinguished by the analyser in any position, it will produce a background illumination in the image which tends to mask the grain contrast effect.

Clearly, one of the main uses of polarized light is to distinguish between areas of varying orientation, since these are revealed as differences of intensity under crossed polars. The technique is, therefore, very useful for studying the effects of deformation, particularly the production of preferred orientation, but information on cleavage faces, twin bands and sub-grain boundaries can also be obtained. If a 'sensitive tint' plate is inserted between the vertical illuminator and the analyser each grain of a sample may be identified by a characteristic colour which changes as the specimen is rotated on the stage. This application is useful in the assessment of the degree of preferred orientation and in recrystallization studies. Other uses of polarized light include distinguishing and identifying phases in multi-phase alloys.

2.3 X-ray and neutron diffraction

The use of diffraction methods is of great importance in the analysis of metal crystals. Not only can they reveal the main features of the lattice structure, i.e. the lattice parameter and type of structure, but also other details such as the arrangement of different kinds of atoms in crystals, the presence of imperfections, the orientation, sub-grain and grain size, the size and density of precipitates and the state of lattice distortion.

X-ray and neutron diffraction techniques can be applied to 'bulk' samples and the basic principles of each method of diffraction will be considered. Electron diffraction also is used widely, but is applied to thin specimens and is carried out generally in conjunction with electron microscopy, so that crystallographic information is obtained from the structural area of interest. Electron microscopy and diffraction are considered in section 2.4.

2.3.1 The principles and methods of x-ray diffraction

The production of x-rays X-rays are a form of electromagnetic radiation differing from light waves ($\lambda = 400$–800 nm (0.4–0.8 μm)) in that they have a shorter wavelength ($\lambda \approx 0.1$ nm). These rays are produced when a metal target is bombarded with fast electrons in a vacuum tube. The radiation emitted, as shown in *Figure 2.3(a)*, can be separated into two components, a continuous spectrum which is spread over a wide range of wavelengths and a superimposed line spectrum characteristic of the metal being bombarded. The energy of the white radiation, as the continuous spectrum is called, increases as the atomic number of the target and approximately as the square of the applied voltage, while the characteristic radiation is excited only when a certain critical voltage is exceeded. The characteristic radiation is produced when the accelerated electrons have sufficient energy to eject one of the inner electrons (1*s* level, for example) from its shell. The vacant 1*s* level is then occupied by one of the other electrons from a higher energy level, and during the transition an emission of x-radiation takes place. If the electron falls from an adjacent shell then the radiation emitted is known as $K\alpha$-radiation, since the vacancy in the first shell (i.e. $n = 1$ termed the K-shell by x-ray crystallographers) is filled by an electron from the second shell (termed the L-shell), and the wavelength can be derived from the relation

$$h\nu = E_L - E_K \quad (2.2)$$

However, if the K-shell vacancy is filled by an electron from an M-shell (i.e. the next highest quantum shell) then $K\beta$ radiation is emitted. *Figure 2.3* shows that, in fact, one cannot be excited without the other, and the characteristic K-radiation emitted from a copper target is in detail composed of a strong $K\alpha$-doublet and a weaker $K\beta$-line.

The absorption of x-rays In transversing a specimen, an x-ray beam loses intensity according to the equation

$$I = I_0 \exp[-\mu x] \quad (2.3)$$

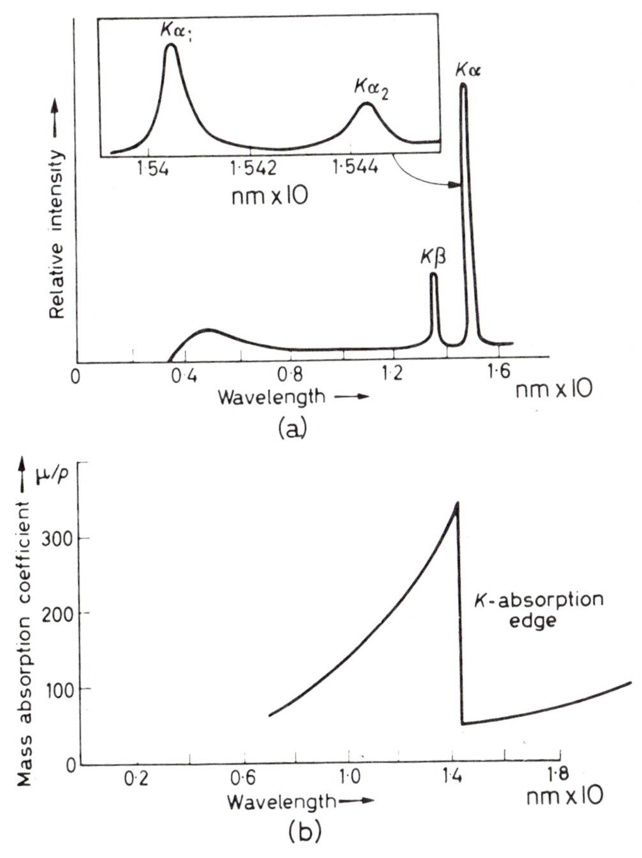

Figure 2.3 (a) Distribution of x-ray intensity from a copper target, and (b) dependence of absorption coefficient on x-ray wavelength for nickel

where I_0 and I are the values of the initial and final intensities respectively, μ is a constant, known as the linear absorption coefficient which depends on the wavelength of the x-rays and the nature of the absorber, and x is the thickness of the specimen*. The variation of absorption with wavelength is of particular interest, as shown in *Figure 2.3(b)* which is the curve for nickel. It varies approximately as λ^3 until a critical value of λ ($=0.148$ nm) is reached, when the absorption decreases precipitously. The critical wavelength λ_K at which this decrease occurs is known as the K absorption edge, and is the value at which the x-ray beam has acquired just sufficient energy to eject an electron from the K-shell of the absorbing material. The value of λ_K is characteristic of the absorbing material, and similar L and M absorption edges occur at higher wavelengths.

This sharp variation in absorption with wavelength has many applications in x-ray practice, but its most common use is in filtering out unwanted $K\beta$-

* This absorption equation is the basis of radiography, since a cavity, crack or similar defect will have a much lower μ-value than the sound metal. Such defects can be detected by the appearance of an intensity difference registered on a photographic film placed behind the x-irradiated object.

radiation. For example, if a thin piece of nickel foil is placed in a beam of x-rays from a copper target, absorption of some of the short wavelength white radiation and most of the $K\beta$-radiation will result, but the strong $K\alpha$-radiation will be only slightly attenuated. This filtered radiation is sufficiently monochromatic for many x-ray techniques, but for more specialized studies when a pure monochromatic beam is required, crystal monochromators are used. The x-ray beam is then reflected from a crystal, such as quartz or lithium fluoride, which is oriented so that only the desired wavelength is reflected according to the Bragg law (*see* below).

Diffraction of x-rays by crystals The phenomena of interference and diffraction are commonplace in the field of light. The standard school physics laboratory experiment is to determine the spacing of a grating, knowing the wavelength of the light impinging on it, by measuring the angles of the diffracted beam. The only conditions imposed on the experiment are that (i) the grating be periodic, and (ii) the wavelength of the light is of the same order of magnitude as the spacing to be determined. This experiment immediately points to the application of x-rays in determining the spacing and inter-atomic distances in crystals, since both are about 0.1–0.4 nm in dimension. Rigorous consideration of diffraction from a crystal in terms of a three-dimensional diffraction grating is complex, but Bragg has simplified the problem by showing that diffraction is equivalent to symmetrical reflection from the various crystal planes, provided certain conditions are fulfilled. *Figure 2.4(a)* shows a beam of x-rays of wavelength λ, impinging at an angle θ on a set of crystal planes of spacing d. The beam reflected at the angle θ can be real only if the rays from each successive plane reinforce each other. For this to be the case, the extra distance a ray, scattered from each successive plane, has to travel, i.e. the path difference, must be equal to an integral number of wavelengths, $n\lambda$. For example, the second ray shown in *Figure 2.4(a)* has to travel farther than the first ray by the distance $PO + OQ$. The condition for reflection and reinforcement is then given by

$$n\lambda = PO + OQ = 2ON \sin\theta = 2d \sin\theta \qquad (2.4)$$

This is the well-known Bragg law and the critical angular values of θ for which the law is satisfied are known as Bragg angles.

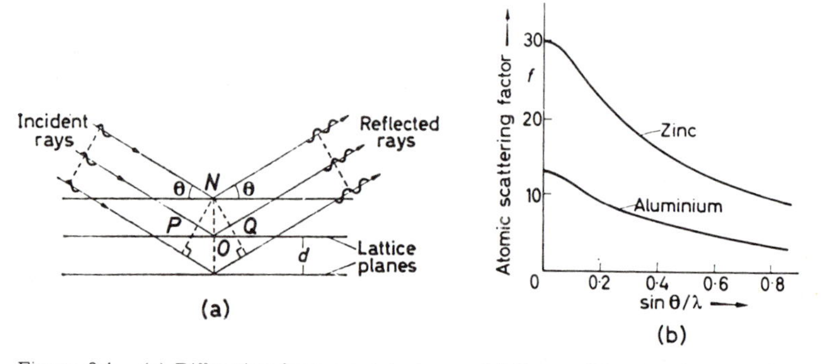

Figure 2.4 (a) Diffraction from crystal planes. (b) Form of the atomic scattering curves for aluminium and zinc

The directions of the reflected beams are determined entirely by the geometry of the lattice, which in turn is governed by the orientation and spacing of the crystal planes. If for a crystal of cubic symmetry we are given the size of the structure cell, a, the angles at which the beam is diffracted from the crystal planes (hkl) can easily be calculated from the interplanar spacing relationship

$$d_{(hkl)} = a/\sqrt{(h^2+k^2+l^2)} \tag{2.5}$$

It is conventional to incorporate the order of reflection, n, with the Miller index, and when this is done the Bragg law becomes

$$\lambda = 2a\sin\theta/\sqrt{(n^2h^2+n^2k^2+n^2l^2)} = 2a\sin\theta/\sqrt{N} \tag{2.6}$$

where N is known as the reflection or line number. To illustrate this let us take as an example the second order reflection from (100) planes. Then, since $n=2$, $h=1$, $k=0$, and $l=0$, this reflection is referred to either as the 200 reflection or as line 4. The lattice planes which give rise to a reflection at the smallest Bragg angle are those which are most widely spaced, i.e. those with a spacing equal to the cell edge, d_{100}. The next planes in order of decreased spacing will be the $\{110\}$ planes for which $d_{110}=a/\sqrt{2}$, while the octahedral planes will have a spacing equal to $a/\sqrt{3}$. The angle at which any of these planes in a crystal reflect an x-ray beam of wavelength λ may be calculated by inserting the appropriate value of d into the Bragg equation.

To ensure that Bragg's law is satisfied and that reflections from various crystal planes can occur, it is necessary to provide a range of either θ or λ values. The various ways in which this can be done leads to the standard methods of x-ray diffraction, namely: (a) the Laue method, (b) the rotating crystal method, and (c) the powder method.

The Laue method In the Laue method a stationary single crystal is bathed in a beam of white radiation. Then, because the specimen is a fixed single crystal, the variable necessary to ensure that the Bragg law is satisfied for all the planes in the crystal has to be provided by the range of wavelengths in the beam, i.e. each set of crystal planes chooses the appropriate λ from the white spectrum to give a Bragg reflection. Radiation from a target metal having a high atomic number (e.g. tungsten) is often used, but almost any form of white radiation is suitable. In the experimental arrangement shown in *Figure 2.5(a)*, either a transmission photograph or back-reflection photograph may be taken, and the pattern of spots which are produced lie on ellipses in the transmission case or hyperbolae in the back-reflection case. All spots on any ellipse or hyperbola are reflections from planes of a single zone (i.e. where all the lattice planes are parallel to a common direction, the zone axis) and, consequently, the Laue pattern is able to indicate the symmetry of the crystal. For example, if the beam is directed along a [111] or [100] direction in the crystal, the Laue pattern will show three-fold or four-fold symmetry respectively. The Laue method is used extensively for the determination of the orientation of single crystals and, while charts are available to facilitate this determination, the method consists essentially of plotting the zones taken from the film on to a stereogram, and comparing the angles between them with a standard projection of that crystal structure. In recent years the use of

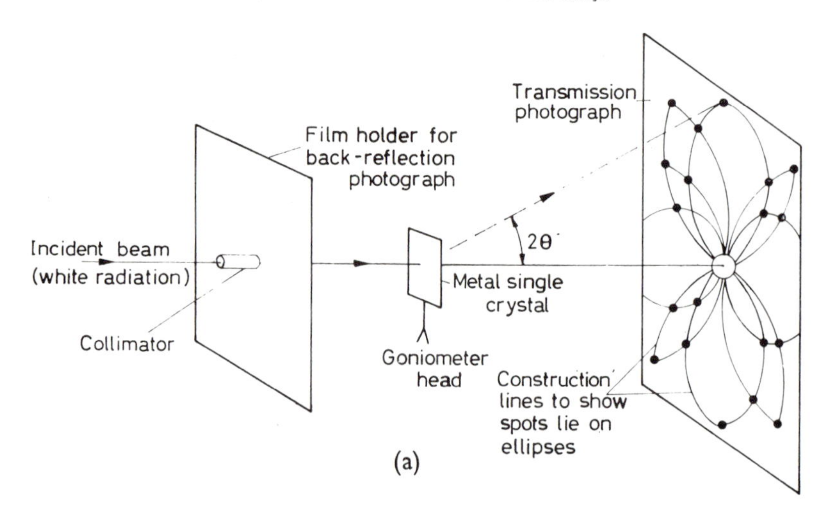

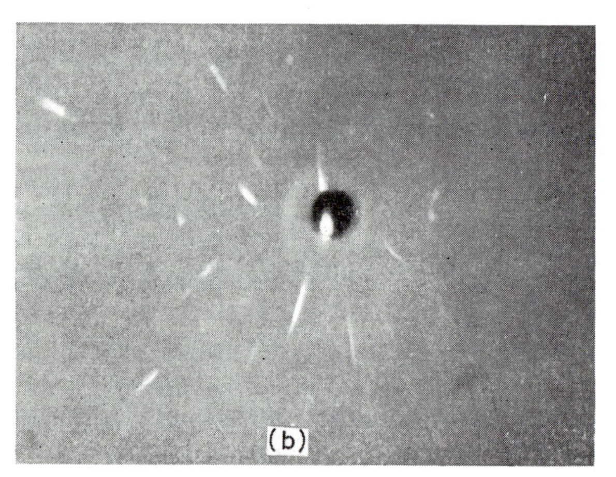

Figure 2.5 (a) Laue method of x-ray diffraction. (b) Asterism on a Laue transmission photograph of deformed zinc (after Cahn, *J. Inst. Metals,* 1949, **77,** 121)

the Laue technique has been extended to the study of imperfections resulting from crystal growth or deformation, because it is found that the Laue spots from perfect crystals are sharp, while those from imperfect or deformed crystals are, as shown in *Figure 2.5(b)*, elongated. This elongated appearance of the diffraction spots is known as asterism and it arises in an analogous way to the reflection of light from curved mirrors.

The rotating crystal method This method utilizes a single crystal which is rotated, or oscillated, in a beam of monochromatic x-rays. In this case, because λ is fixed, it is the variation of θ as the crystal rotates which allows the Bragg law to be satisfied, i.e. different crystal planes are brought into the reflecting position as the specimen rotates. The technique is used principally for the determination of the more complex crystal structures, by studying the blackness and positions of

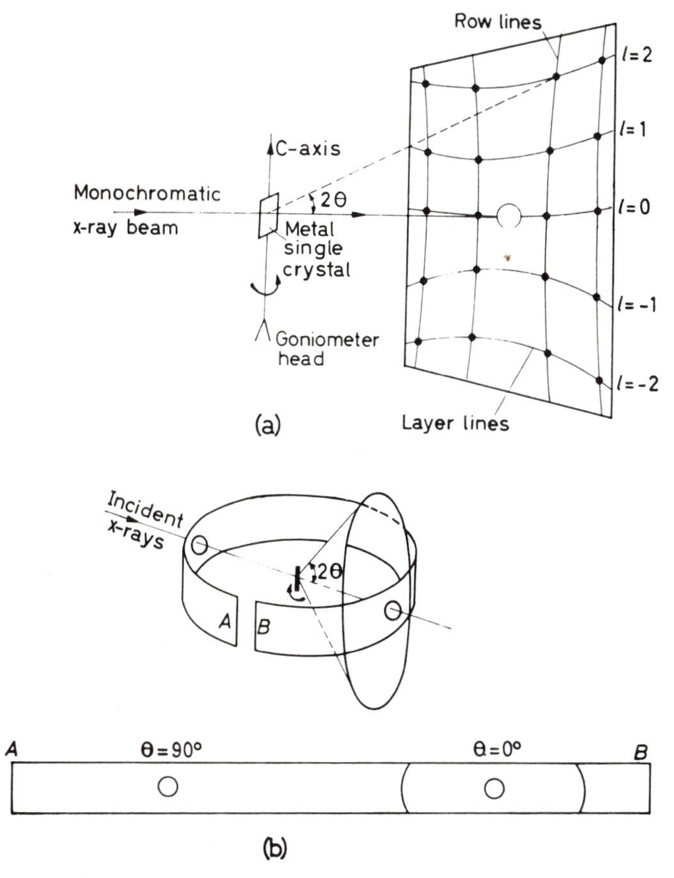

Figure 2.6 (a) Rotating crystal, and (b) powder method of x-ray diffraction

the diffraction spots, and the crystal is usually oriented so that the rotation is about one of the major axes. A typical 'rotating-crystal' film is shown in *Figure 2.6(a)*, from which it can be seen that the reflections are arranged on layer lines and that the spots on the succession of layer lines form a series of row lines. In consequence, all planes parallel to the rotation axis reflect into the zero layer line, in such a way that a photograph taken with the *c*-axis of the crystal as the axis of rotation has the zero layer line made up of reflections from planes of the (*hko*) type. Similarly, the planes of the type (*hkl*) and (*hkl*) will reflect into the first layer line above and below the central one respectively, and it is easy to see that the *c*-parameter of the crystal can be obtained from the distance between the layer lines.

The powder method The powder method, devised independently by Debye and Scherrer, is probably the most generally useful of all the x-ray techniques. It employs monochromatic radiation and a finely powdered, or fine-grained polycrystalline, wire specimen. In this case, θ is the variable, since the collection of randomly oriented crystals will contain sufficient particles with the correct orientation to allow reflections from each of the possible reflecting planes, i.e. the powder pattern results from a series of superimposed rotating crystal patterns. The angle between the direct x-ray beam and the reflected ray is 2θ, and consequently each set of crystal planes gives rise to a cone of reflected rays of semi-angle 2θ, where θ is the Bragg angle for that particular set of reflecting planes producing the cone. Thus, if a film is placed around the specimen, as shown in *Figure 2.6(b)*, the successive diffracted cones, which consist of hundreds of rays from hundreds of grains, intersect the film to produce concentric curves around the entrance and exit holes.

Precise measurement of the pattern of diffraction lines is required for many applications of the powder method, but a good deal of information can readily be obtained merely by inspection. One example of this is in the study of deformed metals, since after deformation the individual spots on the diffraction rings are blurred so much that line broadening occurs, especially at high Bragg angles. On low temperature annealing the cold worked material will tend to recover and this is indicated on the photograph by a sharpening of the broad diffraction lines. At higher annealing temperatures the metal will completely regain its softness by a process known as recrystallization (*see* Chapter 10) and this phenomenon is accompanied by the completion of the line sharpening process. With continued annealing, the grains absorb each other to produce a structure with an overall coarser grain size and, because fewer reflections are available to contribute to the diffraction cones, the lines on the powder photograph take on a spotty appearance. This latter behaviour is sometimes used as a means of determining the grain size of a polycrystalline sample. In practice an x-ray photograph is taken for each of a series of known grain sizes to form a set of standards, and with them an unknown grain size can be determined quite quickly by comparing the corresponding photograph with the set of standards. Yet a third use of the powder method as an inspection technique is in the detection of a preferred orientation of the grains of a polycrystalline aggregate. This is possible because a random orientation of the grains will produce a uniformly intense diffraction ring, while a preferred orientation, or texture, will concentrate the intensity at certain positions on the ring. The details of the texture require considerable interpretation.

Other applications of the powder method depend on the accurate measurement of either line position or line intensity. The arrangement of the diffraction lines in any pattern is characteristic of the material being examined and, consequently, an important practical use of the method is in the identification of unknown phases. Thus, it will be evident that equation 2.6 can indicate the position of the reflected beams, as determined by the size and shape of the unit cell, but not the intensities of the reflected beams. These are determined, not by the size of the unit cell, but by the distribution of atoms within it, and while simple cubic lattices give reflections for every possible value of $(h^2+k^2+l^2)$ all other structures give characteristic absences. It is by studying the indices of the 'absent' reflections that enables the different structures to be distinguished.

In calculating the intensity scattered by a given atomic structure we have first to consider the intensity scattered by one atom, and then go on to consider the contribution from all the other atoms in the particular arrangement which make up that structure. The efficiency of an atom in scattering x-rays is usually denoted by f, the atomic scattering factor, which is the ratio of amplitude scattered by an atom A_a to that by a single electron A_e. If atoms were merely points, their scattering factors would be equal to the number of electrons they contain, i.e. to their atomic numbers, and the relation $I_a = Z^2 . I_e$ would hold since intensity is proportional to the square of amplitude. However, because the size of the atom is comparable with the wavelength of x-rays, scattering from different parts of the atom is not in phase, and the result is that $I_a \leqslant Z^2 . I_e$. The scattering factor, therefore, depends both on angle θ and on the wavelength of x-rays used, as shown in *Figure 2.4(b)*, because the path difference for the individual waves scattered from the various electrons in the atom is zero when $\theta = 0$ and increases with increasing θ. Thus, to consider the intensity scattered by a given structure, it is necessary to sum up the waves which come from all the atoms of one unit cell of that structure, since each wave has a different amplitude and a different phase angle due to the fact that it comes from a different part of the structure. The square of the amplitude of the resultant wave, F, then gives the intensity, and this may be calculated by using the f-values and the atomic co-ordinates of each atom in the unit cell. It can be shown that a general formula for the intensity is

$$\begin{aligned} I \propto |F|^2 = [&f_1 \cos 2\pi(hx_1 + ky_1 + lz_1) \\ &+ f_2 \cos 2\pi(hx_2 + ky_2 + lz_2) + \ldots]^2 \\ &+ [f_1 \sin 2\pi(hx_1 + ky_1 + lz_1) \\ &+ f_2 \sin 2\pi(hx_2 + ky_2 + lz_2) + \ldots]^2 \end{aligned} \tag{2.7}$$

where $x_1, y_1, z_1; x_2, y_2, z_2$, etc., are the co-ordinates of those atoms having scattering factors f_1, f_2, etc., respectively and hkl are the indices of the reflection being computed. For structures having a centre of symmetry, which includes most metals, the expression is much simpler because the sine terms vanish.

This equation may be applied to any structure, but to illustrate its use let us examine a pure metal crystallizing in the b.c.c. structure. From *Figure 1.2(c)* it is clear that the structure has identical atoms (i.e. $f_1 = f_2$) at the co-ordinates (000) and $(\frac{1}{2}\frac{1}{2}\frac{1}{2})$ so that equation 2.7 becomes:

$$I \propto f^2[\cos 2\pi . O + \cos 2\pi(h/2 + k/2 + l/2)]^2 = f^2[1 + \cos \pi(h + k + l)]^2 \tag{2.8}$$

It then follows that I is equal to zero for every reflection having $(h + k + l)$ an odd number. The significance of this is made clear if we consider in a qualitative way the 100 reflection shown in *Figure 2.7(a)*. To describe a reflection as the first-order reflection from (100) planes implies that there is 1λ phase-difference between the rays reflected from planes A and those reflected from planes A'. However, the reflection from the plane B situated half way between A and A' will be $\lambda/2$ out of phase with that from plane A, so that complete cancellation of the 100 reflected ray will occur. The 100 reflection is therefore absent, which agrees with the prediction made from equation 2.8 that the reflection is missing when $(h + k + l)$ is an odd number. A similar analysis shows that the 200 reflection will be present (*Figure 2.7(b)*), since the ray from the B plane is now exactly 1λ out of phase with the rays from A and A'. In consequence, if a diffraction pattern is taken from a material having a b.c.c. structure, because of the rule governing the sum of the

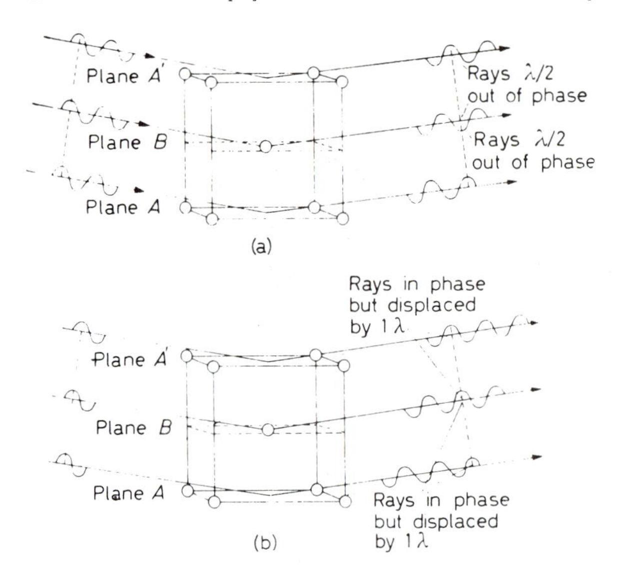

Figure 2.7 (a) (100) reflection from b.c.c. cell showing interference of diffracted rays, (b) (200) reflection showing reinforcement (after Barrett, *Structure of Metals,* 1952, courtesy of McGraw-Hill Book Co.

indices, the film will show diffraction lines almost equally spaced with indices $N=2,(110);4,(200);6,(211);8,(220);\ldots$, as shown in *Figure 2.8(a)*. Application of equation 2.7 to a pure metal with f.c.c. structure shows that 'absent' reflections will occur when the indices of that reflection are mixed, i.e. when they are neither all odd nor all even. Thus, the corresponding diffraction pattern will contain lines according to $N=3,4,8,11,12,16,19,20$, etc; and the characteristic feature of the arrangement is a sequence of two lines close together and one line separated, as shown in *Figure 2.8(b)*.

Equation 2.7 is the basic equation used for determining unknown structures, since the determination of the atomic positions in a crystal is based on this relation between the co-ordinates of an atom in a unit cell and the intensity with which it will scatter x-rays.

The determination of the lattice parameter

Perhaps the most common use of the powder method is in the accurate determination of lattice parameters. From the Bragg law we have the relation $a=\lambda\sqrt{(N)}/2\sin\theta$ which, because both λ and N are known and θ can be measured for the appropriate reflection, can be used to determine the lattice parameter of a material. Several errors are inherent in the method, however, and the most common include shrinkage of the film during processing, eccentricity of the specimen in the camera, and absorption of the x-rays in the sample. These errors affect the high-angle diffraction lines least and, consequently, the most accurate parameter value is given by determining a value of a from each diffraction line,

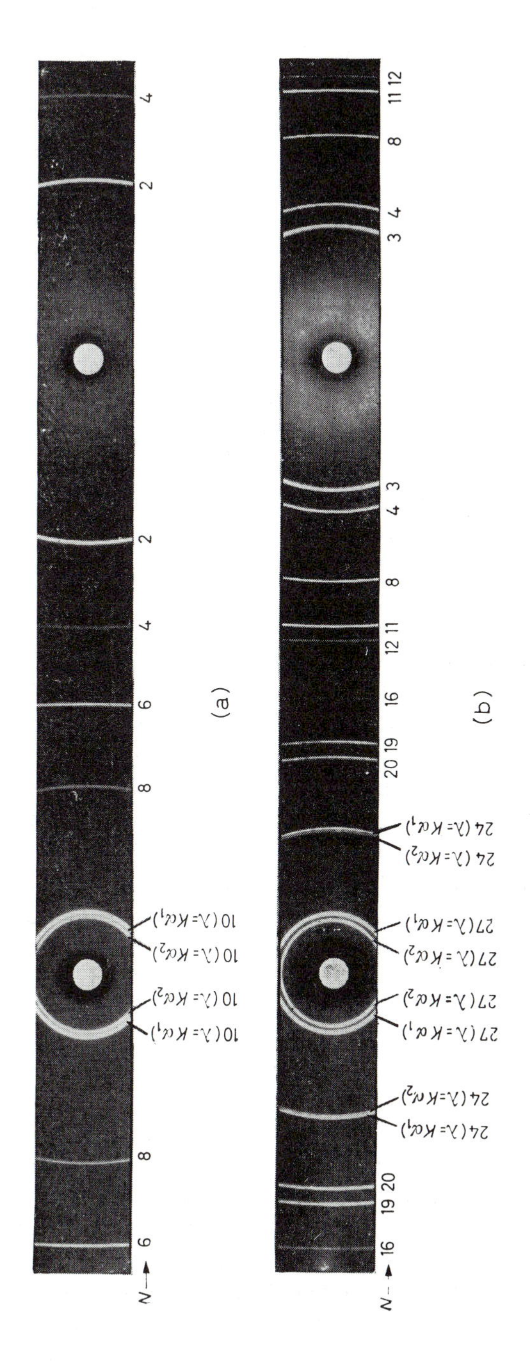

Figure 2.8 Powder photographs taken in a Phillips camera (114mm radius) of (a) iron with cobalt radiation using an iron filter, and (b) aluminium with copper radiation using a nickel filter. The high-angle lines are resolved and the separate reflections for $\lambda = K\alpha_1$ and $\lambda = K\alpha_2$ are observable

plotting it on a graph against an angular function* of the $\cos^2\theta$-type and then extrapolating the curve to $\theta = 90°$.

The determination of precision lattice parameters is of importance in many fields of physical metallurgy, particularly in the study of thermal expansion coefficients, density determinations, the variation of properties with alloy composition, precipitation from solid solution, and thermal stresses. Some of these problems are discussed in other sections, but at this stage it is instructive to consider the application of lattice parameter measurements to the determination of phase boundaries in equilibrium diagrams, since this illustrates the general usefulness of the technique. The diagrams shown in *Figure 2.9(a)* and *(b)* indicate the principle of the method. A variation of alloy composition within the single-phase field, α, produces a variation in the lattice parameter, a, since solute B, which has a different atomic size to the solvent A, is being taken into solution. However, at the phase boundary this variation in a ceases, because at a given temperature the composition of the α-phase remains constant in the two-phase field, and the marked discontinuity in the plot of lattice parameter versus composition indicates the position of the phase boundary at that temperature. The change in solid solubility with temperature may then be obtained, either by taking diffraction photographs in a high temperature camera at various temperatures, or by quenching the powder sample from the high temperature to

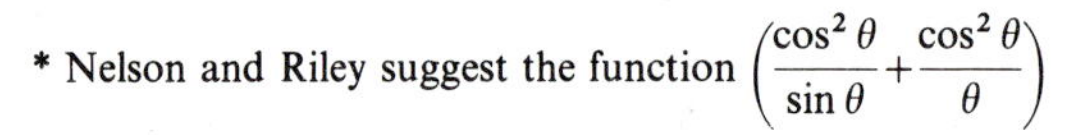

* Nelson and Riley suggest the function $\left(\dfrac{\cos^2\theta}{\sin\theta} + \dfrac{\cos^2\theta}{\theta}\right)$

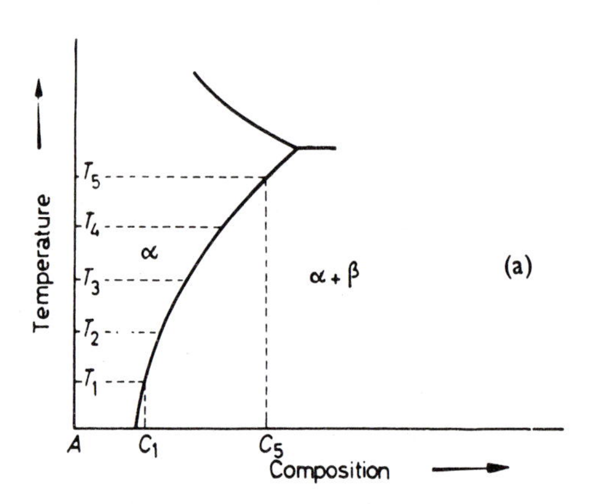

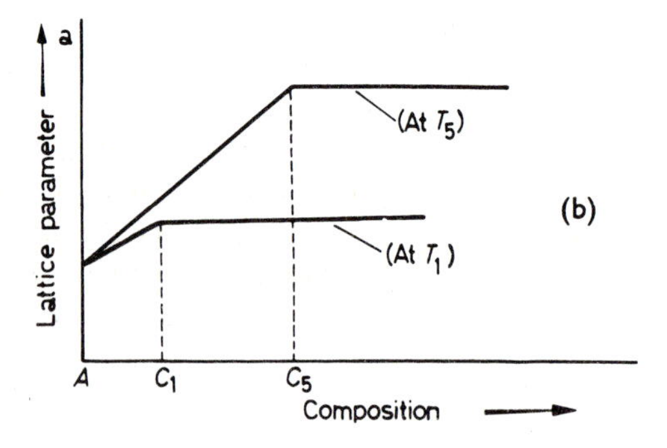

Figure 2.9 (a) and (b) Phase-boundary determination using lattice parameter measurements

room temperature (in order to retain the high temperature state of solid solution down to room temperature) and then taking a powder photograph at room temperature.

Line broadening

Diffraction lines are not always sharp because of various instrumental factors such as slit size, specimen condition, and spread of wavelengths, but in addition the lines may be broadened as a result of lattice strain in the region of the crystal diffracting and also its limited dimension. Strain gives rise to a variation of the interplanar spacing Δd and hence diffraction occurs over a range $\Delta\theta$ and the breadth due to strain is then

$$\beta_s = \eta \tan \theta \tag{2.9}$$

where η is the strain distribution. If the dimension of the crystal diffracting the x-rays is small*, then this also gives rise to an appreciable 'particle-size' broadening given by the Scherrer formula

$$\beta_p = \lambda / t \cos \theta \tag{2.10}$$

where t is the effective particle size. In practice this size is the region over which there is coherent diffraction and is usually defined by boundaries such as dislocation walls. It is possible to separate the two effects by plotting the experimentally measured broadening $\beta \cos \theta/\lambda$ against $\sin \theta/\lambda$, when the intercept gives a measure of t and the slope η.

Direct recording of x-rays

In addition to photographic recording, the diffracted x-ray beam may be detected directly using a counter tube (either Geiger, proportional or scintillation type) with associated electrical circuitry. The geometrical arrangement of such an x-ray spectrometer, or diffractometer, is shown in *Figure 2.10(a)*. A divergent beam of filtered, or monochromatized radiation impinges on the flat face of a powder specimen. This specimen is rotated at precisely one-half of the angular speed of the receiving slit so that a constant angle between the incident and reflected beams is maintained. The receiving slit is mounted in front of the counter on the counter tube arm, and behind it is usually fixed a scatter slit to ensure that the counter receives radiation only from that portion of the specimen illuminated by the primary beam. The intensity diffracted at the various angles is recorded automatically on a chart of the form shown in *Figure 2.10(c)*, and this can quickly be analysed for the appropriate θ and d values. The most widespread application of the technique in the metal industry is in routine chemical analysis, since accurate intensity measurements allow a quantitative estimate of the various elements in the sample to be made. In research, the technique has been applied to problems such as the degree of order in alloys, the density of stacking faults in

* The optical analogue of this effect is the broadening of diffraction lines from a grating with a limited number of lines.

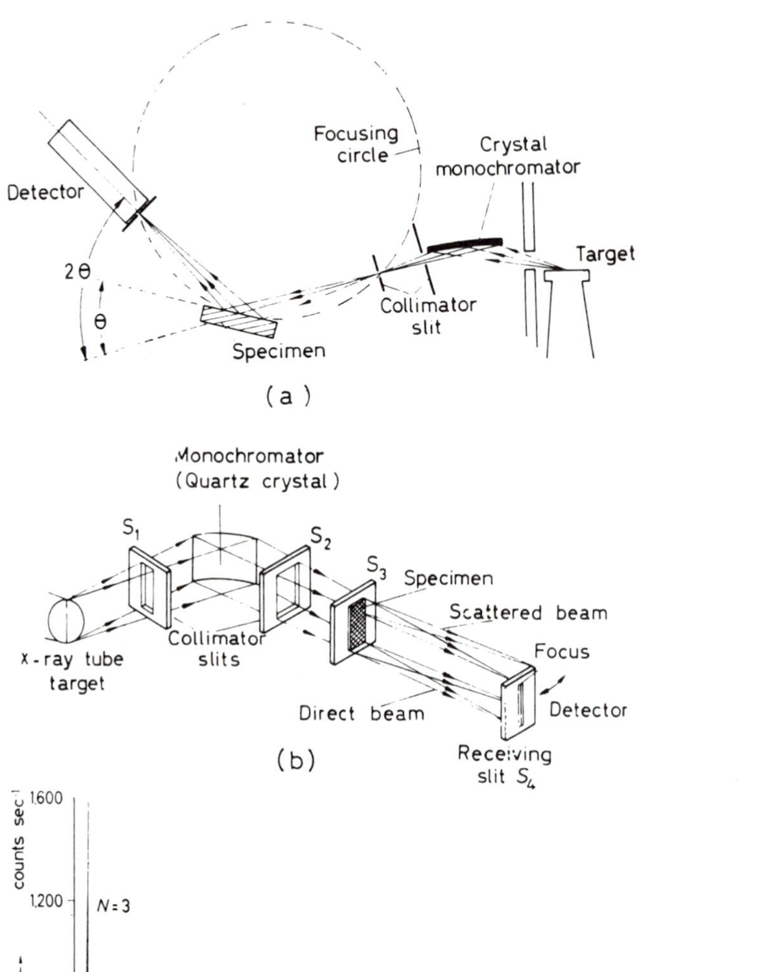

Figure 2.10 Geometry of (a) conventional diffractometer, and (b) small-angle scattering diffractometer (c) Chart record of diffraction pattern from aluminium powder with copper radiation using nickel filter

deformed alloys, elastic constant determination and the study of imperfections.

Direct recording has also been applied to other more specialized x-ray techniques, such as the determination of preferred orientation in textured materials and the scattering of x-rays at small angles. The scattering of intensity into the low angle region ($\varepsilon = 2\theta < 10°$) arises from the presence of inhomogeneities within the material being examined (such as small clusters of solute atoms), where these inhomogeneities have dimensions only 10 to 100 times the wavelength of the incident radiation. The origin of the scattering can be attributed to the differences in electron density between the heterogeneous regions and the surrounding matrix*, so that precipitated particles afford the most common source of scattering; other heterogeneities such as dislocations, vacancies and cavities must also give rise to some small angle scattering, but the intensity of the scattered beam will be much weaker than that from precipitated particles. The experimental arrangement suitable for this type of study is shown in *Figure 2.10(b)*.

Interpretation of much of the small angle scatter data is based on the approximate formula derived by Guinier,

$$I = Mn^2 I_e \exp\left[-4\pi^2 \varepsilon^2 R^2 / 3\lambda^2\right] \tag{2.11}$$

where M is the number of scattering aggregates, or particles, in the sample, n represents the difference in number of electrons between the particle and an equal volume of the surrounding matrix, R is the radius of gyration of the particle, I_e is the intensity scattered by an electron, ε is the angle of scattering and λ is the wavelength of x-rays. From this equation it can be seen that the intensity of small angle scattering is zero if the inhomogeneity, or cluster, has an electron density equivalent to that of the surrounding matrix, even if it has quite different crystal structure. On a plot of $\log_{10} I$ as a function of ε^2, the slope near the origin, $\varepsilon = 0$, is given by

$$P = -(4\pi^2/3\lambda^2) R^2 \log_{10} e \tag{2.12}$$

which for Cu $K\alpha$ radiation gives the radius of gyration of the scattering aggregate to be

$$R = 0.645 \times P^{1/2} \text{ Å} \tag{2.13}$$

It is clear that the technique is ideal for studying regions of the lattice where segregation on too fine a scale to be observable in the optical microscope has occurred, e.g. the early stages of phase precipitation (*see* Chapter 11), and the aggregation of lattice defects (*see* Chapter 9).

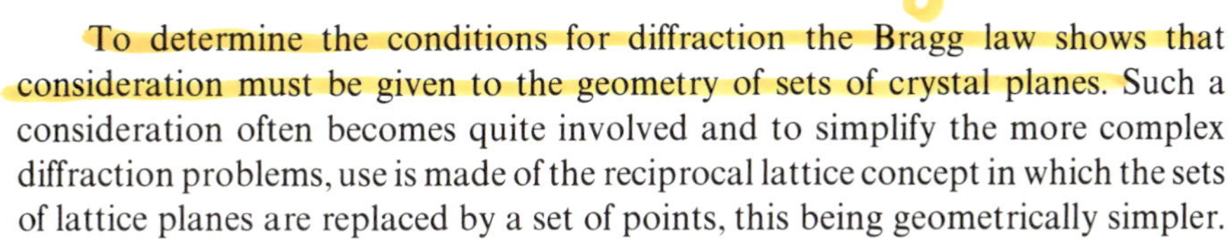

The reciprocal lattice and the reflection sphere

To determine the conditions for diffraction the Bragg law shows that consideration must be given to the geometry of sets of crystal planes. Such a consideration often becomes quite involved and to simplify the more complex diffraction problems, use is made of the reciprocal lattice concept in which the sets of lattice planes are replaced by a set of points, this being geometrically simpler.

The reciprocal lattice is constructed from the real lattice by drawing a line from the origin normal to the lattice plane *hkl* under consideration of length, d^*, equal to the reciprocal of the interplanar spacing d_{hkl}. The construction of part of the reciprocal lattice from a face-centred cubic crystal lattice is shown in *Figure*

* The halo around the moon seen on a clear frosty night is the best example obtained without special apparatus, of the scattering of light at small angles by small particles.

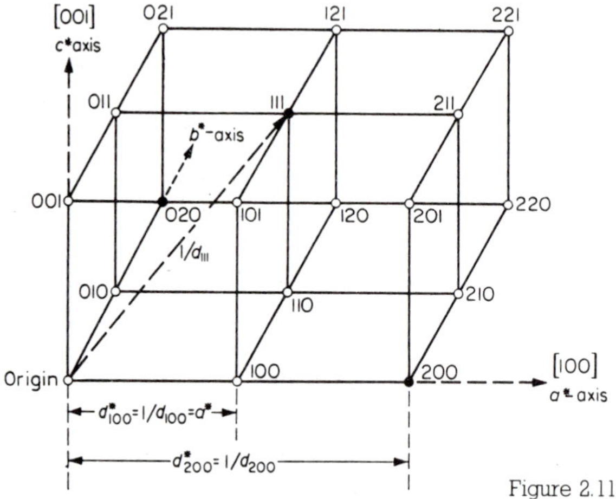

Figure 2.11 F.C.C. reciprocal lattice

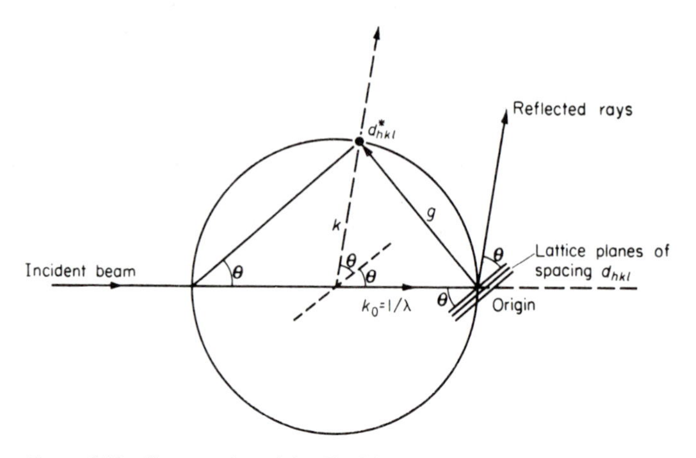

Figure 2.12 Construction of the Ewald reflecting sphere

2.11. Included in the reciprocal lattice are the points which correspond not only to the true lattice planes with Miller indices (*hkl*) but also to the fictitious planes (*nh, nk, nl*) which give possible x-ray reflections. The reciprocal lattice therefore corresponds to the diffraction spectrum possible from a particular crystal lattice and, since a particular lattice type is characterized by 'absent' reflections the corresponding spots in the reciprocal lattice will also be missing. It can be deduced that a f.c.c. Bravais lattice is equivalent to a b.c.c. reciprocal lattice, and vice versa.

A simple geometrical construction using the reciprocal lattice gives the conditions that correspond to Bragg reflection. Thus, if a beam of wavelength λ is incident on the origin of the reciprocal lattice, then a sphere of radius $1/\lambda$ drawn through the origin will intersect those points which correspond to the reflecting planes of a stationary crystal. This can be seen from *Figure 2.12* in which the reflecting plane *AB* has a reciprocal point at d^*. If d^* lies on the surface of the sphere of radius $1/\lambda$ then

$$d^* = 1/d_{hkl} = 2 \sin \theta/\lambda \tag{2.14}$$

and the Bragg law is satisfied; the line joining the origin to the operating reciprocal lattice spot is usually referred to as the g-vector. It will be evident that at any one setting of the crystal, few, if any, points will touch the sphere of reflection. This is the condition for a stationary single crystal and a monochromatic beam of x-rays, when the Bragg law is not obeyed except by chance. To ensure that the Bragg law is satisfied the crystal has to be rotated in the beam, since this corresponds to a rotation of the reciprocal lattice about the origin when each point must pass through the reflection surface. The corresponding reflecting plane reflects twice per revolution.

To illustrate this feature let us re-examine the powder method. In the powder specimen the number of crystals is sufficiently large that all possible orientations are present and in terms of the reciprocal lattice construction we may suppose that the reciprocal lattice is rotated about the origin in all possible directions. The locus of any one lattice point during such a rotation is of course a sphere. This locus-sphere will intersect the sphere of reflection in a small circle about the axis of the incident beam as shown in *Figure 2.13*, and any line joining the centre of the reflection sphere to a point on this small circle is a possible direction for a diffraction maximum. This small circle corresponds to the powder halo discussed previously. From *Figure 2.13* it can be seen that the radius of the sphere describing the locus of the reciprocal lattice point (hkl) is $1/d_{(hkl)}$ and that the angle of deviation of the diffracted beam 2θ is given by the relation

$$(2/\lambda) \sin \theta = 1/d_{(hkl)}$$

which is the Bragg condition.

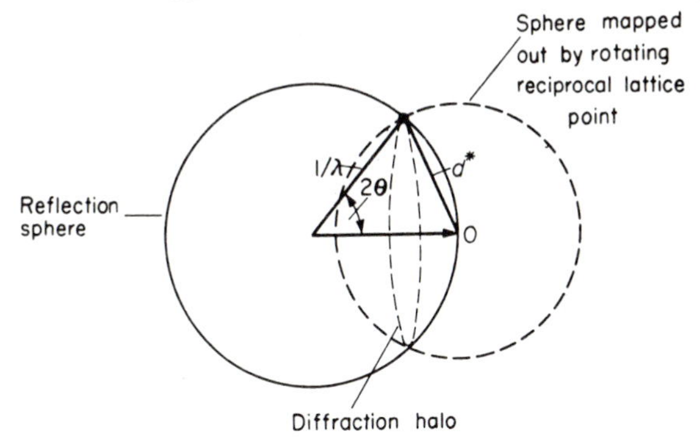

Figure 2.13 Principle of the powder method

X-ray topography

With x-rays it is possible to study individual crystal defects by detecting the differences in intensity diffracted by regions of the crystal near dislocations, for example, and more nearly perfect regions of the crystal. *Figure 2.14(a)* shows the experimental arrangement schematically in which collimated monochromatic $K\alpha$-radiation and photographic recording is used.

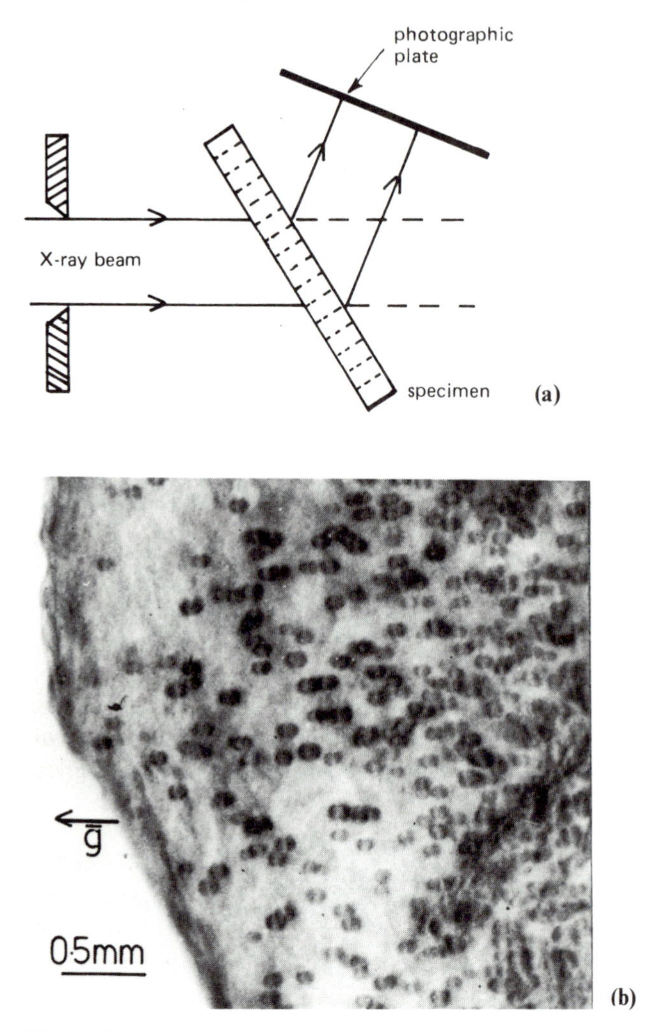

Figure 2.14 (a) Geometry of x-ray topographic technique, (b) topograph from a magnesium single crystal showing dislocation loops. $g = 01\bar{1}0$

Any imperfections give rise to local changes in diffracted or transmitted x-ray intensities and, consequently, dislocations show up as bands of contrast, some 5 to 50 μm wide. No magnification is used in recording the diffraction image, but subsequent magnification of up to 500 times may be achieved with high resolution x-ray emulsions. Large areas of the crystal to thicknesses of 10–100 μm can be mapped using scanning techniques provided the dislocation density is not too high ($\ngtr 10^4$ mm^{-2}).

The x-ray method of detecting lattice defects suffers from the general limitations that the resolution is low and exposure times are long (12 hours) although very high intensity x-ray sources are now available from synchrotrons and are being used increasingly with very short exposure times (~minutes). By comparison, the thin-film electron microscopy method (*see* page 51) is capable of revealing dislocations with a much higher resolution because the dislocation image width is 10 nm or less and magnifications up to 100 000 times are possible. The x-ray method does, however, have the great advantage of being able to reveal dislocations in crystals which are comparatively thick ($\sim$1 mm, cf. $\sim$100 nm (0.1 μm) in foils suitable for transmission electron microscopy). The technique has been used for studying in detail the nature of dislocations in thick single crystals with very low dislocation densities, such as found in semi-conducting materials; *Figure 2.14(b)* shows an example of an x-ray topograph revealing dislocations in magnesium by this technique.

2.3.2 Neutron diffraction

The advent of nuclear piles, or reactors, has stimulated the application of neutron diffraction to those problems of physical metallurgy which cannot be solved satisfactorily by other diffraction techniques. In a conventional pile the fast neutrons produced by fission are slowed down by repeated collisions with a 'moderator' of graphite or heavy water until they are slow enough to produce further fission. If a collimator is inserted into the pile, some of these slow neutrons* will emerge from it in the form of a beam, and the equivalent wavelength λ of this neutron beam of energy E in electron-volts is given by

$$\lambda = (0.081/E)\ \text{Å} \tag{2.15}$$

The equilibrium temperature in a pile is usually in the range 0–100 °C, which corresponds to a peak energy of several hundredths of an electron-volt. The corresponding wavelength of the neutron beam is about 0.15 nm (1.5 Å) and, since this is very similar to the wavelength of x-rays, it is to be expected that thermal neutrons will be diffracted by crystals.

The properties of x-ray and neutron beams differ in many respects. The distribution of energy among the neutrons in the beam approximately follows the Maxwellian curve appropriate to the equilibrium temperature and, consequently, there is nothing which corresponds to characteristic radiation. The neutron beam is analogous to a beam of white x-rays, and as a result it has to be monochromatized before it can be used in neutron crystallography. Then, because only about 1 in 10^3 of the neutrons in the originally weak collimated beam are reflected from the monochromator, it is necessary to employ very wide beams several inches in cross-section to achieve a sufficiently high counting rate in the boron-trifluoride counter detector (photographic detection is possible but not generally useful). In consequence, neutron spectrometers, although similar in principle to x-ray diffractometers, have to be constructed on a massive scale.

* These may be called 'thermal' neutrons because they are in thermal equilibrium with their surroundings.

Neutron beams do, however, have advantages over x-rays or electrons, and one of these is the extremely low absorption of thermal neutrons by most elements. *Table 2.1* shows that even in the most highly absorbent elements (e.g. lithium, boron, cadmium and gadolinium) the mass absorption coefficients are only of the same order as those for most elements for a comparable x-ray wavelength, and for other elements the neutron absorption is very much less indeed. This penetrative property of the neutron beam presents a wide scope for neutron crystallography, since the whole body of a specimen may be examined and not merely its surface. Problems concerned with preferred orientation, internal stresses, cavitation and structural defects are but a few of the possible applications, some of which are discussed more fully later.

TABLE 2.1 X-ray and neutron mass absorption coefficients

Element	*At. No.*	*X-rays* ($\lambda=0.19$ *nm*)	*Neutrons* ($\lambda=0.18$ *nm*)
Li	3	1.5	5.8
B	5	5.8	38.4
C	6	10.7	0.002
Al	13	92.8	0.005
Fe	26	72.8	0.026
Cu	29	98.8	0.03
Ag	47	402	0.3
Cd	48	417	13.0
Gd	61	199	183.0
Au	79	390	0.29
Pb	82	429	0.0006

Another difference is concerned with the intensity of scattering per atom, I_a. For x-rays, where the scattering is by electrons, the intensity I_a increases with atomic number and is proportional to the square of the atomic-form factor. For neutrons, where the scattering is chiefly by the nucleus, I_a appears, to the metallurgist at least, to be quite unpredictable. Not only does the scattering power per atom vary apparently at random from atom to atom, but also from isotope to isotope of the same atom. Moreover, the nuclear component to the scattering does not decrease with increasing angle, as it does with x-rays, because the nucleus which causes the scattering is about 10^{-12} mm in size compared with 10^{-7} mm, which is the size of the electron cloud which scatters x-rays. *Table 2.2* gives some of the scattering amplitudes for x-rays and thermal neutrons.

The fundamental difference in the origin of scattering between x-rays and neutrons affords a method of studying structures, such as hydrides and carbides, which contain both heavy and light atoms. When x-rays are used, the weak intensity contributions of the light atoms are swamped by those from the heavy atoms, but when neutrons are used, the scattering power of all atoms is roughly of the same order. Similarly, structures made up of atoms whose atomic numbers are nearly the same (e.g. iron and cobalt, or copper and zinc), can be studied

TABLE 2.2 Scattering amplitudes for x-rays and thermal neutrons

Element	*At. No.*	*Scattering amplitudes* *X-rays for* $\sin\theta/\lambda=0.5$ $\times 10^{-12}$		*Neutrons** $\times 10^{-12}$
H	1	0.02		−0.4
Li	3	0.28	Li^6	0.7
			Li^7	−0.25
C	6	0.48		0.64
N	7	0.54		0.85
O	8	0.62		0.58
Al	13	1.55		0.35
Ti	22	2.68		−0.38
Fe	26	3.27	Fe^{56}	1.0
			Fe^{57}	0.23
Co	27	3.42		0.28
Cu	29	3.75		0.76
Zn	30	3.92		0.59
Ag	47	6.71	Ag^{107}	0.83
			Ag^{109}	0.43
Au	79	12.37		0.75

* The negative sign indicates the scattered and incident waves are in phase for certain isotopes and hence for certain elements. Usually the scattered wave from an atom is 180° out of phase with the incident wave.

more easily by using neutrons. This aspect is discussed later in relation to the behaviour of ordered alloy phases.

The major contribution to the scattering power arises from the nuclear component, but there is also an electronic (magnetic spin) component to the scattering. This arises from the interaction between the magnetic moment of the neutron and any resultant magnetic moment which the atom might possess. As a result, the neutron diffraction pattern from paramagnetic materials, where the atomic moments are randomly directed (*see* Chapter 5), shows a broad diffuse background, due to incoherent (magnetic) scattering, superimposed on the sharp peaks which arise from coherent (nuclear) scattering. In ferromagnetic metals the atomic moments are in parallel alignment throughout a domain, so that this cause of incoherent scattering is absent. In some materials (e.g. NiO or FeO) an alignment of the spins takes place, but, in this case, the magnetization directions of neighbouring pairs of atoms in the structure are opposed and, in consequence, cancel each other out. For these materials, termed anti-ferromagnetic, there is no net spontaneous magnetization and neutron diffraction is a necessary and important tool for investigating their behaviour (*see* Chapter 5).

2.4 Electron metallography

Equation 2.1 shows that to increase the resolving power of a microscope it is necessary to employ shorter wavelengths. Radiation of wavelengths in the range 250–280 nm can be provided by metal arc lamps but all optical components which transmit the light must be made of quartz because of the high absorption of ultra-violet radiation by glass. A difficulty which exists in the use of such

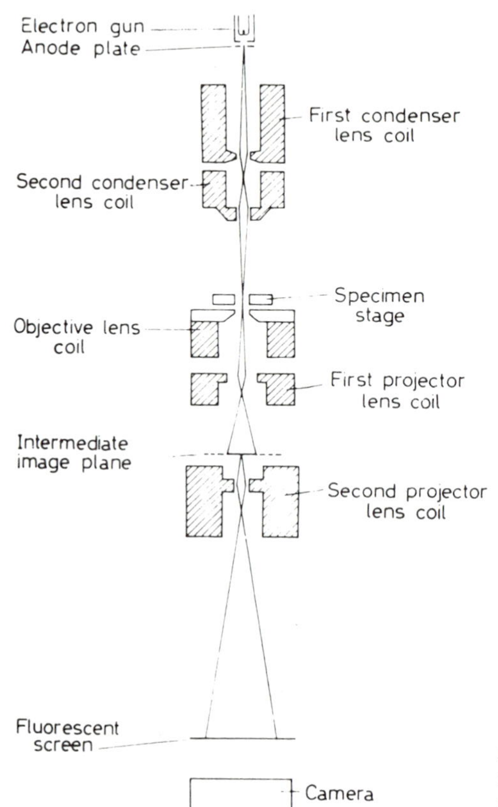

Figure 2.15 Schematic arrangement of a basic electron microscope system

apparatus is in focusing and, as a consequence, in some of the modern equipment the objective may be of the reflecting type, in which case the specimen may be focused in white light and photographed in ultra-violet. However, even with such complex arrangements the resolution can hardly exceed 100 nm, and it is for this reason that the electron microscope has been developed to allow the observation of structures which have dimensions down to as little as 1 nm.

An electron microscope consists of an electron gun and an assembly of lenses all enclosed in an evacuated column, as shown schematically in *Figure 2.15*. The optical arrangement is similar to that of the glass lenses in a projection-type light microscope, although it is customary to use three stages of magnification in the electron microscope. The lenses are usually of the magnetic type, i.e. current-carrying coils which are completely surrounded by a soft iron shroud except for a narrow gap in the bore, energized by d.c. current and, unlike the lenses in a light microscope, which have fixed focal lengths, the focal length can be controlled by regulating the current through the coils of the lens. This facility compensates for the fact that it is difficult to move the large magnetic lenses in the evacuated column of the electron microscope in an analogous manner to the glass lenses in a light microscope.

The condenser lenses are concerned with collimating the electron beam and illuminating the object, which is placed in the bore of the objective lens. The function of the objective lens is to form a magnified image of up to about 40 times in the object plane of the intermediate, or first projector lens. A small part of this image then forms the object for the first projector lens, which gives a second image, again magnified up to about 40 times, in the object plane of the second projector lens. The second projector lens is capable of enlarging this image a further 50 times to form a final image on the fluorescent viewing screen. This image, magnified up to 100 000 times, may also be recorded on a photographic plate placed beneath the viewing screen.

It will be recalled from Chapter 1 that the motion of a stream of electrons can be accounted for by assigning to the radiation a wavelength λ given by the equation $\lambda = \boldsymbol{h}/mv$, where $\boldsymbol{h}$ is Planck's constant and mv is the momentum of the electron. The electron wavelength is inversely proportional to the velocity, and hence to the voltage applied to the electron gun, according to the approximate relation

$$\lambda = \sqrt{(150/\mathrm{V})}\ \text{Å} \tag{2.16}$$

and, since normal operating voltages are between 50 and 100 kV, the value of λ used varies from 0.054 Å to 0.035 Å. With a wavelength of 0.05 Å, if one could obtain a value of $(\mu \sin \alpha)$ for electron lenses comparable to that for optical lenses, i.e. 1.4, it would be possible to see the orbital electrons. However, magnetic lenses are more prone to spherical and chromatic aberration than glass lenses and, in consequence, small apertures, which correspond to α-values of about 0.002 radians, must be used. As a result, the optimum resolution of the electron microscope is limited to less than 1 nm. It will be appreciated, of course, that a variable magnification is possible in the electron microscope without relative movement of the lenses, as in a light microscope, because the depth of focus of each image, being inversely proportional to the square of the numerical aperture, is so great.

Although the examination of metals may be carried out with the electron beam, impinging on the metal surface at a 'glancing incidence', most electron microscopes are aligned for the use of a transmission technique, since added information on the interior of the specimen may be obtained. In consequence, the thickness of the metal specimen has to be limited to below a micrometre, because of the restricted penetration power of the electrons. Three methods now in general use for preparing such thin films are (i) chemical thinning, (ii) electropolishing, and (iii) bombarding with a beam of ions at a potential of about 3 kV. Chemical thinning has the disadvantage of preferentially attacking either the matrix or the precipitated phases, and so the electropolishing technique is used extensively to prepare thin metal foils. Ion beam thinning is quite slow but is the only way of preparing thin ceramic and semiconducting specimens.

2.4.1 Transmission electron microscopy

Transmission electron microscopy provides both image and diffraction information from the same small volume down to 1 μm in diameter. Ray diagrams for the two modes of operation, imaging and diffraction, are shown in

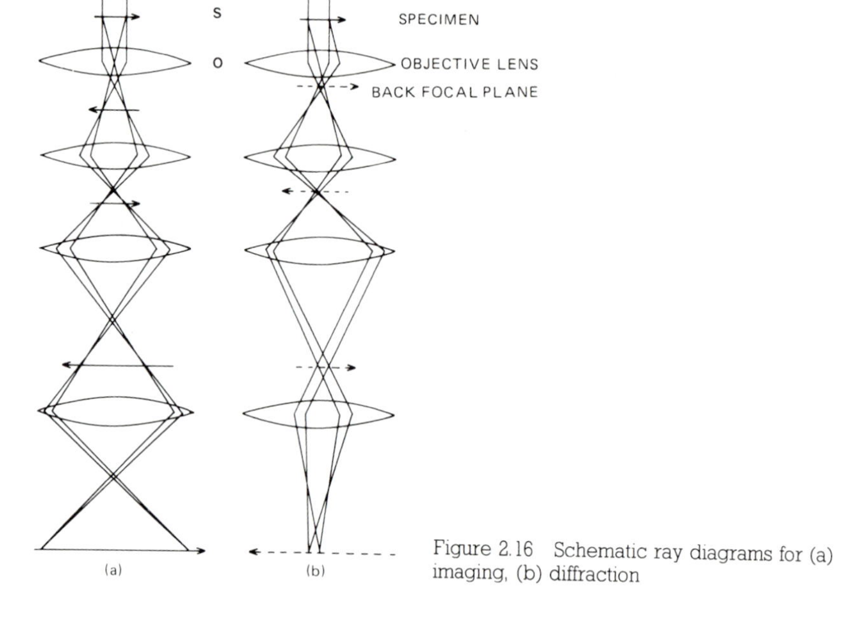

Figure 2.16 Schematic ray diagrams for (a) imaging, (b) diffraction

Figure 2.16(*a*) and (*b*). Diffraction contrast* is the most common technique used and, as shown in *Figure 2.17*(*a*), involves the insertion of an objective aperture in the back focal plane, i.e. in the plane in which the diffraction pattern is formed to select either the directly transmitted beam or a strong diffracted beam. Images obtained in this way cannot possibly contain information concerning the periodicity of the crystal, since this information is contained in the spacing of diffraction maxima and the directions of diffracted beams; information excluded by the objective aperture.

Variations in intensity of the selected beam is the only information in such images since the image is simply a magnified image of the selected spot. Imaging carried out by selecting one beam as in TEM is unusual and images cannot be interpreted simply as high magnification images of periodic objects. In formulating a suitable theory (*see* Chapter 7) it is necessary to consider what factors can influence the intensity of the directly transmitted beam and the intensity of diffracted beams. The obvious factors are (i) local changes in scattering factor, e.g. particles of heavy metal in light metal matrix, (ii) local changes in thickness, and (iii) local changes in orientation of the specimen, or discontinuities in the crystal planes which give rise to the diffracted beams. Fortunately the interpretation of any intensity changes is relatively straightforward if it is assumed that there is only one strong diffracted beam excited. Moreover, since this can be achieved easily experimentally, by orienting the crystal such that strong diffraction occurs from only one set of crystal planes,

* Another imaging mode does allow more than one beam to interfere in the image plane and hence crystal periodicity can be observed; the larger the collection angle, which is generally limited by lens aberrations, the smaller the periodicity that can be resolved. Interpretation of this direct lattice imaging mode, while apparently straightforward is still controversial, and will not be covered here (*see* Chapter 7).

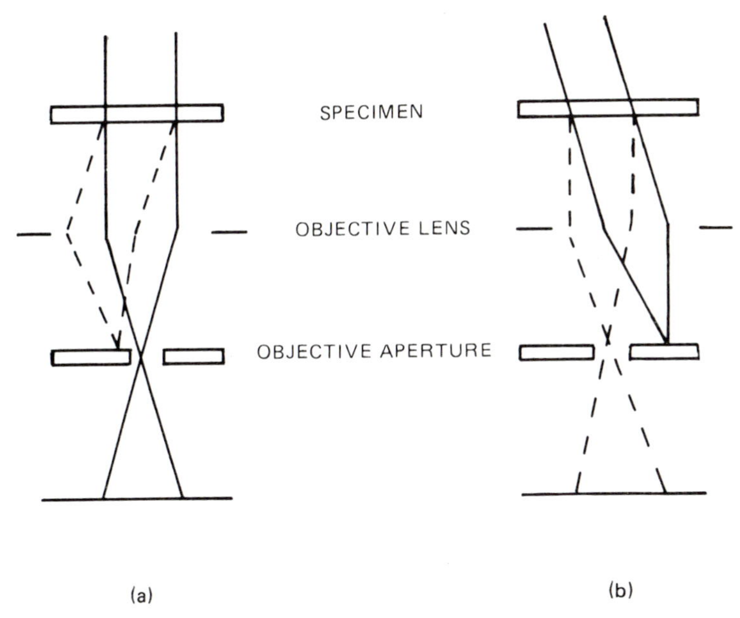

Figure 2.17 Schematic diagram illustrating (a) bright-field, (b) dark-field, image formation

virtually all TEM is carried out with a two-beam condition: a direct and a diffracted beam. When the direct, or transmitted, beam only is allowed to contribute to the final image by inserting a small aperture in the back focal plane to block the strongly diffracted ray, then contrast is shown on a bright background and is known as bright-field imaging. If the diffracted ray only is allowed through the aperture by tilting the incident beam then contrast on a dark background is observed and is known as dark-field imaging. These two arrangements are shown in *Figure 2.17*.

A dislocation or any other crystal defect can be seen in the electron microscope because it locally changes the orientation of the crystal thereby altering the diffracted intensity. This is illustrated in *Figure 2.18*; that part of a grain or crystal which is not oriented at the Bragg angle, i.e. $\theta > \theta_B$, is not strongly diffracting electrons. However, in the vicinity of the dislocation the lattice planes are tilted such that locally the Bragg law is satisfied and strong diffraction arises from near the defect. These diffracted rays are blocked by the objective aperture and prevented from contributing to the final image. The dislocation therefore appears as a dark line (where electrons have been removed) on a bright background in the bright-field picture.

The success of transmission electron microscopy (TEM) is due, to a great extent, to the fact that it is possible to define the diffraction conditions which give rise to the dislocation contrast by obtaining a diffraction pattern from the same small volume of crystal (as small as 1 μm diameter) as that from which the electron micrograph is taken. Thus, it is possible to obtain the crystallographic and diffraction information necessary to interpret electron micrographs. To

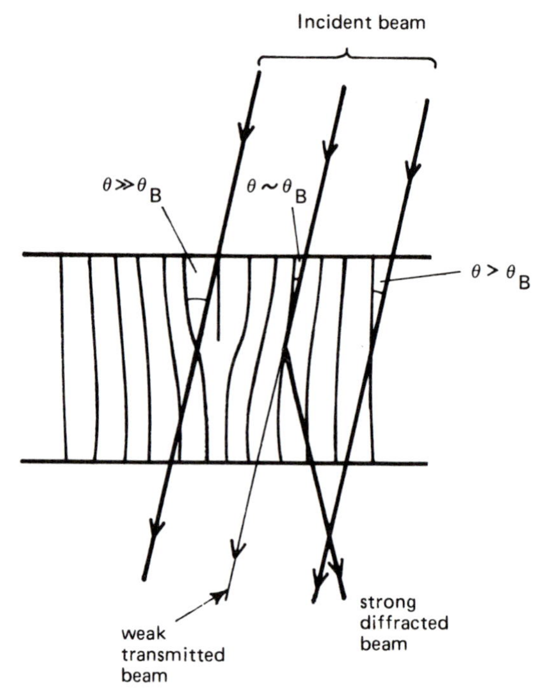

Figure 2.18 Mechanism of diffraction contrast: the planes to the RHS of the dislocation are bent so that they closely approach the Bragg condition and the intensity of the direct beam emerging from the crystal is therefore reduced

obtain a selected area diffraction pattern (SAD) an aperture is inserted in the plane of the first image so that only that part of the specimen which is imaged within the aperture can contribute to the diffraction pattern. The power of the diffraction lens is then reduced so that the back focal plane of the objective is imaged, and then the diffraction pattern, which is focused in this plane, can be seen after the objective aperture is removed.

The usual type of transmission electron diffraction pattern from a single crystal region is a cross-grating pattern of the form shown in *Figure 2.19*. The simple explanation of the pattern can be given by considering the reciprocal lattice and reflecting sphere construction commonly used in x-ray diffraction. In electron diffraction the electron wavelength is extremely short ($\lambda = 0.037$ Å at 100 kV) so that the radius of the Ewald reflecting sphere is about 25 Å^{-1}, which is about 50 times greater than g, the reciprocal lattice vector. Moreover, because λ is small the Bragg angles are also small (about 10^{-2} radians or $\frac{1}{2}^{\circ}$ for low order reflections) and hence the reflection sphere may be considered as almost planar in this vicinity. If the electron beam is closely parallel to a prominent zone axis of the crystal then several reciprocal points (somewhat extended because of the limited thickness of the foil) will intersect the reflecting sphere, and a projection of the prominent zone in the reciprocal lattice is obtained, i.e. the SAD pattern is really a photograph of a reciprocal lattice section. *Figure 2.19* shows some standard

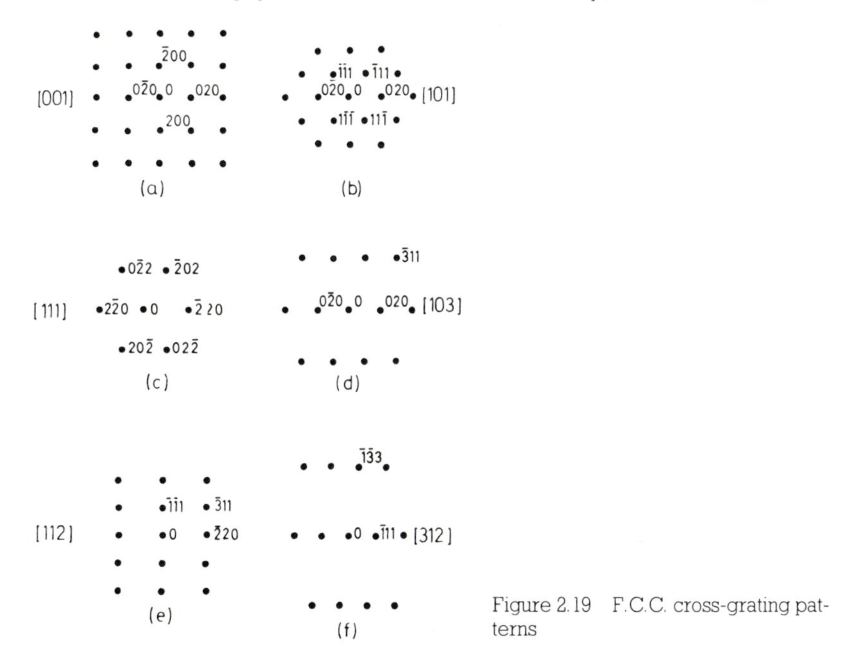

Figure 2.19 F.C.C. cross-grating patterns

cross-grating for face centred cubic crystals. Because the Bragg angle for reflection is small ($\approx \frac{1}{2}°$) only those lattice planes which are almost vertical, i.e. parallel to the direction of the incident electron beam, are capable of Bragg scattering the electrons out of the objective aperture and giving rise to image contrast. Moreover, because the foil is buckled or purposely tilted, only one family of the various sets of approximately vertical lattice planes will diffract strongly and the SAD pattern will then show only the direct beam spot and one strongly diffracted spot (*see* insert *Figure 7.6*). The indices g of the crystal planes (hkl) which are set at the Bragg angle can be obtained from the SAD. Often the planes are near to, but not exactly at the Bragg angle and it is necessary to determine the precise deviation which is usually represented by the parameter s, as shown in the Ewald sphere construction in *Figure 2.20*. The deviation parameter s is determined from Kikuchi lines, observed in diffraction patterns obtained from somewhat thicker areas of the specimen, which form a pair of bright and dark lines associated with each reflection, spaced $|g|$ apart.

The Kikuchi lines arise from inelastically scattered rays, originating at some point P in the specimen (*see Figure 2.21*), being subsequently Bragg diffracted. Thus, for the set of planes in *Figure 2.21*, those electrons travelling in the directions PQ and PR will be Bragg-diffracted at Q and R and give rise to rays in the directions QQ′ and RR′. Since the electrons in the beam RR′ originate from the scattered ray PR, this beam will be less intense than QQ′, which contains electrons scattered through a smaller angle at P. Because P is a spherical source this rediffraction at points such as Q and R gives rise to cones of rays which, when they intersect the film, approximate to straight lines.

The selection of the diffracting conditions used to image the crystal defects can be controlled using Kikuchi lines. Thus the planes (hkl) are at the Bragg angle

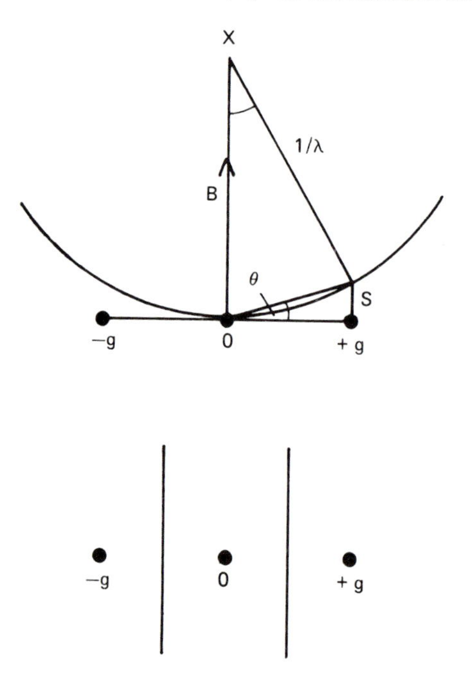

Figure 2.20 Schematic diagram to illustrate the determination of s at the symmetry position, together with associated diffraction pattern

when the corresponding pair of Kikuchi lines passes through 000 and g_{hkl}, i.e. $s = 0$. Tilting of the specimen so that this condition is maintained (which can be done quite simply, using modern double-tilt specimen stages) enables the operator to select a specimen orientation with a close approximation to two-beam conditions. Tilting the specimen to a particular orientation, i.e. electron beam direction, can also be selected using the Kikuchi lines as a 'navigational' aid. The series of Kikuchi lines make up a Kikuchi map, as shown in *Figure 2.21(b)*, which can be used to tilt from one pole to another (as one would use an Underground map).

2.4.2 High-voltage electron microscopy

The most serious limitation of conventional transmission electron microscopes (CTEM) is the limited thickness of specimens examined (50–500 nm). This makes preparation of samples from heavy elements difficult, gives limited containment of particles and other structural features within the specimen, and restricts the study of dynamical processes such as deformation, annealing, etc., within the microscope. However, the usable specimen thickness is a function of the accelerating voltage and can be increased by the use of higher voltages. Because of this, high-voltage microscopes (HVEM) have been built and installed in a number of laboratories.

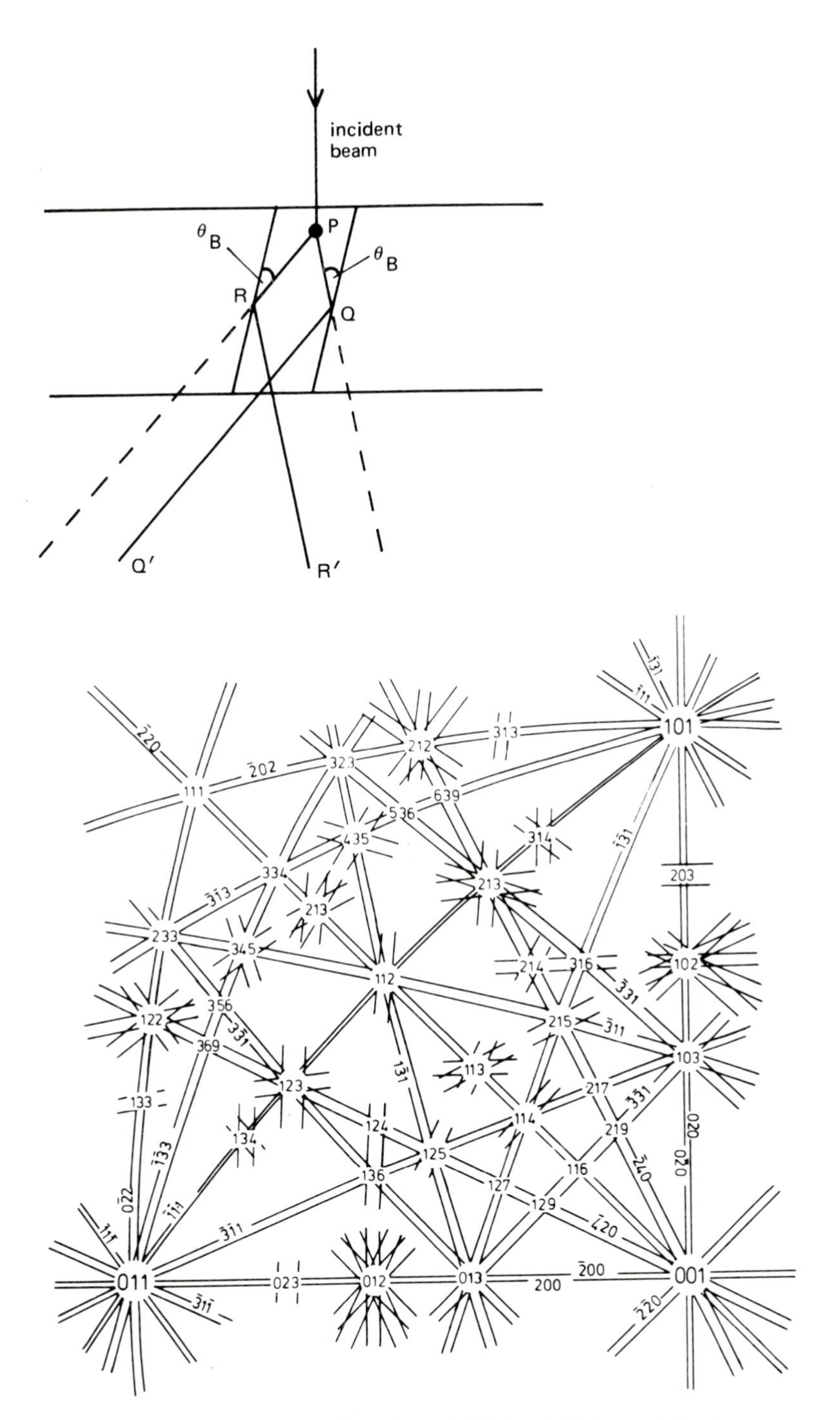

Figure 2.21 Kikuchi lines (a) formation of, (b) from FCC crystal forming a Kikuchi map

The electron wavelength λ decreases rapidly with voltage and at 1000 kV the wavelength $\lambda \approx 0.001$ nm. The decrease in λ produces corresponding decreases in the Bragg angles θ, and hence the Bragg angles at 1000 kV are only about one third of their corresponding values at 100 kV. One consequence of this is that an additional projector lens is usually included in high-voltage microscope. This is often called the diffraction lens and its purpose is to increase the diffraction camera length so that the diffraction spots are more widely spaced on the photographic plate.

The principal advantages of HVEM are: (1) a factor of 5–6 increase in usable foil thickness, (2) a reduced ionization damage rate in ionic, polymer and biological specimens, and the increased space around the specimen. The range of materials is therefore widened and includes (1) materials which are difficult to prepare as thin foils, such as tungsten and uranium, (2) materials which contain large precipitates and are attacked either faster or slower than the matrix materials in the polishing bath, and which consequently give rise to great difficulty at the thicknesses required for 100 kV microscopy, and (3) materials in which the defect being studied is too large to be conveniently included within a 100 kV specimen; these include large voids, precipitates and some dislocation structures such as grain boundaries.

Many processes such as recrystallization, deformation, recovery, martensitic transformation, etc. are dominated by the effects of the specimen surfaces in thin samples and the use of thicker foils enables these phenomena to be studied as they occur in bulk materials. With thicker foils and the increased space inside HVEMs it is possible to design and construct intricate stages which enable the specimen to be cooled, heated, strained and exposed to various chemical environments while it is being looked through. Cooling and heating stages enable a temperature range of $-150\,^{\circ}$C to $1000\,^{\circ}$C to be achieved without difficulty. Such stages have been used to investigate the phenomena mentioned above. Straining stages have been used successfully to study dislocation interactions in different materials. When environmental stages are used the electrons have to penetrate the (usually gaseous) environment without their image information being degraded too much.

A disadvantage of HVEM is that as the beam voltage is raised the energy transferred to the atom by the fast electron increases until it becomes sufficient to eject the atom from its site. The amount of energy transferred from one particle to another in a collision depends on the ratio of the two masses (*see* Chapter 9). Because the electron is very light compared with an atom, the transfer of energy is very inefficient and the electron needs to have several hundred keV before it can transmit the 25 eV or so necessary to displace an atom. To avoid radiation damage it is necessary to keep the beam voltage below the critical displacement value which is ≈ 100 kV for Mg and ≈ 1300 kV for Au. There is much basic scientific interest in radiation damage for technological reasons and the HVEM enables the damage processes to be studied directly.

2.4.3 The scanning electron microscope

The surface structure of a metal can be studied in the TEM by the use of thin

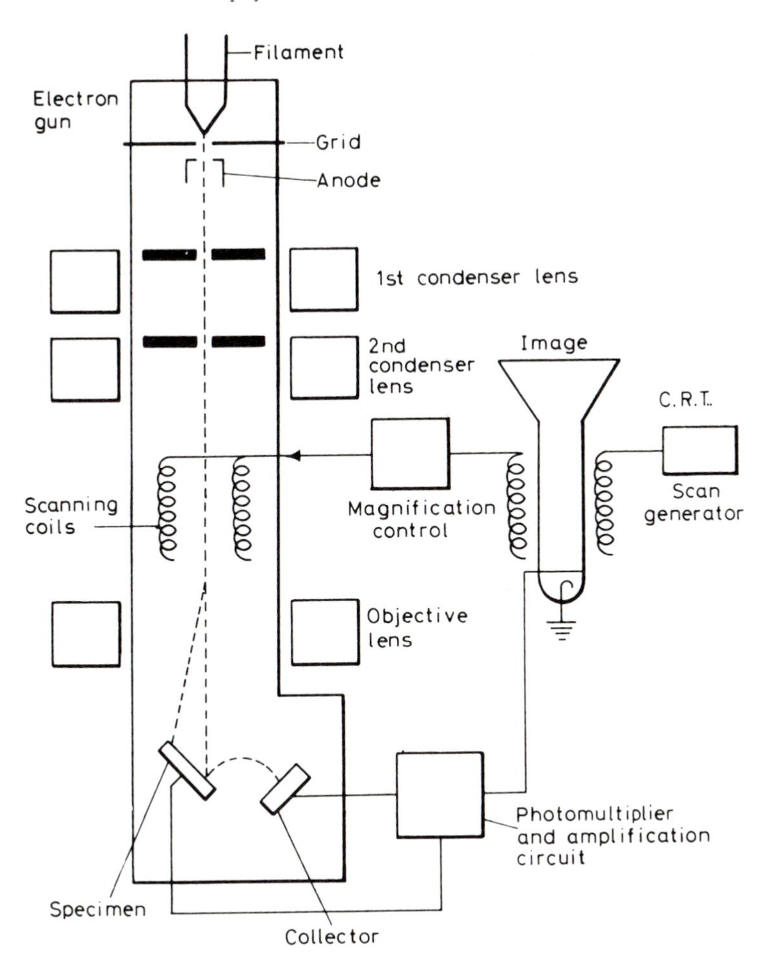

Figure 2.22 Schematic diagram of a scanning electron microscope (Cambridge Instrument Co.)

transparent replicas of the surface topography. Three different types of replica are in use, (i) oxide, (ii) plastic, and (iii) carbon replicas. However, since the development of the scanning electron microscope (SEM) it is very much easier to study the surface structure directly.

A diagram of the SEM is shown in *Figure 2.22*. The electron beam is focused to a spot ≈100 Å diameter and made to scan the surface in a raster. Electrons from the specimen are focused with an electrostatic electrode on to a biased scintillator. The light produced is transmitted via a Perspex light pipe to a photomultiplier and the signal generated is used to modulate the brightness of an oscilloscope spot which traverses a raster in exact synchronism with the electron beam at the specimen surface. The image observed on the oscilloscope screen is similar to the optical image and the specimen is usually tilted towards the collector at a low angle ($<30°$) to the horizontal, for general viewing.

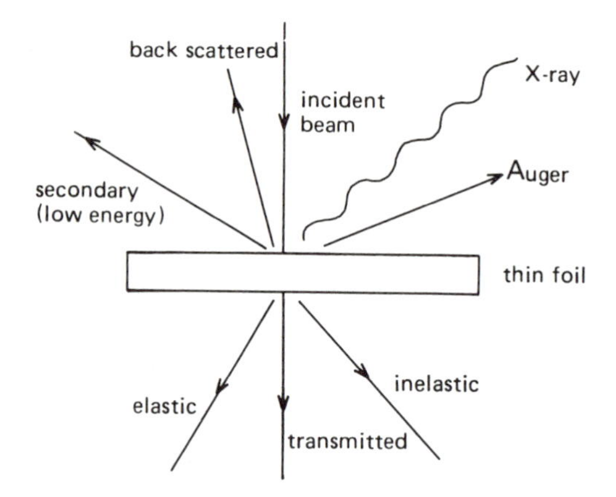

Figure 2.23 Scattering of incident electrons by thin foil. With a bulk specimen the transmitted, elastic and inelastic scattered beams are absorbed

As initially conceived, the SEM used back-scattered electrons (with $E \approx 30\,\mathrm{kV}$ which is the incident energy) and secondary electrons ($E \approx 100\,\mathrm{eV}$) which are ejected from the specimen (*see Figure 2.23*). Since the secondary electrons are of low energy they can be bent round corners and give rise to the topographic contrast. The intensity of back-scattered electrons is proportional to atomic number but contrast from these electrons tends to be swamped because, being of higher energy, they are not so easily collected by the normal collector system used in SEMs. If the secondary electrons are to be collected a positive bias of ≈ 200 V is applied to the grid in front of the detector; if only the back-scattered electrons are to be collected the grid is biased negatively to ≈ 200 V.

Perhaps the most significant development in recent years has been the gathering of information relating to chemical composition. As discussed in section 2.3.1, materials bombarded with high-energy electrons can give rise to the emissions of x-rays characteristic of the material being bombarded. The x-rays emitted when the beam is stopped on a particular region of the specimen may be detected either with a solid state (Li-drifted silicon) detector which produces a voltage pulse proportional to the energy of the incident photons (energy dispersive method) or with an x-ray spectrometer to measure the wavelength and intensity (wavelength dispersive method). The microanalysis of materials is presented in section 2.5. Alternatively, if the beam is scanned as usual and the intensity of the x-ray emission, characteristic of a particular element, is used to modulate the CRT, an image showing the distribution of that element in the sample will result. X-ray images are usually very 'noisy' because the x-ray production efficiency is low, necessitating exposures a thousand times greater than electron images.

Collection of the back-scattered (BS) electrons with a specially located detector on the bottom of the lens system gives rise to some exciting applications and opens up a completely new dimension for SEM from bulk samples. The BS

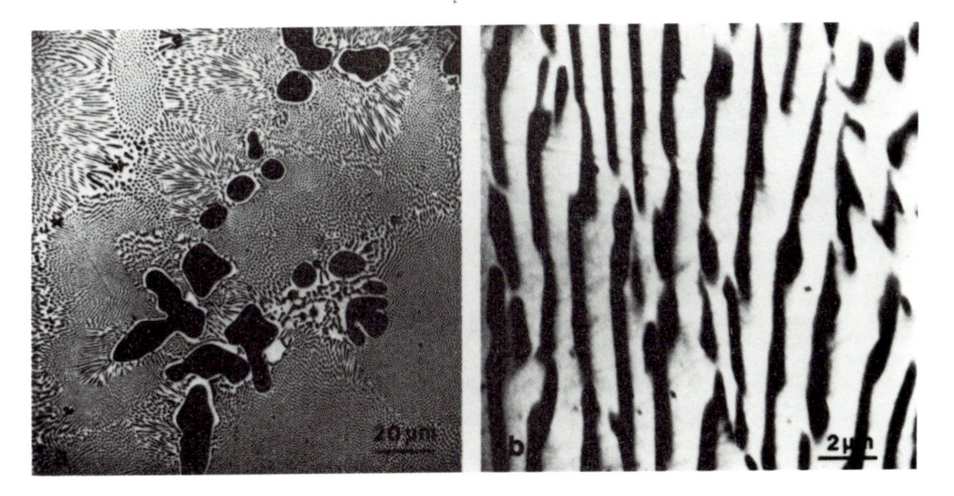

Figure 2.24 Back-scattered electron image by atomic number contrast from Cu–70% Ag alloy showing (a) α-dendrites + eutectic, (b) eutectic (Courtesy, B. W. Hutchinson)

electrons are very sensitive to atomic number and hence are particularly important in showing contrast from changes of composition, as illustrated by the image from a copper alloy in *Figure 2.24*(*a*) and (*b*). This atomic number contrast is particularly effective in studying alloys which normally are difficult to study because they cannot be etched. The intensity of back-scattered electrons is also sensitive to the orientation of the incident beam relative to the crystal. This effect will give rise to 'orientation' contrast from grain to grain in a polycrystalline specimen as the scan crosses several grains. In addition, the effect is also able to provide crystallographic information from bulk specimens by a process known as electron channelling. As the name implies, the electrons are channelled between crystal planes and the amount of channelling per plane depends on its packing and spacing. If the electron beam impinging on a crystal is rocked through a large angle then the amount of channelling will vary with angle and hence the BS image will exhibit contrast in the form of electron channelling patters which can be used to provide crystallographic information. *Figure 2.25* shows the 'orientation' or channelling contrast exhibited by a Fe–3%Si specimen during secondary recrystallization (a process used for transformer lamination production) and the channelling pattern can be analysed to show that the new grain possesses the Goss texture. Electron channelling occurs only in relatively perfect crystals and hence the degradation of electron channelling patterns may be used to monitor the level of plastic strain, for example to map out the plastic zone around a fatigue crack as it develops in an alloy.

The electron beam may also induce electrical effects which are of importance particularly in semiconductor materials. Thus a 30 kV electron beam can generate some thousand excess free electrons and the equivalent number of ions ('holes'), the vast majority of which recombine. In metals this recombination

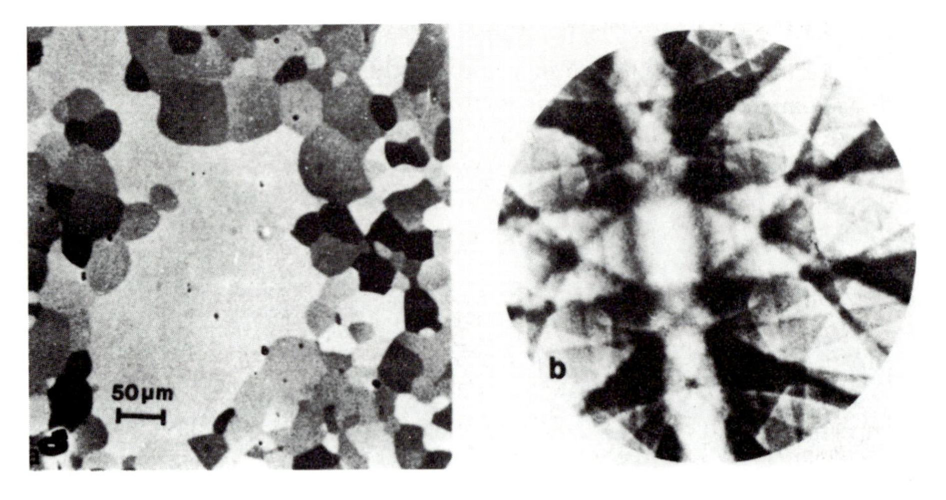

Figure 2.25 (a) Back-scattered electrons image, (b) associated channelling pattern, from secondary recrystallized Fe–3% Si (Courtesy, B. W. Hutchinson)

process is very fast, 10^{-12} s, but in semiconductors may be a few seconds depending on purity. These excess current carriers will have a large effect on the limited conductivity. Also the carriers generated at one point will diffuse towards regions of lower carrier concentration and voltages will be established whenever the carriers encounter regions of different chemical composition, e.g. impurities around dislocations. The conductivity effect can be monitored by applying a potential difference across the specimen from an external battery and using the magnitude of the resulting current to modulate the CRT brightness to give an image of conductivity variation.

The voltage effect arising from different carrier concentrations or from accumulation of charge on an insulator surface or from the application of an external electromotive force can modify the collection of the emitted electrons and hence give rise to voltage contrast. Similarly, a magnetic field arising from ferromagnetic domains, for example, will affect the collection efficiency of emitted electrons and lead to magnetic field contrast.

The secondary electrons, i.e. lightly bound electrons ejected from the specimen, which give topographical information are generated by the incident electrons, by the back-scattered electrons and by x-rays. The resolution is typically ≈ 10 nm at 20 kV for medium atomic weight elements and is limited by spreading of electrons as they penetrate into the specimen. The back-scattered electrons are also influenced by beam spreading and for a material of medium atomic weight the resolution is ≈ 100 nm. The specimen current mode is limited both by spreading of the beam and the noise of electronic amplification to a spatial resolution of 500 nm and somewhat greater values ≈ 1 µm apply to the beam-induced conductivity and x-ray modes. This limitation on spatial resolution is removed in the STEM which uses thin specimens.

2.4.4 Scanning transmission electron microscopy (STEM)

Recently, some of the instrumental features of both CTEM and SEM have been combined in a new type of instrument known as a scanning transmission electron microscope. In STEM, *Figure 2.26(a)*, a small probe is scanned over the specimen and a detector situated below the specimen collects the appropriate transmitted electrons which are used to modulate the intensity of a cathode-ray tube scanned synchronously with the specimen scan. As in SEM the magnification is obtained in STEM by the ratio of the CRT scan to the scan on the specimen instead of using the objective lens (and subsequent lenses) to produce magnified images as in CTEM. The resolution in STEM cannot be better than the size of the incident electron probe and therefore, if a resolution approaching that of a CTEM is to be obtained then it is essential that a high-intensity gun be used so that there are sufficient electrons in such a small probe. Electron sources produced by field emission are appropriate but require ultra-high vacuum for their operation and are therefore expensive. The lack of post-objective lenses can have some disadvantages for studying materials and hence dedicated STEMs have not been so generally used. Instead, STEM attachments are now available which can be fitted to a CTEM so that the microscope can be operated in either mode (*see Figure 2.26(b)*). The CTEM objective lens operates as a condenser lens in STEM and a fine probe is produced at, and scanned across, the surface of the specimen so that the microscope can be operated in the STEM mode. The STEM detector is located usually just above or below the CTEM viewing screen. The lenses below the objective lenses are used simply to transfer the time-dependent signal, which is in the back focal plane of the objective lens, to the detector. These lenses are clearly not functioning as magnifying lenses in the imaging mode – the magnification arising solely from the scan ratio on the CRT and specimen. However, these lenses are used to control the effective size of the STEM detector by altering the camera constant; at small camera constants the detector can collect the direct beam and several diffracted beams but at large camera constants a single beam can be collected.

Diffraction patterns can easily be observed in a CTEM which has a STEM attachment simply by stopping the probe, removing the STEM detector (if it is above the viewing screen), when the diffraction pattern can be observed on the CTEM screen and photographed using the CTEM camera. In a dedicated STEM the diffraction pattern is generated generally by scanning the stationary diffraction pattern formed by the specimen over the STEM detector. A small divergence must be used to obtain good spatial resolution in a diffraction pattern, which is ≈ 2 nm and an improvement of more than two orders of magnitude over that possible in CTEM or HVEM. An example of the spatial resolution of diffraction information available in STEM is shown in *Figure 2.27*. In (a) an area of heavily deformed α-brass imaged in CTEM at 100 kV is shown consisting of small sub-grains, in (b) a SAD pattern obtained using the smallest available aperture, and in (c) a STEM diffraction pattern from a 50 Å diameter region within the area. Clearly, diffraction information from such small areas can be obtained only by STEM microdiffraction.

A major advantage of a STEM microscope over CTEM lies in the ability to

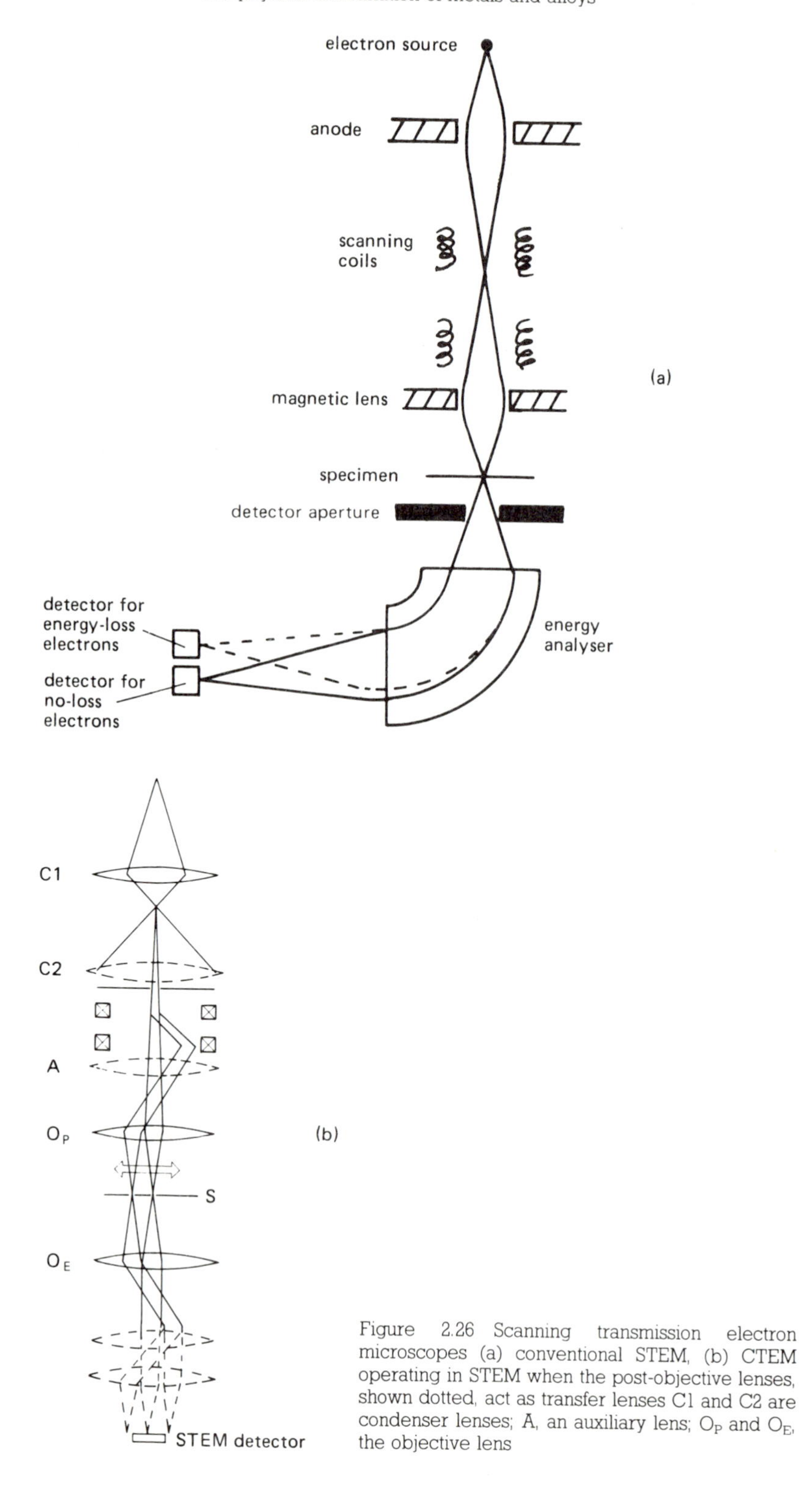

Figure 2.26 Scanning transmission electron microscopes (a) conventional STEM, (b) CTEM operating in STEM when the post-objective lenses, shown dotted, act as transfer lenses C1 and C2 are condenser lenses; A, an auxiliary lens; O_P and O_E, the objective lens

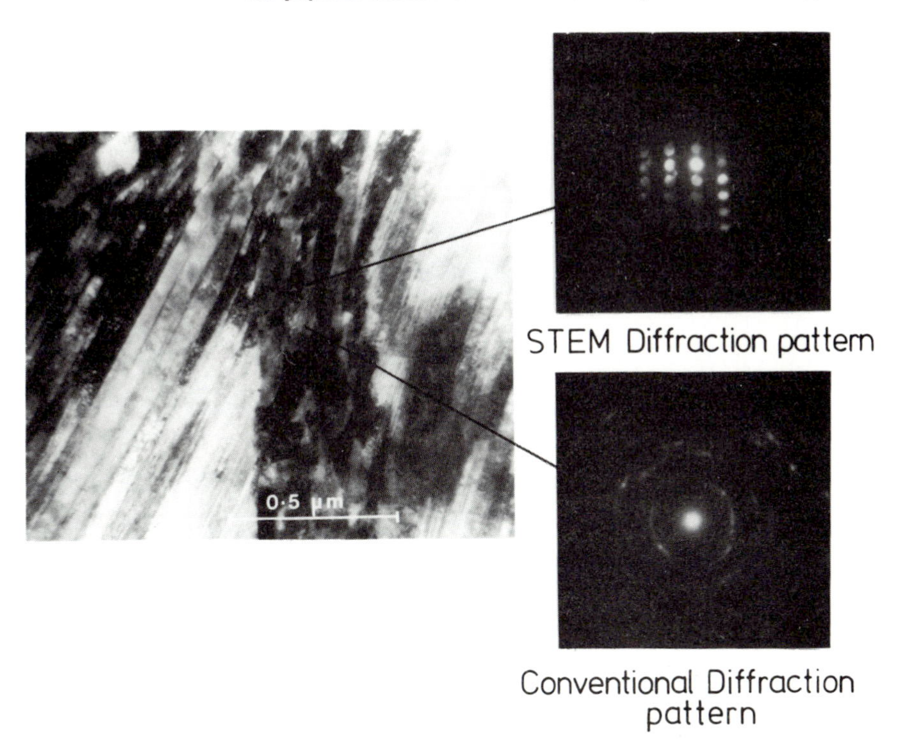

Figure 2.27 Conventional and STEM diffraction pattern from small sub-grains in α-brass

obtain diffraction data from small volumes of crystals. Also since most STEMs are fitted with x-ray detectors and/or electron loss spectrometers it is possible to analyse chemically such small regions simply by stopping the probe on a selected area and collecting the x-rays, and the energy-loss electrons for the light elements. In addition, by deliberately using a large beam divergence, patterns, termed convergent beam diffraction patterns, can be generated which also enable the crystal structure of these very small volumes to be determined. Thus micro-analysis in its broadest sense can be carried out in the STEM mode far more conveniently and with greatly improved spatial accuracy than by any other technique.

2.4.5 Convergent beam diffraction patterns (CBDPs)

A ray diagram illustrating the formation of a convergent beam diffraction pattern is shown in *Figure 2.28*. The discs of intensity which are formed in the back focal plane contain information which is of three types:

1. Fringes within discs formed by strongly diffracted beams. If the crystal is tilted to 2-beam conditions, these fringes can be used to determine the specimen thickness very accurately.

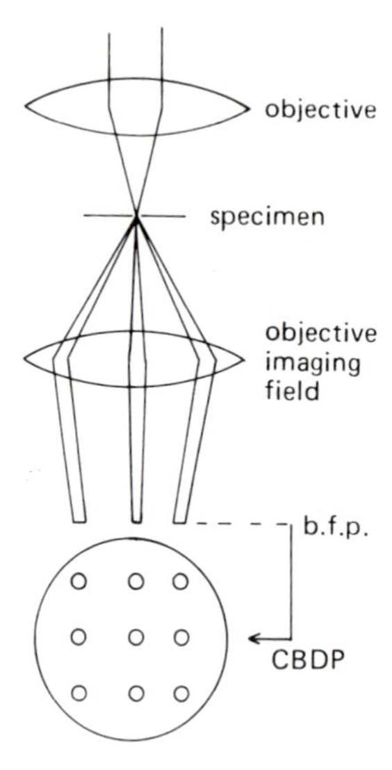

Figure 2.28 Formation of convergent beam diffraction pattern in the back-focal plane of the objective lens

2. High angle information in the form of fine lines (somewhat like Kikuchi lines) which are visible in the direct beam and in the higher order Laue zones (HOLZ). These HOLZ are visible in a pattern covering a large enough angle in reciprocal space. The fine line pattern can be used to measure the lattice parameter to 1 in 10^4. *Figure 2.29(a)* shows an example of HOLZ lines for a silicon crystal centred [111]. Pairing a dark line through the zero-order disc with its corresponding bright line through the higher-order disc allows the lattice parameter to be determined, the distance between the pair being sensitive to the temperature, etc.
3. Detailed structure both within the direct beam and within the diffracted beams which show certain well-defined symmetries when the diffraction pattern is taken precisely along an important zone axis. The patterns can therefore be used to give crystal structure information, particularly the point group and space group. This information, together with the chemical composition from EELS or EDX (see page 68) and the size of the unit cell from the indexed diffraction patterns can be used to define the specific crystal structure, i.e. the atomic positions. *Figure 2.29(b)* indicates the 3-fold symmetry in a CBDP from silicon taken along the [111] axis.

2.5 Microanalysis

Electron probe microanalysis (EPMA) of bulk samples is now a routine technique for obtaining rapid, accurate analysis of alloys. A small electron probe (≈ 100 nm diameter) is used to generate x-rays from a defined area of a polished specimen and the intensity of the various characteristic x-rays measured using either wavelength dispersive spectrometers (WDS) or energy dispersive spectrometers (EDS). Typically the accuracy of the analysis is $\pm 0.1\%$. One of the

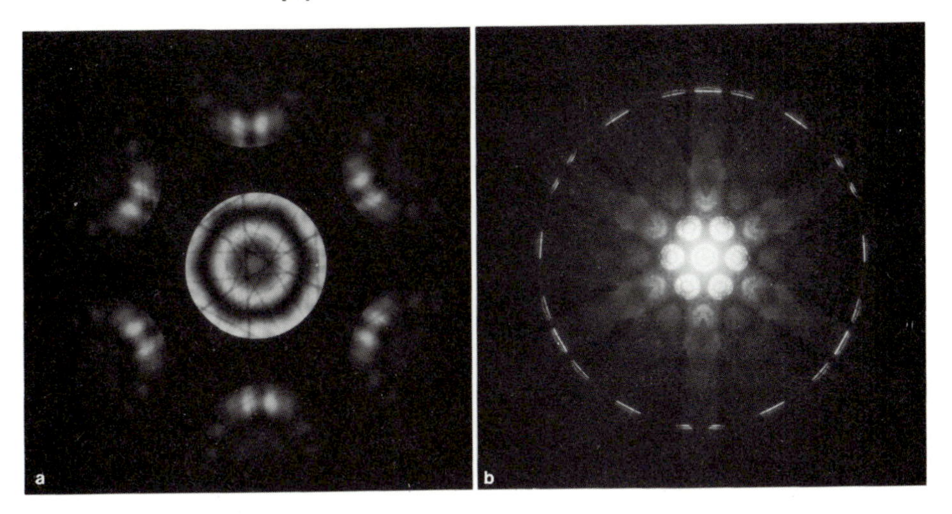

Figure 2.29 <111> CBDPs from Si. (a) Zero layer + HOLZ (Higher Order Laue Zones) in direct beam, (b) zero layer + FOLZ (First Order Laue Zones)

limitations of EPMA of bulk samples is that the volume of the sample which contributes to the x-ray signal is relatively independent of the size of the electron probe, because high-angle elastic scattering of electrons within the sample generates x-rays (*see Figure 2.30*). The consequence of this is that the spatial resolution of EPMA is no better than $\simeq 2\,\mu$m. In the last few years EDX detectors have been interfaced to transmission electron microscopes which are capable of operating in the scanning transmission mode (STEM) with an electron probe as small as 2 nm. The combination of electron transparent samples, in which high-angle elastic scattering is limited, and a small electron probe, leads to a dramatic improvement in the potential spatial resolution of x-ray micro-analysis. In addition, interfacing of energy loss spectrometers has enabled light

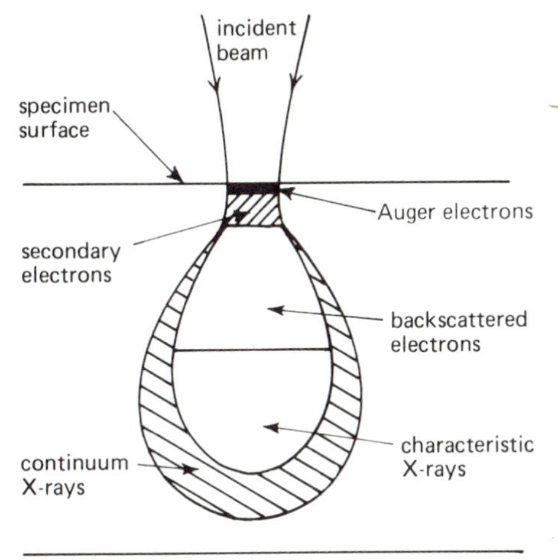

Figure 2.30 Schematic diagram showing the generation of electrons and x-rays within the specimen

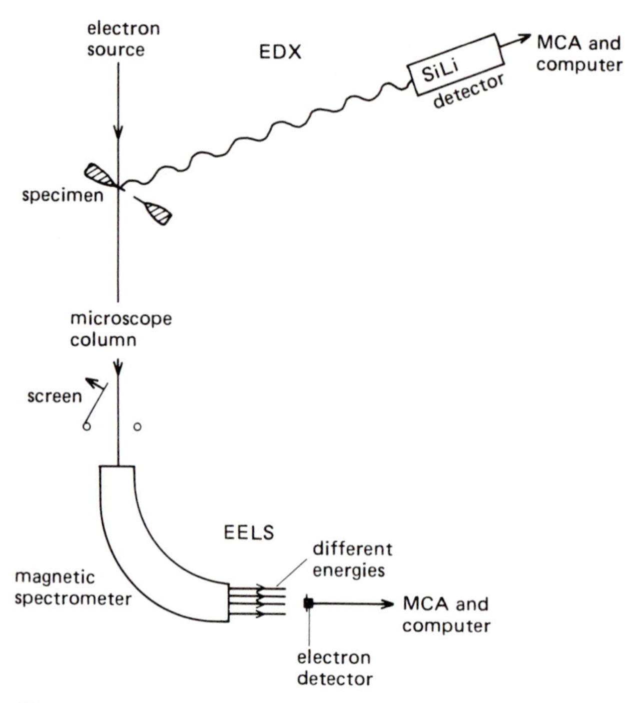

Figure 2.31 Schematic drawing of EDX and EELS in TEM

elements to be detected and measured, so that electron microchemical analysis is now a powerful tool in the characterization of materials (*see Figure 2.31*).

Inelastic scattering

Inelastically scattered electrons are those which lose energy when they interact with matter. The two mechanisms important in chemical analysis are (i) excitation of the electron gas or plasmon scattering, and (ii) single electron scattering.

Plasmon scattering The fast electron excites a ripple in the plasma of free electrons in the solid. The energy of this 'plasmon' depends only on the volume concentration of free electrons n in the solid and is given by $E_p = [ne^2/m]^{1/2}$. Typically E_p, the energy loss suffered by the fast electron is $\approx 15\,\text{eV}$ and the scattering intensity/unit solid angle has an angular half-width given by $\theta_E = E_p/2E_0$, where E_0 is the incident voltage; θ_E is therefore $\approx 10^{-4}$ radians. The energy of the plasmon is converted very quickly into atom vibrations (heat) and the mean free path for plasmon excitation is small, $\approx$50–150 nm.

Single electron scattering Energy may be transferred to single electrons (rather than to the large number $\approx 10^5$, involved in plasmon excitation), by the incident fast electrons. Lightly bound valency electrons may be ejected, and these electrons can be used to form secondary electron images in SEM; a very large number of electrons with energies up to ≈ 50 eV are ejected when a high energy electron beam strikes a solid. The useful collisions are those where the single electron is bound. There is a minimum energy required to remove the single electron, i.e. ionization, but provided the fast electron gives the bound electron more than this minimum amount, it can give the bound electron any amount of energy, up to its own energy (e.g. 100 keV). Thus, instead of the single electron excitation process turning up in the energy loss spectrum of the fast electron as a peak, as happens with plasmon excitation, it turns up as an edge. Typically, the mean free path for inner shell ionization is several micrometres and the energy loss can be several keV. The angular half-width of scattering is given by $\Delta E/2E_0$. Since the energy loss ΔE can vary from ≈ 10 eV to tens of keV the angle can vary upwards from 10^{-4} radians.

X-rays and Auger electrons

A plasmon, once excited, decays to give heat, which is not at all useful. In contrast, an atom which has had an electron removed from it decays in one of two ways, both of which turn out to be very useful in chemical analysis. The first step is the same for both cases. An electron from an outer shell, which therefore has more energy than the removed electron, drops down to fill the hole left by the removal of the bound electron. Its extra energy, equal to the difference in energy between the two levels involved, ΔE and therefore absolutely characteristic of the atom, must be dissipated. This may happen in two ways: (i) by the creation of a photon whose energy, $h\nu$, equals the energy difference ΔE. For electron transition of interest, ΔE, and therefore $h\nu$, is such that the photon is an x-ray (as discussed in section 2.3); (ii) by transferring the energy to a neighbouring electron, which is then ejected from the atom. This is an 'Auger' electron. Its energy when detected will depend on the original energy difference ΔE minus the binding energy of the ejected electron. Thus the energy of the Auger electron depends on three atomic levels rather than two as for emitted photons. The energies of the Auger electrons are sufficiently low that they escape from within only about 5 nm of the surface. This is therefore a surface analysis technique. The ratio of photon–Auger yield is called the fluorescence ratio ω, and depends on the atom and the shells involved. For the K-shell, ω is given by $\omega_k = X_k/(A_k + X_k)$, where X_k and A_k are respectively the number of x-ray photons and Auger electrons emitted. A_k is independent of atomic number Z, and X_k is proportional to Z^4 so that $\omega_k Z^4/(a + Z^4)$, where $a = 1.12 \times 10^6$. Light elements and outer shells (L-lines) have lower yields; for K series transitions ω_k varies from a few per cent for carbon up to $\geqslant 90\%$ for gold.

X-ray and electron energy measurement

With electron beam instrumentation it is required to measure (i) the wavelength or energies of emitted x-rays (WDX and EDX), (ii) the energy losses of the fast electrons (EELS), and (iii) the energies of emitted electrons (AES). Nowadays (i) and (ii) can be carried out on the modern STEM using special detector systems.

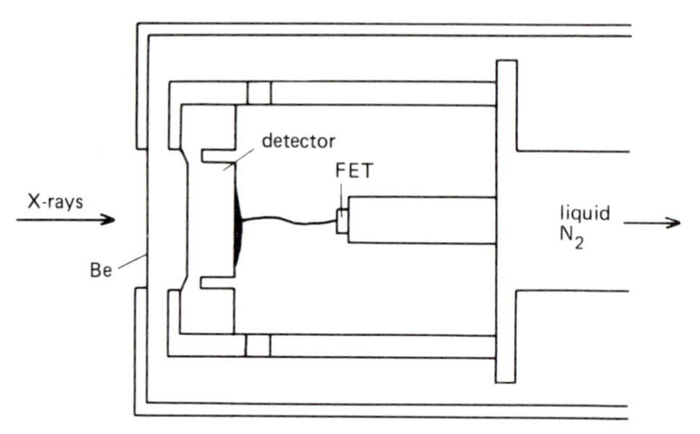

Figure 2.32 Schematic diagram of Si–Li x-ray detector

WDX spectrometer. For a crystal of known d-spacing, x-rays of a specific wavelength, λ, will be diffracted at an angle θ, given by the well-known Bragg equation, $n\lambda = 2d \sin \theta$. Different wavelengths are selected by changing θ and, in order to cover the necessary range of wavelengths, several crystals of different d-spacings can be used successively in a spectrometer. The range of wavelength is 0.1–2.5 nm and the corresponding d-spacing for practicable values of θ, which lie between $\approx 15°$ and $65°$, is achieved by using crystals such as LiF, quartz, mica, etc. In a WDX spectrometer the specimen (which is the x-ray source), a bent crystal of radius $2r$ and the detector all lie on the focusing circle radius r and different wavelength x-rays are collected by the detector by setting the crystal at different angles, θ as discussed on page 41. The operation of the spectrometer is very time consuming since only one particular x-ray wavelength can be focused on to the detector at any one time.

The resolution of WDX spectrometers is controlled in the main by the perfection of the crystal, which influences the range of wavelengths over which the Bragg condition is satisfied, and by the size of the entrance slit to the x-ray detector; taking the resolution ($\Delta\lambda$) to ~ 0.01 Å then $\lambda/\Delta\lambda$ is about 300 which, for a medium atomic weight sample, leads to a peak–background ratio of about 250. The crystal spectrometer normally uses a proportional counter to detect the x-rays, producing an electrical signal by ionization of the gas in the counter. The electrical signal is proportional to the x-ray energy, i.e. inversely proportional to the wavelength. The window of the counter needs to be thin and of low atomic number to minimize x-ray absorption. The output pulse from the counter is amplified and differentiated to produce a short pulse. The time constant of the electrical circuit is of the order of 1 μs which leads to possible count rates of at least 10^5/s.

EDX spectrometer. EDX detectors have developed rapidly in recent years and have replaced WDX detectors on transmission microscopes and are used together with WDX detectors on microprobes and on SEMs. A schematic diagram of a Si–Li detector is shown in *Figure 2.32*. X-rays enter through the thin Be window and produce electron–hole pairs in the Si–Li. Each electron–hole pair requires 3.8 eV, at the operating temperature of the detector, and the number of pairs produced by a photon of energy E_p is thus $E_p/3.8$. The charge produced by a typical x-ray photon is $\approx 10^{-16}$ coulombs, and this is amplified to give a shaped pulse, the height of which is then a measure of the energy of the incident x-ray photon. The data is stored in a multi-channel analyser. Provided that the x-ray photons arrive with a sufficient time interval between them, the energy of each incident photon can be measured and the output presented as an intensity versus energy display. The amplification pulse shaping takes about 50 μs and if a second pulse arrives before the preceding pulse is processed, both pulses are rejected. This results in significant dead time for count rates $\geqslant$4000/s. Since the count rate refers to the count across the whole energy range, typically 0–20 keV, the time for collecting statistically significant data from minor components can be prohibitively long.

The number of electron–hole pairs generated by an x-ray of a given energy is subject to normal statistical fluctuations and this, taken together with electronic noise, limits the energy resolution of a Si–Li detector to about a few hundred eV, which worsens with increase in photon energy. The main advantage of EDX detectors is that simultaneous collection of the whole range of x-rays is possible and the energy characteristics of all the elements (above Na in the periodic table) can be obtained in a matter of seconds. The main disadvantages are the relatively poor resolution, which leads to a peak–background ratio of about 50, and the limited count rate.

The variation in efficiency of a Si–Li detector must be allowed for when quantifying x-ray analysis. At low energies ($\leqslant$1 kV) the x-rays are mostly absorbed in the Be window and at high energies ($\geqslant$20 kV), the x-rays pass through the detector so that the decreasing cross section for electron–hole pair generation results in a reduction in efficiency. The Si–Li detector thus has optimum detection efficiency between about 1 and 20 kV. The equation for detector efficiency is given in terms of I_a, the x-ray intensity arriving at the Be window, and the measured intensity I_m, by

$$I_m = I_a \oint_i^3 \left[\exp\left\{ -\left(\frac{\mu}{\rho}\right)_i \rho_i t_i \right\} \right] \left\{ 1 - \exp\left[-\left(\frac{\mu}{\rho}\right)_{si} \rho_{si} t_{si}^* \right] \right\} \tag{2.17}$$

where $\oint_i^3$ is used to denote the product of the three absorption terms, one each for absorption in the Be window, in the gold contact layer and the silicon dead layer. The final term in curly brackets allows for the production of x-rays not absorbed in the active layer of thickness t_{si}^*. The terms $\exp -(\mu/\rho)$ are mass absorption coefficients of the incident x-rays in the detector.

2.5.1 Electron microanalysis of thin foils

There are several simplifications which arise from the use of thin foils in (S)TEM. The most important of these arises from the fact that the average energy loss which electrons suffer on passing through a thin foil is only about 2%, and this small average loss means that the ionization cross section can be taken as a constant. Thus the number of characteristic x-ray photons generated from a thin sample is given simply by the product of the electron path length and the appropriate cross section Q, i.e. the probability of ejecting the electron, and the fluorescent yield ω. The intensity generated by element A is then given by

$$I_A = iQ\omega n \tag{2.18}$$

where Q is the cross section per cm^2 for the particular ionization event, ω the fluorescent yield, n the number of atoms in the excited volume, and i the current incident on the specimen. Microanalysis is usually carried out under conditions where the current is unknown and interpretation of the analysis simply requires that the ratio of the x-ray intensities from the various elements be obtained. For the simple case of a very thin specimen for which absorption and x-ray fluorescence can be neglected, then the measured x-ray intensity from element A is given by

$$I_A \propto n_A Q_A \omega_A a_A \eta_A \tag{2.19}$$

and for element B by

$$I_B \propto n_B Q_B \omega_B a_B \eta_B \tag{2.20}$$

where n, Q, ω, a and η represent the number of atoms, the ionization cross sections, the fluorescent yields, the fraction of the K line (or L and M) which is collected and the detector efficiencies respectively for elements A and B. Thus in the alloy made up of elements A and B

$$\frac{n_A}{n_B} \propto \frac{I_A Q_B \omega_B a_B \eta_B}{I_B Q_A \omega_A a_A \eta_A} = K_{AB} \frac{I_A}{I_B} \tag{2.21}$$

This equation forms the basis for x-ray microanalysis of thin foils where the constant K_{AB} contains all the factors needed to correct for atomic number differences, and is known as the Z-correction. Thus from the measured intensities, the ratio of the number of atoms A to the number of atoms B, i.e. the concentrations of A and B in the alloy, can be calculated using the computed values for Q, ω, η, etc. A simple spectrum for stoichiometric NiAl is shown in *Figure 2.33* and the values of I_K^{Al} and I_K^{Ni}, obtained after stripping the background, are given in *Table 2.3* together with the final analysis. The absolute accuracy of any x-ray analysis depends either on the accuracy of the constants Q, ω, etc. or on the standards used to calibrate the measured intensities.

TABLE 2.3 Relationships between measured intensities and composition for a NiAl alloy

	Measured intensities	*Cross-section Q, (10^{-24} cm^2)*	*Fluorescent yield ω*	*Detector efficiency η*	*Analysis at %*
NiK$_\alpha$	16 250	297	0.392	0.985	50.6
AlK$_\alpha$	7 981	2935	0.026	0.725	49.4

If the foil is too thick then an absorption correction (A) may have to be made to the measured intensities, since in traversing a given path length to emerge from the surface of the specimen, the x-rays of different energies will be absorbed differently. This correction involves a knowledge of the specimen thickness which has to be determined by one of the various techniques but usually from CBDPs. The attenuation of x-ray photons of an incident beam of intensity, I_0 in a path

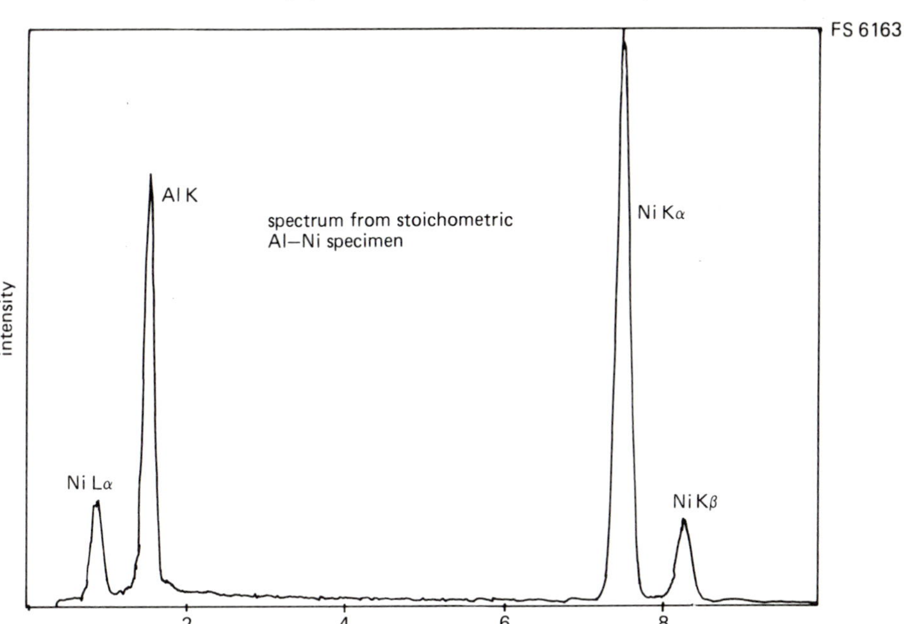

Figure 2.33 EDX spectrum from a stoichiometric Ni–Al specimen

length l in a material of density $\bar{\rho}$ is given by the relation

$$I = I_0 \exp\left\{-\left(\frac{\mu}{\rho}\right)\bar{\rho}l\right\} \tag{2.22}$$

where (μ/ρ) is the mass absorption coefficient of the x-ray in the specimen with units of cm^2/g. Occasionally a fluorescence (F) correction is also needed since elements Z between Ti and Ni are fluoresced by the element Z+2. This 'no-standards' Z(AF) analysis can give an overall accuracy of $\approx 2\%$ and can be carried out on-line with laboratory computers.

2.5.2 Electron energy loss spectroscopy (EELS)

A disadvantage of EDX is that the x-rays from the light elements are absorbed in the detector window. Windowless detectors can be used but have some disadvantages, which have led to the development of EELS.

It is evident that EELS is possible only on transmission specimens. Hence, magnetic prism electron spectrometers have been interfaced to STEMS to collect all the transmitted electrons lying within a cone of width α. The intensity of the various electrons, i.e. those transmitted without loss of energy and those that have been inelastically scattered and lost energy, is then obtained by dispersing the electrons with a magnetic prism which separates out spatially the electrons of different energies (*see* Figures 2.26 and 2.31). Because of the limited dispersive power and the aberrations of the magnetic spectrometer, the resolution depends

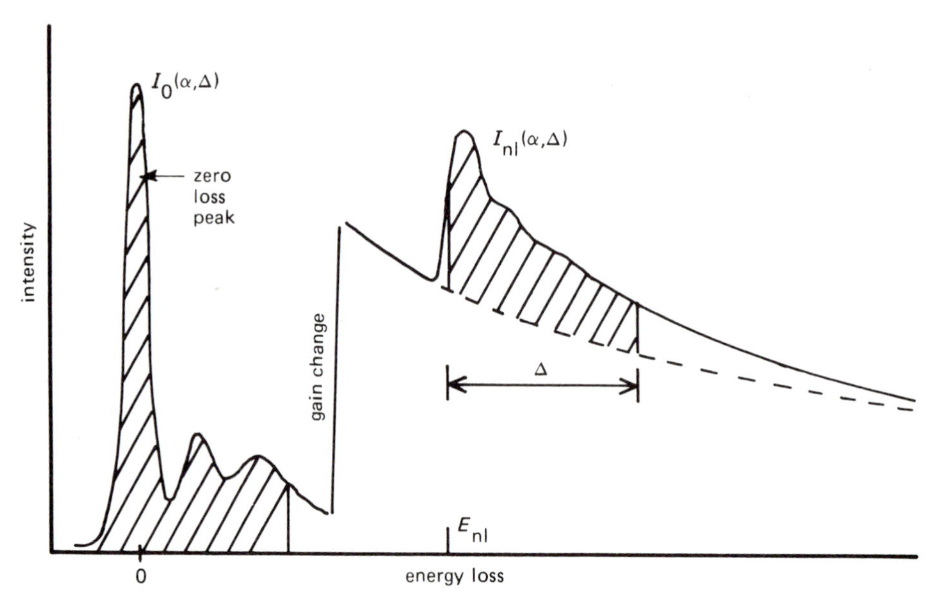

Figure 2.34 Schematic energy-loss spectrum, showing the zero-loss and plasmon regions together with the characteristic ionization edge

on the object size at the spectrometer object plane δ, and on the semi-divergence angle β at the spectrometer object plane.

A typical EELS spectrum illustrated in *Figure 2.34* shows three distinct regions. The zero loss peak is made up from those electrons which have (i) not been scattered by the specimen, (ii) suffered phonon scattering ($\approx 1/40$ eV) and (iii) elastically scattered. The energy width of the zero loss peak is caused by the energy spread of the electron source (up to ≈ 2 eV for a thermionic W filament) and the energy resolution of the spectrometer (typically a few eV). The second region of the spectrum extends up to about 50 eV loss and is associated with plasmon excitations corresponding to electrons which have suffered one, two, or more plasmon interactions. Since the typical mean free path for the generation of a plasmon is about 50 nm, many electrons suffer single plasmon losses and only in specimens which are too thick for electron loss analysis will there be a significant third plasmon peak. The relative size of the plasmon loss peak and the zero loss peak can also be used to measure the foil thickness. Thus the ratio of the probability of exciting a plasmon loss P_1, to not exciting a plasmon, P_0 is given by $P_1/P_0 = t/L$, where t is the thickness, L the mean free path for plasmon excitation, and P_1 and P_0 are given by the relative intensities of the zero loss and the first plasmon peak; if the second plasmon peak is a significant fraction of the first peak this indicates that the specimen will be too thick for accurate microanalysis.

The third region is made up of a continuous background on which the characteristic ionization losses are superimposed. Qualitative elemental analysis can be carried out simply by measuring the energy of the edges and comparing them with tabulated energies. The actual shape of the edge can also help to define the chemical state of the element. Quantitative analysis requires the measurement of the ratios of the intensities of the electrons from elements A and B which have suffered ionization losses. In principle, this allows the ratio of the number of A atoms, N_A and B atoms, N_B, to be obtained simply from the appropriate ionization cross sections, Q_K. Thus the number of A atoms will be given by

$$N_A = (1/Q_K^A)[I_A^K/I_0] \qquad (2.23)$$

and the number of B atoms by a similar expression, so that

$$N_A/N_B = I_K^A Q_K^B / I_K^B Q_K^A \qquad (2.24)$$

where I_K^A is the measured intensity of the K edge for element A, similarly for I_K^B and I_0 is the measured intensity of the zero loss peak. This expression is similar to the thin foil EDX equation.

To obtain I_K the background has first to be removed so that only loss electrons remain. This background is quite complex and arises from the low energy tails associated with lower energy inner shell losses, plasmon losses, and valence electron excitations. An empirical equation of the form AE^{-r} is therefore used to fit the smoothly falling background for 50–100 eV on the high energy (lower loss) side of the edge. This background is then extrapolated below the edge (*see Figure 2.34*) and the intensity above this background is taken as I_K. Because of the presence of other edges there is a maximum energy range over which I_K can be measured which is about 50–100 eV. The value of Q_K must therefore be replaced by $Q_K(\Delta)$ which is a partial cross-section calculated for atomic transition within an energy range Δ of the ionization threshold.

On the above basis $I_K^A(\Delta)$ can be obtained and related to N_A, via the partial cross section $Q_K^A(\Delta)$ but, because not all of the loss electrons are collected by the spectrometer, it is necessary to modify the analysis equation further by use of a double partial cross section $Q(\Delta, \alpha)$, where α is the angular range of scatter at the specimen which is accepted by the spectrometer. Thus

$$N_A = \frac{1}{Q(\Delta, \alpha)} \left[\frac{I_K^A(\Delta, \alpha)}{I_0(\Delta, \alpha)} \right] \qquad (2.25)$$

where $I_K^A(\Delta, \alpha)$ is the area under the edge after stripping up to energy Δ above the edge, and similarly $I_0(\Delta, \alpha)$ is extended up to Δ. Thus analysis of a binary alloy is carried out using the equation

$$\frac{N_A}{N_B} = \frac{Q_K^B(\Delta, \alpha)}{Q_K^A(\Delta, \alpha)} \frac{I_K^A(\Delta, \alpha)}{I_K^B(\Delta, \alpha)} \qquad (2.26)$$

Values of $Q(\Delta, \alpha)$ may be calculated from data in the literature for the specific value of ionization edge, Δ, α and incident accelerating voltage, but give an analysis accurate to only about 5%; a greater accuracy might be possible if standards are used.

2.5.3 Auger electron spectroscopy (AES)

As mentioned on page 67, Auger electrons originate from a surface layer a few atoms thick and therefore AES is a technique to look at the composition of the surface of a solid. It is obviously an important technique for studying

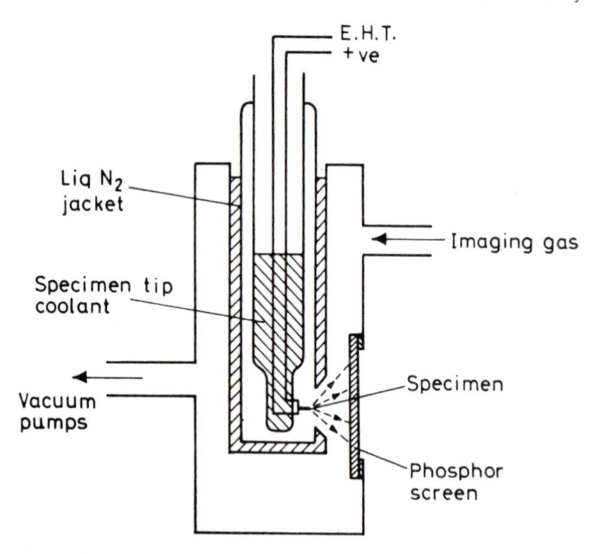

Figure 2.35 Schematic diagram of the microscope chamber

oxidation, catalysis and other surface chemical reactions, but has also been used successfully to determine the chemistry of fractured interfaces and grain boundaries, e.g. temper embrittlement of steels.

The basic instrumentation involves a focusable electron gun, an electron analyser and a sample support and manipulation system, all in an ultra-high vacuum environment to minimize adsorption of gases on to the surface during analysis. Two types of analyser are in used, a cylindrical mirror analyser (CMA) and hemispherical analyser (HSA), both of which are of the energy dispersive type as for EELS, with the difference that the electron energies are much lower, and electrostatic rather than magnetic 'lenses' are used to separate out the electrons of different energies.

In the normal distribution the Auger electron peaks appear small on a large and often sloping background, which gives problems in detecting weak peaks since amplification enlarges the background slope as well as the peak. It is therefore customary to differentiate the spectrum so that the Auger peaks are emphasized as doublet peaks with a positive and negative displacement against a nearly flat background. This is achieved by electronic differentiation by applying a small AC signal of a particular frequency in the detected signal. Chemical analysis through the outer surface layers can be carried out by depth profiling with an argon ion gun.

2.6 Field ion microscopy

Individual vacancies or interstitials cannot be imaged in CTEM but can be seen in a field-ion microscope. A sectional diagram of this instrument is shown in *Figure 2.35*. The specimen takes the form of a short, fine wire, electropolished at one end to a hemispherical tip 100–300 atoms in radius and welded at the other end to a filament between tungsten electrodes cooled in liquid nitrogen. The whole assembly is evacuated to below 10^{-6} mm Hg into which a trace of helium or neon gas is leaked and the specimen electrically charged, relative to the screen, to a positive potential of 5–15 kV. Straight lines of electrostatic force run radially from the tip of the screen and the image of the specimen tip is carried by the gas atoms to the fluorescent screen.

By positively charging the tip of the specimen, the free electron gas at the surface is pulled slightly into the metal, so partly exposing positive charged metal ions on the surface. When one of the gas atoms approaches a surface ion it gives up an electron to the metal and so becomes a positive helium (or neon) ion. It then accelerates down the lines of force radiating from the tip to produce an image on the screen. Since the ion current from any point on the surface depends on the local field strength, the image reveals the detailed field distribution. This field is greatly influenced by the atomic structure of the surface, so that protruding atoms give rise to high local fields and hence bright image points. The ions travel perpendicularly to the local tip surface and hence the magnification is approximately the ratio of screen distance to tip radius, or about 10^6. It is thus possible to obtain a separate and visible image from each ionizing atom in the surface of the tip.

High-melting-point refractory metals can readily be studied, and with specimen coolants and image gases with a lower ionization potential, most metals having a melting point above about 1000 °C can be studied. Present studies include (i) grain boundaries, (ii) precipitates in steels and Ni-base systems, (iii) phase transformations, (iv) ordered alloys, and (v) surface film formation.

2.7 Mechanical properties

Real crystals, however carefully prepared, contain lattice imperfections which profoundly affect those properties sensitive to structure. Careful examination of the mechanical behaviour of metals can give information on the nature of these lattice defects. In some branches of industry the common mechanical tests, such as tensile, hardness, impact, creep and fatigue tests, may be used, not to study the 'defect state', but to check the quality of the product produced against a standard specification, but whatever its purpose the mechanical test is of importance to physical metallurgy and deserves special consideration. The theories underlying the various mechanical properties are given in later chapters.

It is inevitable that a large number of different machines for performing the tests outlined above are in general use. This is because it is often necessary to know the effect of temperature and strain rate at vastly different levels of stress and temperature, depending on the material being tested. Consequently, no attempt is made here to describe the details of the various testing machines.

2.7.1 The tensile test

In a tensile test the ends of a test piece are fixed into grips, one of which is attached to the load measuring device on the tensile machine, and the other to the straining device. The strain is usually applied by means of a motor-driven crosshead and the elongation of the specimen is indicated by its relative movement. The load necessary to cause this elongation may be obtained from the elastic deflection of either a beam or proving ring, which may be measured by using hydraulic, optical or electro-mechanical methods. The latter method (where there is a change in the resistance of strain gauges attached to the beam) is, of course, easily adpted into a system for autographically recording the load-elongation curve.

The load–elongation curves for both polycrystalline mild steel and copper are shown in *Figure 2.36*(*a*) and (*b*). The corresponding stress (load per unit area, P/A) versus strain (change in length per unit length, dl/l) curves may be obtained knowing the dimensions of the test piece. At low stresses the deformation is elastic, reversible and obeys Hooke's law with stress linearly proportional to strain. The proportionality constant connecting stress and strain is known as the elastic modulus and may be either (a) the elastic or Young's modulus, E, (b) the rigidity or shear modulus μ, or (c) the bulk modulus K, depending on whether the strain is tensile, shear or hydrostatic compressive, respectively. Young's modulus, bulk modulus, shear modulus and Poisson's ratio ν, the ratio of lateral contractions to longitudinal extension in uniaxial tension, are related according to

$$K=\frac{E}{2(1-2\nu)},\quad \mu=\frac{E}{2(1+\nu)},\quad E=\frac{9K\mu}{3K+\mu}$$

In general, the elastic limit is an ill-defined stress, but for impure iron and low carbon steels the onset of plastic deformation is denoted by a sudden drop in load indicating both an upper and lower yield point*. This yielding behaviour is characteristic of many metals, particularly those with b.c.c. structure containing small amounts of solute element (*see* Chapter 8). For materials not showing a sharp yield point, a conventional definition of the beginning of plastic flow is the 0.1 per cent proof stress, in which a line is drawn parallel to the elastic portion of the stress–strain curve from the point of 0.1 per cent strain.

For control purposes the tensile test gives valuable information on the ultimate tensile strength (UTS = maximum load/original area) and ductility (percentage reduction in area or percentage elongation) of the material. When it is used as a research technique, however, the exact shape and fine details of the curve, in addition to the way in which the yield stress and fracture stress vary with temperature, alloying additions and grain size, are probably of greater significance.

The increase in stress from the initial yield up to the UTS indicates that the specimen hardens during deformation, i.e. work hardens. On straining beyond the UTS the metal still continues to work harden, but at a rate too small to compensate for the reduction in cross-sectional area of the test piece. The

* Load relaxations are obtained only on 'hard' beam Polanyi-type machines where the beam deflection is small over the working load range. With 'soft' machines, those in which the load measuring device is a soft spring, rapid load variations are not recorded because the extensions required are too large, while in dead loading machines no load relaxations are possible. In these latter machines sudden yielding will show as merely an extension under constant load.

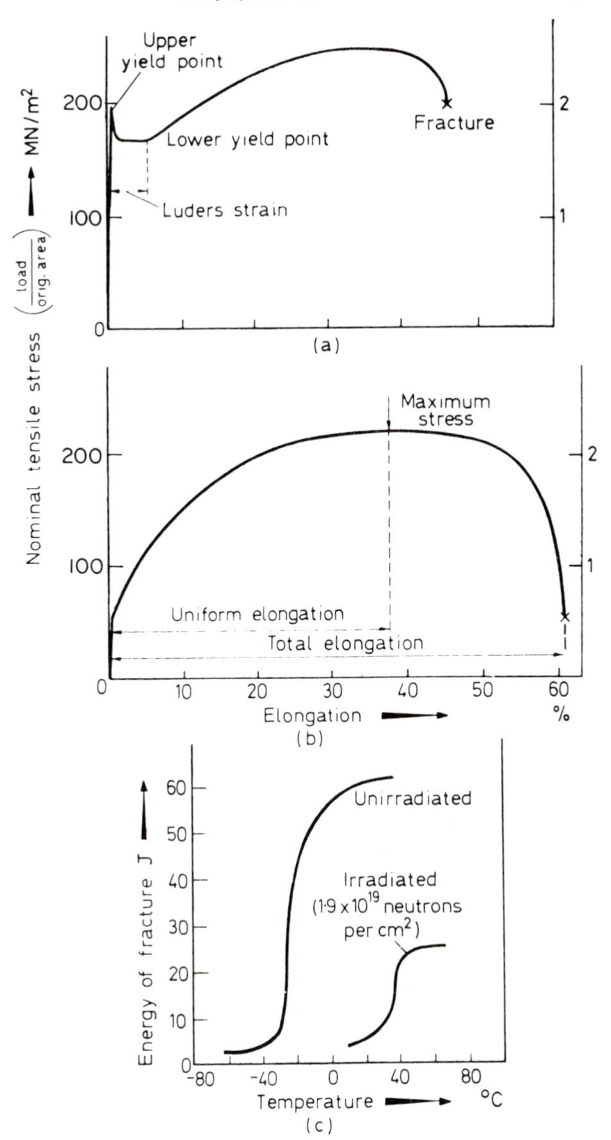

Figure 2.36 Stress–elongation curves for (a) impure iron, (b) copper. (c) Ductile–brittle transition in mild steel (after Churchman, Mogford and Cottrell, *Phil. Mag.* **2,** 1957, 1273)

deformation then becomes unstable, such that as a localized region of the gauge length strains more than the rest, it cannot harden sufficiently to raise the stress for further deformation in this region above that to cause further strain elsewhere. A neck then forms in the gauge length, and further deformation is confined to this region until fracture. Under these conditions, the reduction in area $(A_0 - A_1)/A_0$ where A_0 and A_1 are the initial and final areas of the neck gives a measure of the localized strain, and is a better indication than the strain to fracture measured along the gauge length.

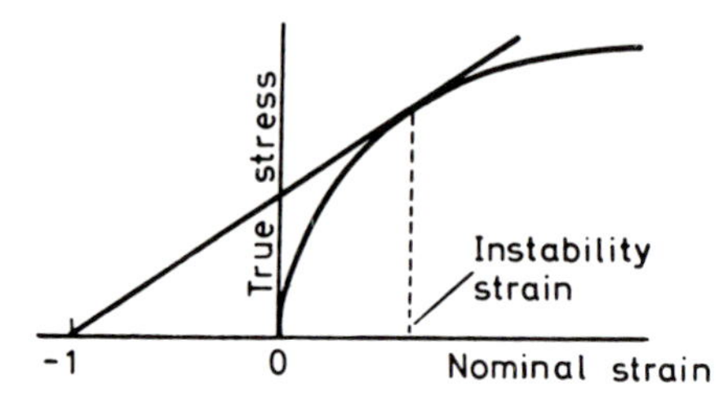

Figure 2.37 Considères construction

True stress–true strain curves are often plotted to show the work hardening and strain behaviour at large strains. The true stress σ is the load P divided by the area A of the specimen at that particular stage of strain and the total true strain in deforming from initial length l_0 to length l_1 is $\varepsilon=\int_{l_0}^{l_1}(\mathrm{d}l/l)=\ln(l_1/l_0)$. The true stress–strain curves often fit the Ludwig relation $\sigma=k\varepsilon^n$ where n is a work hardening coefficient ≈ 0.1–0.5 and k the strength coefficient. Plastic instability, or necking, occurs when an increase in strain produces no increase in load supported by the specimen, i.e. $\mathrm{d}P=0$, and hence since $P=\sigma A$, then

$$\mathrm{d}P=A\,\mathrm{d}\sigma+\sigma\,\mathrm{d}A=0$$

defines the instability condition. During deformation, the specimen volume is essentially constant, i.e. $\mathrm{d}V=0$, and from

$$\mathrm{d}V=\mathrm{d}(lA)=A\,\mathrm{d}l+l\,\mathrm{d}A=0$$

we obtain

$$\frac{\mathrm{d}\sigma}{\sigma}=-\frac{\mathrm{d}A}{A}=\frac{\mathrm{d}l}{l}=\mathrm{d}\varepsilon$$

Thus, necking occurs at a strain at which the slope of the true stress–true strain curve equals the true stress at that strain, i.e. $\mathrm{d}\sigma/\mathrm{d}\varepsilon=\sigma$. Alternatively, since $k\varepsilon^n=\sigma=\mathrm{d}\sigma/\mathrm{d}\varepsilon=nk\varepsilon^{n-1}$ then $\varepsilon=n$ and necking occurs when the true strain equals the strain hardening exponent. The instability condition may also be expressed in terms of the conventional (nominal strain)

$$\frac{\mathrm{d}\sigma}{\mathrm{d}\varepsilon}=\frac{\mathrm{d}\sigma}{\mathrm{d}\varepsilon_n}\frac{\mathrm{d}\varepsilon_n}{\mathrm{d}\varepsilon}=\frac{\mathrm{d}\sigma}{\mathrm{d}\varepsilon_n}\left(\frac{\mathrm{d}l/l_0}{\mathrm{d}l/l}\right)=\frac{\mathrm{d}\sigma}{\mathrm{d}\varepsilon_n}\frac{l}{l_0}=\frac{\mathrm{d}\sigma}{\mathrm{d}\varepsilon_n}(1+\varepsilon_n)=\sigma$$

which allows the instability point to be located using Considères construction (*see Figure 2.37*), by plotting the true stress against nominal strain and drawing the tangent to the curve from $\varepsilon_n=-1$ on the strain axis. The point of contact is the instability stress and the tensile strength is $\sigma/(1+\varepsilon_n)$.

Tensile specimens can also give information on the type of fracture exhibited. Usually in polycrystalline metals transgranular fractures occur, i.e. the fracture surface cuts through the grains, and the 'cup and cone' type of fracture is extremely common in really ductile metals such as copper. In this, the fracture starts at the centre of the necked portion of the test piece and at first grows roughly perpendicular to the tensile axis, so forming the 'cup', but then, as it nears the outer surface, it turns into a 'cone' by fracturing along a surface at about 45° to the tensile axis. In detail the 'cup' itself consists of many irregular surfaces at about 45° to the tensile axis, which gives the fracture a fibrous appearance. Cleavage is also a fairly common type of transgranular fracture, particularly in materials of b.c.c. or c.p.h. structure when tested at low temperatures. The fracture surface follows certain crystal planes, e.g. {100} planes, as is shown by the grains revealing large bright facets, but the surface also appears granular with 'river lines' running across the facets where cleavage planes have been torn apart. Intercrystalline fractures sometimes occur, often without appreciable deformation. This type of fracture is usually caused by a brittle second phase precipitating out around the grain boundaries, as shown by copper containing bismuth or antimony.

2.7.2 Hardness test

The hardness of a metal, defined as the resistance to penetration, gives a conveniently rapid indication of its deformation behaviour. The hardness tester forces a small sphere, pyramid or cone into the surface of the metals by means of a known applied load, and the hardness number (Brinell or Vickers diamond pyramid) is then obtained from the diameter of the impression. The hardness may be related to the yield or tensile strength of the metal, since during the indentation, the material around the impression is plastically deformed to a certain percentage strain. The Vickers hardness number (VPN) is defined as the load divided by the pyramidal area of the indentation, in kg/mm^2, and is about three times the yield stress for materials which do not work harden appreciably. The Brinell hardness number (BHN) is defined as the stress P/A, in kg/mm^2 where P is the load and A the surface area of the spherical cap forming the indentation. Thus

$$\mathrm{BHN} = P/(\tfrac{1}{2}\pi D^2)\{1 - [1 - (d/D)^2]^{1/2}\}$$

where d and D are the indentation and indentor diameters respectively. For consistent results the ratio d/D should be maintained constant and small. Under these conditions soft materials have similar values of BHN and VPN. Hardness testing is of importance in both control work and research, especially where information on brittle materials at elevated temperatures is required.

2.7.3 Impact testing

A material may have a high tensile strength and yet be unsuitable for shock loading conditions. To determine this the impact resistance is usually measured by means of the notched or un-notched Izod or Charpy impact test. In this test a load swings from a given height to strike the specimen, and the energy dissipated in the fracture is measured. The test is particularly useful in showing the decrease in ductility and impact strength of materials of b.c.c. structure at moderately low temperatures. For example, carbon steels have a relatively high ductile–brittle transition temperature (*Figure 2.36*(*c*)) and, consequently, they may be used with safety at sub-zero temperatures only if the transition temperature is lowered by suitable alloying additions or by refining the grain size. Nowadays, increasing importance is given to defining a fracture toughness parameter K_c for an alloy,

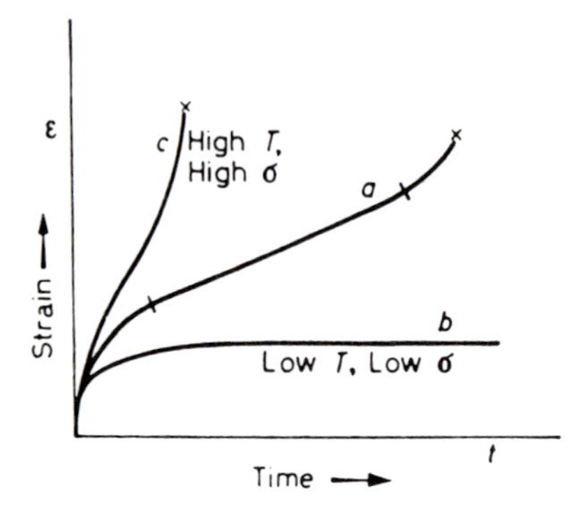

Figure 2.38 Typical creep curves

since many alloys contain small cracks which, when subjected to some critical stress, propagate; K_c defines the critical combination of stress and crack length. Brittle fracture is discussed more fully in Chapter 14.

2.7.4 Creep

Creep is defined as plastic flow under constant stress, and although the majority of tests are carried out under constant load conditions, equipment is available for reducing the loading during the test to compensate for the small reduction in cross-section of the specimen. At relatively high temperatures creep appears to occur at all stress levels, but the creep rate increases with increasing stress at a given temperature. For the accurate assessment of creep properties, it is clear that special attention must be given to the maintenance of the specimen at a constant temperature, and to the measurement of the small dimensional changes involved. This latter precaution is necessary, since in many materials a rise in temperature by a few tens of degrees is sufficient to double the creep rate. *Figure 2.38* curve *a* shows the characteristics of a typical creep curve and following the instantaneous strain caused by the sudden application of the load, the creep process may be divided into three stages, usually termed primary or transient creep, second or steady state creep and tertiary or accelerating creep. The characteristics of the creep curve often vary, however, and the tertiary stage of creep may be advanced or retarded if the temperature and stress at which the test is carried out is high or low respectively (*see Figure 2.38*, curves *b* and *c*). Creep is discussed more fully in Chapter 13.

2.7.5 Fatigue

The fatigue phenomenon is concerned with the premature fracture of metals under repeatedly applied low stresses, and is of importance in many branches of engineering, e.g. aircraft structures. Several different types of testing machines have been constructed in which the stress is applied by bending, torsion, tension or compression, but all involve the same principle of subjecting the material to constant cycles of stress. To express the characteristics of the stress system, three properties are usually quoted: these include (1) the maximum range of stress, (2)

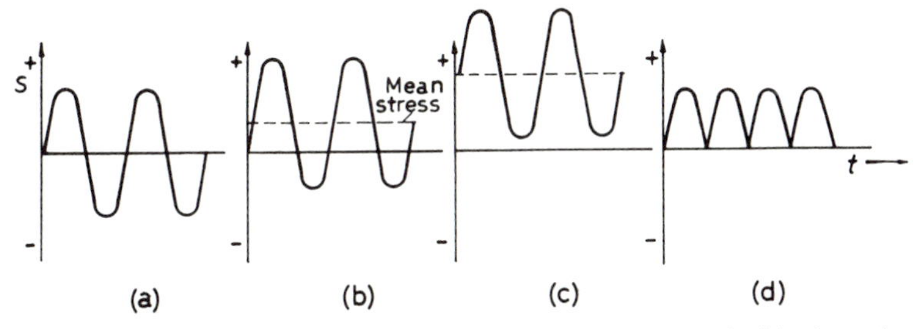

Figure 2.39 Alternative forms of stress cycling: (a) reversed, (b) alternating (mean stress ≠ zero), (c) fluctuating, and (d) repeated

the mean stress, and (3) the time period for the stress cycle. Four different arrangements of the stress cycle are shown in *Figure 2.39*, but the reverse and the repeated cycle test, e.g. 'push-pull', are the most common, since they are the easiest to achieve in the laboratory.

The standard method of studying fatigue is to prepare a large number of specimens free from flaws, and to subject them to tests using a different range of stress, S, on each group of specimens. The number of stress cycles, N, endured by each specimen at a given stress level is recorded and plotted, as shown in *Figure 2.40*. This S–N diagram indicates that some metals can withstand indefinitely the application of a large number of stress reversals, provided the applied stress is below a limiting stress known as the endurance limit. For certain ferrous materials when they are used in the absence of corrosive conditions the assumption of a safe working range of stress seems justified, but for non-ferrous materials and for steels when they are used in corrosive conditions a definite endurance limit cannot be defined. Fatigue is discussed in more detail in Chapter 13.

2.8 Physical properties

A knowledge of the physical properties is important for the correct application of metals. The measurements of these properties is also useful in the study of crystal structure (particularly in the way the structure may vary with

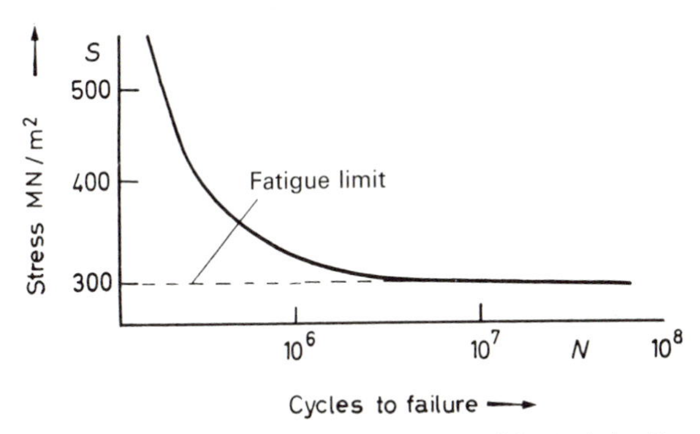

Figure 2.40 S–N curve for a commercial steel in the normalized condition

temperature or composition) and because the properties may throw light on the electronic structure of metals.

2.8.1 Density

This property, defined as the weight per unit volume of a material, increases regularly with increasing atomic numbers in each sub-group. The reciprocal of the density, i.e. $1/\rho$, is the specific volume v, while the product of v and the atomic weight W is known as the atomic volume Ω. The density may be determined by the usual 'immersion' method, but it is instructive to show how x-rays can be used. For example, a powder photograph may give the lattice parameter of an f.c.c. metal, say copper, as 0.36 nm. Then $1/(3.6 \times 10^{-10})^3$ or 2.14×10^{28} cells of this size (0.36 nm) are found in a cube 1 m edge length. The total number of atoms in 1 m^3 is then $4 \times 2.14 \times 10^{28} = 8.56 \times 10^{28}$ since an f.c.c. cell contains 4 atoms. Furthermore, the weight of a copper atom is 63.57 times the weight of a hydrogen atom (which is 1.63×10^{-24} g) so that the weight of 1 m^3 of copper, i.e. the density, is $8.56 \times 10^{28} \times 63.57 \times 1.63 \times 10^{-24} = 8.9\ \mathrm{Mg\,m^{-3}}$.

On alloying, the density of a metal changes. This is because the mass of the solute atom differs from that of the solvent, and also because the lattice parameter usually changes on alloying. The parameter change may often be deduced from Vegard's law, which assumes that the lattice parameter of a solid solution varies linearly with atomic concentration, but numerous deviations from this ideal behaviour do exist.

2.8.2 Thermal properties (dilatometry, specific heat)

The effect of raising the temperature of a metal is to increase the amplitude of vibration of the atoms, which give rise to a thermal expansion of the lattice (*see* Chapter 4). The change in dimensions with temperature is usually expressed in terms of the linear coefficient of expansion α, given by $\alpha = (1/l)(\mathrm{d}l/\mathrm{d}T)$, where l is the original length of the specimen and T is the absolute temperature. Because of the anisotropic nature of metal crystals, the value of α usually varies with the direction of measurement in the crystal, and even in a particular crystallographic direction the dimensional change with temperature may not always be uniform. The change in dimensions of a specimen can be transmitted to a sensitive dial gauge or electrical transducer by means of a silica rod.

Phase changes in the solid state are usually studied by *dilatometry*. When a phase transformation takes place, because the new phase usually occupies a different volume to the old phase, discontinuities are observed in the α v. T curve. Moreover, some of the 'nuclear metals' which exist in many allotropic forms, such as uranium and plutonium, show a negative coefficient of linear expansion along one of the crystallographic axes in certain of their allotropic modifications.

The change in volume of a metal with temperature is important in many metallurgical operations such as casting, welding and heat treatment. Of particular importance is the volume change associated with the melting or, alternatively, the freezing phenomenon since this is responsible for many of the defects, both of a macroscopic and microscopic size, which exist in metal crystals.

Most metals increase their volume by about 3 per cent on melting, although those metals which have crystal structures of lower co-ordination, such as bismuth, antimony or gallium, contract on melting. This volume change is quite small, and while the liquid structure is more open than the solid structure, it is clear that the liquid state resembles the solid state more closely than it does the gaseous phase. For the simple metals the latent heat of melting, which is merely the work done in separating the atoms from the close-packed structure of the solid to the more open liquid structure, is only about one thirtieth of the latent heat of evaporation, while the electrical and thermal conductivities are reduced only to three-quarters to one-half of the solid state values.

The *specific heat* is another thermal property important in the metallurgical operation of casting or heat treatment, since it determines the amount of heat required in the process. Thus, the specific heat (denoted by C_p, when dealing with the specific heat at constant pressure) controls the increase in temperature, dT, produced by the addition of a given quantity of heat, dQ, to one gram of metal, so that

$$dQ = C_p dT$$

When thermal energy is supplied to the metal, it is clear from the above discussion that part of it will be absorbed by the lattice to increase the amplitude of vibration of the ions, but a small part of it will also be absorbed by the electrons at the top of the energy band. A knowledge of the specific heat is, therefore, of importance in understanding both the electronic structure and the lattice structure of metals.

The lattice behaviour will be discussed more fully in Chapter 4, particularly the relation between specific heat and phase transformations, but it is convenient here to discuss briefly the electronic contribution. It emerges from the band theory that the electronic contribution to the specific heat is linear in temperature and that the proportionality constant, η, is given by

$$\eta = \pi^2 \boldsymbol{k} N(E)/3$$

where $\boldsymbol{k}$ is Boltzmann's constant and $N(E)$ is the density of electron states at the Fermi surface. However, since the total specific heat is given by

$$C = C_{\text{lattice}} + C_{\text{electronic}}$$

$$C = \text{const. } T^3 + \eta T \tag{2.27}$$

it is only at low temperatures that the lattice term becomes small. Nevertheless, from the way in which the specific heat varies with temperature at these low temperatures (≈ 4 K), η can be obtained from a plot of C/T versus T^2 by extrapolation to $T = 0$.

2.8.3 Electrical conductivity, superconductivity, semiconductivity

One of the most important electronic properties of metals is the electrical conductivity, κ, and the reciprocal of the conductivity (known as the resistivity, ρ) is defined by the relation, $R = \rho l/A$, where R is the resistance of the specimen, l is the length and A is the cross-sectional area. As discussed in Chapter 1, the

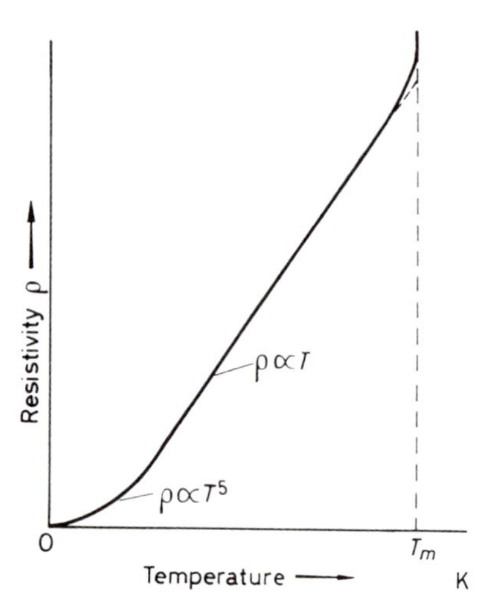

Figure 2.41 Variation of resistivity with temperature

characteristic feature of a metal is its high electrical conductivity which arises from the ease with which the electrons can migrate through the lattice. The high thermal conduction of metals also has a similar explanation, and the Wiedmann–Franz law shows that the ratio of the electrical and thermal conductivities is nearly the same for all metals at the same temperature.

Since conductivity arises from the motion of conduction electrons through the lattice, resistance must be caused by the scattering of electron waves by any kind of irregularity in the lattice arrangement. Irregularities can arise from any one of several sources, such as temperature, alloying, deformation or nuclear irradiation, since all will disturb, to some extent, the periodicity of the lattice. The effect of temperature is particularly important and, as shown in *Figure 2.41*, the resistance increase linearly with temperature above about 100 K up to the melting point. On melting, the resistance increases markedly because of the exceptional disorder of the liquid state. However, for some metals such as bismuth, the resistance actually decreases, owing to the fact that the special zone structure which makes bismuth a poor conductor in the solid state is destroyed on melting.

In most metals the resistance approaches zero at absolute zero, as shown in *Figure 2.41*, but in some metals (e.g. lead, tin and mercury), the resistance suddenly drops to zero at some finite critical temperature above 0 K. Such metals are called superconductors. The critical temperature is different for each metal but is always close to absolute zero; the highest critical temperature known for an element is 8 K for niobium. Superconductivity is discussed in more detail in Chapter 5.

The fundamental difference between a metal and an insulator was outlined in Chapter 1, but it is worth mentioning here that the general behaviour of insulators can be modified either by the application of high temperatures or by the addition of impurities. Clearly, insulators may become conductors at elevated temperatures if the thermal agitation is sufficient to enable electrons to jump the energy gap into the unfilled zone above. Pure Si and Ge are materials with an

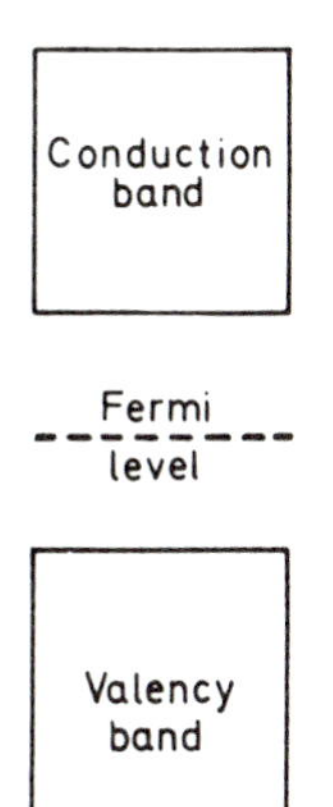

Figure 2.42 Schematic diagram of an intrinsic semiconductor showing the relative positions of the conduction and valency bonds

energy gap small enough to be surmounted by thermal excitation. In such intrinsic semiconductors, as they are called, the current carriers are electrons in the conduction band and holes in the valency band in equal numbers as shown in *Figure 2.42*. The motion of a hole in the valency band is equivalent to the motion of an electron in the opposite direction. Alternatively, conduction may be produced by the presence of impurities which either add a few electrons to an empty zone or remove a few from a full zone. Materials which have their conductivity developed in this way are commonly known as semiconductors. Silicon and germanium containing small amounts of impurity have semiconducting properties at ambient temperatures and, as a consequence, they are frequently used in electronic transistor devices. Silicon normally has completely filled zones, but becomes conducting if some of the silicon atoms, which have four valency electrons, are replaced by phosphorus, arsenic or antimony atoms which have five valency electrons. The extra electrons go into empty zones, and as a result silicon becomes an n-type semiconductor, since conduction occurs by negative carriers. On the other hand, the addition of elements of lower valency than silicon, such as aluminium, removes electrons from the filled zones leaving behind 'holes' in the valency band structure. In this case silicon becomes a p-type semiconductor, since the movement of electrons in one direction of the zone is accompanied by a movement of 'holes' in the other, and consequently they act as if they were positive carriers. The conductivity may be expressed as the product of (i) the number of charge carriers, n, (ii) the charge carried by each (i.e. $e = 1.6 \times 10^{-19}$ C) and (iii) the mobility of the carrier, μ.

The band structure of n- and p-type semiconductors is discussed more fully in Chapter 5.

2.8.4 Magnetic properties

When a metal is placed in a magnetic field of strength H, the field induced in the metal is given by

$$B = H + 4\pi I$$

where I is the intensity of magnetization. The quantity I is a characteristic

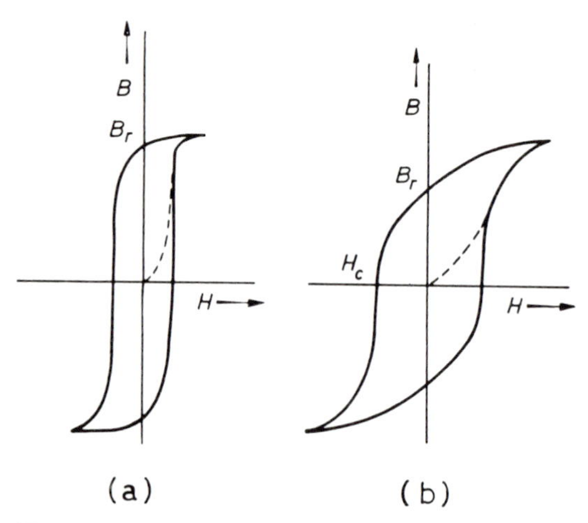

Figure 2.43 *B–H* curves for (a) soft, and (b) hard magnets

property of the metal, and is related to the susceptibility per unit volume of the metal which is defined as

$$\kappa = I/H \tag{2.28}$$

Those metals for which κ is negative, such as copper, silver, gold and bismuth, are repelled by the field and are termed diamagnetic materials. Most metals, however, have positive κ values (i.e. they are attracted by the field) and are either paramagnetic (when κ is small) or ferromagnetic (when κ is very large). Only four pure metals—iron, cobalt and nickel from the transition series, and gadolinium from the rare earth series—are ferromagnetic ($\kappa \approx 1000$) at room temperature, but there are several ferromagnetic alloys and some contain no metals which are themselves ferromagnetic. The Heusler alloy, which contains manganese, copper and aluminium, is one example, and it will be shown in Chapter 5 that ferromagnetism in such alloys is due to the presence of one of the transition metals.

The susceptibility is usually measured by a method which depends upon the fact that when a metal specimen is suspended in a non-uniform transverse magnetic field, a force proportional to $\kappa V . H . \mathrm{d}H/\mathrm{d}x$, where V is the volume of the specimen and $\mathrm{d}H/\mathrm{d}x$ is the field gradient measured transversely to the lines of force, is exerted upon it. This force is easily measured by attaching the specimen to a sensitive balance, and one type commonly used is that designed by Sucksmith. In this balance the distortion of a copper–beryllium ring, caused by the force on the specimen, is measured by means of an optical or electro-mechanical system.

The ability of a ferromagnetic metal to concentrate the lines of force of the applied field is of great practical importance, and while all such materials can be both magnetized and demagnetized, the ease with which this can be achieved usually governs their application in the various branches of engineering. Materials may be generally classified either as magnetically soft (temporary magnets), or as magnetically hard (permanent magnets), and the difference between the two types of magnet may be inferred from *Figure 2.43*. Here, H is the magnetic field necessary to induce a field of strength B inside the material. Upon removal of the field H, a certain residual magnetism B_r, known as the remanence residual, is left in the specimen, and a field H_c, called the coercive force, must be applied in the opposite direction to remove it. A soft magnet is one which is easy both to magnetize and to demagnetize and, as shown in *Figure 2.43*(a), a low value of H is sufficient to induce a large field B in the metal, while only a small field H_c is required to remove it; a hard magnet is a material that is magnetized and demagnetized with difficulty (*Figure 2.43*(b)). Magnetically soft materials are used for transformer laminations and armature stampings where a high permeability and a low hysteresis are desirable; iron–silicon or iron–nickel alloys are commonly used for this purpose. Magnetically hard materials are used for applications where a permanent magnetic field is required, but where electro-magnets cannot be used, such as in electric clocks, meters, etc. Materials commonly used for this purpose include Alnico (Al–Ni–Co) alloys, Cunico (Cu–Ni–Co) alloys, ferrites (barium and strontium), samarium–cobalt alloys ($SmCo_5$ and $Sm_2(CoFeCuZr)_{17}$) and Neomax ($Nd_2Fe_{14}B$). The Alnico's have high remanence but poor coercivities, the ferrites have rather low remanence but good coercivities together with very cheap raw material costs. The rare-earth magnets have a high performance but are rather costly although the Nd-based alloys are cheaper than the Sm-based alloys.

Magnetic measurements are also useful in throwing light on the electronic structure of metals. Thus, it may be shown that in simple metals, such as sodium or copper, the paramagnetic susceptibility per unit volume due to the spin of the electrons in the energy band is given by

$$\chi = 2\mu_B^2 N(E) \tag{2.29}$$

where μ_B is the unit of magnetization known as the 'Bohr magneton' ($\mu_B = e\boldsymbol{h}/4\pi mc$), and $N(E)$ is the density of states at the Fermi surface. In copper the contribution to paramagnetic susceptibility is outweighed by the diamagnetic contribution, as mentioned on page 88. Measurements of this type show that the transition metals, which are either strongly paramagnetic or ferromagnetic, have quite complicated electronic structures. This will be discussed in Chapter 5.

Suggestions for further reading

Andrews, K.W., *Physical Metallurgy, Techniques and Applications*, vols 1 and 2, George Allen & Unwin, 1973

Belk, J.A. (ed.), *Electron Microscopy and Microanalysis of Crystalline Materials*, Applied Science Publishers, 1979

Bowen, D.K. and Hall, C.R., *Microscopy of Materials*, Macmillan, 1975

Cullity, B.D., *Elements of X-ray Diffraction*, Addison-Wesley, 1967

Loretto, M.H., *Electron Beam Analysis of Materials*, Chapman & Hall, 1984

Quantitative Microanalysis with High Spatial Resolution, Metals Society, London, 1981

Chapter 3

Phase diagrams and solidification

3.1 The determination of phase diagrams

Many of the techniques discussed in Chapter 2 are used in the determination of phase diagrams. Particularly important are the metallographic and x-ray analysis techniques, but the physical property measurements are also widely used, e.g. dilatometry, electrical resistivity and thermal analysis.

One of the oldest methods of studying a metal is by observing its behaviour on heating or cooling. This thermal analysis, in addition to indicating the melting point, gives information on any changes of crystal structure, i.e. phase transformations, which exist in the solid state. A pure metal solidifies at a fixed temperature, so that if the temperature of the metal is measured after successive equal intervals of time, a cooling curve of the form shown in *Figure 3.1*(*a*) is obtained. The change of phase from liquid to solid is marked by the evolution of latent heat which, as can be seen from the diagram, results in the appearance of a horizontal portion in the cooling curve. Similarly, if a change of crystal structure takes place in the solid state, the transformation will be accompanied by latent heat, and a discontinuity again appears in the cooling curve. To increase the sensitivity in detecting the transition inverse rate, curves are used in which dt/dT is plotted against T.

The process of solidification occurs by a mechanism of nucleation and growth, i.e. minute nuclei, or seed crystals, are formed in various parts of the melt which then grow at the expense of the surrounding liquid until the whole volume is solid. During the freezing process the nuclei grow more rapidly along certain crystallographic directions, and this results in the formation of long branch-like crystals, known as dendrites, as shown in *Figure 3.1*(*b*). Eventually the outward growth of a dendrite is halted when contact is made with neighbouring growths, and then the remaining liquid freezes in the interstices between the dendrite arms. Every contact surface then becomes the boundary between two crystals, so that each original nucleus produces a crystal or grain of its own, separated from the neighbouring grains by grain boundaries (*Figure 3.1*(*c*)). The grain boundary is simply a narrow transition region, of a few atom diameters width, in which the atoms adjust themselves from the crystal orientation of the one grain to that of the other. Clearly, if the number of nuclei formed on freezing is small then the

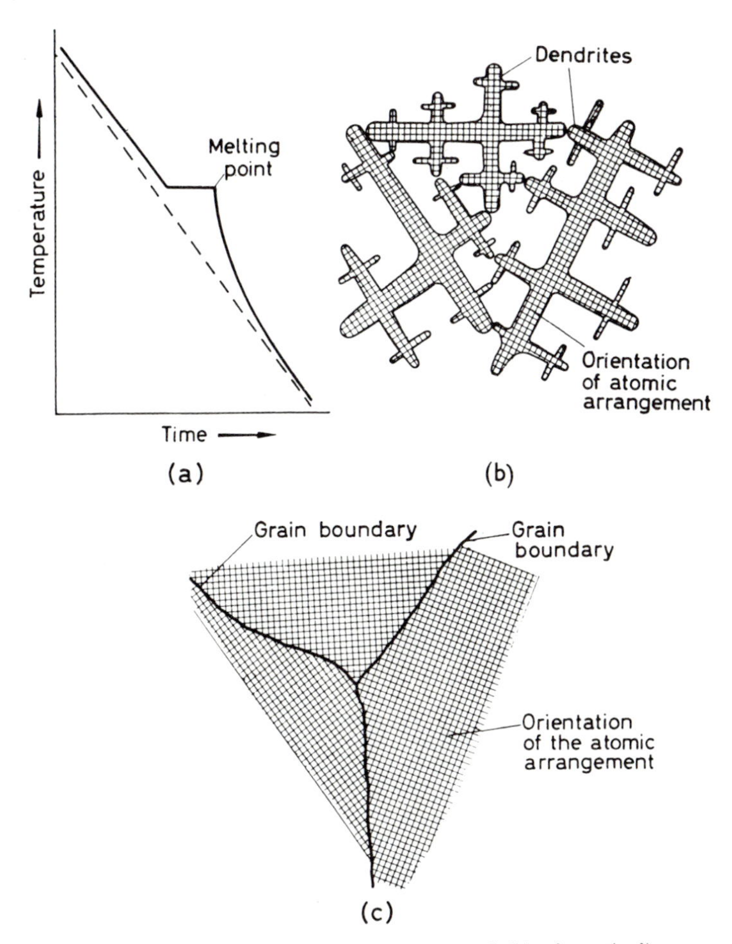

Figure 3.1 (a) Ideal cooling curve for a pure metal, (b) schematic diagram of dendrites interlocking, and (c) formation of a grain from each dendrite

resulting polycrystalline grain size will be large. In the limit, if only one nucleus is allowed to grow then only one grain, i.e. a single crystal, will result. The preparation of metal single crystals formed by such controlled solidification, and by other methods is discussed in section 3.6.

3.2 The equilibrium or phase diagram

In general, an alloy solidifies not at a fixed temperature but over a range of temperature. Moreover, even after becoming solid the alloy may undergo constitutional changes which profoundly affect its properties and use. It is, therefore, of considerable importance to know both the structures and compositions of the phases that will exist in equilibrium in an alloy at a given composition and temperature.

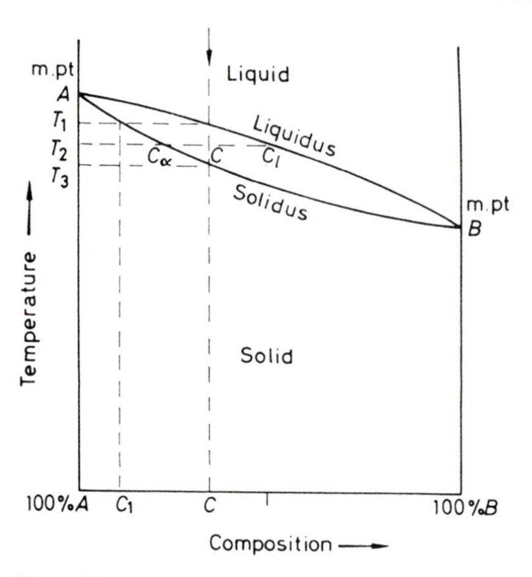

Figure 3.2 Phase diagram showing complete miscibility in the solid state

When dealing with a pure metal, only the effect of temperature on its structure need be considered. However, when a second element is added to the parent metal the problem becomes more complex. Composition is now an additional variable, and the solute atoms may mix with the solvent atoms to form either a primary solid solution where the crystal structure is the same as the parent metal, or an intermediate phase which has a different crystal structure from that of the pure metal. Since the equilibrium structure of an alloy depends on both temperature and composition, it is found convenient to show the existence of a particular phase by graphical representation on a diagram, known as an equilibrium or phase diagram, in which temperature is plotted as the ordinate and composition as the abscissa. In other words, the phase diagram is a map which shows at a glance the phases which exist in equilibrium for any combination of temperature and alloy composition.

In practice, it is found that, depending on the nature of the two metals involved (i.e. the crystal structure, the size of the atoms, the valency), several basic types of binary equilibrium diagram exist. The general features of these basic systems are shown in *Figures 3.2, 3.3* and *3.4*, and they represent the alloying behaviour for the following conditions.

(1) Metal *A* and metal *B* are completely soluble in each other in both the liquid and solid state.

(2) Metal *A* and metal *B* are completely soluble in each other in the liquid state, but completely insoluble in the solid state.

(3) The two metals *A* and *B* are completely soluble in the liquid state, but only partially soluble in the solid state.

Once these simple binary systems are understood, even the most complicated diagram becomes easy to interpret.

3.2.1 Complete solubility in the solid state

The phase diagram for the alloying of two metals having the same crystal structure, similar atomic size and similar melting points is shown in *Figure 3.2*. Common examples include the systems copper–nickel and gold–platinum. The diagram is divided into three separate areas, or phase fields, by two phase boundaries known as the liquidus and solidus. At temperatures above that denoted by the liquidus, alloys of all compositions from pure A to pure B will be liquid, while at temperatures below that denoted by the solidus the alloys are solid. It should be noted that, unlike a pure metal, alloys freeze over a range of temperature, and that the field between the liquidus and solidus curves represents the temperature interval during which the alloys are in a 'pasty' condition. To illustrate this, let us consider the solidification of an alloy of composition C. On cooling, freezing starts at a temperature T_1, when the first solid to form is solid α of composition C_1 given by the solidus curve at the appropriate temperature T_1. It can be seen that the first dendrites to form are richer in A than is the original liquid and, consequently, the remaining melt will be richer in B than the original liquid. On cooling below T_1 the alloy passes into the two-phase field and at any temperature in this range, the composition and proportions of the phases which co-exist can be deduced from the diagram. Thus, at the temperature T_2 the composition of the α-phase is C_α and that of the liquid phase is C_l, while the relative amounts are given by the 'lever rule' as

$$\frac{\text{amount of } \alpha\text{-phase}}{\text{amount of liquid}} = \frac{C_l - C}{C - C_\alpha}$$

With further cooling the relative proportions of the two phases change continuously, with the amount of liquid decreasing progressively to zero at T_3. Furthermore, the composition of each of the two phases varies with temperature in such a way that the composition of the α-phase moves down the solidus and that of the liquid down the liquidus. At T_3 the alloy emerges from the two-phase field consisting entirely of an α-solid solution of solute B in solvent A with a composition C.

From the above discussion, it is clear that the composition of the α-phase deposited becomes progressively richer in B and, consequently, if equilibrium is to be maintained in the alloy, with phase compositions able to re-adjust themselves to produce a uniform composition, extensive diffusion or atomic migration must occur in the solid. However, in practice the cooling rate is often too fast to allow appreciable diffusion, and only partial atomic redistribution takes place. The grains of the α-phase formed, therefore, vary in composition about the average composition C, and the centre of the grain which corresponds to the first α deposited is rich in A, while that at the outside of the grain is correspondingly rich in B. Such a non-uniform α-solid solution is said to be cored, and it will be evident that the faster the rate of cooling the more pronounced will be the degree of coring; coring in chill cast ingots is, therefore, quite extensive.

A cored α-solid solution has inferior properties when compared with a homogeneous alloy, and for this reason the cast alloy is subsequently given a homogenization treatment. One method of achieving a homogeneous alloy is to re-heat the ingot to a temperature just below the solidus temperature where atomic migration is rapid. However, it is more usual to cold-work the billet before annealing, since this treatment has three added advantages over annealing alone. Firstly, the dendrite structure is broken up by the deformation treatment so that *A*-rich areas are brought in contact with *B*-rich areas, thereby reducing the distance over which diffusion must occur. Secondly, imperfections introduced by deformation accelerate the rate of diffusion during a subsequent anneal. Thirdly, deformation followed by annealing often leads to recrystallization during which the cast structure is replaced by a new equiaxed uniform grain structure (*see* Chapter 10).

3.2.2 Complete insolubility in the solid state

When two metals are insoluble in each other in the solid state their alloying behaviour is demonstrated by a simple eutectic system; an approximate example of this system is lead–antimony. The characteristics of this diagram, shown in *Figure 3.3*(*a*), may be summarized as follows: (1) the two liquidus curves *AE* and *EB* meet at a minimum intersection point *E*, called the eutectic point, and (2) all alloys become completely solid at the same temperature, i.e. the solidus line is a horizontal line at the eutectic temperature. The diagram has four phase fields: liquid, (*A* + liquid), (*B* + liquid) and (*A* + *B*), and the microstructure of any alloy in such a system may be deduced after considering the cooling of alloy *C*. On reaching the liquidus, the first solid to be deposited is of composition *A* and with further cooling the liquid continues to deposit solid *A* while the composition of the liquid 'travels down the liquidus'. When the eutectic temperature is reached the structure consists of solid *A*-crystals in a liquid of composition *E*; the relative amounts of liquid and *A*-crystals are given according to the lever rule, by the ratio *FG*/*GE*. Below this temperature the remaining liquid freezes by the solidification of small quantities of *A* and *B* side by side, and this gives rise to a duplex structure characteristic of all eutectic alloys. This duplex structure arises because the initial formation of an *A*-crystal causes the surrounding liquid to be left supersaturated in *B*, with the result that conditions are then favourable for the *B*-crystal to be precipitated next to the *A*-crystal already formed. Further cooling of the alloy from the eutectic temperature does not alter the composition or structure of the phases, since there are no changes in phase composition in the solid state.

It is evident that the phase *A* separates in two different stages, since primary *A* forms while passing through the (*A* + liquid) region, and also during the freezing of the liquid of eutectic composition when some secondary *A* is formed. These different forms of *A* are clearly distinguishable in the optical microscope, as shown in the inset in *Figure 3.3*(*a*), since the phase *A* formed at the eutectic decomposition is often more finely dispersed.

The cooling curve of the alloy *C* is shown in *Figure 3.3*(*b*). A change of slope is evident at the temperature T_1, due to the evolution of heat accompanying the separation of *A*-crystals. The curve is not horizontal, however, because pure *A*-crystals continue to separate out as the temperature falls. In contrast, when the eutectic separates out, the cooling curve does become horizontal and the

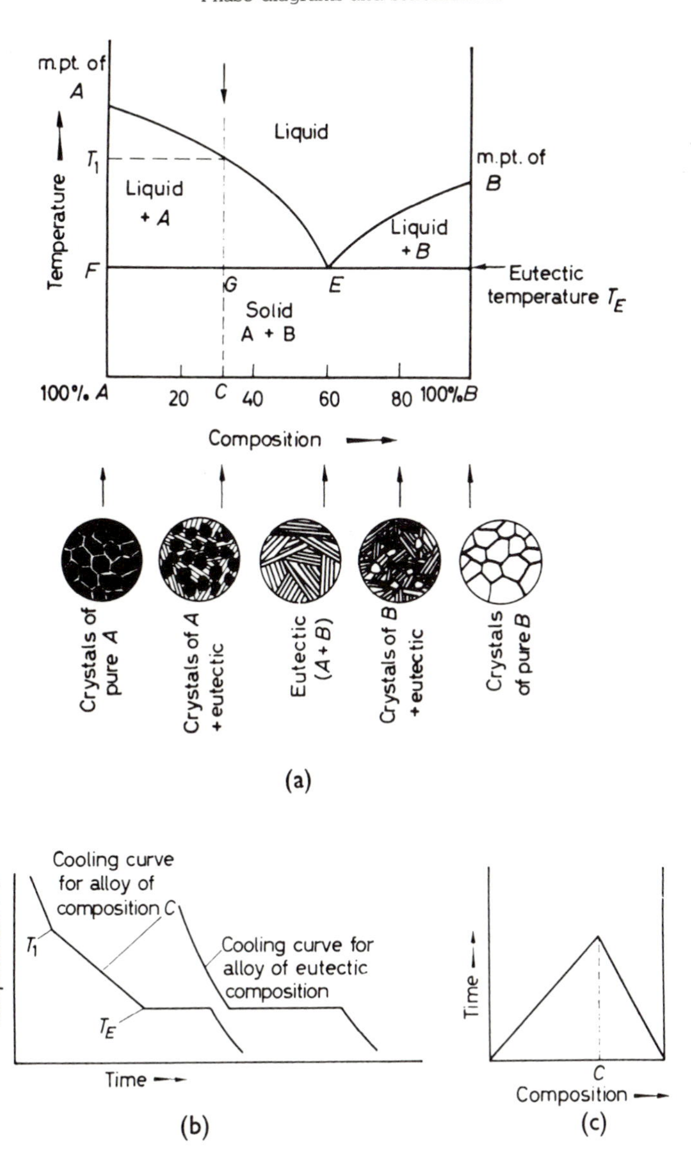

Figure 3.3 (a) Simple eutectic diagram with microstructures of some selected alloys, (b) typical cooling curves of a eutectic and non-eutectic alloy, and (c) dependence of the duration of cooling arrest at eutectic temperature on composition

duration of the arrest is proportional to the amount of eutectic in the alloy. It therefore follows that this type of diagram could be completely determined merely by examining the duration of the arrest on a cooling curve as a function of alloy composition, as illustrated in *Figure 3.3(c)*. However, even in the case of such a simple equilibrium diagram, it is usual to confirm the details by means of metallographic and x-ray examination.

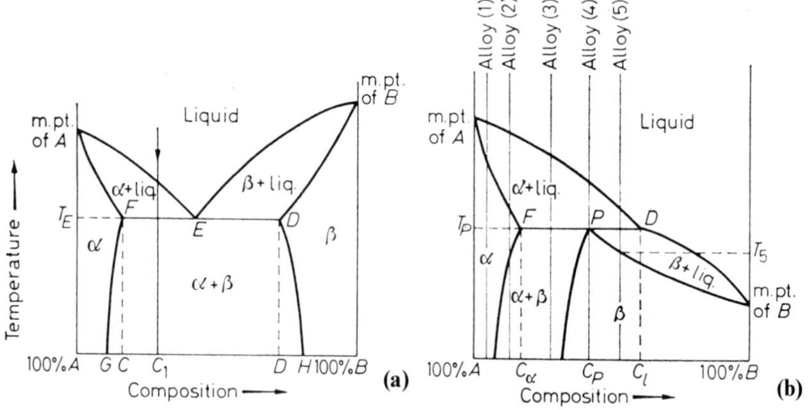

Figure 3.4 Phase diagrams showing (a) eutectic and (b) peritectic, both with limited solid solubility

3.2.3 Partial solubility in the solid state

The next alloy system to be examined is that in which the component metals *A* and *B* are only partially soluble in each other in the solid state. If the two metals have similar melting points the diagram will again exhibit a eutectic, but if the melting points differ widely the resultant diagram is likely to exhibit a peritectic reaction. The form of both of these diagrams is shown in *Figure 3.4(a)* and *(b)* respectively, from which it can be seen that each system is made up of six phase fields: liquid, α, (α + liquid), β, (β + liquid) and ($\alpha + \beta$). The primary solution of *B* in *A* in the solid phase is denoted by the α-solid solution, while that of *A* in *B* is denoted by the β-solid solution. The limit of primary solubility of the one metal in the other is normally greatest at the eutectic, or peritectic, temperature and at lower temperatures the solubility decreases. The extent of the primary solid solution also depends on the characteristics of the two metals *A* and *B* (*see* Chapter 5).

Let us consider first the eutectic-type of diagram shown in *Figure 3.4(a)*; common examples of this system include the copper–silver, lead–tin and aluminium–silicon alloys. Alloys of *B* and *A* up to a solute content *G*, solidify as described for the simple solid solution series. Alloys of composition greater than *H* also behave similarly, and it is clear that unless cooling is extremely slow a cored solid solution will result. Alloys in the composition ranges *G* to *C* or *D* to *H* solidify as α or β solid solutions respectively, but, on further cooling, as the primary solid solubility boundary is crossed, deposit respectively β or α from solution, the amount increasing with decreasing temperature. All alloys with compositions between *C* and *D* solidify in the manner previously described for the simple eutectic with the exception that α-, or β-solid solutions and not pure metals separate from the liquid. An alloy of composition C_1 on cooling down to the eutectic temperature, T_E, consists of liquid and α-phases in the proportions given by the lever rule. As T_E is approached the amount of α increases, and the concentration of the metal *B* in the α-phase also increases. Immediately the temperature of the alloy falls below T_E the remaining liquid phase becomes unstable and freezes as the eutectic mixture of ($\alpha + \beta$). Below T_E, however, i.e. from

T_E to room temperature, the compositions and the relative proportions of the α and β phases, do not remain constant but change as indicated by the sloping primary solid solubility curves of both the α and β phase fields.

If equilibrium is to be maintained during all stages of cooling, extensive atomic migration, or diffusion, is necessary in order that the actual phase compositions can change continuously as indicated by the equilibrium diagram. However, these changes in the solid state take a considerable time to complete, and often they may be prevented, or at least retarded, by quenching or even rapid cooling; a fact which forms the basis of many heat-treatment processes. In contrast, if the cooling rate of an alloy of composition between G and C is sufficiently slow the α-phase will decompose to give more β-phase as the alloy cools down below the temperature T_E. Then, the β-phase precipitates in one or more of three preferential places: (1) at grain boundaries, (2) around dislocations and inclusions, and (3) on specific crystallographic planes. The choice of the particular precipitating site depends on several factors, of which the grain size and rate of nucleation are particularly important. If the grain size is large, the amount of grain boundary area is relatively small, and β-phase deposition within the grains is then more likely; when specific crystallographic planes within a grain are used as precipitation sites, e.g. $\{111\}$ planes, the structure has a mesh-like appearance in the microscope, which is known as Widmanstätten structure.

The peritectic system—Although this type of system, as shown in its simplest form in *Figure 3.4*(*b*), occurs only occasionally in practice (one example is the silver–platinum system) the peritectic reaction is a rather common feature of some of the more complicated diagrams (*see* for example *Figure 3.5*).

The alloys labelled 1 and 2 solidify as discussed previously. Thus, the alloy 1 forms a cored α-solid solution unless cooled extremely slowly, while alloy 2 solidifies first as an α-solid solution, but subsequently the β-phase precipitates out in increasing amounts as the $(\alpha+\beta)$ phase boundary is crossed. The alloy 4, which is of the peritectic composition, begins to solidify by depositing α crystals and continues to do so until the peritectic temperature, T_P, is reached; the structure of the alloy at this stage then consists of α, of composition C_α, and liquid, of composition C_l. Immediately below the temperature T_P, however, a reaction, somewhat like a chemical reaction, takes place according to the relation

$$\alpha(C_\alpha)+\text{liquid}(C_l)\rightarrow\beta(C_P)$$

For this reaction to go to completion the alloy must be of the peritectic composition C_P, when an entirely new single phase β is produced.

When the composition of the alloy is to the left of the peritectic composition C_P, as for example alloy 3, then at T_P it will contain more α than is required by the above reaction. The structure produced on cooling below T_P then consists of $(\beta+\alpha)$ according to the reaction

$$\alpha(C_\alpha)+\text{liquid}(C_l)\rightarrow\beta(C_P)+\text{excess }\alpha(C_\alpha)$$

and the proportions of each phase at a given temperature are given by the lever rule. If the composition of the alloy is to the right of C_P, as for example alloy 5, then at T_P the alloy will contain more liquid than is required for the peritectic reaction to go to completion. Following the peritectic reaction the immediate

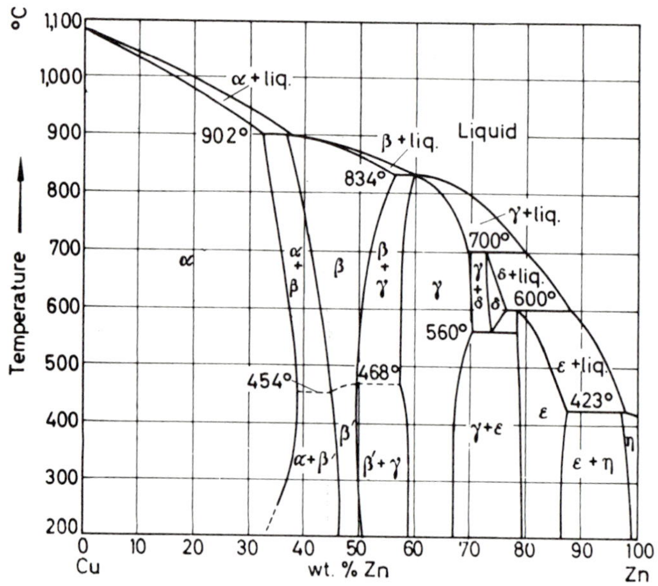

Figure 3.5 (a) Phase diagram for copper–zinc (from G. V. Raynor, *Annot. Equilib. Diag. No. 3*, courtesy of the Institute of Metals)

structure produced is (β + liquid) but on further cooling the composition of each of the two phases varies with temperature in such a way that the composition of the β moves along the line PB and the composition of the liquid moves along the line DB. The relative amount of liquid also decreases progressively until at a temperature T_5 the alloy is completely β. Alloys with composition between D and 100 per cent B will, of course, form a β-solid solution as described previously.

3.2.4 Important phase diagrams

The phase diagrams of most alloy systems contain more than one of the characteristics so far discussed. To illustrate this, let us consider the well-known alloy system of copper and zinc, which gives rise to the important industrial alloys known as the brasses (*Figure 3.5*(*a*)). Copper dissolves up to approximately 40 per cent zinc in primary solid solution, and the cooling of any alloy in this range, which of course will be f.c.c., produces a typical α-solution structure; by contrast the other primary solid solution, the c.p.h. phase, is extremely limited. A special feature of the diagram is that it contains several secondary solid solutions, i.e. phases which occur in the middle of the diagram with a fairly wide range of homogeneity, and it can be seen that each of these intermediate phases is formed during freezing from a peritectic reaction. It is also evident that the δ-phase becomes unstable below 560°C and decomposes by a eutectoid* change into $\gamma + \varepsilon$. The diagram is also notable for the β-phase order–disorder transformation which occurs in alloys of about 50 per cent zinc in the temperature range 450 to 470°C. Above 450°C, the b.c.c. β-phase exists as a disordered solid solution (*see* p.167), but at lower temperatures the zinc atoms are distributed regularly on the b.c.c. lattice and this ordered phase is denoted by β'.

Another interesting diagram, shown in *Figure 3.5*(*b*), is that for the iron–carbon system, which gives rise to the group of alloys known as steels. The

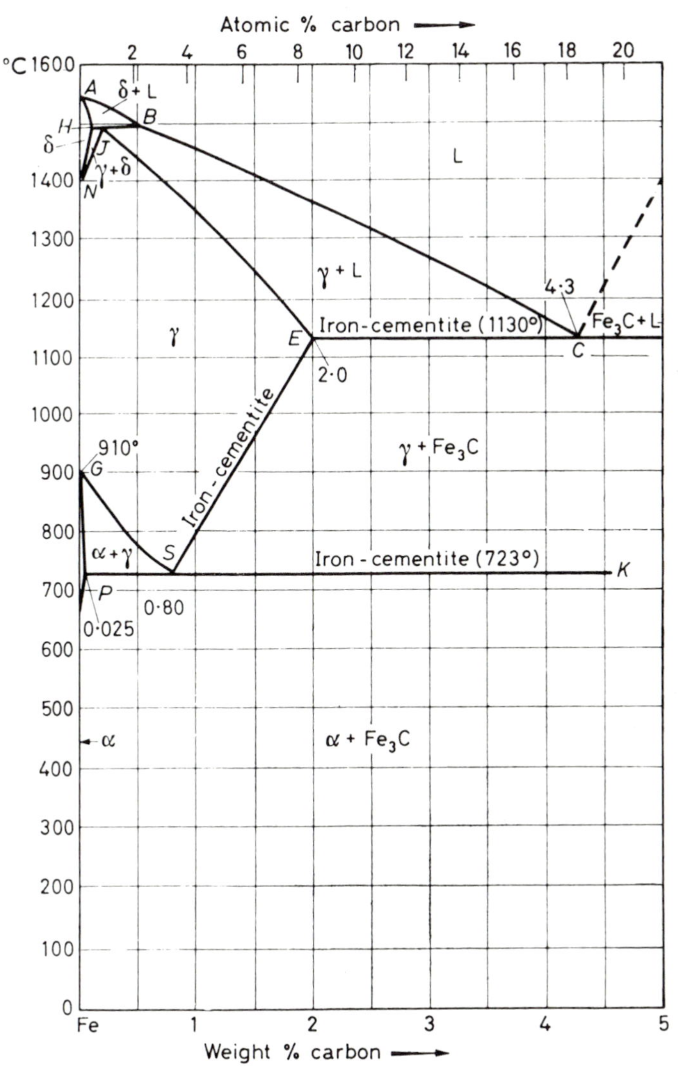

Figure 3.5 (b) Phase diagram for iron–carbon systems (from Mehl and Hagel, *Progress in Metal Physics 6*, courtesy of Pergamon Press)

polymorphic changes which take place in pure iron result in the formation of three different single phase fields in iron-rich alloys, namely: α-iron, or ferrite (b.c.c.), γ-iron, or austenite (f.c.c.) and δ-iron (b.c.c.)*. The temperatures at which these phase changes take place are known as *A*-points as a result of the arrests noted on a cooling curve. In addition to the solution formed with carbon in b.c.c. and f.c.c. iron, a compound known as cementite, Fe_3C, is also formed, and the diagram in the range 0 to 2.0 per cent carbon, where the austenite transforms by a eutectoid reaction to ferrite and cementite, is similar to that of the eutectic diagrams already discussed. The alternate formation of ferrite and cementite in

* Eutectoid and peritectoid reactions occur in the solid state, and are exactly analogous to the eutectic and peritectic changes involved in the solid–liquid transformation.

* β-iron does not exist, but at one time the term was used to denote the non-magnetic form of α-iron which exists above the Curie point.

eutectoid steel, 0.80 wt per cent carbon, gives rise to a finely divided eutectoid structure known as pearlite.

3.2.5 Limitations of phase diagrams

Phase diagrams are extremely useful in the interpretation of alloy structures but are subject to several restrictions. The most important limitation is that the diagram gives information only on the constitution of alloys and not on the structural distribution of the phases. This is unfortunate, since the structural distribution of phases, which depends upon the surface energy between phases and the strain energy caused by a transformation (*see* Chapter 11), plays an important role in the mechanical behaviour of alloys. This is easily understood if we consider a duplex alloy of $(\alpha+\beta)$ containing only a small amount of β-phase. The β-phase may be distributed entirely within the α-grains, in which case the mechanical properties of the alloy would be largely governed by that of the α-phase, but if the β-phase is deposited around the α-phase grain boundaries, then the strength of the alloy will be largely dictated by that of the β-phase.

A second limitation is that the diagrams show only the equilibrium state, whereas alloys in practical use are rarely in equilibrium. By suitable heat treatment, such as quenching, it is often possible to retain the high temperature phase at room temperature, and in some cases to produce structures other than those shown on the equilibrium diagram (e.g. martensite, *see* Chapter 12). Clearly, in the use of such alloys it is important to know at what rate the equilibrium condition will be attained, and the effect of temperature on this rate of approach to equilibrium. In dealing with problems of phase equilibrium the phase rule of Willard Gibbs is extremely useful. This states that for solid and liquid systems in which temperature and composition are the only significant variables $P+F=C+1$, where P is the number of phases in equilibrium with each other, F is the number of degrees of freedom and C is the number of components. For a pure metal $C=1$, so that $F=2-P$ and hence for a two-phase state to coexist in equilibrium $F=0$ and there are no degrees of freedom, i.e. the temperature cannot vary. The solid and liquid phases coexist at a single temperature, i.e. the melting point. In a binary alloy $C=2$ and hence $F=3-P$. A single phase alloy can exist over a range of both temperature and composition. In a two-phase alloy, the solid and liquid phases can exist over a range of temperature, but for three phases to coexist $F=0$ and corresponds to the eutectic or peritectic point.

3.3 Constitutional undercooling

We have seen that small amounts of solute elements profoundly affect the solidification of metals. On freezing, the solute atoms tend to remain in the liquid, rather than solidify with the solvent atoms at a crystal–liquid interface. The solute rejected by the crystals becomes concentrated in the remaining liquid as a result of which the freezing point of the liquid decreases. The liquid just ahead of a freezing solid–liquid interface is, however, more enriched than liquid further away, and hence the composition of the liquid varies with distance from the interface as shown in *Figure 3.6*. There is a corresponding variation with distance

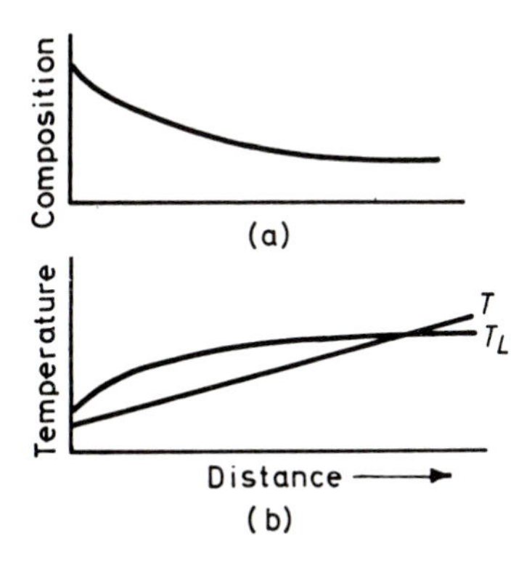

Figure 3.6 Variation with distance from the solid–liquid interface of (a) composition and (b) actual temperature T and freezing temperature T_L

of the temperature at which the liquid will freeze, since liquid close to the interface will have its freezing temperature lowered more than liquid further away. Consequently for a positive temperature gradient, as shown in *Figure 3.6*(*b*), there is a layer of liquid in which the actual temperature T is below the freezing temperature T_L; the liquid is constitutionally undercooled.

With reference to the simple binary shown in *Figure 3.2* a distribution coefficient k can be defined as the ratio of the solute concentration in the solid to that in the liquid with which it is in equilibrium, i.e. c_S/c_L. For an alloy of composition c_0, the first solid to freeze is kc_0 where $k<1$, and the liquid adjacent to the solid becomes richer in solute than c_0. The next solid to freeze will have a higher solute concentration and eventually, for a constant rate of growth of the solid–liquid interface, a steady state is reached for which the solute concentration at the interface reaches a limiting value of c_0/k and decreases exponentially into the liquid to the bulk composition. This profile is illustrated in *Figure 3.6*(*a*) and can be shown by applying Fick's second law of diffusion (*see* page 144) to be

$$c_L = c_0\left[1+\frac{1-k}{k}\exp\left(-\frac{R}{D}x\right)\right]$$

where x is the distance into the liquid ahead of the interface, c_L is the solute concentration in the liquid at the point x, R the rate of solidification, and D the diffusion coefficient of the solute in the liquid. The temperature distribution in the liquid can be calculated if it is assumed that k is constant and the liquidus line is a straight line of slope m. For the curves of *Figure 3.6*(*b*),

$$T = T_0 - mc_0/k + G_L x$$

and

$$T_L = T_0 - mc_0\left[1+\frac{1-k}{k}\exp\left(-\frac{R}{D}x\right)\right]$$

where T_0 is the freezing temperature of the pure solvent metal, T_L the liquidus temperature for the liquid of composition c_L, T the real temperature at any points x, and G_L is the temperature gradient in the liquid. The zone of constitutional undercooling can be eliminated by increasing the temperature gradient, such that

$$\frac{dT}{dx} = G_L > \frac{dT_L}{dx}$$

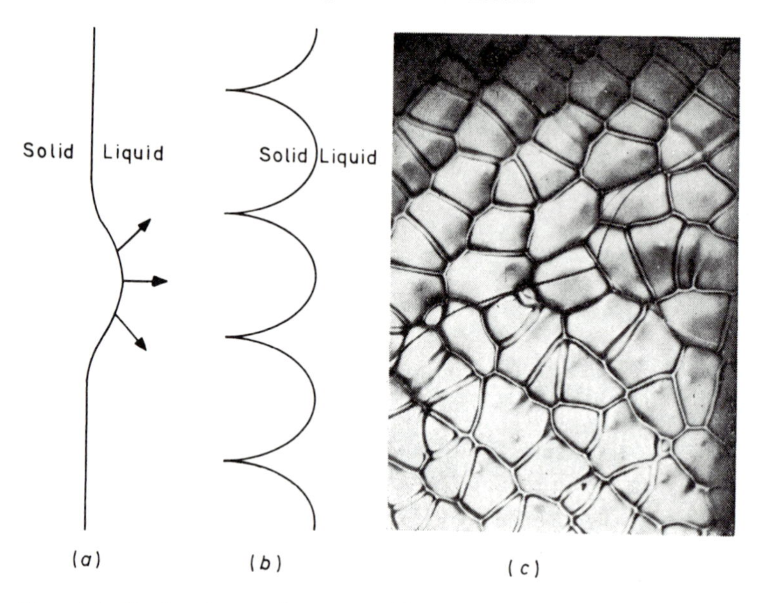

Figure 3.7 The breakdown of a planar solid–liquid interface (a), (b), leading to the formation of a cellular structure of the form shown in (c) for Sn/0.5 at % Sb × 140

Substituting for T_L and putting $[1-(R/D)x]$ for the exponential gives the condition

$$G_L > \frac{mc_0R}{D}\left(\frac{1-k}{k}\right) \tag{3.1}$$

As a result of the zone of undercooled liquid ahead of the interface a macroscopically planar solid–liquid interface becomes unstable and an interface with a cellular morphology develops. Because of the undercooling, the interface projects out into the liquid and the subsequent rapid freezing enables it to grow still more. The initial projecting interface (*see Figure 3.7*) rejects solute into the liquid at its base which thus remains unfrozen. The interface eventually breaks up into dome-shaped growth cells with a profile of the form shown in *Figure 3.7(b)*. The solute-rich liquid in the regions between the cells eventually freezes at a much lower temperature and a crystal with a periodic columnar distribution of solute or impurity is produced. The cells are usually hexagonal in shape about 0.1 to 0.05 mm in diameter, as illustrated in *Figure 3.7(c)* and the cell walls contain a higher concentration of solute than the centre. The growth conditions in the transition from planar to cellular structures is shown in *Figure 3.8*.

It is often found that superimposed on the cellular structure, but coarser in scale, is another form of substructure known as the lineage or macromosaic structure. This structure arises within a given crystal originating from the same nucleus, where because of thermal effects different parts of the crystal vary in orientation by small angles ($\approx 1°$). The lineage boundaries between the mis-oriented regions are roughly parallel to the direction of growth, and are equivalent to an array of edge dislocations (*see* page 211). When the extent of the

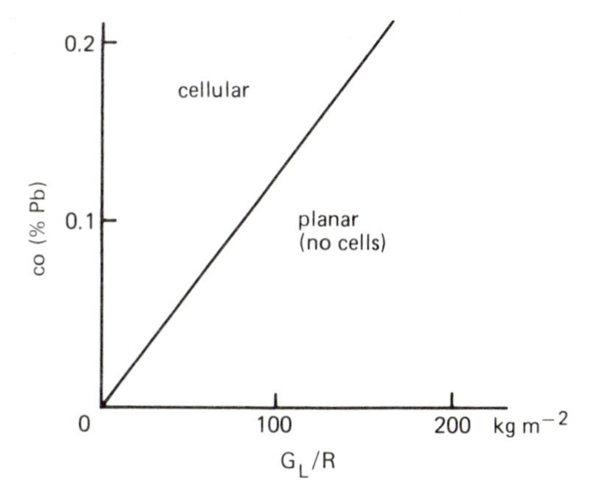

Figure 3.8 Conditions for cellular growth in tin-containing lead

undercooling into the liquid is increased as for example, by reducing the temperature gradient, the cellular structure is not stable and a few cells grow rapidly as cellular dendrites. The branches of the dendrites are interconnected and are an extreme development of the dome-shaped bulges of the cell structure in the direction of rapid growth. The growth of dendrites in a constitutionally undercooled alloy is slower than in a pure metal, because the solute must diffuse away from the dendrite–liquid surface, and is limited to the undercooled zone. From the relationship given in equation 3.1, it is evident that to ensure a planar interface and reduce the tendency cellular and dendritic structures requires high G_L, low R and low c_0.

3.4 Metal structures

When a metal is solidified the grain structure is usually not as uniform as that discussed in section 3.1, the extent of the non-uniformity being dependent on the rate of cooling and other factors. When metal is poured into a metal mould for example, the metal in contact with the mould wall is cooled very rapidly. This high cooling rate at the surface influences the nucleation of crystals to a greater extent than their growth with the result that many fine-grained equiaxed crystals are formed as shown in *Figure 3.9*. With increasing thickness of the chilled zone of

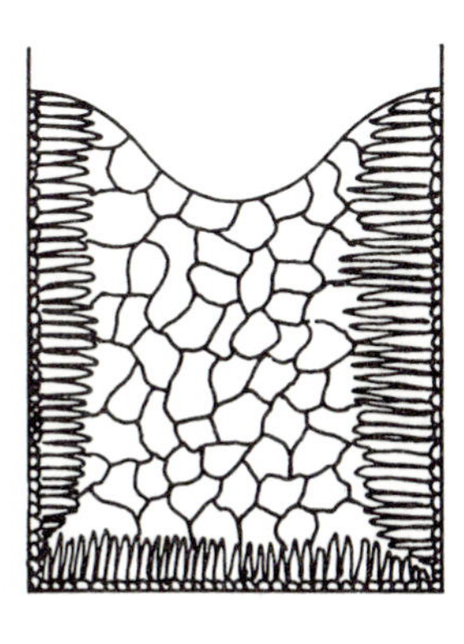

Figure 3.9 Chill-cast ingot structure

crystals, the temperature gradient from the liquid to the mould wall becomes less steep and the rate of cooling decreases. Crystal growth rather than nucleation of new crystals is then favoured and, as a consequence, the crystals at the liquid–metal interface of the chilled zone begin to grow towards the centre of the mould, so forming long columnar crystals. Some crystals have orientations such that their rapid growth direction, e.g. $\langle 100 \rangle$ in cubic crystals, is along the direction of heat flow, and hence they grow more rapidly than their neighbours. During growth these crystals widen and stifle the growth of slower growing crystals from the chilled zone so that this selective growth process gives rise to a preferred orientation in the columnar grains.

In pure metals the columnar zone may extend to the centre of the ingot, but in alloys particularly, the liquid ahead of the growing columnar crystals is constitutionally undercooled and heterogeneous nucleation of new crystals takes place. These nuclei grow into coarse, randomly oriented, equi-axed crystals at the centre of the ingot. As mentioned in the previous section, a lowering of the temperature gradient increases the extent of the constitutional undercooling and hence the possibility of fresh nucleation. This is just the condition that exists in the final stages of solidification in an ingot when the temperature gradients have largely flattened out.

To produce good mechanical properties it is generally necessary to develop a uniform, fine grain size. This may be achieved by reducing the length of the columnar zone and promoting nucleation ahead of the solidifying metal front. To do this, the required amount of constitutional undercooling is introduced as quickly as possible by adding some solute to increase the difference between liquidus and solidus and by reducing the temperature of the melt. In addition, various 'foreign' nucleating agents have been discovered empirically which provide good centres for heterogeneous nucleation, e.g. Ti in aluminium which forms TiN particles, C in magnesium alloys, and Al in steels (*see* page 135).

During solidification of an ingot, segregation always occurs. This may be of two different forms, (i) normal and (ii) inverse segregation. The normal form is the direct result of solute rejection at an advancing interface and increases as the rate of solidification increases and as k (*see* page 101), the distribution coefficient decreases. Inverse segregation is caused by interdendritic flow of enriched liquid to the outer parts of the ingot by the pressure set up by solid contraction and volume shrinkage in this region of the casting. This form of segregation increases with increasing fluidity of the liquid and with a lowering of the freezing rate.

3.5 Zone refining

An important application of the non-equilibrium effect associated with the liquidus–solidus curves has been developed by Pfann, and is known as zone refining. This technique has been applied to the purification of many metals, particularly silicon and germanium for the electronics industry, and depends upon the fact that for an alloy in the (α + liquid) condition (*see Figure 3.2*), where the liquidus temperature falls as the solute content increases, the liquid will always contain more solute than the solid, i.e. the distribution coefficient $\kappa = c_s/c_l$ is less than unity. Thus, if a travelling furnace is passed over an alloy bar so that a narrow molten zone moves along the bar, some solute will be transported from the starting end to the other end. The first solid to freeze is purer than the average by a factor of k, while that which freezes last is correspondingly enriched. By repeatedly passing zones along the bar the impurities in some metals can be reduced well below the limit of detection (e.g. less than 1 part in 10^{10} for germanium).

3.6 Growth of single crystals

Metal specimens containing no grain boundaries, i.e. single crystals, have been grown for many years for research studies. With the development of semiconductors, single crystals of silicon, germanium and various compounds have been grown industrially with a resultant refinement in technique. There are two basic methods of preparation involving (i) solidification from the melt, and (ii) grain growth in the solid state.

In the simplest solidification method the polycrystalline metal to be converted to a single crystal is supported in a horizontal boat of graphite or other non-reacting material and made to freeze progressively from one end by passing an electrical furnace, with a peak temperature set 10°C or so above the melting point, over the boat. Although several nuclei may form during the initial solidification, the sensitivity of the growth rate to orientation usually results in one of the crystals swamping the others and eventually forming the entire growth front. The method is particularly useful for seeding crystals of a pre-determined orientation. The seed crystal of given orientation is placed adjacent to the polycrystalline sample in the boat and the junction melted before commencing the progressive melting and solidification process. Wires may be grown in heat-resistant glass or silica tubes suitably internally coated with Aquadag. A modern development of these methods is the use of a water-cooled copper boat to hold the sample in an evacuated silica tube over which is passed a high frequency heating coil to produce a molten zone.

Most techniques are derived from the Bridgman and Czochralski methods. In the former, a vertical mould tapered at the bottom end is lowered slowly through a tubular furnace with a temperature gradient. In the Czochralski method, often referred to as the crystal pulling technique, a seed crystal is withdrawn slowly from the surface of a molten metal to enable the melt to solidify with the orientation of the seed. The crystal is usually rotated as it is withdrawn to produce a cylindrical crystal. This technique is commonly used for the preparation of Si and Ge crystals when the whole operation is carried out in vacuo.

Crystals may also be prepared vertically by a 'floating' zone technique and has the advantage that the specimen is no longer in contact with any source of contamination. The polycrystalline rod is gripped at the top and bottom in water cooled grips and a small molten zone, produced by a high frequency coil or electron bombardment from a circular filament, is passed up the sample. Many high melting point metals such as W, Mo, and Ta, can be produced by this

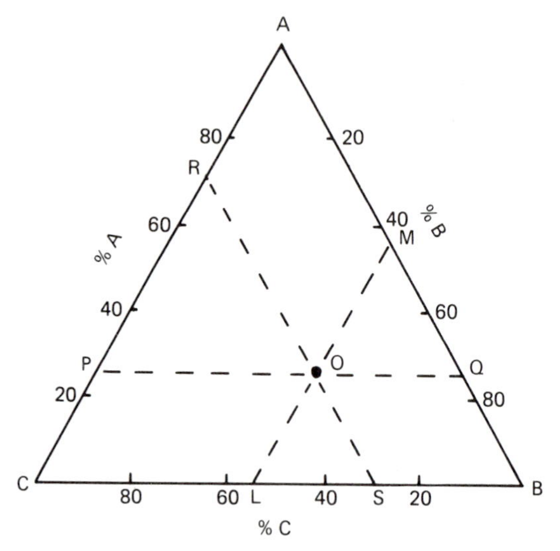

Figure 3.10 Triangular representation of a ternary alloy system ABC

method in a purified state, since a zone refining action is also introduced.

Methods involving grain growth in the solid state depend on the annealing of deformed samples. In the strain anneal technique a fine grained polycrystalline metal is critically strained approximately 1–2 per cent elongation in tension and subsequently annealed in a moving gradient furnace with a peak temperature set below the melting point or transformation temperature. By lightly straining, very few recrystallization nuclei are formed (*see* page 367) and during the annealing treatment one favourable nucleus grows more rapidly than the others and consumes the other potential nuclei. The method has been applied to high stacking fault energy metals such as Al and silicon-iron. Single crystals are difficult to grow in metals with low stacking fault energy such as Au and Ag because of the ease of formation of annealing twins (*see* page 367) which lead to multiple orientations. Hexagonal metals are also difficult to prepare because deformation twins formed during deformation act as effective nucleation sites.

3.7 Ternary equilibrium diagrams

In considering equilibrium diagrams for ternary systems three independent variables have to be specified, i.e. two to define the composition and the third to define temperature. Consequently a 3-D model is required and an equilateral triangle ABC, with a temperature axis perpendicular to it is used. The corners of the triangle represent the pure metals, the sides the three appropriate binaries and a point inside the triangle represents a ternary alloy composition. For the ternary alloy O in *Figure 3.10*, the concentration of the components is given by $C_A = PC$, $C_C = RA$ and $C_B = RP$ and since $C_A + C_B + C_C = 1$, then $PB + RA + RP = 1$. From the triangular representation ABC all alloys lying on a line (i) parallel to a side must have the same composition of the component opposite to the base, e.g. C_A is

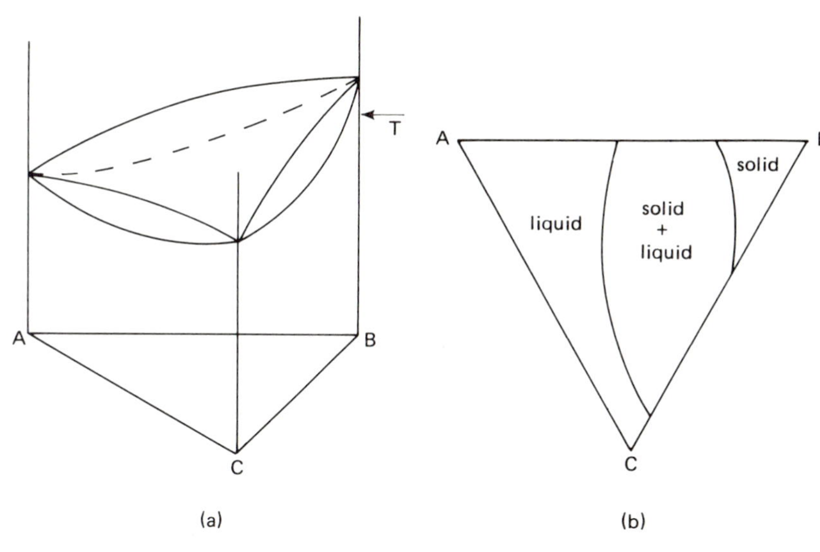

Figure 3.11 (a) Ternary diagram for a system with complete solid solubility

Figure 3.11 (b) Horizontal section at temperature T through the diagram shown in (a)

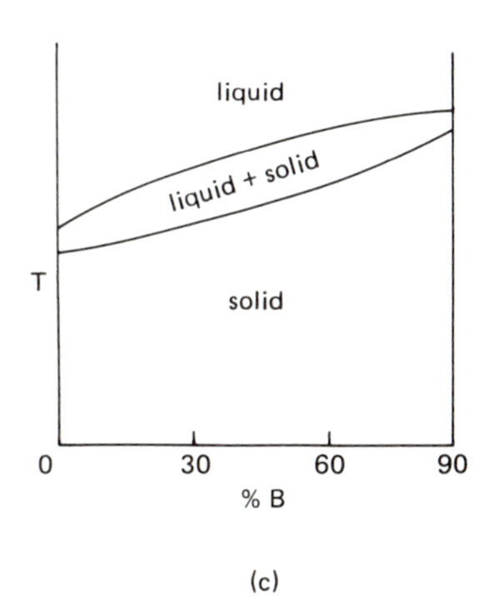

Figure 3.11 (c) Vertical section through diagram shown in (a)

constant along PQ, and (ii) through one corner have a constant proportion of the other two components.

3.7.1 Ternary diagram for complete solid solubility

This is shown in *Figure 3.11(a)*. The binaries along each side are similar to that of *Figure 3.2*. The liquidus and solidus lines of the binaries become liquidus and solidus surfaces in the ternary system, the (solid + liquid) region forming a convex-lens shape and the solid phase becoming a volume in the ternary bounded by a surface. An example of this sytem is Ag–Au–Pd.

A series of isothermal or horizontal sections can be examined over the complete temperature range to give a full representation of the equilibrium

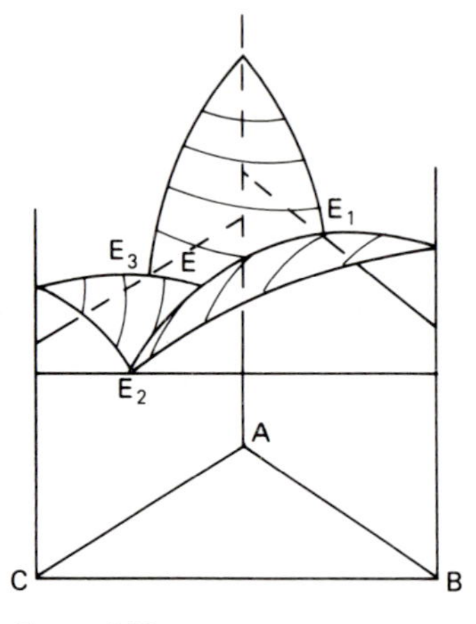

Figure 3.12 (a) Schematic ternary system for complete immiscibility in solid state

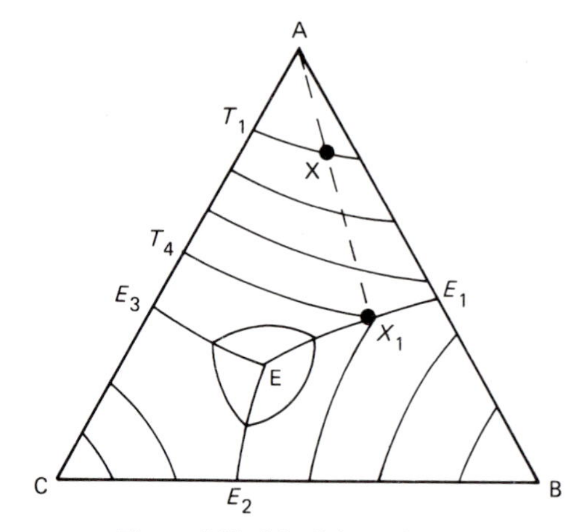

Figure 3.12 (b) Schematic ternary system for projection of liquidus surface on to base

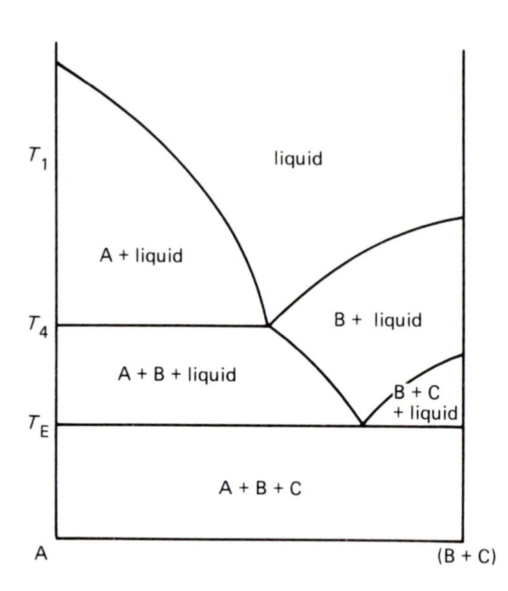

Figure 3.12 (c) Schematic ternary system for a vertical section

diagram. For the simple diagram of *Figure 3.11(a)* horizontal sections from room temperature up to the solidus surface show no variation, merely solid phase. At temperature T, the horizontal section cuts through the (solid + liquid) phase field and looks like *Figure 3.11(b)*; alloys can be liquid, solid or (solid + liquid) depending on the composition. The liquidus surface varies in extent with temperature increasing with decreasing temperature and can be shown as a 'contour' line on the diagram.

Vertical sections may also be useful, particularly if they are taken either (i) parallel to one side of the base triangle, i.e. at constant proportion of one of the components (*see Figure 3.11(c)*, or (ii) along a line through a corner of the triangle.

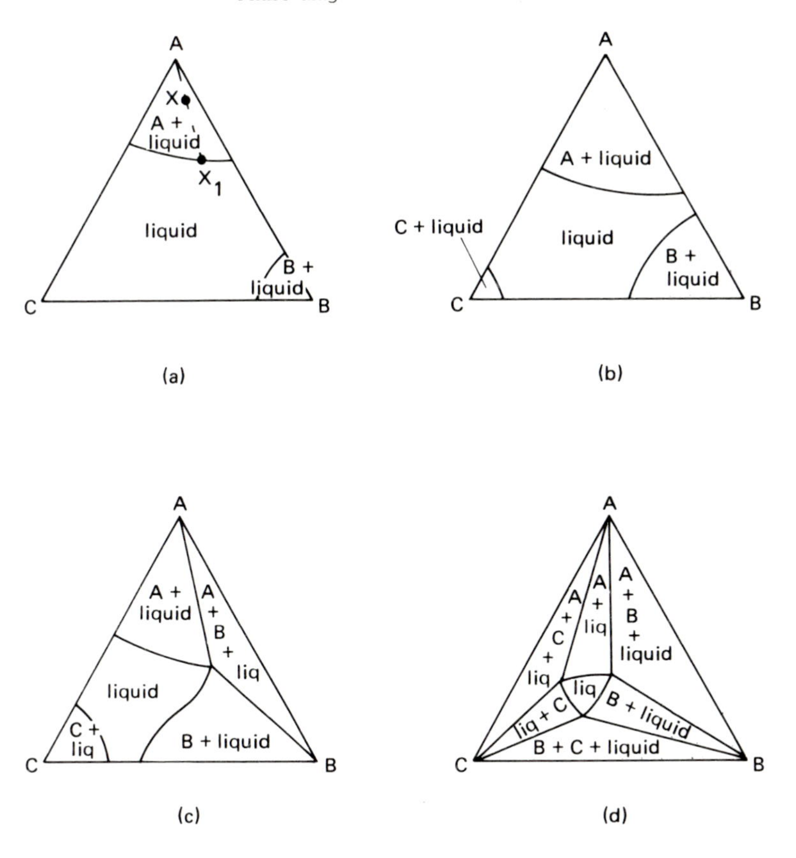

Figure 3.13 Horizontal sections at four temperatures above T_E through the ternary diagram in Figure 3.12 (a)

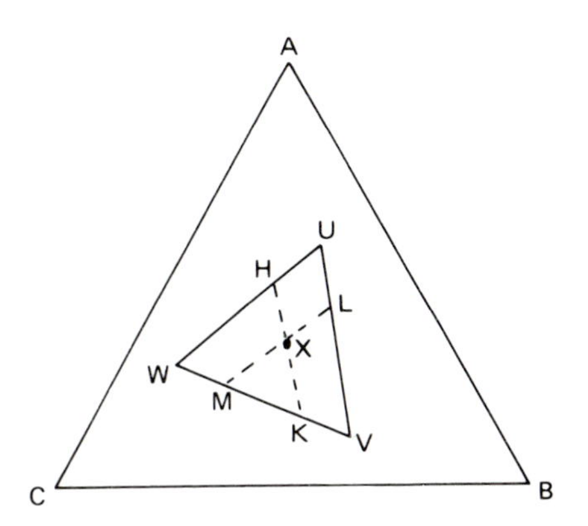

Figure 3.14 Constitution in a three-phase field UVW

3.7.2 Ternary eutectic

For the case where there is complete immiscibility in the solid state, e.g. Pb–Bi–Sn, the three binary eutectics are constructed on the sides of the diagrams, *Figure 3.12(a)*. The liquidus of the ternary forms three separate smooth liquidus surfaces, extending inwards from the melting point of each pure metal and sloping downwards to form three valleys which meet at E, the ternary eutectic point. A projection of this liquidus surface on to the base is shown in *Figure 3.12(b)*.

Let us consider the solidification of an alloy X which starts at the temperature T_1 when solid of pure A separates out and the liquid becomes richer in B and C. The ratio of B to C in the liquid remains constant as the temperature falls to T_4, when the liquid composition is given by the point X_1 in the valley between the two liquidus surfaces. This liquid is in equilibrium with both solid A (when solid/liquid $= XX_1/AX$) and pure B. On further cooling the composition of the liquid follows the valley X_1E and in the secondary stage of freezing pure B separates out as well as A. The freezing behaviour for B-rich or C-rich alloys is similar and the liquid phase eventually reaches either the valley E_1E or E_2E, depositing either A and C or B and C before finally reaching the ternary eutectic

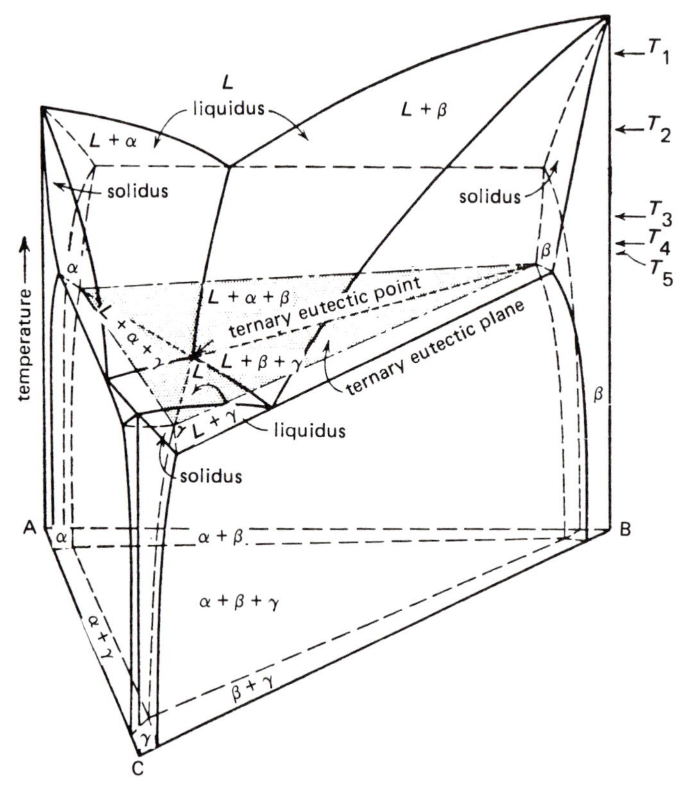

Figure 3.15 A ternary phase diagram with solid solutions (after Rhines, *Phase Diagrams in Metallurgy*)

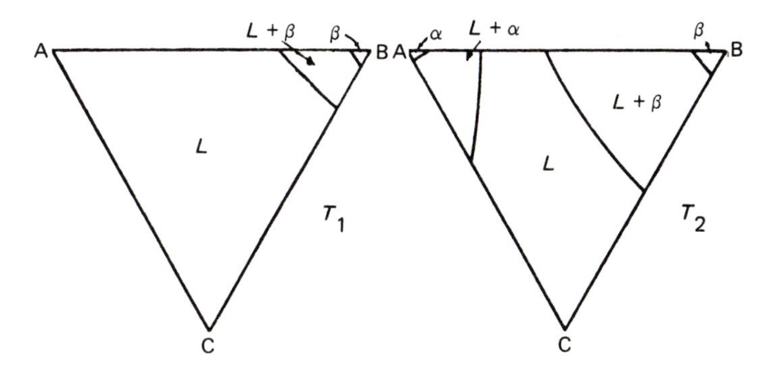

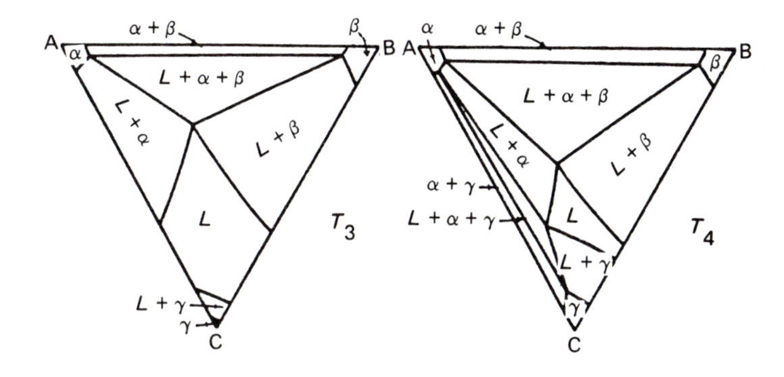

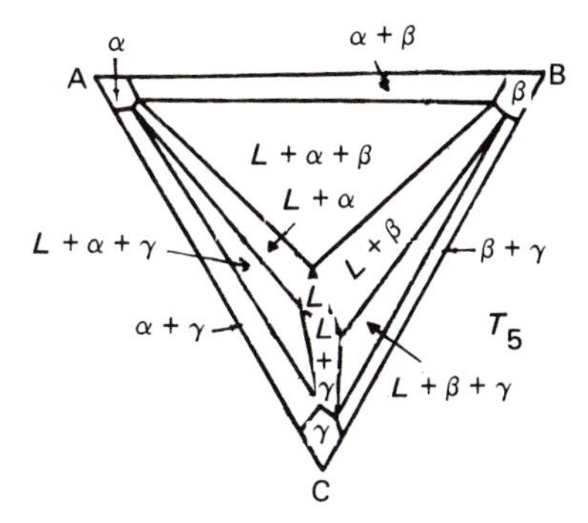

Figure 3.16 Horizontal sections of five temperatures above T_E through the ternary in Figure 3.15

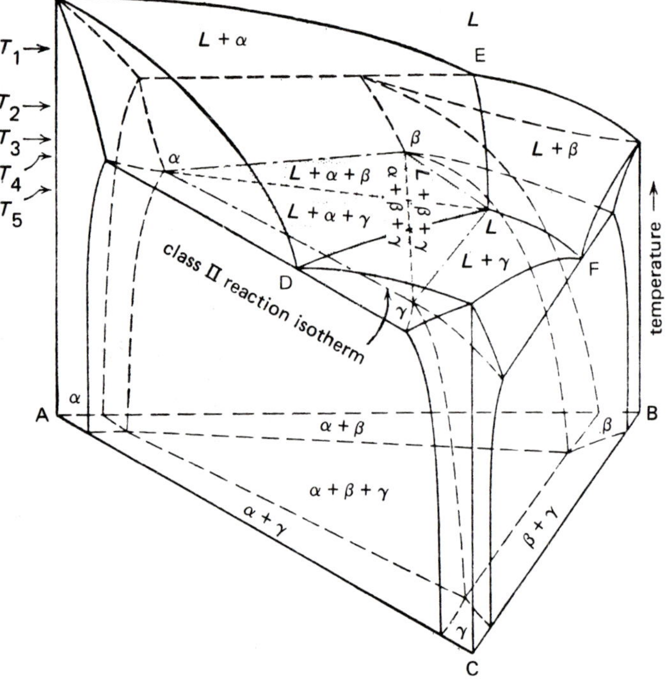

Figure 3.17 Ternary diagram with a peritectic (after Rhines, *Phase Diagrams in Metallurgy*)

point E. Here the liquid is in equilibrium with A, B and C, and at T_E the three-phase mixture freezes into the ternary eutectic structure. Every ternary alloy will completely solidify at the same temperature T_E and so the solidus or tertiary surface is a horizontal plane across the diagram; this can be seen in the vertical section shown in *Figure 3.12(c)*. The compositions and proportions of the phases present in an alloy at a given temperature can be found from horizontal sections. Four such sections at temperatures above T_E are shown in *Figure 3.13*. From these sections it can be seen that boundaries between single and two-phase regions are curved, between two- and three-phase regions, straight lines, and three-phase regions are bounded by three straight lines. In a three-phase field the compositions of the three phases are given by the corners of the triangle, e.g. UVW in *Figure 3.14*. For any alloy in this field only the proportions of U, V and W change and an alloy X has the proportions $U:V:W$ given by $HK:WH:VK$.

3.7.3 Ternary diagrams with solid solutions

If primary solid solubility exists, then instead of pure A, B, or C separating out, the solid solutions α, β or γ form. In *Figure 3.15* the liquidus surface has the same form as before, but beneath it there is a new phase boundary surface near each corner of the diagram, representing the limit of solid solubility in each of the pure metals A, B and C.

Horizontal sections above T_E are shown in *Figure 3.16* and below T_E the horizontal section is contained in the base triangle ABC of *Figure 3.15*. It is

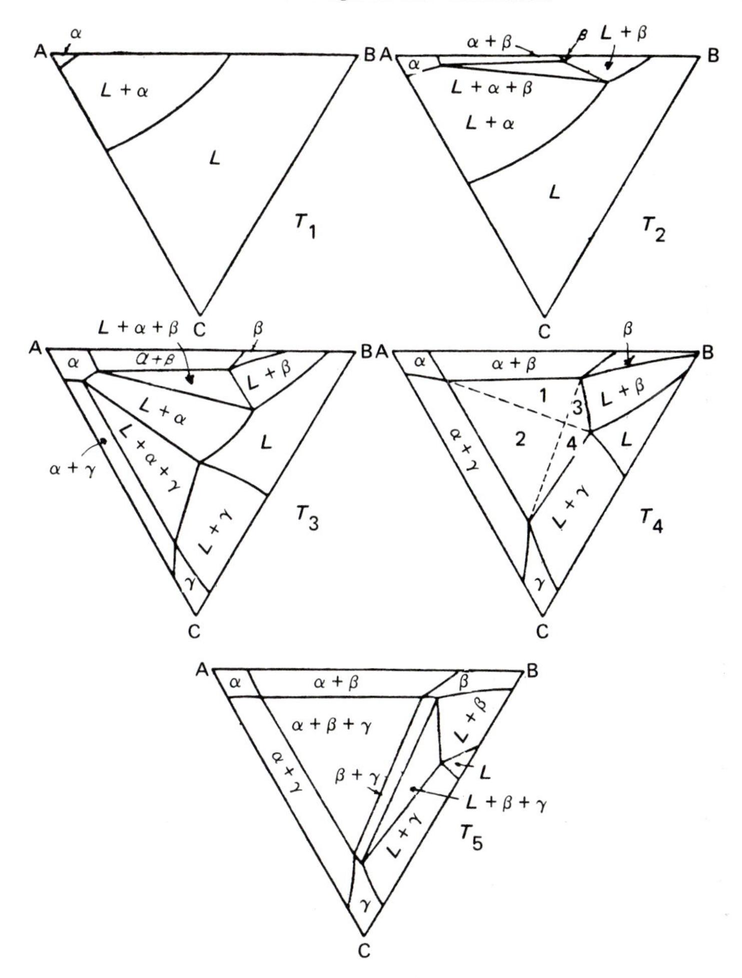

Figure 3.18 Horizontal sections at five temperatures through the diagram in Figure 3.17

readily seen that alloys with compositions near the corners solidify in one stage to primary solid solutions α, β or γ, while alloys near the sides of the diagram do not undergo a tertiary stage of solidification, since no ternary eutectic (which is now a phase mixture of α, β and γ) is formed.

3.7.4 Ternary diagrams with a peritectic

There are no new principles involved if other reactions are introduced, but a peritectic system AB is included in *Figure 3.17* to illustrate the way in which new features may be considered. The curve EL running into the body of the ternary system represents the path of the liquid composition taking part in the $L+\alpha \rightarrow \beta$ peritectic reaction. The curve DL represents the liquid composition for the eutectic reaction, $L \rightarrow \alpha+\gamma$. The intersection point L is the peritectic point where $L+\alpha \rightarrow \beta+\gamma$. From L the eutectic valley LF runs to the lower

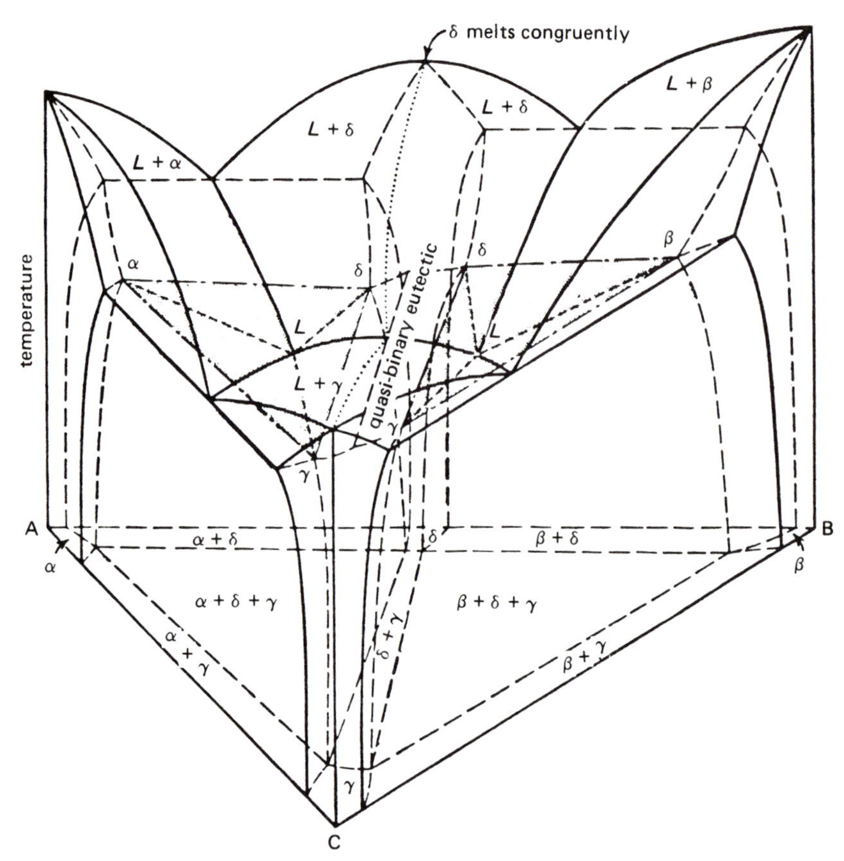

Figure 3.19 Ternary system containing an intermediate phase δ which forms a quasi-binary with component C (after Rhines, *Phase Diagrams in Metallurgy*)

eutectic where $L \rightarrow \beta + \gamma$. At the peritectic temperature four phases L, α, β and γ, with compositions lying at the corners of the shaded trapezium, coexist; the trapezium is termed the peritectic plane.

Horizontal sections for five temperatures T_1 to T_5 are given in *Figure 3.18* and it is instructive to follow the solidification of four alloys lying in the four different sections of the trapezium (*see* isotherm T_4). Alloy 1 starts to solidify with a primary separation of α followed by a secondary deposition of $\alpha + \beta$. On cooling through the four-phase reaction temperature $L + \alpha + \beta \rightarrow \alpha + \beta + \gamma$. For alloy 2, primary α also separates initially but the secondary deposition is $\alpha + \gamma$ and in the four-phase reaction $L + \alpha + \gamma \rightarrow \alpha + \beta + \gamma$. For alloys 3 and 4, the primary α is quite small and the secondary deposition is $\alpha + \beta$ and $\alpha + \gamma$, respectively. Both alloys will decrease the α and liquid content to form $\beta + \gamma$ in the four-phase reaction and $\beta + \gamma$ will continue to form at lower temperatures until the liquid is consumed. It must be remembered however, that peritectic alloys do not often solidify under equilibrium conditions because of the envelopment of the α-phase by reaction products which hinder diffusion. An excess amount of α is generally present and also liquid after the four-phase reaction.

3.7.5 Ternary systems containing intermetallic phases

More complex ternary systems can often be broken down into simpler, basic forms in certain regions of the diagram. A common example of this is of a system containing an intermetallic phase. *Figure 3.19* shows such a phase δ in the binary system AB and this forms a quasi-binary with the component C; a vertical section along δC is equivalent to a binary eutectic with solid solubility. This quasi-binary divides the ternary diagram into two independent regions which are easily seen from a horizontal section; for the solid state this section is included in the base of the triangle ABC of *Figure 3.19*.

Suggestions for further reading

Hume-Rothery, W., Smallman, R.E. and Haworth, C., *Structures of Metals and Alloys*, Institute of Metals, London, 1969

Rhines, F.N., *Phase Diagrams in Metallurgy*, McGraw-Hill, 1965

West, D.R.F., *Ternary Equilibrium Diagrams*, MacMillan

Winegard, W.C., *An Introduction to the Solidification of Metals*, Institute of Metals, 1964

Chapter 4

Thermodynamics of crystals

4.1 Introduction

So far, we have built up a working model of the structure of a metal and outlined the various experimental techniques used to determine (a) the metallic structure, (b) modifications to the structure (e.g. by alloying), and (c) the property changes which accompany such structural modifications. Before dealing with the behaviour and theory of metallic properties in detail, it is necessary to develop our understanding of why and how structural changes occur. For example, why metals such as iron, tin and titanium change their structure on heating, why all metals melt, and why the temperature of such phase transformations can be different on heating from that on cooling. It is clear that temperature is an important metallurgical variable and elementary reasoning suggests that if any metallic system, whether pure metal or alloy, undergoes a structural change at a temperature T K, then in that system a particular structural form must become unstable at that temperature. Consequently, to understand such transformations it is necessary to examine the effect of temperature on the structural stability of metal crystals.

4.2 The effect of temperature on metal crystals

If we consider a metal at absolute zero temperature, the ions sit in a potential well of depth E_{r_0} below the energy of a free atom (*Figure 4.1*). The effect of raising the temperature of the crystal is to cause the ions to oscillate in this asymmetrical potential well about their mean positions. As a consequence, this motion causes the energy of the system to rise, and it can be seen from *Figure 4.1* that the rise in energy increases with increasing amplitude of vibration. The increasing amplitude of vibration also causes an expansion of the crystal, since as a result of the sharp rise in energy below r_0 the ions as they vibrate to and fro do not approach much closer than the equilibrium separation, r_0, but separate more widely when moving apart. When the distance r is such that the atoms are no longer interacting, the metal is transformed to the gaseous phase, and the energy to bring this about is the energy of evaporation.

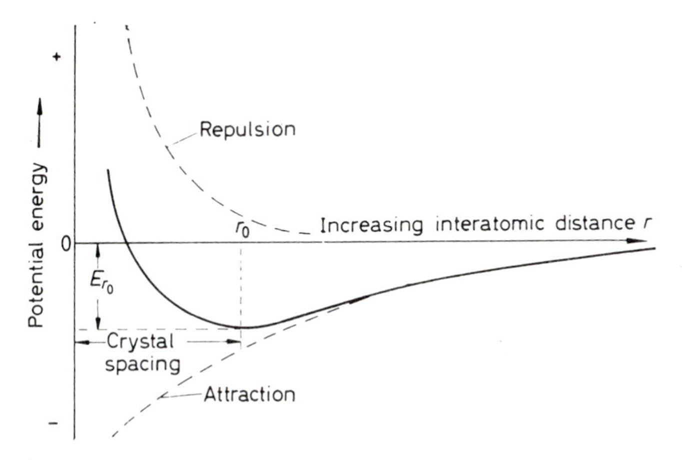

Figure 4.1 Variation in potential energy with interatomic distance

The specific heat of a metal (*see* Chapter 2) is due almost entirely to the vibrational motion of the ions. However, a part of the specific heat is due to the motion of the free electrons, but this part is very small and therefore usually neglected. Nevertheless, the electronic specific heat does become important at high temperatures, especially in transition metals which have electrons in incomplete shells. This will be dealt with later, but for the moment will be neglected.

The classical theory of specific heat assumes that an atom can oscillate in any one of three directions, and hence a crystal of N atoms can vibrate in $3N$ independent normal modes, each with its characteristic frequency. Furthermore, the mean energy of each normal mode will be $\mathbf{k}T$, so that the total vibrational thermal energy of the metal is $E=3N\mathbf{k}T$. Now in solid and liquid metals, the volume changes on heating are very small (e.g. the volume of most, but not all, metals increases by only 3 per cent on melting) and, consequently, it is customary to consider the specific heat at constant volume. If N, the number of atoms in the crystal, is equal to the number of atoms in a gram atom (i.e. Avogadro's number), the heat capacity per gram atom, i.e. the atomic heat, at constant volume is given by

$$C_v=\left(\frac{\mathrm{d}Q}{\mathrm{d}T}\right)_v=\frac{\mathrm{d}E}{\mathrm{d}T}=3N\mathbf{k}=24.95\ \mathrm{J/K} \tag{4.1}$$

where

$$\frac{\mathrm{d}Q}{\mathrm{d}T}=\frac{\mathrm{d}E}{\mathrm{d}T}$$

when $\mathrm{d}V$ is zero. In practice, of course, when the specific heat is experimentally determined, it is the specific heat at constant pressure, C_p, which is measured, not C_v, and this is given by

$$C_p=\left(\frac{\mathrm{d}E+P\,\mathrm{d}V}{\mathrm{d}T}\right)_p=\frac{\mathrm{d}H}{\mathrm{d}T}$$

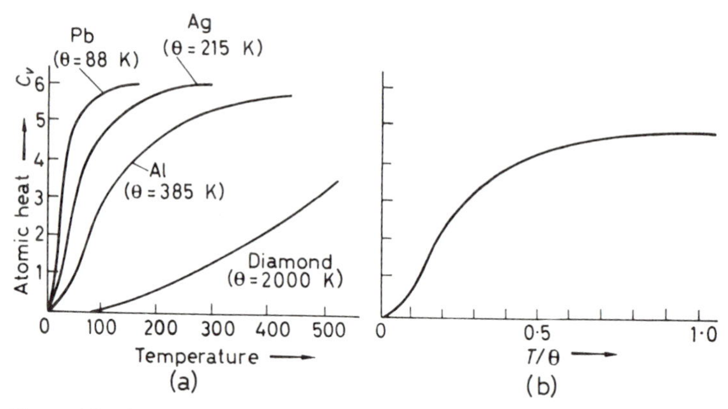

Figure 4.2 The variation of atomic heat with temperature

where $H = E + PV$ is known as the heat content or enthalpy. C_p is greater than C_v by a few per cent because some work is done against interatomic forces when the crystal expands, and it can be shown that

$$C_p - C_v = 9\alpha^2 VT/\beta$$

where α is the coefficient of linear thermal expansion, V is the volume per gram-atom and β is the compressibility.

Dulong and Petit were the first to point out that the specific heat of most materials, when determined at sufficiently high temperatures and corrected to apply to constant volume, is approximately equal to $3\boldsymbol{R}$, where $\boldsymbol{R}$ is the gas constant. However, deviations from the 'classical' value of the atomic heat occur at low temperatures, as shown in *Figure 4.2(a)*. This deviation is readily accounted for by the quantum theory, since the vibrational energy must then be quantized in multiples of $\boldsymbol{h}\nu$, where $\boldsymbol{h}$ is Planck's constant and ν is the characteristic frequency of the normal mode of vibration.

According to the quantum theory, the mean energy of a normal mode of the crystal is

$$E(\nu) = \tfrac{1}{2}\boldsymbol{h}\nu + \{\boldsymbol{h}\nu/\exp(\boldsymbol{h}\nu/\boldsymbol{k}T) - 1\} \tag{4.2}$$

where $\frac{1}{2}\boldsymbol{h}\nu$ represents the energy a vibrator will have at the absolute zero of temperature, i.e. the zero-point energy; clearly equation 4.2 tends to the classical value of $\boldsymbol{k}T$ at high temperatures. Using the assumption made by Einstein (1907) that all vibrations have the same frequency (i.e. all atoms vibrate independently), the heat capacity is

$$C_v = (\mathrm{d}E/\mathrm{d}T)_v = 3N\boldsymbol{k}(\boldsymbol{h}\nu/\boldsymbol{k}T)^2[\exp(\boldsymbol{h}\nu/\boldsymbol{k}T)/\{\exp(\boldsymbol{h}\nu/\boldsymbol{k}T) - 1\}^2] \tag{4.3}$$

This equation is rarely written in such a form because most materials have different values of ν. It is more usual to express ν as an equivalent temperature defined by $\Theta_E = \boldsymbol{h}\nu/\boldsymbol{k}$, where Θ_E is known as the Einstein characteristic temperature. Consequently, when C_v is plotted against T/Θ_E, the specific heat curves of all pure metals coincide and the value approaches zero at very low temperatures and rises to the classical value of $3N\boldsymbol{k}=3R\simeq 25.2$ J/g at high temperatures.

Einstein's formula for the specific heat is in good agreement with experiment for $T\gtrsim\Theta_E$, but is poor for low temperatures where the practical curve falls off less rapidly than that given by the Einstein relationship. However, the discrepancy can be accounted for, as shown by Debye, by taking account of the fact that the atomic vibrations are not independent of each other. This modification to the theory gives rise to a Debye characteristic temperature Θ_D, which is defined by

$$\boldsymbol{k}\Theta_D=\boldsymbol{h}\nu_D$$

where ν_D is Debye's maximum frequency. *Figure 4.2(b)* shows the atomic heat curves of *Figure 4.2(a)* plotted against T/Θ_D; in most metals for low temperatures $(T/\Theta_D\ll 1)$ a T^3 law is obeyed, but at high temperatures the free electrons make a contribution to the atomic heat which is proportional to T and this causes a rise of C above the classical value (*see* Chapter 2).

4.3 The specific heat curve and transformations

The specific heat of a metal varies smoothly with temperature, as shown in *Figure 4.2(a)* provided that no phase change occurs. On the other hand, if the metal undergoes a structural transformation the specific heat curve exhibits a discontinuity, as shown in *Figure 4.3*. If the phase change occurs at a fixed temperature, the metal undergoes what is known as a first-order transformation; for example, the α to γ, γ to δ and δ to liquid phase changes in iron shown in *Figure 4.3(a)*. At the transformation temperature the latent heat is absorbed without a rise in temperature, so that the specific heat ($\mathrm{d}Q/\mathrm{d}T$) at the transformation temperature is infinite. In some cases, known as transformations of the second order, the phase transition occurs over a range of temperature (e.g. the order–disorder transformation in alloys), and is associated with a specific heat peak of the form shown in *Figure 4.3(b)*. Obviously the narrower the temperature range T_1-T_c, the sharper is the specific heat peak, and in the limit when the total change occurs at a single temperature, i.e. $T_1=T_c$, the specific heat becomes infinite and equal to the latent heat of transformation. A second order

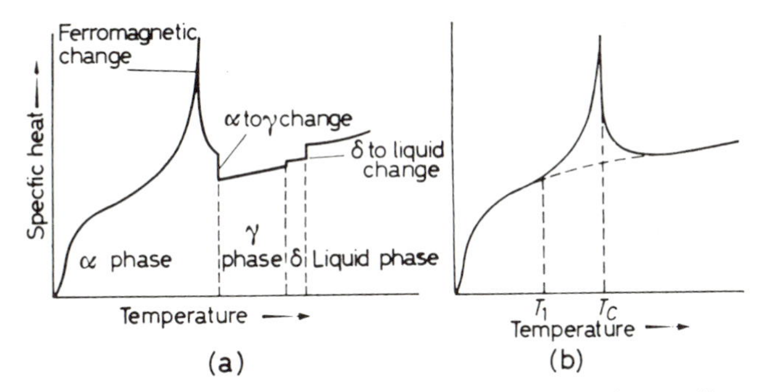

Figure 4.3 The effect of solid state transformations on the specific heat–temperature curve

transformation also occurs in iron (*see Figure 4.3(a)*), and in this case is due to a change in ferromagnetic properties with temperature.

4.4 Heat content, entropy and free energy

Every material in a given state has a characteristic heat content or enthalpy, H. The rate of change of heat content with temperature is equal to the specific heat of the material measured at constant pressure, C_p (*see* section 4.2). Then, since

$$(\mathrm{d}H/\mathrm{d}T)_p = C_p$$

by integration

$$H = H_0 + \int_0^T C_p \,\mathrm{d}T \tag{4.4}$$

where H is the absolute enthalpy of the material at T K, H_0, the integration constant, is the enthalpy at absolute zero, and $\int_0^T C_p \,\mathrm{d}T$ can be taken to be the amount of energy in thermal vibrations of the atoms. If the metal exhibits a transformation at a temperature T_t, then*

$$H = H_0 + \int_0^{T_t} C'_p \,\mathrm{d}T + L_t + \int_{T_t}^{T} C''_p \,\mathrm{d}T$$

where C'_p and C''_p are the specific heat values for the two phases, and L_t is the latent heat of transformation. In general, when a reaction occurs in a system it is accompanied by a change in heat content usually termed 'the heat of reaction', ΔH.

The heat content, H, is often termed a state property, which by definition means that it is uniquely defined by the state of the system. Thus if, after a cyclic series of changes, the final state of a system is the same as its initial state, the heat content must have reverted to its original value. That is, the change of any individual state property, e.g. pressure, temperature, volume, mass, etc, summed over the whole cycle must be zero; in the case under discussion

$$\oint \mathrm{d}H = 0$$

For any given reaction a knowledge of the quantity H is important, but it does not provide a criterion for equilibrium, nor does it determine when a phase

* In practice, it is not possible to obtain absolute values of enthalpies since values of H_0 are unobtainable. This is of little consequence, however, since a convenient zero of heat content may be defined; this is usually taken to be the heat content of pure elements in equilibrium at 25°C. Thus, for an element we have

$$H = H_{298} + \int_{298}^{T} C_p \,\mathrm{d}T = \int_{298}^{T} C_p \,\mathrm{d}T$$

change occurs, as shown by the occurrence of both exothermic and endothermic reactions. To provide this criterion it is also necessary to consider a second important state property known as the entropy, S. We shall see later that S may be regarded as a measure of the state of disorder of the structure, but for the present it is sufficient to treat it as a defined quantity arising naturally from classical thermodynamics. It may be shown that for any material passed though a complete cycle of events

$$\oint \frac{dQ}{T} = 0$$

where dQ is the heat exchanged between the system and its surroundings during each infinitesimal step and T is the temperature at which the transfer takes place. However, for this equation to be true, the heat must be exchanged isothermally and reversibly, i.e. in very small steps, so that the system never deviates far from equilibrium, otherwise the sum $\oint dQ/T$ will be greater than zero.

It will be seen that the function dQ/T behaves as a state property, and it is therefore convenient to define a quantity S such that $dS = dQ/T$, so that $\oint dS = 0$; entropy so defined is then a state property. At constant pressure, $dQ = dH$ and consequently

$$dS = dQ/T = C_p dT/T$$

which by integration gives

$$S = S_0 + \int_0^T (C_p/T)\,dT = S_0 + \int_0^T C_p\,d(\ln T) \tag{4.5}$$

where S is the entropy at T K usually measured in joules per degree per gram. The integration constant S_0 represents the entropy at absolute zero, which for an ordered crystalline substance is taken to be zero; this is often quoted as the third law of thermodynamics. Clearly, any reaction or transformation within a system will be associated with a characteristic entropy change given by

$$dS = S_\beta - S_\alpha$$

where dS is the entropy of transformation and S_β and S_α are the entropy values of the new phase β and the old phase α, respectively. It is a consequence of this that any irreversible change which takes place in a system (e.g. the combustion of a metal), must be accompanied by an increase in the total entropy of the system. This is commonly known as the second law of thermodynamics.

The quantity entropy could serve as a criterion for deciding the equilibrium state of a system, but it is much more convenient to work in terms of energy. Accordingly, it is more common to deal with the quantity TS, which has the units of energy, rather than just S, and to separate the total energy of the system H into two components according to the relation

$$H = G + TS$$

where G is that part of the energy of the system which causes the process to occur and TS is the energy requirement resulting from the structural change involved. The term G is known as Gibbs' free energy and is defined by the equation

$$G = H - TS \tag{4.6}$$

It is, of course, a state property, since it is a function entirely of state properties, so that every material in a given state will have a characteristic value of G. Moreover, the change of free energy accompanying a change represents the 'driving force' of the change, and in terms of the quantities discussed above is given by the expression

$$dG = dH - T\,dS \equiv dE + P\,dV - T\,dS$$

It may be shown that all spontaneous changes in a system must be accompanied by a reduction of the total free energy of that system, and thus for a change to occur the free energy change ΔG must be negative. It also follows that the equilibrium condition of a reaction will correspond to the state where $dG = 0$, i.e. zero driving force.

For solids and liquids at atmospheric pressure the volume change accompanying changes of state is very small and hence $P\,dV$ is also very small. It is therefore reasonable to neglect this term in the free energy equation and use as the criterion of equilibrium $dE - T\,dS = 0$. This is equivalent to defining the quantity $E - TS$ to be a minimum in the equilibrium state, for by differentiation

$$\begin{aligned} d(E - TS) &= dE - T\,dS - S\,dT \\ &= dE - T\,dS \text{ (since } T \text{ is constant)} \\ &= 0 \text{ for the equilibrium state} \end{aligned}$$

The quantity $E - TS$ thus defines the equilibrium state at constant temperature and volume, and is given the symbol F, the Helmholtz free energy ($F = E - TS$), to distinguish it from the Gibbs free energy ($G = H - TS$). In considering changes in the solid state it is thus a reasonable approximation to use F in place of G. The enthalpy H is the sum of the internal and external energies which reduces to $H \simeq E$ when the external energy PV is neglected.

4.5 The statistical nature of entropy

Since metals are always used at temperatures above 0 K, the relation $G = H - TS$ indicates the important role that entropy plays in determining the stability of phases. The equation represents to adopt a configuration of low enthalpy or energy, H, on the one hand and the tendency to adopt a high entropy, S, on the other. The phase which actually occurs is the one which satisfies both of these tendencies best at the temperature concerned.

When the temperature is high the quantity TS is large, so that the entropy of the phases is an important factor. In physical terms the entropy of a system may be considered as defining the degree of randomness or disorder in the atomic arrangements. For example, at 0 K each atom is positioned on its lattice in a regular geometrical array so that for a pure metal where every lattice site is identical and where there is no thermal vibration of the atoms, there is no degree of disorder, and the entropy is zero, i.e. $S_0 = 0$. As the temperature is raised each atom vibrates about its lattice site, and its true position is less definite.

Consequently, a certain degree of disorder is introduced and the entropy of the system increases with temperature.

A second example, of some importance in metallurgical practice, is the increase in entropy associated with the formation of a disordered solid solution from the pure components. This arises because over and above the entropies of the pure components A and B, the solution of B in A has an extra entropy due to the numerous ways in which the two kinds of atoms can be arranged amongst each other. This entropy of disorder or mixing is of the form shown in *Figure 4.13(a)*. As a measure of the disorder of a given state we can, purely from statistics, consider W the number of distributions which belong to that state. Thus, if the crystal contains N sites, n of which contain A-atoms and $(N-n)$ contain B-atoms, it can be shown that the total number of ways of distributing the A and B atoms on the N sites is given by

$$W = \frac{N!}{n!(N-n)!}$$

This is a measure of the extra disorder of solution, since $W=1$ for the pure state of the crystal because there is only one way of distributing N indistinguishable pure A or pure B atoms on the N sites.

The quantity W behaves like entropy in that the assembly with the maximum number of distributions is the most random one, and is, therefore, the most probable. Moreover, when a system changes its configuration it does so from one of low probability to one of high probability, and the configuration with maximum probability is most likely the equilibrium one. In order that the thermodynamic and statistical definitions of entropy be in agreement the quantity, W, which is a measure of the configurational probability of the system, is not used directly, but in the form

$$S = \boldsymbol{k} \ln W \tag{4.7}$$

where $\boldsymbol{k}$ is Boltzmann's constant. From this equation it can be seen that entropy is a property which measures the probability of a configuration, and that the greater the probability the greater is the entropy. Substituting for W in the statistical equation of entropy and using Stirling's approximation* we obtain

$$\begin{aligned} S &= \boldsymbol{k} \ln [N!/n!(N-n)!] \\ &= \boldsymbol{k}[N \ln N - n \ln n - (N-n) \ln (N-n)] \end{aligned}$$

for the entropy of disorder or mixing. The form of this entropy is shown in *Figure 4.13(a)*, where $c = n/N$ is the atomic concentration of A in the solution. It is of particular interest to note the sharp increase in entropy for the addition of only a small amount of solute. This fact accounts for the difficulty of producing really pure metals, since the entropy factor, $-T\,\mathrm{d}S$, associated with impurity addition, usually outweighs the energy term, $\mathrm{d}H$, so that the free energy of the material is almost certainly lowered by contamination.

* Stirling's theorem states that if N is large

$$\ln N! = N \ln N - N$$

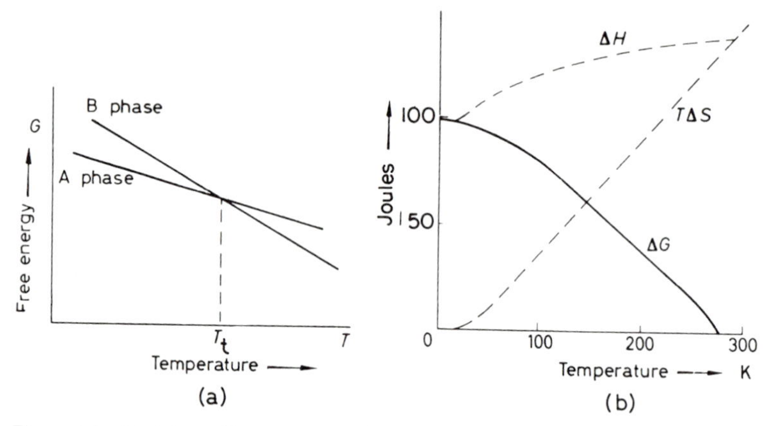

Figure 4.4 Variation of free energy with temperature. (a) Hypothetical. (b) For tin (after Lumsden, *Thermodynamics of Alloys*, by courtesy of the Institute of Metals)

Yet a third example is the ability of a material to change its phase. This phenomenon is discussed in some detail in section 4.7, but it is worth pointing out here that the phase which appears at the high temperatures is that which has the largest specific heat. The reason for this is because the phase with the largest specific heat has a larger number of ways of absorbing heat and has a greater degree of randomness and entropy.

4.6 Free energy of transformation

If a metal undergoes a structural change from phase α to phase β at a temperature T_t then it does so because above this temperature the free energy of the β phase, G_β, becomes lower than the free energy of the α phase, G_α. For this to occur the free energy curves must vary with temperature in the manner shown in *Figure 4.4(a)*. It can be seen that at T_t the free energy of the α-phase is equal to that of the β-phase so that ΔG is zero; T_t is, therefore, the equilibrium transformation point.

To illustrate this further, let us consider a simple example of the transformation of a pure metal, tin, according to the reaction

Sn (grey)$\rightleftharpoons$Sn (white)

The free energy of this reaction varies with temperature as a result of the temperature dependence of the free energy of each form. Thus, we can write, at any temperature (T K)

$$\Delta G = G\,(\text{white}) - G\,(\text{grey})$$

$$= \Delta H - T\Delta S$$

$$= \Delta H_{298} + \int_{298}^{T} \Delta C_p \, \mathrm{d}T - T\int_{0}^{T} \Delta C_p \, \mathrm{d}(\ln T)$$

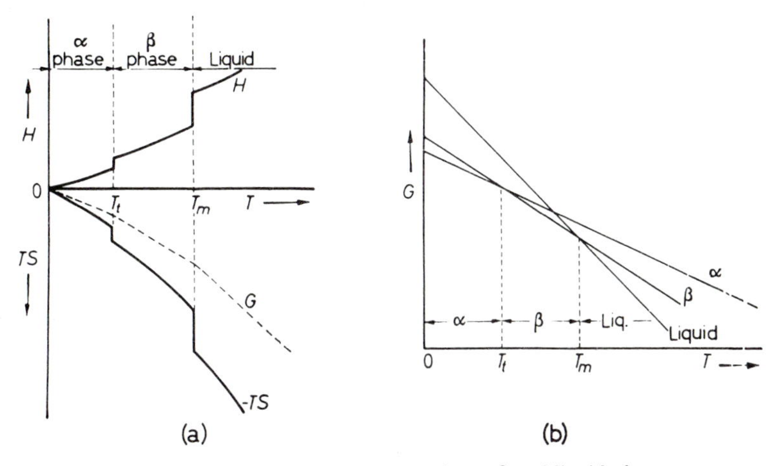

Figure 4.5 Free energy–temperature curves for α, β and liquid phases

where ΔH_{298} is the heat of transformation at 298 K, and ΔC_p is the difference in specific heat between white and grey tin, respectively. The value of ΔG at any temperature can be calculated simply from the heat of transformation at a single temperature and the specific heat values of the two phases over the range 0–T K. The values of heat content, entropy and free energy of the grey to white tin transformation, obtained in this way are shown in *Figure 4.4(b)*. From the diagram it is seen that $\Delta G = 0$ at 292 K, so that this temperature is the equilibrium transformation point. Above this critical temperature, ΔG for the reaction becomes negative, so that the tin will exist in the white form as this gives a lower free energy.

It is important to realize that although the value of ΔG indicates at what temperature a transformation should take place, the actual transformation also depends on the kinetics of the process and the creation of nuclei. Thus, in practice, both forms of tin can exist below the equilibrium transformation point for long periods. This feature of a transformation is discussed in section 4.9.

4.7 The variation of free energy with temperature, and polymorphism

The above example is a simple illustration of the application of thermodynamic principles to a single transformation. Frequently, it is necessary to consider the possibility of successive transformations occurring in a given temperature range, and in this case, it is more instructive to examine the way in which the absolute value of the free energy of a crystal varies with temperature. This is best seen by reference to *Figure 4.5(a)*, where H and $-TS$ are plotted as a function of temperature. At the transformation temperature, T_t, the change in heat content ΔH is equal to the latent heat L, while the entropy change ΔS is equal to L/T_t. In consequence, a plot of $G = H - TS$ shows no sharp discontinuities at T_t (since $\Delta H = T_t \Delta S$) or T_m but merely a discontinuity of slope. A plot of G versus temperature for each of the three phases considered, α, β and liquid, would then be of the form shown in *Figure 4.5(b)**. In any temperature range the phase with the lowest free energy is the stable phase.

The relative magnitude of the free energy value governs the stability of any phase, and from *Figure 4.5* it can be seen that the free energy G at any temperature is in turn governed by two factors: (i) the value of G at 0 K, G_0, and (ii) the slope of the G versus T curve, i.e. the temperature dependence of free energy. Both of these terms are influenced by the vibrational frequency, and consequently the specific heat of the atoms, as can be shown mathematically. For example, if the temperature of the system is raised from T to $T+\mathrm{d}T$ the change in free energy of the system $\mathrm{d}G$ is

$$\begin{aligned} \mathrm{d}G &= \mathrm{d}H - T\,\mathrm{d}S - S\,\mathrm{d}T \\ &= C_p\,\mathrm{d}T - T(C_p\,\mathrm{d}T/T) - S\,\mathrm{d}T = -S\,\mathrm{d}T \end{aligned}$$

so that the free energy of the system at a temperature T is

$$G = G_0 - \int_0^T S\,\mathrm{d}T$$

At the absolute zero of temperature, the free energy G_0 is equal to H_0, and then

$$G = H_0 - \int_0^T S\,\mathrm{d}T$$

which if S is replaced by $\int_0^T (C_p/T)\,\mathrm{d}T$ becomes

$$G = H_0 - \int_0^T \left[\int_0^T (C_p/T)\,\mathrm{d}T \right] \mathrm{d}T \tag{4.8}$$

Equation 4.8 indicates that the free energy of a given phase decreases more rapidly with rise in temperature the larger its specific heat. The intersection of the free energy–temperature curves, shown in *Figure 4.4(a)* and *4.5(b)*, therefore takes place because the low temperature phase has a smaller specific heat than the higher temperature phase.

At low temperatures the second term in equation 4.8 is relatively unimportant, and the phase that is stable is the one which has the lowest value of H_0, i.e. the most close-packed phase which is associated with a strong bonding of the atoms. However, the more strongly bound the phase, the higher is its elastic constant, the higher the vibrational frequency, and consequently the smaller the specific heat (*see Figure 4.2(a)*). Thus, the more weakly bound structure, i.e. the phase with the higher H_0 at low temperature, is likely to appear as the stable phase at higher temperatures. This is because the second term in equation 4.8 now becomes important and G decreases more rapidly with increasing temperature for the phase with the largest value of $\int (C_p/T)\,\mathrm{d}T$. From *Figure 4.2(b)* it is clear that a large $\int (C_p/T)\,\mathrm{d}T$ is associated with a low characteristic temperature and

* Although high temperature phases cannot exist at low temperature, it is theoretically possible to calculate H and S from a knowledge of the characteristic temperature Θ obtained from the specific heat–temperature curve.

hence, with a low vibrational frequency such as is displayed by a metal with a more open structure and small elastic strength. In general, therefore, when phase changes occur the more close-packed structure usually exists at the low temperatures and the more open structures at the high temperatures. From this viewpoint a liquid, which possesses no long-range structure, has a higher entropy than any solid phase so that ultimately all metals must melt at a sufficiently high temperature, i.e. when the TS term outweighs the H term in the free energy equation (*see Figure 4.5*).

The sequence of phase changes in such metals as titanium, zirconium, etc., is in agreement with this predicted transition and, moreover, the alkali metals, lithium and sodium, which are normally b.c.c. at ordinary temperatures, can be transformed to f.c.c. at sub-zero temperatures. It is interesting to note that iron, being b.c.c. (α-iron) even at low temperatures and f.c.c. (γ-iron) at high temperatures, is an exception to this rule. In this case, the stability of the b.c.c. structure is thought to be associated with its ferromagnetic properties. By having a b.c.c. structure the interatomic distances are of the correct value for the exchange interaction to allow the electrons to adopt parallel spins (this is a condition for magnetism, *see* Chapter 5). While this state is one of low entropy it is also one of minimum internal energy, and in the lower temperature ranges this is the factor which governs the phase stability, so that the b.c.c. structure is preferred.

Iron is also of interest because the b.c.c. structure, which is replaced by the f.c.c. structure at temperatures above 910 °C, reappears as the δ-phase above 1400°C. This behaviour is attributed to the large electronic specific heat of iron which is a characteristic feature of most transition metals. Thus, the Debye characteristic temperature of γ-iron is lower than that of α-iron and this is mainly responsible for the α to γ transformation. However, the electronic specific heat of the α-phase becomes greater than that of the γ-phase above about 300 °C and eventually at higher temperatures becomes sufficient to bring about the return to the b.c.c. structure at 1400 °C.

4.8 Thermodynamics of lattice defects

As outlined in Chapter 1, a knowledge and understanding of the defect state is of considerable importance to physical metallurgy. In the present section it is of interest to consider whether such defects are in thermodynamic equilibrium in the lattice. First, let us consider a dislocation. Clearly, a crystal containing a dislocation is less ordered and, therefore, has a greater configurational entropy than a perfect crystal. On the other hand, the distortion associated with such a line defect gives rise to a lattice strain energy which raises the internal energy of the crystal. It is therefore necessary to decide which term predominates to see if a dislocation can exist as a thermodynamically stable lattice defect.

It is evident that a configurational entropy arises from the number of ways the dislocation can be arranged in the crystal. If the dislocation line is straight and of length L, the number of places for the dislocation is $3A/a^2$, where A is area of the crystal normal to the line of the dislocation and a is the crystal spacing. From the Boltzmann formula, equation 4.7, the contribution to the free energy at a temperature T, $(-TS)$, is

$$G_{(\text{disl.})} = -(a/L)kT \ln(3A/a^2)$$

per atom plane threaded by the dislocation. The contribution to the free energy from this source is extremely small ($\approx 10^{-6}\,\boldsymbol{k}T$), but a higher configurational entropy exists if the dislocation line is flexible, since the line can take many more forms in the crystal. However, even in this case the contribution to G must be less than $-\boldsymbol{k}T \ln Z$, where Z is the number of neighbouring sites into which the path of the dislocation line may deviate in each atomic plane through which it passes. The total contribution to the free energy per atom plane from the configurational entropy is, therefore, less than $2\boldsymbol{k}T$. A second contribution to the free energy may arise from the effect that a dislocation line has on the vibrational frequency of a crystal, since this frequency depends on elastic strain. An estimate of this vibrational contribution has been made and shown to be about $-3\boldsymbol{k}T$ per atom plane. The total free energy contribution arising from a dislocation line in a crystal is then about $-5\boldsymbol{k}T$ for each atomic plane through which it threads. At room temperature this is only about $\frac{1}{8}$ eV and, since the strain energy contribution is positive and equal to several electron volts per atomic plane threaded by the dislocation (*see* Chapter 6), the free energy is raised by the strain energy more than it is lowered by the contribution from the increase of entropy. It is evident, therefore, that a dislocation, although geometrically feasible, cannot exist as a thermodynamically stable lattice defect.

The strain energy of a dislocation is so large that even at high temperatures (according to the approximate calculations discussed above) it outweighs the negative contributions to the free energy. On the other hand, a similar analysis for the other common lattice defects, i.e. vacancies, which we might previously have thought to be unstable lattice defects, shows that they are, in fact, thermodynamically stable in certain temperature ranges. We know, for example, that a crystal is in equilibrium at a temperature T when its free energy $G = H - TS$ is a minimum and at 0 K this condition reduces to $G = H$, so that the stable state, i.e. that of minimum energy, is the one which is most perfect. At higher temperatures, however, defects may be thermodynamically stable since the increased energy associated with the introduction of defects may be off-set by the entropy increase associated with the increased irregularity or disorder of the lattice, i.e. $\mathrm{d}G = \mathrm{d}H - T\,\mathrm{d}S$ is negative. This is certainly the case for lattice vacancies.

It is of particular interest to estimate the number of lattice vacancies in a crystal in equilibrium at a temperature T. Thus, if E_f is the energy required to form one such defect (usually expressed in electron volts per atom) by removing an atom from the lattice and depositing it in a normal site on the crystal surface, the total energy increase resulting from the formation of n such defects is nE_f. The accompanying entropy increase may be calculated using the relation $S = \boldsymbol{k} \ln W$, knowing the number of ways of distributing n defects and N atoms on $N+n$ lattice sites, i.e. $(N+n)!/n!N!$ Then the free energy, G, or strictly F of a crystal of n defects, relative to the free energy of the perfect crystal, is

$$F = nE_f - \boldsymbol{k}T \ln[(N+n)!/n!N!] \qquad (4.9)$$

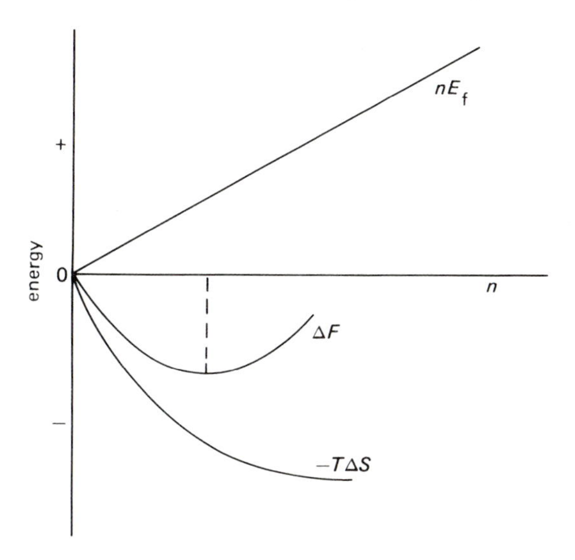

Figure 4.6 Variation of the energy of a crystal with addition of n vacancies

which by the use of Stirling's theorem simplifies to

$$F = nE_f - \boldsymbol{k}T[(N+n)\ln(N+n) - n\ln n - N\ln N]$$

The equilibrium value of n is that for which $\mathrm{d}F/\mathrm{d}n = 0$, which defines the state of minimum free energy as shown in *Figure 4.6**. Thus, differentiating equation 4.9, gives

$$0 = E_f - \boldsymbol{k}T[\ln(N+n) - \ln n] = E_f - \boldsymbol{k}T\ln[(N+n)/n]$$

so that

$$\frac{n}{N+n} = \exp[-E_f/\boldsymbol{k}T]$$

Usually N is very large compared with n, so that the expression can be taken to give the atomic concentration, c, of lattice vacancies, $n/N = \exp[-E_f/\boldsymbol{k}T]$. A more rigorous calculation of the concentration of vacancies in thermal equilibrium in a perfect lattice shows that although c is principally governed by the Boltzmann factor $\exp[-E_f/\boldsymbol{k}T]$, the effect of the vacancy on the vibrational properties of the lattice also leads to an entropy term which is independent of temperature and usually written as $\exp[S_f/\boldsymbol{k}]$. The fractional concentration may thus be written

$$c = n/N = \exp[S_f/\boldsymbol{k}]\exp[-E_f/\boldsymbol{k}T] = A\exp[-E_f/\boldsymbol{k}T] \qquad (4.10)$$

The value of the entropy term is not accurately known but it is usually taken to be within a factor of ten of the value 10; for simplicity we will take it to be unity.

The equilibrium number of vacancies rises rapidly with increasing tempera-

* $\mathrm{d}F/\mathrm{d}n$ or $\mathrm{d}G/\mathrm{d}n$ is known as the chemical potential as discussed on page 140.

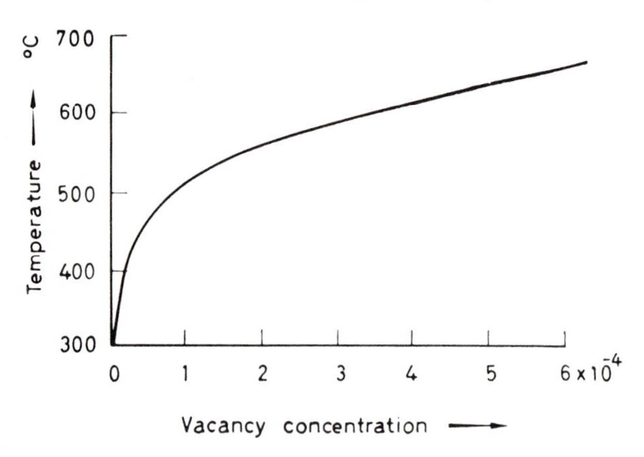

Figure 4.7 Equilibrium concentration of vacancies as a function of temperature for aluminium (after Bradshaw and Pearson, *Phil. Mag.* 1957, **2,** 570)

ture, owing to the exponential form of the expression, and for most common metals has a value of about 10^{-4} near the melting point (*see Figure 4.7*). For example, $\boldsymbol{k}T$ at room temperature (300 K) is $\approx 1/40$ electron volt, and for aluminium $E_f = 0.7$ eV, so that at 900 K we have

$$c = \exp\left[-\frac{7}{10} \times \frac{40}{1} \times \frac{300}{900}\right] = \exp[-9.3] = 10^{-[9.3/2.3]} \approx 10^{-4}$$

As the temperature is lowered, c should decrease in order to maintain equilibrium and to do this the vacancies must migrate to positions in the lattice where they can be annihilated; these lattice positions are usually known as 'vacancy sinks' and include such places as the free surface, grain boundaries and dislocations.

Clearly the surface of a sample or the grain boundary interface are a considerable distance, in atomic terms, from the centre of a grain and so the dislocations in the body of the grain or crystal are the most efficient 'sink' for vacancies. Vacancies are annihilated at the edge of the extra half-plane of atoms of the dislocation, as shown in *Figure 4.8*. This causes the dislocation to climb, as discussed in Chapter 6. The process whereby vacancies are annihilated at vacancy sinks such as surfaces, grain boundaries and dislocations, to satisfy the thermodynamic equilibrium concentration at a given temperature is, of course, reversible. When a metal is heated the equilibrium concentration increases and to produce this additional concentration the surfaces, grain boundaries and dislocations in the crystal reverse their role and act as vacancy sources and emit vacancies; the extra half-plane of atoms in *Figure 4.8* climbs in the opposite sense (*see Figure 4.8*(*c*) *and* (*d*)).

However, below a certain temperature, the migration of vacancies will be too slow for equilibrium to be maintained, and at the lower temperatures a concentration of vacancies in excess of the equilibrium number will be retained in the lattice. Moreover, if the cooling rate of the metal or alloy is particularly rapid, as, for example, in quenching, the vast majority of the vacancies which exist at high temperatures can be 'frozen-in' (*see* Chapter 9). Such vacancies are of considerable importance in governing the kinetics of many physical processes.

The energy of formation of an interstitial atom is much higher than that for a vacancy and of the order of 4 eV. At temperatures just below the melting point the concentration of such point defects is only about 10^{-15} and, therefore, interstitials are of little consequence in normal behaviour of metals and alloys. They are however, of considerable importance in the behaviour of solids which have been subjected to irradiation by high energy particles (*see* Chapter 9).

4.9 The rate of reaction

The occurrence of a non-equilibrium number of lattice vacancies at certain temperatures is further evidence that in practice, metals and alloys often exist in a state which is not that of minimum free energy. Another example already mentioned is the formation of cored solid solutions. In industrial practice such a metastable equilibrium is the rule rather than the exception and, in fact, is the basis of quenching as a standard heat treatment operation (*see* precipitation hardening and the decomposition of austenite, Chapters 11 and 12). The equilibrium state can, of course be approached more readily as the temperature of the system is raised, but it is necessary to examine those factors which prevent such a speedy reaction at the lower temperature.

Let us consider the non-equilibrium concentration of vacancies discussed above. To maintain equilibrium as the temperature is lowered, the vacancies must migrate through the lattice to vacancy sinks. Such a migration process is not

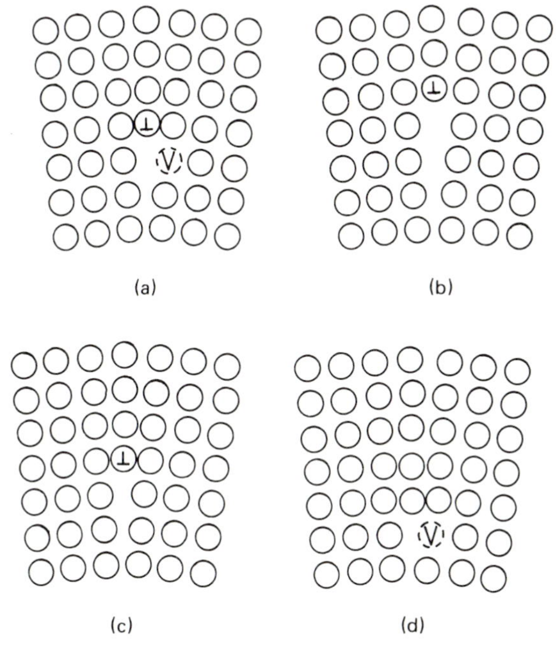

Figure 4.8 Climb of a dislocation, (a)–(b) to annihilate, (c)–(d) to create, a vacancy

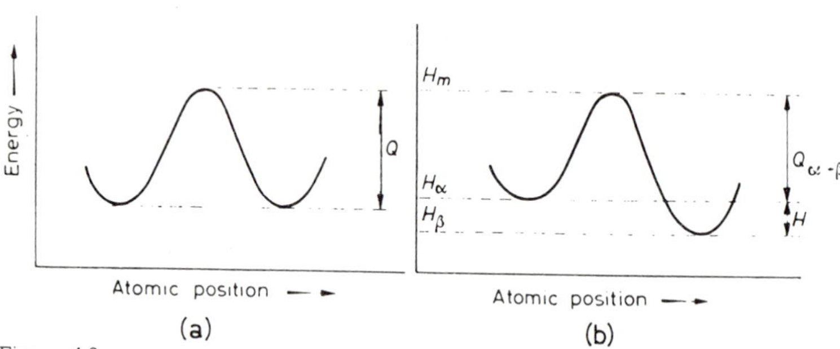

Figure 4.9

straightforward, however, since, owing to the periodic nature of the lattice, the internal energy of an atom as a function of its position is as shown in *Figure 4.9(a)*. For the atom or vacancy to move from one lattice position to the next it must obtain the necessary energy to surmount the energy barrier, Q. The rate of such a reaction will, therefore, depend on (i) the magnitude of activation energy barrier, Q, and (ii) the number of atoms, n, that possess the required energy Q, at any instant.. The first factor depends on the bonding of the structure, and is characteristic of a given material. The second factor concerns the way in which the energy is distributed in the alloy since, owing to thermal vibrations, at any given temperature, some atoms will have exceptionally high energies and others exceptionally low. The probability of an atom having sufficient energy to jump the barrier is given, from the Maxwell–Boltzmann distribution law, as proportional to $\exp[-Q/\boldsymbol{k}T]$ where $\boldsymbol{k}$ is Boltzmann's constant, T is the temperature and Q is usually expressed* as the energy per atom in electron volts. The number of times a vacancy makes a jump from A to B in unit time is therefore proportional to $\exp[-Q/\boldsymbol{k}T]$, and the rate of reaction is given by

$$\text{Rate} = A\exp[-Q/\boldsymbol{k}T] \tag{4.11}$$

where A is a constant involving n and ν, the frequency of vibration. This formula is of great importance in physical metallurgy since it describes the kinetics of many processes, and it also shows that temperature has a profound effect on reaction rate. To determine Q experimentally, the reaction velocity is measured at different temperatures and, since

$$\ln(\text{Rate}) = \ln A - Q/\boldsymbol{k}T$$

the slope of the ln (rate) versus $1/T$ curve gives $Q/\boldsymbol{k}$. In the above example, the value of Q may be identified with the activation energy, E_m, necessary for the movement of vacancies (*see* Chapter 9).

In deriving equation 4.11, usually called an Arrhenius equation after the Swedish chemist who first studied reaction kinetics, no account is taken of the entropy of activation, i.e. the change in entropy as a result of the transition. In considering a general reaction the probability expression should be written in terms of the free energy of activation per atom F or G rather than just the internal

* Q may also be given as the energy in J mol^{-1} in which case the rate equation becomes

$$\text{Rate of reaction} = A\exp[-Q/\boldsymbol{R}T]$$

where $\boldsymbol{R} = \boldsymbol{k}N$ is the gas constant, i.e. 8.314 J mol^{-1} K^{-1}.

energy or enthalpy, the rate equation then becomes

$$\text{Rate} = A \exp[-F/\boldsymbol{k}T] = A \exp[S/\boldsymbol{k}] \exp[-E/\boldsymbol{k}T]$$

The slope of the ln (rate) versus $1/T$ curve then gives the temperature dependence of the reaction rate, which is governed by the activation energy or enthalpy, and the magnitude of the intercept on the ln (rate) axis depends on the temperature-independent terms and include the frequency factor and the entropy term.

4.10 The mechanism of phase changes

We have seen that the application of thermodynamic concepts readily explains the changes which occur in metals with changes in temperature. It now remains to outline the mechanism whereby these transformations take place. Since almost any alteration in the solid state involves a redistribution of the atoms in that solid, the kinetics of the change necessarily depend upon the rate of atomic migration. The transport of atoms through the crystal is more generally termed diffusion, and it is evident from the previous section that this can occur more easily with the aid of vacancies, since the basic act of diffusion is the movement of an atom to an empty adjacent atomic site.

Let us consider that during a phase change an atom is moved from an α-phase lattice site to a more favourable β-phase lattice site. The energy of the atom should vary with distance as shown in *Figure 4.9(b)*, where the potential barrier which has to be overcome arises from the interatomic forces between the moving atom and the group of atoms which adjoin it and the new site. Only those atoms with an energy greater than Q are able to make the jump, where $Q_{\alpha\rightarrow\beta} = H_m - H_\alpha$ and $Q_{\beta\rightarrow\alpha} = H_m - H_\beta$ are the activation enthalpies for heating and cooling respectively. Consequently, the reaction rate is given by inserting the appropriate activation energy (enthalpy) in the rate equation and the net energy release, $H = H_\alpha - H_\beta$, is the heat of reaction.

During the transformation it is not necessary for the entire system to go from α to β at one jump and, in fact, if this were necessary, phase changes would practically never occur. Instead, most phase changes occur by a process of nucleation and growth (cf. solidification, Chapter 3). Chance thermal fluctuations provide a small number of atoms with sufficient activation energy to break away from the matrix (the old structure) and form a small nucleus of the new phase, which then grows at the expense of the matrix until the whole structure is transformed. By this mechanism, the amount of material in the intermediate configuration of higher free energy is kept to a minimum, as it is localized into atomically thin layers at the interface between the phases. Because of this mechanism of transformation the factors which determine the rate of phase change are: (i) the rate of nucleation, N (i.e. the number of nuclei formed in unit volume in unit time), and (ii) the rate of growth, G (i.e. the rate of increase in radius with time). Both processes require activation energies, which in general are not equal, but the values are much smaller than that needed to change the whole lattice from α to β in one operation.

Even with such an economical process as nucleation and growth transformation difficulties occur and it is common to find that the transformation temperature, even under the best experimental conditions, is slightly higher on heating than on cooling. This sluggishness of the transformation is known as hysteresis, and is attributed to the difficulties of nucleation, since diffusion, which controls the growth process, is usually high at temperatures near the transformation temperature and is, therefore, not rate controlling. Perhaps the simplest phase change to indicate this is the solidification of a liquid metal.

The transformation temperature, as shown on the equilibrium diagram, represents the point at which the free energy of the solid phase is equal to that of the liquid phase. Thus, we may consider the transition, as given in a phase diagram, to occur when the bulk or chemical free energy change, ΔG_v, is infinitesimally small and negative, i.e. when a small but positive driving force exists. However, such a definition ignores the process whereby the bulk liquid is transformed to bulk solid, i.e. nucleation and growth. When the nucleus is formed the atoms which make up the interface between the new and old phase occupy positions of compromise between the old and new structure, and as a result these atoms have rather higher energies than the other atoms. Thus, there will always be a positive free energy term opposing the transformation as a result of the energy required to create the surface of interface. Consequently, the transformation will occur only when the sum $\Delta G_v + \Delta G_s$ becomes negative, where ΔG_s arises from the surface energy of solid–liquid interface. Normally, for the bulk phase change, the number of atoms which form the interface is small and ΔG_s compared with ΔG_v can be ignored. However, during nucleation ΔG_v is small, since it is proportional to the amount transformed, and ΔG_s, the extra free energy of the boundary atoms, becomes important due to the large surface area to volume ratio of small nuclei. Therefore before transformation can take place the negative term ΔG_v must be greater than the positive term ΔG_s and, since ΔG_v is zero at the equilibrium freezing point, it follows that undercooling must result.

Homogeneous nucleation

Dealing with this problem quantitatively, since ΔG_v depends on the volume of the nucleus and ΔG_s is proportional to its surface area, we can write for a spherical nucleus of radius r,

$$\Delta G = (-4\pi r^3 \Delta G_v/3) + 4\pi r^2 \gamma \quad (4.12)$$

where ΔG_v is the bulk free energy change involved in the formation of the nucleus of unit volume and γ is the surface free energy of unit area. When the nuclei are small the positive surface energy term predominates, while when they are large the negative volume term predominates, so that the change in free energy as a function of nucleus size is as shown in *Figure 4.10(a)*. This indicates that a critical nucleus size exists below which the free energy increases as the nucleus grows, and above which further growth can proceed with a lowering of free energy; ΔG_{max} may be considered as the energy or work of nucleation W. Both r_c and W mat be calculated since $\mathrm{d}\Delta G/\mathrm{d}r = -4\pi r^2 \Delta G_v + 8\pi r\gamma = 0$ when $r = r_c$ and thus

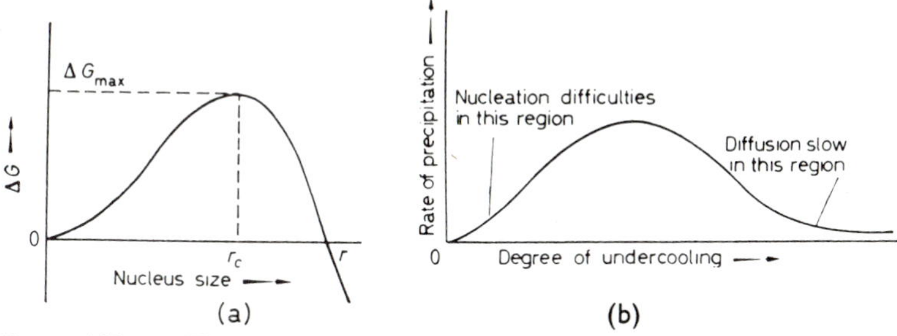

Figure 4.10 (a) Effect of nucleus size on the free energy of nucleus formation. (b) Effect of undercooling on the rate of precipitation

$r_c = 2\gamma/\Delta G_v$. Substituting in equation 4.12 for r_c gives

$$W = 16\pi\gamma^3/3\,\Delta G_v^2 \tag{4.13}$$

The surface energy factor γ is not strongly dependent on temperature but the chemical free energy and, therefore, the greater the degree of undercooling or super-saturation, the greater is the release of chemical free energy and the smaller the critical nucleus size and energy of nucleation. This can be shown analytically since $\Delta G_v = \Delta H - T\,\Delta S$, and at $T = T_e$, $\Delta G_v = 0$, so that $\Delta H = T_e \Delta S$. It therefore follows that

$$\Delta G_v = (T_e - T)\Delta S = \Delta T\,\Delta S$$

and because $\Delta G_v \propto \Delta T$, then

$$W \propto \gamma^3/\Delta T^2 \tag{4.14}$$

Consequently, since nuclei are formed by thermal fluctuations, the probability of forming a smaller nucleus is greatly improved, and the rate of nucleation increases according to

$$\begin{aligned}\text{Rate} &= A\exp[-Q/\boldsymbol{k}T]\exp[-\Delta G_{\max}/\boldsymbol{k}T] \\ &= A\exp[-(Q+\Delta G_{\max})/\boldsymbol{k}T]\end{aligned} \tag{4.15}$$

The term $\exp[-Q/\boldsymbol{k}T]$ is introduced to allow for the fact that rate of nucleus formation is in the limit controlled by the rate of atomic migration. Clearly, with very extensive degrees of undercooling, when $\Delta G_{\max} \ll Q$, the rate of nucleation approaches $\exp[-Q/\boldsymbol{k}T]$ and, because of the slowness of atomic mobility, this becomes small at low temperature, *Figure 4.10(b)*. While this range of conditions can be reached for liquid glasses the nucleation of liquid metals normally occurs at temperatures before this condition is reached. (By splat cooling, small droplets of the metal are cooled very rapidly (10^5 K s^{-1}) and an amorphous solid may be produced.) Nevertheless, the principles are of importance in metallurgy since in the isothermal transformation of eutectoid steel, for example, the rate of transformation initially increases and then decreases with lowering of the transformation temperature (*see* TTT curves, Chapter 12).

Heterogeneous nucleation

In practice homogeneous nucleation rarely takes place and heterogeneous nucleation occurs either on the mould walls or on insoluble impurity particles.

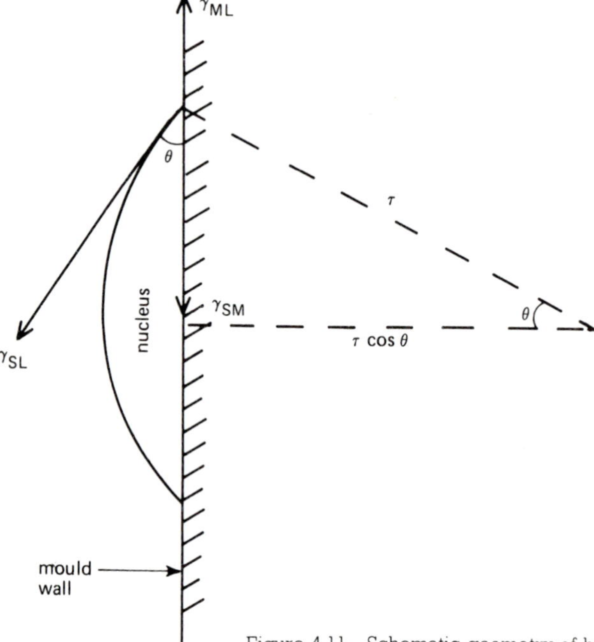

Figure 4.11 Schematic geometry of heterogenous nucleation

From equation 4.14 it is evident that a reduction in the interfacial energy γ would facilitate nucleation at small values of ΔT. *Figure 4.11* shows how this occurs at a mould wall or pre-existing solid particle, where the nucleus has the shape of a spherical cap to minimize the energy and the 'wetting' angle θ is given by the balance of the interfacial tensions in the plane of the mould wall, i.e. $\cos\theta = (\gamma_{ML} - \gamma_{SM})/\gamma_{SL}$.

The formation of the nucleus is associated with an excess free energy given by

$$\Delta G = -V\,\Delta G_v + A_{SL}\gamma_{SL} + A_{SM}\gamma_{SM} - A_{SM}\gamma_{ML}$$

$$= \frac{\pi}{3}(2 - 3\cos\theta + \cos^3\theta)r^3\,\Delta G_v + 2\pi(1-\cos\theta)r^2\gamma_{SL}$$

$$+ \pi r^2 \sin^2\theta(\gamma_{SM} - \gamma_{LM}) \tag{4.16}$$

Differentiation of this expression for the maximum, i.e. $\mathrm{d}\Delta G/\mathrm{d}r = 0$ gives $r_c = 2\gamma_{SL}/\Delta G_v$ and

$$W = -(16\pi\gamma^3/3\,\Delta G_v^2[(1-\cos\theta)^2(2+\cos\theta)/4] \tag{4.17}$$

or

$$W_{(\text{heterogeneous})} = W_{(\text{homogeneous})}[S(\theta)]$$

The shape factor $S(\theta) \leqslant 1$ is dependent on the value of θ and the work of nucleation is therefore less for heterogeneous nucleation. When $\theta = 180°$, no wetting occurs and there is no reduction in W; when $\theta \to 0°$ there is complete wetting and $W \to 0$; and when $0 < \theta < 180°$ there is some wetting and W is reduced.

Nucleation in solids

When the transformation takes place in the solid state, i.e. between two solid phases, a second factor giving rise to hysteresis operates. The new phase usually has a different parameter and crystal structure from the old so that the transformation is accompanied by dimensional changes. However, the changes in volume and shape cannot occur freely because of the rigidity of the surrounding matrix, and elastic strains are induced. The strain energy and surface energy created by the nuclei of the new phase are positive contributions to the free energy and so tend to oppose the transition.

The total free energy change is

$$\Delta G = -V\Delta G_v + A\gamma + V\Delta G_s \tag{4.18}$$

where A is the area of interface between the two phases and γ the interfacial energy per unit area, and ΔG_s is the misfit strain energy per unit volume of new phase. For a spherical nucleus of the second phase

$$\Delta G = -\tfrac{4}{3}\pi r^3(\Delta G_v - \Delta G_s) + 4\pi r^2\gamma \tag{4.19}$$

and the misfit strain energy reduces the effective driving force for the transformation. Differentiation of equation 4.19 gives

$$r_c = -2\gamma/(\Delta G_v - \Delta G_s), \quad \text{and} \quad W = 16\pi\gamma^3/3(\Delta G_v - \Delta G_s)^2$$

The value of γ can vary widely from a few mJ/m^2 to several hundred mJ/m^2 depending on the coherency of the interface. A coherent interface is formed when the two crystals have a good 'match' and the two lattices are continuous across the interface. This happens when the interfacial plane has the same atomic configuration in both phases, e.g. (111) in f.c.c. and (0001) in h.c.p. When the 'match' at the interface is not perfect it is still possible to maintain coherency by straining one or both lattices, as shown in *Figure 4.12(a)*. These coherency strains increase the energy and for large misfits it becomes energetically more favourable to form a semi-coherent interface in which the mismatch is periodically taken up by misfit dislocations and the coherency strains can then be relieved by a cross-grid of dislocations in the interface plane, the spacing of which depends on the Burgers vector b of the dislocation and the misfit ε, i.e. b/ε. The interfacial energy for semi-coherent interfaces arises from the change in composition across the interface or chemical contribution as for fully coherent interfaces, plus the energy of the dislocations (*see* Chapter 6). The energy of a semi-coherent interface is 200–500 mJ/m^2 and increases with decreasing dislocation spacing until the dislocation strain fields overlap. When this occurs the discrete nature of the dislocations is lost and the interface becomes incoherent. The incoherent interface is somewhat similar to a high-angle grain boundary (*see Figure 3.1*) with its energy of 0.5 to 1 J/m^2 relatively independent of the orientation.

The surface and strain energy effects discussed above play an important role in phase separation. When there is coherence in the atomic structure across the interface between precipitate and matrix the surface energy term is small, and it is the strain energy factor which controls the shape of the particle. A plate-shaped particle is associated with the least strain energy, while a spherical shaped particle

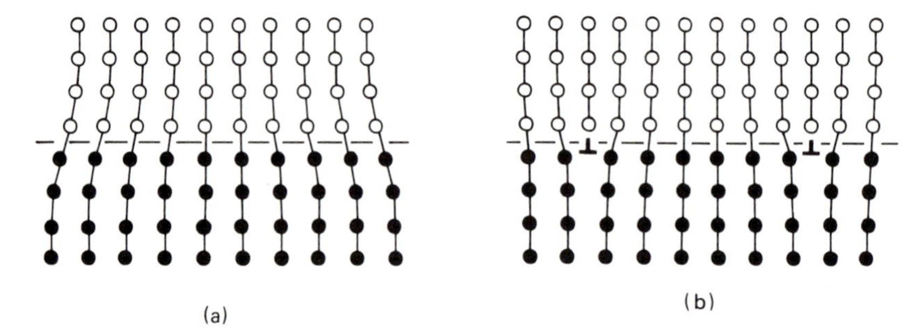

Figure 4.12 Schematic representation of interface structures (a) coherent and (b) incoherent

is associated with maximum strain energy but the minimum surface energy (cf. *Figures 11.4* and *11.6*). On the other hand, surface energy determines the crystallographic plane of the matrix on which a plate-like precipitate forms. Thus, the habit plane is the one which allows the planes at the interface to fit together with the minimum of disregistry; the frequent occurrence of the Widmanstätten structure may be explained on this basis. It is also observed that precipitation occurs most readily in regions of the lattice which are somewhat disarranged, e.g. at grain boundaries, inclusions, dislocations or other positions of high internal stress caused by plastic deformation (*see Figure 11.12*). Such regions have an unusually high free energy and necessarily are the first areas to become unstable during the transformation. Also, new phases can form there with a minimum increase in surface energy. This behaviour is considered again in Chapter 11.

4.11 The equilibrium diagram

The basic features of the various equilibrium diagrams are readily accounted for in terms of the thermodynamic principles already outlined. The diagram merely represents on one plot the composition limits of the stable phases determined from the relative positions of the free energy curves at various temperatures. We have seen in section 4.5 that the entropy of the system varies with composition as shown in *Figure 4.13*(*a*) and, consequently, the corresponding free energy versus composition curve is of the form shown in *Figure 4.13*(*b*), (*c*) or (*d*) depending on whether the solid solution is ideal or deviates from ideal behaviour. The variations of enthalpy with composition, or heat of mixing, is linear for an ideal solid solution, but if A atoms prefer to be in the vicinity of B atoms rather than A atoms, and B atoms behave similarly, the enthalpy will be lowered by alloying (*Figure 4.13*(*c*). A positive deviation occurs when A and B atoms prefer like atoms as neighbours and the free energy curve takes the form shown in *Figure 4.13*(*d*). In diagrams (*b*) and (*c*) the curvature $\mathrm{d}G^2/\mathrm{d}c^2$ is everywhere positive whereas in (*d*) there are two minima and a region of negative

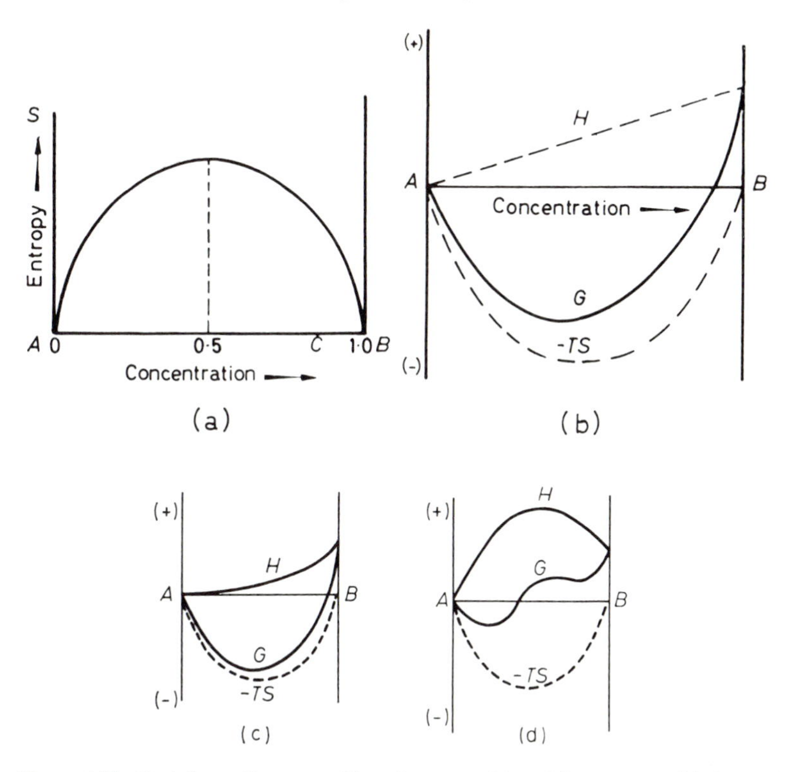

Figure 4.13 Variation with composition of entropy (a) and free energy (b) (for an ideal solid solution) and non-ideal solid solutions (c) and (d)

curvature between points of inflexion* given by $dG^2/dc^2 = 0$. A free energy curve for which d^2G/dc^2 is positive, i.e. simple U-shaped, gives rise to a homogeneous solution. When a region of negative curvature exists, the stable state is a phase mixture rather than a homogeneous solid solution, as shown in *Figure 4.14(a)*. An alloy of composition c has a lower free energy G_c when it exists as a mixture of A-rich phase (α_1) of composition c_A and B-rich phase (α_2) of composition c_B in the proportions given by the lever rule, i.e. $\alpha_1/\alpha_2 = (c_B - c)/(c - c_A)$. Alloys with composition $c < c_A$ or $c > c_B$ exist as homogeneous solid solutions and are denoted by phases, α_1 and α_2 respectively. Partial miscibility in the solid state can also occur when the crystal structures of the component metals are different. The free energy curve then takes the form shown in *Figure 4.14(b)*, the phase being denoted by α and β.

To illustrate how the equilibrium diagram is derived, let us examine the relative positions of the free energy curves at various temperatures for a simple binary system with complete miscibility in both the liquid and solid state, e.g.

* The composition at which $d^2G/dc^2 = 0$ varies with temperature and the corresponding temperature–composition curves are called spinodal lines.

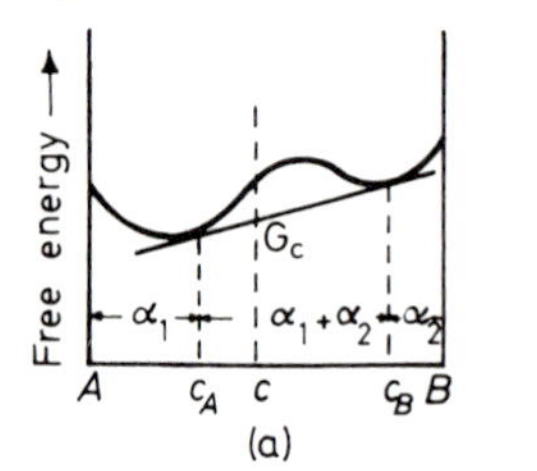

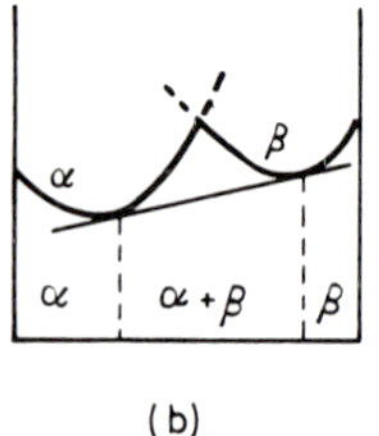

Figure 4.14 Showing extent of phase fields at a given temperature

Cu–Ni. Only two phases, solid and liquid, appear and at a high temperature T_1, i.e. above the melting point, the free energy curve of the solid relative to that of the liquid is as shown in *Figure 4.15*(*a*). The liquid phase has a lower free energy than the solid phase and, therefore, at this temperature all alloys of the system are liquid. As the temperature is lowered the absolute value of the free energy for both phases increases but, because that of the solid phase changes less rapidly (*see* section 4.7), the free energy value for the liquid approaches that of the solid. Hence, at a temperature T_2 (*Figure 4.15*(*b*)) the liquid is in equilibrium with solid of composition equal to A; this represents the melting point of pure A. At still lower temperatures more and more of the alloy becomes stable in the solid phase and, as shown in *Figure 4.15*(*c*), the solid phase is more stable than the liquid phase provided an alloy composition c_1 is not exceeded. Between c_1 and c_2 the phase mixture $(L+S)$ is more stable than either the liquid or solid phase alone. This is because the free energy of such a phase mixture, as given by the appropriate point on the common tangent drawn between the curves for the α and liquid phases, is lower than that for either the α or liquid phase. The temperature T_4 (*Figure 4.15*(*d*)) represents the lowest temperature at which the liquid is in equilibrium with any solid, i.e. the melting point of B. At all lower temperatures the free energy curve of the solid phase is lower than that of the liquid phase, which means the alloy exists as a solid across the complex composition range.

In more complex systems, the relative positions of the free energy curves of several phases (liquid, solid α, β and γ phases) must be considered. *Figure 4.16* shows the relative positions of the free energy curves to be expected for a simple peritectic alloy system. From *Figure 4.16*(*b*) it is clear that at T_2, alloys of composition c_1 to c_2 prefer to exist as a phase mixture (α + liquid), because this has a lower free energy than any of the other phases alone. A similar situation occurs at T_3 (*Figure 4.16*(*c*), between compositions c_3 to c_4 and c_5 to c_6.

4.11.1 Chemical potential

The concept of chemical potential $\mu = (\mathrm{d}G/\mathrm{d}n)$ is used to denote the rate of change of free energy with n, such that the total free energy of a system made up of

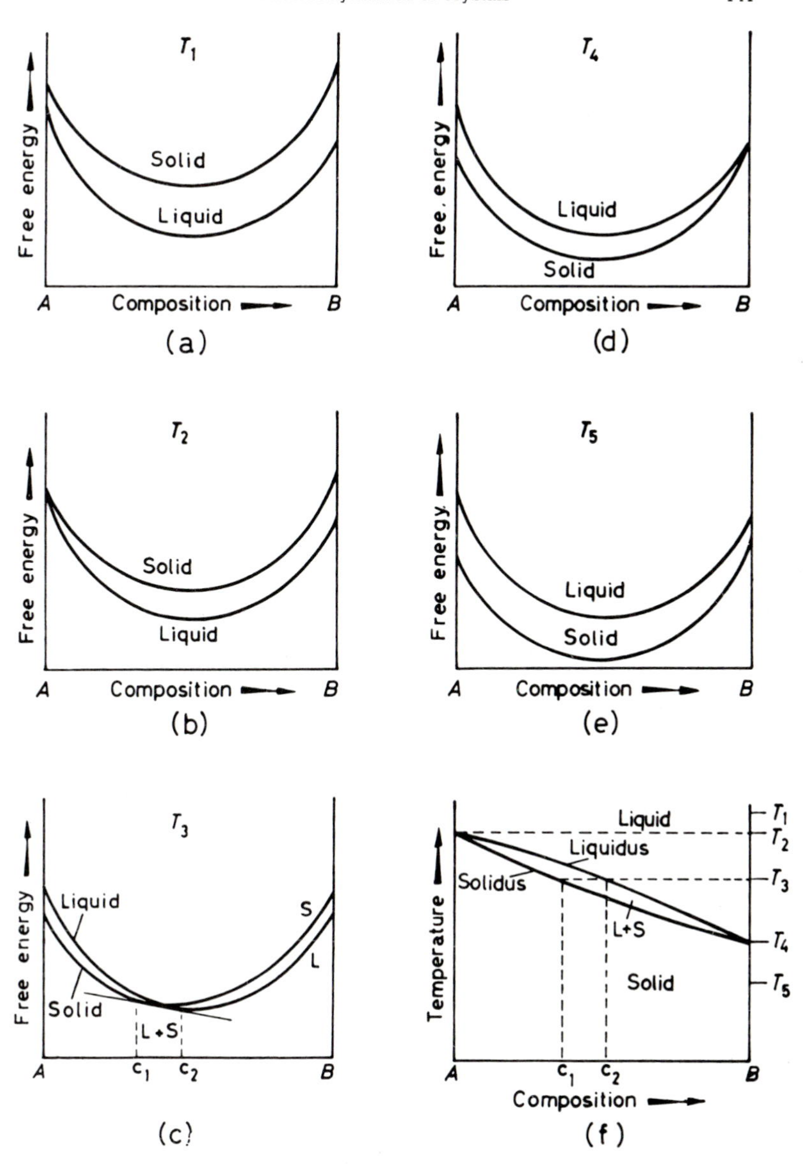

Figure 4.15 (a)–(e) Free energy–composition curves at temperatures T_1, T_2, T_3, ... for the solid and liquid phases of the copper–nickel type of diagram (f)

n_A moles of component A, n_B moles of component B and n_C moles of component C, is

$$G = \mu_A n_A + \mu_B n_B + \mu_C n_C$$

Hence, μ_A is the chemical potential of A in a particular phase with the other variables P, T, n_B and n_C held constant. For equilibrium between phases, say α, β, γ in a system, the chemical potential of any component in all the phases is the same, i.e. $\mu_A^\alpha = \mu_A^\beta = \mu_A^\gamma$.

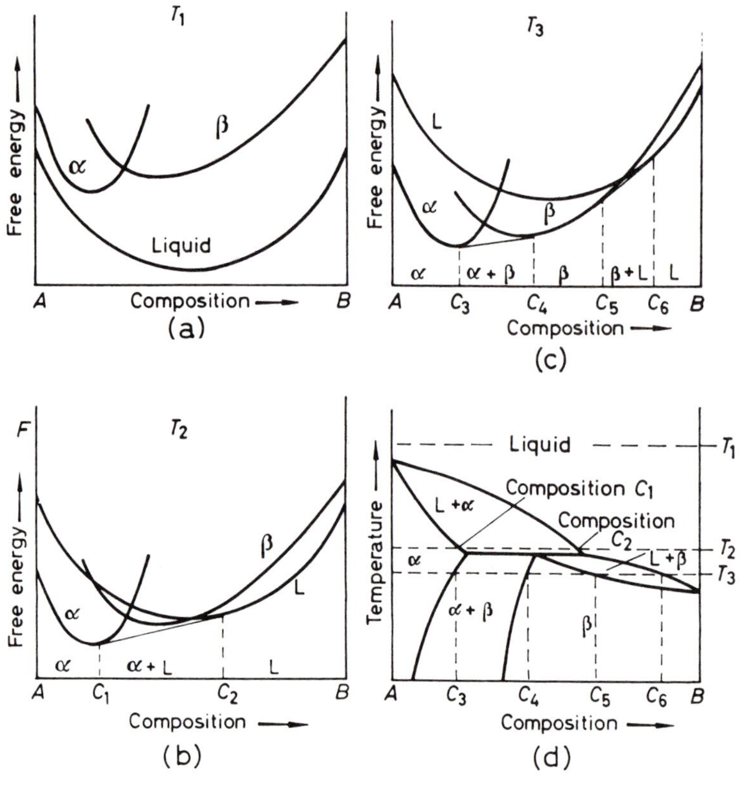

Figure 4.16 (a)–(c) Free energy–composition curves for the solid and liquid phases of a peritectic type of diagram (d). (After Cottrell, *Theoretical and Structural Metallurgy* (1954), courtesy of Prof. Cottrell and Edward Arnold Ltd)

4.12 Diffusion

Some knowledge of diffusion is essential in understanding the behaviour of metals and alloys particularly at elevated temperatures. A few examples include such commercially important processes as annealing, heat treatment, the age hardening of alloys, sintering, surface hardening of steel, oxidation and creep. Apart from the specialized diffusion processes, such as grain boundary diffusion and diffusion down dislocation channels, a distinction is frequently drawn between diffusion in pure metals, homogeneous alloys and inhomogeneous alloys. In a pure material self-diffusion can be observed by using radioactive tracer atoms. In a homogeneous alloy diffusion of each component can also be measured by a tracer method, but in an inhomogeneous alloy, diffusion can be determined by chemical analysis merely from the broadening of the interface between the two metals as a function of time. Inhomogeneous alloys are common in metallurgical practice, e.g. cored solid solutions, and in such cases diffusion always occurs in such a way as to produce a macroscopic flow of solute atoms down the concentration gradient. Thus, if a bar of an alloy, along which there is a concentration gradient (*Figure 4.17*) is heated for a few hours at a temperature

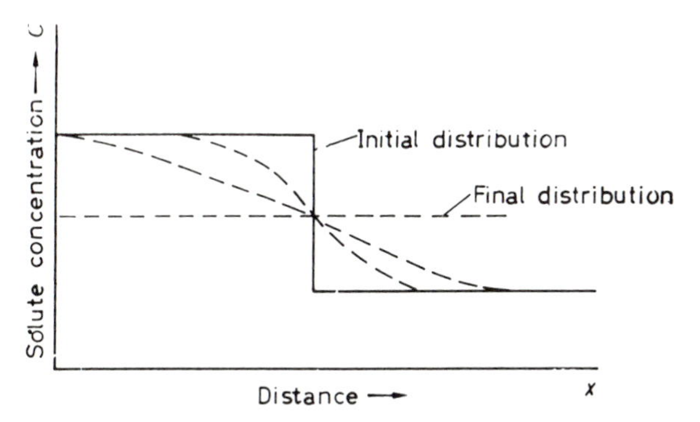

Figure 4.17 Effect of diffusion on the distribution of solute in an alloy

where atomic migration is fast, i.e. near the melting point, the solute atoms are redistributed until the bar becomes uniform in composition. This occurs even though the individual atomic movements are random, simply because there are more solute atoms to move down the concentration gradient than there are to move up. This fact forms the basis of Fick's law of diffusion which is

$$\mathrm{d}n/\mathrm{d}t = -D\,\mathrm{d}c/\mathrm{d}x \tag{4.20}$$

Here the number of atoms diffusing in unit time across unit area through a unit concentration gradient is known as the diffusivity or diffusion coefficient,* D. It is usually expressed as units of $\mathrm{cm^2\,s^{-1}}$ or $\mathrm{m^2\,s^{-1}}$ and depends on the concentration and temperature of the alloy.

Diffusion laws

To illustrate these we may consider the flow of atoms in one direction x, by taking two atomic planes A and B of unit area separated by a distance b, as shown in *Figure 4.18*. If c_1 and c_2 are the concentrations of diffusing atoms in these two planes ($c_1 < c_2$) the corresponding number of such atoms in the respective planes is $n_1 = c_1 b$ and $n_2 = c_2 b$. If the probability that any one jump in the $+x$ direction is p_x, then the number of jumps per unit time made by one atom is $p_x v$, where v is the mean frequency with which an atom leaves a site irrespective of directions. The number of diffusin atoms leaving A and arriving at B in unit time is $(p_x v c_1 b)$ and the number making the reverse transition is $(p_x v c_2 b)$ so that the net gain of atoms at B is

$$p_x v b(c_1 - c_2) = J_x$$

with J_x the flux of diffusing atoms. Setting $c_1 - c_2 = -b(\mathrm{d}c/\mathrm{d}x)$ this flux becomes

$$J_x = -p_x v_v b^2(\mathrm{d}c/\mathrm{d}x) = -\tfrac{1}{2} v b^2(\mathrm{d}c/\mathrm{d}x) = -D(\mathrm{d}c/\mathrm{d}x) \tag{4.21}$$

In cubic lattices, diffusion is isotropic and hence all six orthogonal directions are equally likely so that $p_x = \frac{1}{6}$. For simple cubic structures $b = a$ and thus

* The conduction of heat in a still medium also follows the same laws as diffusion.

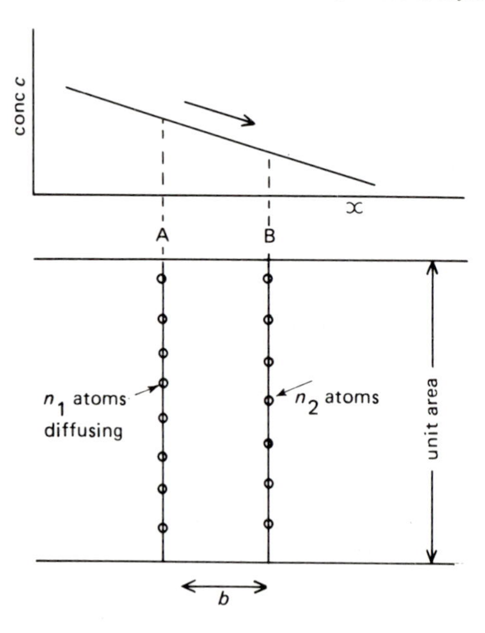

Figure 4.18 Diffusion of atoms down a concentration gradient

$$D_x = D_y = D_z = \tfrac{1}{6}\nu a^2 = D \tag{4.22}$$

whereas in f.c.c. structures $b = a/\sqrt{2}$ and $D = \frac{1}{12}\nu a^2$, and in b.c.c. structures $D = \frac{1}{24}\nu a^2$.

Fick's first law only applies if a steady state exists in which the concentration at every point is invariant, i.e. $(dc/dt) = 0$ for all x. To deal with non-stationary flow in which the concentration at a point changes with time, we take two planes A and B, as before, separated by unit distance and consider the rate of increase of the number of atoms (dc/dt) in a unit volume of the specimen; this is equal to the difference between the flux into and that out of the volume element. The flux across one plane is J_x and across the other $(J_x + 1)\,dJ/dx$ the difference being $-(dJ/dx)$. We thus obtain Fick's second law of diffusion

$$\frac{dc}{dt} = -\frac{dJ_x}{dx} = \frac{d}{dx}\left(D_x \frac{dc}{dx}\right) \tag{4.23}$$

When D is independent of concentration this reduces to

$$\frac{dc_x}{dt} = -D_x \frac{d^2c}{dx^2} \tag{4.24}$$

and in three dimensions becomes

$$\frac{dc}{dt} = \frac{d}{dx}\left(D_x \frac{dc}{dx}\right) + \frac{d}{dy}\left(D_y \frac{dc}{dy}\right) + \frac{d}{dz}\left(D_z \frac{dc}{dz}\right)$$

An illustration of the use of the diffusion equations is the behaviour of a diffusion couple, where there is a sharp interface between pure metal and an alloy. *Figure 4.17* can be used for this example and as the solute moves from alloy to the pure metal the way in which the concentration varies is shown by the dotted lines.

The solution to Fick's second law is given by

$$c=\frac{c_0}{2}\left[1-\frac{2}{\sqrt{\pi}}\int_0^{x/[2\sqrt{(Dt)}]}\exp(-y^2)\,\mathrm{d}y\right] \tag{4.25}$$

where c_0 is the initial solute concentration in the alloy and c is the concentration at a time t at a distance x from the interface. The integral term is known as the Gauss error function (erf (y)) and as $y \to \infty$, erf $(y) \to 1$. It will be noted that at the interface where $x=0$, then $c=c_0/2$, and in those regions where the curvature $\partial^2 c/\partial x^2$ is positive the concentration rises, in those regions where the curvature is negative the concentration falls, and where the curvature is zero the concentration remains constant.

This particular example is important because it can be used to model the depth of diffusion after time t, e.g. in the case-hardening of steel, providing the concentration profile of the carbon after a carburizing time t, or dopant in silicon. Starting with a constant composition at the surface, the value of x where the concentration falls to half the initial value, i.e. $1-\mathrm{erf}(y)=\frac{1}{2}$, is given by $x=\sqrt{(Dt)}$. Thus knowing D at a given temperature the time to produce a given depth of diffusion can be estimated.

The diffusion equations developed above can also be transformed to apply to particular diffusion geometries. If the concentration gradient has spherical symmetry about a point, c varies with the radial distance r and, for constant D,

$$\frac{\mathrm{d}c}{\mathrm{d}t}=D\left(\frac{\mathrm{d}^2 c}{\mathrm{d}r^2}+\frac{2}{r}\frac{\mathrm{d}c}{\mathrm{d}r}\right) \tag{4.26}$$

When the diffusion field has radial symmetry about a cylindrical axis, the equation becomes

$$\frac{\mathrm{d}c}{\mathrm{d}t}=D\left(\frac{\mathrm{d}^2 c}{\mathrm{d}r^2}+\frac{1}{r}\frac{\mathrm{d}c}{\mathrm{d}r}\right) \tag{4.27}$$

and the steady state condition $(\mathrm{d}c/\mathrm{d}t)=0$ is given by

$$\frac{\mathrm{d}^2 c}{\mathrm{d}r^2}+\frac{1}{r}\frac{\mathrm{d}c}{\mathrm{d}r}=0 \tag{4.28}$$

which has a solution $c=A\ln r+B$. The constants A and B may be found by introducing the appropriate boundary conditions and for $c=c_0$ at $r=r_0$ and $c=c_1$ at $r=r_1$ the solution becomes

$$c=\frac{c_0\ln(r_1/r)+c_1\ln(r/r_0)}{\ln(r_1/r_0)}$$

The flux through any shell of radius r is $-2\pi rD(\mathrm{d}c/\mathrm{d}r)$ or

$$J=-\frac{2\pi D}{\ln(r_1/r_0)}(c_1-c_0) \tag{4.29}$$

Diffusion equations are of importance in many diverse problems and in Chapter 9 are applied to the diffusion of vacancies from dislocation loops and the sintering of voids.

4.12.1 The mechanisms of diffusion

The transport of atoms through the lattice may conceivably occur in many ways. The term 'interstitial diffusion' describes the situation when the moving atom does not lie on the crystal lattice, but instead occupies an interstitial position. Such a process is likely in interstitial alloys where the migrating atom is very small, e.g. carbon, nitrogen or hydrogen in iron. In this case, the diffusion process for the atoms to move from one interstitial position to the next in a perfect lattice is not defect-controlled. A possible variant of this type of diffusion has been suggested for substitutional solutions in which the diffusing atoms are only temporarily interstitial and are in dynamic equilibrium with others in substitutional positions. However, the energy to form such an interstitial is many times that to produce a vacancy and, consequently, the most likely mechanism is that of the continual migration of vacancies. With vacancy diffusion, the probability that an atom may jump to the next site will depend on: (i) the probability that the site is vacant (which in turn is proportional to the fraction of vacancies in the crystal), and (ii) the probability that it has the required activation energy to make the transition. For self-diffusion where no complications exist, the diffusion coefficient is therefore given by

$$D = \tfrac{1}{6}a^2 f \nu \exp[(S_f + S_m)/k] \exp[-E_f/\boldsymbol{k}T] \exp[-E_m/\boldsymbol{k}T]$$
$$= D_0 \exp[-(E_f + E_m)/\boldsymbol{k}T] \qquad (4.30)$$

The factor f appearing in D_0 is known as a correlation factor and arises from the fact that any particular diffusion jump is influenced by the direction of the previous jump. Thus when an atom and a vacancy exchange places in the lattice there is a greater probability of the atom returning to its original site than moving to another site, because of the presence there of a vacancy; f is 0.80 and 0.78 for f.c.c. and b.c.c. lattices respectively. Values for E_f and E_m are discussed in Chapter 9, E_f is the energy of formation of a vacancy, E_m the energy of migration, and the sum of the two energies, $Q = E_f + E_m$, is the activation energy for self-diffusion,* E_d.

In alloys, the problem is not so simple and it is found that the self-diffusion energy is smaller than in pure metals. This observation has led to the suggestion that in alloys the vacancies associate preferentially with solute atoms in solution; the binding of vacancies to the impurity atoms increases the effective vacancy concentration near those atoms so that the mean jump rate of the solute atoms is much increased. This association helps the solute atom on its way through the lattice, but, conversely, the speed of vacancy migration is reduced because it lingers in the neighbourhood of the solute atoms, as shown in *Figure 4.19*. The phenomenon of association is of fundamental importance in all kinetic studies since the mobility of a vacancy through the lattice to a vacancy sink will be governed by its ability to escape from the impurity atoms which trap it. This problem is considered again in Chapter 9.

When considering diffusion in alloys it is important to realise that in a binary solution of A and B the diffusion coefficients D_A and D_B are generally not equal.

* The entropy factor $\exp[(S_f + S_m)/k]$ is usually taken to be unity.

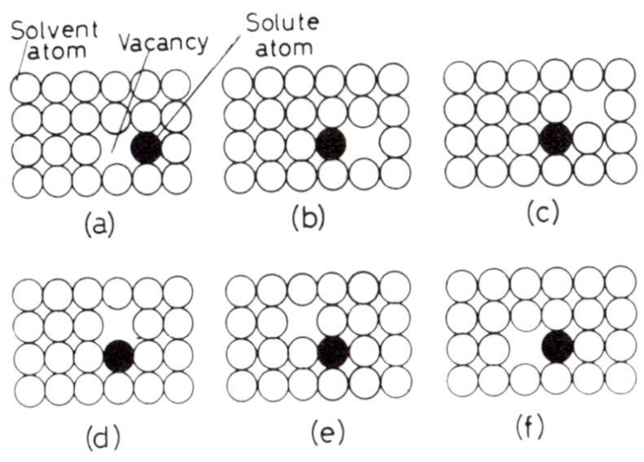

Figure 4.19 Solute atom–vacancy association during diffusion

This inequality of diffusion was first demonstrated by Kirkendall using an α-brass/copper couple (*Figure 4.20*). He noted that if the position of the interfaces of the couple were marked (e.g. with fine W or Mo wires), during diffusion the markers move towards each other, showing that the zinc atoms diffuse out of the alloy more rapidly than copper atoms diffuse in. This being the case, it is not surprising that several workers have shown that porosity develops in such systems on that side of the interface from which there is a net loss of atoms.

The Kirkendall effect is of considerable theoretical importance since it confirms the vacancy mechanism of diffusion. This is because the observations cannot easily be accounted for by any other postulated mechanisms of diffusion, such as direct place-exchange, i.e. where neighbouring atoms merely change place with each other. The Kirkendall effect is readily explained in terms of vacancies since the lattice defect may interchange places more frequently with one atom than the other. The effect is also of some practical importance, especially in the fields of metal-to-metal bonding, sintering and creep.

4.12.2 Factors affecting diffusion

The two most important factors affecting the diffusion coefficient D are temperature and composition. Because of the activation energy term the rate of diffusion increases with temperature according to equation 4.30 above, while

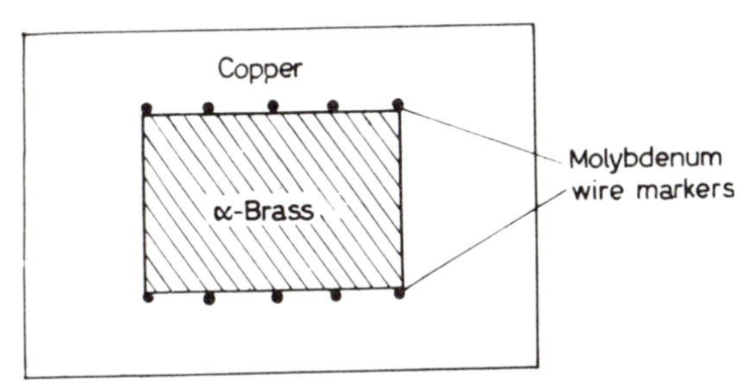

Figure 4.20 α-brass–copper couple for demonstrating the Kirkendall effect

each of the quantities D, D_0 and Q varies with concentration; for a metal at high temperatures $Q \approx 20RT_m$, D_0 is 10^{-5} to $10^{-3}\,m^2\,s^{-1}$, and $D \simeq 10^{-12}\,m^2\,s^{-1}$. Because of this variation of diffusion coefficient with concentration, the most reliable investigations into the effect of other variables necessarily concern self-diffusion in pure metals.

Diffusion is a structure-sensitive property and, therefore, D is expected to increase with increasing lattice irregularity. In general, this is found experimentally. In metals quenched from a high temperature the excess vacancy concentration $\approx 10^9$ leads to enhanced diffusion at low temperatures since $D = D_0 c_v \exp(-E_m/kT)$. Grain boundaries and dislocations are particularly important in this respect and produce enhanced diffusion. Diffusion is faster in the cold worked state than in the annealed state, although recrystallization may take place and tend to mask the effect. The enhanced transport of material along dislocation channels has been demonstrated in aluminium where voids connected to a free surface by dislocations anneal out at appreciably higher rates than isolated voids. Measurements show that surface and grain boundary forms of diffusion also obey Arrhenius equations, with lower activation energies than for volume diffusion, i.e. $Q_{vol} \geqslant 2Q_{g.b} \geqslant 2Q_{surface}$. This behaviour is understandable in view of the progressively more open atomic structure found at grain boundaries and external surfaces. It will be remembered however, that the relative importance of the various forms of diffusion does not entirely depend on the relative activation energy or diffusion coefficient values. The amount of material transported by any diffusion process is given by Fick's law and for a given composition gradient also depends on the effective area through which the atoms diffuse. Consequently, since the surface area (or grain boundary area) to volume ratio of any polycrystalline solid is usually very small, it is only in particular phenomena, e.g. sintering, oxidation, etc., that grain boundaries and surfaces become important. It is also apparent that grain boundary diffusion becomes more competitive the finer the grain size, and the lower the temperature. The latter feature follows from the lower activation energy which makes it less sensitive to temperature change. As the temperature is lowered, the diffusion rate along grain boundaries (and also surfaces) decreases less rapidly than the diffusion rate through the lattice. The importance of grain boundary diffusion and dislocation pipe diffusion is discussed again in Chapter 13 in relation to deformation at elevated temperatures, and is demonstrated convincingly on the deformation maps (*see Figure 13.10*) where the creep field is extended to lower temperatures when grain boundary (coble creep) rather than lattice diffusion (Herring–Nabarro creep) operates.

Because of the strong binding between atoms, pressure has little or no effect but it is observed that with extremely high pressure on soft metals, e.g. sodium, an increase in Q may result. The rate of diffusion also increases with decreasing density of atomic packing. For example, self-diffusion is slower in f.c.c. iron or thallium than in b.c.c. iron or thallium when the results are compared by extrapolation to the transformation temperature. This is further emphasized by the anisotropic nature of D in metals of open structure. Bismuth (rhombohedral) is an example of a metal in which D varies by 10^6 for different directions in the lattice; in cubic crystals D is isotropic.

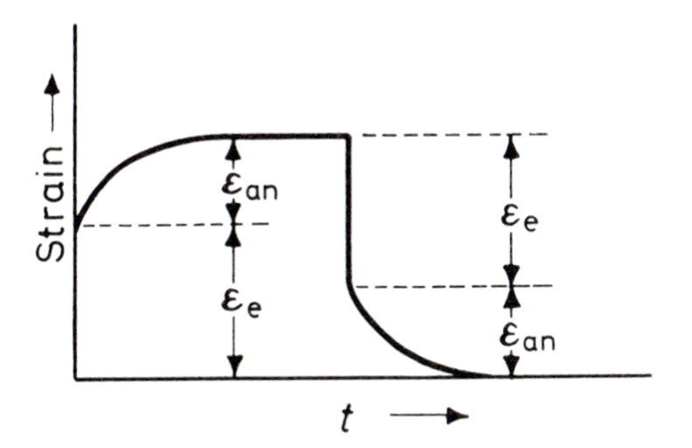

Figure 4.21

4.13 Anelasticity and internal friction

For an elastic solid it is generally assumed that stress and strain are directly proportional to one another, but in practice the elastic strain is usually dependent on time as well as stress so that the strain lags behind the stress; this is an anelastic effect. On applying a stress at a level below the conventional elastic limit, a specimen will show an initial elastic strain ε_e followed by a gradual increase in strain until it reaches an essentially constant value, $\varepsilon_e+\varepsilon_{an}$ as shown in *Figure 4.21*. When the stress is removed the strain will decrease, but a small amount remains which decreases slowly with time. At any time t the decreasing anelastic strain is given by the relation $\varepsilon=\varepsilon_{an}\exp(-t/\tau)$ where τ is known as the relaxation time, and is the time taken for the anelastic strain to decrease to $1/e\simeq 36.79$ per cent of its initial value. Clearly, if τ is large, the strain relaxes very slowly, while if small the strain relaxes quickly.

In materials under cyclic loading this anelastic effect leads to a decay in amplitude of vibration and therefore a dissipation of energy by internal friction. Internal friction is defined in several different but related ways. Perhaps the most common uses the logarithmic decrement $\delta=\ln(A_n/A_{n+1})$, the natural logarithm of successive amplitudes of vibration. In a forced vibration experiment near a resonance, the factor $(\omega_2-\omega_1)/\omega_0$ is often used, where ω_1 and ω_2 are the frequencies on the two sides of the resonant frequency ω_0 at which the amplitude of oscillation is $1/\sqrt{2}$ of the resonant amplitude. Also used is the specific damping capacity $\Delta E/E$, where ΔE is the energy dissipated per cycle of vibrational energy E, i.e. the area contained in a stress–strain loop. Yet another method uses the phase angle α by which the strain lags behind the stress, and if the damping is small it can be shown that

$$\tan\alpha=\frac{\delta}{\pi}=\frac{1}{2\pi}\frac{\Delta E}{E}=\frac{\omega_2-\omega_1}{\omega_0}=Q^{-1}$$

By analogy with damping in electrical systems $\tan\alpha$ is often written equal to Q^{-1}.

There are many causes of internal friction arising from the fact that the migration of atoms, lattice defects and thermal energy are all time dependent processes. The latter gives rise to thermoelasticity and occurs when an elastic stress is applied to a specimen too fast for the specimen to exchange heat with its surroundings and so cools slightly. As the sample warms back to the surrounding

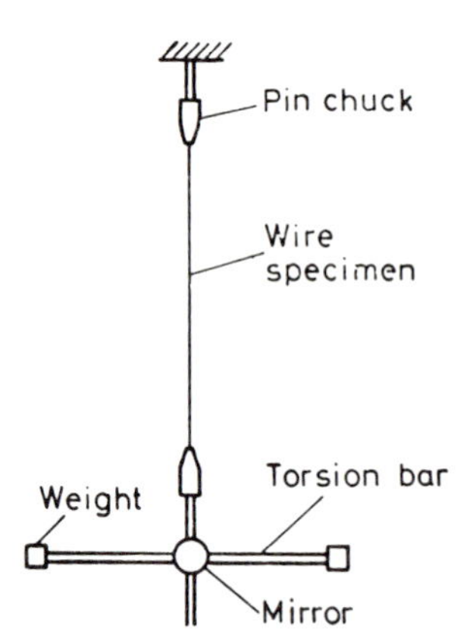

Figure 4.22 Schematic diagram of a Kê torsion pendulum

temperature it expands thermally, and hence the dilatation strain continues to increase after the stress has become constant.

The diffusion of atoms can also give rise to anelastic effects in an analogous way to the diffusion of thermal energy gives thermoelastic effects. A particular example is the stress induced diffusion of carbon or nitrogen in iron. A carbon atom occupies the interstitial site along one of the cell edges (*see Figure 5.6*) slightly distorting the lattice tetragonally. Thus when iron is stretched by a mechanical stress, the crystal axis oriented in the direction of the stress develops favoured sites for the occupation of the interstitial atoms relative to the other two axes. Then if the stress is oscillated, such that first one axis and then another is stretched, the carbon atoms will want to jump from one favoured site to the other. Mechanical work is therefore done repeatedly, dissipating the vibrational energy and damping out the mechanical oscillations. The maximum energy is dissipated when the time per cycle is of the same order as the time required for the diffusional jump of the carbon atom.

The simplest and most convenient way of studying this form of internal friction is by means of a Kê torsion pendulum, shown schematically in *Figure 4.22*. The specimen can be oscillated at a given frequency by adjusting the moment of inertia of the torsion bar. The energy loss per cycle $\Delta E/E$ varies smoothly with the frequency according to the relation

$$\frac{\Delta E}{E}=2\left(\frac{\Delta E}{E}\right)_{\max}\left[\frac{\omega\tau}{1+(\omega\tau)^2}\right]$$

and has a maximum value when the angular frequency of the pendulum equals the relaxation time of the process; at low temperatures around room temperature this is interstitial diffusion. In practice, it is difficult to vary the angular frequency over a wide range and thus it is easier to keep ω constant and vary the relaxation time. Since the migration of atoms depends strongly on temperature according to an Arrhenius type equation, the relaxation time $\tau_1=1/\omega_1$ and the peak occurs at a temperature T_1. For a different frequency value ω_2 the peak occurs at a different temperature T_2, and so on (*see Figure 4.23*). It is thus possible to ascribe an activation energy ΔH for the internal process producing the damping by plotting $\ln\tau$ versus $1/T$, or from the relation

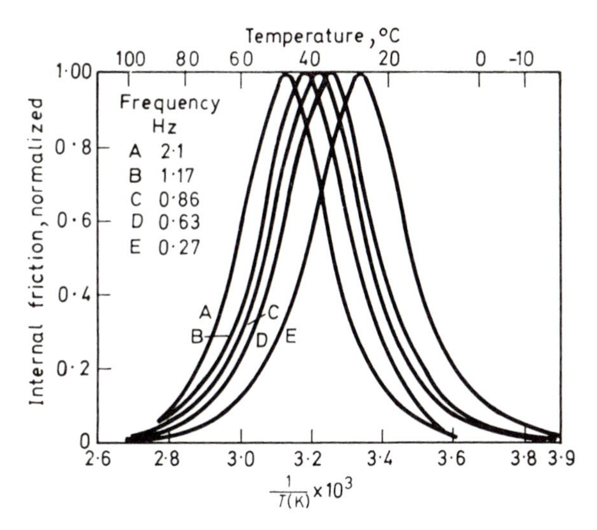

Figure 4.23 Internal friction as a function of temperature for Fe with C in solid solution at five different pendulum frequencies (From Wert and Zener, *Phys. Rev.*, 1949, **76,** 1169, by permission of the American Institute of Physics)

$$\Delta H = R \frac{\ln(\omega_2/\omega_1)}{\dfrac{1}{T_1} - \dfrac{1}{T_2}}$$

In the case of iron the activation energy is found to coincide with that for the diffusion of carbon in iron. Similar studies have been made for other metals. In addition, if the relaxation time is τ the mean time an atom stays in an interstitial position is $\frac{3}{2}\tau$, and from the relation $D = \frac{1}{24}a^2 v$ for b.c.c. lattices derived previously (page 146), the diffusion coefficient may be calculated directly from

$$D = \frac{1}{36}\left(\frac{a^2}{\tau}\right)$$

Many other forms of internal friction exist in metals arising from different relaxation processes to those discussed above, and hence occurring in different frequency and temperature regions. One important source of internal friction is that due to stress relaxation across grain boundaries. The occurrence of a strong internal friction peak due to grain boundary relaxation was first demonstrated on polycrystalline aluminium at 300 °C by Kê and has since been found in numerous other metals. It indicates that grain boundaries behave in a somewhat viscous manner at elevated temperatures and grain boundary sliding can be detected at very low stresses by internal friction studies. The grain boundary sliding velocity produced by a shear stress τ is given by $v = \tau d/\eta$ and its measurement gives values of the viscosity η which extrapolate to that of the liquid at the melting point, assuming the boundary thickness to be $d \simeq 5$ Å.

Movement of low energy twin boundaries in crystals, domain boundaries in ferromagnetic materials and dislocation bowing and unpinning all give rise to internal friction and damping.

Suggestions for further reading

Christian, J.W., *The Theory of Transformations in Metals and Alloys*, Pergamon Press, 1965

Hume-Rothery, W., Smallman, R.E. and Haworth, C., *The Structure of Metals and Alloys*, Institute of Metals, London, 1969

Kubaschewski, O. and Evans, E.Ll., *Metallurgical Thermochemistry*, Pergamon Press, 1958

Lumsden, J., *Thermodynamics of Alloys*, Institute of Metals, London, 1952

Porter, D.A. and Easterling, K.E., *Phase Transformations in Metals and Alloys*, Van Nostrand Reinhold, 1981

Shewmon, P.G., *Diffusion in Solids*, McGraw-Hill, 1963

Chapter 5

The structure of alloys

5.1 Introduction

When a metal B is alloyed to a metal A several different structures and atomic arrangements may be obtained in the alloy, depending upon the relative amounts of the component metals and upon the temperature of the alloy. Thus, if the two types of atoms behave as if they were similar and become homogeneously dispersed amongst each other, a solid solution of the type shown in *Figure 1.8(a)* will be formed. However, in only a few alloy systems does the solid solution exist over the entire composition range from pure A to pure B, one example being the copper–nickel system. More usually, the second element enters into solid solution only to a limited extent and, in this case, a primary solid solution is formed which has the same crystal structure as the parent metal (*see* for example the copper–zinc system, *Figure 3.5(a)*. Then at higher concentrations of the second element, new phases, generally termed intermediate phases, are formed in which the crystal structure usually differs from that of the parent metals. These intermediate phases are also called secondary solid solutions if they exist over wide ranges of composition, or intermetallic compounds if the range of homogeneity is small.

5.2 Primary substitutional solid solutions

As a result of a comparison of the solubilities of various solute elements in the noble metals, copper, silver and gold, several general rules* governing the extent of the primary solid solutions have been formulated. Extension of these experimental observations to solvents from other groups such as magnesium and iron show that, in general, these rules form a useful basis for predicting alloying behaviour. In brief the rules are as follows:

(1) *The atomic size factor*—If the atomic diameter of the solute atom differs by more than 15 per cent from that of the solvent atom the extent of the primary solid solution is small. In such cases it is said that the size-factor is unfavourable for extensive solid solution.

* These are usually called the Hume-Rothery rules because it was chiefly Hume-Rothery and his colleagues who formulated them.

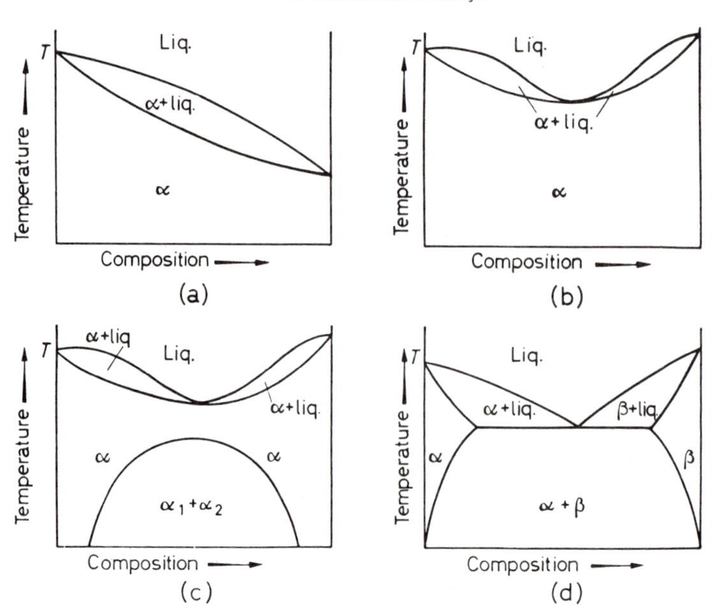

Figure 5.1 Effect of size factor on the form of the equilibrium diagram; examples include (a) Cu–Ni, Au–Pt, (b) Ni–Pt, (c) Au–Ni, and (d) Cu–Ag

(2) *The electrochemical effect*—The more electropositive the one component and the more electronegative the other, the greater is the tendency for the two elements to form compounds rather than extensive solid solutions.

(3) *Relative valency effect*—A metal of higher valency is more likely to dissolve to a large extent in one of lower valency than vice versa.

5.2.1 The size factor effect

Clearly, two metals are able to form a continuous range of solid solutions only if they have the same crystal structure, e.g. copper and nickel both have f.c.c. structures. However, even when the crystal structures of the two elements are the same, the extent of the primary solubility is limited if the atomic size of the two metals, usually taken as the closest distance of approach of atoms in the crystal of the pure metal, is unfavourable. This is demonstrated in *Figure 5.1* for alloy systems where rules 2 and 3 have been observed, i.e. the electrochemical properties of the two elements are similar and the solute is dissolved in a metal of lower valency. As the size difference between the atoms of the two component metals A and B approaches 15 per cent, the equilibrium diagram changes from that of the copper–nickel type to one of a eutectic system with limited primary solid solubility.

The size-factor effect is due to the distortion produced in the parent lattice around the dissolved misfitting solute atom. In these localized regions the interatomic distance will differ from that given by the minimum in the $E-r$ curve of *Figure 4.1*, so that the internal energy and hence the free energy, G, of the system is raised. In the limit when the lattice distortion is greater than some critical value the primary solid solution becomes thermodynamically unstable relative to some other phase.

The size factor rule may be deduced semi-quantitatively from elasticity considerations, since this allows a calculation of the strain energy ($E_s = 8\pi\mu r_0^3\varepsilon^2$) associated with the solution of a solute atom in an alloy of concentration, c, i.e. the heat of solution, to be made. According to this theory, in which atoms are regarded as rigid spheres and electronic considerations are ignored, the height of the dome-shaped two-phase region on the temperature–composition diagram (*Figure 5.1*(*c*)) increases with increasing misfit, and the maximum at $c = \frac{1}{2}$ is given by

$$T = 2\mu\Omega\varepsilon^2/\boldsymbol{k} \tag{5.1}$$

where μ is the shear modulus of the alloy, Ω its atomic volume, $\boldsymbol{k}$ is Boltzmann's constant and ε is the misfit of the solute atom in the solvent which is equal to $(r_1 - r_o/r_o)$ where r_1 and r_o are the atomic radii of the solute and solvent atoms respectively. If there is to be no complete solid solution at any temperature, then T must be equal to the melting point T_m of the alloy, which gives the condition for limited primary solid solubility to be when

$$|\varepsilon| > (\boldsymbol{k}T_m/2\mu\Omega)^{1/2} \tag{5.2}$$

For most metals, $\boldsymbol{k}T_m/\mu\Omega$ is about 0.04 and substitution in the above relationship shows that the limiting misfit value $|\varepsilon|$ is 14 per cent.

5.2.2 The electrochemical effect

This effect is best demonstrated by reference to the alloying behaviour of an electropositive solvent with solutes of increasing electronegativity. The electronegativity of elements in the Periodic Table increases from left to right in any period and from bottom to top in any group. Thus, if magnesium is alloyed with elements of Group IV the compounds formed, Mg_2(Si, Sn or Pb), become more stable in the order lead, tin, silicon, as shown by their melting points, 550, 778 and 1085 °C respectively. In accordance with rule 2 the extent of the primary solid solution is small ($\approx$7.75 atomic per cent, 3.35 atomic per cent, and negligible, respectively at the eutectic temperature) and also decreases in the order lead, tin, silicon. Similar effects are also observed with elements of Group V, which includes the elements bismuth, antimony and arsenic, when the compounds Mg_3 (Bi, Sb or As)$_2$ are formed.

The importance of compound formation in controlling the extent of the primary solid solution can be appreciated by reference to *Figure 5.2*, where the curves represent the free-energy versus composition relationship between the α-phase and compound at a temperature T. It is clear from *Figure 5.2*(*a*) that at this temperature the α-phase is stable up to a composition c_1, above which the phase mixture (α + compound) has the lower free energy. When the compound becomes more stable, as shown in *Figure 5.2*(*b*), the solid solubility decreases, and correspondingly the phase mixture is now stable over a greater composition range which extends from c_3 to c_4.

The above example is an illustration of a more general principle that the

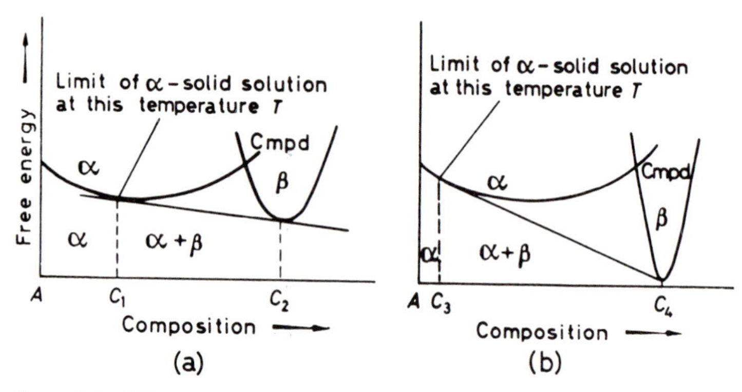

Figure 5.2 Influence of compound stability on the solubility limit of the α-phase at a given temperature

solubility of a phase decreases with increasing stability, and may also be used to show that the concentration of solute in solution increases as the radius of curvature of the precipitate particle decreases. Small precipitate particles are less stable than large particles and the variation of solubility with particle size is recognized in classical thermodynamics by the Thomson–Freundlich equation

$$\ln [c(r)/c] = 2\gamma\Omega/kTr$$

where $c(r)$ is the concentration of solute in equilibrium with small particles of radius r, c the equilibrium concentration, γ the precipitate/matrix interfacial energy and Ω the atomic volume (*see* Chapters 9 and 11).

5.2.3 The relative valency effect

This is a general rule for alloys of the univalent metals, copper, silver and gold, with those of higher valency. Thus, for example, copper will dissolve approximately 40 per cent zinc in solid solution but the solution of copper in zinc is limited. For solvent elements of higher valencies the application is not so general, and in fact exceptions, such as that exhibited by the magnesium–indium system, occur.

5.3 The form of the liquidus and solidus curves

The fact that the primary solid solubility is affected by the three factors discussed above, implies that the form of the liquidus and solidus curves will also be influenced. This may be demonstrated if we alter any one of the three alloying factors while keeping the other two constant. For example, if we take copper or silver as solvents and add to them elements which have favourable size factors, i.e. those elements which follow copper or silver in the Periodic Table, it is found that a definite valency effect exists. In each series, increasing the valency of the solute results in a more restricted solid solution and a steeper fall in both the liquidus and solidus curves. These observations, an example of which is shown in *Figure*

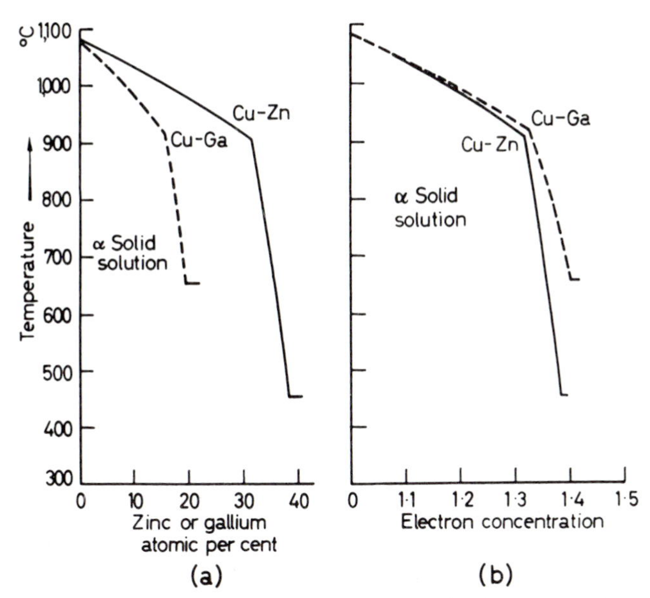

Figure 5.3 Solidus and solid solubility curves for the copper–zinc and copper–gallium systems (after Hume-Rothery, Smallman and Howath, *Structure of Metals and Alloys,* by courtesy of the Institute of Metals)

5.3(a), suggest that if the curves were plotted in terms of the product of the atomic concentration of solute and valency they would superimpose. The equivalent composition used in practice is the electron concentration, or the ratio of the number of valency electrons to atoms in the alloy*, i.e. the e/a ratio, and *Figure 5.3(b)* shows the data for the copper–zinc and copper–gallium systems plotted in this way against the e/a ratio. A comparison of *Figure 5.3(a)* and *(b)* clearly demonstrates the important role played by electronic structure in alloying.

When all the factors affecting alloying are operative the situation is less clear, but an examination of many systems suggests that the depression of the liquidus and solidus curves increases both with increasing difference in size factor and with increasing difference in valency between the solute and solvent. Moreover, the solidus curves appear to be much more affected than the liquidus curves. With magnesium as solvent, although the tendency to compound formation tends to complicate the observations, it is nevertheless still possible to see the effect of valency. The addition of solute elements such as cadmium ($Z=2$), aluminium ($Z=3$), or tin ($Z=4$), causes both the liquidus and solidus curves to fall more and more steeply with increasing valency.

5.4 The primary solid solubility boundary

It is not yet possible to predict the exact form of the α-solid solubility boundary, but in general terms the boundary may be such that the range of

* For example, a copper–zinc alloy containing 40 atomic per cent zinc has an e/a ratio of 1.4, i.e. for every 100 atoms, 60 are copper each contributing one valency electron and 40 are zinc each contributing 2 valency electrons, so that $e/a = (60 \times 1 + 40 \times 2)/100 = 1.4$.

primary solid solution either (a) increases, or (b) decreases with rise of temperature. Both forms arise as a result of the increase in entropy which occurs when solute atoms are added to a solvent. It will be remembered that this entropy of mixing is a measure of the extra disorder of the solution compared to the pure metal, and takes the form shown in *Figure 4.13(a)*.

The most common form of phase boundary is that indicating that the solution of one metal in another increases with rise in temperature. This follows from thermodynamic reasoning since increasing the temperature favours the structure of highest entropy (because of the $-TS$ term in the relation $G=H-TS$) and in alloy systems of the simple eutectic type an α-solid solution has a higher entropy than a phase mixture $(\alpha+\beta)$. Thus, if the alloy exists as a phase mixture $(\alpha+\beta)$ at the lower temperatures, it does so because the value of H happens to be less for the mixture than for the homogeneous solution at that composition. However, because of its greater entropy term, the solution gradually becomes preferred at high temperatures. In more complex alloy systems, particularly those containing intermediate phases of the secondary solid solution type (e.g. copper–zinc, copper–gallium, copper–aluminium, etc.), the range of primary solid solution decreases with rise in temperature. This is because the β-phase, like the α-phase, is a disordered solid solution. However, since it occurs at a higher composition, it is evident from *Figure 4.13(a)* that it has a higher entropy of mixing, and consequently its free energy will fall more rapidly with rise in temperature. This is shown schematically in *Figure 5.4*. The point of contact on the free energy curve of the α-phase, determined by drawing the common tangent to the α and β curves, governs the solubility c at a given temperature T. The steep fall with temperature of this common tangent automatically gives rise to a decreasing solubility limit.

As discussed in the previous section, the electron concentration is an important factor controlling the slope of the liquidus and solidus curves, and consequently this factor must also be reflected in the composition limit of the α-phase. In support of this it is observed that in many alloys of copper or silver the f.c.c. α-solid solution reaches the limit of its solubility at an electron to atom ratio of about 1.4. The divalent elements zinc, cadmium and mercury have solubilities of approximately 40 atomic per cent* (e.g. copper–zinc, silver–cadmium, silver–mercury), the trivalent elements approximately 20 atomic per cent (e.g. copper–aluminium, copper–gallium, silver–aluminium, silver–indium) and the tetravalent elements about 13 per cent (e.g. copper–germanium, copper–silicon, silver–tin), respectively. The valency factor, therefore, has the same influence on these primary solubility values as it does on the liquidus and solidus curves. In all the above examples the solute and solvent atoms have favourable size factors, but if this is not the case the solubility limit is less than that given by an electron-to-atom ratio of 1.4; tin is on the borderline of size factor favourability and hence dissolves in copper only up to 9.2 atomic per cent. It is also found that alloys with gold as solvent show much lower electron–atom ratios at saturation even when

* When using this rule in practice remember that x atomic per cent is equal to $[xW_B/xW_B+(100-x)W_A\times 100]$ weight per cent, where W_B is the atomic weight of the solute and W_A that of the solvent.

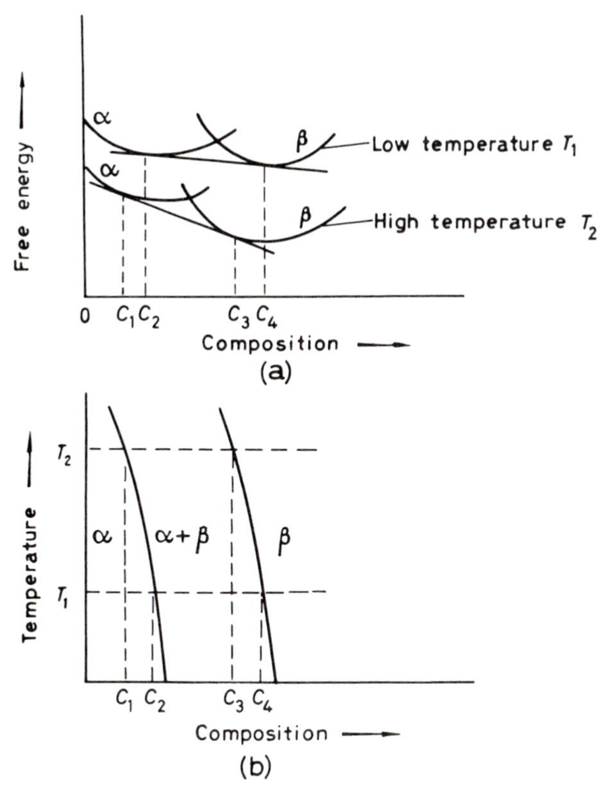

Figure 5.4 (a) The effect of temperature on the relative positions of the α-and β-phase free energy curves for an alloy system having a primary solid solubility of the form shown in (b)

the size factor is favourable; the value for gold–cadmium is 1.33, and that for gold–tin is 1.205.

The limit of solubility has been explained by Jones in terms of the Brillouin zone structure of alloy phases. It is assumed that the density of states–energy curve for the two phases, α (the close packed phase) and β (the more open phase), is of the form shown in *Figure 5.5(a)* where the $N(E)$ curve deviates from the parabolic relationship as the Fermi surface approaches the zone boundary. As the solute is added to the solvent lattice and more electrons are put into the zone, the top of the Fermi level moves towards A, i.e. where the density of states is high and the total energy E for a given electron concentration is low. Above this point the number of available energy levels decreases so markedly that the introduction of a few more electrons per atom causes a sharp increase in energy. Thus, just above this critical point the α structure becomes unstable relative to the alternative β structure which can accommodate the electrons within a smaller energy range, i.e. the energy of the Fermi level is lower if the β-phase curve is followed rather than the α-phase curve. The composition for which E_{max} reaches the point E_A is therefore a critical one, since the alloy will adopt that phase which has the lowest energy. It can be shown that this point corresponds to an electron-to-atom ratio of approximately 1.4, which appears to show successfully why the

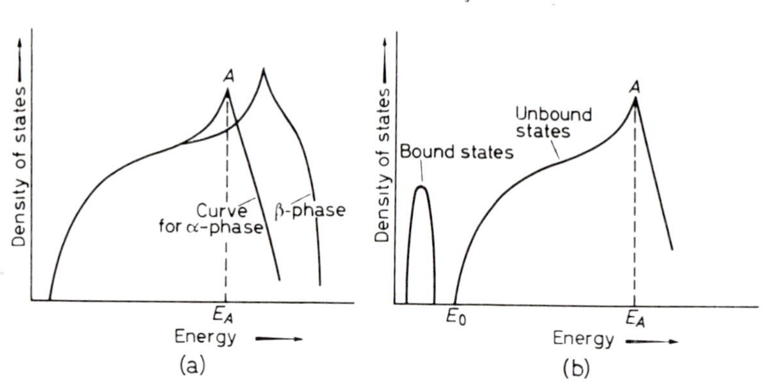

Figure 5.5 Density of states–energy curves (after Raynor, in *Structure of Metals*, courtesy of the Institute of Metals)

solubilities of many elements in copper or silver occur up to concentrations of 1.4 electrons per atom but no further. If it is assumed that the electrons are free and the Fermi surface approximates to a sphere when it touches the zone boundary then, since $E = h^2/2m\lambda^2$ and the Fermi energy $E_F = (h^2/8m)(3N/\pi V)^{2/3}$ the electron wavelength at the Fermi level is

$$\lambda = 2(\pi V/3N)^{1/3}$$

For the Fermi surface touching at the nearest point, the wavelength is the Bragg wavelength $\lambda = 2d$ for waves normal to the reflecting planes responsible for the zone boundary. In f.c.c. crystals these are the $\{111\}$ planes for which $d = a/\sqrt{3}$ where the lattice parameter a is given by $(4V/N_0)^{1/3}$ with N_0 the number of atoms in a volume V. Hence,

$$\lambda = (2/\sqrt{3})(4V/N_0)^{1/3}$$

and eliminating λ between these equations, gives $N/N_0 = \pi\sqrt{3}/4 \simeq 1.36$ for the electron concentration above which the f.c.c. α-phase should become unstable. For b.c.c. crystal structure the $\{110\}$ planes are responsible for the zone boundary and $N/N_0 \simeq 1.48$ for the β-phase. The lower solubilities in gold could be accounted for by the fact that the Fermi surface is slightly distorted from the spherical shape (*see* below) in such a way that it will need less solute metal than in the case of copper or silver to raise the electron energy level to the critical point E_A.

A similar reasoning to that of Jones may be used to account for the occurrence of the electron compounds (*see* section 5.6.3), at 3/2, 21/13 and 7/4 electrons per atom, respectively.

While there is little doubt that the energy band picture of a metal is justified, many quantitative details are in doubt. Work by Friedel, for example, suggests that when a polyvalent metal such as aluminium dissolves in copper it does not exist as an Al^{3+} ion with three valency electrons per atom belonging to the lattice as a whole, but instead contributes only one valency electron to the first allowed energy band for the alloy. The remainder are effectively tied to the solute atom, and may be considered to occupy what are known as 'bound states'. On this picture, therefore, the valency band for the alloy still contains only one electron per atom, as for the solvent copper itself, and consequently the dependence of solid solubility on electron concentration is a little more difficult to understand. A more detailed consideration of the theory, however, shows that in general, it is the states of lowest energy in the band system for the solvent that give rise to the bound states in the alloy. This is schematically shown in *Figure 5.5(b)*, where it is the areas beneath both curves which now correspond to a total of 2 electrons per atom. Thus, for every bound state introduced by a solute atom, the area beneath the curve for the unbound states between E_0 and E_A in *Figure 5.5(b)* will be correspondingly decreased by the equivalent of one energy level per atom. Consequently, the Fermi level, although corresponding to effectively only one electron per atom in the unbound energy levels, will nevertheless be shifted progressively toward E_A, so that an explanation such as that used by Jones can still be used.

Nowadays there are several ways of determining the shape of the Fermi surface from physical property measurements (*see* Chapter 2). One method known as the anomalous skin effect, measures the surface resistance to a high-frequency current at low temperatures. Another, known as the De Haas–Van Alphen effect, measures the variation of the magnetic suscepibility with the magnetizing field strength. In both experiments, the contour of the Fermi surface is mapped out by measuring the physical property as a function of the orientation in a single crystal of the material. Using the anomalous skin effect, Pippard finds that the Fermi surface in copper is distorted from the spherical shape, thus tending to invalidate the free electron model which assumes a spherical Fermi surface. However, work by Cohen and Heine indicates that the Fermi surface becomes more nearly spherical in copper alloys than in pure copper, owing to the reduction by alloying of the energy gaps across the zone surfaces. Thus in the region of electron concentration where the α-phase becomes unstable relative to the β-phase, it is quite possible that the free electron model is valid.

To summarize, it would appear that the precise details of how the energy levels are occupied by electrons remain, especially in alloys, incompletely understood. Nevertheless, the application of the simple theory of quasi-free electrons is useful, since it allows a general interpretation of metallic behaviour (*see also* Chapter 1). As for the details, it is not surprising that modifications to the simple theory are necessary as more work, of both a computational and experimental nature, is done.

5.5 Interstitial solid solutions

Interstitial solid solutions are formed when the solute atoms can fit into the interstices of the lattice of the solvent. However, an examination of the common crystal lattices shows that the size of the available interstices is restricted, and consequently only the small atoms, such as hydrogen, boron, carbon or nitrogen, with atomic radii very much less than one nanometre form such solutions. The most common examples occur in the transition elements and in particular the solution of carbon or nitrogen in iron is of great practical importance. In f.c.c. iron

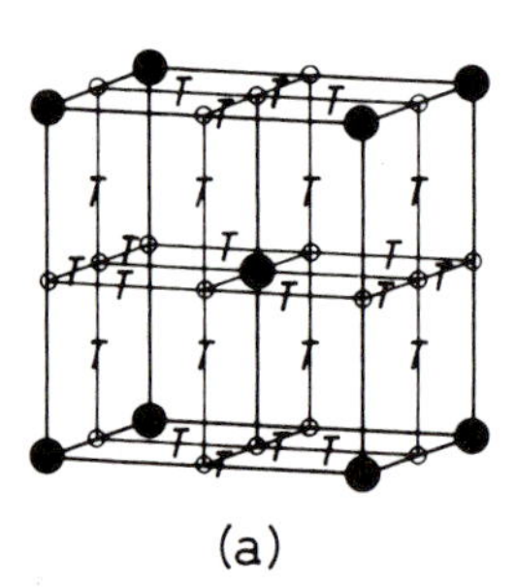

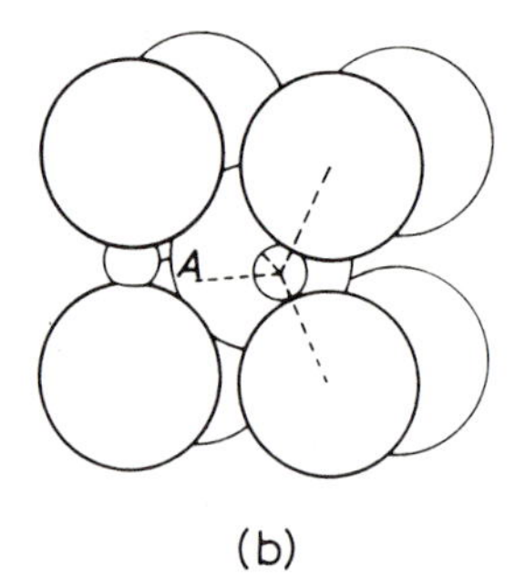

Figure 5.6 (a) Body centred cubic lattice showing the relative positions of the main lattice sites, the octahedral interstices marked O, and the tetrahedral interstices marked *T*. (b) Structure cell of iron showing the distortions produced by the two different interstitial sites. Only three of the iron atoms surrounding the octahedral sites are shown, the fourth, centred at *A*, has been omitted for clarity (after Williamson and Smallman, *Acta Cryst.* 1953, **6,** 361)

(austenite) the largest interstice or 'hole' is at the centre of the unit cell (co-ordinates $\frac{1}{2}, \frac{1}{2}, \frac{1}{2}$) where there is space for an atom of radius 52 pm (0.52 Å), i.e. $0.41r$ if r is the radius of the solvent atom. A carbon atom (80 pm (0.8 Å) diameter) or a nitrogen atom (70 pm (0.7 Å) diameter) therefore expands the lattice on solution, but nevertheless dissolves in quantities up to 1.7 weight per cent and 2.8 weight per cent respectively. Although the b.c.c. lattice is the more open structure the largest interstice is smaller than that in the f.c.c. In b.c.c. iron (ferrite) the largest hole is at the position ($\frac{1}{2}, \frac{1}{4}, 0$) and is a tetrahedral site where four iron atoms are situated symmetrically around it; this can accommodate an atom of radius 36 pm, i.e. $0.29r$, as shown in *Figure 5.6(a)*. However, internal friction and x-ray diffraction experiments show that the carbon or nitrogen atoms do not use this site, but instead occupy a smaller site which can accommodate an atom only $0.154r$, or 19 pm. This position ($0, 0, \frac{1}{2}$) at the mid-points of the cell edges is known as the octahedral site since, as can be seen from *Figure 5.6(b)*, it has a distorted octahedral symmetry for which two of the iron atoms are nearer to the centre of the site than the other four nearest neughbours. The reason for the interstitial atoms preferring this small site is thought to be due to the elastic properties of the b.c.c. lattice. The two iron atoms which lie above and below the interstice, and which are responsible for the smallness of the hole, can be pushed away more easily than the four atoms around the larger interstice. As a result, the solution of carbon in α-iron is extremely limited (0.02 weight per cent) and the structure becomes distorted into a body-centred tetragonal lattice; the c axis for each interstitial site is, however, disordered, so that this gives rise to a structure which is statistically cubic. The body-centred tetragonal structure forms the basis of martensite (an extremely hard metastable constituent of steel), since the quenching treatment given to steel retains the carbon in supersaturated solution (*see* Chapter 12).

5.6 Intermediate phases

The phases which form in the intermediate composition regions of the equilibrium diagram may be either (1) electrochemical or full-zone compounds, (2) size-factor compounds, or (3) electron compounds. The term 'compound' still persists even though many of these phases do not obey the valency laws of chemistry and often exist over a wide composition range. No sharp distinction exists between these three different types of compound and, as we shall see later, many factors may be involved in their formation, so that the characteristics of the phases observed are those which arise from the resultant of these various factors.

5.6.1 Electrochemical compounds

We have already seen that a strong tendency for compound formation exists when one element is electropositive and the other is electronegative. The magnesium-based compounds are probably the most common examples having the formula Mg_2(Pb, Sn, Ge or Si). These have many features in common with salt-like compounds since their compositions satisfy the chemical valency laws, their range of solubility is small, and usually they have high melting points. Moreover, many of these types of compounds have crystal structures identical with definite chemical compounds such as sodium chloride, NaCl, or calcium fluoride, CaF_2. In this respect the Mg_2X series are anti-isomorphous with the CaF_2 fluorspar structure, i.e. the magnesium metal atoms are in the position of the non-metallic fluoride atoms and the metalloid atoms such as tin or silicon take up the position of the metal atoms in calcium fluoride.

Even though these compounds obey all the chemical principles that we are familiar with, they may in fact often be considered as special electron compounds. For example, the first Brillouin zone of the CaF_2 structure is completely filled at 8/3 electrons per atom, which significantly is exactly that supplied by the compound Mg_2Pb, Sn,..., etc. Justification for calling these full-zone compounds is also furnished by electrical conductivity measurements. Contrary to the behaviour of salt-like compounds which exhibit low conductivity even in the liquid state, the compound Mg_2Pb shows the normal conduction (which indicates the possibility of zone overlapping) while Mg_2Sn behaves like a semiconductor (indicating that a small energy gap exists between the first and second Brillouin zones).

In general, it is probable that both concepts are necessary to describe the complete situation. As we shall see in Section 5.6.3, with increasing electrochemical factor even true electron compounds begin to show some of the properties associated with chemical compounds, and the atoms in the lattice take up ordered arrangements.

5.6.2 Size-factor compounds

When the atomic diameters of the two elements differ only slightly, electron compounds are formed, as discussed in the next section. However, when the difference in atomic diameter is appreciable, definite size-factor compounds are formed which may be of the (a) interstitial, or (b) substitutional type.

A consideration of several interstitial solid solutions has shown that if the interstitial atom has an atomic radius 0.41 times that of the metal atom then it can fit into the largest available lattice interstice without distortion. When the ratio of the radius of the interstitial atom to that of the metal atom is greater than 0.41 but less than 0.59, interstitial compounds are formed; hydrides, borides, carbides and nitrides of the transition metals are common examples. These compounds usually take up a simple structure of either the cubic or hexagonal type, with the metal atoms occupying the normal lattice sites and the non-metal atoms occupying the interstices. In general, the phases occur over a range of composition which is often centred about a simple formula such as M_2X and MX. Common examples are carbides and nitrides of titanium, zirconium, hafnium, vanadium, niobium and tantalum, all of which crystallize in the NaCl structure. It is clear, therefore, that these phases do not form merely as a result of the small atom fitting into the interstices of the solvent lattice, since vanadium, niobium and tantalum are b.c.c., while titanium, zirconium and hafnium are c.p.h. By changing their structure to f.c.c. the transition metals allow the interstitial atom not only a larger 'hole' but also six metallic neighbours. The formation of bonds in three directions at right angles, such as occurs in the sodium chloride arrangement, imparts a condition of great stability to these MX carbides.

When the ratio $r_{(\mathrm{interstitial})}$ to $r_{(\mathrm{metal})}$ exceeds 0.59 the distortion becomes appreciable, and consequently more complicated crystal structures are formed. Thus, iron nitride, where $r_N/r_{Fe}=0.56$, takes up a structure in which nitrogen lies at the centre of six atoms as suggested above, while iron carbide, i.e. cementite, Fe_3C, for which the ratio is 0.63, takes up a more complex structure.

For intermediate atomic size difference, i.e. about 20 to 30 per cent, an efficient packing of the atoms can be achieved if the crystal structure common to the Laves phases is adopted. These phases, classified by Laves and his co-workers, have the formula AB_2 and each A atom has 12 B neighbours and 4 A neighbours, while each B atom is surrounded by six like and six unlike atoms. The average co-ordination number of the structure (13.33) is higher, therefore, than that achieved by the packing of atoms of equal size. These phases crystallize in one of three closely related structures which are isomorphous with the compounds $MgCu_2$ (cubic), $MgNi_2$ (hexagonal) or $MgZn_2$ (hexagonal). The secret of the close relationship between these structures is that the small atoms are arranged on a space lattice of tetrahedra.

The different ways of joining such tetrahedra account for the different structures. This may be demonstrated by an examination of the $MgCu_2$ structure. The small B atoms lie at the corners of tetrahedra which are joined point-to-point throughout space, as shown in *Figure 5.7(a)*. Such an arrangement provides large holes of the type shown in *Figure 5.7(b)* and these are best filled when the atomic ratio $r_{(\mathrm{large})}/r_{(\mathrm{small})}=1.225$. The complete cubic structure of $MgCu_2$ is shown in *Figure 5.7(c)*. The $MgZn_2$ structure is hexagonal, and in this case the tetrahedra are joined alternately point-to-point and base-to-base in long chains to form a Wurtzite type of structure. The $MgNi_2$ structure is also hexagonal and although very complex it is essentially a mixture of both the $MgCu_2$ and $MgNi_2$ types.

The range of homogeneity of these phases is narrow. This limited range of

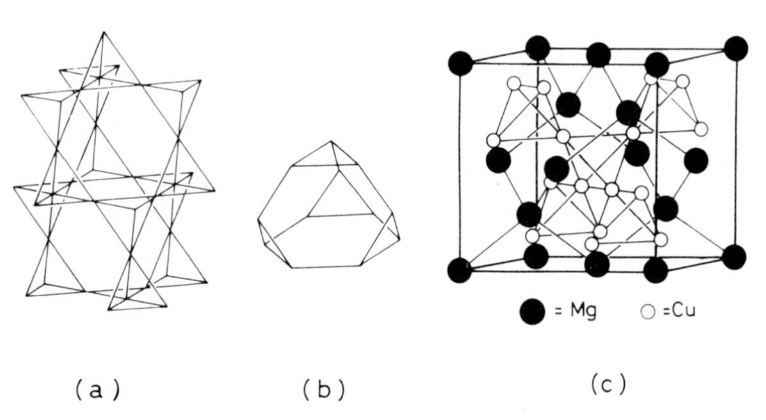

Figure 5.7 (a) Framework of the $MgCu_2$ structure. (b) Shape of hole in which large Mg atom is accommodated. (c) Complete $MgCu_2$ structure (after Hume-Rothery, Smallman and Howath, *Structure of Metals and Alloys*, courtesy of the Institute of Metals)

homogeneity is not due to any ionic nature of the compound, since ionic compounds usually have low co-ordination numbers whereas Laves phases have high co-ordination numbers, but because of the stringent geometrical conditions governing the structure. However, even though the chief reason for their existence is that the ratio of the radius of the large atom to that of the small is about 1.2, there are indications that electronic factors may play some small part. For example, provided the initial size-factor condition is satisfied then if the e/a ratio is high (e.g. 2), there is a tendency for compounds to crystallize in the $MgZn_2$ structure, while if the e/a ratio is low (e.g. 4/3), then there is a tendency for the $MgCu_2$ type of structure to be formed. This electronic feature is demonstrated in the magnesium–nickel–zinc ternary system. Thus, even though the binary systems contain both the $MgZn_2$ and $MgNi_2$ phases the ternary compound MgNiZn has the $MgCu_2$ structure, presumably because its e/a ratio is 4/3. *Table 5.1* shows a few common examples of each type of Laves structure, from which it is evident that there is also a general tendency for transition metals to be involved.

TABLE 5.1 Compounds which exist in a Laves phase structure

$MgCu_2$ type	*$MgNi_2$ type*	*$MgZn_2$ type*	
$AgBe_2$			
$BiAu_2$	$BaMg_2$		
$NbCo_2$	Nb (Mn or Fe)$_2$	$NbCo_2$	with excess of B metal
$TaCo_2$	$TaMn_2$	$TaCo_2$	
Ti (Be, Co, or Cr)$_2$	Ti (Mn or Fe)$_2$	$TiCo_2$	
U (A., Co, Fe or Mn)$_2$	UNi_2	$ZrFe_2$	
Zr (Co, Fe or W)$_2$	Zr (Cr, Ir, Mn, Re, Ru, Os, or V)$_2$		

5.6.3 Electron compounds

An examination of the alloys of copper, silver and gold with the B sub-group metals shows that their equilibrium diagrams have many similarities. In general, all possess the sequence α, β, γ, ε of structurally similar phases, and while each phase does not occur at the same composition when this is measured in weight per cent or atomic per cent, they do so if composition is expressed in terms of electron concentration. Hume-Rothery and his co-workers have pointed out that the e/a ratio is not only important in governing the limit of the α-solid solution but also in controlling the formation of certain intermediate phases; for this reason they have been termed ‘electron compounds’. Structural determination of many intermediate phases, by Westgren, has confirmed this hypothesis and has allowed a general classification of such compounds to be made.

In terms of those phases observed in the copper–zinc system (*see* page 98) β-phases are found at an e/a ratio of 3/2 and these phases are often either disordered b.c.c. in structure or odered CsCl-type, β'. In the copper–aluminium system for example, the β-structure is found at Cu_3Al, where the three valency electrons from the aluminium and the one from each copper atom make up a ratio of 6 electrons to 4 atoms, i.e. $e/a = 3/2$. Similarly, in the copper–tin system the β-phase occurs at Cu_5Sn with 9 electrons to 6 atoms giving the governing e/a ratio. The γ-brass phase, Cu_5Zn_8, has a complex cubic (52 atoms per unit cell) structure, and is characterized by an e/a ratio of 21/13, while the ε-brass phase, $CuZn_3$, has a cp.h. structure and is governed by an e/a ratio of 7/4. A list of some of these structurally analogous phases is given in *Table 5.2.*

TABLE 5.2 Some selected structurally analogous phases

Electron–atom ratio 3:2			*Electron–atom ratio 21:13*	*Electron–atom ratio 7:4*
β-brass (b.c.c.)	β-manganese (*complex cubic*)	(c.p.h.)	γ-*brass* (*complex cubic*)	ε-*brass* (*c.p.h.*)
(Cu, Ag or Au)Zn		AgZn	(Cu, Ag or Au)(Zn or Cd)$_8$	(Cu, Ag or Au)(Zn or Cd)$_3$
CuBe	(Ag or Au)$_3$Al	AgCd		
	Cu_5Si		Cu_9Al_4	Cu_3Sn
(Ag or Au)Mg	$CoZn_3$	Ag_3Al		Cu_3Si
(Ag or Au)Cd		Au_5A_u	$Cu_{31}Sn_8$	Ag_5Al_3
(Cu or Ag)$_3$Al				
(Cu_5Sn or Si)			(Fe, Co, Ni, Pd or Pt)$_5$Zn$_{21}$	
(Fe, Co or Ni)Al				

A close examination of this table shows that some of these phases, e.g. Cu_5Si and Ag_3Al, exist in different structural forms for the same e/a ratio. Thus, Ag_3Al is basically a 3/2 b.c.c. phase, but it only exists as such at high temperatures; at intermediate temperatures it is c.p.h. and at low temperatures β-Mn. It is also noticeable that to conform with the appropriate electron-to-atom ratio the transition metals are credited with zero valency. The basis for this may be found in their electronic structure which is characterized by an incomplete d-band below an occupied outermost s-band. The nickel atom, for example, has an electronic structure denoted by (2) (8) (16) (2), i.e. two electrons in the first quantum shell, eight in the second, sixteen in the third and two in the fourth shells, and while this indicates that the free atom has two valency electrons, it also shows two electrons missing from the third quantum shell. Thus, if nickel contributes valency electrons, it also absorbs an equal number from other atoms to fill up the third quantum shell so that the net effect is zero.

Without doubt the electron concentration is the most important single factor which governs these compounds. However, as for the other intermediate phases, a closer examination shows that the interplay of all factors must be taken into account, and the easily classified compounds are merely those in which a clearly recognizable factor predominates. A consideration of the 3/2 electron compounds, for example, shows that several secondary factors are also of importance to their formation. These can be listed as follows.

(i) *Atomic size*—In general the b.c.c. 3/2 compounds are only formed if the size factor is less than $\pm 18\%$. Moreover, with increasing size factor difference, the range of homogeneity is displaced towards a lower e/a concentration.

(ii) *Valency of the solute*—An increase in the valency of the solute tends to favour c.p.h. and β-Mn structures at the expense of the b.c.c. structure. However, the size factor principle interferes even in this generalization, since large size-factor differences favour the b.c.c. lattice at the expense of the other structures. In the copper–tin system, for example, because tin has a valency of 4, a c.p.h. or β-Mn structure might be expected whereas, because tin has a borderline size-factor, the 3/2 compound has a b.c.c. structure.

(iii) *Electrochemical factor*—The electrochemical factor which predominates in compounds such as Mg_2Sn is also present to some degree in the electron compounds. Thus, the tendency for these phases in copper, silver and gold alloys to become ordered solid solutions increases in the sequence copper$\rightarrow$silver$\rightarrow$gold. A high electrochemical factor leads to ordering up to the melting point, and the liquidus curve rises to a maximum in a similar manner to that shown by electrochemical compounds such as Mg_2Sn; a good example is exhibited by the gold–magnesium system.

(iv) *Temperature*—An increase in temperature favours the b.c.c. structure in preference to the c.p.h. or β-Mn structure.

5.7 Order–disorder phenomena

A substitutional solid solution can be one of two types, either ordered in which the A and B atoms are arranged in a regular pattern, or disordered in which the distribution of the A and B atoms is random. From the previous section it is clear that the necessary condition for the formation of a superlattice, i.e. an ordered solid solution is that dissimilar atoms must attract each other more than similar atoms. In addition, the alloy must exist at or near a composition which can be expressed by a simple formula such as AB, A_3B or AB_3.

5.7.1 Examples of ordered structures

CuZn While the disordered solution is b.c.c. with equal probabilities of having copper or zinc atoms at each lattice point, the ordered lattice has copper atoms and zinc atoms segregated to cube corners $(0, 0, 0)$ and centres $(\frac{1}{2}, \frac{1}{2}, \frac{1}{2})$, respectively. The superlattice in the β-phase therefore takes up the CsCl structures as illustrated in *Figure 5.8(a)*. Other examples of the same type, which may be considered as being made up of two interpenetrating simple cubic lattices, are Ag (Mg, Zn or Cd), AuNi, NiAl, FeAl and FeCo.

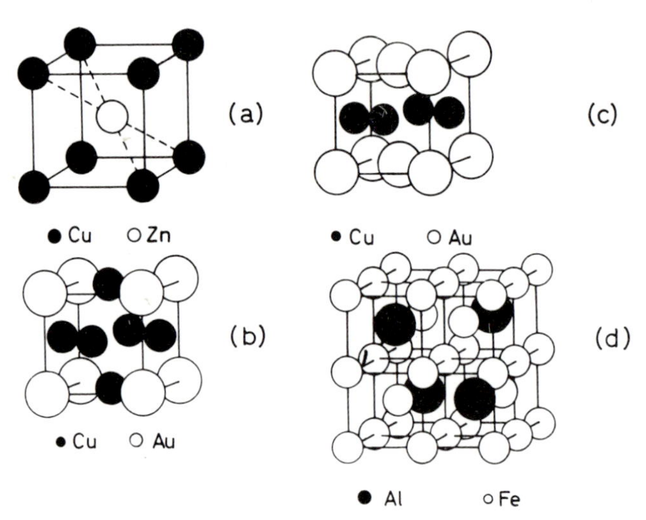

Figure 5.8 Examples of ordered structures, (a) CuZn, (b) Cu_3Au, (c) CuAu, (d) Fe_3Al

$AuCu_3$ This structure, which occurs less frequently than the β-brass type, is based on the f.c.c. structure with copper atoms at the centres of the faces $(0, \frac{1}{2}, \frac{1}{2})$ and gold atoms at the corners (0, 0, 0), as shown in *Figure 5.8*(*b*). Other examples include $PtCu_3$, (Fe or Mn) Ni_3, and $(MnFe)Ni_3$.

AuCu The AuCu structure shown in *Figure 5.8*(*c*) is also based on the f.c.c. lattice, but in this case alternate (001) layers are made up of copper and gold atoms respectively. Hence, because the atomic sizes of copper and gold differ, the lattice is distorted into a tetragonal structure having an axial ratio $c/a = 0.93$.

Fe_3Al Like FeAl, the Fe_3Al structure is based on the b.c.c. lattice but, as shown in *Figure 5.8*(*d*), eight simple cells are necessary to describe the complete ordered arrangement. In this structure any individual atom is surrounded by the maximum number of unlike atoms and the aluminium atoms are arranged tetrahedrally in the cell.

Mg_3Cd This ordered structure is based on the c.p.h. lattice. Other examples are $MgCd_3$ and Ni_3Sn.

Ordered structures can occur not only in binary alloys but also in ternary and quaternary alloys. The Heusler alloy, Cu_2MnAl, which is ferromagnetic when in the ordered condition, has a structure which is based on Fe_3Al with manganese and aluminium atoms taking alternate body centring positions respectively. In fact many alloys with important magnetic properties have ordered lattices, and some (antiferromagnetic) have a form of superlattice which is based on the magnetic moment of the atoms and hence is shown up not by x-ray but by neutron diffraction (*see* section 5.7.3).

Another important structure which occurs in certain alloys is the defect

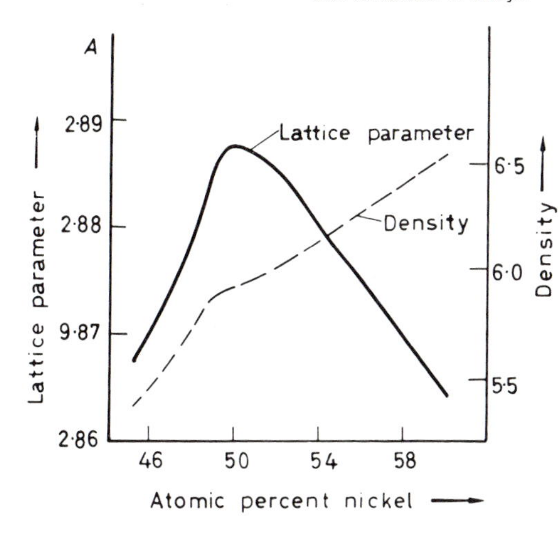

Figure 5.9 Variation of lattice parameter and density in the b.c.c. β-phase NiAl (after Bradley and Taylor, *Proc. Phys. Soc.* 1937, **159,** 56)

lattice (e.g. the b.c.c. NiAl phase). This ordered 3/2 electron compound has a wide range of homogeneity about the 50/50 composition and it is found that when the nickel content is reduced below 50 atomic per cent some of the lattice sites vacated by the nickel atoms are not taken up by the aluminium atoms so that the structure as a whole contains a large number of lattice vacancies. The influence of such point defects on the physical properties is shown in *Figure 5.9*, where an anomalous decrease in both density and lattice parameter is observed. The reason for the thermodynamic stability of these vacancies is that the true factor governing the formation of 3/2 compounds is the number of electrons per unit cell and not the number of electrons per atom: in general these two ratios are identical. In the compound NiAl, as the composition deviates from stoichiometry towards pure aluminium, the electron to atom ratio becomes greater than 3/2, but to prevent the compound becoming unstable the lattice takes up a certain proportion of vacancies to maintain the number of electrons per unit cell at a constant value of 3. Such defects obviously increase the entropy of the alloy, but the fact that these phases are stable at low temperatures, where the entropy factor is unimportant, demonstrates that their stability is due to a lowering of internal energy.

5.7.2 Long- and short-range order

The discussion so far suggests that in an ordered alloy the lattice may be regarded as being made up of two or more interpenetrating sub-lattices, each containing different arrangements of atoms. Moreover, the term 'superlattice' would imply that such a coherent atomic scheme extends over large distances, i.e. the crystal possesses long-range order. Such a perfect arrangement can exist only at low temperatures, since the entropy of an ordered structure is much lower than that of a disordered one, and with increasing temperature the degree of long-range order, S, decreases until at a critical temperature T_c it becomes zero; the general form of the curve is as shown in *Figure 5.10*. Partially ordered structures are achieved by the formation of small regions (domains) of order, each of which

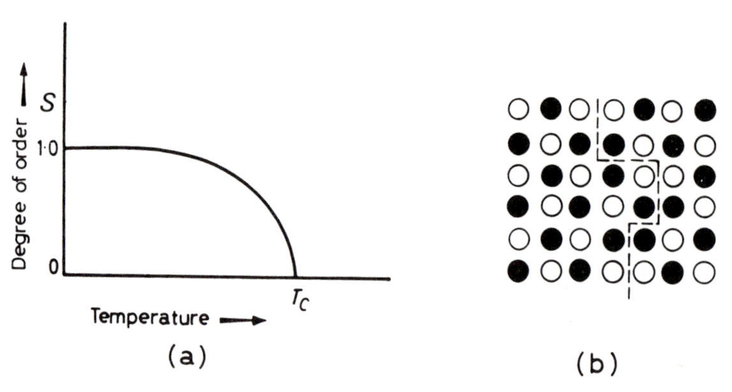

Figure 5.10 (a) Influence of temperature on the degree of order; (b) an antiphase domain boundary

are separated from each other by domain or anti-phase domain boundaries, across which the order changes phase (*Figure 5.10(b)*). However, even when long-range order is destroyed, the tendency for unlike atoms to be neighbours still exists, and short-range order results above T_c. The transition from complete disorder to complete order is a nucleation and growth process and may be likened to the annealing of a cold worked structure (*see* Chapter 10). At high temperatures well above T_c, there are more than the random number of AB atom pairs, and with the lowering of temperature small nuclei of order continually form and disperse in an otherwise disordered matrix. As the temperature, and hence thermal agitation, is lowered these regions of order become more extensive, until at T_c they begin to link together and the alloy consists of an interlocking mesh of small ordered regions. Below T_c these domains absorb each other (cf. grain growth, *see* Chapter 10) as a result of antiphase domain boundary mobility until long-range order is established.

Some order–disorder alloys can be retained in a state of disorder by quenching to room temperature while in others, e.g. β-brass, the ordering process occurs almost instantaneously. Clearly, changes in the degree of order will depend on atomic migration, so that the rate of approach to the equilibrium configuration will be governed by an exponential factor of the usual form, i.e. Rate$=A\,\mathrm{e}^{-Q/RT}$. However, Bragg has pointed out that the ease with which interlocking domains can absorb each other to develop a scheme of long-range order, will also depend on the number of possible ordered schemes the alloy possesses. Thus, in β-brass (*see Figure 5.8(a)*) only two different schemes of order are possible, while in f.c.c. lattices such as Cu_3Au (*Figure 5.8(b)*) four different schemes are possible and the approach to complete order is less rapid.

5.7.3 The detection of order

The determination of an ordered superlattice is usually done by means of the x-ray powder technique. In a disordered solution every plane of atoms is statistically identical and, as discussed in Chapter 2, there are reflections missing in the powder pattern of the material. In an ordered lattice, on the other hand,

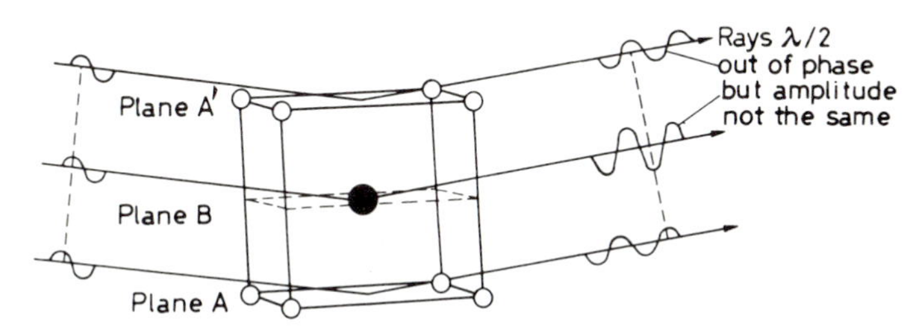

Figure 5.11 Formation of a weak 100 reflection from an ordered lattice by the interference of diffracted rays of unequal amplitude

alternate planes become A-rich and B-rich respectively so that these 'absent' reflections are no longer missing but appear as extra superlattice lines. This can be seen from *Figure 5.11*; while the diffracted rays from the A planes are completely out of phase with those from the B planes their intensities are not identical, so that a weak reflection results.

Application of the structure factor equation indicates that the intensity of the superlattice lines is proportional to $|F^2|=S^2(f_A-f_B)^2$, from which it can be seen that in the fully disordered alloy, where $S=0$, the superlattice lines must vanish. In some alloys such as copper–gold, the scattering factor difference (f_A-f_B) is appreciable and the superlattice lines are, therefore, quite intense and easily detectable. In other alloys, however, such as iron–cobalt, nickel–manganese, copper–zinc, the term (f_A-f_B) is negligible for x-rays and the superlattice lines are very weak; in copper–zinc, for example, the ratio of the intensity of the superlattice lines to that of the main lines is only about 1:3500. In some cases special x-ray techniques can enhance this intensity ratio; one method is to use an x-ray wavelength near to the absorption edge when an anomalous depression of the f-factor occurs which is greater for one element than for the other. As a result, the difference between f_A and f_B is increased. A more general technique, however, is to use neutron diffraction since the scattering factors for neighbouring elements in the Periodic Table can be substantially different. Conversely, as *Table 2.2* indicates, neutron diffraction is unable to show the existence of superlattice lines in Cu_3Au, because the scattering amplitudes of copper and gold for neutrons are approximately the same, although x-rays show them up quite clearly.

Sharp superlattice lines are observed as long as order persists over lattice regions of about 10^{-3} mm, large enough to give coherent x-ray reflections. When long-range order is not complete the superlattice lines become broadened, and an estimate of the domain size can be obtained from a measurement of the line breadth, as discussed in Chapter 2. *Figure 5.12* shows variation of order S and domain size as determined from the intensity and breadth of powder diffraction lines. The domain sizes determined from the Scherrer line broadening formula are in very good agreement with those observed by TEM. Short-range order is much more difficult to detect but nowadays direct measuring devices allow weak x-ray intensities to be measured more accurately, and as a result considerable information on the nature of short-range order has been obtained by studying the intensity of the diffuse background between the main lattice lines.

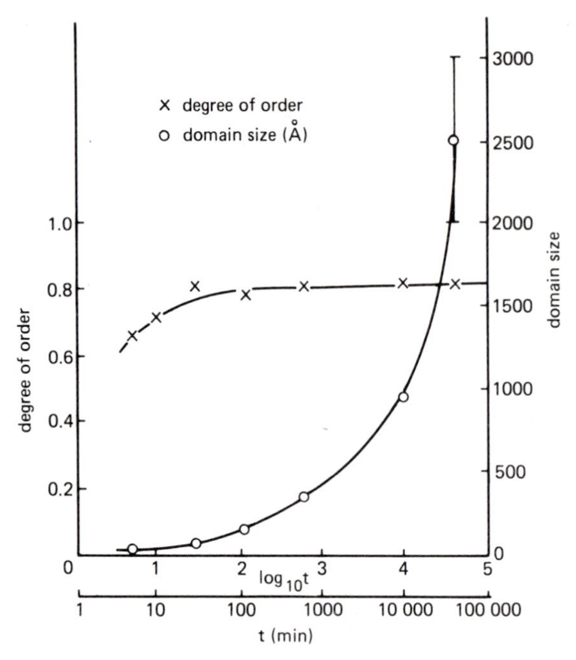

Figure 5.12 Degree of order and domain size during isothermal annealing at 350°C after quenching from 465°C (after Morris, Besag and Smallman, *Phil. Mag.* **29**, 43, (1974). Courtesy Taylor and Francis)

High-resolution transmission microscopy of thin metal foils allows the structure of domains to be examined directly. The alloy CuAu is of particular interest, since it has a face-centred tetragonal structure, often referred to as CuAu 1 (*see Figure 5.8*(*c*)), below 380 °C, but between 380 °C and the disordering temperature of 410 °C it has the CuAu 11 structures shown in *Figure 5.13*. The (002) planes are again alternately gold and copper, but half-way along the *a*-axis of the unit cell the copper atoms switch to gold planes and vice versa. The spacing between such periodic anti-phase domain boundaries is 5 unit cells or about 2 nm, so that the domains are easily resolvable in TEM, as seen in *Figure 5.14*(*a*). The isolated domain boundaries in the simpler superlattice structures such as CuAu 1, although not in this case periodic, can also be revealed by electron microscope, and an example is shown in *Figure 5.14*(*b*). Apart from static observations of these superlattice structures, annealing experiments inside the microscope also allow the effect of temperature on the structure to be examined directly. Such observations have shown that the transition from CuAu 1 to CuAu 11 takes place, as predicted, by the nucleation and growth of anti-phase domains.

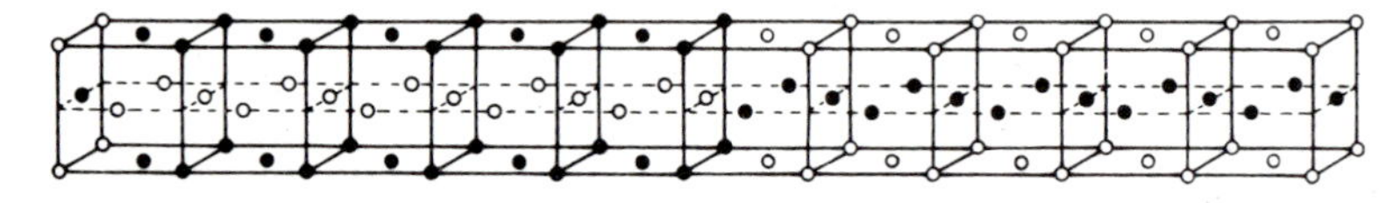

Figure 5.13 One unit cell of the orthorhombic superlattice of CuAu, i.e. CuAu 11 (from *J. Inst. Metals,* 1958–9, **87,** 419, by courtesy of the Institute of Metals)

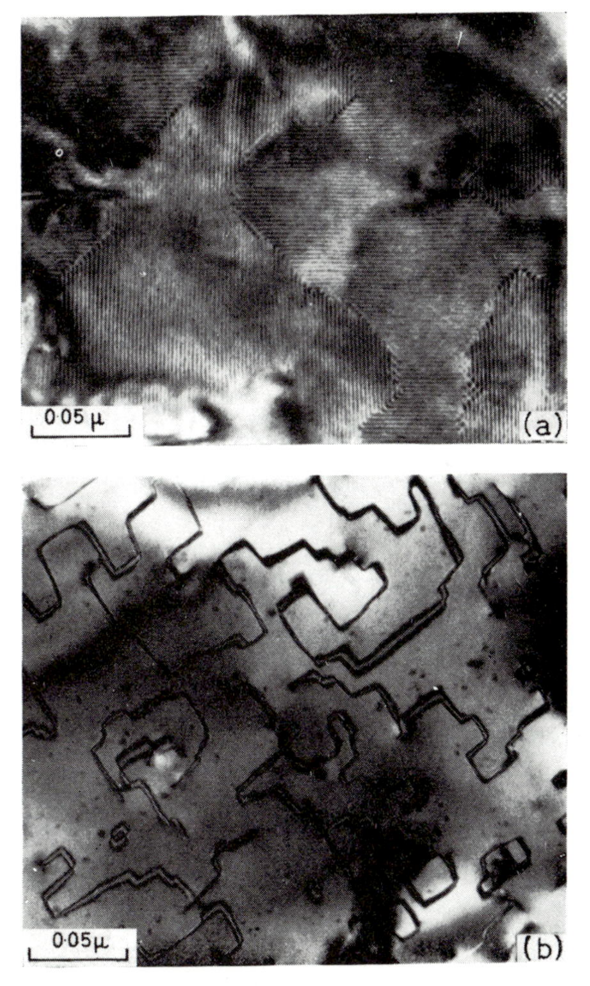

Figure 5.14 Electron micrographs of (a) CuAu 11, and (b) CuAu 1 (from Pashley and Presland, *J. Inst. Metals,* 1958–9, **87,** 419, courtesy of the Institute of Metals)

5.7.4 The influence of ordering on properties

Specific heat The order–disorder transformation has a marked effect on the specific heat, since energy is necessary to change atoms from one configuration to another. However, because the change in lattice arrangement takes place over a range of temperature, the specific heat versus temperature curve will be of the form shown in *Figure 4.3*(*b*). In practice the excess specific heat, above that given by Dulong and Petit's law, does not fall sharply to zero at T_c owing to the existence of short-range order, which also requires extra energy to destroy it as the temperature is increased above T_c.

Electrical resistivity As discussed in both Chapters 1 and 2 any form of disorder in a metallic structure, e.g. impurities, dislocations or point defects, will make a

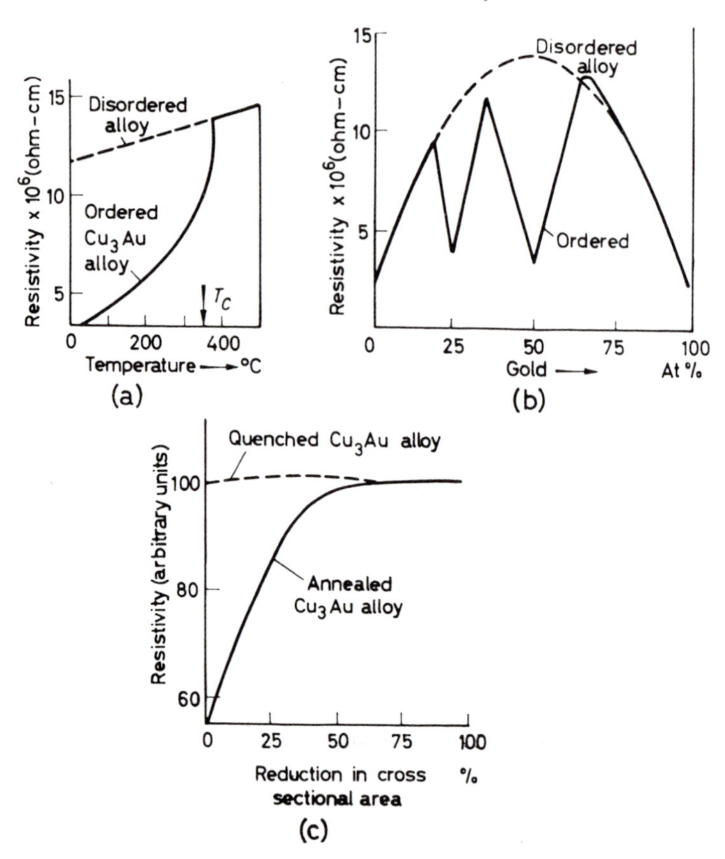

Figure 5.15 Effect of (a) temperature, (b) composition, and (c) deformation on the resistivity of copper–gold alloys (after Barrett, *Structure of Metals*, 1952, courtesy of McGraw-Hill Book Co.

large contribution to the electrical resistance. Accordingly, superlattices below T_e have a low electrical resistance, but on raising the temperature the resistivity increases, as shown in *Figure 5.15(a)* for ordered Cu_3Au. The influence of order on resistivity is further demonstrated by the measurement of resistivity as a function of composition in the copper–gold alloy system. As shown in *Figure 5.15(b)*, at composition near Cu_3Au and CuAu, where ordering is most complete, the resistivity is extremely low, while away from these stoichiometric compositions the resistivity increases; the quenched (disordered) alloys given by the dotted curve, also have high resistivity values.

Mechanical properties The mechanical properties are altered when ordering occurs. The change in yield stress is not directly related to the degree or ordering however, and in fact Cu_3Au crystals have a lower yield stress when well ordered than when only partially ordered. Experiments show that such effects can be accounted for if the maximum strength as a result of ordering is associated with critical domain size. In the alloy Cu_3Au, the maximum yield strength is exhibited by quenched samples after an annealing treatment of 5 min at 350 °C which gives a domain size of 6 nm (*see Figure 5.12*). However, if the alloy is well ordered and the domain size larger, the hardening is insignificant. In some alloys such as CuAu or CuPt, ordering produces a change of crystal structure and the resultant lattice strains can also lead to hardening. Interpretation of such effects in terms of dislocation is discussed in Chapters 8 and 10.

Thermal agitation is the most common means of destroying long-range order, but other methods, e.g. deformation, are equally effective. *Figure 5.15(c)* shows that cold work has a negligible effect upon the resistivity of the quenched (disordered) alloy but considerable influence on the well annealed (ordered) alloy. Irradiation by neutrons or electrons also markedly affects the ordering (*see* Chapter 9).

Magnetic properties The order–disorder phenomenon is of considerable importance in the application of magnetic materials. The kind and degree of order affects the magnetic hardness, since small ordered regions in an otherwise disordered lattice induce strains which affect the mobility of magnetic domain bouundaries. To understand such behaviour more fully it is necessary to examine magnetic materials in greater detail.

5.8 The magnetic properties of metals and alloys

As outlined in Chapter 2, three types of magnetic materials are commonly known, dia-, para- and ferromagnetic. Interest in magnetism is large not only because of the practical importance, but also because it throws light on the complex electronic structure of the rare-earth and transition elements. As a consequence, it is customary nowadays to speak of five rather than three kinds of magnetism, since to the above list have been added antiferromagnetism and ferrimagnetism.

5.8.1 Dia- and paramagnetism

Diamagnetism is a universal property of the atom since it arises from the motion of electrons in their orbits around the nucleus. Electrons moving in this way represent electrical circuits and it follows from Lenz's law that this motion is altered by an applied field in such a manner as to set up a repulsive force. The diamagnetic contribution from the valency electrons is small, but from a closed shell it is proportional to the number of electrons in it and to the square of the radius of the 'orbit'. In many metals this diamagnetic effect is outweighed by a paramagnetic contribution, the origin of which is to be found in the electron spin. Each electron behaves like a small magnet and in a magnetic field can take up one of two orientations, either along the field or in the other opposite direction, depending on the direction of the electron spin. Accordingly, the energy of the electron is either decreased or increased and may be represented conveniently by the band theory. Thus, if we regard the band of energy levels as split into two halves, as shown in *Figure 5.19(a)*, each half associated with electrons of opposite spin, it follows that in the presence of the field, some of the electrons will transfer their allegiance from one band to the other until the Fermi energy level is the same in both. It is clear, therefore, that in this state there will be a larger number of electrons which have their energy lowered by the field than have their energy raised. This condition defines paramagnetism, since there will be an excess of unpaired spins which give rise to a resultant magnetic moment.

It is evident that an insulator will not be paramagnetic since the bands are full and the lowered half-band cannot accommodate those electrons which wish to 'spill-over' from the raised half-band. On the other hand it is not true, as one might expect, that conductors are always paramagnetic. This follows because in some elements the natural diamagnetic contribution outweighs the paramagnetic contribution; in copper, for example, the newly filled *d*-shell gives rise to a larger diamagnetic contribution.

5.8.2 Ferromagnetism

The theory of ferromagnetism is difficult and at present not completely understood. Nevertheless, from the electron theory of metals it is possible to build up a band picture of ferromagnetic materials which goes a long way to explain not only their ferromagnetic properties, but also the associated high resistivity and electronic specific heat of these metals compared to copper. In recent years considerable experimental work has been done on the electronic behaviour of the transition elements, and this suggests that the electronic structure of iron is somewhat different to that of cobalt and nickel. For this reason it will be convenient to consider initially the well-established features of the subject and then to consider in a separate section the more speculative aspects of the electronic structure of these transition metals.

Ferromagnetism, like paramagnetism, has its origin in the electron spin. In ferromagnetic materials, however, permanent magnetism is obtained and this indicates that there is a tendency for electron spins to remain aligned in one direction even when the field has been removed (*see Figure 5.19(b)*. In terms of the band structure this means that the half-band associated with one spin is automatically lowered when the vacant levels at its top are filled by electrons from the top of the other; the change in potential energy associated with this transfer is known as the exchange energy. Thus, while it is energetically favourable for a condition in which all the spins are in the same direction, an opposing factor is the Pauli exclusion principle, because if the spins are aligned in a single direction many of the electrons will have to go into higher quantum states with a resultant increase in kinetic energy. In consequence, the conditions for ferromagnetism are stringent, and only electrons from partially filled *d* or *f* levels can take part. This condition arises because only these levels have (1) vacant levels available for occupation, and (2) a high density of states which is necessary if the increase in kinetic energy accompanying the alignment of spins is to be smaller than the decrease in exchange energy. Both of these conditions are fulfilled in the transition and rare-earth metals, but of all the metals in the long periods only the elements iron, cobalt and nickel are ferromagnetic at room temperature, gadolinium just above RT ($T_c \approx 16$°C) and the majority are in fact strongly paramagnetic. This observation has led to the conclusion that the exchange interactions are most favourable, at least for the iron group of metals, when the ratio of the atomic radius to the radius of the unfilled shell, i.e. the *d*-shell, is

TABLE 5.3 Radius (Å) of electronic orbits of atoms of transition metals of first long period (after Slater '*Quantum Theory of Matter*')

Element	*3d*	*4s*	*Atomic radius in metal*
Sc	0.61	1.80	1.60
Ti	0.55	1.66	1.47
V	0.49	1.52	1.36
Cr	0.45	1.41	1.28
Mn	0.42	1.31	1.28
Fe	0.39	1.22	1.28
Co	0.36	1.14	1.25
Ni	0.34	1.07	1.25
Cu	0.32	1.03	1.28

somewhat greater than 3 (*see Table 5.3*). As a result of this condition it is hardly surprising that there are a relatively large number of ferromagnetic alloys and compounds, even though the base elements themselves are not ferromagnetic.

In ferromagnetic metals the strong interaction results in the electron spins being spontaneously aligned, even in the absence of an applied field. However, a specimen of iron can exist in an unmagnetized condition because such an alignment is limited to small regions, or domains, which statistically oppose each other. These domains are distinct from the grains of a polycrystalline metal and in general there are many domains in a single grain, as shown in *Figure 5.16*. Under the application of a magnetic field the favourably oriented domains grow at the expense of the others by the migration of the domain boundaries until the whole specimen appears fully magnetized. At high field strengths it is also possible for unfavourably oriented domains to 'snap-over' into more favourable orientations quite suddenly, and this process, which can often be heard using sensitive equipment, is known as the Barkhausen effect.

The state in which all the electron spins are in complete alignment is possible only at low temperatures. As the temperature is raised the saturation magnetization is reduced, falling slowly at first and then increasingly rapidly, until a critical temperature, known as the Curie temperature is reached. Above this temperature, T_c, the specimen is no longer ferromagnetic, but becomes paramagnetic, and for the metals iron, cobalt, and nickel this transition occurs at 780°, 1075° and 365 °C respectively. Such a co-operative process may be readily understood from

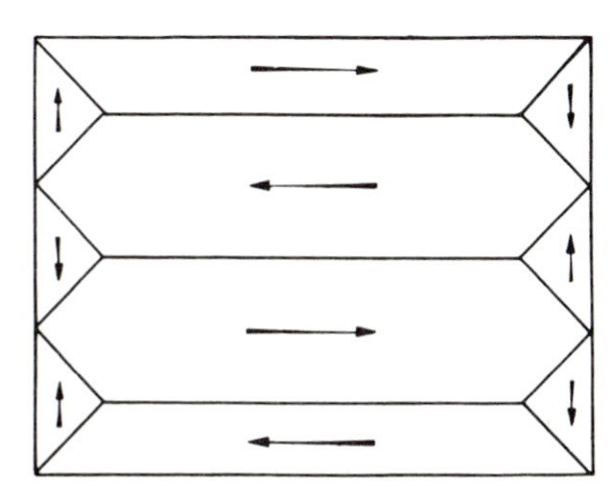

Figure 5.16 Simple domain structure in a ferromagnetic material. The arrows indicate the direction of magnetization in the domains

thermodynamic reasoning, since the additional entropy associated with the disorder of the electron spins makes the disordered (paramagnetic) state thermodynamically more stable at high temperatures. This behaviour is similar to that shown by materials which undergo the order–disorder transformation and, as a consequence, ferromagnetic metals exhibit a specific heat peak of the form previously shown (*see Figure 4.3*(*b*)).

A ferromagnetic crystal in its natural state has a domain structure. From *Figure 5.16* it is clear that by dividing itself into domains the crystal is able to eliminate those magnetic poles which would otherwise occur at the surface. The width of the domain boundary or Bloch wall is not necessarily small, however, and in most materials is of the order of 100 atoms in thickness. By having a wide boundary the electron spins in neighbouring atoms are more nearly parallel, which is a condition required to minimize the exchange energy. On the other hand, within any one domain the direction of magnetization is parallel to a direction of easy magnetization, i.e. $\langle 100 \rangle$ in iron, $\langle 111 \rangle$ in nickel and $\langle 001 \rangle$ in cobalt, and as one passes across a boundary the direction of magnetization rotates away from one direction of easy magnetization to another. To minimize this magnetically disturbed region the crystal will try to adopt a boundary which is as thin as possible. Consequently, the boundary width adopted is one of compromise between the two opposing effects, and the material may be considered to possess a magnetic interfacial or surface energy.

5.8.3 Magnetic alloys

The work done in moving a domain boundary depends on the energy of the boundary, which in turn depends on the magnetic anisotropy. The ease of magnetization also depends on the state of internal strain in the material and the presence of impurities. Both these latter factors affect the magnetic 'hardness' through the phenomenon of magnetostriction, i.e. the lattice constants are slightly altered by the magnetization so that a directive influence is put upon the orientation of magnetization of the domains. Materials with internal stresses are hard to magnetize or demagnetize, while materials free from stresses are magnetically soft. Hence, since internal stresses are also responsible for mechanical hardness, the principle which governs the design of magnetic alloys is to make permanent magnetic materials as mechanically hard and soft magnets as mechanically soft as possible.

In the development of magnetically soft materials it is found that those elements which form interstitial solid solutions with iron are those which broaden the hysteresis loop most markedly. For this reason, it is common to remove such impurities from transformer iron by vacuum melting or hydrogen annealing. However, such processes are expensive and, consequently, alloys are frequently used as 'soft' magnets, particularly iron–silicon and iron–nickel alloys (because silicon and nickel both reduce the amount of carbon in solution). The role of Si is to form a γ-loop (*see* Chapter 12) and hence remove transformation strains and also to improve orientation control. In the production of iron–silicon alloys the factors which are controlled include the grain size, the orientation difference from one grain to the next, and the presence of non-magnetic inclusions, since all are major sources of coercive force. The coercive force increases with decreasing grain size because the domain pattern in the neighbourhood of a grain boundary is complicated owing to the orientation difference between two adjacent grains. Complex domain patterns can also arise at the free surface of the metal unless these are parallel to a direction of easy magnetization. Accordingly, to minimize the coercive force, rolling and annealing schedules are adopted to produce a preferred oriented material with a strong 'cube-texture', i.e. one with two $\langle 100 \rangle$ directions in the plane of the sheet (*see* Chapter 10). This procedure is extremely important, since transformer material is used in the form of thin sheets to minimize eddy-current losses. Fe–Si–B in the amorphous state is finding increasing application in transformers.

The iron–nickel series, Permalloys, present many interesting alloys and are used chiefly in communication engineering where a high permeability is a necessary condition. The alloys in the range 40 to 55 per cent nickel are characterized by a high permeability and at low field strengths this may be as high as 15 000 compared with 500 for annealed iron. The 50 per cent alloy, Hypernik, may have a permeability which reaches a value of 70 000, but the highest initial and maximum permeability occurs in the composition range of the $FeNi_3$ superlattice, provided the ordering phenomenon is suppressed. An interesting development in this field is in the heat treatment of the alloys while in a strong magnetic field. By such a treatment the permeability of 'Permalloy 65' has been increased to about 260 000. This effect is thought to be due to the fact that during alignment of the domains, plastic deformation is possible and magnetostrictive strains may be relieved.

In the development of magnetically hard materials the principle is to obtain, by alloying and heat treatment, a matrix containing finely divided particles of a second phase. These fine precipitates, usually differing in lattice parameter from the matrix, set up coherency strains in the lattice which affect the domain boundary movement. Alloys of copper–nickel–iron, copper–nickel–cobalt and aluminium–nickel–cobalt are of this type. An important advance in this field is to make the particle size of the alloy so small, i.e. less than a hundred nonometres diameter, that each grain contains only a single domain. Then magnetization can occur only by the rotation of the direction of magnetization *en bloc*. Alnico alloys containing 6–12% Al, 14–25% Ni, 0–35% Co, 0–8% Ti, 0–6% Cu in 40–70% Fe depend on this feature and are the most commercially important permanent magnet materials. They are precipitation hardened alloys and are heat-treated to produce rod-like precipitates (30 nm × 100 nm) lying along $\langle 100 \rangle$ in the b.c.c. matrix. During magnetic annealing the rods form along the $\langle 100 \rangle$ axis nearest to the direction of the field, when the remanence and coercivity are markedly increased. Sm_2 (Co, Fe, Cu, Zr)$_{17}$ alloys also rely on the pinning of magnetic domains by fine precipitates. Clear correlation exists between mechanical hardness and intrinsic coercivity. $SmCo_5$ magnets depend on the very high magnetocrystalline anisotropy of this compound and the individual grains are single domain particles. The big advantage of these magnets over the Alnicos is their much higher coercivities.

The Heusler alloys, copper–manganese–aluminium, are of particular interest because they are made up from non-ferromagnetic metals and yet exhibit ferromagnetic properties. The magnetism in this group of alloys is associated with the compound Cu_2MnAl, evidently because of the presence of manganese atoms. The compound has the Fe_3Al-type superlattice when quenched from 800 °C, and in this state is ferromagnetic, but when the alloy is slowly cooled it has a γ-brass structure and is non-magnetic, presumably because the correct exchange forces arise from the lattice rearrangement on ordering. A similar behaviour is found in both the copper–manganese–gallium and the copper–manganese–indium systems.

The order–disorder phenomenon is also of magnetic importance in many other systems. As discussed previously, when ordering is accompanied by a structural change, i.e. cubic to tetragonal, coherency strains are set up which often lead to magnetic hardness. In FePt, for example, extremely high coercive forces are produced by rapid cooling. However, because the change in mechanical properties accompanying the transformation is found to be small, it has been suggested that the hard magnetic properties in this alloy are due to the small particle-size effect, which arises from the finely laminated state of the structure.

5.8.4 Anti-ferromagnetism and ferrimagnetism

Apart from the more usual dia-, para- and ferromagnetic materials, there are certain substances which are termed anti-ferromagnetic; in these the net moments of neighbouring atoms are aligned in opposite directions, i.e. anti-parallel. Many oxides and chlorides of the transition metals are examples including both chromium and α-manganese, and also manganese–copper alloys. Some of the relevant features of anti-ferromagnetism are similar in many respects to ferromagnetism, and are summarized as follows.

(1) In general, the magnetization directions are aligned parallel or anti-parallel to crystallographic axes, e.g. in MnI and CoO the moment of the Mn^{2+} and Co^{2+} ions are aligned along a cube edge of the unit cell. The common directions are termed directions of anti-ferromagnetism.
(2) The degree of long-range anti-ferromagnetic ordering progressively decreases with increasing temperature and becomes zero at a critical temperature, T_n, known as the Néel temperature; this is the anti-ferromagnetic equivalent of the Curie temperature.
(3) An anti-ferromagnetic domain is a region in which there is only one common direction of anti-ferromagnetism; this is probably affected by lattice defects and strain.

The most characteristic property of an anti-ferromagnetic material is that its susceptibility χ shows a maximum as a function of temperature, as shown in *Figure 5.17*(*a*). As the temperature is raised from 0 K the interaction which leads to anti-parallel spin alignment becomes less effective, until at T_n the spins are 'free'. Above this temperature the material is paramagnetic so that the susceptibility decreases with increasing temperature in the usual way, according to the Curie–Weiss Law, $\chi = \text{const.}/(T+\theta)$, where θ is the Néel temperature.

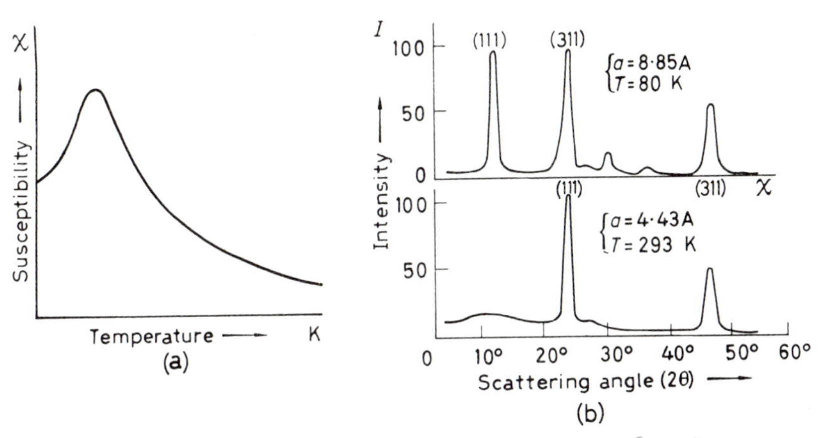

Figure 5.17 (a) Variation of magnetic susceptibility with temperature for an antiferro-magnetic material; (b) neutron diffraction pattern from the antiferro-magnetic powder MnO above and below the critical temperature for ordering (after Shull and Smart, *Phys. Rev.* 1949, **76,** 1256)

Similar characteristic features are shown in the resistivity curves due to scattering as a result of spin disorder. However, the application of neutron diffraction techniques provides a more direct method of studying anti-ferromagnetic structures, as well as giving the magnetic moments associated with the ions of the metal. As outlined in Chapter 2, there is a magnetic scattering of neutrons in the case of certain magnetic atoms, and owing to the different scattering amplitude of the parallel and anti-parallel atoms, the possibility arises of the existence of superlattice lines in the anti-ferromagnetic state. In manganese oxide MnO, for example, the parameter of the magnetic unit cell is 0.885 nm, whereas the chemical unit cell (NaCl structure) is half this value, 0.443 nm. This atomic arrangement is analogous to the structure of an ordered alloy and the existence of magnetic superlattice lines below the Néel point (122 K) has been observed, as shown in *Figure 5.17(b)*.

Some magnetic materials have properties which are intermediate between those of antiferromagnetic and ferromagnetic. This arises if the moments in one direction are unequal in magnitude to those in the other, as for example in magnetite, Fe_3O_4, where the ferrous and ferric ions of the $FeOFe_2O_3$ compound occupy their own particular sites. Néel has called this state *ferrimagnetism* and the corresponding materials are termed ferrites. Such materials are of importance in the field of electrical engineering because they are ferromagnetic without being appreciably conducting; eddy current troubles in transformers are, therefore, not so great. Strontium ferrite is extensively used in applications such as electric motors, because of these properties and low material costs.

5.9 The electronic structure of the transition metals

In building up a picture of the periodic system of the elements, it was noted that the *d*-states of the free atoms of the elements were not occupied after the *p*-

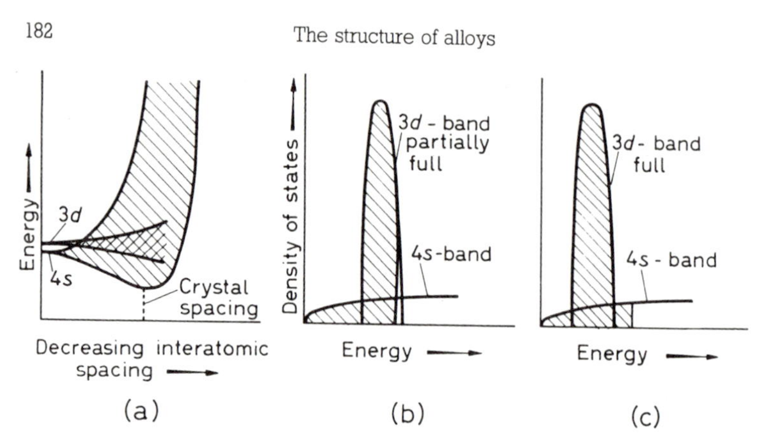

Figure 5.18 (a) Broadening of 3*d* and 4*s* levels in the transition metals, (b) filling of the 3*d* (ten-electron) band, and 4*s* (two-electron) band in nickel, (c) filling of 3*d* and 4*s* bands in copper

states having the same principal quantum number. The reason for this is that *d*-states penetrate deeply into the atom core, and hence they have a higher energy than *s*- and *p*-states with a greater principal quantum number. The elements of the short periods of the Periodic Table do not, therefore, have electrons in *d*-states. With increasing atomic number the energy of outer *s*- and *p*-states will, however, eventually have a higher energy than the unfilled *d*-states of lower principal quantum number. When this happens the atoms will, with increasing atomic number, be able to include electrons in *d*-states. This state of affairs begins at scandium and continues until copper is reached, when all possible *d*-states for principal quantum number 3 have been filled. The same process, but introducing 4*d* and 5*d* electrons rather than 3*d* electrons, takes place in the second and third long periods, and these elements as a group are termed the transition metals.

When transition metal atoms come together to form the metal, the energy levels corresponding to the *d*-states of the free atom broaden out into a band of electronic levels. Because *d*-electrons have atomic orbits which lie closer to the core of the atom than the outer *s*- and *p*-electrons, there is less interaction between *d*-electrons from adjacent atoms in the metallic structure, and hence the *d*-band does not broaden out to anything like the same extent as the *s*- and *p*-bands (*Figure 5.18(a)*). Moreover, since there are less possible *d*-states contributed by each atom of a transition metal, and since the band width of the *d*-states is narrow, then it follows that the *d*-band must have a very high density of states, as shown in *Figure 5.18(b)*. *Figure 5.18(c)* shows that copper, which is not a transition metal, has a completely filled *d*-band.

In the case of a metal in which there is only a small interaction between electrons in identical states in adjacent atoms of the structure, the band formed from such states has its top and bottom respectively raised and lowered by approximately the same amount with respect to the discrete energy level of the state in the free atoms, as shown in *Figure 5.18(a)*. It follows, therefore, that if the band is only partially filled, then the energy of the electrons in the narrow band will have a lower energy than in the free atom. The existence of such a band, which is the d-band in the transition elements, confers a great stability on the early members of the transition series which have less than 5 d-electrons per atom. This great stability can readily be seen by examining the melting points of the early transition elements of the first long periods:

Melting	Sc	Ti	V	Cr	Mn	Fe	Co	Ni
point (°C)	1400	1800	1710	1830	1260	1535	1490	1452

The progressive increase in melting point from scandium to chromium is well marked. A similar process occurs in the second and third long periods, and the metals molybdenum and tungsten which correspond to chromium in the first long period have melting points of 2600 °C and 3380 °C respectively, which makes these two of the most stable metals that exist.

As the number of d-electrons increases towards 10 per atom and at the same time because the atomic number, and hence the charge on the ion, is increasing, the d-electrons in the metal are drawn more tightly to the core, as can be seen in *Table 5.3*. Thus the d-electrons in the later transition elements can be considered to be approaching a state in which they are largely localized around the ions in the metal lattice. In consequence, the d-band is very narrow, and would be reduced to a discrete level in the limiting case if all the d-electrons were completely localized. The existence of a narrow d-band with a very high density of states in the later transition metals has many interesting consequences. One of the most far-reaching is the onset of ferromagnetism in iron, cobalt and nickel, as discussed in section 5.8.

In a d-band which is nearly filled it is possible, providing the atomic spacing in the metal lies within a certain range of values, for the electrons in the d-band to reduce their total energy by filling one half of the band with electrons of one spin, and leaving twice as many unfilled states in the other sub-band containing electrons of opposite spin. When this occurs the material contains more d-electrons of one spin than the opposite spin, and since an electron with unpaired spins contributes a magnetic moment, the material becomes ferromagnetic. Ferromagnetism of this type occurs only in metals of the first long period since the $3d$-electrons have a more localized character than the $4d$ and $5d$-electrons. Metals towards the ends of the second and third long period also have an unfavourable interatomic distance thus Ni is strongly ferromagnetic, Pd strongly paramagnetic and Pt weakly paramagnetic.

In nickel the experimental value of the saturation magnetic moment is 0.6 Bohr magneton per atom, which, because of the way the Bohr magneton is defined, means that in the ferromagnetic state the number of electrons of one type of spin must exceed that of the other type to the extent of 0.6 per atom. Then, since there is a total of ten electrons in the $4s$ and $3d$ bands, the observed saturation magnetic moment can be accounted for if the number of electrons per atom in the d-band is 9.4, and the number in the s-band 0.6. The change in the distribution of electrons between the $4s$ and $3d$ bands as the nickel becomes ferromagnetic can then be pictured as shown in *Figure 5.19*.

By assuming such a d-band structure it is possible to account for some of the physical properties of nickel alloys. For example, nickel and copper form a

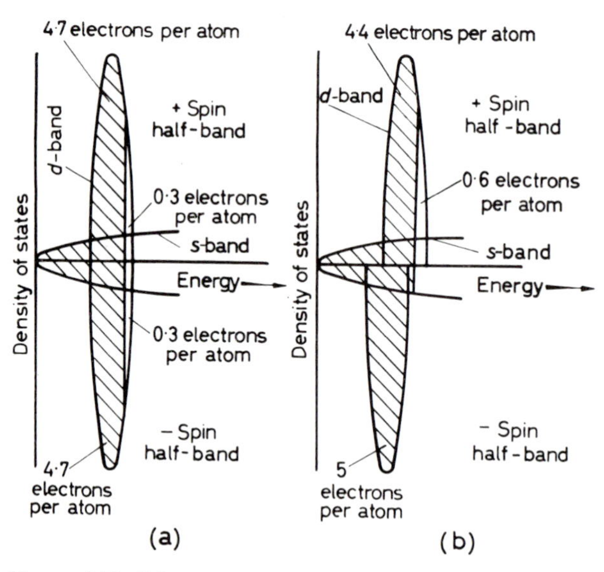

Figure 5.19 Schematic representation of (a) paramagnetic nickel, (b) ferromagnetic nickel (after G. V. Raynor, *Structure of Metals*, by courtesy of the Institution of Metallurgists)

complete series of solid solutions and experimentally it is found that the addition of copper to nickel causes the ferromagnetism to disappear at 60 atomic per cent. This change in magnetic properties is due to the fact that copper contributes one electron per atom to the band, so that the *d*-band will be full at a composition of 60 atomic per cent copper. By a similar reasoning it can be shown that zinc will be twice as effective as copper in reducing the magnetic moment of nickel.

The observed value of the saturation magnetic moment for cobalt, 1.7 Bohr magnetons per atom, may be accounted for in a similar way. In this case the *d*-band would have 1.7 vacant energy levels per atom, with 0.7 electrons per atom in the *s*-band. When iron is examined, however, it becomes clear that some modifications to the simple picture applied to nickel and cobalt must be made. The saturation magnetic moment in iron is 2.2 Bohr magnetons per atom and, consequently, if we were to arrange for 2.2 vacant energy levels in the *d*-band, there would be 7.8 electrons per atom in this band, which leaves only 0.2 electrons per atom in the *s*-band. Even allowing for the contribution of the *d*-band to cohesion, this relatively small number of *s*-electrons is incompatible with the high melting point of iron. In consequence, it is now believed that there are more electrons contributing to the bonding as in nickel and cobalt.

One suggestion is that in iron there are about 0.6 electrons per atom in the *s*-band, and that the form of the *d*-band is such that, in the ferromagnetic state, neither that half of the band which contains electrons of positive spin, nor that half which contains electrons of negative spin are full. The saturation magnetic moment of 2.2 Bohr magnetons can then be accounted for if the 'positive' half-band, which as shown in *Figure 5.19* could accommodate 5 electrons per atomic full, contains only 4.8 electrons per atom, and the 'negative' half-band 2.6

electrons per atom. Another suggestion, based on the results of some x-ray and neutron diffraction experiments, is that two of the eight electrons in 3*d* and 4*s* bands of iron are in 'bound' states, and, while making no contribution to bonding and conduction, are responsible for the magnetic properties; the remaining six electrons contribute to the Fermi surface and are thought to account for the bonding and conduction. Thus, the two electrons in 'bound' states with aligned spins contribute an effective magnetic moment of 2 Bohr magnetons per atom, and the remaining 0.2 Bohr magnetons per atom is considered to be due to other small contributions such as orbital motion.

The influence of alloying elements on the magnetic properties of iron would also suggest that the form of the electronic structure which exists in iron must be very stable. Thus, unlike the alloying behaviour of nickel, when solute elements are dissolved in iron to form substitutional solid solutions, the saturation magnetic moment is generally reduced by 2.2 Bohr magnetons for each atomic substitution, i.e. the value remains at 2.2 Bohr magnetons per atom of iron present in this alloy.

Apart from producing the strong magnetic properties, great strength and high melting point, the *d*-band is also responsible for the poor electrical conductivity and high electronic specific heat of the transition metals. When an electron is scattered by a lattice irregularity it jumps into a different quantum state, and it will be evident that the more vacant quantum states there are available in the same energy range, the more likely will be the electron to deflect at the irregularity. The high resistivities of the transition metals may, therefore, be explained by the ease with which electrons can be deflected into vacant *d*-states. Phonon-assisted *s*–*d* scattering gives rise to the non-linear variation of ρ with temperature observed at high temperatures. The high electronic specific heat is also due to the high density of states in the unfilled *d*-band, since this gives rise to a considerable number of electrons at the top of the Fermi distribution which can be excited by thermal activation. In copper, of course, there are no unfilled levels at the top of the *d*-band into which electrons can go, and consequently both the electronic specific heat and electrical resistance is low.

The shape of the density of states curve of the *d*-band is more complicated than that shown in *Figures 5.18* and *5.19*. Much work is being done on the detailed structure of the *d*-band in transition metals, and a fuller picture will emerge. However, work so far indicates that elements are much more individualistic in alloys.

5.10 Semiconductors

In Chapter 2 it was pointed out that Si and Ge are common intrinsic semiconductors for which the current carriers are electrons in the conduction band and holes in the valency band, in equal numbers (*see Figure 2.42*). This band structure can be modified by the addition of impurities having a valency greater or less than four. The metals then become extrinsic semiconductors.

A pentavalent impurity which donates conduction electrons without producing holes in the valency band is called a donor. The spare electrons of the

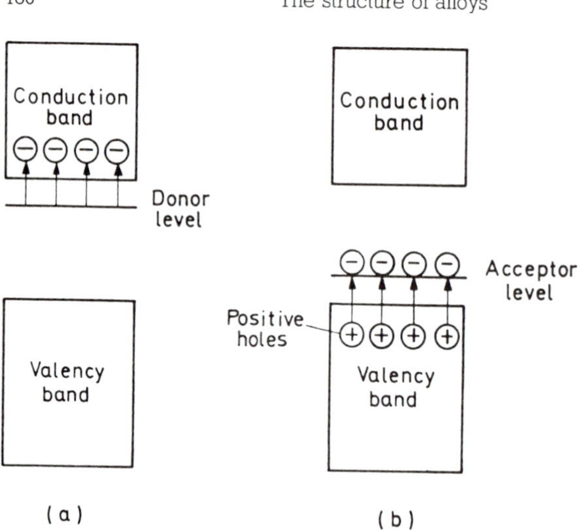

Figure 5.20 Schematic energy band structure of (a) *n*-type and (b) *p*-type semiconductor

impurity-atoms are bound in the vicinity of the impurity atoms in energy levels known as the donor levels, which are near the conduction band. If the impurity exists in an otherwise intrinsic semiconductor the number of electrons in the conduction band become greater than the number of holes in the valency band and hence, the electrons are the majority carriers and the holes the minority carriers. Such a material is an *n*-type extrinsic semiconductor (*see Figure 5.20(a)*).

Trivalent impurities in Si or Ge show the opposite behaviour leaving an empty electron state, or hole, in the valency band. If the hole separates from the so-called acceptor atom an electron is excited from the valency band to an acceptor level $\Delta E \approx 0.01$ eV. Thus, with impurity elements such as Al, Ga or In creating holes in the valency band in addition to those created thermally, the majority carriers are holes and the semiconductor is of the *p*-type extrinsic form (*see Figure 5.20(b)*). For a semiconductor where both electrons and holes carry current the conductivity is given by

$$K = n_e e \mu_e + n_h e \mu_h$$

where n_e and n_h are respectively the volume concentration of electrons and holes, and μ_e and μ_h the mobility of the carriers, i.e. electrons and holes.

Semiconductor materials are extensively used in electronic devices such as the *p–n* rectifying junction, transistor (a double-junction device) and the tunnel diode. Semiconductor regions of either *p* or *n*-type can be produced by carefully controlling the distribution and impurity content of Si or Ge single crystals, and the boundary between *p* and *n*-type extrinsic semiconductor materials is called a *p–n* junction. Such a junction conducts a large current when the voltage is applied in one direction, but only a very small current when the voltage is reversed. The action of a *p–n* junction as a rectifier is shown schematically in *Figure 5.21*. The junction presents no barrier to the flow of minority carriers from either side, but since the concentration of minority carriers is low, it is the flow of majority

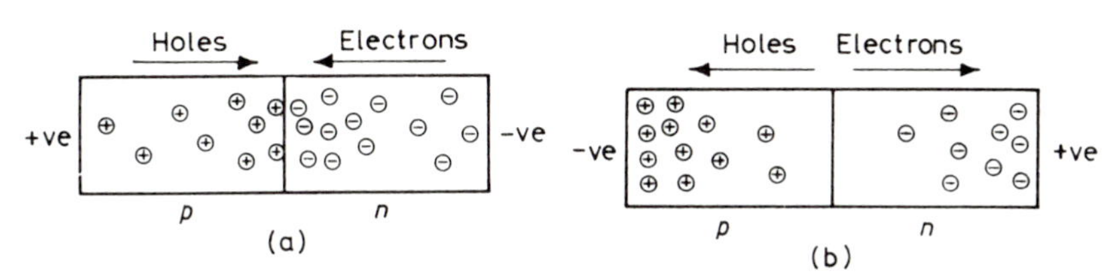

Figure 5.21 Schematic illustration of *p-n* junction rectification with (a) forward bias and (b) reverse bias

carriers which must be considered. When the junction is biased in the forward direction, i.e. *n*-type made negative and the *p*-type positive, the energy barrier opposing the flow of majority carriers from both sides of the junction is reduced. Excess majority carriers enter the *p* and *n* regions, and these recombine continuously at or near the junction to allow large currents to flow. When the junction is reversed biased the energy barrier opposing the flow of majority carriers is raised, few carriers move and little current flows.

A transistor is essentially a single crystal with two *p–n* junctions arranged back to back to give either a *p–n–p* or *n–p–n* two-junction device. For a *p–n–p* device the main current flow is provided by the positive holes, while for a *n–p–n* device the electrons carry the current. Connections are made to the individual regions of the *p–n–p* device, designated emitter, base and collector respectively, as shown in *Figure 5.22*, and the base is made slightly negative and the collector more negative relative to the emitter. The emitter–base junction is therefore forward biased and a strong current of holes passes through the junction into the *n*-layer which because it is thin (10^{-2} mm) largely reach the collector base junction without recombining with electrons. The collector–base junction is reversed biased and the junction is no barrier to the passage of holes; the current through the second junction is thus controlled by the current through the first junction. A small increase in voltage across the emitter–base junction produces a large injection of holes into the base and a large increase in current in the collector, to give the amplifying action of the transistor.

Many varied semiconductor materials such as InSb and GaAs have been developed apart from Si and Ge. However, in all cases very high purity and crystal perfection is necessary for efficient semiconducting operations and to produce the material, zone-refining techniques are used before growing the single

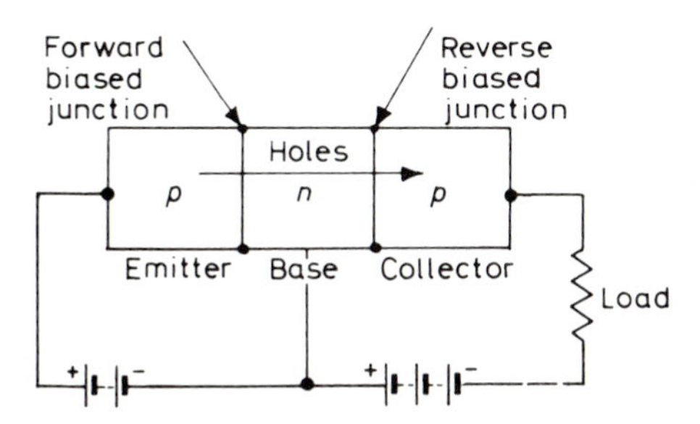

Figure 5.22 Schematic diagram of a *p-n-p* transistor

crystal perfection are necessary for efficient semiconducting operations and, to semiconductor integrated circuits are extensively used in microelectronic equipment and these are produced by vapour deposition through masks on to a single Si-slice, followed by diffusion of the deposits into the base crystal.

5.11 Superconductivity

At low temperatures (<20 K) some metals have zero electrical resistivity and become superconductors (*see* page 86). This superconductivity disappears if the temperature of the metal is raised above a critical temperature T_c, if a sufficiently strong magnetic field is applied or when a high current density flows. The critical field strength H_c, current density J_c and temperature T_c are interdependent. *Figure 5.23* shows the dependence of H_c on temperature for a number of metals; metals with high T_c and H_c values, which include the transition elements, are known as hard superconductors, those with low values such as Al, Zn, Cd, Hg, white-Sn are soft superconductors. The curves are roughly parabolic and approximate to the relation $H_c = H_0[1-(T/T_c)^2]$ where H_0 is the critical field at 0 K; H_0 is about 1.6×10^5 A/m for Nb.

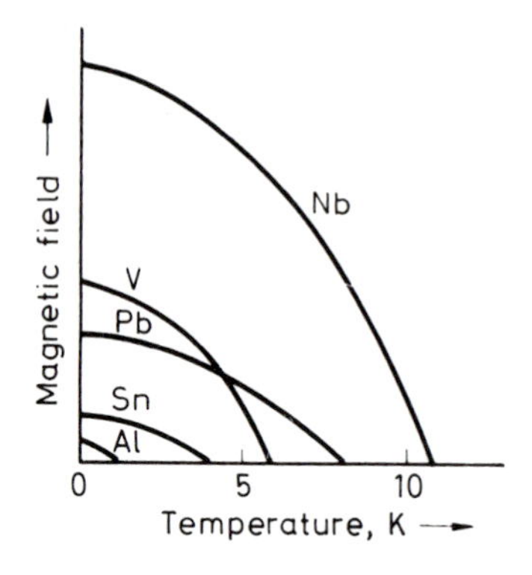

Figure 5.23 Variation of critical field H_c as a function of temperature for several pure metal superconductors

Superconductivity arises from conduction electron–electron attraction resulting from a distortion of the lattice through which the electrons are travelling; this is clearly a weak interaction since for most metals it is destroyed by thermal activation at very low temperatures. As the electron moves through the lattice it attracts nearby positive ions thereby locally causing a slightly higher positive charge density. A nearby electron may in turn be attracted by the net positive charge, the magnitude of the attraction depending on the electron density, ionic charge and lattice vibrational frequencies such that under favourable conditions the effect is slightly stronger than the electrostatic repulsion between electrons. The importance of the lattice ions in superconductivity is supported by the observation that different isotopes of the same metal (e.g. Sn and Hg) have different T_c values proportional to $M^{-1/2}$, where M is the atomic mass of the isotope. Since both the frequency of atomic vibrations and the velocity of elastic waves also varies as $M^{-1/2}$, the interaction between electrons and lattice vibrations, i.e. electron–phonon interaction, must be at least one cause of superconductivity.

The theory of superconductivity indicates that the electron–electron attraction is strongest between electrons in pairs, such that the resultant momentum of each pair is exactly the same and the individual electrons of each pair have opposite spin. With this particular form of ordering the total electron energy (i.e. kinetic and interaction) is lowered and effectively introduces a finite energy gap between this organized state and the usual more excited state of motion. The gap corresponds to a thin shell at the Fermi surface, but does not produce an insulator or semiconductor, because the application of an electric field causes the whole Fermi distribution together with gap, to drift to an unsymmetrical position, so causing a current to flow. This current remains even when the electric field is removed, since the scattering which is necessary to alter the displaced Fermi distribution is suppressed.

At 0 K all the electrons are in paired states but as the temperature is raised, pairs are broken by thermal activation giving rise to a number of normal electrons in equilibrium with the superconducting pairs. With increasing temperature the number of broken pairs increases until at T_c they are finally eliminated together with the energy gap; the superconducting state then reverts to the normal conducting state. The superconductivity transition is a second-order transformation and a plot of C/T as a function of T^2 deviates from the linear behaviour exhibited by normal conducting metals (page 86) the electronic contribution being zero at 0 K. The main theory of superconductivity, due to Bardeen, Cooper and Schrieffer (BCS) attempts to relate T_c to the strength of the interaction potential, the density of states at the Fermi surface and to the average frequency of lattice vibration involved in the scattering, and provides some explanation for the variation of T_c with the e/a ratio for a wide range of alloys, as shown in *Figure 5.24*. The main effect is attributable to the change in density of states with e/a ratio. Superconductivity is thus favoured in compounds of polyvalent atoms with crystal structures having a high density of states at the Fermi surface. Compounds with high T_c values, such as Nb_3Sn (18.1 K), Nb_3Al (17.5 K), V_3Si (17.0 K), V_3Ga (16.8 K), all crystallize with the β-tungsten structure and have e/a ratio close to 4.7; T_c is very sensitive to the degree of order and to deviation from the stoichiometric ratio, so values probably correspond to the non-stoichiometric condition.

The magnetic behaviour of superconductivity is as remarkable as the corresponding electrical behaviour, as shown in *Figure 5.25* by the Meissner effect for an ideal (structurally perfect) superconductor. It is observed for a specimen placed in a magnetic field ($H < H_c$) which is then cooled down below T_c, that magnetic lines of force are pushed out. The specimen is a perfect diamagnetic material with zero inductance as well as zero resistance. Such a material is termed an ideal type I superconductor. An ideal type II superconductor behaves similarly at low field strengths, with $H < H_{c1} < H_c$, but then allows a gradual penetration of the field returning to the normal state when penetration is complete at $H > H_{c2} > H_c$. In detail, the field actually penetrates to a small extent in type I superconductors when it is below H_c and in type II superconductors

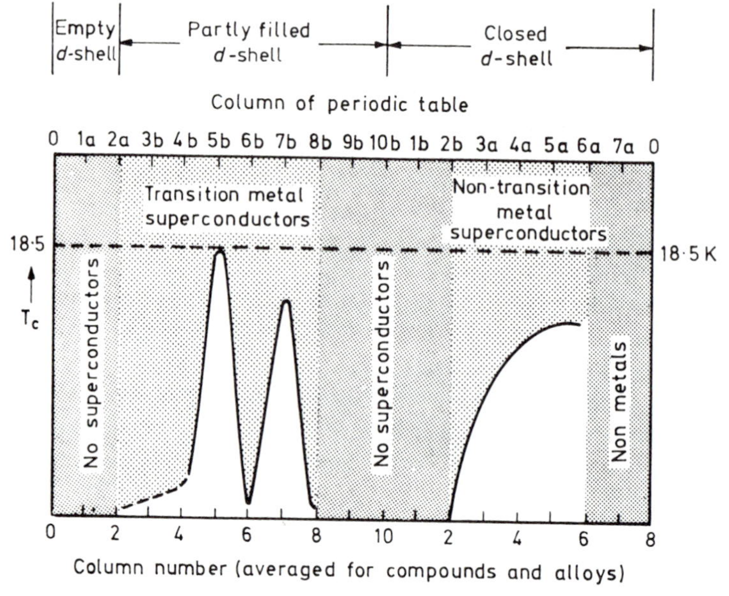

Figure 5.24 The variation of T_c with position in the periodic table. (From Matthias, *Progress in Low Temperature Physics*, Vol. II, p. 138, ed by C. J. Gorter, 1957, courtesy of North-Holland Publishing Co.)

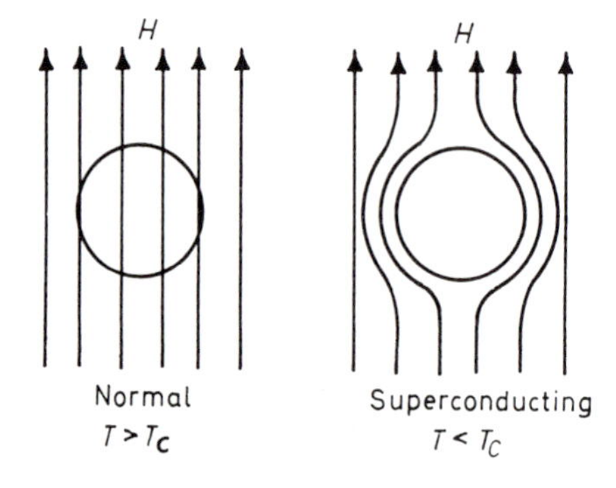

Figure 5.25 The Meissner effect; shown by the expulsion of magnetic flux when the specimen becomes superconducting

when H is below H_{c1}, and decays away at a penetration depth $\approx 10^{-4}$–10^{-5} mm.

The observation of the Meissner effect in type I superconductors implies that the surface between the normal and superconducting phases has an effective positive energy. In the absence of this surface energy, the specimen would break-up into separate fine regions of superconducting and normal material to reduce the work done in the expulsion of the magnetic flux. A negative surface energy exists between the normal and superconducting phases in a type II superconductor and hence the superconductor exists naturally in a state of finely separated superconducting and normal regions. By adopting a 'mixed state' of normal and superconducting regions the volume of interface is maximized while at the same time keeping the volume of normal conduction as small as possible. The structure of the mixed state is believed to consist of lines of normal phases parallel to the applied field through which the field lines run, embedded in a

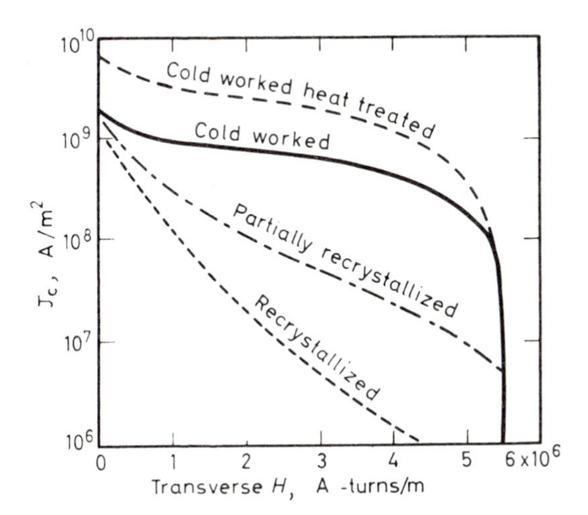

Figure 5.26 The effect of processing on the J_c v. H curve of an Nb-25% Zr alloy wire which produces a fine precipitate and raises J_c (From Rose, Shepard and Wulff, *Structure and Properties of Materials*, 1966, courtesy of John Wiley and Sons)

superconducting matrix. The field falls off with distances from the centre of each line over the characteristic distance λ, and vortices or whirlpools of supercurrents flow around each line; the flux line, together with its current vortex, is called a fluxoid. At H_{c1}, fluxoids appear in the specimen and increase in number as the magnetic field is raised. At H_{c2}, the fluxoids completely fill the cross-section of the sample and type II superconductivity disappears. Type II superconductors are of particular interest because of their high critical fields which makes them potentially useful for the construction of high field electromagnets and solenoids. To produce a magnetic field of ≈ 10 T with a conventional solenoid would cost more than ten times that of a superconducting solenoid wound with Nb_3Sn wire. By embedding Nb wire in a bronze matrix it is possible to form channels of Nb_3Sn by interdiffusion. The conventional installation would require considerable power, cooling water and space, whereas the superconducting solenoid occupies little space, has no steady-state power consumption and uses relatively little liquid helium. It is necessary, however, for the material to carry useful currents without resistance in such high fields, which is not usually the case in annealed homogeneous type II superconductors. Fortunately, the critical current density is extremely sensitive to microstructure and is markedly increased by precipitation hardening, cold work, radiation damage, etc., because the lattice defects introduced pin the fluxoids and tend to immobilize them. *Figure 5.26* shows the influence of metallurgical treatment on the critical current density.

Suggestions for further reading

Barrett, C.S. and Massalski, T.B., *Structure of Metals*, McGraw-Hill, 1980

Dew-Hughes, D., *Metallurgy of Superconductors*, Academic Press (1979)

Dugdale, J.S., *The Electrical Properties of Metals and Alloys*, Edward Arnold, 1977

Hume-Rothery, W., Smallman, R.E. and Haworth, C., *The Structure of Metals and Alloys*, Institute of Metals Monograph and Report series, London, 1969

Kittel, C., *Introduction to Solid State Physics*, Wiley, 1953

Mott, N.F. and Jones, H., *The Theory of the Properties of Metals and Alloys*, Oxford University Press, 1936

Ziman, J.M. (ed.), *The Physics of Metals, 1 Electrons*, Cambridge University Press, 1969

Chapter 6

Dislocations in crystals

6.1 Elastic and plastic deformation

It is well known that metals deform both elastically and plastically. Elastic deformation takes place at low stresses and has three main characteristics, namely (i) it is reversible, (ii) stress and strain are linearly proportional to each other according to Hooke's Law and (iii) it is usually small (i.e. <1 per cent elastic strain).

The stress at a point in a body is usually defined by considering an infinitesimal cube surrounding that point and the forces applied to the faces of the cube by the surrounding material. These forces may be resolved into components parallel to the cube edges and when divided by the area of a face give the nine stress components shown in *Figure 6.1(a)*. A given component σ_{ij} is the force acting in the j-direction per unit area of face normal to the i-direction. Clearly, when $i=j$ we have normal stress components (e.g. σ_{xx}) which may be either tensile (conventionally positive) or compressive (negative), and when $i \neq j$ (e.g. σ_{xy}) the stress components are shear. These shear stresses exert couples on the cube and to prevent rotation of the cube the couples on opposite faces must balance and hence $\sigma_{ij}=\sigma_{ji}$. Thus, stress has only six independent components. In this book limited use is made of the stress notation* given above. It is found more convenient to denote the shear stresses by τ and the normal stresses by σ and, since σ_{xx}, for example, is clearly a normal stress, the second suffix can be dropped for these stresses, as shown in *Figure 6.1(b)*.

When a body is strained, small elements in that body are displaced. If the initial position of an element is defined by its coordinates (x, y, z) and its final position by $(x+u,\ y+v,\ z+w)$ then the displacement is (u, v, w). If this displacement is constant for all elements in the body, no strain is involved, only a rigid translation. For a body to be under a condition of strain the displacements

* The nine components of stress σ_{ij} form a second rank tensor usually written

$$\begin{matrix} \sigma_{xx} & \sigma_{xy} & \sigma_{xz} \\ \sigma_{yx} & \sigma_{yy} & \sigma_{yz} \\ \sigma_{zx} & \sigma_{zy} & \sigma_{zz} \end{matrix}$$

and is known as the stress tensor.

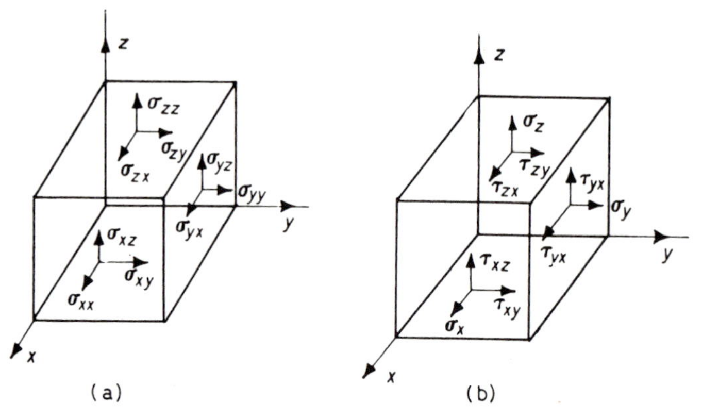

Figure 6.1 Normal and shear stress components

must vary from element to element. A uniform strain is produced when the displacements are linearly proportional to distance. In one dimension then $u = ex$ where $e = du/dx$ is the coefficient of proportionality or nominal tensile strain. For a three-dimensional uniform strain, each of the three components u, v, w is made a linear function in terms of the initial elemental coordinates, viz.:

$$u = e_{xx}x + e_{xy}y + e_{xz}z$$

$$v = e_{yx}x + e_{yy}y + e_{yz}z$$

$$w = e_{zx}x + e_{zy}y + e_{zz}z$$

The strains $e_{xx} = du/dx$, $e_{yy} = dv/dy$, $e_{zz} = dw/dz$ are the tensile strains along the x, y and z axes respectively. The strains e_{xy}, e_{yz}, etc, produce shear strains and in some cases a rigid body rotation. The rotation produces no strain and can be allowed for by rotating the reference axes (*see Figure 6.2*). In general therefore, $e_{ij} = \varepsilon_{ij} + \omega_{ij}$ with ε_{ij} the strain components and ω_{ij} the rotation components. If, however, the shear strain is defined as the angle of shear, this is twice the corresponding shear strain component, i.e. $\gamma = 2\varepsilon_{ij}$. The strain tensor, like the stress tensor, has nine components which are usually written as:

$$\begin{matrix} \varepsilon_{xx} & \varepsilon_{xy} & \varepsilon_{xz} \\ \varepsilon_{yx} & \varepsilon_{yy} & \varepsilon_{yz} \\ \varepsilon_{zx} & \varepsilon_{zy} & \varepsilon_{zz} \end{matrix} \quad \text{or} \quad \begin{matrix} \varepsilon_x & \tfrac{1}{2}\gamma_{xy} & \tfrac{1}{2}\gamma_{xz} \\ \tfrac{1}{2}\gamma_{yx} & \varepsilon_y & \tfrac{1}{2}\gamma_{yz} \\ \tfrac{1}{2}\gamma_{zx} & \tfrac{1}{2}\gamma_{zy} & \varepsilon_z \end{matrix}$$

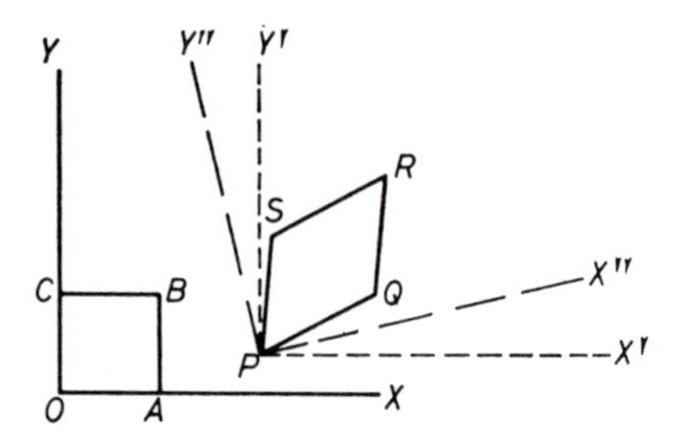

Figure 6.2 Deformation of a square *OABC* to a parallelogram *PQRS* involving (i) a rigid body translation *OP* allowed for by redefining new axes *X'Y'*, (ii) a rigid body rotation allowed for by rotating the axes to *X''Y''*, and (iii) a change of shape involving both tensile and shear strains

where ε_x, etc. are tensile strains and γ_{xy}, etc. are shear strains. All the simple types of strain can be produced from the strain tensor by setting some of the components equal to zero. For example, a pure dilatation, i.e. change of volume without change of shape, is obtained when $\varepsilon_x=\varepsilon_y=\varepsilon_z$ and all other components are zero. Another example is a uniaxial tensile test when the tensile strain along the x-axis is simply $e=\varepsilon_x$. However, because of the strains introduced by lateral contraction, $\varepsilon_y=-ve$ and $\varepsilon_z=-ve$ where v is Poisson's ratio; all other components of the strain tensor are zero.

At small elastic deformations, the stress is linearly proportional to the strain. This is Hooke's law and in its simplest form relates the uniaxial stress to the uniaxial strain by means of the modulus of elasticity. For a general situation, it is necessary to write Hooke's law as a linear relationship between the six stress components and the six strain components, viz.:

$$\sigma_x=c_{11}\varepsilon_x+c_{12}\varepsilon_y+c_{13}\varepsilon_z+c_{14}\gamma_{yz}+c_{15}\gamma_{zx}+c_{16}\gamma_{xy}$$

$$\sigma_y=c_{21}\varepsilon_x+c_{22}\varepsilon_y+c_{23}\varepsilon_z+c_{24}\gamma_{yz}+c_{25}\gamma_{zx}+c_{26}\gamma_{xy}$$

$$\sigma_z=c_{31}\varepsilon_x+c_{32}\varepsilon_y+c_{33}\varepsilon_z+c_{34}\gamma_{yz}+c_{35}\gamma_{zx}+c_{36}\gamma_{xy}$$

$$\tau_{yz}=c_{41}\varepsilon_x+c_{42}\varepsilon_y+c_{43}\varepsilon_z+c_{44}\gamma_{yz}+c_{45}\gamma_{zx}+c_{46}\gamma_{xy}$$

$$\tau_{zx}=c_{51}\varepsilon_x+c_{52}\varepsilon_y+c_{53}\varepsilon_z+c_{54}\gamma_{yz}+c_{55}\gamma_{zx}+c_{56}\gamma_{xy}$$

$$\tau_{xy}=c_{61}\varepsilon_x+c_{62}\varepsilon_y+c_{63}\varepsilon_z+c_{64}\gamma_{yz}+c_{65}\gamma_{zx}+c_{66}\gamma_{xy}$$

The constants $c_{11}, c_{12}, \ldots, c_{ij}$ are called the elastic stiffness constants*.

Taking account of the symmetry of the crystal, many of these elastic constants are equal or become zero. Thus in cubic crystals there are only three independent elastic constants c_{11}, c_{12} and c_{44} for the three independent modes of deformation. These include the application of (i) a hydrostatic stress p to produce a dilatation Θ given by

$$p=-\tfrac{1}{3}(c_{11}+2c_{12})\Theta=-\kappa$$

where κ is the bulk modulus, (ii) a shear stress on a cube face in the direction of the cube axis defining the shear modulus $\mu=c_{44}$, and (iii) a rotation about a cubic axis defining a shear modulus $\mu_1=\tfrac{1}{2}(c_{11}-c_{12})$. The ratio μ/μ_1 is the elastic anisotropy factor and in elastically isotropic crystals it is unity with $2c_{44}=c_{11}-c_{12}$; the constants are all inter-related with $c_{11}=\kappa+\tfrac{4}{3}\mu$, $c_{12}=\kappa-\tfrac{2}{3}\mu$ and $c_{44}=\mu$. The bulk modulus κ, shear modulus μ and Young's modulus E are inter-related by the equations given in Chapter 2.

Table 6.1 shows that most metals are far from isotropic and, in fact, only tungsten is isotropic; the alkali metals and β-compounds are mostly anisotropic. Generally, $2c_{44}>(c_{11}-c_{12})$ and hence, for most elastically anisotropic metals E is maximum in the $\langle 111\rangle$ and minimum in the $\langle 100\rangle$ directions. Molybdenum and niobium are unusual in having the reverse anisotropy when E is greatest along $\langle 100\rangle$ directions. Most commercial materials are polycrystalline, and consequently they have approximately isotropic properties. For such materials

* Alternatively the strain may be related to the stress, e.g. $\varepsilon_x=s_{11}\sigma_x+s_{12}\sigma_y+s_{13}\sigma_z+\ldots$, in which case the constants $s_{11}, s_{12}, \ldots, s_{ij}$ are called elastic compliances.

TABLE 6.1 Elastic constants of cubic crystals (GN/m^2)

Metal	c_{11}	c_{12}	c_{44}	$2c_{44}/(c_{11}-c_{12})$
Na	006.0	004.6	005.9	8.5
K	004.6	003.7	002.6	5.8
Fe	237.0	141.0	116.0	2.4
W	501.0	198.0	151.0	1.0
Mo	460.0	179.0	109.0	0.77
Al	108.0	62.0	28.0	1.2
Cu	170.0	121.0	75.0	3.3
Ag	120.0	90.0	43.0	2.9
Au	186.0	157.0	42.0	3.9
Ni	250.0	160.0	118.0	2.6
β-brass	129.1	109.7	82.4	8.5

the modulus value is usually independent of the direction of measurement because the value observed is an average for all directions, in the various crystals of the specimen. However, if during manufacture a preferred orientation of the grains in the polycrystalline specimen occurs, the material will behave, to some extent, like a single crystal and some 'directionality' will occur.

The limit of the elastic range cannot be defined exactly but may be considered to be that value of the stress below which the amount of plasticity (irreversible deformation) is negligible, and above which the amount of plastic deformation is far greater than the elastic deformation. If we consider the deformation of a metal in a tensile test, one or other of two types of curve may be obtained (*Figure 2.36*). *Figure 2.36*(*a*) shows the stress–strain curve characteristic of iron, from which it can be seen that plastic deformation begins abruptly at A and continues initially with no increase in stress. The point A is known as the yield point and the stress at which it occurs is the yield stress. *Figure 2.36*(*b*) shows a stress–strain curve characteristic of copper, from which it will be noticed that the transition to the plastic range is gradual. No abrupt yielding takes place and in this case the stress required to start macroscopic plastic flow is known as the flow stress.

Once the yield or flow stress has been exceeded plastic or permanent deformation occurs, and this is found to take place by one of two simple processes, slip or glide and twinning. During slip, shown in *Figure 6.3*(*a*), the top half of the crystal moves over the bottom half along certain crystallographic planes, known as slip planes, in such a way that the atoms move forward by a whole number of lattice vectors; as a result the continuity of the lattice is maintained. During twinning (*Figure 6.3*(*b*) the atomic movements are not whole lattice vectors, and the lattice generated in the deformed region, although the same as the parent lattice, is oriented in a twin relationship to it. It will also be observed that in contrast to slip, the sheared region in twinning occurs over many atom planes, the atoms in each plane being moved forward by the same amount relative to those of the plane below them.

6.1.1 Resolved shear stress

All working processes such as rolling, extrusion, forging etc., cause plastic deformation and, consequently, these operations will involve the processes of slip

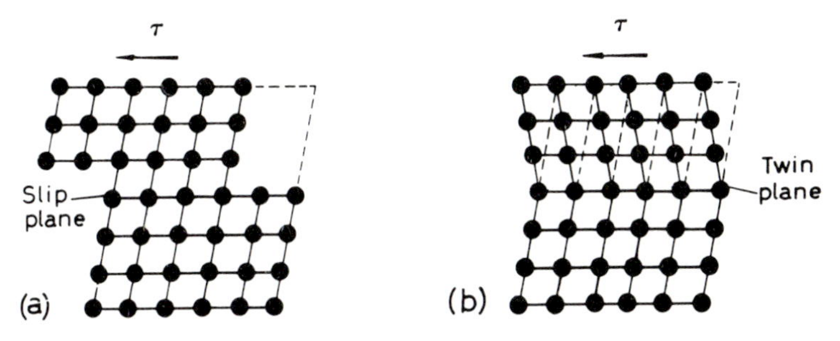

Figure 6.3 Slip and twinning in a crystal

or twinning outlined above. The stress system applied during these working operations is often quite complex, but for plastic deformation to occur the presence of a shear stress is essential. The importance of shear stresses becomes clear when it is realized that these stresses arise in most processes and tests even when the applied stress itself is not a pure shear stress. This may be illustrated by examining a cylindrical crystal of area A in a conventional tensile test under a uniaxial load P. In such a test, slip occurs on the slip plane, shown shaded in *Figure 6.4*, the area of which is $A/\cos\phi$, where ϕ is the angle between the normal to the plane OH and the axis of tension. The applied force P is spread over this plane and may be resolved into a force normal to the plane along OH, $P\cos\phi$, and a force along OS, $P\sin\phi$. Here, OS is the line of greatest slope in the slip plane and the force $P\sin\phi$ is a shear force. It follows that the applied stress (force/area) is made up of two stresses, a normal stress $(P/A)\cos^2\phi$ tending to pull the atoms apart, and a shear stress $(P/A)\cos\phi\sin\phi$ trying to slide the atoms over each other.

In general, slip does not take place down the line of greatest slope unless this happens to coincide with the crystallographic slip direction. It is necessary, therefore, to know the resolved shear stress on the slip plane and in the slip direction. Now, if OT is taken to represent the slip direction the resolved shear stress will be given by

$$\sigma = P\cos\phi\sin\phi\cos\chi/A$$

where χ is the angle between OS and OT. Usually this formula is written more simply as

$$\sigma = P\cos\phi\cos\lambda/A \qquad (6.1)$$

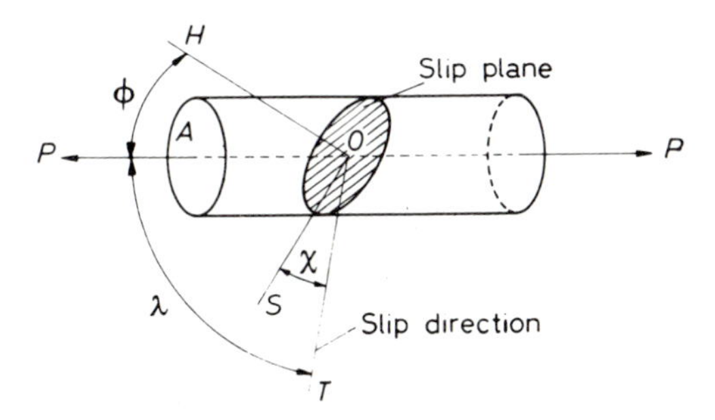

Figure 6.4 Relation between the slip plane, slip direction and the axis of tension for a cylindrical crystal

where λ is the angle between the slip direction OT and the axis of tension. It can be seen that the resolved shear stress has a maximum value when the slip plane is inclined at 45° to the tensile axis, and becomes smaller for angles either greater than or less than 45°. When the slip plane becomes more nearly perpendicular to the tensile axis ($\phi > 45°$) it is easy to imagine that the applied stress has a greater tendency to pull the atoms apart than to slide them. When the slip plane becomes more nearly parallel to the tensile axis ($\phi < 45°$) the shear stress is again small but in this case it is because the area of the slip plane, $A/\cos\phi$, is correspondingly large.

A consideration of the tensile test in this way shows that it is shear stresses which lead to plastic deformation, and for this reason the mechanical behaviour exhibited by a material will depend, to some extent, on the type of test applied. For example, a ductile material can be fractured without displaying its plastic properties if tested in *a state of hydrostatic or triaxial tension*, since under these conditions the resolved shear stress on any plane is zero. Conversely, materials which normally exhibit a tendency to brittle behaviour in a tensile test will show ductility if tested under conditions of high shear stresses and low tension stresses. In commercial practice, extrusion approximates closely to a system of hydrostatic pressure, and it is common for normally brittle materials to exhibit some ductility when deformed in this way, e.g. when extruded.

6.1.2 The relation of slip to crystal structure

An understanding of the fundamental nature of plastic deformation processes is provided by experiments on single crystals only, because if a polycrystalline sample is used the result obtained is the average behaviour of all the differently oriented grains in the material. Such experiments with single crystals show that, although the resolved shear stress is a maximum along lines of greatest slope in planes at 45° to the tensile axis, slip occurs preferentially along certain crystal planes and directions. Three well-established laws governing the slip behaviour exist, namely: (i) the direction of slip is almost always that along which the atoms are most closely packed, (ii) slip usually occurs on the most closely packed plane, and (iii) from a given set of slip planes and directions, the crystal operates on that system (plane and direction) for which the resolved shear stress is largest. The slip behaviour observed in f.c.c. metals shows the general applicability of these laws, since slip occurs along $\langle 110 \rangle$ directions in $\{111\}$ planes. In c.p.h. metals slip occurs along $\langle 11\bar{2}0 \rangle$ directions, since these are invariably the closest packed, but the active slip plane depends on the value of the axial ratio. Thus, for the metals cadmium and zinc, c/a is 1.886 and 1.856 respectively, the planes of greatest atomic density are the $\{0001\}$ basal planes and slip takes place on these planes. When the axial ratio is appreciably smaller than the ideal value of $c/a = 1.633$ the basal plane is not so closely packed, nor so widely spaced, as in cadmium and zinc, and other slip planes operate. In zirconium ($c/a = 1.589$) and titanium ($c/a = 1.587$), for example, slip takes place on the $\{10\bar{1}0\}$ prism planes at room temperature and on the $\{10\bar{1}1\}$ pyramidal planes at higher temperatures. In magnesium the axial ratio ($c/a = 1.624$) approximates to the ideal value, and although only basal slip occurs at room temperature, at temperatures above 225 °C slip on the $\{10\bar{1}1\}$ planes has also been observed. B.c.c. metals have a single well-defined close packed $\langle 111 \rangle$ direction but several planes of equally high density of packing, i.e. $\{112\}$, $\{110\}$ and $\{123\}$. The choice of slip plane in these metals is often influenced by temperature and a preference is shown for $\{112\}$ below $T_m/4$, $\{110\}$ from $T_m/4$ to $T_m/2$ and $\{123\}$ at high temperatures, where T_m is the melting point. Iron often slips on all the slip planes at once in a common $\langle 111 \rangle$ slip direction, so that a slip line, i.e. the line of intersection of a slip plane with the outer surface of a crystal, takes on a wavy appearance.

6.1.3 Law of critical resolved shear stress

This law states that slip takes place along a given slip plane and direction when the shear stress reaches a critical value. In most crystals the high symmetry of atomic arrangement provides several crystallographic equivalent planes and directions for slip (i.e. c.p.h. crystals have three systems made up of 1 plane containing 3 directions, f.c.c. crystals have twelve systems made up of 4 planes each with 3 directions, while b.c.c. crystals have many systems) and in such cases slip occurs first on that plane and along that direction for which the maximum stress acts (law iii above). This is most easily demonstrated by testing in tension a series of zinc single crystals. Then, because zinc is c.p.h. in structure only one plane is available for the slip process and the resultant stress–strain curve will depend on the inclination of this plane to the tensile axis. The value of the angle ϕ is determined by chance during the process of single crystal growth, and consequently all crystals will have different values of ϕ, and the corresponding stress–strain curves will have different values of the flow stress as shown in *Figure 6.5(a)*. However, because of the criterion of a critical resolved shear stress, a plot of resolved shear stress (i.e. the stress on the glide plane in the glide direction) versus strain should be a common curve, within experimental error, for all the specimens. This plot is shown in *Figure 6.5(b)*.

The importance of a critical shear stress may be demonstrated further by taking the crystal which has its basal plane oriented perpendicular to the tensile

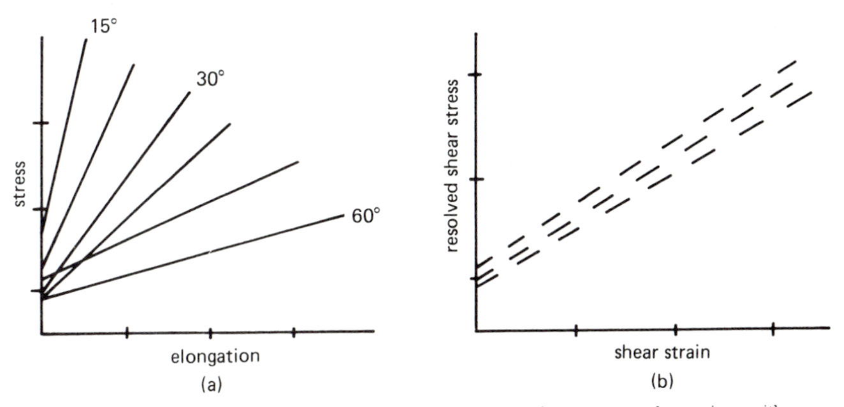

Figure 6.5 Schematic representation of (a) variation of stress v. elongation with orientation of basal plane, (b) constancy of resolved shear stress

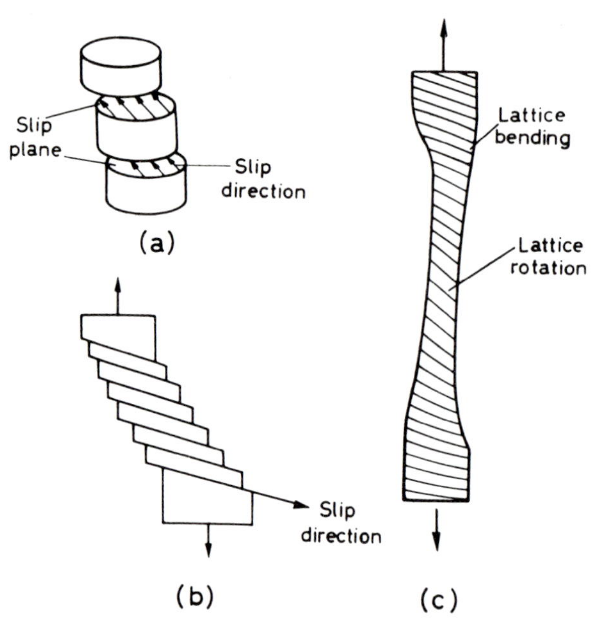

Figure 6.6 (a) and (b) show the slip process in an unconstrained single crystal; (c) illustrates the plastic bending in a crystal gripped at its ends

axis, i.e. $\phi = 0°$, and subjecting it to a bend test. In contrast to its tensile behaviour, where it is brittle it will now appear ductile, since the shear stress on the slip plane is only zero for a tensile test and not for a bend test. On the other hand, if we take the crystal with its basal plane oriented parallel to the tensile axis, i.e. $\phi = 90°$, this specimen will appear brittle whatever stress system is applied to it. For this crystal, although the shear force is large, owing to the large area of the slip plane, $A/\cos\phi$, the resolved shear stress is always very small and insufficient to cause deformation by slipping.

6.1.4 A multiple slip

The fact that slip bands, each consisting of many slip lines, are observed on the surface of deformed crystals shows that deformation is inhomogeneous, with extensive slip occurring on certain planes, while the crystal planes lying between them remain practically undeformed. *Figure 6.6*(*a*) and (*b*) shows such a crystal in which the set of planes shear over each other in the slip direction. In a tensile test, however, the ends of a crystal are not free to move 'sideways' relative to each other, since they are constrained by the grips of the tensile machine. In this case, the central portion of the crystal is altered in orientation, and rotation of both the slip plane and slip direction into the axis of tension occurs, as shown in *Figure 6.6*(*c*). This behaviour is more conveniently demonstrated on a stereographic projection of the crystal by considering the rotation of the tensile axis relative to the crystal rather than vice versa. This is illustrated in *Figure 6.7*(*a*) for the

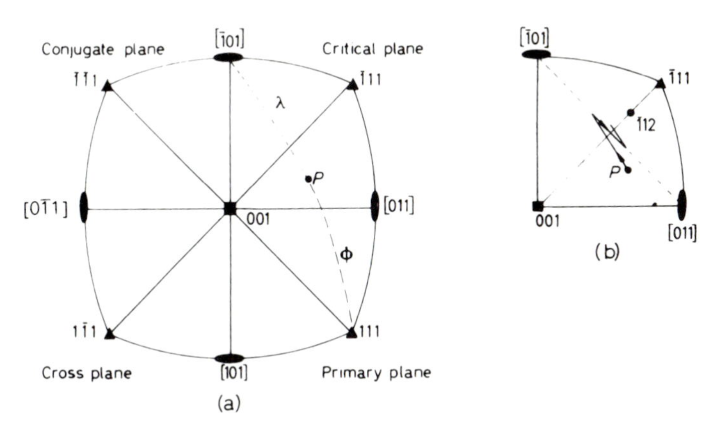

Figure 6.7 Stereographic representation of (a) slip systems in f.c.c. crystals, and (b) overshooting of the primary slip system

deformation of a crystal with f.c.c. structure. The tensile axis, P, is shown in the unit triangle and the angles between P and $[\bar{1}01]$, and P and 111 are equal to λ and ϕ respectively. The active slip system is the (111) plane and the $[\bar{1}01]$ direction, and as deformation proceeds the change in orientation is represented by the point, P, moving along the zone, shown broken in *Figure 6.7(a)*, towards $[\bar{1}01]$, i.e. λ decreasing and ϕ increasing.

As slip occurs on the one system, the primary system, the slip plane rotates away from its position of maximum resolved shear stress until the orientation of the crystal reaches the $[001]$–$[\bar{1}11]$ symmetry line. Beyond this point, slip should occur equally on both the primary system and a second system (the conjugate system) $(\bar{1}\bar{1}1)$ $[011]$, since these two systems receive equal components of shear stress. Subsequently, during the process of multiple or duplex slip the lattice will rotate so as to keep equal stresses on the two active systems, and the tensile axis moves along the symmetry line towards $[\bar{1}12]$. This behaviour agrees with early observations on virgin crystals of aluminium and copper, but not with those made on certain alloys, or pure metal crystals given special treatments (e.g. quenched from a high temperature or irradiated with neutrons). Results from the latter show that the crystal continues to slip on the primary system after the orientation has reached the symmetry line, causing the orientation to overshoot this line, i.e. to continue moving towards $[\bar{1}01]$, in the direction of primary slip. After a certain amount of this additional primary slip the conjugate system suddenly operates, and further slip concentrates itself on this sytem, followed by overshooting in the opposite direction. This behaviour, shown in *Figure 6.7(b)*, is understandable when it is remembered that slip on the conjugate system must intersect that on the primary system, and to do this is presumably more difficult than to 'fit' a new slip plane in the relatively undeformed region between those planes on which slip has already taken place. As we shall see later, the intersection process is more difficult in materials which have a low stacking fault energy, e.g. α-brass.

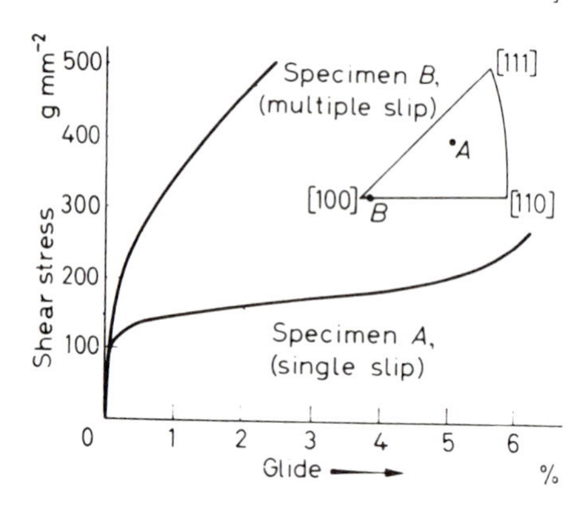

Figure 6.8 Stress–strain curves on aluminium deformed by single and multiple slip (after Lücke and Lange, *Z. Metallk.* 1950, **41,** 65)

6.1.5 The relation between work hardening and slip

The curves of *Figure 2.36* show that following the yield phenomenon a continual rise in stress is required to continue deformation, i.e. the flow stress of a deformed metal increases with the amount of strain. This resistance of the metal to further plastic flow as the deformation proceeds is known as work hardening. The degree of work hardening varies for metals of different crystal structure, and is low in hexagonal metal crystals such as zinc or cadmium, which usually slip on one family of planes only. The cubic crystals harden rapidly on working but even in this case when slip is restricted to one slip system (*see* the curve for specimen *A*, *Figure 6.8*) the coefficient of hardening, defined as the slope of the plastic portion of the stress–strain curve, is small. Thus, this type of hardening, like overshoot, must be associated with the interaction which results from slip on intersecting families of planes. This interaction will be dealt with more fully in Chapter 10 on work hardening (page 340).

6.2 Dislocations in crystals

Figure 6.3 indicates that a crystal changes its shape during deformation by the slipping of atomic layers over one another. This theoretical strength of crystals was first calculated by Frenkel for the simple rectangular-type lattice shown in *Figure 6.9* with *a* the spacing between the planes. The shearing force

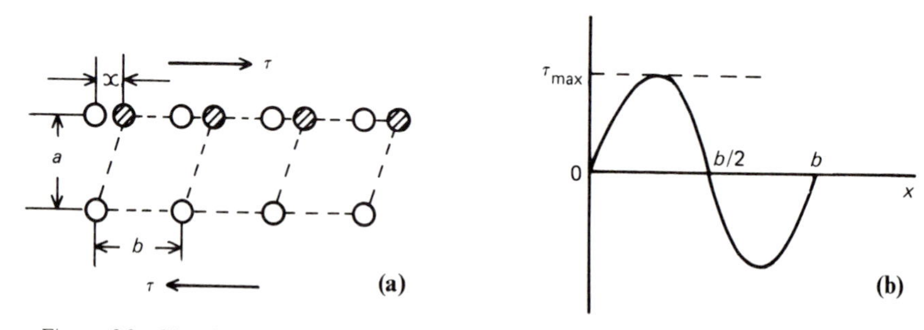

Figure 6.9 Slip of crystal planes (a) atomic model, (b) stress v. displacement curve

required to move a plane of atoms over the plane below will be periodic, since for displacements $x < b/2$, where b is the spacing of atoms in the shear direction, the lattice resists the applied stress, but for $x > b/2$ the lattice forces assist the applied stress. The simplest function with these properties is a sinusoidal relation of the form

$$\tau = \tau_m \sin(2\pi x/b) \simeq \tau_m 2\pi x/b$$

where τ_m is the maximum stress at a displacement $x = b/4$. For small displacements the elastic strain given by x/a is equal to τ/μ from Hooke's law, where μ is the shear modulus, so that

$$\tau_m = (\mu/2\pi)b/a \tag{6.2}$$

and since $b \simeq a$, the theoretical strength of a perfect crystal is of the order of $\mu/10$. More refined calculations show that τ_m is about $\mu/30$, that is, perfect single crystals should be rather strong and difficult to deform. A striking experimental property of single crystals, however, is their softness, which shows that the critical shear stress to produce slip is very small (about $10^{-5}\,\mu$ or $\approx 50\,\mathrm{g\,mm^{-2}}$). This discrepancy between the theoretical and experimental strength of crystals may be resolved if it is assumed that slip occurs by a nucleation and growth process in which complete atomic planes do not slip over each other as rigid bodies, but instead that slip starts at a localized region in the lattice and then spreads gradually over the remainder of the plane as shown in *Figure 6.10(a)*. Such a process would appear to be logical, since the slipping of atomic planes over each other like playing cards necessitates a rigid coupling of the atoms, and this in turn would imply an infinite value of Young's modulus in the direction of slip or, alternatively, that the force on the plane is uniformly distributed with thermal vibrations playing no role.

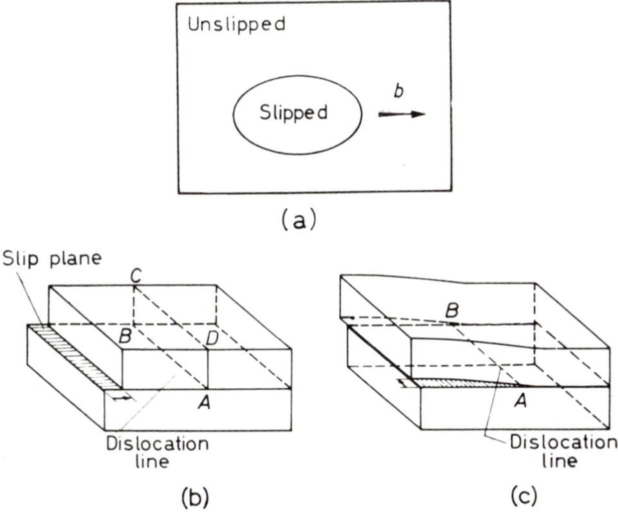

Figure 6.10 Schematic representation of (a) dislocation loop, (b) edge dislocation, and (c) screw dislocation

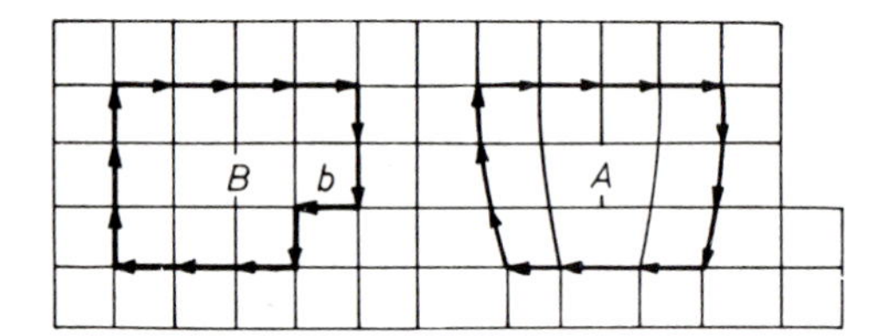

Figure 6.11 A Burgers circuit round a dislocation *A* fails to close when repeated in a perfect lattice *B* unless completed by a closure vector equal to the Burgers vector *b*

In general, therefore, the slip plane may be divided into two regions, one where slip has occurred and the other which remains unslipped. Between the slipped and unslipped regions the lattice structures will be dislocated and, consequently, the boundary is referred to as a dislocation line, or dislocation. Three simple properties of a dislocation are immediately apparent, namely: (i) it is a line discontinuity, (ii) it forms a closed loop in the interior of the crystal or emerges at the surface, and (iii) the difference in the amount of slip across the dislocation line is constant. This latter property is probably the most important, since a dislocation is characterized by the magnitude and direction of the slip movement associated with it. This is called the Burgers vector, *b*, which for any given dislocation line is the same all along its length. The Burgers vector is defined by constructing a Burgers circuit as shown in *Figure 6.11.* A sequence of lattice vectors is taken to form a closed clockwise circuit around the dislocation. The same sequence of vectors is then taken in the perfect lattice, when it is found that the circuit fails to close. The closure vector FS defines *b* for the dislocation. With this FS/RH (right hand) convention it is necessary to choose one direction along the dislocation line as positive. If this direction is reversed the vector *b* is also reversed.

6.2.1 Edge and screw dislocations

It is evident from *Figure 6.10*(*a*) that some sections of the dislocation line are perpendicular to *b*, others are parallel to *b* while the remainder lie at an angle to *b*. This variation in the orientation of the line with respect to the Burgers vector gives rise to a difference in the structure of the dislocation. *Figure 6.10*(*b*) shows a schematic diagram of a dislocation line AB. Here the slip plane is shown shaded and the top left-hand side of the crystal has moved by one atomic spacing in the slip direction, i.e. perpendicular to AB. This slip movement has not been transmitted to the whole of the slip plane, however, but terminates at the dislocation line AB. It is clear, therefore, that there must be an extra half-plane of atoms ABCD terminating on the dislocation line. In this example, the dislocation line is normal to the slip direction and as a result it is called an edge dislocation. In contrast, when the line of the dislocations is parallel to the slip direction the dislocation line is known as a screw dislocation. This type of dislocation is shown schematically in *Figure 6.10*(*c*) where, in this case, AB is the direction of slip. From the diagram shown in *Figure 6.10*(*a*) it is evident that the dislocation line is rarely pure edge or pure screw, but it is convenient to think of these ideal dislocations since any dislocation can be resolved into edge and screw components.

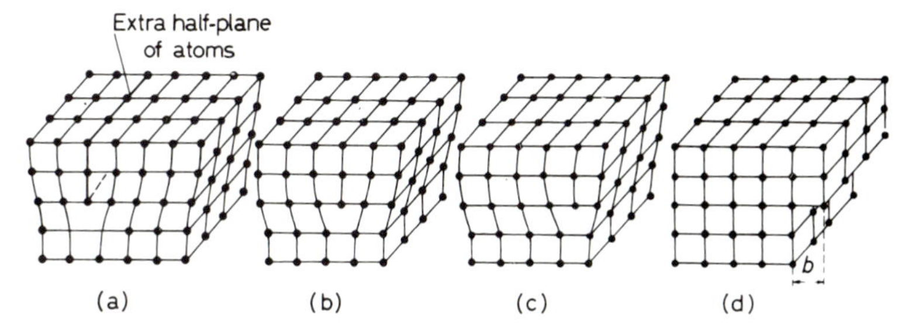

Figure 6.12 Slip caused by the movement of an edge dislocation

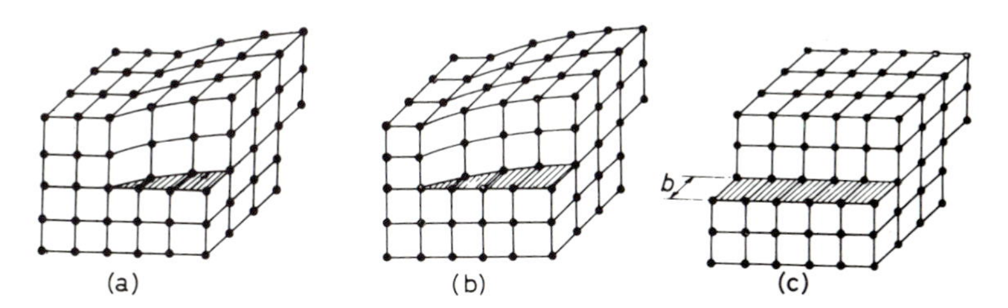

Figure 6.13 Slip caused by the movement of a screw dislocation

6.2.2 The mechanism of slip and climb

The atomic structure of an edge dislocation is shown in *Figure 6.12(a)*. Here the extra half-plane of atoms is above the slip plane of the crystal, and consequently the dislocation is called a positive edge dislocation and is often denoted by the symbol $\perp$. When the half-plane is below the slip plane it is termed a negative dislocation. If the resolved shear stress on the slip plane is τ and the Burgers vector of the dislocation b, the force on the dislocation, i.e. force per unit length of dislocation, is $F=\tau b$. This can be seen by reference to *Figure 6.10(b)* if the crystal is of side L. The force on the top face (stress $\times$ area) is $\tau \times L^2$. Thus, when the two halves of the crystal have slipped the relative amount b, the work done by the applied stress (force $\times$ distance) is $\tau L^2 b$. On the other hand, the work done in moving the dislocation (total force on dislocation $FL \times$ distance moved) is FL^2, so that equating the work done gives F (force per unit length of dislocation) $=\tau b$. *Figure 6.12* indicates how slip is propagated by the movement of a dislocation under the action of such a force. The extra half-plane moves to the right until it produces the slip step shown at the surface of the crystal; the same shear will be produced by a negative dislocation moving from the right to left*. The slip process as a result of a screw dislocation is shown in *Figure 6.13*. It must be realized, however, that the dislocation line is more usually of the form shown in *Figure 6.10(a)*, and that slip occurs by the movement of all parts of the dislocation loop, i.e. edge, screw and mixed components, as shown in *Figure 6.14*.

A dislocation is able to glide in that slip plane which contains both the line of the dislocation and its Burgers vector. The edge dislocation is confined to glide in

* An obvious analogy to the slip process is the 'shunting' movement of a line of railway trucks or the movement of a caterpillar in the garden.

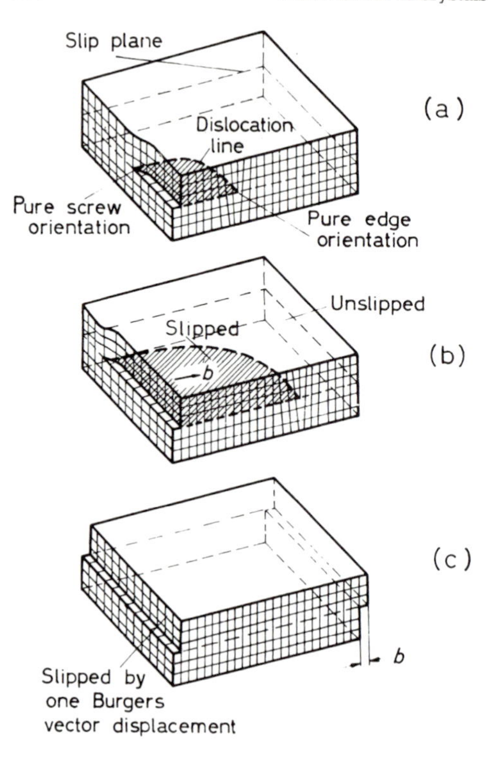

Figure 6.14 Process of slip by the expansion of a dislocation loop in the slip plane

one plane only. An important difference between the motion of a screw dislocation and that of an edge dislocation arises from the fact that the screw dislocation is cylindrically symmetrical about its axis with its b parallel to this axis. To a screw dislocation all crystal planes passing through the axis look the same and, therefore, the motion of the screw dislocation is not restricted to a single slip plane, as is the case for a gliding edge dislocation. The process whereby a screw dislocation glides into another slip plane having a slip direction in common with the original slip plane, as shown in *Figure 6.15(a)*, is called cross-slip. Usually, the cross slip plane is also a close-packed plane, e.g. $\{111\}$ in f.c.c. crystals, and is suitable stressed.

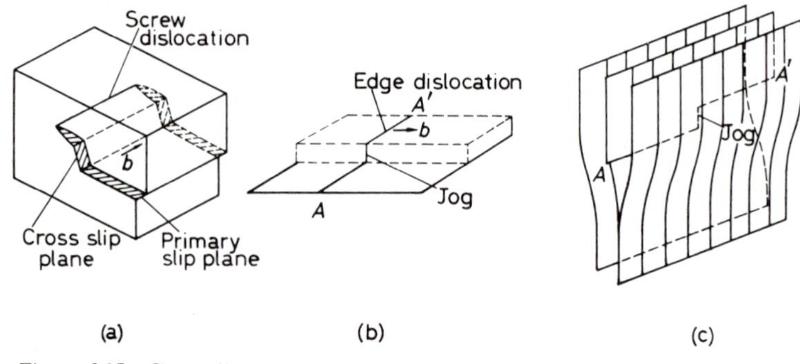

Figure 6.15 Cross-slip (a), and climb (b) and (c), in a crystal

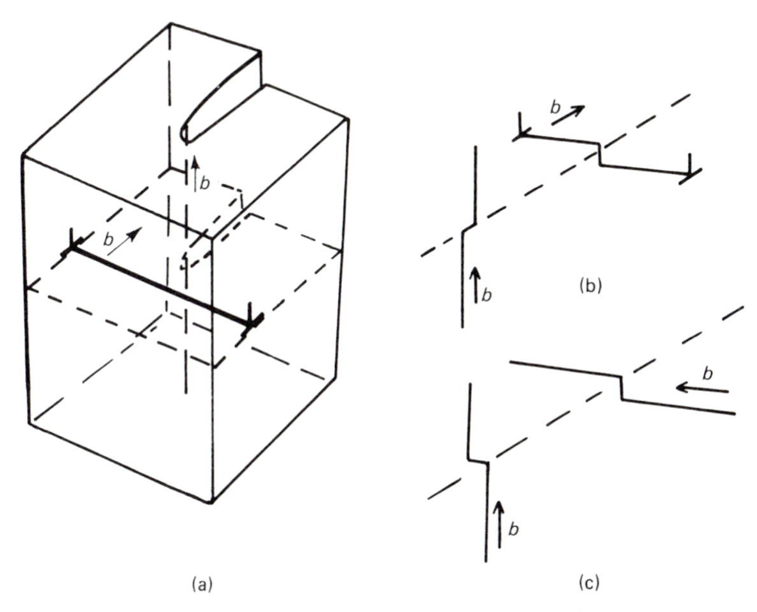

Figure 6.16 Dislocation intersections (a)–(b) screw-edge, (c) screw–screw

The mechanism of slip illustrated above shows that the glide motion of an edge dislocation is restricted, since it can only glide in that slip plane which contains both the dislocation line and its Burgers vector; this is a conservative motion. However, movement of the dislocation line in a direction normal to the slip plane can occur under certain conditions; this is called dislocation climb. To move the extra half-plane either up or down, as is required for climb, requires mass transport by diffusion and is a non-conservative motion (*see* page 131). For example, if vacancies diffuse to the dislocation line it climbs up and the extra half-plane will shorten. However, since the vacancies will not necessarily arrive at the dislocation at the same instant, or uniformly, the dislocation climbs one atom at a time. Moreover, in general, the dislocation will not lie in a particular slip plane but instead some sections will lie in one plane and other sections in parallel neighbouring planes. Where the dislocation jumps from one plane to another is known as jog, and from the diagrams of *Figure 6.15*(*b*) and (*c*) it is evident that a jog in a dislocation may be regarded as a short length of dislocation not lying in the same slip plane as the main dislocation but having the same Burgers vector.

If the jog is formed by thermal activation the number of jogs per unit length of a dislocation which exists in equilibrium at a given temperature will be given by the exponential relation

$$n = A \exp[-E_j/\boldsymbol{k}T]$$

where E_j is the energy required to form a jog. Jogs may also be present, however, because the moving dislocation has cut through intersecting dislocations, i.e. forest dislocations, during its glide motion, and in the lower range of temperature this process will be the major source of jogs. Two examples of jogs formed from the crossings of dislocations are shown in *Figure 6.16*. *Figure 6.16*(*a*) shows a crystal containing a screw dislocation running from top to bottom which has the effect of 'ramping' all the planes in the crystal. If an edge dislocation moves through the crystal on a horizontal plane then the screw dislocation becomes jogged as the top half of the crystal is sheared relative to the bottom. In addition, the screw dislocation becomes jogged since one part has to take the upper ramp and the other part the lower ramp. The result is shown schematically in *Figure 6.16*(*b*). *Figure 6.16*(*c*) shows the situation for a moving screw cutting through the vertical screw; the jog formed in each dislocation is edge in character since it is perpendicular to its Burgers vector which lies along the screw axis.

A jog in an edge dislocation will not impede the motion of the dislocation in its slip plane because it can, in general, move with the main dislocation line by glide, not in the same slip plane (*see Figure 6.16*(*b*)) but in an intersecting slip plane that contains the line of the jog and the Burgers vectors. In the case of a jog in a screw dislocation the situation is not so clear, since there are two ways in which the jog can move. Since the jog is merely a small piece of edge dislocation it may move sideways, i.e. conservatively, along the screw dislocation and attach itself to an edge component of the dislocation line. Conversely, the jog may be dragged along with the screw dislocations. This latter process requires the jog to climb and, because it is a non-conservative process, must give rise to the creation of a row of point defects, i.e. either vacancies or interstitials depending on which way the jog is forced to climb. Clearly, such a movement is difficult but, nevertheless, may be necessary to give the dislocation sufficient mobility. The 'frictional' drag of jogs will make a small contribution to the work hardening of metals (*see* Chapter 10).

Apart from elementary jogs, or those having a height equal to one atomic plane spacing, it is possible to have composite jogs where the jog height is several atomic plane spacings. Such jogs can be produced, for example, by part of a screw dislocation cross-slipping from the primary plane to the cross-slip plane, as shown in *Figure 6.17*(*a*). In this case, as the screw dislocation glides forward it trails the composite jog behind, since it acts as a frictional drag. As a result, two parallel dislocations of opposite sign are created in the wake of the moving screw, as shown in *Figure 6.17*(*b*); this arrangement is called a dislocation dipole. Dipoles formed as debris behind moving dislocations are frequently seen in electron micrographs taken from deformed crystals and examples are shown in *Figures 9.18* and *10.5*.

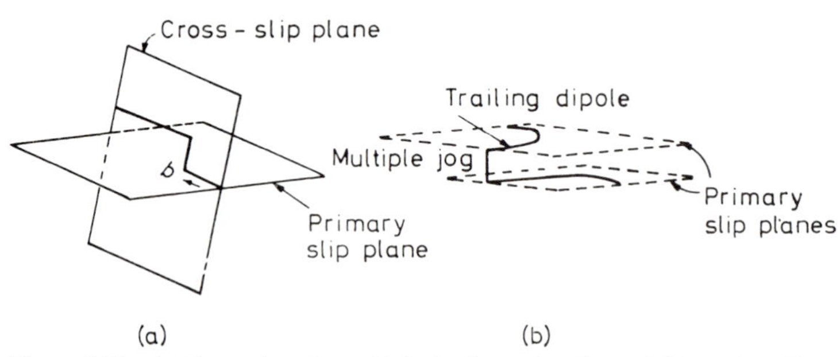

Figure 6.17 (a) Formation of a multiple jog by cross-slip, and (b) motion of jog to produce a dipole

6.2.3 Elastic properties of dislocations

That there is distortion around a dislocation line is evident from an examination of *Figures 6.12* and *6.13*. At the centre of the dislocation the strains are too large to be treated by elasticity theory, but beyond a distance r_0, equal to a few atom spacings, Hooke's law can be applied. It is therefore necessary to define a core to the dislocation at a cut-off radius $r_0(\approx b)$ inside which elasticity theory is no longer applicable. A screw dislocation can then be considered as a cylindrical shell of length l and radius r contained in an elastically isotropic medium (*Figure 6.18*). A discontinuity in displacement exists only in the z-direction, i.e. parallel to the dislocation*, such that $u=v=0$, $w=b$. The elastic strain thus has to accommodate a displacement $w=b$ around a length $2\pi r$. In an elastically isotropic crystal the accommodation must occur equally all round the shell and indicates the simple relation $w=b\theta/2\pi$ in polar (r, θ, z) co-ordinates. The corresponding (shear strain $\gamma_{\mathrm{h}z}(=\gamma_{z\theta})=b/2\pi r$ and shear stress $\tau_{\theta z}(=\tau_{z\theta})=ub/2\pi r$ *which acts on the end faces of the cylinder with* σ_{rr} and $\tau_{r\theta}$ equal to zero. Alternatively, the stresses are given in cartesian co-ordinates (x, y, z)

$$\tau_{xz}(=\tau_{zx})=-\mu by/2\pi(x^2+y^2)$$

$$\tau_{yz}(=\tau_{zy})=-\mu bx/2\pi(x^2+y^2) \qquad (6.3)$$

with all other stresses equal to zero. The field of a screw dislocation is therefore purely one of shear, having radial symmetry with no dependence on θ. This mathematical description is related to the structure of a screw which has no extra half-plane of atoms and cannot be identified with a particular slip plane.

An edge dislocation has a more complicated stress and strain field than a screw. The distortion associated with the edge dislocation is one of plane strain, since there are no displacements along the z-axis, i.e. $w=0$. In plane deformation the only stresses to be determined are the normal stresses σ_x, σ_y along the x- and y-axes respectively, and the shear stress τ_{xy} which acts in the direction of the y-axis on planes perpendicular to the x-axis. The third normal stress $\sigma_z=v(\sigma_x+\sigma_y)$ where v is Poisson's ratio, and the other shear stresses τ_{yz} and τ_{zx} are zero. In polar co-ordinates r, θ and z, the stresses are σ_{rr}, $\sigma_{\theta\theta}$, and $\tau_{r\theta}$.

Even in the case of the edge dislocation the displacement b has to be accommodated round a ring of length $2\pi r$, so that the strains and the stresses must contain a term in $b/2\pi r$. Moreover, because the atoms in the region $0<\theta<\pi$ are under compression and for $\pi<\theta<2\pi$ in tension, the strain field must be of the form $(b/2\pi r)f(\theta)$, where $f(\theta)$ is a function such as $\sin\theta$ which changes sign when θ

* This is termed anti-plane strain.

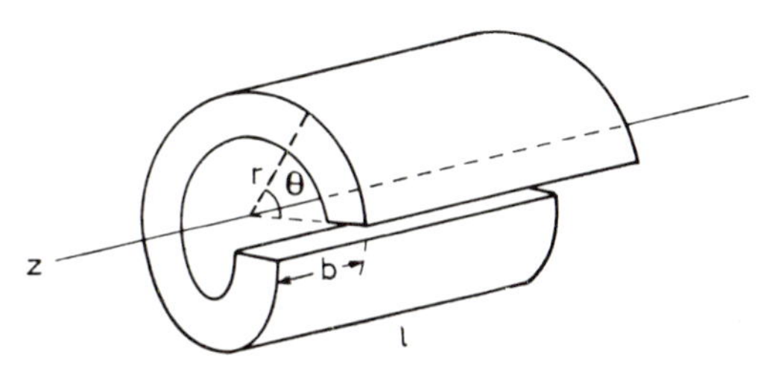

Figure 6.18 Screw dislocation in an elastic continuum

changes from 0 to 2π. It can be shown that the stresses are given by

$$\sigma_{rr}=\sigma_{\theta\theta}=-D\sin\theta/r;\quad \sigma_{r\theta}=D\cos\theta/r;$$

$$\sigma_x=-D\frac{y(3x^2+y^2)}{(x^2+y^2)^2};\quad \sigma_y=D\frac{y(x^2-y^2)}{(x^2+y^2)^2};\quad \tau_{xy}=D\frac{x(x^2-y^2)}{(x^2+y^2)^2}$$

where $D=\mu b/2\pi(1-\nu)$. These equations show that the stresses around dislocations fall off as $1/r$ and hence the stress field is long range in nature.

Strain energy of a dislocation

A dislocation is a line defect extending over large distances in the crystal and, because of its great length, it possesses a total strain energy. An estimate of the elastic strain energy of screw dislocation can be obtained by taking the strain energy (i.e. $\frac{1}{2}\times$stress$\times$strain per unit volume) in an annular ring around the dislocation of radius r and thickness $\mathrm{d}r$ to be $\frac{1}{2}\times(\mu b/2\pi r)\times(b/2\pi r)\times 2\pi r\,\mathrm{d}r$. The total strain energy per unit length of dislocation is then obtained by integrating from r_0 the core radius, to r the outer radius of the stress field, and is

$$E=\frac{\mu b^2}{4\pi}\int_{r_0}^{r}\frac{\mathrm{d}r}{r}=\frac{\mu b^2}{4\pi}\ln\left[\frac{r}{r_0}\right] \qquad (6.4)$$

With an edge dislocation this energy is modified by the term $(1-\nu)$ and hence is about 50 per cent greater than a screw. For a unit dislocation in a typical crystal $r_0\simeq 2.5\times10^{-7}$ mm, $r\simeq 2.5\times10^{-3}$ mm, $\ln[r/r_0]=9.2$, so that the energy is approximately μb^2 per unit length of dislocation, which for copper [taking $\mu=4\times10^{10}$ N/m^2, $b=2.5\times10^{-10}$ m, and $1\,\mathrm{eV}=1.6\times10^{-19}$ J] is about 4 electron volts for every atom plane threaded by the dislocation*. If the reader prefers to think in terms of one metre of dislocation line, then this length is associated with about 2×10^{10} electron volts. We shall see later that heavily deformed metals contain approximately 10^{16} m/m^3 of dislocation line which leads to a large amount of energy stored in the lattice (i.e. ≈ 4 J/g for Cu). Clearly, because of this high line energy a dislocation line will always tend to shorten its length as much as possible, and from this point of view it may be considered to possess a line tension, $T\approx\alpha\mu b^2$, analogous to the surface energy of a soap film, where $\alpha\approx\frac{1}{2}$.

Interaction of dislocations

The strain field around a dislocation, because of its long-range nature, is also important in influencing the behaviour of other dislocations in the crystal. Thus, it is not difficult to imagine that a positive dislocation will attract a negative dislocation lying on the same slip plane in order that their respective strain fields should cancel. Moreover, as a general rule it can be said that the dislocations in a crystal will interact with each other to take up positions of minimum energy to reduce the total strain energy of the lattice.

* The energy of the core must be added to this estimate. The core energy is about $\mu b^2/10$ or $\frac{1}{2}$ eV per atom length.

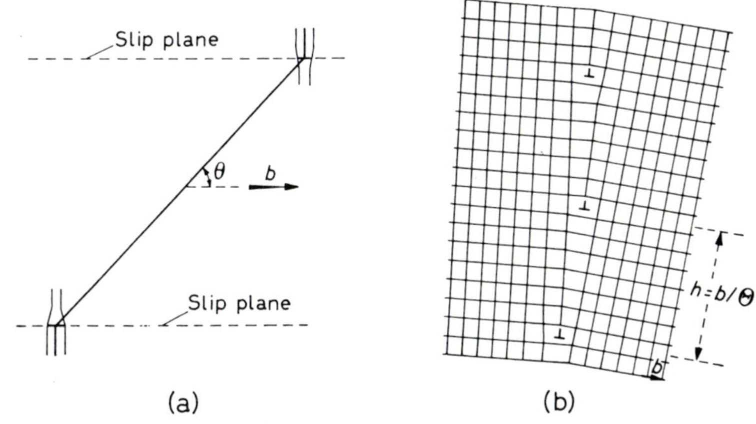

Figure 6.19 Interaction between dislocations not on the same slip plane: (a) unlike dislocation, (b) like dislocations. The arrangement in (b) constitutes a small angle boundary

Two dislocations of the same sign will repel each other, because the strain energy of two dislocations on moving apart would be $2 \times b^2$ whereas if they combined to form one dislocation of Burgers vector $2b$, the strain energy would then be $(2b)^2 = 4b^2$; a force of repulsion exists between them. The force is, by definition, equal to the change of energy with position (dE/dr) and for screw dislocations is simply $F = \mu b^2/2\pi r$ where r is the distance between the two dislocations. Since the stress field around screw dislocations has cylindrical symmetry the force of interaction depends only on the distance apart, and theabove expression for F applies equally well to parallel screw dislocations on neighbouring slip planes. For parallel edge dislocations the force–distance relationship is not quite so simple. When the two edge dislocations lie in the same slip plane the relation is similar to that for two screws and has the form $F = \mu b^2/(1-\nu)2\pi r$, but for edge dislocations with the same Burgers vector but not on the same slip plane the force also depends on the angle θ between the Burgers vector and the vector distance joining the two dislocations (*Figure 6.19(a)*). Taylor's rule still applies and edge dislocations of the same sign repel and of opposite sign attract along the line between them, but the component of force in the direction of slip, which governs the motion of a dislocation, varies with the angle θ. With unlike dislocations an attractive force is experienced for $\theta > 45°$ but a repulsive force for $\theta < 45°$, and in equilibrium the dislocations remain at an angle of 45° to each other. For like dislocations the converse applies and the position $\theta = 45°$ is now one of unstable equilibrium. Thus, edge dislocations which have the same Burgers vector but which do not lie on the same slip plane will be in equilibrium when $\theta = 90°$, and consequently they will arrange themselves in a plane normal to the slip plane, one above the other a distance h apart. Such a wall of dislocations constitutes a small-angle grain boundary as shown in *Figure 6.19(b)* where the angle across the boundary is given by $\Theta = b/h$. This type of dislocation array is also called a sub-grain or polygonization boundary, and is important in the annealing of deformed metals.

By this arrangement the long range stresses from the individual dislocations are cancelled out beyond a distance of the order of h from the boundary. It then follows that the energy of the crystal boundary will be given approximately by the sum of the individual energies, each equal to $\{\mu b^2/4\pi(1-\nu)\ln(h/r_0)$ per unit length. There are $1/h$ or θ/b dislocations in a unit length, vertically, and hence, in terms of the misorientation across the boundary $\theta = b/h$, the energy $\gamma_{g.b}$ per unit area of boundary is

$$\gamma_{gb} = \frac{\mu b^2}{4\pi(1-\nu)}\ln\left(\frac{h}{r_0}\right)\times\frac{\theta}{b} = \left[\frac{\mu b}{4\pi(1-\nu)}\right]\theta\ln\left(\frac{b}{\theta r}\right) = E_0\theta[A - \ln\theta] \qquad (6.5)$$

where $E_0 = \mu b/4\pi(1-\nu)$ and $A = \ln(b/r_0)$; this is known as the Read–Shockley formula. Values from it give good agreement with experimental estimates even up to relatively large angles. For $\theta \sim 25^\circ$, $\gamma_{gb} \sim \mu b/25$ or $\sim 0.4\,\mathrm{J/m^2}$, which surprisingly is close to the value for a general large-angle grain boundary.

6.2.4 Imperfect dislocations

The Burgers vector is the most important property of a dislocation, since it defines the atomic displacement produced as the dislocation moves through the lattice. Its value is governed by the crystal structure because during slip it is necessary to retain an identical and mechanically stable lattice structure both before and after the passage of the dislocation. Clearly, such a stable lattice arrangement is assured if the dislocation has a Burgers vector equal to one lattice vector and, since the energy of a dislocation depends on the square of the Burgers vector, such a unit, or perfect, dislocation has the lowest energy when its Burgers vector is equal to the shortest available lattice vector. This vector, by definition, is parallel to the direction of closest packing in the structure, which agrees with experimental observations of the slip direction.

The Burgers vector is conveniently specified by its components along the principal crystal axes. In the f.c.c. lattice the shortest lattice vector is associated with slip from a cube corner to a face centre, and has components $a/2, a/2, 0$. This is usually written $\frac{1}{2}a[110]$, where a is the lattice parameter and $[110]$ is the slip direction. The magnitude of the vector, or the strength of the dislocation, is then given by $\sqrt{\{a^2(1^2+1^2+0^2)/4\}} = a/\sqrt{2}$. The corresponding slip vectors for the b.c.c. and c.p.h. structures are $b = \frac{1}{2}a[111]$ and $b = \frac{1}{3}a[11\bar{2}0]$ respectively.

In certain other processes of crystal plasticity the Burgers vector of the dislocation may not be equal to a whole lattice vector; the dislocation is then referred to as an imperfect, partial, or half-dislocation. One obvious example where partial dislocations are important is in the dislocation mechanism of twinning, since it is clear from *Figure 6.3(b)* that the vectors describing the atom movements are not whole lattice vectors. This mechanism is described in more detail in Chapter 8. Partial dislocations may also be important during slip where conditions are such that it is easier for a pair of dislocations, each associated with less distortion than one unit dislocation, to pass through the lattice. The pair of partial dislocations must glide together as a unit dislocation in order to maintain the lattice identity, and such a combination is referred to as an extended dislocation. The lattice left in the wake of the first moving partial dislocation is not identical with that ahead of it, and the lattice disorder or stacking fault which exists across the slip plane is not corrected until after the passage of the second half of the extended dislocation. The reader will appreciate this process more fully after reading the next section.

6.3 Dislocations in close-packed crystals

6.3.1 Extended dislocations

The relationship between the two close-packed structures c.p.h. and f.c.c., has been discussed in Chapter 1, where it was seen that both structures may be built up from stacking close-packed planes of spheres. The shortest lattice vector in the f.c.c. structure joins a cube corner atom to a neighbouring face centre atom and defines the observed slip direction; one such slip vector $\frac{1}{2}a[10\bar{1}]$ is shown as b_1 in *Figure 6.20(a)* which is for glide in the (111) plane. However, an atom which

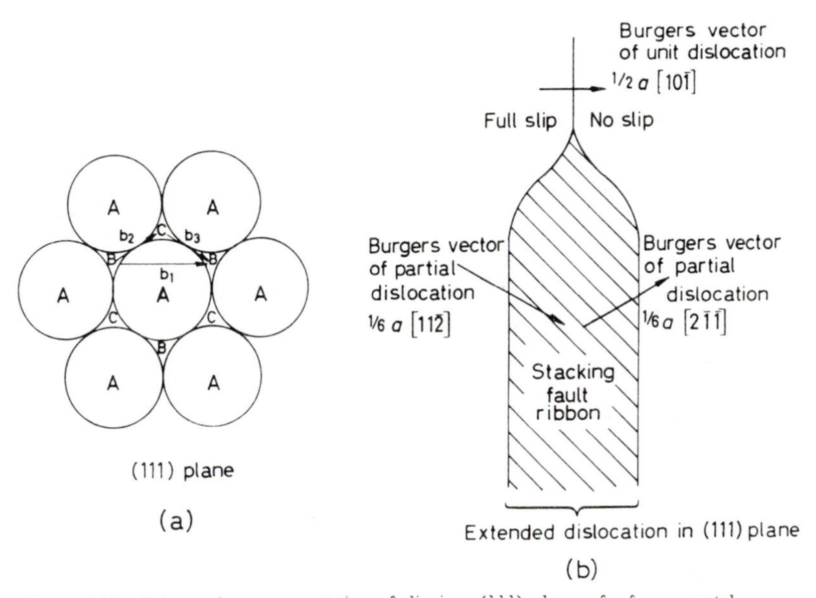

Figure 6.20 Schematic representation of slip in a (111) plane of a f.c.c. crystal

sits in a B position on top of the A plane would move most easily initially towards a C position and, consequently, to produce a macroscopical slip movement along $[10\bar{1}]$ the atoms might be expected to take a zig-zag path of the type B→C→B following the vectors $b_2=\frac{1}{6}a[2\bar{1}\bar{1}]$ and $b_3=\frac{1}{6}a[11\bar{2}]$ alternately. It will be evident, of course, that during the initial part of the slip process when the atoms change from B positions to C positions, a stacking fault in the (111) layers is produced and the stacking sequence changes from ABCABC... to ABCACABC.... During the second part of the slip process the correct stacking sequence is restored.

To describe the atom movements during slip, discussed above, Heidenreich and Shockley have pointed out that the unit dislocation must dissociate into two half dislocations*, which for the case of glide in the (111) plane would be according to the reaction:

$$\tfrac{1}{2}a[10\bar{1}] \rightarrow \tfrac{1}{6}a[2\bar{1}\bar{1}] + \tfrac{1}{6}a[11\bar{2}]$$

Such a dissociation process is (i) algebraically correct, since the sum of the Burgers vector components of the two partial dislocations, i.e. $a/6[2+1]$, $a/6[\bar{1}+1]$, $a/6[\bar{1}+\bar{2}]$, are equal to the components of the Burgers vector of the unit dislocation, i.e. $a/2$, 0, $\bar{a}/2$, and (ii) energetically favourable, since the sum of the strain energy values for the pair of half dislocations is less than the strain energy value of the single unit dislocation, where the initial dislocation energy is proportional to b_1^2 $(=a^2/2)$ and the energy of the resultant partials to $b_2^2+b_3^2=a^2/3$. These half dislocations, or Shockley partial dislocations, repel each other by a force that is approximately $F=(\mu b_2 b_3 \cos 60)/2\pi d$, and separate as shown in *Figure 6.20*(*b*). A sheet of stacking faults is then formed in the slip plane between the partials, and it is the creation of this faulted region, which has a higher energy than the normal lattice, that prevents the partials from separating too far. Thus, if γ J/m^2 is the energy per unit area of the fault, the force per unit length exerted on the dislocations by the fault is γ N/m and the equilibrium separation d is given by equating the repulsive force F between the two half dislocations, $\mu a^2/24\pi d$, to the force exerted by the fault, γ. The equilibrium separation of two partial dislocations is then given by

$$d=\mu a^2/24\pi\gamma \tag{6.6}$$

from which it can be seen that the width of the stacking fault 'ribbon' is inversely proportional to the value of the stacking fault energy γ and also depends on the value of the shear modulus μ.

Figure 6.21 shows that the undissociated edge dislocation has its extra half plane corrugated (*Figure 6.21*(*a*)) which may be considered as two $(10\bar{1})$ planes displaced relative to each other and labelled *a* and *b* in *Figure 6.21*(*b*). On dissociation, planes *a* and *b* are separated by a region of crystal in which across the slip plane the atoms are in the wrong sites (*see Figure 6.21*(*c*)). Thus the high strain energy along a line through the crystal associated with an undissociated dislocation is spread over a plane in the crystal for a dissociated dislocation (*see Figure 6.21*(*d*)) thereby lowering its energy.

A direct estimate of γ can be made from the observation of extended dislocations in the electron microscope and from observations on other stacking fault defects (*see* Chapter 9). Such measurements show that the stacking fault energy for pure f.c.c. metals ranges from about 16 mJ/m^2 for silver to $\approx$200 mJ/m^2 for nickel, with gold $\approx$30, copper $\approx$40 and aluminium 135 mJ/m^2 respectively. Since stacking faults are coherent interfaces or boundaries they have energies considerably lower than non-coherent interfaces such as free surfaces for

* The correct indices for the vectors involved in such dislocation reactions can be obtained from *Figure 6.27*.

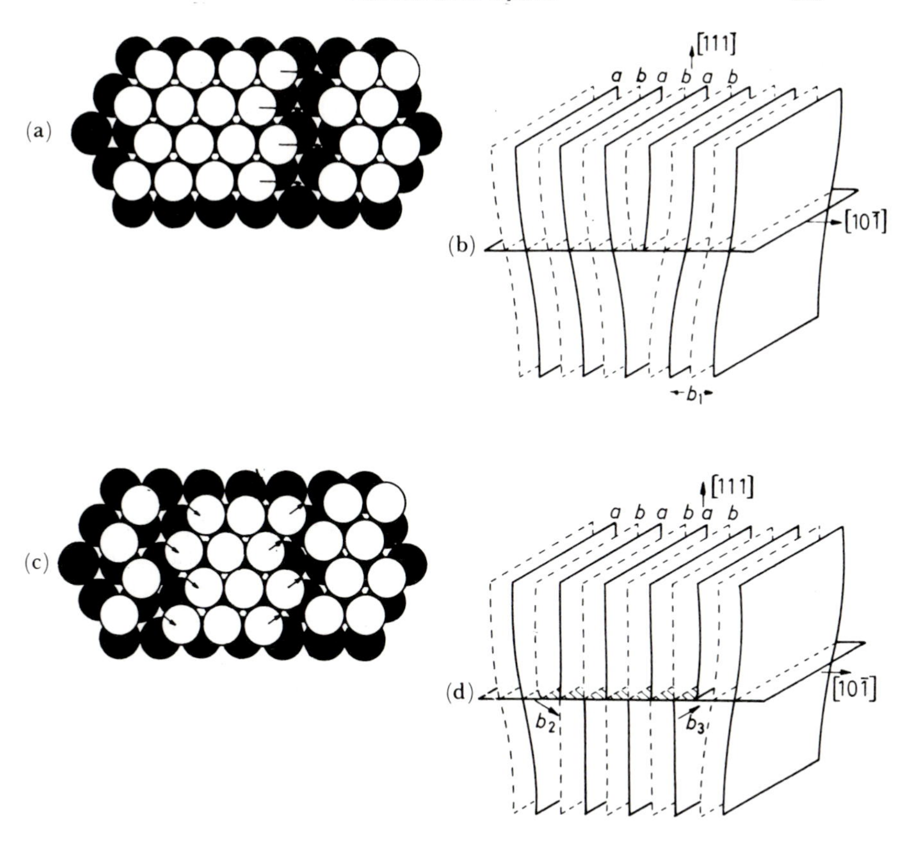

Figure 6.21 Edge dislocation structure in the f.c.c. lattice (a) and (b) undissociated, (c) and (d) dissociated; (a) and (c) are viewed normal to the (111) plane (From Hume-Rothery, Smallman and Haworth, *Structure of Metals and Alloys*, Monograph No. 1, 1969, courtesy of the Institute of Metals)

which $\gamma_s \approx \mu b/8 \approx 1.5\,\text{J/m}^2$ and grain boundaries for which $\gamma_{gb} \approx \gamma_s/3 \approx 0.5\,\text{J/m}^2$.

The energy of a stacking fault can be estimated from twin boundary energies since a stacking fault ABCBCABC may be regarded as two overlapping twin boundaries CBC and BCB across which the next nearest neighbouring planes are wrongly stacked. In f.c.c. crystals any sequence of three atomic planes not in the ABC or CBA order is a stacking violation and is accompanied by an increased energy contribution. A twin has one pair of second nearest neighbour planes in the wrong sequence, two third neighbours, one fourth neighbour and so on; an intrinsic stacking fault two second nearest neighbours, three third and no fourth nearest neighbour violations. Thus, if next-next nearest neighbour interactions are considered to make a relatively small contribution to the energy then an approximate relation $\gamma \simeq 2\gamma_T$ is expected. The free energies of interfaces can be determined from the equilibrium form of the triple junction where three interfaces, such as surfaces, grain boundaries or twins, meet. For the case of a

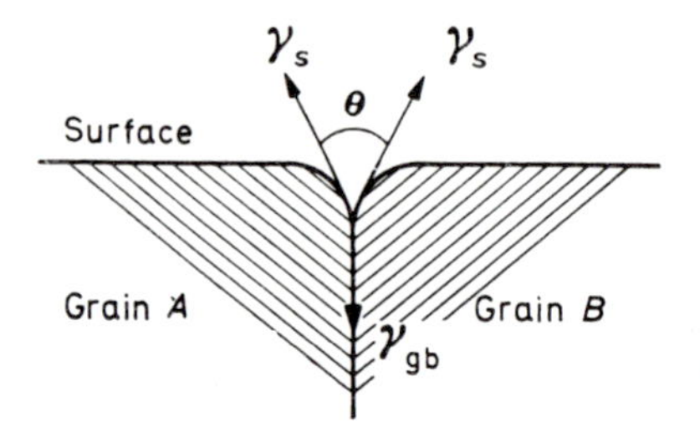

Figure 6.22 Grain boundary/surface triple function

grain boundary intersecting a free surface, shown in *Figure 6.22*,

$$\gamma_{gb} = 2\gamma_s \cos \theta/2$$

and hence γ_{gb} can be obtained by measuring the dihedral angle θ and knowing γ_s. Similarly, measurements can be made of the ratio of twin boundary energy to the average grain boundary energy and, knowing either γ_s or γ_{gb} gives an estimate of γ_T and hence γ. The frequency of occurrence of annealing twins generally confirms the above classification of stacking fault energy and it is interesting to note that in aluminium, a metal with a relatively high value of γ, annealing twins are rarely, if ever, observed, while they are seen in copper which has a lower stacking fault energy. Electron microscope measurements of γ show that the stacking fault energy is lowered by solid solution alloying and is influenced by those factors which affect the limit of primary solubility.

On alloying, the free energies of the α-phase and its neighbouring phase become more nearly equal, i.e. the stability of the α-phase is decreased relative to some other phase, and hence can more readily tolerate mis-stacking.

Figure 6.23 shows the reduction of γ for copper with addition of solutes such as Zn, Al, Si and Ge, and is consistent with the observation that annealing twins occur more frequently in α-brass or Cu–Si than pure copper. Substituting the appropriate values for μ, a and γ in equation 6.6 indicates that in silver and copper the partials are separated to about 12 and 6 atom spacings respectively. For nickel the width is about $2b$ since although nickel has a high γ its shear modulus is also very high. In contrast, aluminium has a lower $\gamma \approx 135\,\text{mJ/m}^2$ but also a considerably lower value for μ and hence the partial separation is limited to about $1b$ and may be considered to be unextended. By alloying, the important

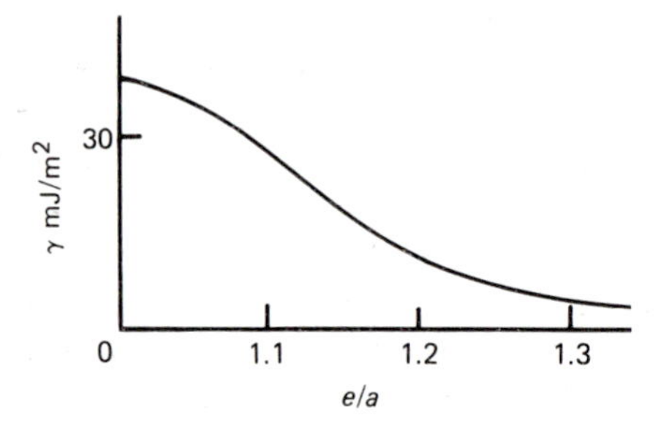

Figure 6.23 Decrease in stacking fault energy γ for copper with alloying addition (e/a)

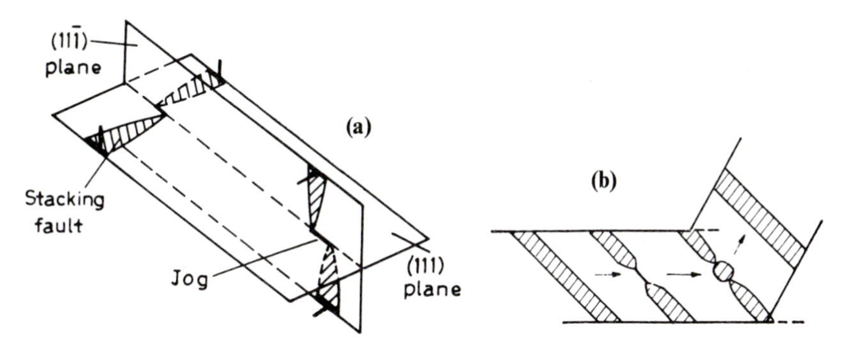

Figure 6.24 (a) The crossing of extended dislocations. (b) Various stages in the cross-slip of a dissociated screw dislocation

dimensionless parameter $\gamma/\mu b$ can be significantly altered and very wide dislocations produced, as found in the brasses, bronzes and austenitic stainless steels. However, no matter how narrow or how wide the partials are separated the two half dislocations are bound together by the stacking fault, and consequently, they must move together as a unit across the slip plane.

The width of the stacking fault ribbon is of importance in many aspects of plasticity because in some stage of deformation it becomes necessary for dislocations to intersect each other; the difficulty which dislocations have in intersecting each other gives rise to one source of work-hardening. With extended dislocations the intersecting process is particularly difficult since the crossing of stacking faults would lead to a complex fault in the plane of intersection. The complexity may be reduced, however, if the half-dislocations coalesce at the crossing point, so that they intersect as perfect dislocations; the partials then being constricted together at their jogs, as shown in *Figure 6.24(a)*. In low stacking fault energy materials the jogs themselves, particularly multiple jogs, may be dissociated. The structure of such dissociate jogs is discussed in Chapter 10.

The width of the stacking fault ribbon is also important to the phenomenon of cross-slip in which a dislocation changes from one slip plane to another intersecting slip plane. As discussed previously, for glide to occur the slip plane must contain both the Burgers vector and the line of the dislocation, and, consequently, for cross-slip to take place a dislocation must be in an exact screw orientation. If the dislocation is extended, however, the partials have first to be brought together to form an unextended dislocation as shown in *Figure 6.24(b)*, before the dislocation can spread into the cross-slip plane. The constriction process will be aided by thermal activation and hence the cross-slip tendency increases with increasing temperature. The constriction process is also more difficult the wider the separation of the partials. In aluminium, where the dislocations are relatively unextended, the frequent occurrence of cross-slip is expected, but for low stacking fault energy metals, e.g. copper or gold, the activation energy for the process will be high. Nevertheless, cross-slip may still occur in those regions where a high concentration of stress exists, as for example when dislocations pile up against some obstacle, where the width of the extended dislocation may be reduced below the equilibrium separation. Often in a piled-up group of dislocations in f.c.c. crystals, screw dislocations escape from the group by gliding in an intersecting plane, but then after moving a certain distance in this cross-slip plane they act in a glide plane parallel to the original slip plane; this phenomenon is often termed double cross-slip.

6.3.2 Sessile dislocations

The Shockley partial dislocation has its Burgers vector lying in the plane of the fault and hence is glissile. Some dislocations however, have their Burgers vector not lying in the plane of the fault with which they are associated, and are incapable of gliding, i.e. they are sessile. The simplest of these, the Frank sessile dislocation loop, is shown in *Figure 6.25*(*a*). This dislocation is believed to form as a result of the collapse of the lattice surrounding a cavity which has been produced by the aggregation of vacancies on to a (111) plane (*see* Chapter 9). As shown in *Figure 6.25*(*a*), if the vacancies aggregate on the central A-plane the adjoining parts of the neighbouring B and C planes collapse to fit in close-packed formation. The Burgers vector of the dislocation line bounding the collapsed sheet is normal to the plane with $b=\frac{1}{3}a[111]$, where a is the lattice parameter, and such a dislocation is sessile since it encloses an area of stacking fault which cannot move with the dislocation. A Frank sessile dislocation loop can also be produced by inserting an extra layer of atoms between two normal planes of atoms, as occurs when interstitial atoms aggregate following high energy particle irradiation (*see* Chapter 9). For the loop formed from vacancies the stacking sequence changes from the normal ABCABCA... to ABCBCA..., whereas inserting a layer of atoms, e.g. an A-layer between B and C, the sequence becomes ABCABACA.... The former type of fault with one violation in the stacking sequence is called an intrinsic fault, the latter with two violations is called an extrinsic fault. The stacking sequence violations are conveniently shown by using the symbol Δ to denote any normal stacking sequence AB, BC, CA but ∇ for the

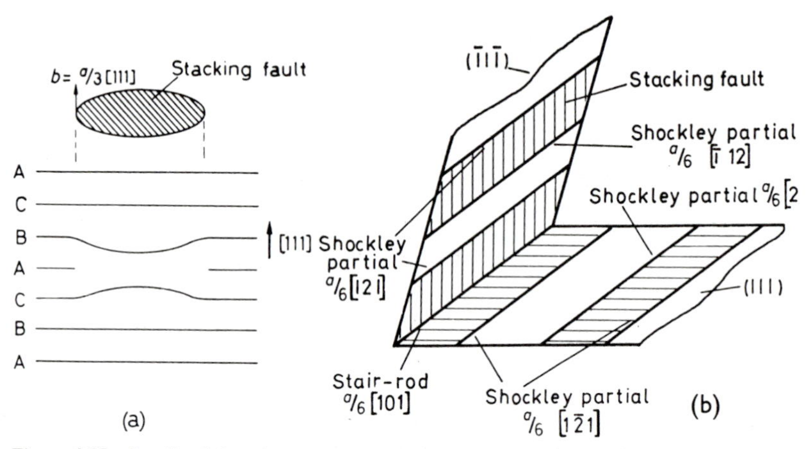

Figure 6.25 Sessile dislocations: (a) a Frank sessile dislocation; (b) stair-rod dislocation as part of a Lomer–Cottrell barrier

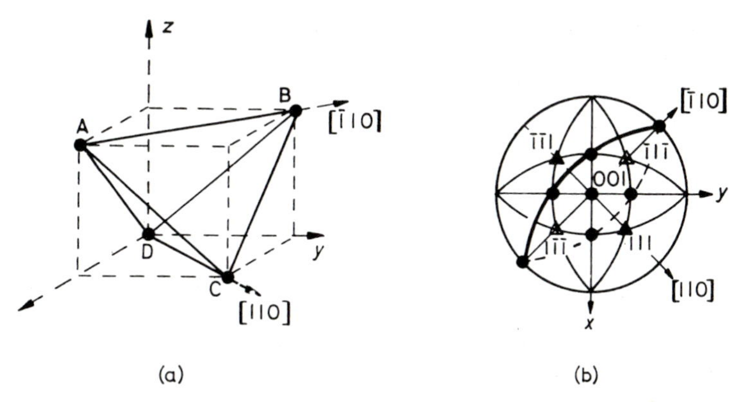

Figure 6.26 (a) Construction and (b) orientation of the Thompson tetrahedron ABCD. The slip directions in a given {111} plane may be obtained from the trace of that plane as shown for the (111) plane in (b)

reverse sequence AC, CB, BA. The normal f.c.c. stacking sequence is then given by ΔΔΔΔ..., the intrinsic fault by ΔΔ∇ΔΔ... and the extrinsic fault by ΔΔ∇∇ΔΔ.... The reader may verify that the fault discussed in the previous section is also an intrinsic fault, and that a series of intrinsic stacking faults on neighbouring planes gives rise to a twinned structure ABCABACBA or ΔΔΔΔ∇∇∇∇Δ. Electron micrographs of Frank sessile dislocation loops are shown in *Figures 9.7* and *9.8*.

Another common obstacle is that formed between extended dislocations on intersecting {111} slip planes, as shown in *Figure 6.25(b)*. Here, the combination of the leading partial dislocation lying in the (111) plane with that which lies in the ($\bar{1}1\bar{1}$) plane forms another partial dislocation, often referred to as a 'stair-rod' dislocation, at the junction of the two stacking fault ribbons by the reaction

$$\tfrac{1}{6}a[\bar{1}12]+\tfrac{1}{6}a[2\bar{1}\bar{1}]\rightarrow\tfrac{1}{6}a[101]$$

The indices for this reaction can be obtained from *Figure 6.27*, and it is seen that there is a reduction in energy from $[(a^2/6)+(a^2/6)]$ to $a^2/18$. This triangular group of partial dislocations, which bounds the wedge-shaped stacking fault ribbon lying in a $\langle 101\rangle$ direction, is obviously incapable of gliding and such an obstacle, first considered by Lomer and Cottrell, is known as a Lomer–Cottrell barrier.

6.3.3 The Thompson reference tetrahedron

All the dislocations common to the f.c.c. structure, which we have discussed in the previous sections, can be represented conveniently by means of the Thompson reference tetrahedron (*Figure 6.26(a)*), formed by joining the three nearest face-centering atoms to the origin D. Here ABCD is made up of four {111} planes (111), ($\bar{1}\bar{1}1$), ($\bar{1}1\bar{1}$) and ($1\bar{1}\bar{1}$) as shown by the stereogram given in *Figure 6.26(b)*, and the edges AB, BC, CA... correspond to the $\langle 110\rangle$ directions in these planes. Then, if the mid-points of the faces are labelled, α, β, γ, δ, as shown in *Figure 6.27(a)*, all the dislocation Burgers vectors are represented. Thus, the edges (AB, BC...) correspond to the normal slip vectors, $\tfrac{1}{2}a\langle 110\rangle$. The half-disloca-

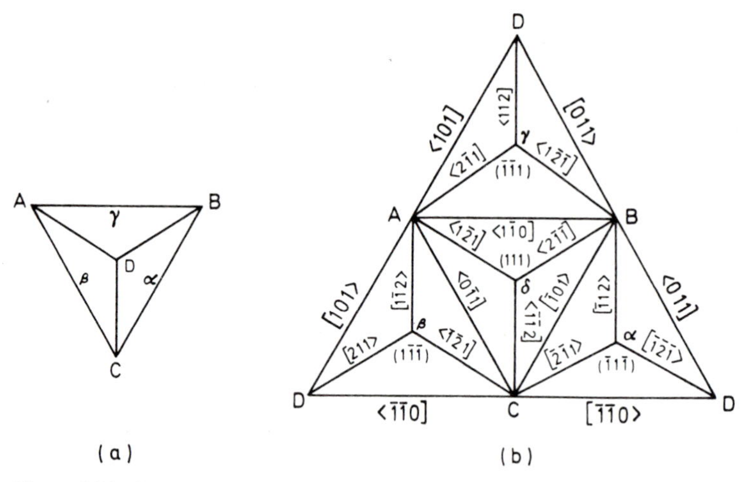

Figure 6.27 A Thompson tetrahedron (a) closed and (b) opened out. In (b) the notation [110> is used in place of the usual notation [110] to indicate the sense of the vector direction

tions, or Shockley partials, into which these are dissociated have Burgers vectors of the $\frac{1}{6}a\langle 112\rangle$ type and are represented by the Roman–Greek symbols Aγ, Bγ, Dγ, Aδ, Bδ, etc, or Greek–Roman symbols γA, γB, γD, δA, δB, etc. The dissociation reaction given in the first reaction on page 214 is then simply written

$$\mathrm{BC} \rightarrow \mathrm{B}\delta + \delta\mathrm{C}$$

and there are six such dissociation reactions in each of the four $\{111\}$ planes. It is conventional to view the slip plane from outside the tetrahedron along the positive direction of the unit dislocation BC, and on dissociation to produce an intrinsic stacking fault arrangement; the Roman–Greek partial Bδ is on the right and the Greek–Roman partial δC on the left. A screw dislocation with Burgers vector BC which is normally dissociated in the δ-plane is capable of cross-slipping into the α-plane by first constricting $\mathrm{B}\delta + \delta\mathrm{C} \rightarrow \mathrm{BC}$ and then re-dissociating in the α-plane $\mathrm{BC} \rightarrow \mathrm{B}\alpha + \alpha\mathrm{C}$.

The Frank partial dislocation has a Burgers vector perpendicular to the (111) plane on which it lies and is represented by Aα, Bβ, Cγ, Dδ, αA, etc. Such a dislocation Aα is capable of combining with a Shockley partial in the α-plane, e.g. αC to form a perfect dislocation AC. Moreover, when the Frank dislocation lies along a $\langle 110\rangle$ direction it can reduce its energy by dissociating on an intersecting $\{111\}$ plane, forming a stair-rod at the junction of the two $\{111\}$ planes, e.g. $\mathrm{A}\alpha \rightarrow \mathrm{A}\delta + \delta\alpha$ when the Frank dislocation lies along $[\bar{1}01]$ common to both α and δ-planes. The stair-rod dislocation formed at the apex of a Lomer–Cottrell barrier can also be represented by the Thompson notation. As an example, let us take the interaction between dislocations on the δ and α-planes. Two unit dislocations BA and DB respectively are dissociated according to

$$\mathrm{BA} \rightarrow \mathrm{B}\delta + \delta\mathrm{A} \text{ (on the } \delta\text{-plane)}$$

and $\mathrm{DB} \rightarrow \mathrm{D}\alpha + \alpha\mathrm{B}$ (on the α-plane) and when the two Shockley partials αB and Bδ interact, a stair-rod dislocation $\alpha\delta = a/6[101]$ is formed. This low energy dislocation is pure edge and therefore sessile. If the other pair of partials interact then the resultant Burgers vector is $(\delta\mathrm{A} + \mathrm{D}\alpha) = a/3[101]$and of higher energy. This vector is written in Thompson's notation as $\delta\mathrm{D}/\mathrm{A}\alpha$ and is a vector equal to twice the length joining the mid-points of δA and Dα. The stair-rod formed in the above reaction lies at an acute bend, it is also possible to have stair-rods at obtuse bends in which case the Burgers vector is either $a/6\langle 110\rangle$, $a/3\langle 100\rangle$ or $a/6\langle 301\rangle$.

6.4 Dislocations in hexagonal structures

In a h.c.p. structure with axial ratio c/a, the most closely packed plane of atoms is the basal plane (0001) and the most closely packed directions $\langle 11\bar{2}0\rangle$. The smallest unit lattice vector is a, but to indicate the direction of the vector $\langle u, v, w\rangle$ in Miller–Bravais indices, it is written as $a/3\langle 11\bar{2}0\rangle$ where the magnitude of the vector in terms of the lattice parameters is given by $a[3(u^2 + uv + v^2) + (c/a)^2 w^2]^{1/2}$. The usual slip dislocation therefore has a Burgers vector $a/3\langle 11\bar{2}0\rangle$ and glides in the (0001) plane. This slip vector is $\langle a/3, a/3, \bar{2}(a/3), 0\rangle$ and has no component along the c-axis and so can be written without difficulty as $a/3\langle 11\bar{2}0\rangle$. However, when the vector has a component along the c-axis, as for example $\langle a/3, a/3, \bar{2}(a/3), 3c\rangle$, difficulty arises and to avoid confusion the vectors are referred to unit distances (a, a, a, c) along the respective axes, e.g. $\frac{1}{3}\langle 11\bar{2}0\rangle$ and $\frac{1}{3}\langle 11\bar{2}3\rangle$. Other dislocations can be represented in a notation similar to that for the f.c.c. structure, but using a double-tetrahedron or bi-pyramid instead of the single tetrahedron previously adopted, as shown in *Figure 6.28*. An examination leads to the following simple types of dislocation:

1. Six perfect dislocations with Burgers vectors in the basal plane along the sides of the triangular base ABC. They are AB, BC, CA, BA, CB and AC and are denoted by a or $\frac{1}{3}\langle 11\bar{2}0\rangle$.

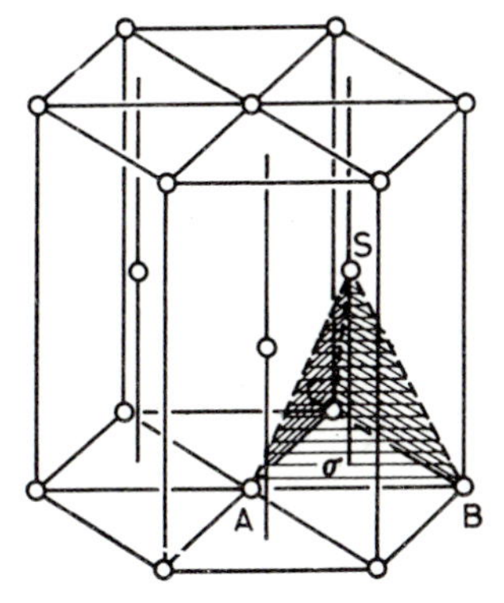

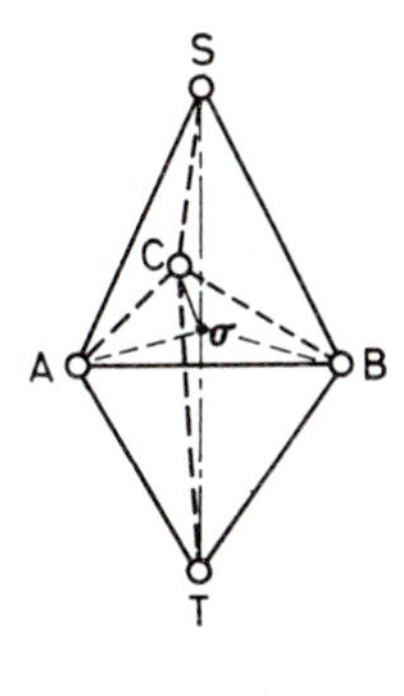

Figure 6.28 Burgers vectors in the h.c.p. lattice (after Berghezan, Fourdeux and Amelinckx, *Acta. Met.*, 1961, **9**, 464, courtesy of Pergamon Press)

2. Six partial dislocations with Burgers vectors in the basal plane represented by the vectors Aσ, Bσ, Cσ and their negatives. These dislocations arise from dissociation reactions of the type

   $$\mathrm{AB} \rightarrow \mathrm{A}\sigma + \sigma\mathrm{B}$$

   and may also be written as p or $\frac{1}{3}\langle 10\bar{1}0\rangle$.
3. Two perfect dislocations perpendicular to the basal plane represented by the vectors ST and TS of magnitude equal to the cell height c or $\langle 0001\rangle$.
4. Partial dislocations perpendicular to the basal plane represented by the vectors σS, σT, Sσ, Tσ of magnitude $c/2$ or $\frac{1}{2}\langle 0001\rangle$.
5. Twelve perfect dislocations of the type $\frac{1}{3}\langle 11\bar{2}3\rangle$ with a Burgers vector represented by SA/TB which is a vector equal to twice the join of the mid-points of SA and TB. These dislocations are more simply referred to as $(c+a)$ dislocations.
6. Twelve partial dislocations, which are a combination of the partial basal and non-basal dislocations, and represented by vectors AS, BS, CS, AT, BT and CT or simply $(c/2)+p$ equal to $\frac{1}{6}\langle 20\bar{2}3\rangle$. Although these vectors represent a displacement from one atomic site to another the resultant dislocations are imperfect because the two sites are not identical.

The energies of the different dislocations are given in a relative scale in *Table 6.2*, assuming c/a is ideal.

TABLE 6.2 Dislocations in the c.p. hexagonal structures

Type	*AB, BC*	*Aσ, Bσ*	*ST, TS*	*σS, σT*	*AS, BS*	*SA/TB*
Vector	$\frac{1}{3}\langle 11\bar{2}0\rangle$	$\frac{1}{3}\langle 10\bar{1}0\rangle$	$\langle 0001\rangle$	$\frac{1}{2}\langle 0001\rangle$	$\frac{1}{6}\langle 20\bar{2}3\rangle$	$\frac{1}{3}\langle 11\bar{2}3\rangle$
Energy	a^2	$a^2/3$	$c^2=8a^2/3$	$2a^2/3$	a^2	$11a^2/3$

There are many similarities between the dislocations in the c.p.h. and f.c.c. structure and thus it is not necessary to discuss them in great detail. It is, however, of interest to consider the two basic processes of glide and climb.

Dislocation glide

A perfect slip dislocation in the basal plane $\mathrm{AB}=\frac{1}{3}[\bar{1}2\bar{1}0]$ may dissociate into two Shockley partial dislocations separating a ribbon of intrinsic stacking fault which violates the two next nearest neighbours in the stacking sequence. There are actually two possible slip sequences, either a B-layer slides over an A-layer, i.e. Aσ followed by σB (*see Figure 6.29(a)*) or an A-layer slides over a B-layer by the passage of a σB partial followed by an Aσ (*see Figure 6.29(b)*). The dissociation given by

$$\mathrm{AB} \rightarrow \mathrm{A}\sigma + \sigma\mathrm{B}$$

may be written in Miller–Bravais indices as

$$\tfrac{1}{3}[\bar{1}2\bar{1}0] \rightarrow \tfrac{1}{3}[01\bar{1}0] + \tfrac{1}{3}[\bar{1}100]$$

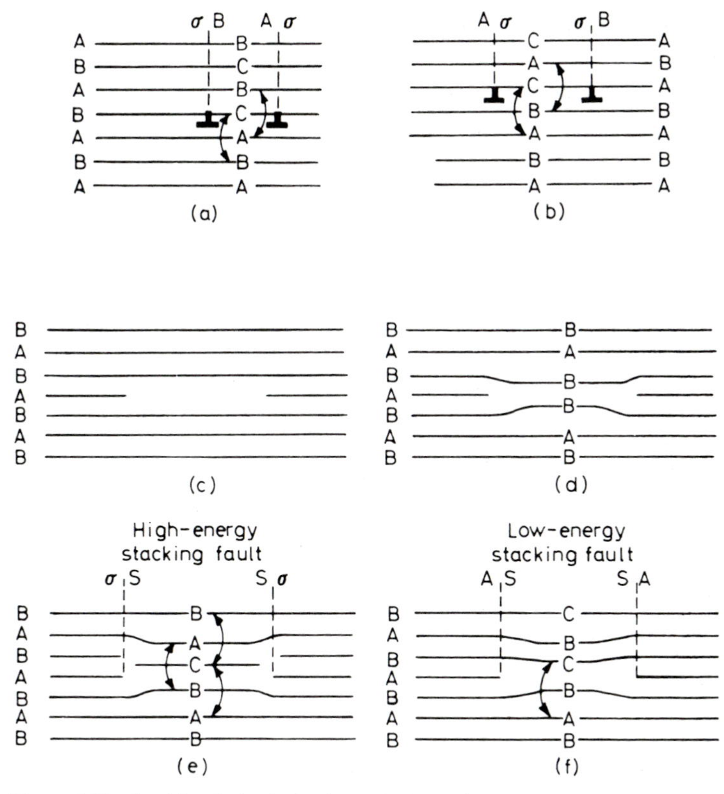

Figure 6.29 Stacking faults in the h.c.p. lattice (after Partridge, *Met. Reviews*, 1967, **118,** 169 by courtesy of the American Society for Metals)

This reaction is similar to that in the f.c.c. lattice and the width of the ribbon is again inversely proportional to the stacking fault energy γ. Dislocations dissociated in the basal planes have been observed in cobalt, which undergoes a phase transformation and for which γ is considered to be low ($\approx 25\,\text{mJ/m}^2$). For the other common c.p.h. metals Zn, Cd, Mg, Ti, Be, etc. γ is high [250–300 mJ/m^2]. No measurements of intrinsic faults with two next nearest neighbour violations have been made, but intrinsic faults with one next nearest neighbour violation (*see below* and Chapter 9) have been measured and show that Mg $\approx 125\,\text{mJ/m}^2$, Zn $\approx 140\,\text{mJ/m}^2$, and Cd ≈ 150–$175\,\text{mJ/m}^2$. It is thus reasonable to conclude that intrinsic faults associated with Shockley partials have somewhat higher energy. Dislocations in these metals are therefore not very widely dissociated. A screw dislocation lying along a $[\bar{1}2\bar{1}0]$ direction is capable of gliding in three different glide planes but the small extension in the basal plane will be sufficient to make basal glide easier than in either the pyramidal $(10\bar{1}1)$ or prismatic $(10\bar{1}0)$ glide, *see Figure 7.1*. Pyramidal and prismatic glide will be more favoured at high temperatures in metals with high stacking fault energy when thermal activation aids the constriction of the dissociated dislocations. It is also observed that non-basal slip is favoured in metals with $c/a < 1.633$.

Dislocation climb

Stacking faults may be produced in hexagonal lattices by the aggregation of point defects. If vacancies aggregate as a platelet, as shown in *Figure 6.29*(*c*), the resultant collapse of the disc-shaped cavity (*Figure 6.29*(*d*)) would bring two similar layers into contact. This is a situation incompatible with the close-packing and suggests that simple Frank dislocations are energetically unfavourable in c.p.h. lattices. This unfavourable situation can be removed by either one of two mechanisms as shown in *Figure 6.29*(*e*) and (*f*). In *Figure 6.29*(*e*) the B-layer is converted to a C-position by passing a pair of equal and opposite partial dislocations (dipole) over adjacent slip planes. The Burgers vector of the dislocation loop will be of the σS type and the energy of the fault, which is extrinsic, will be high because of the three next nearest neighbour violations. In *Figure 6.29*(*f*) the loop is swept by a Aσ-type partial dislocation which changes the stacking of all the layers above the loop according to the rule $A \rightarrow B \rightarrow C \rightarrow A$. The Burgers vector of the loop is of the type AS, and from the dislocation reaction $A\sigma + \sigma S \rightarrow AS$ or $\frac{1}{3}[10\bar{1}0] + \frac{1}{2}[0001] \rightarrow \frac{1}{6}[20\bar{2}3]$ and the associated stacking fault, which is intrinsic, will have a lower energy because there is only one next-nearest neighbour violation in the stacking sequence. As in f.c.c. metals interstitials may be aggregated into platelets on close-packed planes and the resultant structure, shown in *Figure 6.30*(*a*), is a dislocation loop with Burgers vector Sσ, containing a high energy stacking fault. This high energy fault can be changed to one with lower energy by having the loop swept by a partial as shown in *Figure 6.30*(*b*).

All these faulted dislocation loops are capable of climbing by the addition or removal of point defects to the dislocation line. The shrinkage and growth of vacancy loops has been studied in some detail in Zn, Mg and Cd and examples, together with the climb analysis are discussed in Chapter 9.

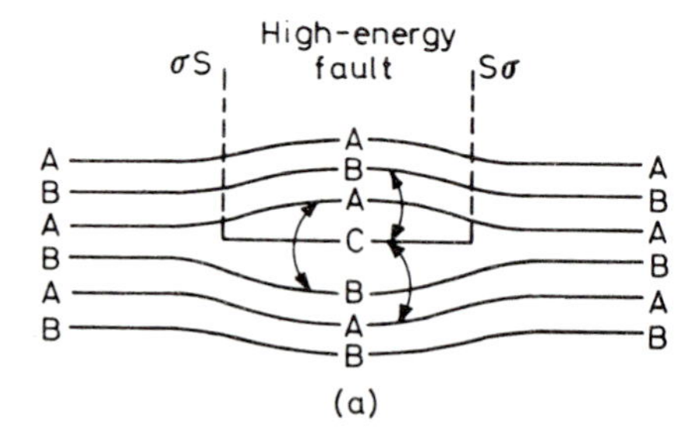

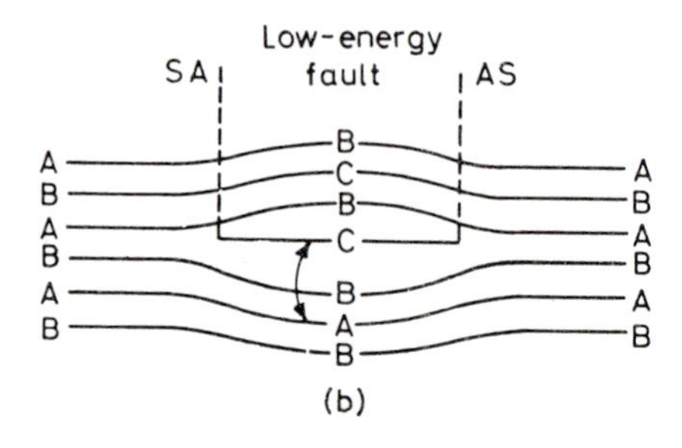

Figure 6.30 Dislocation loop formed by aggregation of interstitials in a h.c.p. lattice with (a) high-energy and (b) low-energy stacking fault

6.5 Dislocations in the b.c.c. lattice

The shortest lattice vector in the b.c.c. lattice is $\frac{1}{2}a[111]$, which joins an atom at a cube corner to the one at the centre of the cube; this is the observed slip direction. The slip plane most commonly observed is (110) which, as shown in *Figure 6.31*, has a distorted close-packed structure. The (110) planes are packed in an ABABAB sequence and three $\{110\}$ type planes intersect along a $\langle 111 \rangle$ direction. It therefore follows that screw dislocations are capable of moving in any of the 3 $\{110\}$ planes and for this reason the slip lines are often wavy and ill-defined. By analogy with the f.c.c. structure it is seen that in moving the B-layer along the $[\bar{1}\bar{1}1]$ direction it is easier to shear in the directions indicated by the three vectors b_1, b_2 and b_3. These three vectors define a possible dissociation reaction

$$\frac{a}{2}[\bar{1}\bar{1}1] \rightarrow \frac{a}{8}[\bar{1}\bar{1}0] + \frac{a}{4}[\bar{1}\bar{1}2] + \frac{a}{8}[\bar{1}\bar{1}0]$$

The stacking fault energy of pure b.c.c. metals is considered to be very high, however, and hence no faults have been observed directly. Because of the stacking sequence ABABAB of the (110) planes the formation of a Frank partial dislocation in the b.c.c. structure gives rise to a situation similar to that for c.p.h. structure, i.e. the aggregation of vacancies or interstitials will bring either two A-layers or two B-layers into contact with each other. The correct stacking sequence can be restored by shearing the planes to produce perfect dislocations $a/2[111]$ or $a/2[11\bar{1}]$.

Slip has also been observed on planes indexed as (112) and (123) planes, and although some workers attribute this latter observation to varying amounts of slip on different (110) planes, there is evidence to indicate that (112) and (123)

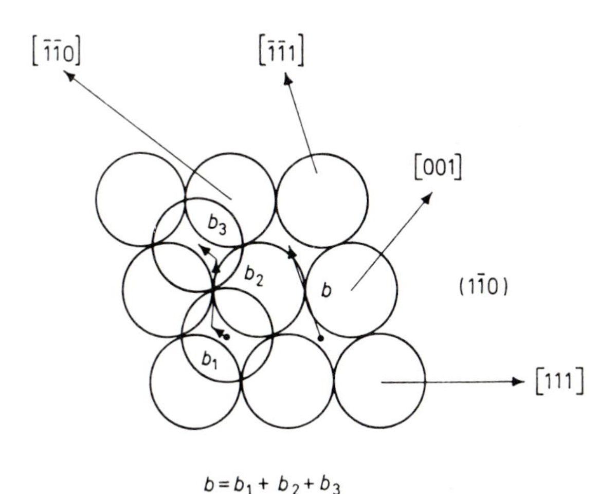

Figure 6.31 The (110) plane of the b.c.c. lattice (after Weertman, *Introduction to Dislocations*, courtesy of Collier-Macmillan International)

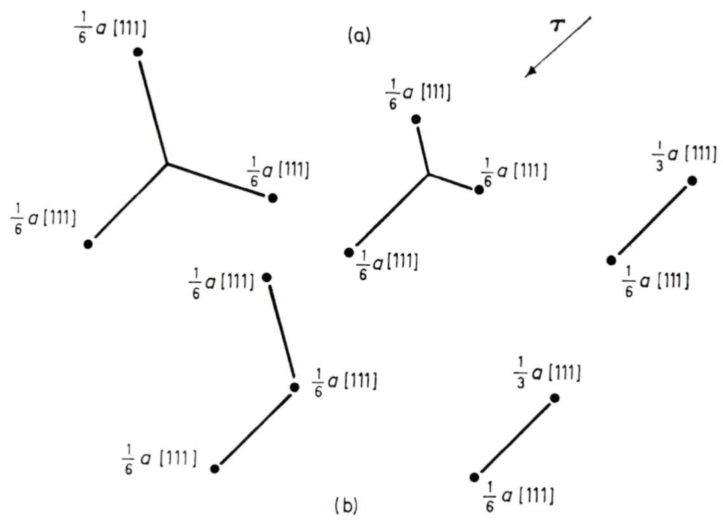

Figure 6.32 Dissociated $a/2$ [111] dislocation in the b.c.c. lattice (after Mitchell, Foxall and Hirsch, *Phil. Mag.*, 8(1963) 1895, courtesy of Taylor and Francis)

are definite slip planes. The packing of atoms in a (112) plane conforms to a rectangular pattern, the rows and columns parallel to the $[1\bar{1}0]$ and $[11\bar{1}]$ directions respectively, with the closest distance of approach along the $[11\bar{1}]$ direction. The stacking sequence of the (112) planes is ABCDEFAB... and the spacing between the planes $a/\sqrt{6}$. It has often been suggested that the unit dislocation can dissociate in the (112) plane according to the reaction

$$\frac{a}{2}[11\bar{1}] \rightarrow \frac{a}{3}[11\bar{1}] + \frac{a}{6}[11\bar{1}]$$

because the homogeneous shear necessary to twin the structure is $1/\sqrt{2}$ in a $\langle 111 \rangle$ on a (112) and this shear can be produced by a displacement $a/6[11\bar{1}]$ on every successive (112) plane. It is therefore believed that twinning takes place by the movement of partial dislocations. However, it is generally recognized that the stacking fault energy is very high in b.c.c. metals so that dissociation must be limited. Moreover, because the Burgers vectors of the partial dislocations are parallel, it is not possible to separate the partials by an applied stress unless one of them is anchored by some obstacle in the crystal.

When the dislocation line lies along the $[11\bar{1}]$ direction it is capable of dissociating in any of the three $\{112\}$ planes, i.e. (112), $(\bar{1}21)$ and $(2\bar{1}1)$, which intersect along $[11\bar{1}]$. Furthermore, the $a/2[11\bar{1}]$ screw dislocation could dissociate according to

$$\frac{a}{2}[11\bar{1}] \rightarrow \frac{a}{6}[11\bar{1}] + \frac{a}{6}[11\bar{1}] + \frac{a}{6}[11\bar{1}]$$

to form the symmetrical fault shown in *Figure 6.32*. Sleeswyk has shown, however, that the symmetrical configuration should be unstable, and the equilibrium configuration is one partial dislocation at the intersection of two

{112} planes and the other two lying equidistant, one in each of the other two planes. At larger stresses this unsymmetrical configuration can be broken up and the partial dislocations induced to move on three neighbouring parallel planes, to produce a three-layer twin (*see* Chapter 8). In recent years an asymmetry of slip has been confirmed in many b.c.c. single crystals, i.e. the preferred slip plane may differ in tension and compression. A yield stress asymmetry has also been noted. Such asymmetry implies a limitation of the Schmid law and has been related to asymmetric glide resistance of screw dislocations arising from their 'core' structure.

An alternative dissociation of the slip dislocation proposed by Cottrell is

$$\frac{a}{2}[111] \rightarrow \frac{a}{3}[112] + \frac{a}{6}[11\bar{1}]$$

The dissociation results in a twinning dislocation $a/6[11\bar{1}]$ lying in the (112) plane and a $a/3[112]$ partial dislocation with Burgers vector normal to the twin fault and hence is sessile. There is no reduction in energy by this reaction and is therefore not likely to occur except under favourable stress conditions, as discussed in Chapter 8.

Another unit dislocation can exist in the b.c.c. structure, namely $a[001]$, but it will normally be immobile. This dislocation can form at the intersection of normal slip bands by the reaction,

$$\frac{a}{2}[\bar{1}\bar{1}1] + \frac{a}{2}[111] \rightarrow a[001]$$

with a reduction of strain energy from $3a^2/2$ to a^2. The new $a[001]$ dislocation lies in the (001) plane and is pure edge in character and may be considered as a wedge, one lattice constant thick, inserted between the (001) and hence has been considered as a crack nucleus (*see* Chapter 14). $a[001]$ dislocations can also form in networks of $a/2\langle 111\rangle$ type dislocations.

6.6 Dislocations in ordered structures

A unit dislocation in a disordered alloy becomes an imperfect dislocation when the alloy orders and may be considered as a partial-dislocation in the superlattice with its attached anti-phase boundary interface, as shown in *Figure 6.33(a)*. Thus, when this dislocation moves through the lattice it will completely destroy the order across its slip plane. However, in an ordered alloy any given atom prefers to have unlike atoms as its neighbours, and consequently such a process of slip would require a very high stress. To move a dislocation against the force γ exerted on it by the fault requires a shear stress

$$\tau = \gamma/b \tag{6.7}$$

where b is the Burgers vector; in β-brass where γ is about 0.07 N/m this stress

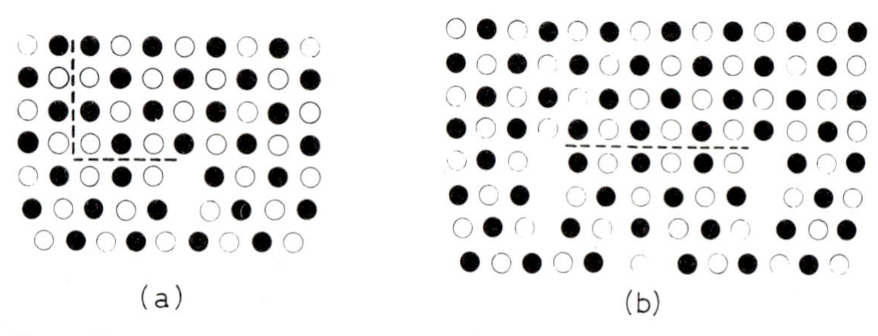

Figure 6.33 Dislocations in ordered structures

is 300 MN/m^2. In practice the critical shear stress of β-brass is an order of magnitude less than this value, and thus one must conclude that slip occurs by an easier process than the movement of unit dislocations. In consequence, by analogy with the slip process in f.c.c. crystals, where the leading partial dislocation of an extended dislocation trails a stacking fault, it is believed that the dislocations which cause slip in an ordered lattice are not single dislocations but coupled pairs of dislocations, as shown in *Figure 6.33(b)*. The first dislocation of the pair, on moving across the slip plane, destroys the order and the second half of the couple completely restores it again, the third dislocation destroys it once more, and so on. In β-brass and similar weakly ordered alloys such as AgMg and FeCo the crystal structure is ordered b.c.c. (or CsCl-type) and, consequently, deformation is believed to occur by the movement of coupled pairs of $\frac{1}{2}a[111]$-type dislocations. The combined slip vector of the coupled pair of dislocations, sometimes called a super-dislocation, is then equivalent to $a[111]$, and, since this vector connects like atoms in the structure, long-range order will be maintained.

The separation of the super-partial dislocations may be calculated, as for Shockley partials, by equating the repulsive force between the two like $a/2\langle 111\rangle$ dislocations to the surface tension of the anti-phase boundary. The values obtained for β-brass and FeCo are about 70 and 50 nm respectively, and thus superdislocations can be detected in the electron microscope using the weak beam technique (*see* Chapter 7 and also *Figure 8.18*). The separation is inversely proportional to the square of the ordering parameter and super-dislocation pairs $\approx$12.5 nm width have been observed more readily in partly ordered FeCo ($S=0.59$).

In alloys with high ordering energies the antiphase boundaries associated with super-dislocations cannot be tolerated and dislocations with a Burgers vector equal to the unit lattice vector $a\langle 100\rangle$ operate to produce slip in $\langle 100\rangle$ directions. The extreme case of this is in ionic-bonded crystals such as CsBr, but strongly ordered intermetallic compounds such as NiAl are also observed to slip in the $\langle 100\rangle$ direction with dislocations having $b=a\langle 100\rangle$.

Ordered alloys of Cu_3Au also give rise to super dislocations with $b=(a/2)\langle 110\rangle$ and, as for the f.c.c. structure, each $a/2\langle 110\rangle$ dislocation may dissociate into two Shockley partials with vector $a/6\langle 112\rangle$ separated by a stacking fault, as shown in *Figure 6.34*. In alloys such as Cu_3Au, Ni_3Mn, etc, the stacking fault ribbon is too small to be observed experimentally but super dislocations have been observed (*see Figure 8.18*). It is evident, however, that the

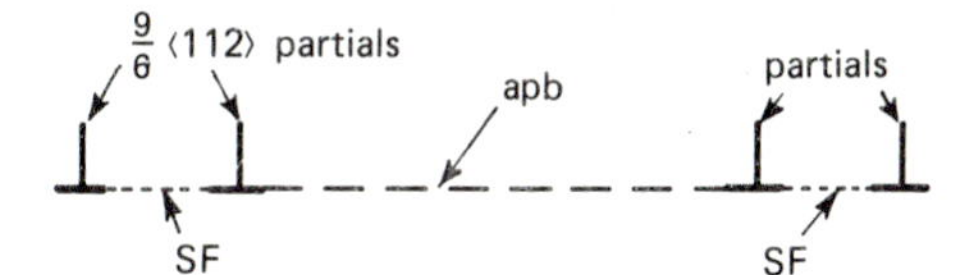

Figure 6.34 Schematic representation of superdislocation structure

cross slip of these super dislocations will be an extremely difficult process. This can lead to a high work-hardening rate in these alloys, as discussed in Chapter 10.

In an alloy possessing short-range order, slip will not occur by the motion of superdislocations since there are no long-range faults to couple the dislocations together in pairs. However, because the distribution of neighbouring atoms is not random the passage of a dislocation will destroy the short-range order between the atoms, across the slip plane. As before, the stress to do this will be large but in this case there is no mechanism, such as coupling two dislocations together, to make the process easier. The fact that, for instance, a crystal of $AuCu_3$ in the quenched state (short-range order) has nearly double the yield strength of the annealed state (long-range order) may be explained on this basis. The maximum strength is exhibited by a partially ordered alloy with a critical domain size of about 6 nm. The transition from deformation by unit dislocations in the disordered state to deformation by super dislocations in the ordered condition gives rise to a peak in the flow stress with change in degree of order (*see* Chapter 8).

Suggestions for further reading

Cottrell, A.H., *The Theory of Crystal Dislocations*, Blackie and Sons, 1964

Freidel, J., *Dislocations*, Pergamon Press, 1964

Hirsch, P.B. (ed), *The Physics of Metals*, 2. Defects, Cambridge University Press, 1975

Hirth, J.P. and Lathe, J., *Theory of Dislocations*, McGraw-Hill, 1984

Nabarro, F.R.N., *Theory of Crystal Dislocations*, Oxford University Press, 1967

'Dislocations and Properties of Real Metals,' Conf. Metals Society, 1984.

Chapter 7

Observation of crystal defects

7.1 Introduction

Much of the earlier evidence for dislocations was indirect, and the density and arrangement of dislocations was inferred either from the measurement of physical properties or from the character of x-ray reflections. One of the most reliable methods is to measure calorimetrically the energy stored in a metal as a result of cold work and to compare the value obtained with the energy of an array of dislocations. For example, the magnitude of the stored energy for cold worked copper is found to be about 40 MJ/m^3 and, since the energy of one metre length of a single dislocation is about 5×10^{-9} J a density of dislocations equal to 8×10^{15} m/m^3, is necessary to account for the observed value. The same result may be obtained using x-rays, since the lattice stored energy can be determined from the powder pattern of a deformed metal if the line broadening due to strain is measured (*see* page 41, Chapter 2).

These methods, and many others, depend to some extent on the assumption that the arrangement of dislocations is random. However, if n dislocations pile up against a barrier, as shown in *Figure 7.2(a)* the group behaves like a super-dislocation of Burgers vector nb. The strain energy is then proportional to $(nb)^2$ and not nb^2, so that fewer dislocations will give the same stored energy. Nevertheless, if the sub-grain size is independently determined, a correction can be applied and the number of dislocations in the pile-up, n, also deduced. The sizes of, and the strains in, sub-grains produced by cold work can be measured using an x-ray micro-beam Laue technique. It is also possible from the micro-beam results alone to make an estimate of the density of dislocations in the sub-grains and in their boundaries.

7.2 Crystal growth

It is found that crystals can grow from the vapour phase almost as fast as molecules strike the surface, even under conditions of low supersaturation. This behaviour is contrary to theoretical expectation because thermal agitation should easily disrupt small groups of atoms in mildly supersaturated vapours,

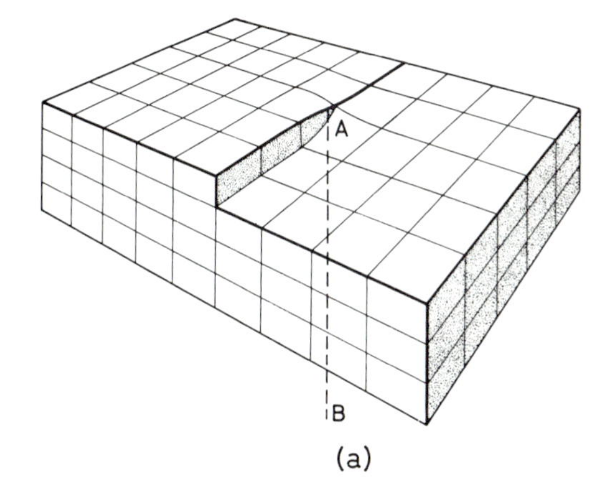

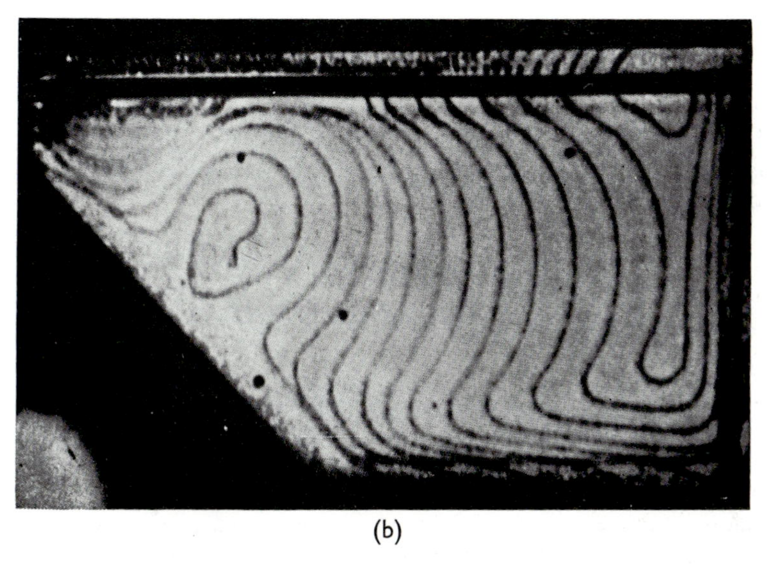

(b)

Figure 7.1 (a) A step on a crystal face caused by a screw dislocation AB emerging at the surface; (b) growth spiral on the surface of a silver crystal (after Forty, *Proc. Roy. Soc.* 1953, **A217,** 262)

with the result that the probability of atoms condensing on a smooth crystal surface and aggregating to form a nucleus for the next layer is small. To explain the observations, Frank suggested that the growth of real crystals does not occur by a layer-upon-layer process but instead is catalysed at places where dislocation lines meet their surfaces.

The ideal condition for continuous and rapid growth of crystals is that the growing faces shall always contain a growth step or terrace upon which atoms can deposit. This is ensured if screw dislocations end at these faces as shown in *Figure 7.1(a)* at the point A. Growth can then occur by the condensation of atoms along the step, so that a new layer is added as the step rotates about its point of emergence. The important feature is that the step never disappears no matter how much the crystal continues to grow. However, since that part of the step near A requires less atoms to make one revolution than the outlying parts of the step, the centre rotates faster than the outside and the step winds up into a 'growth-spiral'.

This idea has been confirmed by observations on many crystals using electron microscopy, multiple-beam interferometry and other optical methods. Step heights which correspond in some cases to a unit Burgers vector and in other cases to multiple vectors have been seen in the electron microscope, and an example found on a crystal of silver is shown in *Figure 7.1*(*b*).

7.3 Direct observations of dislocations

7.3.1 Etch pits

Since dislocations are regions of high energy, their presence can be revealed by the use of an etchant which chemically attacks their sites preferentially. This method has been applied successfully in studying metals, alloys and compounds,

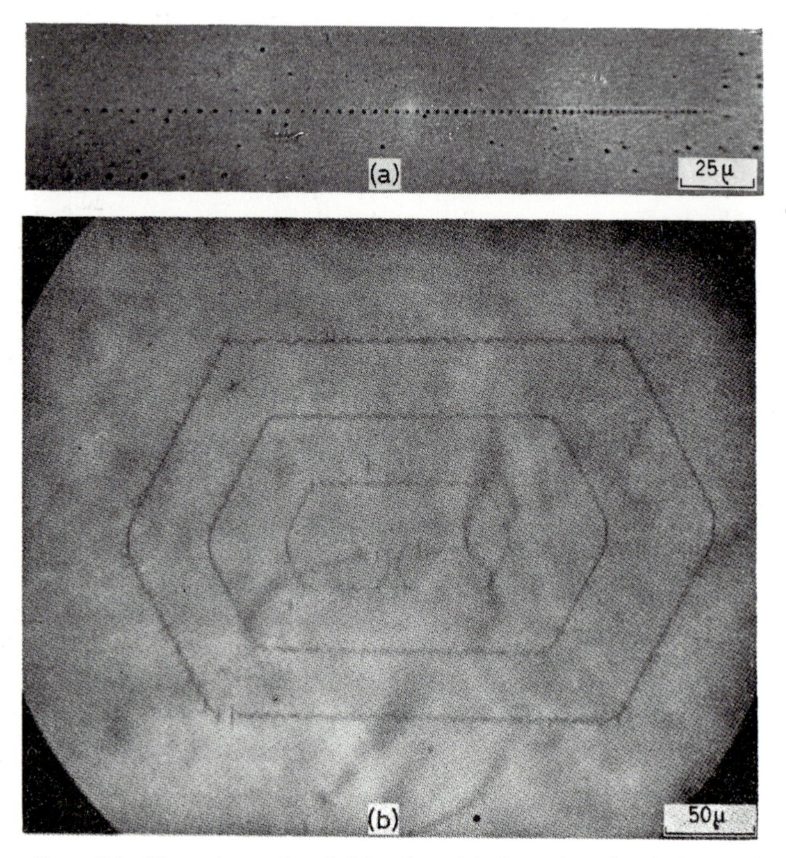

Figure 7.2 Direct observation of dislocations: (a) pile-up in a zinc single crystal (after Gilman, *J. Metals,* Aug. 1956, p. 1000). (b) Frank–Read source in silicon (after Dash, in *Dislocations and Mechanical Properties of Crystals,* 1957, by courtesy of John Wiley & Sons)

and there are many fine examples in existence of etch-pit patterns showing small angle boundaries and pile-ups. *Figure 7.2(a)* shows an etch-pit pattern from an array of piled-up dislocations in a zinc crystal. The dislocations are much closer together at the head of the pile-up, and an analysis of the array, made by Gilman, shows that their spacing is in reasonable agreement with the theory of Eshelby, Frank and Nabarro who have shown that the number of dislocations n that can be packed into a length $L$of slip plane is $n = 2L\tau/\mu b$ where τ is the applied stress. The main disadvantage of the technique is its inability to reveal networks or other arrangements in the interior of the crystal, although some information can be obtained by taking sections through the crystal. Its use is also limited to materials with low dislocation contents ($< 10^4\,mm^{-2}$) because of the limited resolution. In recent years it has been successfully used to determine the velocity v of dislocations as a function of temperature and stress by measuring the distance travelled by a dislocation after the application of a stress for a known time (*see* Chapter 8).

7.3.2 Dislocation decoration

It is well known that there is a tendency for solute atoms to segregate to grain boundaries and, since these may be considered as made up of dislocations, it is clear that particular arrangements of dislocations and sub-boundaries can be revealed by preferential precipitation. Most of the studies in metals have been carried out on aluminium–copper alloys, to reveal the dislocations at the surface, but recently several decoration techniques have been devised to reveal internal structures. The original experiments were made by Hedges and Mitchell in which they made visible the dislocations in AgBr crystals with photographic silver. After a critical annealing treatment and exposure to light, the colloidal silver separates along dislocation lines. The technique has since been extended to other halides, and to silicon where the decoration is produced by diffusing copper into the crystal at 900°C so that on cooling the crystal to room temperature, the copper precipitates. When the silicon crystal is examined optically, using infra-red illumination, the dislocation-free areas transmit the infra-red radiation, but the dislocations decorated with copper are opaque. A fine example of dislocations observed using this technique is shown in *Figure 7.2(b)*.

The technique of dislocation decoration has the advantage of revealing internal dislocation networks but, when used to study the effect of cold work on the dislocation arrangement, suffers the disadvantage of requiring some high temperature heat treatment during which the dislocation configuration may become modified.

7.3.3 Electron microscopy

Direct lattice resolution

In most metals the spacing of the lattice planes is about 0.3 nm, which is now well within the resolving power of the modern 100 kV electron microscope. It is possible, therefore, to observe directly the lattice structure in a metal. However,

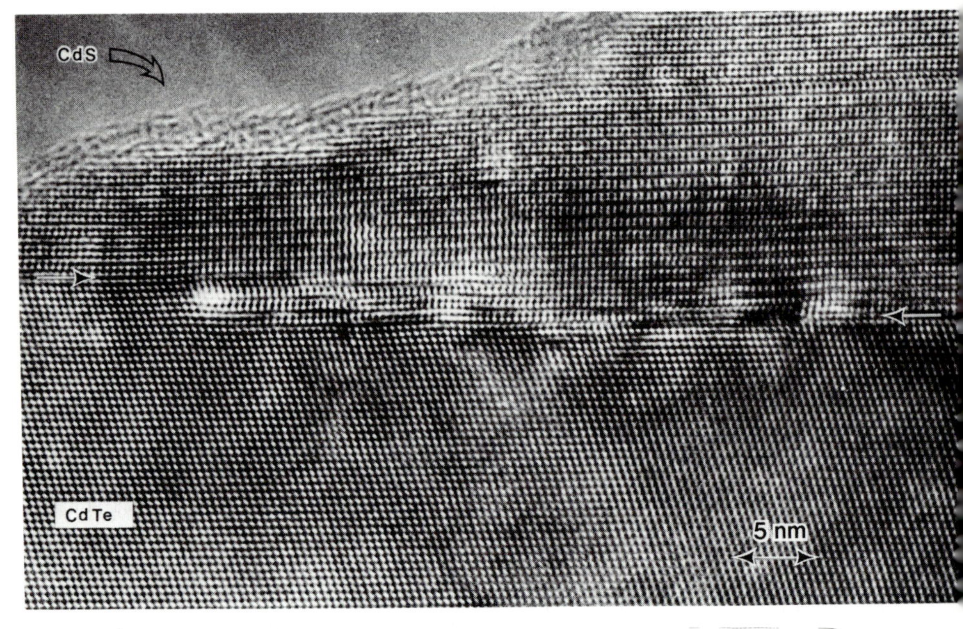

Figure 7.3 Direct lattice resolution micrograph of the interface between CdS and CdTe showing non-uniformity and interface structure (Courtesy R. Sinclair and T. Yamashita)

there are stringent conditions of operation and the technique is largely for research applications. Basically, if the transmitted beam and (at least) one diffracted beam from a very thin specimen are allowed through the objective aperture a periodic fringe pattern will be formed by phase interference between the beams; the larger the collection angle, which is generally limited by lens aberrations, the smaller the periodicity that can be resolved. But, because phase contrast is being used, the precise plane which is being imaged by the objective lens can give rise to images that look different, e.g. atoms rows which normally appear white and tunnels dark will reverse in contrast on going through focus, or as the thickness changes locally. To interpret the images correctly, precise control of the experimental conditions and a comparison of the direct lattice image with computed images is essential. *Figure 7.3* shows a lattice image of CdS grown on CdTe.

In many applications such as precipitate/matrix interfaces, grain boundaries where two crystals meet, or epitaxial growth in semiconductor technology, two lattices of slightly different spacing or orientation overlap and the resultant image shows a lattice of a larger spacing D. These Moiré patterns may be useful in structural analysis and of particular use are the (i) parallel Moiré with spacing $D=d_1d_2/(d_1-d_2)$ where d_1 and d_2 are the spacing of the individual lattices, and (ii) rotation Moiré with spacing $D=d/\alpha$, where d is the spacing of the two lattices which are rotated relative to each other by an angle α. Dislocations present in either of the two lattices giving rise to D will appear in the larger Moiré pattern.

Thin film transmission microscopy

The most notable advance in the direct observation of dislocations in metal crystals has been made by the application of transmission techniques to thin metal specimens. Although the metal lattice is not resolved when examined by TEM, dislocations are revealed in the electron optical image by diffraction contrast, as discussed in Chapter 2.

The technique has been used widely because the dislocation arrangements inside the specimen can be studied. It is possible, therefore, to investigate the effects of plastic deformation, irradiation, heat-treatment, etc on the dislocation distribution and to record the movement of dislocations by taking ciné films of the images on the fluorescent screen of the electron microscope. One disadvantage of the technique is that the metals have to be thinned before examination and, because the surface-to-volume ratio of the resultant specimen is high, it is possible that some re-arrangement of dislocations may occur.

A theory of image contrast has been developed which agrees well with experimental observations. The basic idea is that the presence of a defect in the lattice causes displacements of the atoms from their position in the perfect crystal and these lead to phase changes in the electron waves scattered by the atoms so that the amplitude diffracted by a crystal is altered. The image seen in the microscope represents the electron intensity distribution at the lower surface of the specimen. This intensity distribution has been calculated by the dynamical theory which considers the coupling between the diffracted and direct beams but it is possible to obtain an explanation of many observed contrast effects using the simpler (kinematical) theory in which the interactions between the transmitted and scattered waves are neglected. Thus if an electron wave, represented by the function $\exp(2\pi i k_0 . r)$ where k_0 is the wave vector of magnitude $1/\lambda$, is incident on an atom at position r there will be an elastically scattered wave $\exp(2\pi i k_1 . r)$ with a phase difference equal to $2\pi r(k_1 - k_0)$ when k_1 is the wave vector of the diffracted wave. If the crystal is not oriented exactly at the Bragg angle the reciprocal lattice point will be either inside or outside the reflecting sphere and the phase difference is then $2\pi r(g+s)$ where g is the reciprocal lattice vector of the lattice plane giving rise to reflection, and s is the vector indicating the deviation of the reciprocal lattice point from the reflection sphere (*see Figure 2.20*). To obtain the total scattered amplitude from a crystal it is necessary to sum all the scattered amplitudes from all the atoms in the crystal, i.e. take account of all the different path lengths for rays scattered by different atoms. Since most of the intensity is concentrated near the reciprocal lattice point it is only necessary to calculate the amplitude diffracted by a column of crystal in the direction of the diffracted beam and not the whole crystal, as shown in *Figure 7.4*. The amplitude of the diffracted beam ϕ_g for an incident amplitude $\phi_0 = 1$, is then

$$\phi_g = (\pi i/\xi_g) \int_0^t \exp[-2\pi i(g+s).r\,\mathrm{d}r$$

and since $r.s$ is small and $g.r$ is an integer this reduces to

$$\phi_g = (\pi i/\xi_g) \int_0^t \exp[-2\pi i s.r]\,\mathrm{d}r = (\pi i/\xi_g) \int_0^t \exp[-2\pi i s z]\,\mathrm{d}z$$

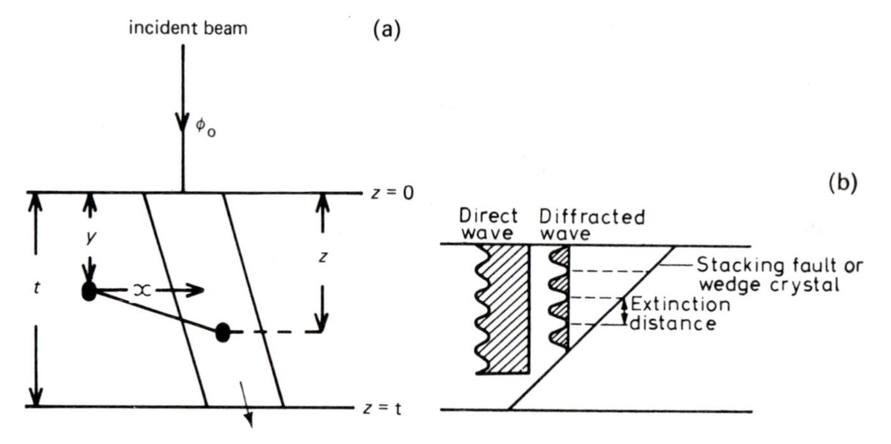

Figure 7.4 (a) Column approximation used to calculate the amplitude of the diffracted beam ϕ_g on the bottom surface of the crystal. The dislocation is at a depth y and a distance x from the column. (b) Variation of intensity with depth in a crystal

where z is taken along the column.

The intensity from such a column is

$$|\phi_g|^2 = I_g = [\pi^2/\xi_g^2](\sin^2 \pi t s/(\pi s)^2)$$

from which it is evident that the diffracted intensity oscillates with depth z in the crystal with a periodicity equal to $1/s$. The maximum wavelength of this oscillation is known as the extinction* distance ξ_g since the diffracted intensity is essentially extinguished at such positions in the crystal. This sinusoidal variation of intensity gives rise to fringes in the electron-optical image of boundaries and defects inclined to the foil surface, e.g. a stacking fault on an inclined plane is generally visible on an electron micrograph as a set of parallel fringes running parallel to the intersection of the fault plane with the plane of the foil (*see Figure 7.4(b) and 7.9*).

In an imperfect crystal atoms are displaced from their true lattice positions. Consequently, if an atom at r_n is displaced by a vector $\boldsymbol{R}$, the amplitude of the wave diffracted by the atom is multiplied by an additional phase factor $\exp[2\pi i(k_1 - k_0).\boldsymbol{R}]$. Then, since $(k_1 - k_0) = g + s$ the resultant amplitude is

$$\phi_g = (\pi i/\xi_g)\int_0^t \exp[-2\pi i(g+s).(r+\boldsymbol{R})]\,\mathrm{d}r$$

If we neglect $s.\boldsymbol{R}$ which is small in comparison with $g.\boldsymbol{R}$, and $g.r$ which gives an integer, then in terms of the column approximation

$$\phi_g = (\pi i/\xi_g)\int_0^t \exp(-2\pi i s z)\exp(-2\pi i g.\boldsymbol{R})\,\mathrm{d}z$$

The amplitude, and hence the intensity, therefore may differ from that scattered by a perfect crystal, depending on whether the phase factor $\alpha = 2\pi g.\boldsymbol{R}$ is finite or not, and image contrast is obtained when $g.\boldsymbol{R} \neq 0$.

* $\xi_g = \pi V \cos\theta/\lambda F$ where V is the volume of the unit cell, θ the Bragg angle and F the structure factor.

Contrast from crystals

In general, crystals observed in the microscope appear light because of the good transmission of electrons. In detail, however, the foils are usually slightly buckled so that the orientation of the crystal relative to the electron beam varies from place to place, and if one part of the crystal is oriented at the Bragg angle, strong diffraction occurs. Such a local area of the crystal then appears dark under bright-field illuminations, and is known as a bend or extinction contour, *see Figure 7.5(b)*. If the specimen is tilted while under observation, the angular

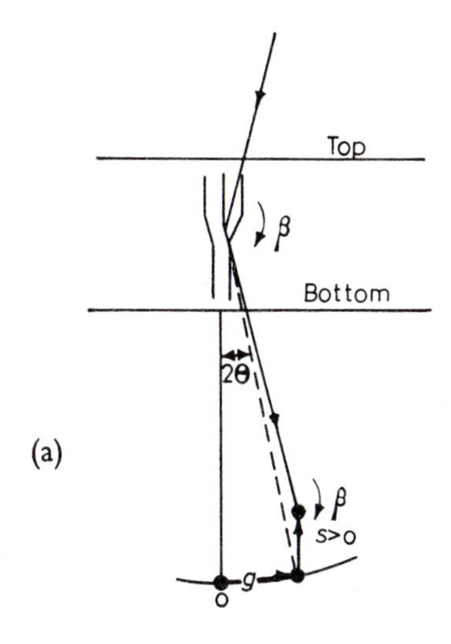

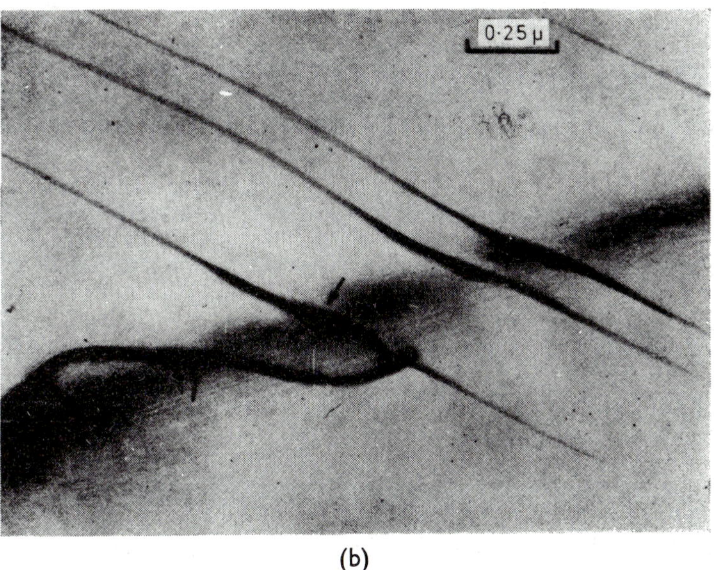

(b)

Figure 7.5 (a) Schematic diagram showing the dependence of the position of the dislocation image on diffraction conditions. (b) Micrograph showing the change in image position on crossing an extinction contour (after Hirsch, Howie and Whelan, *Phil. Trans. Roy. Soc.* A252 (1960) 499, courtesy of the Royal Society)

conditions for strong Bragg diffraction are altered, and the extinction contours, which appear as thick dark lines, can be made to move across the specimen. To interpret micrographs correctly, it is essential to know the correct sense of both g and s. The g-vector is the line joining the origin of the diffraction pattern to the strong diffraction spot and it is essential that its sense is correct with respect to the micrograph, i.e. to allow for any image inversion or rotation by the electron optics. The sign of s can be determined from the position of the Kikuchi lines with respect to the diffraction spots, as discussed in Chapter 2.

Dislocations

Image contrast from imperfections arises from the additional phase factor $\alpha = 2\pi g \cdot \boldsymbol{R}$ in the equation for the diffraction of electrons by crystals. In the case of dislocations the displacement vector $\boldsymbol{R}$ is essentially equal to b the Burgers vector of the dislocation, since atoms near the core of the dislocation are displaced parallel to b. In physical terms, it is easily seen that if a crystal, oriented off the Bragg condition, i.e. $s \neq 0$, contains a dislocation then on one side of the dislocation core the lattice planes are tilted into the reflecting position, and on the other side of the dislocation the crystal is tilted away from the reflecting position. On the side of the dislocation in the reflecting position the transmitted intensity, i.e. passing through the objective aperture, will be less and hence the dislocation will appear as a line in dark contrast. It follows that the image of the dislocation will lie slightly to one or other side of the dislocation core, depending on the sign of $((g \cdot b)s$. This is shown in *Figure 7.5*(a) for the case where the crystal is oriented in such a way that the incident beam makes an angle greater than the Bragg angle with the reflecting planes, i.e. $s > 0$. The image occurs on that side of the dislocation where the lattice rotation brings the crystal into the Bragg position, i.e. rotates the reciprocal lattice point on to the reflection sphere. Clearly, if the diffracting conditions change, i.e. g or s change sign, then the image will be displaced to the other wide of the dislocation core; this effect is shown in *Figure 7.5*(b) when the dislocation crosses a bend contour.

The phase angle introduced by a lattice defect is zero when $g \cdot \boldsymbol{R} = 0$, and hence there is no contrast, i.e. the defect is invisible when this condition is satisfied. Since the scalar product $g \cdot \boldsymbol{R}$ is equal to $g\boldsymbol{R} \cos \theta$, where θ is the angle between g and $\boldsymbol{R}$, then $g \cdot \boldsymbol{R} = 0$ when the displacement vector $\boldsymbol{R}$ is normal to g, i.e. parallel to the reflecting plane producing the image. If we think of the lattice planes which reflect the electrons as mirrors, it is easy to understand that no contrast results when $g \cdot \boldsymbol{R} = 0$, because the displacement vector $\boldsymbol{R}$ merely moves the reflecting planes parallel to themselves without altering the intensity scattered from them; only displacements which have a component perpendicular to the reflecting plane, i.e. tilting the planes, will produce contrast.

A screw dislocation only produces atomic displacements in the direction of its Burgers vector, and hence because $\boldsymbol{R} = b$ such a dislocation will be completely 'invisible' when b lies in the reflecting plane producing the image. A pure edge dislocation, however, produces some minor atomic displacements perpendicular to b, as discussed in Chapter 6 and the displacements give rise to a slight curvature of the lattice planes. An edge dislocation is therefore not completely invisible when b lies in the reflecting planes, but usually shows some evidence of faint residual contrast. In general, however, a dislocation goes out of contrast when the reflecting plane operating contains its Burgers vector, and this fact is commonly used to determine the Burgers vector. To establish b uniquely, it is necessary to tilt the foil so that the dislocation disappears on at least two different reflections. The Burgers vector must then be parallel to the direction which is common to these two reflecting planes. The magnitude of b is usually the repeat distance in this direction.

The use of the $g \cdot b = 0$ criterion is illustrated in *Figure 7.6*. The helices shown in this micrograph have formed by the condensation of vacancies on to screw dislocations (*see* Chapter 9) having their Burgers vector b parallel to the axis of the helix. Comparison of the two pictures in (*a*) and (*b*) shows that the effect of tilting the specimen, and hence changing the reflecting plane, is to make the long helix B in (*a*) disappear in (*b*). In detail, the foil has a [001] orientation and the long screws lying in this plane are $a/2[\bar{1}10]$ and $a/2[110]$. In *Figure 7.6*(*a*) the insert shows the 020 reflection is operating and so $g \cdot b \neq 0$ for either A or B, but in (*b*) the insert shows that the $2\bar{2}0$ reflection is operating and the dislocation B is invisible since its Burgers vector b is normal to the g-vector, i.e. $g \cdot b = 2\bar{2}0 \cdot \frac{1}{2}[110] = (\frac{1}{2} \times 1 \times 2) + (\frac{1}{2} \times 1 \times \bar{2}) + 0 = 0$ for the dislocation B, and is therefore invisible.

Stacking faults

Contrast at a stacking fault arises because such a defect displaces the reflecting planes relative to each other, above and below the fault plane, as illustrated in *Figure 7.7*. In general, the contrast from a stacking fault will not be uniformly bright or dark as would be the case if it were parallel to the foil surface, but in the form of interference fringes running parallel to the intersection of the foil surface with the plane containing the fault. These appear because the diffracted intensity oscillates with depth in the crystal as discussed. The stacking fault displacement vector $\boldsymbol{R}$, defined as the shear parallel to the fault of the portion of crystal below the fault relative to that above the fault which is as fixed, gives rise to a phase difference $\alpha = 2\pi g \cdot \boldsymbol{R}$ in the electron waves diffracted from either side of the fault. It then follows that stacking fault contrast is absent with reflections for which $\alpha = 2\pi$, i.e. for which $g \cdot \boldsymbol{R} = n$. This is equivalent to the $g \cdot b = 0$ criterion for dislocations and can be used to deduce $\boldsymbol{R}$.

The invisibility of stacking fault contrast when $g \cdot \boldsymbol{R} = 0$ is exactly analogous to that of a dislocation when $g \cdot b = 0$ namely that the displacement vector is parallel to the reflecting planes. The invisibility when $g \cdot \boldsymbol{R} = 1, 2, 3, \ldots$ occurs because in these cases the vector $\boldsymbol{R}$ moves the imaging reflecting planes normal to themselves by a distance equal to a multiple of the spacing between the planes. From *Figure 7.7* it can be seen that for this condition the reflecting planes are once again in register on either side of the fault, and, as a consequence, there is no interference between waves from the crystal above and below the fault.

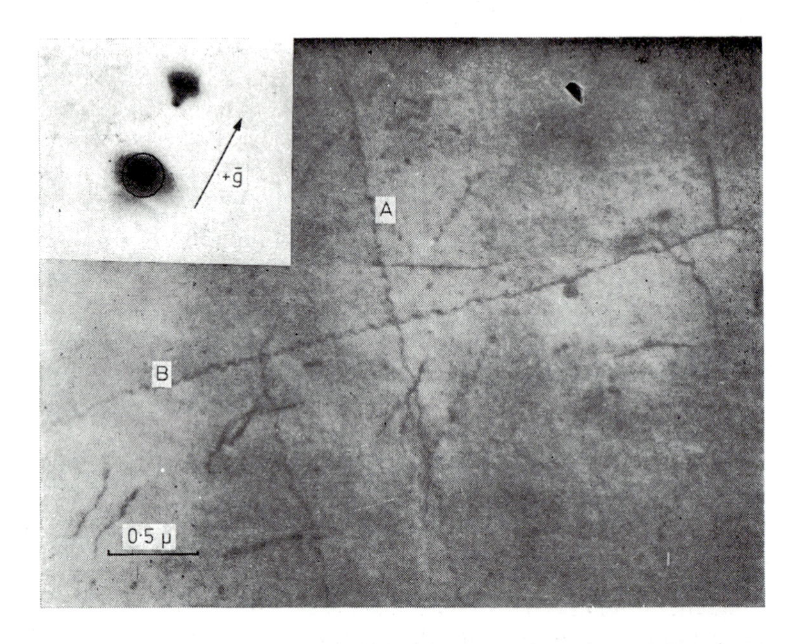

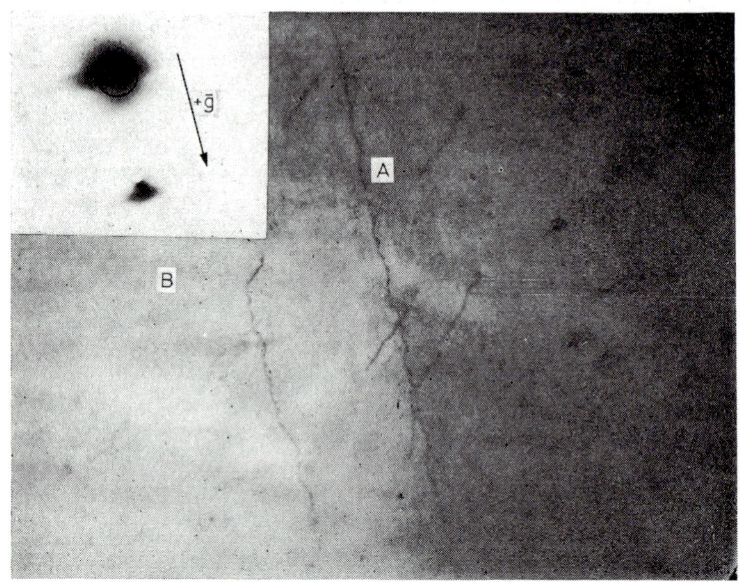

Figure 7.6 (a) Application of the $g.\ b = 0$ criterion. The effect of changing the diffraction condition (see diffraction pattern inserts) makes the long helical dislocation B in (a) disappear in (b) (after Hirsch, Howie and Whelan, *Phil. Trans. Roy. Soc.*, A252 (1960) 499, courtesy of the Royal Society)

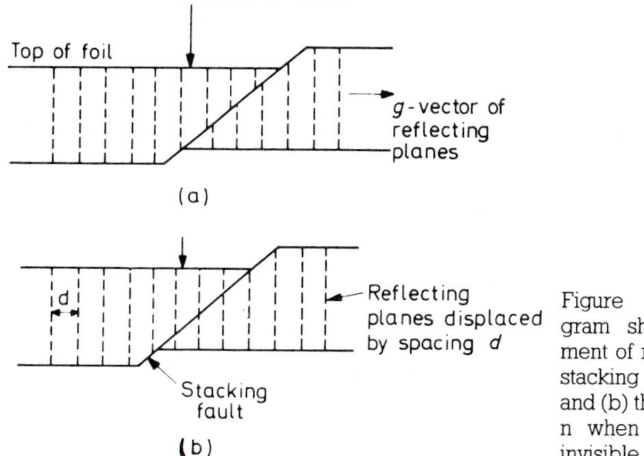

Figure 7.7 Schematic diagram showing (a) displacement of reflecting planes by a stacking fault when $g.\,\boldsymbol{R} = 0$, and (b) the condition for $g.\,\boldsymbol{R} = n$ when the fault would be invisible

Limitations to the kinematical approach to diffraction contrast

There are three major limitations to the use of kinematical theory. These may be summarized as follows:

1. The equations are only valid when the crystal is oriented far from the exact Bragg condition, i.e. when s, the deviation parameter from the exact Bragg reflection g, is large. The kinematical theory gives $I \alpha \sin^2 \pi ts/(\pi s)^2$ with the constant of proportionality π^2/ξ_g^2 and when $s=0$ this gives a diffracted intensity greater than the incident intensities for $t > \xi_g/\pi$. This is physically impossible and limits the experimental conditions under which the results of the theory can be applied.
2. The theory is only strictly applicable for foils whose thickness is less than about half an extinction distance ($\frac{1}{2}\xi_g$), since the incident electron beam on each plane is assumed to be unity, taking no account of the wave scattered out. Normally $\frac{1}{2}\xi_g \ll$ foil thickness, thus invalidating the treatment.
3. No account is taken of absorption, which prevents the explanation of such useful effects as the anomalously low absorption near to the Bragg condition and the asymmetry of dark-field and bright-field images observed experimentally.

Dynamical theory

The object of the dynamical theory is to take into account the interactions between the diffracted and transmitted waves. Again only two beams are considered, viz., the transmitted and one diffracted beam, and experimentally it is usual to orient the specimen in a double tilt stage so that only one strong diffracted beam is excited. The electron wave function is then considered to be made up of two plane waves—an incident or transmitted wave and a reflected or diffracted wave

$$\psi(r) = \phi_0 \exp(2\pi i k_0 . r) + \phi_g \exp(2\pi i k_1 . r)$$

The two waves can be considered to propagate together down a column through the crystal since the Bragg angle is small. Moreover, the amplitudes ϕ_0 and ϕ_g of the two waves are continually changing with depth z in the column because of the reflection of electrons from one wave to another. This is described by a pair of coupled first-order differential equations linking the wave amplitudes ϕ_0 and ϕ_g, viz.

$$\frac{\mathrm{d}\phi_0}{\mathrm{d}z} = \frac{\pi i}{\xi_0}\phi_0 + \frac{\pi i}{\xi_g}\phi_g \exp(2\pi isz)$$

$$\frac{\mathrm{d}\phi_g}{\mathrm{d}z} = \frac{\pi i}{\xi_0}\phi_g + \frac{\pi i}{\xi_g}\phi_g \exp(-2\pi isz)$$

Displacement of an atom $\boldsymbol{R}$ causes a phase change $\alpha = 2\pi g . \boldsymbol{R}$ in the scattered wave, as before, and the differential equations describing the dynamical equilibrium between incident and diffracted waves is given by

$$\frac{\mathrm{d}\phi_g}{\mathrm{d}z} = \frac{i\pi}{\xi_g}\phi_0 \exp(-i\alpha) + i\left(\frac{\pi}{\xi_0} + 2\pi s\right)\phi_g$$

$$\frac{\mathrm{d}\phi_0}{\mathrm{d}z} = \frac{i\pi}{\xi_g}\phi_g \exp(i\alpha) + \frac{i\pi}{\xi_0}\phi_0$$

The first of these describes the change in reflected amplitude ϕ_g because electrons are reflected from the transmitted wave. This change is proportional to ϕ_0, the transmitted wave amplitude, and contains the phase factor. The second equation describes the reflection in the reverse direction and consequently has a phase factor of opposite sign. The quantities ξ_g and ξ_0 are the extinction distance and effective refractive index, respectively.

These equations may be transformed to a more convenient form for numerical calculations by putting

$$\phi_g' = \phi_g \exp(i\pi z/\xi_0)\exp(i\alpha)$$

$$\phi_0' = \phi_0 \exp(-i\pi z/\xi_0)$$

We then obtain (dropping the primes)

$$\frac{\mathrm{d}\phi_0}{\mathrm{d}z} = \frac{\pi i}{\xi_g}\phi_g$$

$$\frac{\mathrm{d}\phi_g}{\mathrm{d}z} = \frac{\pi i}{\xi_g}\phi_0 + 2\pi i\phi_g\left(s + g . \frac{\mathrm{d}\boldsymbol{R}}{\mathrm{d}z}\right)$$

These equations show that the effect of a displacement $\boldsymbol{R}$ is to modify s locally, by an amount proportional to the derivative of the displacement, i.e. $\mathrm{d}\boldsymbol{R}/\mathrm{d}z$ which is the variation of displacement with depth z in the crystal. This was noted in the kinematical theory where $\mathrm{d}\boldsymbol{R}/\mathrm{d}z$ is equivalent to a local tilt of the lattice planes. Absorption in the crystal is accounted for by modifying the parameters ξ_0 and ξ_g and the coupled differential equations integrated numerically starting at the top of the foil ($z = 0$) with boundary conditions $\phi_0 = 1$, $\phi_g = 0$ and finishing at the lower surface ($z = t$). The variation of the intensities

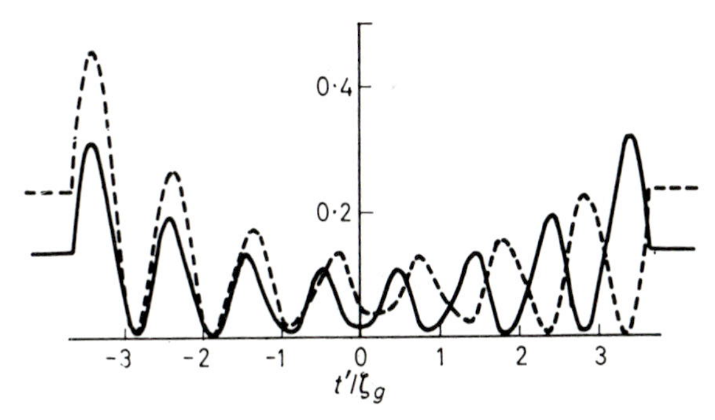

Figure 7.8 Computed intensity profiles about the foil centre for a stacking fault with $\alpha=+2\pi/3$. The full curve is the B.F. and the broken curve the D.F. image. (From Hirsch, Howie and others, *Electron Microscopy of Thin Crystals*, Butterworths, London, 1965)

$|\phi_0|^2$ and $|\phi_g|^2$ for different positions of the column in the crystal, relative to the defect then gives the bright and dark-field images respectively. *Figure 7.8* shows the bright- and dark-field intensity profiles from a stacking fault on an inclined plane, in full and broken lines respectively.

For cases of simple dislocation geometry, e.g. a straight dislocation passing obliquely through the foil, it is possible to use fewer numerical integrations by taking suitable linear combinations of these. The output of the computation may then be presented as a picture produced either on an oscilloscope or by using various combinations of overprinted line printer characters which when photographically reduced resemble a half-tone picture. A wide variety of defects have been computed, some of which are summarized below:

Dislocations In elastically isotropic crystals, perfect screw dislocations show no contrast if the conditions $g \cdot b=0$ is satisfied. Similarly an edge dislocation will be invisible if $g \cdot b=0$ and if $g \cdot b\times \boldsymbol{u}=0$ where $\boldsymbol{u}$ is a unit vector along the dislocation line and $b\times \boldsymbol{u}$ describes the secondary displacements associated with an edge dislocations normal to the dislocations line and b. The computations also show that for mixed dislocation and edge dislocation for which $g \cdot b\times \boldsymbol{u}<0.64$ the contrast produced will be so weak as to render the dislocation virtually invisible. At higher values of $g \cdot b\times \boldsymbol{u}$ some contrast is expected. In addition, when the crystal becomes significantly anisotropic residual contrast can be observed even for $g \cdot b=0$.

The image of a dislocation lies to one side of the core, the side being determined by $(g \cdot b)s$. Thus the image of a dislocation loop lies totally outside the core when (using the appropriate convention) $(g \cdot b)s$ is positive and inside when $(g \cdot b)s$ is negative. Vacancy and interstitial loops can thus be distinguished by examining their size after changing from $+g$ to $-g$, since these loops differ only in the sign of b (*see* Chapter 9).

Partial dislocations Partials for which $g \cdot b=\pm\frac{1}{3}$ (e.g. partial $a/6[\bar{1}\bar{1}2]$ on (111) observed with 200 reflection) will be invisible at both small and large deviations from the Bragg condition. Partials examined under conditions for which

$g.b = \pm\frac{2}{3}$ (i.e. partial $a/6[\bar{2}11]$ on (111) with 200 reflection) are visible except at large deviations from the Bragg condition. A partial dislocation lying above a similar stacking fault is visible for $g.b = \pm\frac{1}{3}$ and invisible for $g.b = \pm\frac{2}{3}$.

Stacking faults For stacking faults running from top to bottom of the foil, the bright-field image is symmetrical about the centre, whereas the dark-field image is asymmetrical (*see Figure 7.8*). The top of the foil can thus be determined from the non-complementary nature of the fringes by comparing bright- and dark-field images. Moreover, the intensity of the first fringe is determined by the sign of the phase-factor α, such that when α is positive the first fringe is bright (corresponding to a higher transmitted intensity) and vice-versa on a positive photographic print.

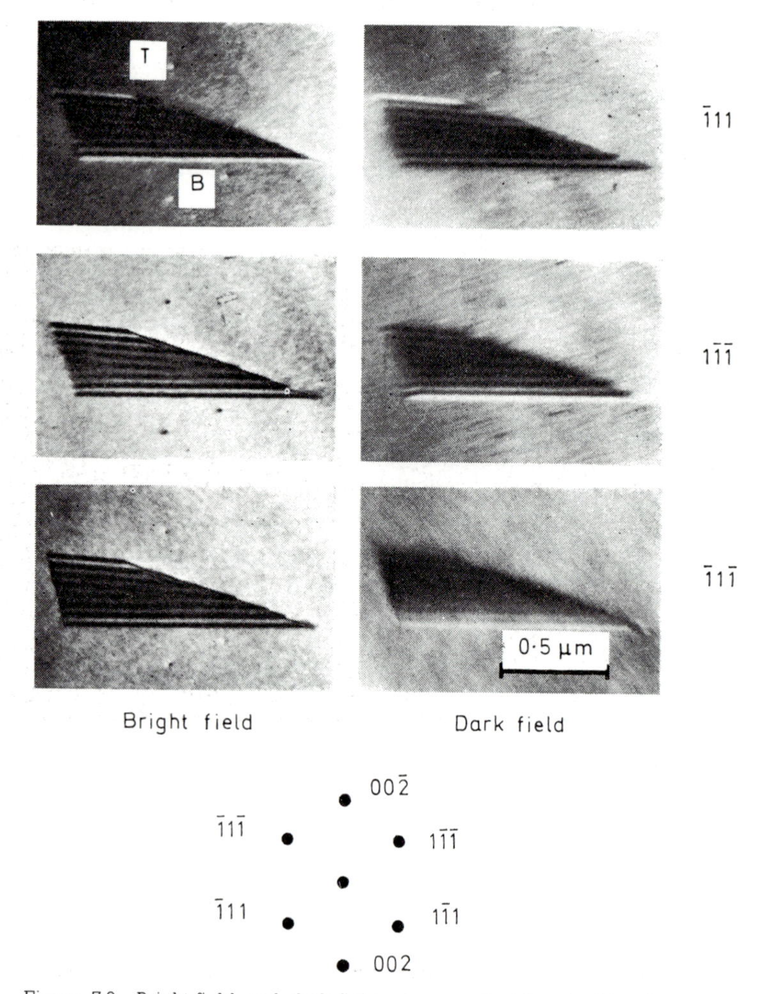

Figure 7.9 Bright-field and dark-field micrographs of a stacking fault in a copper–aluminium alloy. The nature of the first fringe and associated operating diffraction vector indicates that the fault is intrinsic in nature (after Howie and Valdre, *Phil. Mag.*, **8**(1963)1981, courtesy of Taylor and Francis)

It is thus possible to distinguish between intrinsic and extrinsic faults and an example is shown in *Figure 7.9* for an intrinsic fault on (111). The foil orientation is [110] and the non-complementary nature of the first fringe between B.F. and D.F. indicates the top of the foil, marked T. Furthermore, from the B.F. images the first fringe is bright with $\bar{1}11$, and dark with $1\bar{1}\bar{1}$ and $\bar{1}1\bar{1}$.

7.3.4 Weak-beam microscopy

One of the limiting factors in the analysis of defects is the fact that dislocation images have widths of $\xi_g/3$, i.e. typically >10.0 nm. It therefore follows that dislocations closer together than about 20.0 nm are not generally resolved. With normal imaging techniques, the detail that can be observed is limited to a value about fifty to a hundred times greater than the resolution of the microscope. This limitation can be overcome by application of the weak-beam technique in which crystals are imaged in dark-field using a very large deviation parameter s. Under these conditions the background intensity is very low so that weak images are seen in very high contrast and the dislocation images are narrow (≈ 1.5 nm) as shown in *Figure 7.10*. At the large value of s used in weak-beam, the effective extinction

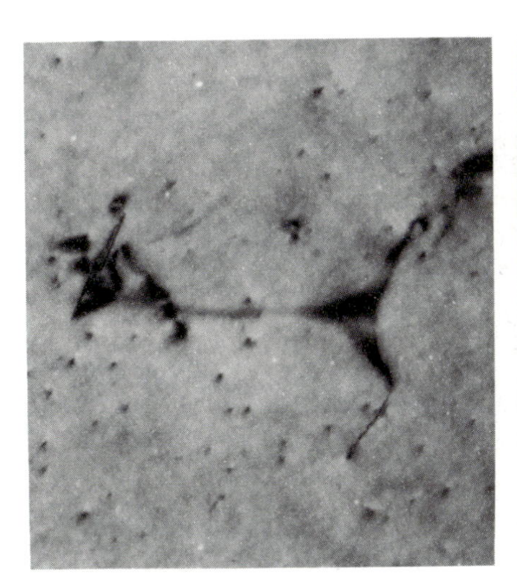

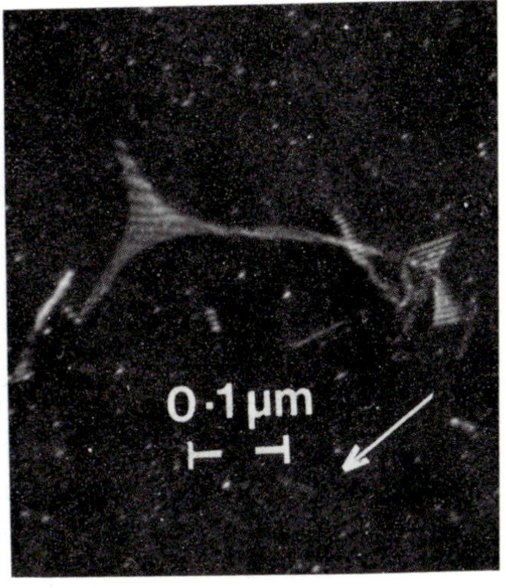

Figure 7.10 Symmetrical node in 21 Cr/14 Ni stainless steel with $\gamma = 18 \pm 4$ mJ/m^2, (a) BF with $g = 11\bar{1}$, (b) weak beam with $g(5g)$

distance is reduced significantly and is given by

$$\xi_g^{\text{eff}} = \xi_g/(1+w^2)^{1/2}$$

where $w = s\xi_g$ and ξ_g is the normal two-beam extinction distance. Thus, if ξ_g is 40.0 nm and s made about 0.2 nm^{-1}, a lower limit for satisfactory weak beam micrographs, then $\xi_g^{\text{eff}} \approx 5.0$ nm and dislocation image widths of $(\xi_g^{\text{eff}}/3) \simeq 1.5$ nm are produced.

In weak beam, when s is set very large the transfer of energy from the direct to the diffracted beam is very small, i.e. the crystal is a long way from the Bragg condition and there is negligible diffraction. Moreover, it is only very near the core of the dislocation that the crystal planes are sufficiently bent to cause the Bragg condition to be locally satisfied, i.e. $g.(dR/dZ)$ be large enough to satisfy the condition $[s+g.(dR/dZ)]=0$. Therefore diffraction takes place from only a small volume near the centre of the dislocation, giving rise to narrow images. The absolute intensity of these images is very small even though the signal-to-background ratio is high, hence long exposures are necessary to record them.

7.4 Arrangements of dislocations in crystals

The density and distribution of dislocations obtained by the present-day direct methods of determination are in good agreement with those deduced previously from the character of x-ray reflections.

Annealed crystals The term 'mosaic' crystal was introduced to account for the high observed integrated x-ray intensities reflected from a real crystal as compared to that calculated for an ideally perfect lattice. The 'mosaic' may be regarded as an assembly of crystallites, or sub-grains, bounded by dislocation walls in such a way that each crystallite is slightly misoriented from its neighbouring crystallite. With this model the x-ray reflections occur over a wider angular range and, as a consequence, an increased integrated intensity results. Since only one dislocation is required to misorient one crystallite from another, the structure is a network of the type shown in *Figure 7.11*(*a*) and (*b*) and the dislocation density is given by $\rho=1/l^2$, where l is the linear dimension of a crystallite; for $l=10^{-4}$ cm the density of dislocations is 10^8 cm^{-2}.

Observations carried out on annealed crystals of silver, alkali halides, silicon and germanium using decoration techniques show that the dislocations form beautiful networks, many of which are arranged in two-dimensional sub-grain boundaries. The mesh size of the network is usually a few micrometres so that the corresponding dislocation density is 10^6–10^8 cm^{-2}, although under special conditions of growth, ρ in silicon and germanium can be as low as 10 to 10^4 cm^{-2}. For metals, the electron microscope indicates that the dislocation density ρ is also about 10^6–10^8 lines cm^{-2}.

Worked crystals Many of the studies of dislocation arrangements have been concentrated on the f.c.c. metals, particularly aluminium, copper and α-brass. They show that there is a gradual transition in the dislocation arrangement as one changes from α-brass to copper to aluminium. This variation in arrangement may be related to the stacking fault energy of the metals concerned through the parameter $\gamma/\mu b$. Thus 70/30 brass and some stainless steels have a low value of γ which gives rise to wide extended dislocations. Thus cross-slip is so difficult at

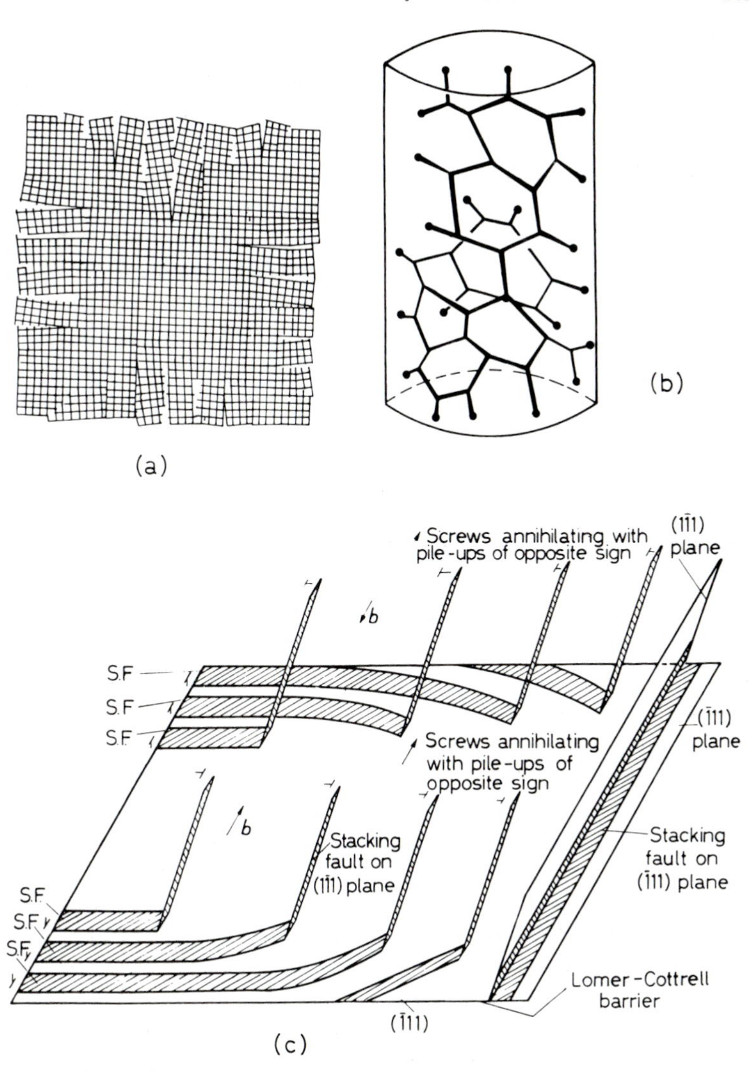

Figure 7.11 (a) A mosaic structure (after Buerger, *Z.Krist.* 1934, **89,** 195), (b) dislocation network (after Cottrell, *Deformations of Solids at High Rates of Strain,* courtesy of the Institution of Mechanical Engineers), and (c) formation of a sub-boundary by cross-slip (after Seeger, in *Dislocations and Mechanical Properties of Crystals,* 1957, courtesy of John Wiley & Sons)

reasonable stress levels that both the screw and edge dislocations are confined to the slip planes. The dislocations are observed to remain in planar arrays and networks on the slip planes and no regular cell walls are formed. This behaviour is reflected in the high density of dislocations (10^9 to 10^{10} lines mm^{-2}) obtained after only moderate strains. However, for copper, which has an intermediate value of γ of about 40 mJ/m^2 the dislocation arrangement is quite different. Cross-slip is possible, but only in regions of high stress, so that the dislocations arrange themselves in complex three-dimensional networks at low deformations, but cross-slip to form poorly developed sub-boundaries at higher deformations (*see Figure 10.5*).

In aluminium, for which the stacking fault energy is about 135 mJ/m^2 the dislocations are relatively unextended and, consequently, cross-slip is so easy that the screw dislocations are able to arrange themselves into almost perfect sub-boundaries. The formation of sub-boundaries by the cross-slip of screw dislocations is shown in *Figure 7.11*(*c*) and is clearly easier for undissociated dislocations. Observations with the electron microscope show that the average sub-grain size is about one micron and that the misorientation across a sub-boundary is about $1\frac{1}{2}^\circ$. The sub-boundaries separate relatively perfect crystallites so that the overall dislocation density is low, and even after a large amount of cold work the dislocation density is not much greater than 10^8 mm^{-2}.

It is now possible to observe stacking faults and widely dissociated dislocations directly in various crystals, and an estimate of γ can be made from such observations. Equilibrium stacking fault configurations such as those observed in dislocation networks lying in $\{111\}$ planes in f.c.c. crystals are used generally to avoid local stress effects. The formation of a dislocation network is discussed in Chapter 10 (*see Figure 10.4*). However, when (using the Thompson notation) an extended dislocation on the α-plane with Burgers vector DC$\rightarrow$ Dα+αC intersects a dislocation on the γ-plane with vector CB$\rightarrow$Cγ+γB, the dislocations constrict at the intersection and then CB dissociates in the α-plane to Cα+αB. The resultant dislocation network takes the form shown in *Figure 7.12*

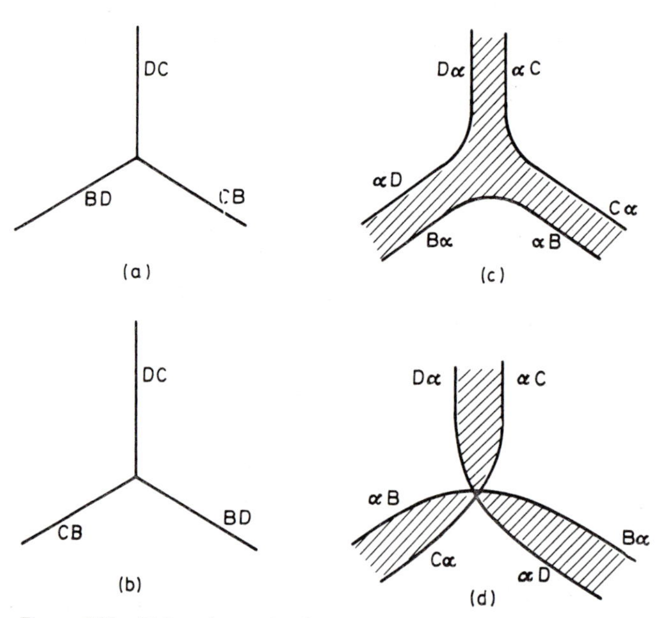

Figure 7.12 Dislocation nodes forming part of a general network in the α-plane, (a) and (b) undissociated, (c) and (d) dissociated; (c) is an extended node and (d) a contracted node

with extended and contracted nodes; when the common partials of two extended dislocations lie on the same side of the node an extended node is formed but when they lie on opposite sides, the node is contracted. From a measurement of either the radius of curvature R or width W of an extended node it is possible to obtain γ. This follows because the force per unit length tending to straighten the dislocation line at the node is $F = T/R$ (where T is the line tension and R the radius of curvature) and is balanced by the opposing force per unit length γ due to the stacking fault tending to contract the node. In equilibrium, the radius of curvature of the node is given by $R = T/\gamma \simeq \mu b^2/k\gamma$ where k depends on the character of the dislocation at the node. Until recently such measurements have been limited to low fault energy materials for which $\gamma/\mu b < 3 \times 10^{-2}$. However, weak-beam microscopy has extended the range of observation quite considerably, as is evident from the dislocation nodes shown in *Figure 7.10*. Weak-beam measurements of the separation of partials in isolated dislocations also give reliable values of γ. The high stacking fault energy metals, such as Ni, Al and c.p.h. metals, cannot be measured by this technique and alternative methods (*see* Chapter 9) are used.

7.5 Origin of dislocations

It was shown in Chapter 4 that the dislocation is a thermodynamically unstable lattice defect. It is, nevertheless, an essential lattice imperfection not only for the propagation of plastic deformation, but also for certain processes of crystal growth and chemical catalysis on crystal faces. Thus even in very carefully prepared crystals of Si the dislocation density $\rho \approx 10^2$–10^4/cm^2 and for most metal crystals $\rho \approx 10^6$–10^8/cm^2. These dislocations may arise in a variety of ways during solidification and cooling. We have already seen that a high concentration of vacancies exists in all metals at elevated temperatures, and during non-equilibrium cooling, as occurs in practice, the excess concentration of point defects can give rise to dislocations (*see* Chapter 9). Dislocations may also be created heterogeneously by thermal or mechanical stresses at stress concentrations near particles, twins, cracks or surface flaws. A striking example is the punching of dislocation loops from inclusions as a result of the stresses produced by the differential contraction between a particle and the matrix during cooling. The stresses in the vicinity of the particle are likely to be compressive and hence to relieve the stress requires the movement of material away from the particle/matrix interface. This can be achieved by the creation of interstitial loops which

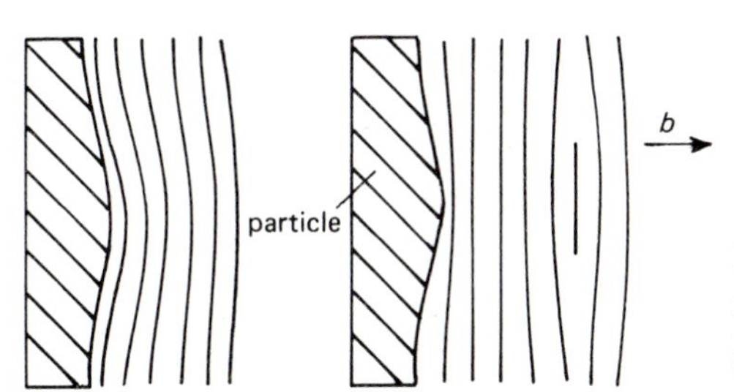

Figure 7.13 Schematic representation of 'punching' of prismatic interstitial dislocation loops from a particle with larger specific volume than the matrix

glide outward from the particle on the surface of a glide cylinder or prism, the axis of which is defined by the Burgers vector of the prismatic dislocation loop (*see Figure 7.13*). Conversely, if the inclusion contracts more than the matrix, a series of prismatic loops equivalent to vacancy platelets would be punched out. This process of prismatic punching at carbide particles in iron can be seen in *Figure 11.12(b)*.

During deformation, the dislocation density increases with increasing strain, reaching values of $\rho \approx 10^{11}$–$10^{12}/\text{cm}^2$. One mechanism whereby ρ increases has already been considered in the discussion of jog intersection and non-conservative motion. Further mechanisms of multiplication are discussed in Chapter 8.

Suggestions for further reading

Hirsch, P., Howie, A., Nicholson, R.B., Pashley, D. and Whelan, M.J., *Electron Microscopy of Thin Crystals*, Butterworths, 1965

Loretto, M.H. and Smallman, R.E., *Defect Analysis in Electron Microscopy*, Chapman and Hall, 1975

Thomas, G. and Goringe, M.J., *Transmission Electron Microscopy of Materials*, John Wiley, 1979

Loretto, M.H., *Electron Beam Analysis of Materials*, Chapman and Hall, 1984

Chapter 8

Deformation of metals and alloys

8.1 Dislocation mobility

The ease with which metal crystals can be plastically deformed at stresses many orders of magnitude less than the theoretical strength ($\tau_t = \mu b/2\pi a$) is quite remarkable. This is due to the most important single property of a dislocation, its mobility. *Figure 8.1*(*a*) shows that as a dislocation glides through the lattice it moves from one symmetrical lattice position to another and at each position the dislocation is in neutral equilibrium, because the atomic forces acting on it from each side are balanced. As the dislocation moves from these symmetrical lattice positions some imbalance of atomic forces does exist, and an applied stress is required to overcome this lattice friction. As shown in *Figure 8.1*(*b*), an intermediate displacement of the dislocation also leads to an approximately balanced force system. Therefore in metals this friction is quite small, particularly in wide dislocations where the transition from the slipped to the unslipped region is spread over several atom distances in the crystal (*see Figure 6.21*), because the wider the dislocation the more nearly do the atomic forces acting on the dislocation from either side cancel. The lattice friction does depend rather sensitively on the width w and has been shown by Peierls and Nabarro to be given by

$$\tau \simeq \mu \exp[-2\pi w/b]$$

for the shear of a rectangular lattice of interplanar spacing a with $w = \mu b/2\pi(1-\nu)\tau_t = a(1-\nu)$. The friction stress is therefore often referred to as the Peierls–Nabarro stress. The two opposing factors affecting w are (i) the elastic energy of the crystal, which is reduced by spreading out the elastic strains, and (ii) the misfit energy, which depends on the number of misaligned atoms across the slip plane. Metals with close-packed structures have extended dislocations and hence w is large. Moreover, the close-packed planes are widely spaced with weak alignment forces between them, i.e. have a small b/a factor. These metals have highly mobile dislocations and are intrinsically soft. In contrast, directional bonding in crystals tends to produce narrow dislocations, which leads to intrinsic hardness and brittleness. Extreme examples are ionic and ceramic crystals and the covalent materials such as diamond and silicon.

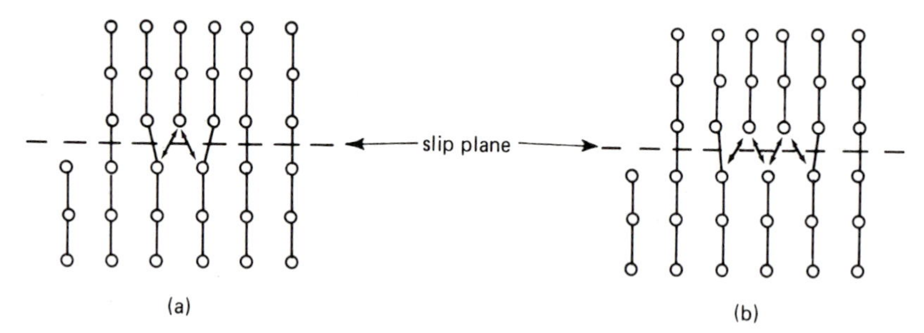

Figure 8.1 Diagram showing structure of edge dislocation during gliding from (a) equilibrium to (b) metastable position

In b.c.c. metals a high lattice friction to the movement of a dislocation may arise from the dissociation of a dislocation on several planes. As discussed in Chapter 6, when a screw dislocation with Burgers vector $\frac{1}{2}a[111]$ lies along a symmetry direction it can dissociate on three crystallographically equivalent planes. If such a dissociation occurs, it will be necessary to constrict the dislocation before it can glide in any one of the slip planes. This constriction will be more difficult to make as the temperature is lowered so that the large temperature dependence of the yield stress in b.c.c. metals, shown in *Figure 8.3(a)* and also *Figure 8.14*, may be due partly to this effect. In f.c.c. metals the dislocations lie on $\{111\}$ planes, and although a dislocation will dissociate in any given (111) plane, there is no direction in the slip plane along which the dislocation could also dissociate on other planes; the temperature-dependence of the yield stress is small as shown in *Figure 8.3(a)*. In c.p.h. metals the dissociated dislocations moving in the basal plane will also have a small Peierls force and be glissile with low temperature-dependence. However, screw dislocations moving on non-basal planes, i.e. prismatic and pyramidal planes, may have a high Peierls force because they are able to extend in the basal plane as shown in *Figure 8.2*. Hence, constrictions will once again have to be made before the screw

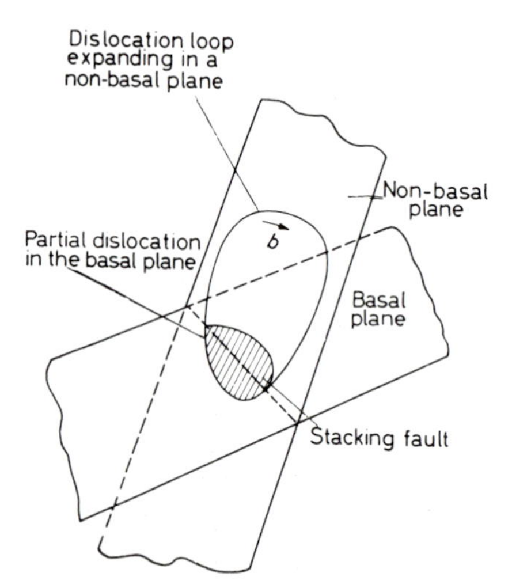

Figure 8.2 Dissociation in the basal plane of a screw dislocation moving on a non-basal glide plane

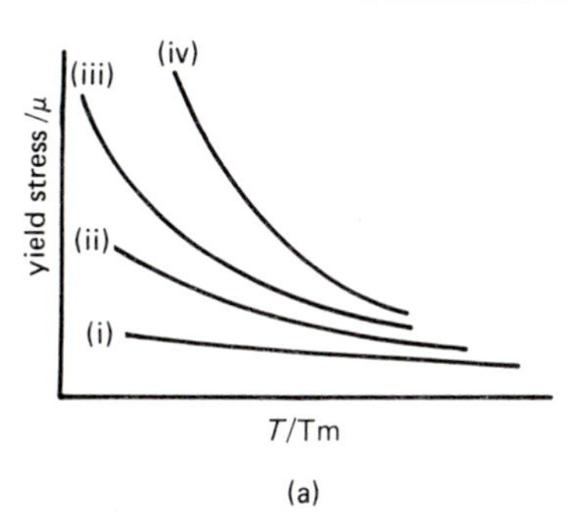

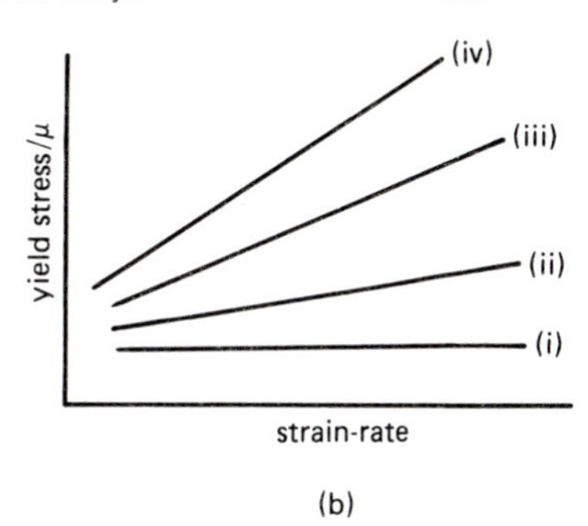

Figure 8.3 Variation of yield stress with (a) temperature, (b) strain-rate, for crystals with (i) FCC (ii) BCC (iii) ionic-bonded (iv) co-valent-bonded structure

dislocations can advance on non-basal planes. This effect contributes to the high critical shear stress and strong temperature-dependence of non-basal glide observed in this crystal system.

The high Peierls–Nabarro stress, which is associated with materials with narrow dislocations, gives rise to a short-range barrier to dislocation motion. Such barriers are effective only over an atomic spacing or so, hence thermal

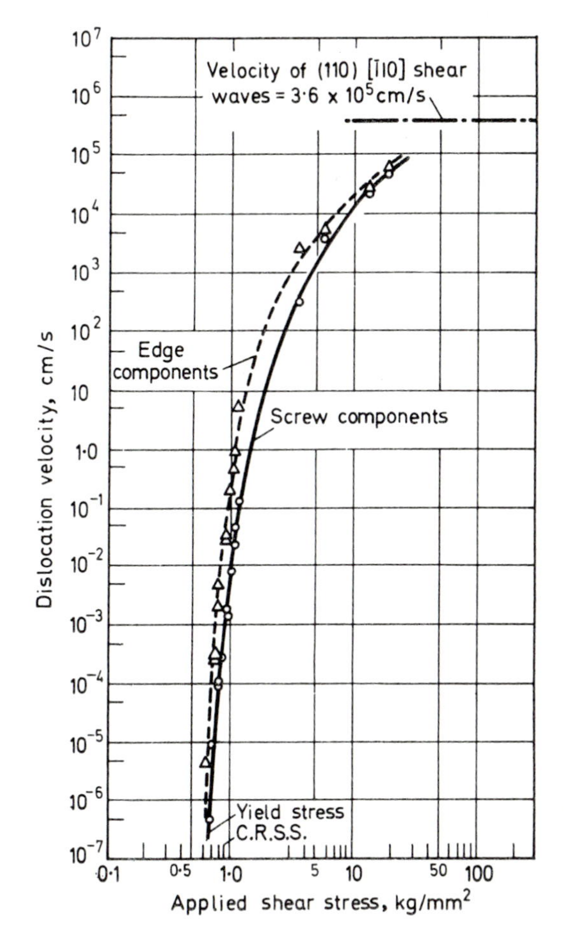

Figure 8.4 Stress dependence of the velocity of edge and screw dislocations in lithium fluoride (from Johnston and Gilman, *J. Appl. Phys.*, 1959, **30**, 129, courtesy of the American Institute of Physics)

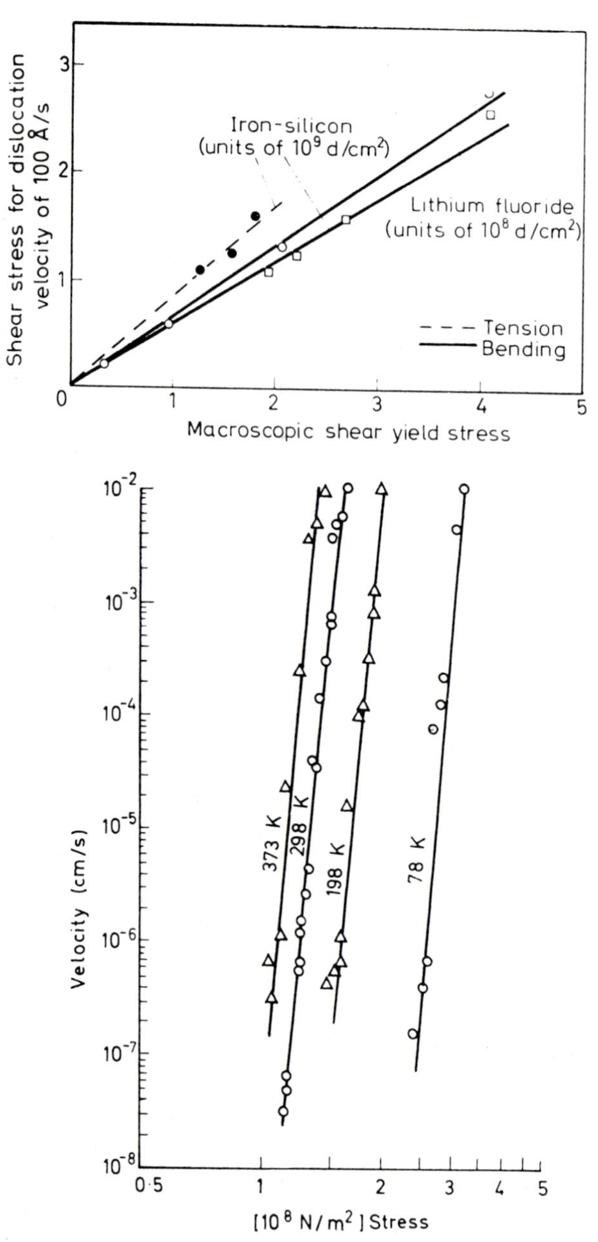

Figure 8.5 (a) Correlation between stress to cause dislocation motion and the macro-yield stresses of crystals. (b) Edge dislocation motions in Fe–3% Si crystals (after Stein and Low, *J. Appl. Physics, 1960,* **30,** 392, courtesy of the American Institute of Physics)

activation is able to aid the applied stress in overcoming them. Thermal activation helps a portion of the dislocation to cross the barrier (*see Figure 8.14*) after which glide then proceeds by the sideways movement of kinks. Materials with narrow dislocations therefore exhibit a significant temperature-sensitivity;

intrinsically hard materials rapidly lose their strength with increasing temperature, as shown schematically in *Figure 8.3(a)*. In this diagram the yield stress/modulus ratio is plotted against T/T_m to remove the effect of modulus which decreases with temperature. *Figure 8.3(b)* shows that materials which exhibit a strong temperature-dependent yield stress also exhibit a high strain-rate sensitivity, i.e. the higher the imposed strain rate the higher the yield stress. This arises because thermal activation is less effective at the faster rate of deformation.

Direct measurements of dislocation velocity v have now been made in some crystals by means of the etch pitting technique; the results of such an experiment are shown in *Figure 8.4*. Edge dislocations move faster than screws, because of the frictional drag of jogs on screws, and the velocity of both varies rapidly with applied stress τ according to an empirical relation of the form $v=(\tau/\tau_0)^n$ where τ_0 is the stress for unit speed and n is an index which varies for different materials. At high stresses the velocity may approach the speed of elastic waves $\approx 10^3$ m/s. The index n is usually low (<10) for intrinsically hard, covalent crystals such as Ge, ≈ 40 for b.c.c. crystals, and high (≈ 200) for intrinsically soft f.c.c. crystals. It is observed that a critical applied stress is required to start the dislocations moving and denotes the onset of microplasticity. A macroscopic tensile test is a relatively insensitive measure of the onset of plastic deformation and the yield or flow stress measured in such a test is related not to the initial motion of an individual dislocation, but to the motion of a number of dislocations at some finite velocity, e.g. ~ 10 nm/s as shown in *Figure 8.5(a)*. Decreasing the temperature of the test or increasing the strain-rate increases the stress level required to produce the same finite velocity (*see Figure 8.5(b)*. This observation is consistent with the increase in yield stress with decreasing temperature or increasing strain-rate. Most metals and alloys are hardened by cold working or by placing obstacles, e.g. precipitates, in the path of moving dislocations to hinder their motion. Such strengthening mechanisms increase the stress necessary to produce a given finite dislocation velocity in a similar way to that found by lowering the temperature.

8.2 Dislocation source operation

When a stress is applied to a metal, the specimen plastically deforms at a rate governed by the strain rate of the deformation process, e.g. tensile testing, rolling, etc., and the strain rate imposes a particular velocity on the mobile dislocation population. In a crystal of dimensions $L_1 \times L_2 \times 1$ cm shown in *Figure 8.6*, a dislocation with velocity v moves through the crystal in time $t=L_1/v$ and produces a shear strain b/L_2, i.e. the strain rate is bv/L_1L_2. If the density of glissible dislocations is ρ, the total number of dislocations which become mobile in the crystal is $\rho L_1 L_2$ and the overall strain rate is thus given by

$$\dot{\gamma}=\frac{b}{L_2}\frac{v}{L_1}\rho L_1 L_2=\rho b v$$

At conventional strain rates, e.g. $1\ \mathrm{s}^{-1}$, the dislocations would be moving at quite moderate speeds of a few cm/s if the mobile density $\approx 10^7/\mathrm{cm}^2$. During high-

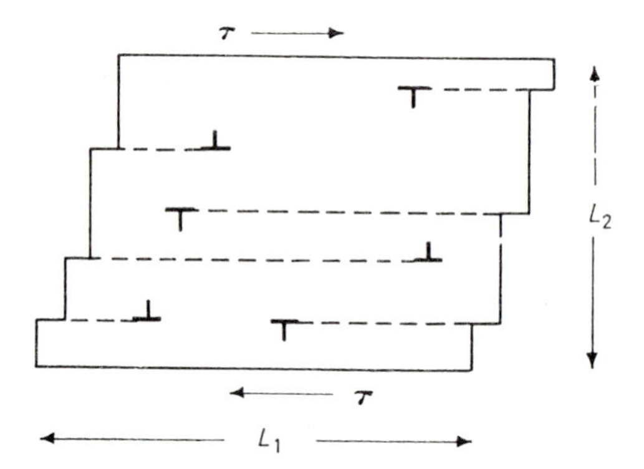

Figure 8.6 Shear produced by gliding dislocations

speed deformation the velocity approaches the limiting velocity. The shear strain produced by these dislocations is given by

$$\gamma = \rho b \bar{x}$$

where x is the average distance a dislocation moves. If the distance $x \simeq 10^{-4}$ cm (the size of an average sub-grain) the maximum strain produced by $\rho \approx 10^{7}$ is about $(10^{7} \times 3 \times 10^{-8} \times 10^{-4})$ which is only a fraction of 1 per cent. In practice, shear strains >100 per cent can be achieved and hence to produce these large strains many more dislocations than the original in-grown dislocations are required. To account for the increase in number of mobile dislocations during straining the concept of a dislocation source has been introduced. The simplest type of source is that due to Frank and Read and accounts for the regenerative multiplication of dislocations. A modified form of the Frank–Read source is the multiple cross-glide source, first proposed by Koehler, which as the name implies depends on the cross-slip of screw dislocations and is therefore more common in metals of intermediate and high stacking fault energy.

Figure 8.7 shows a Frank–Read source consisting of a dislocation line fixed at the nodes A and B (fixed, for example, because the other dislocations that join the nodes do not lie in slip planes). Because of its high elastic energy (≈ 4 eV per atom plane threaded by a dislocation) the dislocation possesses a line tension tending to make it shorten its length as much as possible (position 1, *Figure 8.7*). This line tension T is roughly equal to $\alpha \mu b^2$ where μ is the shear modulus, b the Burgers vector and α a constant usually taken to be about $\frac{1}{2}$. Under an applied stress the dislocation line will bow out, decreasing its radius of curvature until it reaches an equilibrium position in which the line tension balances the force due to the applied stress. Increasing the applied stress causes the line to decrease its radius of curvature further until it becomes semicircular (position 2). Beyond this point it has no equilibrium position so it will expand rapidly, rotating about the nodes and taking up the succession of forms indicated by 3, 4 and 5. Between stages 4 and 5 the two parts of the loop below AB meet and annihilate each other to form a complete dislocation loop, which expands into the slip plane and a new line source between A and B. The sequence is then repeated and one unit of slip is produced by each loop that is generated.

To operate the Frank–Read source the force applied must be sufficient to

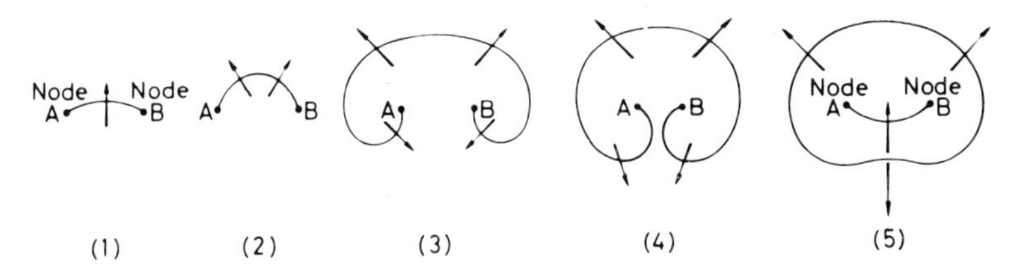

Figure 8.7 Successive stages in the operation of a Frank–Read source. The plane of the paper is assumed to be the slip plane

overcome the restoring force on the dislocation line due to its line tension. Referring to *Figure 8.8*, this would be $2T\,\mathrm{d}\theta/2 > \tau bl\,\mathrm{d}\theta/2$, and if $T \sim \mu b^2/2$ the stress to do this is about $\mu b/l$ where μ and b have their usual meaning and l is the length of the Frank–Read source; the substitution of typical values ($\mu = 4 \times 10^{10}\,\mathrm{N\,m^{-2}}$, $b = 2.5 \times 10^{-10}$ mm, and $l = 10^{-6}$ mm) into this estimate shows that a critical shear stress of about $100\,\mathrm{g\,mm^{-2}}$ is required. This value is somewhat less than, but of the same order as that observed for the yield stress of virgin pure metal single crystals. Another source mechanism involves multiple cross-slip as shown in *Figure 8.9*. It depends on the Frank–Read principle but does not require a dislocation segment to be anchored by nodes. Thus, if part of a moving screw dislocation undergoes double cross-slip (*see* page 206) the two pieces of edge dislocation on the cross-slip plane effectively act as anchoring points for a new source. The loop expanding on the slip plane parallel to the original plane may operate as a Frank–Read source and any loops produced may in turn cross-slip and become a source. This process therefore not only increases the number of dislocations on the original slip plane but also causes the slip band to widen.

The concept of the dislocation source accounts for the observation of slip bands on the surface of deformed metals. The amount of slip produced by the passage of a single dislocation is too small to be observable as a slip line or band under the light microscope. To be resolved it must be at least 300 nm in height and hence ≈ 1000 dislocations must have operated in a given slip band. Moreover, in general, the slip band has considerable width, which tends to support the operation of the cross-glide source as the predominant mechanism of dislocation multiplication during straining.

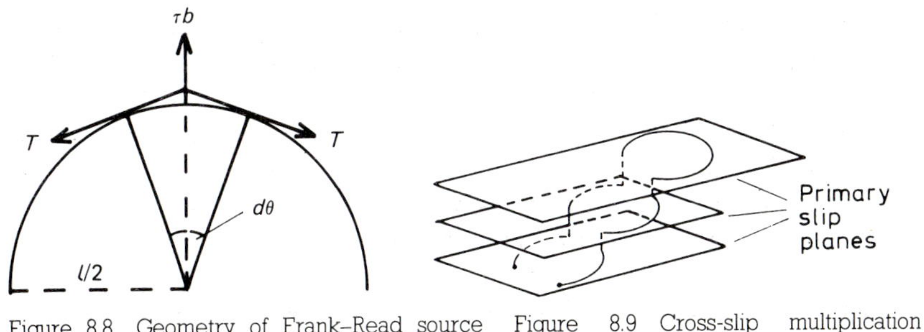

Figure 8.8 Geometry of Frank–Read source used to calculate the stress to operate

Figure 8.9 Cross-slip multiplication source

8.3 Yielding and dislocation multiplication

In some metals, the onset of macroscopic plastic flow begins in an abrupt manner with a yield drop in which the applied stress falls, during yielding, from an upper yield point to a lower yield point. Such yield behaviour is commonly found in intrinsically hard pure materials, e.g. Si, Ge, etc, which have a low dislocation density and a high Peierls–Nabarro stress (*see* section 8.1). When these materials are pulled in a tensile test the overall strain rate $\dot{\gamma}$ imposed on the specimen by the machine has to be matched by the motion of dislocations according to the relation $\dot{\gamma}=\rho bv$. However, because ρ is small the individual dislocations are forced to move at a high speed v, which is only attained at a high stress level (the upper yield stress) because of the large Peierls–Nabarro stress. As the dislocations glide at these high speeds, rapid multiplication occurs and the mobile dislocation density increases rapidly. Because of the increased value of the term ρv, a lower average velocity of dislocations is then required to maintain a constant strain rate, which means a lower glide stress. The stress that can be supported by the specimen thus drops during initial yielding to the lower yield point, and does not rise again until the dislocation–dislocation interactions caused by the increased ρ produce a significant work-hardening.

In the common f.c.c. metals the Peierls stress is quite small and the stress to move a dislocation is almost independent of velocity up to high speeds. If such metals are to show a multiplication type yield point, the density of mobile dislocations must be reduced virtually to zero. This can be achieved as shown in *Figure 8.10* by the tensile testing of whisker crystals which are very perfect. Yielding begins at the stress required to create dislocations in the perfect lattice, and the upper yield stress approaches the theoretical yield strength. Following multiplication, the stress for glide of these dislocations is several orders of magnitude lower.

B.C.C. transition metals such as iron are intermediate in their plastic behaviour between the f.c.c. metals and diamond cubic Si and Ge, as discussed earlier. Because of the significant Peierls stress these b.c.c. metals are capable of exhibiting a sharp yield point even when the initial mobile dislocation density is not zero, as shown by the calculated curves of *Figure 8.11*. However, in practice, the dislocation density of well-annealed pure metals is about 10^4 mm/mm^3 and too high for any significant yield drop due to dislocation multiplication.

Yield points are observed in both b.c.c. and f.c.c. metals when they contain sufficient impurity atoms to immobilize the grown-in dislocation density. The behaviour of dislocation lines to which impurity atoms have segregated is thus of importance in understanding and explaining the phenomenon of the sharp yield point in real metals. The best known example of sharp yielding is that exhibited by soft iron or mild steel where the effect is due to traces (<0.1 per cent) of carbon and nitrogen in the α-iron lattice. The theory of the yield phenomenon is not limited to iron, however, but applies to all metals, whether single or polycrystal. It is true the b.c.c. metals show the phenomenon strongly, but the f.c.c. and c.p.h. metals may also exhibit yielding effects, often to a lesser degree, under appropriate conditions, e.g. dislocation density, impurity (*see Figure 8.13*).

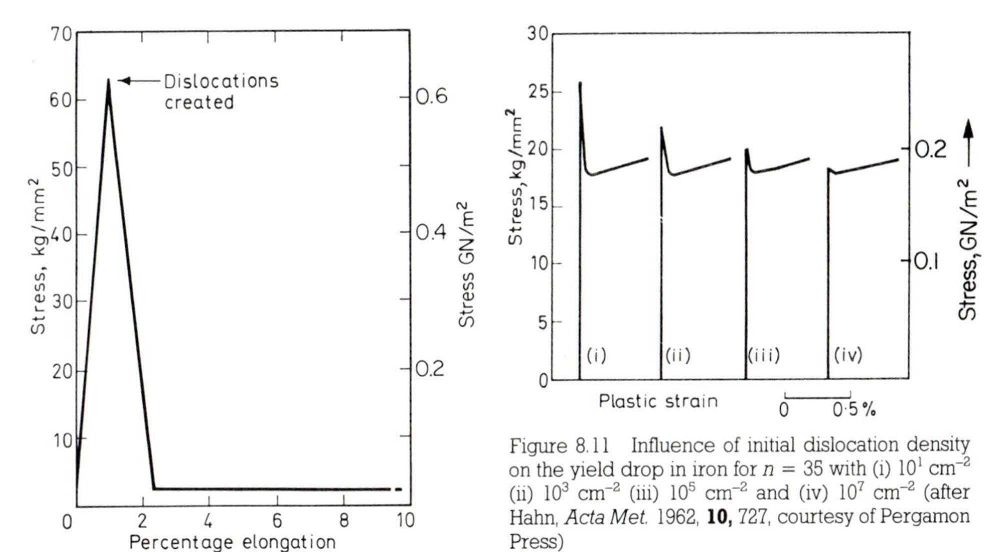

Figure 8.11 Influence of initial dislocation density on the yield drop in iron for $n = 35$ with (i) 10^1 cm^{-2} (ii) 10^3 cm^{-2} (iii) 10^5 cm^{-2} and (iv) 10^7 cm^{-2} (after Hahn, *Acta Met.* 1962, **10,** 727, courtesy of Pergamon Press)

Figure 8.10 Yield point in a copper whisker

8.4 The yield point and related effects

The main characteristics of the yield phenomenon in iron may be summarized as follows.

The yield point A specimen of iron during tensile deformation (*Figure 8.12*(*a*) Curve 1) behaves elastically up to a certain high load A, known as the upper yield point, and then it suddenly yields plastically. The important feature to notice from this curve is that the stress required to maintain plastic flow immediately after yielding has started is lower than that required to start it, as shown by the fall in load from A to B (the lower yield point). A yield point elongation to C then occurs after which the specimen work hardens and the curve rises steadily and smoothly.

Overstraining The yield point can be removed temporarily by applying a small preliminary plastic strain to the specimen. Thus, if after reaching the point D, for example, the specimen is unloaded and a second test is made fairly soon afterwards, a stress–strain curve of type 2 will be obtained. The specimen deforms elastically up to the unloading point, D, and the absence of a yield point at the beginning of plastic flow is characteristic of a specimen in an overstrained condition.

Strain-age hardening If a specimen which has been overstrained to remove the yield point is allowed to rest, or age, before retesting, the yield point returns as

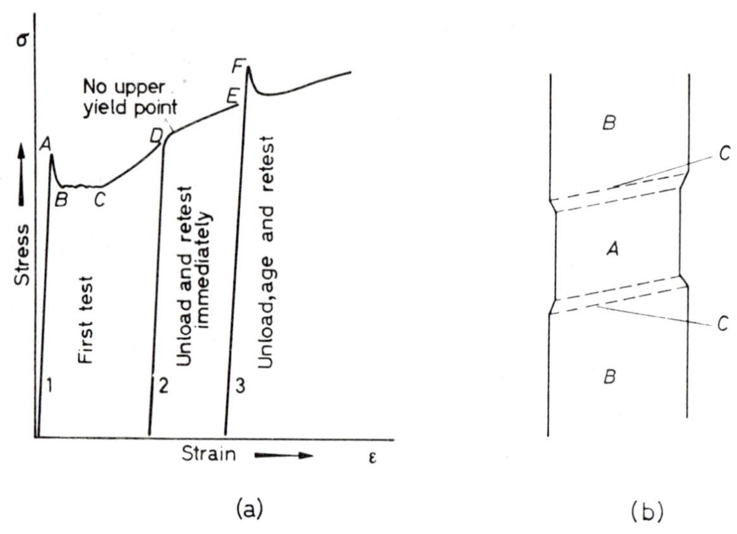

Figure 8.12 Schematic representation of (a) strain ageing and (b) Lüders band formation

shown in *Figure 8.12(a)* curve 3. This process, which is accompanied by hardening (as shown by the increased stress, EF, to initiate yielding) is known as strain ageing or more specifically strain-age hardening. In iron, strain ageing is slow at room temperature but is greatly speeded up by annealing at a higher temperature. Thus, a strong yield point returns after an ageing treatment of only a few seconds at 200°C, but the same yield point will take many hours to develop if ageing is carried out at room temperature.

Blue-brittleness As the temperature of the test is raised, a range is reached where the yield point becomes less pronounced and the slope of the plastic part of the curve becomes steeper and sometimes serrated. Consequently, the specimen rapidly reaches a high stress and failure occurs at a low elongation. This effect is known as blue-brittleness. In iron these effects appear at about 150 °C (at which temperature the specimen acquires a blue tinge due to superficial oxidation), but in very fast strain-rate tests blue-brittleness may occur at higher temperatures, while in slow strain-rate testing the range of temperature in which it occurs may be depressed to around room temperature.

Lüders band formation Closely related to the yield point is the formation of Lüders bands. These bands are markings on the surface of the specimen which distinguish those parts of the specimen that have yielded, A, from those which have not, B. Arrival at the upper yield point is indicated by the formation of one or more of these bands and as the specimen passes through the stage of the yield point elongation these bands spread along the specimen and coalesce until the entire gauge length has been covered. At this stage the whole of the material within the gauge length has been overstrained, and the yield point elongation is complete. The growth of a Lüders band is shown diagrammatically in *Figure 8.12(b)*. It should be noted that the band is a macroscopic band crossing all the grains in the cross-section of a polycrystalline specimen, and thus the edges of the band are not necessarily the traces of individual slip planes. A second point to observe is that the rate of plastic flow in the edges of a band can be very high even in an apparently slow test; this is because the zones, marked C in *Figure 8.12(b)*, are very narrow compared with the gauge length.

These Lüders bands frequently occur in drawing and stamping operations when the surface markings in relief are called stretcher strains. These markings are unsightly in appearance and have to be avoided on many finished products. The remedy consists in overstraining the sheet prior to pressing operations, by means of a temper roll, or roller levelling, pass so that the yield phenomenon is eliminated. It is essential, once this operation has been performed, to carry out pressing before the sheet has time to strain-age; the use of a 'non-ageing' steel is an alternative remedy.

8.4.1 Evidence for the influence of impurity atoms

Experimental work has shown that solute atoms in small concentrations are responsible for the above effects. The first clue, which indicates the effect that traces of carbon or nitrogen have on the properties of α-iron, is given by the grain growth behaviour of ferrite. If a piece of commercial iron is given a strain-anneal treatment, it is found that a state of exaggerated grain growth is difficult to develop, whereas after decarburization by wet-hydrogen annealing, grain growth occurs much faster and single crystals can often be produced. These observations contain a hint concerning the role played by the carbon or nitrogen atoms, since the grain boundaries must be anchored when the interstitial elements are present, but become mobile when they are absent. A similar mechanism must be responsible for the yield phenomenon since the yield point can be removed by annealing at 700 °C in wet-hydrogen atmosphere, and cannot subsequently be restored by any strain-ageing treatment. Conversely, exposing the decarburized specimen to an atmosphere of dry hydrogen containing a trace of hydrocarbon at 700 °C for as little as one minute restores the yield point and its related phenomenon.

The carbon and nitrogen atoms can also be removed from solution in other ways: for example, by adding to the iron such elements as molybdenum, manganese, chromium, vanadium, niobium or titanium which have a strong affinity for forming carbides or nitrides in steels. For this reason, these elements are particularly effective in removing the yield point and producing a non-strain ageing steel.

8.4.2 The formation of 'atmospheres' of solute atoms round dislocations

The distorted atomic structure around a dislocation indicates that there is a large strain energy associated with it. Around an edge dislocation, for example, the atoms above the slip plane are compressed whilst those below are extended. A crystal will try to reduce this strain energy in the core of the dislocation by rearranging the structure there. The widening of dislocations in f.c.c. metals by dissociation is one example of this, but there are other methods, particularly important for narrow dislocations. One of these is that those dislocations with large slip vectors (i.e. $b >$ atomic spacing) may become hollow, as happens in silicon carbide. However, such a process is unlikely for single dislocations in metal crystals, although it may happen when a number of dislocations are piled up so closely by a concentrating stress that they begin to coalesce, and recent theories of brittle fracture are based on this idea (*see* Chapter 14). The most common method of reducing the energy is by the segregation of impurity atoms, for example carbon or nitrogen in iron, to the structure around the dislocations, where they reduce not only the strain energy of the dislocations but also the strain energy associated with the solute atoms.

The strain energy due to the distortion of a solute atom can be relieved if it fits into a structural region where the local lattice parameter approximates to that of the natural lattice parameter of the solute. Such a condition will be brought about by the segregation of solute atoms to the dislocations, with large substitutional atoms taking up lattice positions in the expanded region, and small ones in the compressed region; small interstitial atoms will tend to segregate to interstitial sites below the half-plane. Thus, where both dislocations and solute atoms are present in the lattice, interactions of the stress field can occur, resulting in a lowering of the strain energy of the system. This provides a driving force tending to attract solute atoms to dislocations and if the necessary time for diffusion is allowed, a solute atom 'atmosphere' will form around each dislocation. The precipitation of carbon on dislocations in α-iron is shown in the micrograph of *Figure 11.12*.

8.4.3 The effect of atmospheres on plastic flow

In a solid solution alloy the most stable configuration will be one in which the solute atoms are arranged in atmospheres around the dislocations. Under an applied stress the dislocations will try to move out of their atmosphere, but the atmospheres exert a restraining force tending to pull them back again. The simplest situation for the onset of plastic flow arises if the dislocations can be unpinned from their atmospheres. The upper yield stress is then reached when the anchoring force of the atmosphere is overcome, and once the dislocations have broken away from their atmospheres they move through a relatively 'clean' piece of lattice. The applied force needed to overcome the anchoring effect of the atmosphere is thus larger than the force needed to keep the dislocations in motion once they have been unpinned, and a sudden drop in stress must then occur at the transition from the anchored to the free state; the mobile dislocation density increases very rapidly. The lower yield stress is therefore the stress at which freed dislocations continue to move and produce plastic flow. The overstrained condition corresponds to a case where the dislocations that have been freed from their atmosphere and brought to rest by unloading the specimen, are set in motion again by reloading the specimen before the atmospheres have had time to re-establish themselves by diffusion of the solute atoms. If, however, time is allowed for diffusion to take place, new atmospheres have time to re-form and the original yield-point characteristics should re-appear. This is the strain-aged condition.

Any metal containing dissolved impurity atoms, which produce lattice distortions that can interact with the dislocation stress field, should, if the above explanation is correct, show yield-point effects. This has been shown to be the case in all the common metal structures provided the appropriate solute elements are present, and correct testing procedure adopted. The effect is particularly strong in the b.c.c. metals and has been observed in α-iron, molybdenum, niobium, vanadium and β-brass each containing a strongly interacting interstitial solute element. The hexagonal metals, e.g. cadmium and zinc, can also show the phenomenon provided interstitial nitrogen atoms are added. The copper-based and aluminium-based f.c.c. alloys also exhibit yielding behaviour but often to a lesser degree. In this case it is substitutional atoms, e.g. zinc in α-brass and copper in aluminium alloys, which are responsible for the phenomenon.

The true upper yield point is normally not observed in practice, because non-axial loading, scratches, inclusions, etc, provide stress concentrations that set off local yielding prematurely; the pre-yield micro-strain, i.e. deviations from linearity before the yield point is reached, observed in some tests is evidence of this. However, by taking exceptional precautions in testing procedure, upper yield points far above the usual value can be obtained (e.g. twice the lower yield stress). The upper yield point in conventional experiments on polycrystalline materials is the stress at which prematurely yielded zones can trigger yield in adjacent grains. As more and more grains are triggered the yield zones spread across the specimen and form a Lüders band. Yield at the lower yield point is essentially the same process but occurs at a lower stress, because so many places exist along the front of a fully developed Lüders band where triggering can take place, and because the change in cross-section and orientation intensifies the stress at the Lüders front.

The propagation of yield is thought to occur when a dislocation source operates and releases an avalanche of dislocations into its slip plane which eventually pile up at a grain boundary or other obstacle. The stress concentration at the head of the pile-up acts with the applied stress on the dislocations of the next grain and unpins the nearest source, so that the process is repeated in the next grain. The applied shear stress σ_y at which yielding propagates is given by

$$\sigma_y = \sigma_i + (\sigma_c r^{1/2}) d^{-1/2} \tag{8.1}$$

where r is the distance from the pile-up to the nearest source, $2d$ is the grain diameter and σ_c is the stress required to operate a source which involves unpinning a dislocation τ_c at that temperature. Equation 8.1 reduces to the Hall–Petch equation $\sigma_y = \sigma_i + k_y d^{-1/2}$, where σ_i is the 'friction' stress term, and k_y the grain size dependence parameter ($= m^2 \tau_c r^{1/2}$) discussed in section 8.9.

8.5 The interaction of solute atoms with dislocations

Although it is evident that the yield phenomenon in metals is due to the segregation of solute atoms to dislocations, one question which immediately arises is, why is the pinning of dislocations particularly effective in α-iron but not in metals such as copper or aluminium? This difference in behaviour becomes even more difficult to understand when it is realized that even in essentially pure metals ($>99.99\%$ pure) there are sufficient solute atoms available to allow at least one impurity atom to segregate on each atom plane of every dislocation in the lattice. For example, let us consider a block of copper 1 cm^3 in volume threaded by the normal density of dislocations expected for such a metal, 10^8 lines cm^{-2}. Thus, of the total number of atoms in the block, 6×10^{22}, there are about $10^8 \times (4 \times 10^7)$ or 4×10^{15} lying on dislocation lines, so that the total impurity content required to produce one solute atom on each atom plane of every dislocation is only about $(4 \times 10^{15})/(6 \times 10^{22})$ or 1.5×10^{-5} atomic per cent. To resolve such apparent difficulties it is necessary to examine more closely the ways in which dislocations can interact with solute atoms.

Four main types of interaction have been recognized: elastic, chemical, electrical and geometrical. Of these, the most important is usually the elastic interaction, and so it is dealt with more fully than the others.

Elastic interaction

As pointed out in Chapter 6, the stress field of an edge dislocation possesses both hydrostatic components and shear components, whilst that of a screw dislocation is purely one of shear. Stresses in a lattice can also arise from distortions produced by dissolved solute atoms. Where the distortions are symmetrical the stress fields will contain only hydrostatic components, whereas non-symmetrical distortions give rise to both hydrostatic and shear stresses. Where both dislocations and solute atoms are present in the lattice, interaction of the 4 stress fields can occur, resulting in a lowering of the strain energy of the system.

The strain energy due to distortion can be relieved in the large lattice parameter of the upper region and the small lattice parameter of the lower region can be adopted naturally. Such a condition can be brought about by the segregation of solute atoms to the dislocation, with large atoms taking up positions in the expanded region and small ones in the compressed region. This is an example of a lowering of strain energy produced by the interaction of hydrostatic components of dislocations and solute atom stress fields. It will be clear, of course, that shear stress interactions may also occur in cases where the solute atom distortions are non-symmetrical, and such distortions are often produced by interstitial impurity atoms.

One of the reasons why iron containing carbon or nitrogen shows very marked yield point effects now becomes apparent, since there is a strong interaction, of both a hydrostatic and shear nature, between these solute atoms and the dislocations. The solute atoms occupy interstitial sites in the lattice and produce large tetragonal distortions as well as large volume expansions. Consequently, they can interact with both shear and hydrostatic stresses and can lock screw as well as edge dislocations. Strong yielding behaviour is also expected in other b.c.c. metals, provided they contain interstitial solute elements. On the other hand, in the case of f.c.c. metals the arrangement of lattice positions around either interstitial or substitutional sites is too symmetrical to allow a solute atom to produce an asymmetrical distortion, and the atmosphere locking of screw dislocations, which requires a shear stress interaction, would appear to be impossible. Then by this argument, since the screw dislocations are not locked, a

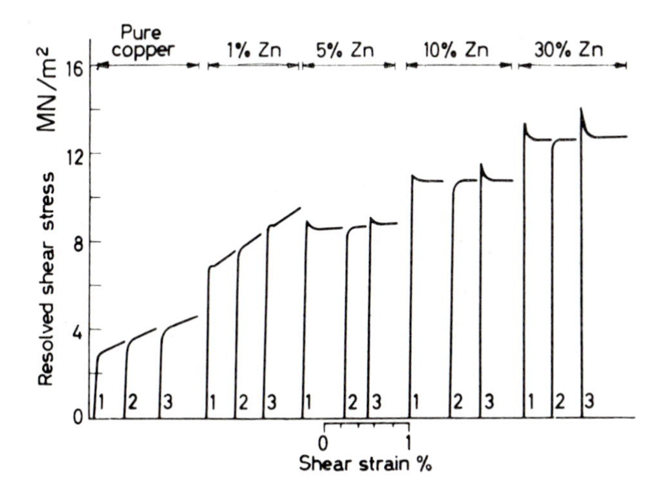

Figure 8.13 Stress–strain curves on copper and brass crystals grown in argon, all strained at room temperature. Curve 1—first loading; curve 2—immediately afterwards; curve 3—after 2 hours at 200°C (after Ardley and Cottrell, *Proc. Roy. Soc.* 1953, **A219,** 328)

drop in stress at the yield point should not be observed. Nevertheless, yield points are observed in f.c.c. materials and one reason for this is, as we have seen in Chapter 6, that unit dislocations in close-packed planes of f.c.c. metals dissociate into pairs of partial dislocations which must glide together because they are elastically coupled by a stacking fault. Moreover, since their Burgers vectors intersect at 120° there is no orientation of the line of the pair for which both can be pure screws. At least one of them must have a substantial edge component, and a locking of this edge component by hydrostatic interactions should cause a locking of the pair although it will undoubtedly be weaker. The yield phenomenon observed in copper containing small amounts of zinc (*see Figure 8.13*), and in aluminium containing copper, magnesium or zinc supports this prediction.

In its quantitative form the theory of solute atom locking has been applied to the formation of an atmosphere around an edge dislocation due to hydrostatic interaction. Since hydrostatic stresses are scalar quantities, no knowledge is required in this case of the orientation of the dislocation with respect to the interacting solute atom, but it is necessary in calculating shear stresses interactions*. Cottrell and Bilby have shown that if the introduction of a solute atom causes a volume change Δv at some point in the lattice where the hydrostatic pressure of the stress field is p, the interaction energy is

$$V = p\,\Delta v = K\Theta\,\Delta v$$

* To a first approximation a solute atom does not interact with a screw dislocation since there is no dilatation around the screw; a second order dilatation exists however, which gives rise to a non-zero interaction falling off with distance from the dislocation according to $1/r^2$. In real crystals, anisotropic elasticity will lead to first order size effects even with screw dislocations and hence a substantial interaction is to be expected.

where K is the bulk modulus and Θ is the local dilatation strain. The dilatation strain at a point (R, θ) from a positive edge dislocation is $b(1-2\nu)\sin\theta/2\pi R(1-\nu)$, and substituting $K=2\mu(1+\nu)/3(1-2\nu)$, where μ is the shear modulus and ν Poisson's ratio, we get the expression

$$V_{(R,\theta)} = b(1+\nu)\mu\,\Delta v\sin\theta/3\pi R(1-\nu) = A\sin\theta/\mathrm{R} \tag{8.2}$$

This is the interaction energy at a point whose polar co-ordinates with respect to the centre of the dislocation are R and θ. We notice that V is positive on the upper side $(0<\theta<\pi)$ of the dislocation for a large atom $(\Delta v>0)$, and negative on the lower side, which agrees with the qualitative picture of a large atom being repelled from the compressed region and attracted into the expanded one.

It is expected that the site for the strongest binding energy V_{max} will be at a point $\theta=3\pi/2$, $R=r_0\simeq b$; and using known values of μ, ν and Δv in equation 8.2 we obtain $A\simeq 3\times 10^{-29}$ N m^2 and $V_{max}\simeq 1$ eV for carbon or nitrogen in α-iron. This value is almost certainly too high because of the limitations of the interaction energy equation in describing conditions near the centre of a dislocation, and a more realistic value obtained from experiment (e.g. internal friction experiments) is $V_{max}\simeq\frac{1}{2}$ to $\frac{3}{4}$ eV. For a substitutional solute atom such as zinc in copper Δv is not only smaller but also easier to calculate from lattice parameter measurements. Thus, if r and $r(1+\varepsilon)$ are the atomic radii of the solvent and solute respectively, where ε is the misfit value, the volume change Δv is $4\pi r^3\varepsilon$ and equation 8.2 becomes

$$V = 4(1+\nu)\mu b\varepsilon r^3\sin\theta/3(1-\nu)R = A\sin\theta/R$$

Taking the known values $\mu=40\,\mathrm{GN/m^2}$ $\nu=0.36$, $b=2.55\times 10^{-10}$ m, $r_0 b/2$ and $\varepsilon=0.06$, we find $A\simeq 5\times 10^{-30}$ N m^2 which gives a much lower binding energy, $V_{max}=\frac{1}{8}$ eV.

Condensed and dilute atmospheres The yield phenomenon is particularly strong in iron because an additional effect is important; this concerns the type of atmosphere a dislocation gathers round itself. During the strain-ageing process migration of the solute atoms to the dislocation occurs and two important cases arise. First, if all the sites at the centre of the dislocation become occupied the atmosphere is then said to be condensed; each atom plane threaded by the dislocation contains one solute atom at the position of maximum binding together with a diffuse cloud of other solute atoms further out. If, on the other hand, equilibrium is established before all the sites at the centre are saturated, a steady state must be reached in which the probability of solute atoms leaving the centre can equal the probability of their entering it. The steady state distribution of solute atoms around the dislocations is then given by the relation

$$c_{(R,\theta)} = c_0\exp\left[V_{(R,\theta)}/\boldsymbol{k}T\right] \tag{8.3}$$

where c_0 is the concentration far from a dislocation, $\boldsymbol{k}$ is Boltzmann's constant, T

is the absolute temperature and c the local impurity concentration at a point near the dislocation where the binding energy is V. This is known as the dilute or Maxwellian atmosphere. Clearly, the form of an atmosphere will be governed by the concentration of solute atoms at the sites of maximum binding energy, V_{max} and for a given alloy (i.e. c_0 and V_{max} fixed) this concentration will be

$$c_{V_{max}} = c_0 \exp(V_{max}/\boldsymbol{k}T) \tag{8.4}$$

as long as $c_{V_{max}}$ is less than unity. The value of $c_{V_{max}}$ depends only on the temperature, and as the temperature is lowered $c_{V_{max}}$ will eventually rise to unity. By definition the atmosphere will then have passed from a dilute to a condensed state. The temperature at which this occurs is known as the condensation temperature T_c, and can be obtained by substituting the value $c_{V_{max}} = 1$ in equation 8.4 when

$$T_c = V_{max}/\boldsymbol{k} \ln(1/c_0) \tag{8.5}$$

Substituting the value of V_{max} for iron, i.e. $\frac{1}{2}$ eV in equation 8.5, we find that only a very small concentration of carbon or nitrogen is necessary to give a condensed atmosphere at room temperature, and with the usual concentration strong yielding behaviour is expected up to temperatures of about 400°C. In the f.c.c. lattice, although the locking between a solute atom and a dislocation is likely to be weaker, condensed atmospheres are still possible if this weakness can be compensated for by sufficiently increasing the concentration of the solution. This may be why examples of yielding in f.c.c. materials have been mainly obtained from alloys. Solid solution alloys of aluminium usually contain less than 0.1 atomic per cent of solute element, and these show yielding in single crystals only at low temperature (e.g. liquid nitrogen temperature, -196 °C) whereas super-saturated alloys show evidence of strong yielding even in polycrystals at room temperature; copper dissolved in aluminium has a misfit value $\varepsilon \simeq 0.12$ which corresponds to $V_{max} = \frac{1}{4}$ eV, and from equation 8.5 it can be shown that a 0.1 atomic per cent alloy has a condensation temperature $T_c = 250$ K. Copper-based alloys, on the other hand, usually form extensive solid solutions, and, consequently, concentrated alloys may exhibit strong yielding phenomena (*see Figure 8.13*). The best known example is α-brass and, because $V_{max} \simeq \frac{1}{8}$ eV, a dilute alloy containing 1 atomic per cent zinc has a condensation temperature ($T_c \simeq 300$ K) at room temperature. At low zinc concentrations (1–10 per cent) the yield point in brass is probably solely due to the segregation of zinc atoms to dislocations. At higher concentrations, however, it may also be due to short-range order which gives rise to a geometrical interaction (*see* page 272) rather than an elastic interaction.

Snoek ordering Another type of interaction of particular importance in b.c.c. metals is that due to local ordering of the interstitial impurity atoms. From *Figure 5.6* it is evident that the interstitial atoms in α-iron can occupy any of the three types of octahedral sites, each of which gives rise to a local distortion of the lattice to a tetragonal configuration. It was pointed out by Snoek, however, that the application of a unidirectional stress to the iron specimen would favour the occupation of those octahedral sites for which the tetragonality coincided with the over-all plastic strain. In this way he was able to explain the internal friction relaxation peak which occurs near room temperature at frequencies about one cycle per second. Applying this idea to the interaction between solute atoms and dislocations, Schoeck and Seeger have suggested that in the stress field of a dislocation a similar state of order in the distribution of carbon atoms exists. As a consequence, before the dislocation can be moved, the state of local order must be destroyed and this, naturally, leads to a higher yield stress. A more detailed examination shows that on this model the yield stress is approximately independent of temperature, and for this reason it is thought that this form of locking operates above room temperature.

Inhomogeneity interaction

A different type of elastic interaction can exist which arises from the different elastic properties of the solvent matrix and the region near a solute. Such an inhomogeneity interaction has been analysed both for a rigid and soft spherical regions; the former corresponds to a relatively hard impurity atom and the latter to a vacant lattice site. The results indicate that the interaction energy VB/r^2 where B is a constant involving elastic constants and atomic size. It is generally believed that the inhomogeneity effect dominates the size effect for vacancy-dislocation interactions (*see* Chapter 9).

8.6 The variation of yield stress with temperature

Before a dislocation source anchored by solute atoms can operate, the dislocation line constituting the source has to be unpinned by pulling it away from its atmosphere. The force to do this is expected to be extremely sensitive to temperature; from the theory of atmospheres two factors are of particular importance. First it is clear that the theoretical break-away stress changes with the nature of the dislocation atmosphere; as the distribution of solute atoms changes from condensed to dilute the pinning stress decreases. The second factor concerns the importance of thermal fluctuations in helping a dislocation to escape from its atmosphere. The binding of a solute atom to a dislocation is short range in nature, and is effective only over an atomic distance or so (*Figure 8.14*). Moreover, the dislocation line is flexible and this enables yielding to begin by throwing forward a small length of dislocation line, only a few atomic spacings long, beyond the position marked x_2 in *Figure 8.14*. The applied stress then separates the rest of the dislocation line from its anchorage by pulling the sides of this loop outward along the dislocation line, i.e. by double kink movement. Thus, when a stress σ, which is less than σ_{max}, is applied to the dislocation it moves

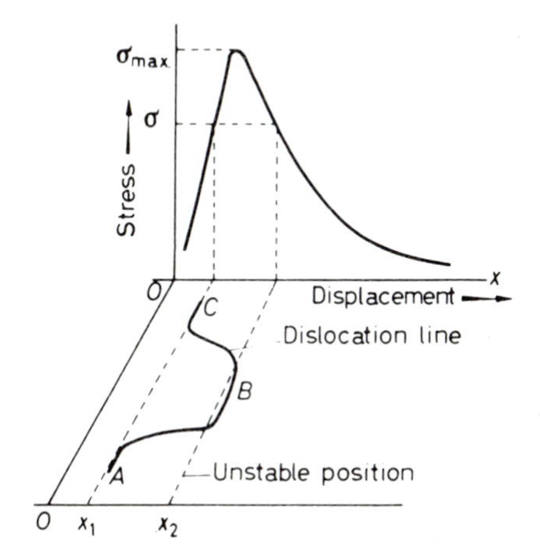

Figure 8.14 Stress–displacement curve for the breakaway of a dislocation from its atmosphere (after Cottrell, *Deformation at High Rates of Strain*, courtesy of the Institution of Mechanical Engineers)

forward to the position of stable equilibrium marked x_1, and even though the applied stress is less than σ_{max} the dislocation can break away from its anchorage provided a thermal stress fluctuation is able to throw a loop forward beyond x_2. This is because the position x_2 is one of unstable equilibrium beyond which the applied stress always exceeds the effective internal stress pulling the dislocation back to its anchorage.

Although the theory of atmospheres predicts that the yield point should depend sensitively on temperature (*see Figure 8.15(a)*), recent experiments have shown that k_y, the grain-size dependence parameter in the Hall–Petch equation, in most annealed b.c.c. metals is almost independent of temperature down to the range (<100 K) where twinning occurs, and that practically all the large temperature-dependence is due to σ_i (*see Figure 8.15(b)*). To explain this observation it is argued that when locked dislocations exist initially in the material, yielding starts by unpinning them if they are weakly locked (this corresponds to the condition envisaged by Cottrell–Bilby), but if they are strongly locked it starts instead by the creation of new dislocations at points of stress concentration, for example at or near grain boundaries. The latter is an athermal process and thus k_y is almost independent of temperature. Because of the rapid diffusion of interstitial elements the conventional annealing and normalizing treatments should commonly produce strong locking. In support of this theory, it is observed that k_y is dependent on temperature in the very early stages of ageing following either straining or quenching but on subsequent ageing k_y becomes temperature-independent. The interpretation of k_y therefore depends on the degree of ageing.

Direct observations of crystals that have yielded show that the majority of the strongly anchored dislocations remain locked and do not participate in the yielding phenomenon. Thus large numbers of dislocations are generated during yielding by some other mechanism than breaking away from Cottrell atmospheres, and the rapid dislocation multiplication, which can take place at the high stress levels, is now considered the most likely possibility. Prolonged

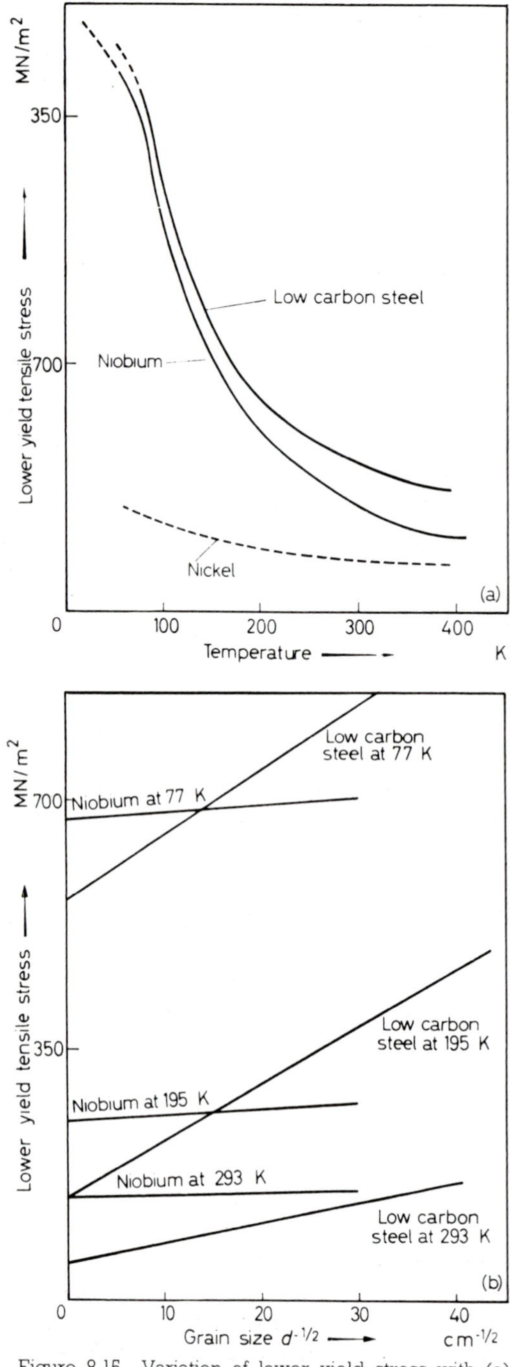

Figure 8.15 Variation of lower yield stress with (a) temperature and (b) grain size, for low carbon steel and niobium; the curve for nickel is shown in (a) for comparison (after Adams, Roberts and Smallman, *Acta Metall.*, 1960, **8**, 328, Hull and Mogford, *Phil. Mag.* 1958, **3**, 1213)

ageing tends to produce coarse precipitates along the dislocation line (*see Figure 11.12*) and unpinning by bowing out between them should easily occur before grain boundary creation. This unpinning process would also give k_y independent of temperature.

8.7 Other types of solute atom-dislocation interaction

Chemical interaction

We have seen in Chapter 6 that when a slip dislocation in the f.c.c. lattice dissociates into a pair of half-dislocations the latter are separated by a stacking fault, i.e. by a layer about two atoms in thickness where the crystal structure locally changes to c.p.h. From alloy chemistry and thermodynamic considerations we would, therefore, expect that when solute atoms segregate to such an extended dislocation, the concentration of solute atoms in the faulted layer would be different from that which exists in the surrounding f.c.c. phase. Moreover, since the stacking fault energy varies with alloy composition, after re-distribution the fault width may be expected to alter. This heterogeneous distribution of solute atoms around a dislocation may be regarded as a chemical interaction with the dislocation and, since it was first suggested by Suzuki, it is often referred to as Suzuki locking.

Experimental verification of this effect is difficult to obtain because the locking force is estimated to be about one-tenth of that due to Cottrell locking, with which it must be associated. Suzuki locking does, however, differ in some respects from atmosphere locking. For example, being temperature independent it could be important at high temperatures where Cottrell locking dies away. Furthermore, the interaction is independent of dislocation orientation and should be equally strong for both edge and screw dislocations; this is another reason why yield points are observed in alloys with close-packed structures.

Electrical interaction

Two solute atoms which dissolve in a short metal, e.g. copper, with the same degree of lattice misfit ε, may still harden the solid solution to a different extent if their valencies are not the same. This suggests that a solute atom may interact electrically with a dislocation, since the excess charge round a solute atom relative to that round a solvent atom depends on the difference in valency. However, this small excess charge which exists round a solute atom can only interact with a dislocation if the latter also has an electrical field associated with it. The electrical field around a dislocation is thought to arise from the lattice distortion associated with it, because there the dilatation strain will locally change the energy level of the Fermi surface. However, because the hydrostatic pressure varies round an edge dislocation, the Fermi level will be different above and below the slip plane

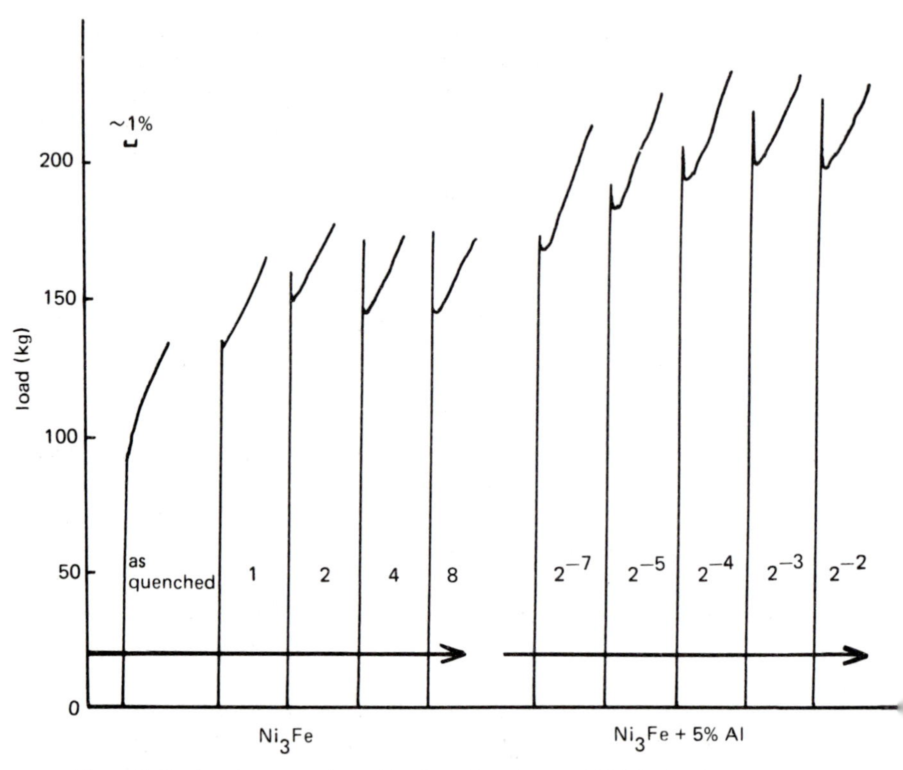

Figure 8.16 Development of a yield point with ageing at 490°C for the times indicated, (a) Ni_3Fe, (b) Ni_3Fe + 5%Al; the tests are at room temperature

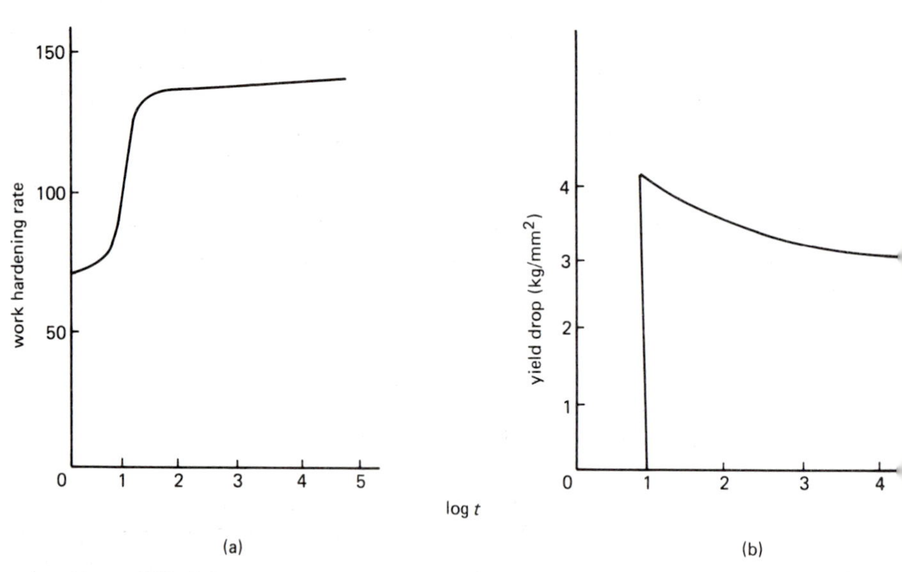

Figure 8.17 Influence of ageing at 350°C on (a) the work-hardening rate, and (b) the development of a yield point in Cu_3Au

and, as the free electrons cannot permit such a state of affairs to exist, some of the electrons will move from the compressed side of the dislocation to the expanded side. This movement will give rise to the formation of an electrical dipole.

Calculations show that the interaction which arises from this effect is small and, in copper at least, is several times smaller than the elastic interaction, e.g. for the solution of zinc in copper it is only 1/50 eV.

Geometrical interaction

This interaction is concerned with the presence of a dislocation in an ordered alloy. The structure of superdislocations separating an anti-phase boundary (APB) has been considered in Chapter 6. (*see Figure 6.33*).

Discontinuous yield points have been observed in a wide variety of A_3B-type alloys. *Figure 8.16* shows the development of the yield point in Ni_3Fe on ageing and the way it is increased by the addition of Al which speeds up the kinetics of ordering. The link between yield points and superdislocations is confirmed by the observation that the yield drop and work-hardening rate increase in the same way with increase of order. In Cu_3Au, for example, a transition from groups of single dislocations to more randomly arranged superdislocation pairs takes place at $\sim S = 0.7$ and this coincides with the onset of a large yield drop and rapid rise in work hardening (*see Figure 8.17*).

Sharp yielding may be explained by at least three mechanisms namely, (i) the stress to operate pre-existing single dislocation sources is much higher than the stress to operate subsequently formed superdislocation sources, (ii) cross-slip of one of the superdislocation pair on to the cube plane to lower the APB energy effectively pinning the pair, and (iii) dislocation locking by re-arrangement of the APB on ageing; the shear APB between a pair of superdislocations is likely to be energetically unstable since there are many like bonds across the interface, and thermal activation will modify this sharp interface by atomic rearrangement. This APB-locking model will give rise to sharp yielding because the energy required by the lead dislocation in creating sharp APB is greater than the energy released by the trailing dislocation initially moving across diffuse APB. Experimental evidence favours the APB-model and weak-beam electron microscopy shows that the superdislocation separation for a shear APB corresponds to an energy of 48 ± 5 mJ/m^2, whereas a larger dislocation separation corresponding to an APB energy of 25 ± 3 mJ/m^2 was observed for a strained and aged Cu_3Au (*see Figure 8.18*).

8.8 The kinetics of strain ageing

Under a force F an atom migrating by thermal agitation acquires a steady drift velocity $v = DF/kT$ (in addition to its random diffusion movements) in the direction of the F, where D is the coefficient of diffusion. The force attracting a solute atom to a dislocation is the gradient of the interaction energy dV/dr and

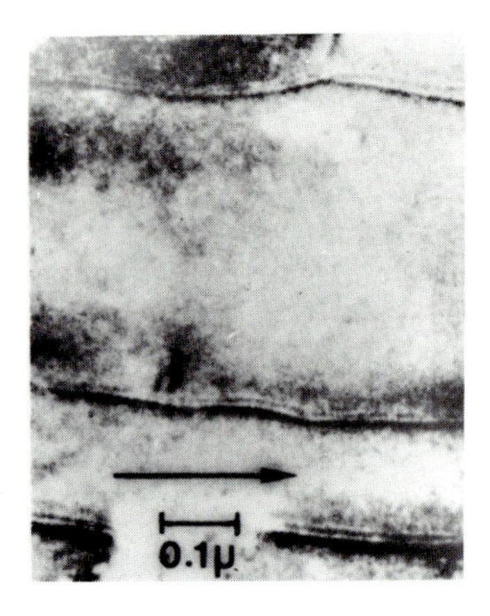

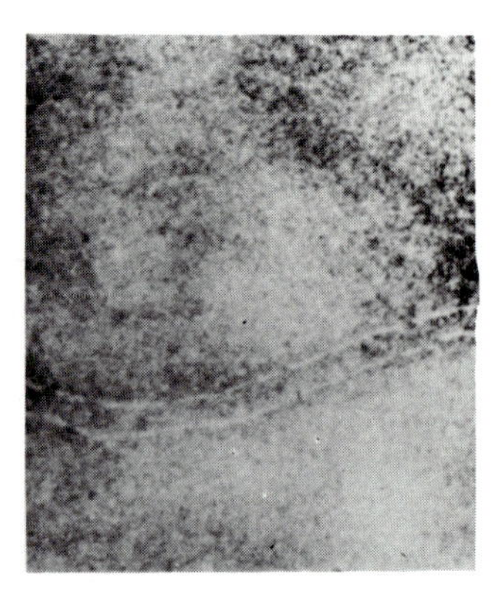

Figure 8.18 Weak-beam micrographs showing separation of superdislocation partials in Cu_3Au (a) as deformed, (b) after ageing at 225°C (after Morris and Smallman)

hence $v = (D/\boldsymbol{k}T)(A/r^2)$. Thus atoms originally at a distance r from the dislocation reach it in a time given approximately by

$$t = r/v = r^3\boldsymbol{k}T/AD$$

After this time t the number of atoms to reach unit length of dislocation is

$$n(t) = \pi r^2 c_0 = \pi c_0[(AD/\boldsymbol{k}T)t]^{2/3}$$

where c_0 is the solute concentration in uniform solution in terms of the number of atoms per unit volume. If ρ is the density of dislocations (cm/cm^3) and f the fraction of the original solute which has segregated to the dislocation in time t then,

$$f = \pi\rho[(AD/\boldsymbol{k}T)t]^{2/3} \tag{8.7}$$

This expression is valid for the early stages of ageing, and may be modified to fit the later stages by allowing for the reduction in the matrix concentration as ageing proceeds, such that the rate of flow is proportional to the amount left in the matrix,

$$\mathrm{d}f/\mathrm{d}t = \pi\rho(AD/\boldsymbol{k}T)^{2/3}(2/3)t^{-1/3}(1-f)$$

which when integrated gives

$$f = 1 - \exp\{-\pi\rho[(AD/\boldsymbol{k}T)t]^{2/3}\} \tag{8.8}$$

This reduces to the simpler equation 8.7 when the exponent is small, and is found to be in good agreement with the process of segregation and precipitation on

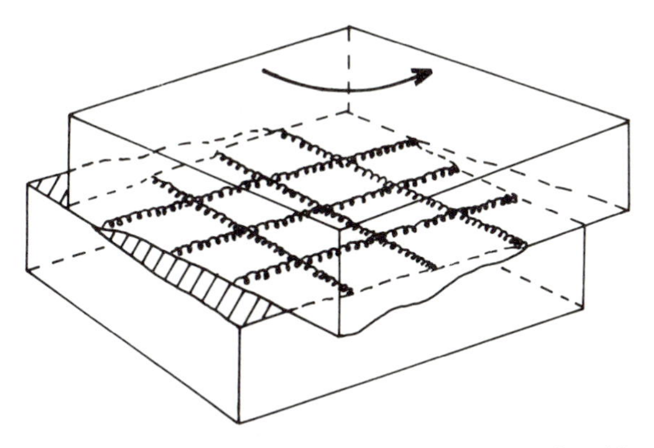

Figure 8.19 Representation of a twist boundary produced by cross-grid of screw dislocations

dislocations in several b.c.c. metals. For carbon in α-Fe, Harper determined the fraction of solute atom still in solution using an internal friction technique and showed that $\log(1-f)$ is proportional to $t^{2/3}$; the slope of the line is $\pi\rho(AD/\boldsymbol{k}T)$ and evaluation of this slope at a series of temperatures allows the activation energy for the process to be determined from an Arrhenius plot. The value obtained for α-iron is 84 kJ/mol which is close to that for the diffusion of carbon in ferrite.

8.9 Influence of grain boundaries on the plastic properties of metals

It might be thought that when a stress is applied to a polycrystalline metal, every grain in the sample deforms as if it were an unconstrained single crystal. This is not the case, however, and the fact that the aggregate does not deform in this manner is indicated by the high yield stress of polycrystals compared with that of single crystals. This increased strength of polycrystals immediately poses the question—is the hardness of a grain caused by the presence of the grain boundary or by the orientation difference of the neighbouring grains? It is now believed that the latter is the case but that the structure of the grain boundary itself may be of importance in special circumstances such as when brittle films, due to bismuth in copper or cementite in steel, form around the grains or when the grains slip past each other along their boundaries during high-temperature creep (*see* Chapter 13).

In a previous chapter we have seen that a small-angle tilt boundary can be described adequately by a vertical wall of dislocations. Rotation of one crystal relative to another, i.e. a twist boundary, can be produced by a crossed grid of two sets of screw dislocations as shown in *Figure 8.19*. These boundaries are of a

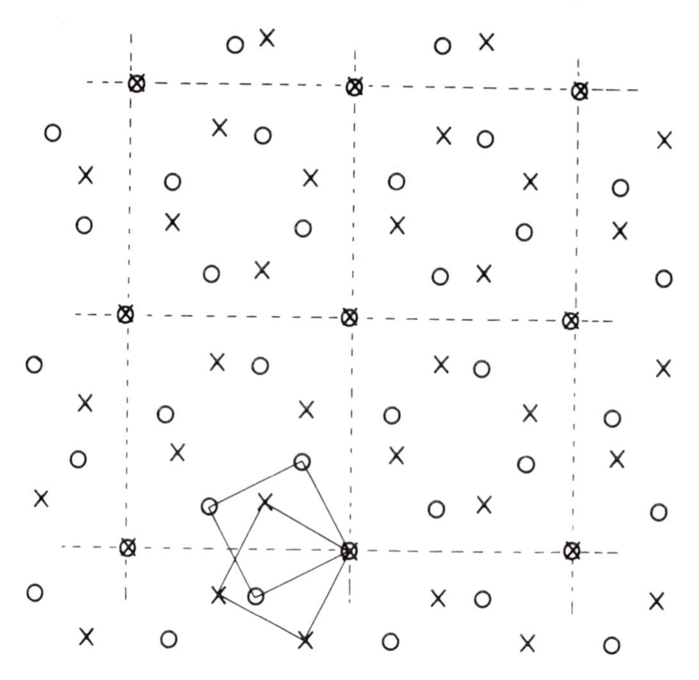

Figure 8.20 Two-dimensional section of a CSL with Σ5 36.9° [100] twist orientation (courtesy P. Goodhew)

particularly simple kind separating two crystals which have a small difference in orientation, whereas a general grain boundary usually separates crystals which differ in orientation by large angles. In this case, the boundary has five degrees of freedom, three of which arise from the fact that the adjoining crystals may be rotated with respect to each other about the three perpendicular axes, and the other two from the degree of freedom of the orientation of the boundary surface itself with respect to the crystals. Nevertheless, such a grain boundary may be described by an arrangement of dislocations, but their arrangement will be complex and the individual dislocations are not easily recognized or analysed.

The simplest extension of the dislocation model for low-angle boundaries to high-angle grain boundaries, is to consider that there are islands of good atomic fit surrounded by non-coherent regions. In a low-angle boundary the 'good fit' is perfect crystal and the 'bad fit' is accommodated by lattice dislocations, whereas for high-angle boundaries the 'good fit' could be an interfacial structure with low energy and the bad fit accommodated by dislocations which are not necessarily lattice dislocations; these dislocations are often termed intrinsic secondary grain boundary dislocations (gbds) and are essential to maintain the boundary at that misorientation.

The regions of good fit are sometimes described by the coincident site lattice (CSL) model, with its development to include the displacement shift complete (DSC) lattice. A CSL is a three-dimensional superlattice on which a fraction 1/Σ

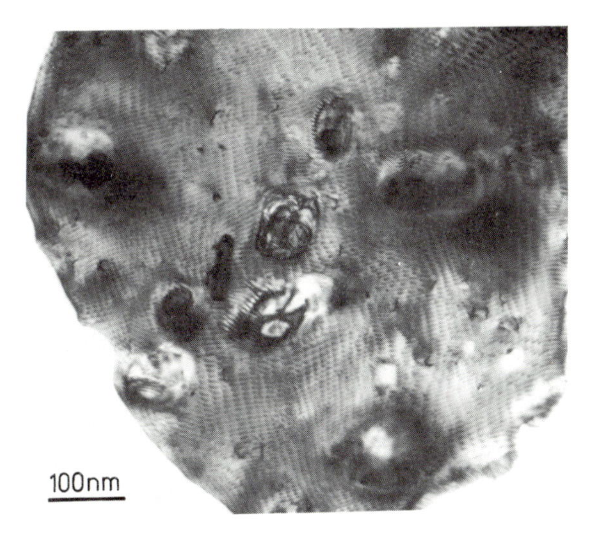

Figure 8.21 Networks of secondary gbds in a gold bicrystal with misorientation near $\Sigma5$ (Courtesy P. Goodhew)

of the lattice points in both crystal lattices lie; for the simple structures there will be many such CSLs, each existing at a particular misorientation. One CSL is illustrated in *Figure 8.20* but it must be remembered that the CSL is three-dimensional, infinite and inter-penetrates both crystals; it does not in itself define an interface. However, an interface is likely to have a low energy if it lies between two crystals oriented such that they share a high proportion of lattice sites, i.e. preferred misorientations will be those with CSLs having low Σ values. Such misorientations can be predicted from the expression

$$\theta = 2\tan^{-1}\left(\frac{b}{a}\sqrt{N}\right)$$

where b and a are integers and $N = h^2 + k^2 + l^2$; the Σ value is then given by $a^2 + Nb^2$, divided by 2 until an odd number is obtained.

The CSL model can only be used to describe certain specific boundary misorientations but it can be extended to other misorientations by allowing the presence of arrays of dislocations which act to preserve a special orientation between them. Such intrinsic secondary dislocations must conserve the boundary structure and, generally, will have Burgers vectors smaller than those of the lattice dislocations. *Figure 8.21* shows an example of $a/10\langle 310\rangle$ screw dislocations in a twist grain boundary $\Delta\theta = 0.7°$ from the special $\Sigma = 5$ boundary at $36.9°$ [100] in gold.

When a polycrystalline specimen is examined in TEM other structural features apart from intrinsic gbds may be observed in a grain boundary, such as 'extrinsic' dislocations which have probably run-in from a neighbouring grain, and interface ledges or steps which curve the boundary. At low temperatures the run-in lattice dislocation tends to retain its character while trapped in the interface, whereas at high temperatures it may dissociate into several intrinsic gbds resulting in a small change in misorientation across the boundary. The analysis of gbds in TEM is not easy, but information about them will eventually further our understanding of important boundary phenomena, e.g. migration of boundaries during recrystallization and grain growth, the sliding of grains during creep and superplastic flow and the way grain boundaries act as sources and sinks for point defects.

A large-angle grain boundary is often regarded simply as a surface, about 2 atoms thick, in which the atomic arrangement is irregular. Field-ion microscopy observations confirm this description. As a dislocation approaches such a boundary a short-ranged force will be exerted upon it by the boundary itself and by the neighbouring crystal. However, because the atoms in the boundary are already so disarranged that their misfit energy will remain relatively unaffected by introducing into the boundary the further irregularity which exists at the centre of the dislocation, the force from the boundary is an attraction and the dislocation will be attracted into the near side of the boundary. On the other hand, the force exerted on the dislocation by the crystal on the opposite side of the boundary is a repulsion, and this arises from the fact that the direction and plane of slip change at the grain boundary due to the orientation difference of the neighbouring crystals.

The importance of the orientation change across a grain boundary to the process of slip has been demonstrated by experiments on bi-crystals, where it is found that as the orientation difference between the two crystals decreases, the direction and plane of slip being the same in both crystals, the mechanical properties change progressively from polycrystalline to single crystal behaviour. Other experiments on 'bamboo'-type specimens, i.e. where the grain boundaries are parallel to each other and all perpendicular to the axis of tension, are also informative. Initially, deformation occurs by slip only in those grains most favourably oriented, but later spreads to all the other grains as those grains which are deformed first, work harden. It is then found that each grain contains wedge-shaped areas near the grain boundary, as shown in *Figure 8.22(a)*, where slip does not operate, which indicates that the continuance of slip from one grain to the next is difficult. From these observations it is natural to enquire what happens in a completely polycrystalline metal where the slip planes must in all cases make contact with a grain boundary. It will be clear that the polycrystalline aggregate must be stronger because, unlike the deformation of bamboo-type samples where it is not necessary to raise the stress sufficiently high to operate those slip planes which made contact with a grain boundary, all the slip planes within any grain of a polycrystalline aggregate make contact with a grain boundary, but, nevertheless, have to be operated. The importance of the grain size on strength is emphasized by *Figure 8.15(b)*, which shows the variation in lower yield stress, σ_y, with grain diameter, $2d$, for both low-carbon steel and commercially pure Nb. The smaller the grain size the higher the yield strength according to a relation of the form

$$\sigma_y = \sigma_i + kd^{-1/2} \tag{8.9}$$

where σ_i is a lattice friction stress and k a constant usually denoted k_y to indicate yielding. Because of the difficulties experienced by a dislocation in

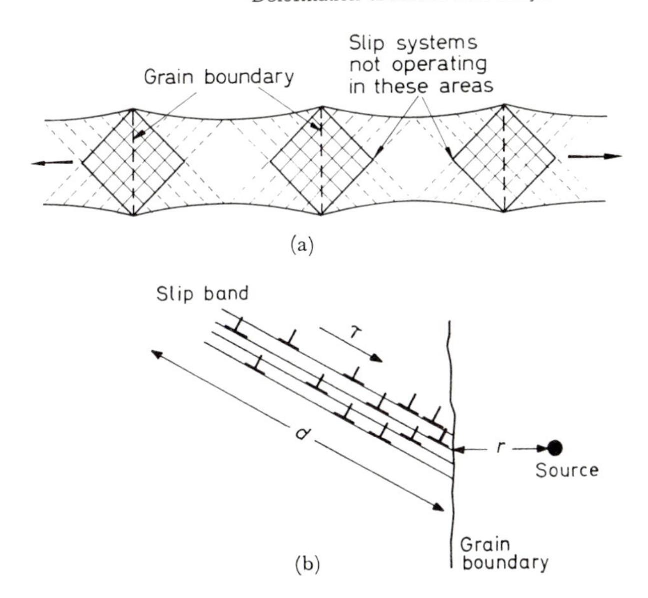

Figure 8.22 (a) Grain-boundary blocking of slip. (b) Blocking of a slip band by a grain boundary

moving from one grain to another, the process of slip in a polycrystalline aggregate does not spread to each grain by forcing a dislocation through the boundary. Instead, the slip band which is held up at the boundary gives rise to a stress concentration at the head of the pile-up group of dislocations which acts with the applied stress and is sufficient to trigger off sources in neighbouring grains. If τ_i is the stress a slip band could sustain if there were no resistance to slip across the grain boundary, i.e. the friction stress, and τ the higher stress sustained by a slip band in a polycrystal, then $(\tau - \tau_i)$ represents the resistance offered by the boundary, which reaches a limiting value when slip is induced in the next grain. The influence of grain size can be explained if the length of the slip band is proportional to d as shown in *Figure 8.22(b)*. Thus, since the stress concentration a short distance r from the end of the slip band is proportional to $(d/4r)^{1/2}$, the maximum shear stress at a distance r ahead of a slip band carrying an applied stress τ in a polycrystal is given by $(\tau - \tau_i)[d/4r]^{1/2}$ and lies in the plane of the slip band. If this maximum stress has to reach a value τ_{max} to operate a new source at a distance r then

$$(\tau - \tau_i)[d/4r]^{1/2} = \tau_{max}$$

or rearranging

$$\tau = \tau_i + (\tau_{max} 2r^{1/2})d^{-1/2}$$

which may be written as

$$\tau = \tau_i + k_s d^{-1/2}$$

It then follows that the tensile flow curve of a polycrystal is given by

$$\sigma = m(\tau_i + k_s d^{-1/2}) \tag{8.10}$$

where m is the orientation factor relating the applied tensile stress σ to the shear stress, i.e. $\sigma = m\tau$. For a single crystal the m-factor has a minimum value of 2 as discussed, but in polycrystals deformation occurs in less favourably oriented grains and sometimes (e.g. hexagonal, intermetallics, etc,) on 'hard' systems, and so the m-factor is significantly higher. From equation 8.9 it can be seen that $\sigma_i = m\tau_i$ and $k = mk_s$.

While there is an orientation factor on a macroscopic scale in developing the critical shear stress within the various grains of a polycrystal, so there is a local orientation factor in operating a dislocation source ahead of a blocked slip band. The slip plane of the sources will not, in general, lie in the plane of maximum shear stress, and hence τ_{max} will need to be such that the shear stress τ_c required to operate the new source must be generated in the slip plane of the source. In general, the local orientation factor dealing with the orientation relationship of adjacent grains will differ from the macroscopic factor of slip plane orientation relative to the axis of stress, so that $\tau_{max} = \frac{1}{2}m'\tau_c$. For simplicity, however, it will be assumed $m' = m$ and hence the parameter k in the Petch equation is given by $k = m^2\tau_c r^{1/2}$.

It is clear from the above treatment that the parameter k depends essentially on two main factors. The first is the stress to operate a source dislocation, and this depends on the extent to which the dislocations are anchored or locked by impurity atoms. Strong locking implies a large τ_c and hence a large k; the reverse is true for weak locking. The second factor is contained in the parameter m which depends on the number of available slip systems. A multiplicity of slip systems enhances the possibility for plastic deformation and so implies a small k. A limited number of slip systems available would imply a large value of k. It then follows, as shown in *Figure 8.23*, that for (i) f.c.c. metals, which have weakly locked dislocations and a multiplicity of slip systems, k will generally be small, i.e. there is only a small grain size dependence of the flow stress, (ii) c.p.h. metals, k will be large because of the limited slip systems, and (iii) b.c.c. metals, because of the strong locking, k will be large.

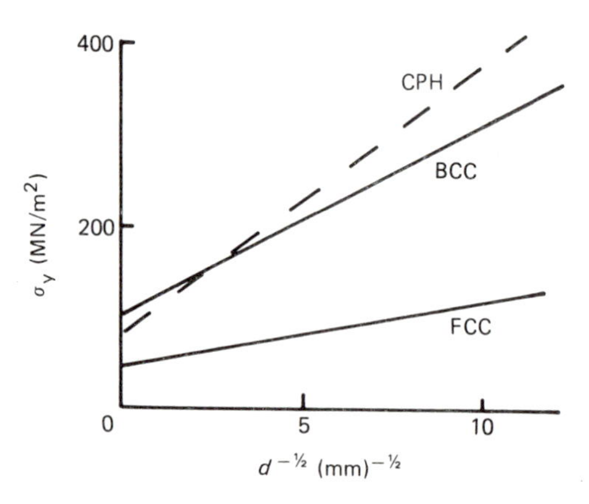

Figure 8.23 Schematic diagram showing the grain size-dependence of the yield stress for crystals of different crystal structure

Each grain does not deform as a single crystal in simple slip, since, if this were so, different grains would then deform in different directions with the result that voids would be created at the grain boundaries. Except in high temperature creep, where the grains slide past each other along their boundaries (*see* Chapter 13), this does not happen and each grain deforms in coherence with its neighbouring grains. However, the fact that the continuity of the metal is maintained during plastic deformation must mean that each grain is deformed into a shape that is dictated by the deformation of its neighbours. Such behaviour will, of course, require the operation of several slip systems, and Von Mises has shown that to allow this unrestricted change of shape of a grain requires at least five independent shear modes. The deformation of metal crystals with cubic structure easily satisfies this condition so that the polycrystals of these metals usually exhibit considerable ductility, and the stress–strain curve generally lies close to that of single crystals of extreme orientations deforming under multiple slip conditions. The hexagonal metals do, however, show striking differences between their single crystal and polycrystalline behaviour. This is because single crystals of these metals deform by a process of basal plane slip, but the three shear systems (two independent) which operate do not provide enough independent shear mechanisms to allow unrestricted changes of shape in polycrystals. Consequently, to prevent gaps opening up at grain boundaries during the deformation of polycrystals, some additional shear mechanisms, such as non-basal slip and mechanical twinning, must operate. Hence, because the resolved stress for non-basal slip and twinning is greater than that for basal-plane slip, yielding in a polycrystal is prevented until the applied stress is high enough to deform by these mechanisms. The work hardening behaviour of polycrystalline aggregates is discussed in Chapter 10.

8.10 Mechanical twinning

Mechanical twinning plays only a minor part in the deformation of the common metals such as copper or aluminium, and its study has consequently been neglected. Nevertheless, twinning does occur in all the common crystal structures under some conditions of deformation. *Table 8.1* shows the appropriate twinning elements for the common structures.

TABLE 8.1 Twinning elements for some common metals

Structure	*Plane*	*Direction*	*Metals*
c.p.h.	$\{10\bar{1}2\}$	$\langle 10\bar{1}\bar{1}\rangle$	Zn, Cd, Be, Mg
b.c.c.	$\{112\}$	$\langle 111\rangle$	Fe, β-brass, W, Ta, Nb, V, Cr, Mo
f.c.c.	$\{111\}$	$\langle 112\rangle$	Cu, Ag, Au, Ag-Au, Cu-Al
Tetragonal	$\{331\}$	—	Sn
Rhombohedral	$\{001\}$	—	Bi, As, Sb

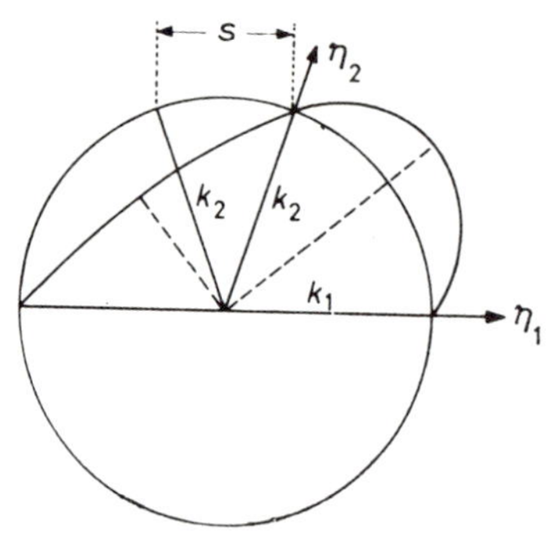

Figure 8.24 Crystallography of twinning

8.10.1 Twinning crystallography

The geometrical aspects of twinning can be represented with the aid of a unit sphere shown in *Figure 8.24*. The twinning plane k_1 intersects the plane of the drawing in the shear direction η_1. On twinning the unit sphere is distorted to an ellipsoid of equal volume, and the shear plane k_1 remains unchanged during twinning, while all other planes become tilted. Distortion of planes occurs in all cases except k_1 and k_2. The shear strain, s, at unit distance from the twinning plane is related to the angle between k_1 and k_2. Thus the amount of shear is fixed by the crystallographic nature of the two undistorted planes. In the b.c.c. lattice, the two undistorted planes are the (112) and $(11\bar{2})$ planes, displacement occurring in a [111] direction a distance of 0.707 lattice vectors. The twinning elements are thus:

k_1	k_2	η_1	η_2	*Shear*
(112)	$(11\bar{2})$	$[11\bar{1}]$	[111]	0.707

where k_1 and k_2 denote the first and second undistorted planes respectively, and η_1 and η_2 denote directions lying in k_1 and k_2, respectively, perpendicular to the line of intersection of these planes. k_1 is also called the composition or twinning plane, while η_1 is called the shear direction. The twins consists of regions of crystal in which a particular set of {112} planes (the k_1 set of planes) is homogeneously sheared by 0.707 in a ⟨111⟩ direction (the η_1 direction). direction). The same atomic arrangement may be visualized by a shear of 1.414 in the *reverse* ⟨111⟩ direction, but this larger shear has never been observed.

8.10.2 Nucleation and growth

During the development of mechanical twins, thin lamellae appear very quickly (≈speed of sound) and these thicken with increasing stress by the steady movement of the twin interface. New twins are usually formed in bursts and are sometimes accompanied by a sharp audible click which coincides with the appearance of irregularities in the stress–strain curve, as shown in *Figure 8.25*. The rapid production of clicks is responsible for the so-called twinning cry (e.g. in tin).

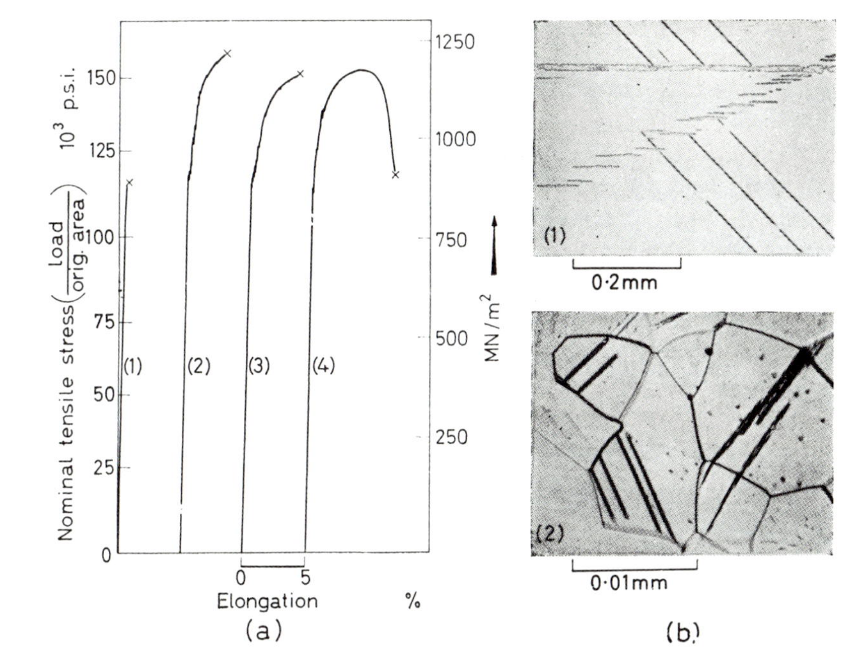

Figure 8.25 (a) Effect of grain size on the stress–strain curves of specimens of niobium extended at a rate of $2.02 \times 10^{-4}s^{-1}$ at 20 K; (1) grain size $2d = 1.414$ mm, (2) grain size $2d = 0.312$ mm, (3) grain size $2d = 0.0951$ mm, (4) grain size $2d = 0.0476$ mm. (b) Deformation twins in specimen 1 and specimen 3 extended to fracture. Etched in 95% HNO_3 + 5% HF (after Adams, Roberts and Smallman, *Acta Metall.* 1960, **8,** 332)

Although most metals show a general reluctance to twin, when tested under suitable conditions they can usually be made to do so. As mentioned in Chapter 6, the shear process involved in twinning must occur by the movement of partial dislocations and, consequently, the stress to cause twinning will depend not only on the line tension of the source dislocation, as in the case of slip, but also on the surface tension of the twin boundary. The stress to cause twinning is, therefore, usually greater than that required for slip and at room temperature deformation will nearly always occur by slip in preference to twinning. As the deformation temperature is lowered the critical shear stress for slip increases, and then, because the general stress level will be high, the process of deformation twinning is more likely.

Twinning is most easily achieved in metals of c.p.h. structure where, because of the limited number of slip systems, twinning is an essential and unavoidable mechanism of deformation in polycrystalline specimens (*see* previous section), but in single crystals the orientation of the specimen, the stress level and the temperature of deformation are all important factors in the twinning process. In metals of the b.c.c. structure twinning may be induced by impact at room temperature or with more normal strain rates at low temperature where the critical shear stress for slip is very high (*see Figure 8.15* also *Figure 8.3*). In contrast, only a few f.c.c. metals have been made to twin, even at low temperatures.

In zinc single crystals it is observed that there is no well-defined critical resolved shear stress for twinning such as exists for slip, and that a very high stress indeed is necessary to nucleate twins. In most crystals, slip usually occurs first and twin nuclei are then created by means of the very high stress concentration which exists at dislocation pile-ups. Once formed, the twins can propagate provided the resolved shear stress is higher than a critical value, because the stress to propagate a twin is much lower than that to nucleate it. This can be demonstrated by deforming a crystal oriented in such a way that basal slip is excluded, i.e. when the basal planes are nearly parallel to the specimen axis. Even in such an oriented crystal it is found that the stress to cause twinning is higher than that for slip on non-basal planes. In this case, non-basal slip occurs first so that when a dislocation pile-up arises and a twin is formed, the applied stress is so high that an avalanche or burst of twins results. It is interesting to note, however, that twins may be nucleated artificially at a lower stress by indenting the specimen during the test with a sharp pin, and these twins will grow if the applied stress is above the twin-propagation stress.

It is also believed that in the b.c.c. metals twin nucleation is more difficult than twin propagation. One possible mechanism is that nucleation is brought about by the stress concentration at the head of a piled-up array of dislocations produced by a burst of slip as a Frank–Read source operates. Such a behaviour is favoured by impact loading, and it is well known that twin lamellae known as Neumann bands are produced this way in α-iron at room temperature. At normal strain rates, however, it should be easier to produce a slip burst suitable for twin nucleation in a material with strongly locked dislocations, i.e. one with a large k value (as defined by equation 8.9) than one in which the dislocation locking is relatively slight (small k values). In this context it is interesting to note that both niobium and tantalum have a small k value and, although they can be made to twin, do so with reluctance compared, for example, with α-iron.

In all the b.c.c. metals the flow stress increases so rapidly with decreasing temperature (*see Figure 8.15*), that even with moderate strain rates (10^{-4} s^{-1}) α-iron will twin at 77 K while niobium with its smaller value of k twins at 20 K. The type of stress–strain behaviour for niobium is shown in *Figure 8.25*(*a*), where three distinct stages can be noted. The pattern of behaviour is characterized by (i) small amounts of slip interspersed between extensive bursts of twinning in the early stages of deformation, (ii) a preponderance of slip with only occasional twinning as deformation is continued, and (iii) an ability (in contrast to the behaviour of specimens tested at 77 K) to work harden whilst deforming by slip. These observations are again in accord with the model of twinning discussed above. Twins, once formed, may themselves act as barriers, allowing further dislocation pile-up and further twin nucleation. After a time, however, most of the Frank–Read sources will have been released from their impurity atmospheres, and the generation of slip dislocations will then no longer occur in sharp bursts so that twin nucleation is suppressed. The action of twins as barriers to slip dislocations could presumably account for the rapid work-hardening observed at 20 K.

F.c.c. metals do not readily deform by twinning but it can occur in certain orientations at low temperatures, and even at 0°C, in favourably oriented crystals of silver. Deformation twins have also been observed in single crystals of copper–aluminium at room temperature. The apparent restriction of twinning to certain orientations and low temperatures may be ascribed to the high shear stress attained in tests on crystals with these orientations, since the stress necessary to produce twinning is high. The exact mechanism for this twinning is not known, except that it must occur by the propagation of a half-dislocation and its associated stacking fault across each plane of a set of parallel (111) planes. However, the stacking faults are wider than the equilibrium separation of the partials, so that the twin process must involve one of the half-dislocations being anchored while the other half separates. Such a process would then produce a wide mono-layer of twin and for the twin to thicken, the half-dislocation must also climb on to successive twin planes, as below for b.c.c. iron.

8.10.3 The effect of impurities on twinning

It is well-established that solid solution alloying favours twinning in f.c.c. metals. For example, silver–gold alloys twin far more readily than the pure metals. Attempts have been made to correlate this effect with stacking fault energy and it has been shown that the twinning stress of copper-based alloys increases with increasing stacking fault energy. Twinning is also favoured by solid solution alloying in b.c.c. metals, and alloys of Mo–Re, W–Re and Nb–V readily twin at room temperature. In this case it has been suggested that the lattice frictional stress is increased and the ability to cross-slip reduced by alloying, thereby confining slip dislocations to bands where stress multiplication conducive to twin nucleation occurs. With a dispersed second phase, twinning generally proceeds unimpeded unless the dispersion takes the form of solute atom clusters. The situation with interstitial impurities is far less readily understood, for increased purity has been shown to favour twinning in some cases and inhibit it in others.

8.10.4 The effect of prestrain on twinning

It was noted as long ago as 1926 that twinning can be suppressed in some metals by a certain amount of prestrain. The effect has since been shown to be virtually universal and it has been demonstrated that the ability to twin may be restored by an ageing treatment. It has been suggested that the effect may be due to the differing dislocation distribution produced under different conditions. For example, niobium will normally twin at $-196°C$, when a heterogeneous arrangement of elongated screw dislocations capable of creating the necessary stress concentrations are formed. Room temperature prestrain, however, inhibits twin formation as the regular network of dislocations produced provides more mobile dislocations and homogenizes the deformation.

8.10.5 Dislocation mechanism of twinning

In contrast to slip, the shear involved in the twinning process is homogeneous throughout the entire twinning region, and each atom plane parallel to the twinning plane moves over the one below it by only a fraction of a lattice spacing in the twinning direction. Nevertheless, mechanical twinning is thought to take place by a dislocation mechanism for the same reasons as slip (*see* Chapter 6), but the dislocations that cause twinning are partial and not unit dislocations. From the crystallography of the process it can be shown that twinning in the c.p.h. lattice, in addition to a simple shear on the twinning plane, must be accompanied by a localized rearrangement of the atoms, and furthermore, only in the b.c.c. lattice does the process of twinning consist of a simple shear on the twinning plane (e.g. a twinned structure in this lattice can be produced by a shear of $1/\sqrt{2}$ in a $\langle 111 \rangle$ direction on a $\{112\}$ plane). Thus, to illustrate the application of the dislocation theory to the problem of twinning, the relatively simple case of twinning in the b.c.c. lattice will be considered.

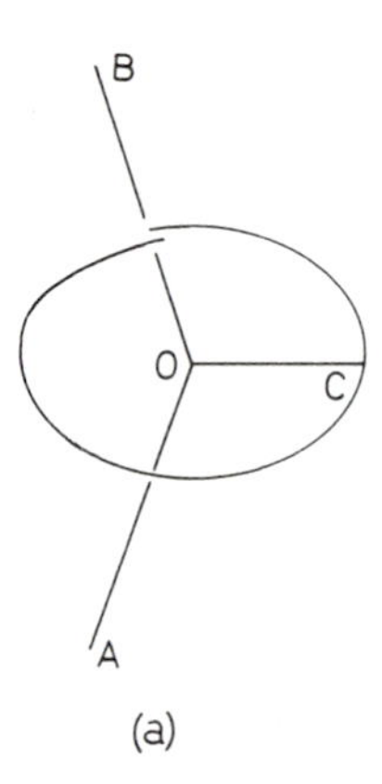

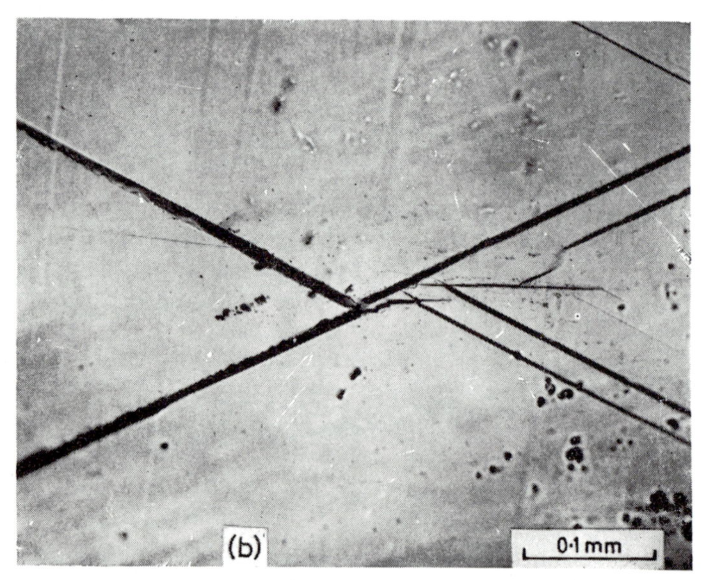

Figure 8.26 (a) Diagram illustrating the pole mechanism of twinning. (b) The formation of a crack at a twin intersection in silicon–iron (after Hull, *Acta Metall.* 1960, **8,** 11)

An examination of *Figure 6.3* shows that the main problem facing any theory of twinning is to explain how twinning develops homogeneously through successive planes of the lattice. Clearly this could be accomplished either by a twinning (partial) dislocation on every plane, or alternatively by the movement of a single dislocation successively from plane to plane. One suggestion, similar in principle to the crystal growth mechanism, is the pole-mechanism proposed by Cottrell and Bilby illustrated in *Figure 8.26(a)*. Here, OA, OB, and OC are dislocation lines. The twinning dislocation is OC, which produces the correct shear as it sweeps through the twin plane about its point of emergence O, and OA and OB form the pole dislocation, being partly or wholly of screw character with a pitch equal to the spacing of the twinning layers. The twinning dislocation rotates round the pole dislocation and in doing so, not only produces a monolayer sheet of twinned crystal, but also climbs up the 'pole' to the next layer. The process is repeated and a thick layer of twin is built up.

The dislocation reaction involved is as follows. The line AOB represents a unit dislocation with a Burgers vector $\frac{1}{2}a[111]$ and that part OB of the line lies in the (112) plane. Then, under the action of stress dissociation of this dislocation can occur according to the reaction

$$\tfrac{1}{2}a[111] \rightarrow \tfrac{1}{3}a[112] + \tfrac{1}{6}a[11\bar{1}]$$

The dislocation with vector $\frac{1}{6}a[11\bar{1}]$ forms a line OC lying in one of the other $\{112\}$ twin planes (e.g. the $(\bar{1}21)$ plane) and produces the correct twinning shear. The line OB is left with a Burgers vector $\frac{1}{3}a[112]$ which is of pure edge type and sessile in the (112) plane.

A dislocation mechanism for the nucleation of thin twins in the b.c.c. lattice based on the dissociation of $\frac{1}{2}a\langle 111\rangle$ dislocations of pure screw orientation has also been proposed. Under an applied stress of the order of the twinning stress the $\frac{1}{2}a\langle 111\rangle$ dislocation may dissociate into three $\frac{1}{6}a\langle 111\rangle$ twinning dislocations, one on each of three neighbouring planes to form a three-layer twin. This elementary twin can easily grow in the composition plane, and if during growth it intersects an $\frac{1}{2}a\langle 111\rangle$ screw dislocation it can then develop in thickness by the pole mechanism.

8.10.6 Twinning and fracture

It has been suggested that a twin, like a grain boundary, may present a strong barrier to slip and that a crack can be initiated by the pile-up of slip dislocations at the twin interface (*see Figure 14.2*). In addition, cracks may be initiated by the intersection of twins, and examples are common in molybdenum, silicon–iron (b.c.c.) and zinc (c.p.h.). *Figure 8.26(b)* shows a very good example of crack nucleation in 3 per cent silicon–iron; the crack has formed along an $\{001\}$ cleavage plane at the intersection of two $\{112\}$ twins, and part of the crack has developed along one of the twins in a zig-zag manner while still retaining $\{001\}$ cleavage facets.

In tests at low temperature on b.c.c. and c.p.h. metals both twinning and fracture readily occur, and this has led to two conflicting views. First, that twins are nucleated by the high stress concentrations associated with fracture, and secondly, that the formation of twins actually initiates the fracture. It is probable that both effects occur. However, there are many observations which demonstrate that twinning is not always essential to brittle fracture, and the fact that in b.c.c. metals the onset of twinning and cleavage occurs under similar conditions is probably explained by the close dependence of both phenomena on the strength of the dislocation locking. It is probably a fair generalization to say that slip is the fundamental process in metals and that both twinning and fracture occur as a result of slip. Under certain restricted conditions, however, it is possible for twins to result in failure directly.

The theory of fracture is discussed more fully in Chapter 14.

Suggestions for further reading

Honeycombe, R.W.K., *Plastic Deformation of Metals*, Edward Arnold, 1984

Hirth, J.P. and Lothe, J., *Theory of Dislocations*, McGraw-Hill, 1984

Cottrell, A.H., *The Mechanical Properties of Matter*, John Wiley & Son, 1964

Hull, D., *Introduction to Dislocations*, Pergamon Press, 1975

Dislocations and Properties of Real Materials, Metals Society, 1984

Chapter 9

Point defects in crystals

9.1 Introduction

Of the various lattice defects which can exist in a metal, the vacancy is the only species that is ever present in appreciable concentrations in thermodynamic equilibrium. The equilibrium concentration increases exponentially with rise in temperature as shown in *Figure 4.7* and, as a consequence, a knowledge of their behaviour is essential for understanding the deformation properties of metals particularly at elevated temperatures. The everyday industrial processes of annealing, homogenization, precipitation, sintering, surface hardening, as well as oxidation and creep, all involve, to varying degrees, the transport of atoms through the lattice with the help of vacancies. Similarly, vacancies enable dislocations to climb, since to move the extra half-plane of a dislocation up or down requires the mass transport of atoms. This mechanism is extremely important in the process of polygonization and recovery (*see* Chapter 10) and, moreover, such a mechanism enables dislocations to climb over obstacles lying in their slip plane, which illustrates one way in which metals can soften and lose their resistance to creep at high temperatures.

The role of vacancies at high temperatures, where an appreciable concentration is thermally maintained, is therefore well established. However, their importance in influencing the behaviour of metals at lower temperatures, i.e. around room temperature, is not so fully understood. It is for this reason that experiments on crystals made defective by quenching or irradiation now play a role in fundamental studies of metals far greater than their apparent ultimate importance.

9.2 The production of vacancies

There are five main methods of introducing point defects into a metal in excess of the equilibrium concentration (we are here chiefly concerned with vacancies rather than interstitials):

1. A rapid quenching from a high temperature can be effective in retaining the concentration of vacancies which exists in thermal equilibrium at that

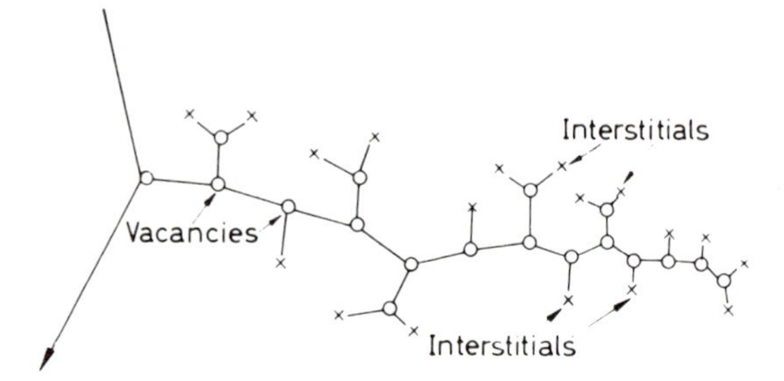

Figure 9.1 Formation of vacancies and interstitials due to particle bombardment (after Cottrell, 46th Thomas Hawksley Memorial Lecture, *Proc. Inst. Mech. Engrs,* 1959, No. 14, courtesy of the Institution of Mechanical Engineers)

temperatures (this can be as high as 1 in 10^4). Quenching is one of the basic treatments for precipitation hardening alloys and the retained vacancies so produced can play an important role in the precipitation process.

2. Excess lattice vacancies are formed when intermetallic compounds such as the transition metal aluminides (e.g. NiAl, CoAl, FeAl) deviate from stoichiometry.
3. Bombardment of the metal with high energy nuclear particles can produce both vacancies and interstitials (*see Figure 9.1*).
4. Point defects can also be formed during plastic deformation at jogs produced on dislocation lines by intersecting dislocations.
5. Oxidation of some metals, e.g. Zn, Cd, Mg, Cu, Ni, is accompanied by the injection of vacancies into the metal lattice (*see* Chapter 15).

The production of lattice defects by quenching, deviations from stoichiometry, and cold work, have been dealt with in previous chapters. The production of point defects by particle bombardment, and oxidation will now be outlined, together with a few additional remarks on the formation of vacancies during cold work.

9.2.1 Vacancy production by bombardment with high-energy particles

There are many different kinds of high energy radiation (e.g. neutrons, electrons, α-particles, protons, deuterons, uranium fission fragments, γ-rays, x-rays) and all of them are capable of producing some form of 'radiation damage' in the materials they irradiate. While all are of importance to some aspects of the solid state, of particular interest to metallurgists is the behaviour of metals and alloys under irradiation in a nuclear reactor. This is because the neutrons produced in a reactor by a fission reaction have extremely high energies of about two million electron volts (i.e. 2 MeV), and being electrically uncharged, and consequently unaffected by the electrical fields surrounding an atomic nucleus, they can travel large distances through a metallic lattice. The resultant damage is therefore not localized, for example at the surface, but is homogeneously distributed throughout the solid.

The fast neutrons (they are given this name because 2 MeV corresponds to a velocity of 2×10^7 m s^{-1}) are slowed down, in order to produce further fission, by the moderator in the pile until they are in thermal equilibrium with their surroundings. The neutrons in a pile will, therefore, have a spectrum of energies which ranges from about 1/40 eV at room temperature (thermal neutrons) to 2 MeV (fast neutrons). However, when a non-fissile metal is placed in a reactor and irradiated most of the damage is caused by the fast neutrons colliding with the atomic nuclei of the metal.

The nucleus of an atom has a small diameter (e.g. 10^{-14} m), and consequently the largest area, or cross-section, which it presents to the neutron for collision is also small. The unit of cross-section is a barn, i.e. 10^{-28} m^2, so that in a material with a cross-section of 1 barn, an average of 10^9 neutrons would have to pass through an atom (cross-sectional area 10^{-19} m^2) for one to hit the nucleus. Conversely, the mean free path between collisions is about 10^9 atom spacings or about 0.3 m. If a metal such as copper (cross-section, 4 barns) were irradiated for 1 day (10^5 s) in a neutron flux of 10^{17} m^{-2} s^{-1}, the number of neutrons passing through unit area, i.e. the integrated flux, would be 10^{22} n.m^{-2} and the chance of a given atom being hit (=integrated flux × cross-section) would be 4×10^{-6}, i.e. about 1 atom in 250 000 would have its nucleus struck.

For most metals the collision between an atomic nucleus and a neutron (or other fast particle of mass m) is usually purely elastic, and the struck atom mass M will have equal probability of receiving any kinetic energy between zero and the maximum $E_{max} = 4E_n Mm/(M+m)^2$, where E_n is the energy of the fast neutron. Thus, the most energetic neutrons can impart an energy of as much as 200 000 eV, to a copper atom initially at rest. Such an atom, called a primary 'knock-on', will do much further damage on its subsequent passage through the lattice, often producing secondary and tertiary knock-on atoms, so that severe local damage results. The neutron, of course, also continues its passage through the lattice producing further primary displacements until the energy transferred in collisions is less than the energy E_d (≈ 25 eV for copper) necessary to displace an atom from its lattice site.

The exact nature of the radiation damage is not yet completely understood, but consists largely of interstitials*, i.e. atoms knocked into interstitial positions in the lattice, and vacancies, i.e. the holes they leave behind. The damaged region, estimated to contain about 60 000 atoms, is expected to be originally pear-shaped in form, having the vacancies at the centre and the interstitials towards the outside. Such a displacement spike or cascade of displaced atoms is shown schematically in *Figure 9.1*. The number of vacancy-interstitial pairs produced by one primary knock-on is given by $n \simeq E_{max}/4E_d$, and for copper is about 1000. Owing to the thermal motion of the atoms in the lattice, appreciable self-annealing of the damage will take place at all except the lowest temperatures, with most of the vacancies and interstitials annihilating each other by re-combination. However, it is expected that some of the interstitials will escape from the surface of the cascade leaving a corresponding number of vacancies in the centre. If this

* The formation energy of an interstitial copper atom is about 4 eV.

number is assumed to be 100, the local concentration will be 100/60 000 or $\approx 2 \times 10^{-3}$.

Another manifestation of radiation damage concerns the dispersal of the energy of the stopped atom into the vibrational energy of the lattice. The energy is deposited in a small region, and for a very short time the metal may be regarded as locally heated. To distinguish this damage from the 'displacement spike', where the energy is sufficient to displace atoms, this heat-affected zone has been called a 'thermal spike'. To raise the temperature of a metal by 1000 °C requires about $3R \times 4.2$ kJ/mol or about 0.25 eV per atom. Consequently, a 25 eV thermal spike could heat about 100 atoms of copper to the melting point, which corresponds to a spherical region of radius about 0.75 nm. It is very doubtful if melting actually takes place, because the duration of the heat pulse is only about 10^{-11} to 10^{-12} s. However, it is not clear to what extent the heat produced gives rise to an annealing of the primary damage, or causes additional quenching damage (e.g. retention of high-temperature phases).

Slow neutrons give rise to transmutation products. Of particular importance is the production of the noble gas elements, e.g. krypton and xenon produced by fission in U and Pu, and helium in the light elements B, Li, Be and Mg. These transmuted atoms can cause severe radiation damage in two ways. First the inert gas atoms are almost insoluble and hence in association with vacancies collect into gas bubbles which swell and crack the material. Secondly, these atoms are often created with very high energies, e.g. as α-particles or fission fragments, and act as primary sources of knock-on damage. The fission of uranium into two new elements is the extreme example when the fission fragments are thrown apart with kinetic energy ≈ 100 MeV. However, because the fragments carry a large charge their range is short and the damage restricted to the fissile material itself, or in materials which are in close proximity. Heavy ions can be accelerated to kilovolt energies in accelerators to produce heavy ion bombardment of materials being tested for reactor application. These moving particles have a short range and the damage is localized.

9.2.2 Vacancies produced by cold work

Several experiments, such as the change of electrical resistivity with deformation, show that vacancies are produced by cold work and that the concentration c depends on the amount of plastic strain ε according to the approximate relation $c = 10^{-4}\,\varepsilon$. These vacancies are presumed to form at jogs produced on dislocation lines by intersecting dislocations, since it is observed that when the deformation is restricted to single slip the vacancy concentration (as shown by the change in electrical resistivity or density) is relatively small, while during double slip the resistance increases and the density decreases.

The formation of point defects by the non-conservative motion of jogs in screw dislocations has already been discussed in Chapter 6. The detailed structure of jogs is discussed in Chapter 10 when considering work hardening. It is pointed out that the jog as well as the glissile dislocation may be dissociated, when it can adopt a complex arrangement of Shockley partials and stair-rods, as shown in *Figure 10.6*. When the jog is moving in the direction to produce

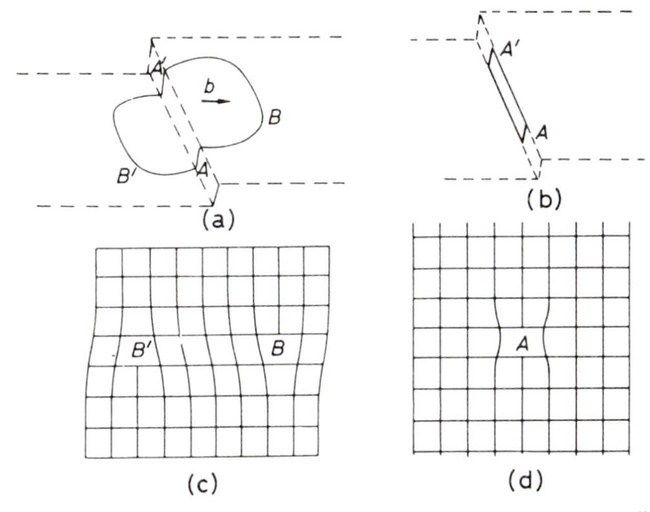

Figure 9.2 Two dislocations of opposite sign on neighbouring parallel glide planes (a), recombining to form a row of vacancies as shown in (b). The lattice arrangement is shown in (c) and (d)

vacancies, the sessile stair-rod dislocation ($\alpha\delta$) is at the back of the dissociated jog configuration and when interstitials are produced the stair-rod is at the front. Thus during the motion of a vacancy-jog the glissile Shockley partial dislocation ($B\delta$) is pushed away from the stair-rod dislocation, while during the motion of an interstitial jog the Shockley partial will be pushed towards the stair-rod. It then follows that the activation energy to constrict an interstitial-jog can be lowered to zero by stress, but the energy to constrict a vacancy-jog must be supplied almost entirely by thermal activation. Except at high temperatures, the vacancy-jog moves non-conservatively to produce a row of vacancies. Interstitial-jogs, however, can move conservatively, but the lower the temperature the higher will be the stress required. The motion of jogs in edge dislocations has not been considered in detail, since most of these are glissile. Nevertheless, a row of point defects may be formed by jogs in edge dislocations in the following way. Imagine a dislocation loop which is expanding in its slip plane under the action of the applied stress becoming jogged as shown in *Figure 9.2(a)*. Then during the unloading stage of deformation the loop contracts and a stage is reached at which a line of point defects can form (*Figure 9.2(b)*). This process is equivalent to the coalescence of two edge dislocation lines of opposite sign on neighbouring slip planes as shown in *Figure 9.2(c)* and *(d)*. The unloading stage of deformation may, therefore, be specially important in producing vacancies, and, because the opportunity for such a process occurring is far greater in fatigue deformation than in uni-directional stressing, it is possible that a high vacancy concentration is maintained in the lattice during fatigue.

9.2.3 Vacancies produced by oxidation

The growth of an oxide film on a metallic surface involves mass transport of either metal ions leading to oxide growth at the oxide–oxygen interface, or oxygen ions leading to growth at the oxide–metal interface. The importance of point defects in the oxide on these diffusion processes is well understood but the inter-relationship between point defects in the oxide and in the metal has only recently been recognized. Experimental results on the epitaxial relationships between metal and oxide show that the oxide–metal interface is not generally sufficiently disordered to act as an infinite source or sink of point defects and provided that point defects are not created or destroyed by co-operative movements of the oxide, the flow of point defects through the oxide must be continuous across the oxide–metal interface. If it is assumed that vacancies are the predominant point defects in the metal then the transfer of metal atoms into the oxide and their subsequent diffusion must leave behind vacant lattice sites in the metal, conversely the substitutional diffusion of oxygen through the oxide and its eventual incorporation into the metal near the interface must result in a flow of vacancies from metal to oxide. Thus an oxide growing by cation diffusion will produce vacancies in the metal whereas an oxide growing by anion diffusion will continually absorb metallic vacancies. A common example of the former type of growth is ZnO, which is known to depart from stoichiometry with an excess of zinc interstitials. The equilibrium equation for the defect products is

$$2Zn_i^{++} + 4e^- + O_2 \rightleftharpoons 2ZnO$$

the reaction being exothermic with 350 kJ released per mole of ZnO formed. Oxidation has the effect of driving the reaction to the right, and hence the concentration of zinc interstitials in the oxide will be reduced. However, the equilibrium concentration of interstitials in the oxide will be restored by the formation of interstitial/vacancy pairs at the oxide/metal interface, with zinc interstitials migrating into the oxide and vacancies migrating into the zinc metal.

9.3 The effect of vacancies on the physical and mechanical properties

The introduction of vacancies into a metal by any of the above methods produces changes in both its physical and mechanical properties. The electrical resistivity is one of the simplest properties which can be employed to investigate changes in vacancy concentration, and consequently is widely used. The increase in resistivity following quenching ($\Delta\rho_0$) may be described by the equation

$$\Delta\rho_0 = A \exp[-E_f/\boldsymbol{k}T_Q] \tag{9.1}$$

where A is a constant involving the entropy of formation, E_f is the formation energy of a vacancy and T_Q the quenching temperature. Thus by measuring the resistivity after quenching from different temperatures enables E_f to be estimated from a plot of $\Delta\rho_0$ versus $1/T_Q$ (*see Table 9.1*). The corresponding vacancy concentration can be obtained if the resistivity of a vacancy is known; for copper this is estimated to be 1.5 μΩ cm per 1 at % vacancies. The formation of vacancies in a crystal produces an increase in the specimen dimensions owing to the removal of atoms to the specimen surface. The concentration of point defects can

TABLE 9.1 Values of vacancy formation (E_f) and migration (E_m) energies in electron volts for some common metals together with the self-diffusion energy (E_D)

Energy	*Cu*	*Ag*	*Au*	*Al*	*Ni*	*Pt*	*Mg*	*Fe*	*W*	*NiAl*
E_f	1.0–1.1	1.1	0.98	0.76	1.4	1.4	0.9	2.13	3.3	1.05
E_m	1.0–1.1	0.83	0.83	0.62	1.5	1.1	0.5	0.76	1.9	2.4
E_D	2.0–2.2	1.93	1.81	1.38	2.9	2.5	1.4	2.89	5.2	3.45

thus be estimated by measuring the specimen length change $\Delta L/L$ and comparing it with the lattice parameter change $\Delta a/a$ as measured by x-rays, i.e. $C_v = 3(\Delta L/L - \Delta a/a)$. This follows because the macroscopic property measures both average dilatation of interatomic spacing due to the presence of the vacancies and the dimensional change due to the increase in the number of lattice sites, whereas the x-rays only measure the dilatation. The sign of $(\Delta L/L - \Delta a/a)$ is positive for vacancy formation and negative for interstitial formation.

At elevated temperatures the very high equilibrium concentration of vacancies which exists in the lattice gives rise to the possible formation of di-vacancy and even tri-vacancy complexes depending on the value of the appropriate binding energy. For equilibrium between single and di-vacancies, then the vacancy concentration is given by

$$c_v = c_{1v} + 2c_{2v}$$

and the di-vacancy concentration by

$$c_{2v} = Azc_{1v}^2 \exp[B_2/\boldsymbol{k}T]$$

where A is a constant involving the entropy of formation of di-vacancies, B_2 the binding energy for vacancy pairs and z a configurational factor which is 6 for close-packed metals and 4 for b.c.c. metals. The binding energy B_2 has been estimated for close packed metals to be in the range $0.1 < B_2 < 0.5$ eV, and for Al $\simeq 0.17$, Au $\simeq 0.3$ and Ag $\simeq 0.38$ eV. The ratio of di-vacancies to single vacancies c_{2v}/c_{1v} depends on the concentration of single vacancies and also on the binding energy; this ratio increases as the temperature is lowered during quenching.

The 'annealing-out' of the retained vacancies occurs by the migration of vacancies to sinks where they are annihilated. The average number of atomic jumps made before annihilation is given by

$$n = Azvt \exp[-E_m/\boldsymbol{k}T_a]$$

where A is a constant (≈ 1) involving the entropy of migration, z the co-ordination around a vacancy, v the Debye frequency ($\approx 10^{13}$/s), t the annealing time at the ageing temperature T_a and E_m the migration energy of a vacancy. The activation energy, E_m, for the movement of vacancies can be obtained by measuring the rate of annealing of the vacancies at a series of annealing temperatures. The rate of annealing is inversely proportional to the time to reach a certain value of 'annealed-out' resistivity. Thus, $1/t_1 = A\exp[-E_m/\boldsymbol{k}T_1]$ and $1/t_2 = A\exp[-E_m/\boldsymbol{k}T_2]$ and by eliminating A we obtain $\ln(t_2/t_1) = E_m[(1/T_2) - (1/T_1)]/\boldsymbol{k}$ where E_m is the only unknown in the expression. It is also possible to measure E_m from the change in annealing rate immediately before and after an abrupt change in annealing temperature. The energy necessary for the formation of the vacancies E_f, can be obtained from a knowledge of the number of vacancies retained after quenching from different temperatures, or indirectly from the energy of self-diffusion, $E_D = E_f + E_m$, since for diffusion it is necessary both to form vacancies and to move them. Values of E_f and E_m for some of the common metals are given in *Table 9.1*. The migration of di-vacancies is an easier process and the activation energy for migration is somewhat lower than E_m for single vacancies.

Before the direct observation of quenched-in lattice defects by transmission microscopy, the processes involved in annealing-out of the vacancy super-saturation had to be deduced from resistivity kinetics. The values of E_m given in *Table 9.1* indicate that vacancies and di-vacancies are mobile at temperatures in the vicinity of room temperatures; Al anneals slightly below room temperature and Cu, Ag and Au slightly above. The annealing behaviour depends on the quenching temperature, quenching rate and binding energy. During the iso-chronal annealing of aluminium quenched from $T_Q \lesssim 470°C$ it has been observed that the extra resistivity anneals out in one stage at about room temperature with an activation energy of ~ 0.45 eV. This value is lower than E_m for single vacancies and thus the annealing process is thought to involve the migration of single vacancies, di-vacancies and possible tri-vacancies. The average number of atomic jumps n involved before annihilation is 10^{10}, and hence the distance travelled by a vacancy defect to a sink ($\sqrt{n} \times b$) is about 30 μm if a random walk process is assumed. This distance is too small for the vacancy defects to be annihilated only at free surfaces or grain boundaries, and hence it was deduced that existing dislocations are the most likely point defect sink. For high quenching tempera-tures the annealing process takes place in two stages, as shown in *Figure 9.3*; stage I near room temperature with an activation energy ≈ 0.58 eV and $n \approx 10^4$, and stage II in the range 140–200 °C with an activation energy of ~ 1.3 eV. The value of n for the first stage indicates that the vacancy defects, mainly single vacancies, migrate an average distance ≈ 30 nm (300 Å) (which is less than the existing dislocation sink distance) to form defects, probably large clusters, which are more stable than isolated vacancies. The second stage of recovery then involves the disappearance of the resistivity associated with these stable clustered defects, and usually takes place at a temperature where self-diffusion is rapid.

For gold quenched from below 750 °C, the resistivity anneals in a simple

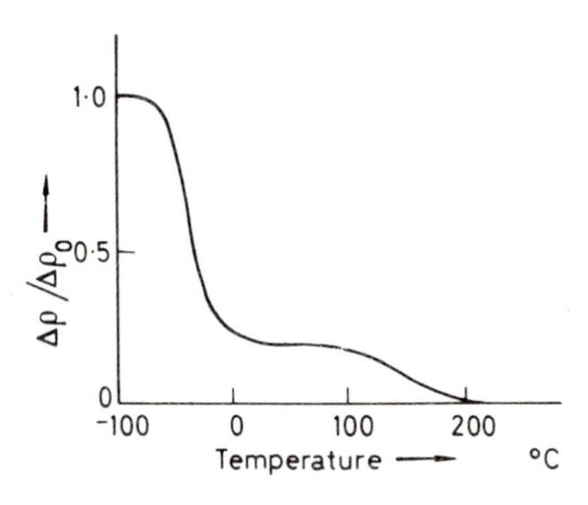

Figure 9.3 Variation of quenched-in resistivity with temperature of annealing for aluminium (after Panseri and Federighi, *Phil. Mag.* 1958, **3**, 1223)

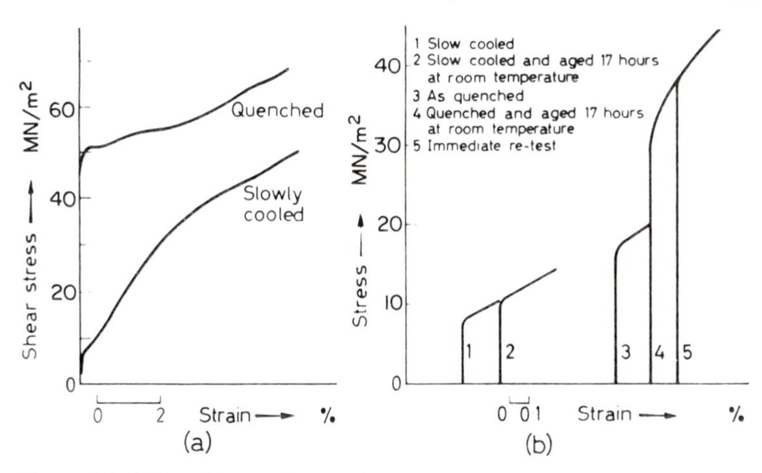

Figure 9.4 Effect of quenching on the stress–strain curves from (a) aluminium (after Maddin and Cottrell, *Phil. Mag.* 1955, **66,** 735), and (b) gold (after Adams and Smallman, unpublished)

manner with an activation energy of ≈ 0.82 eV from which it was deduced that the annealing process involves single vacancies to fixed sinks such as dislocations and grain boundaries. For high quenching temperatures the major defect present after quenching is the di-vacancy and the resistivity annealing process during stage I involves interaction between di-vacancies to form higher vacancy complexes which are relatively immobile and thus become the sinks for further vacancy aggregation. Unlike aluminium, however, the clustered vacancy defects formed in stage I are extremely stable and stage II does not take place until well above the temperature range where self-diffusion is rapid; the resistivity associated with the stable clustered defects anneals at about 650 °C in gold and even higher in silver.

Apart from their effect on resistivity, vacancies produced by both quenching and irradiation lead, as shown by internal friction experiments, to a decrease of mechanical damping due to dislocations. Large changes in the mechanical properties of metal can also be produced by vacancies and, as shown in *Figure 9.4*, the shape of the stress–strain curve is very dependent on the rate of cooling and a large increase in the yield stress may occur after quenching. This increase in yield stress, which can also be produced by irradiation (*Figure 9.5*) or fatigue (*Figure 13.14*) is associated with the appearance of marked coarsening of the slip lines and an increased tendency to overshoot.

While much valuable information has been gained about the energies of formation and movement of vacancies from studies of changes in the physical properties produced by vacancies, these do not indicate by what mechanism the vacancies are annealed out, nor the mechanism by which they produce hardening. The drop in stress on yielding shown in *Figures 9.4* and *9.5* indicates that dislocations are initially locked by jogs (or perhaps atmospheres when the vacancies may condense to form small cavities) formed by the migration of point defects to them. However, in addition to a yield phenomenon there is also a general hardening of the lattice, as shown by the raised level of the stress-strain

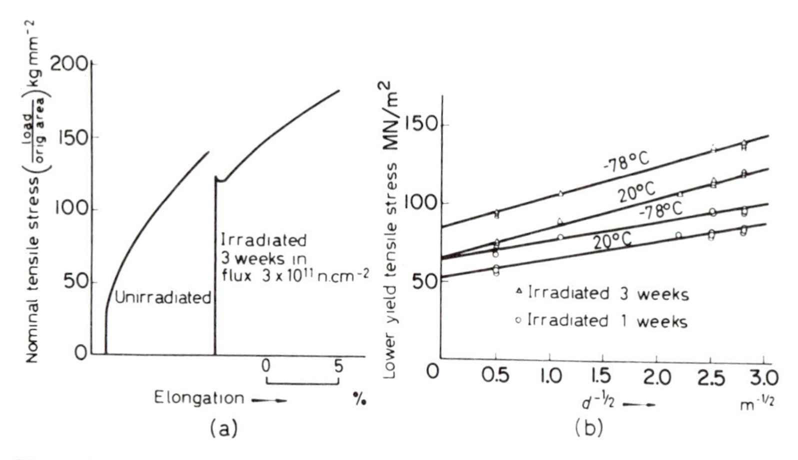

Figure 9.5 (a) Stress–strain curves for unirradiated and irradiated fine-grained polycrystalline copper, tested at 20°C; (b) variation of yield stress with grain size and neutron dose (after Adams and Higgins, *Phil. Mag.* 1959, **4,** 777)

curve. Thus, the yield plateaux may be due not only to the unlocking of dislocations from their sources but also the movement of dislocations past small obstacles in the lattice. The existence of the latter mode of hardening, often termed friction hardening, has been demonstrated in irradiated copper by observations of the dependence of the lower yield stress on grain size. The results, reproduced in *Figure 9.5(b)*, show that the relation $\sigma_y = \sigma_i + k_y d^{-1/2}$, which is a general relation describing the propagation of yielding in materials is obeyed and that two sources of hardening exist—a lattice friction hardening, σ_i, and a source hardening, k_y. In fatigue the friction hardening manifests itself by the gradual increase in the flow stress during continued cycling (*see Figure 13.14*). It is also evident from *Figure 9.5* that the flow stress of a material containing vacancy defects varies sensitively with temperature. This large temperature-dependence of flow stress has also been attributed to a lattice friction contribution. For copper, the temperature-dependence of the lower yield stress, taken as the ratio of the yield stress at 293 K to that at 78 K, is about 0.75 for fatigue, quench, and irradiation hardening, but only 0.90 for unidirectional work hardening.

Friction hardening could arise from the formation of clustered vacancy defects in the lattice which hinder the motion of glide dislocations. The most direct experimental evidence for the formation of both jogs and clusters of point defects in the form of small voids, dislocation loops and stacking fault tetrahedra has been obtained in electron transmission microscope experiments. These observations are discussed in section 9.5.

9.4 The nucleation of point defect clusters

The concentration of vacancies in equilibrium at room temperature is only about 10^{-13}, and, consequently, if a concentration of about 10^{-4} is produced by quenching, irradiation or fatigue, the excess vacancies must be removed from solution. This will occur if the vacancies migrate to the free surface or to a grain boundary. Alternatively, the vacancies might reach other permanent sinks, such as dislocations, and be annihilated there; the most likely sinks are dislocations and during the annihilation process the dislocations will climb. Clearly, the movement of a dislocation in this manner will be favoured by a high supersaturation of vacancies, and the dislocation may be thought of as being influenced by an effective stress somewhat analogous to osmotic pressure. The magnitude of this stress may be estimated from the chemical potential, if we let $\mathrm{d}F$ represent the change of free energy when $\mathrm{d}n$ vacancies are added to the system. Then, from Chapter 4,

$$\begin{aligned} \mathrm{d}F/\mathrm{d}n &= E_f + \boldsymbol{k}T \ln (n/N) \\ &= -\boldsymbol{k}T \ln c_0 + \boldsymbol{k}T \ln c \\ &= \boldsymbol{k}T \ln (c/c_0) \end{aligned}$$

where c is the actual concentration and c_0 the equilibrium concentration of vacancies. This may be rewritten as

$$\mathrm{d}F/\mathrm{d}V = \text{Energy/volume} = \text{stress} = (\boldsymbol{k}T/b^3)[\ln (c/c_0)] \tag{9.2}$$

where $\mathrm{d}V$ is the volume associated with $\mathrm{d}n$ vacancies and b^3 is the volume of one vacancy. Inserting typical values, $\boldsymbol{k}T \simeq 1/40$ eV at room temperature, $b = 0.25$ nm, shows $\boldsymbol{k}T/b^3 \simeq 150$ MN/m^2. Thus, even a moderate 1 per cent supersaturation of vacancies, i.e. when $(c/c_0) = 1.01$ and $\ln (c/c_0) = 0.01$, introduces a chemical stress σ_c equivalent to 1.5 MN/m^2.

The equilibrium concentration of vacancies at a temperature T_2 will be given by $c_2 = \exp[-E_f/\boldsymbol{k}T_2]$ and at T_1 by $c_1 = \exp[-E_f/\boldsymbol{k}T_1]$. Then, since

$$\ln (c_2/c_1) = (E_f/\boldsymbol{k})\left[\frac{1}{T_1} - \frac{1}{T_2}\right]$$

the chemical stress produced by quenching a metal from a high temperature T_2 to a low temperature T_1 is

$$\sigma_c = (\boldsymbol{k}T/b^3) \ln (c_2/c_1) = (E_f/b^3)\left[1 - \frac{T_1}{T_2}\right]$$

For aluminium, E_f is about 0.7 eV so that quenching from 900 K to 300 K produces a chemical stress of about 3 GN/m^2. This stress is extremely high, several times the theoretical yield stress, and must be relieved in some way. Migration of vacancies to grain boundaries and dislocations will occur, of course, but it is not surprising that the point defects form additional vacancy sinks by the spontaneous nucleation of dislocation loops, and other stable lattice defects.

9.5 Electron microscope observations of vacancy defects

9.5.1 Quenching

Dislocation loops

The first direct observations of the clustering of vacancies in quenched metals were made on quenched aluminium. These original observations indicated that specimens of aluminium, quenched from near the melting point into water, contain many dislocation loops (*Figure 9.6*(*a*)); the average number of loops is 10^{12} mm^{-3}, their average diameter is 30 nm and the density of dislocations in the loops is about 10^8 mm^{-2}. Each loop arises from a collapsed disc of vacancies and thus the initial concentration of dispersed vacancies must have been 10^{-4}; a value in agreement with the number of vacancies estimated from resistivity measurements.

To reduce their energy, many of the loops lying on $\{111\}$ planes take up regular crystallographic forms with their edges parallel to the $\langle 110 \rangle$ directions in the loop plane. The loops shown in *Figure 9.6* are not Frank sessile dislocations as expected from the collapse of a vacancy disc, but prismatic dislocations since no fringe contrast, of the type arising from stacking faults, can be seen within the defects. The dislocation loop of Burgers vector $b = \frac{1}{3}a[111]$, formed by the condensation of a single sheet of vacancies on a (111) plane, encloses an area of stacking fault which cannot move with the dislocation, and in a metal of high stacking fault energy, when the loop encloses a region of high energy, subsequent changes may be expected to eliminate the fault. Kuhlmann-Wilsdorf earlier had suggested that if the area of the loop is swept with a Shockley partial dislocation $\frac{1}{6}a[11\bar{2}]$, the Frank sessile may be transformed into a total prismatic dislocation, according to the reaction

$$\frac{1}{3}a[111] + \frac{1}{6}a[11\bar{2}] \rightarrow \frac{1}{2}a[110] \tag{9.4}$$

This reaction is more easily followed with the aid of the Thompson tetrahedron (*Figure 6.27*), when reaction 9.4 can be rewritten as

$$A\alpha + \alpha D \rightarrow AD$$

Physically, this means that the disc of vacancies aggregated on a (111) plane of a metal with high stacking fault energy, besides collapsing, also undergoes a shear movement. The dislocation loops shown in *Figure 9.6* are therefore, unit dislocations with their Burgers vector $\frac{1}{2}a[110]$ inclinded at an angle to the original (111) plane. A prismatic dislocation loop lies on the surface of a cylinder, the cross-section of which is determined by the dislocation loop, and the axis of which is parallel to the [110] direction. Such a dislocation is not sessile, and under the action of a shear stress it is capable of movement by prismatic slip in the [110] direction. This form of slip was actually seen in the electron microscope during the initial observations.

The distribution of loops is not uniform throughout a specimen, and is particularly influenced by the existence of vacancy sinks. For example, around a grain boundary a denuded zone is found, as shown in *Figure 9.6*(*b*), because the boundary acts as a sink for vacancies, and hence the degree of supersaturation in this region is always too low to allow the formation of loops. A denuded zone is also found in the region of a dislocation line (*Figure 9.6*(*c*)). This may be taken as evidence that a dislocation is also a sink for vacancies, and that the complex configuration of the dislocation is the result of climb of various sections.

To unfault the Frank sessile dislocation loop it is necessary to nucleate a Shockley partial dislocation somewhere in the loop which, on expanding, removes the faults and reacts with the Frank dislocation to produce a new (unit)

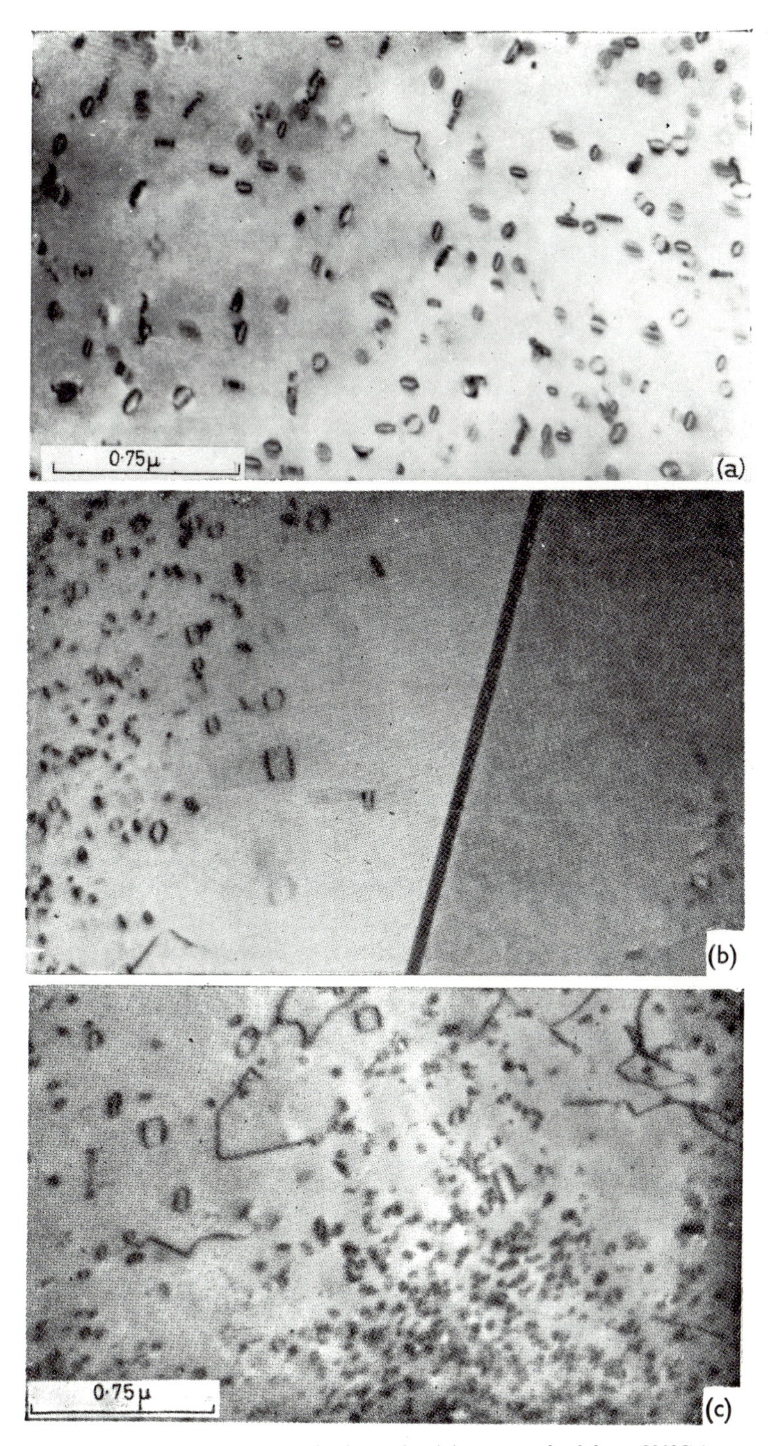

Figure 9.6 Electron micrographs from aluminium quenched from 630°C into water showing (a) prismatic loops lying on {111} planes, (b) a zone denuded of loops in the region of a grain boundary, and (c) a zone denuded of loops in the region of a dislocation (after Hersch, Silcox, Smallman and Westmacott, *Phil. Mag.* 1958, **3**, 897. Courtesy Taylor and Francis)

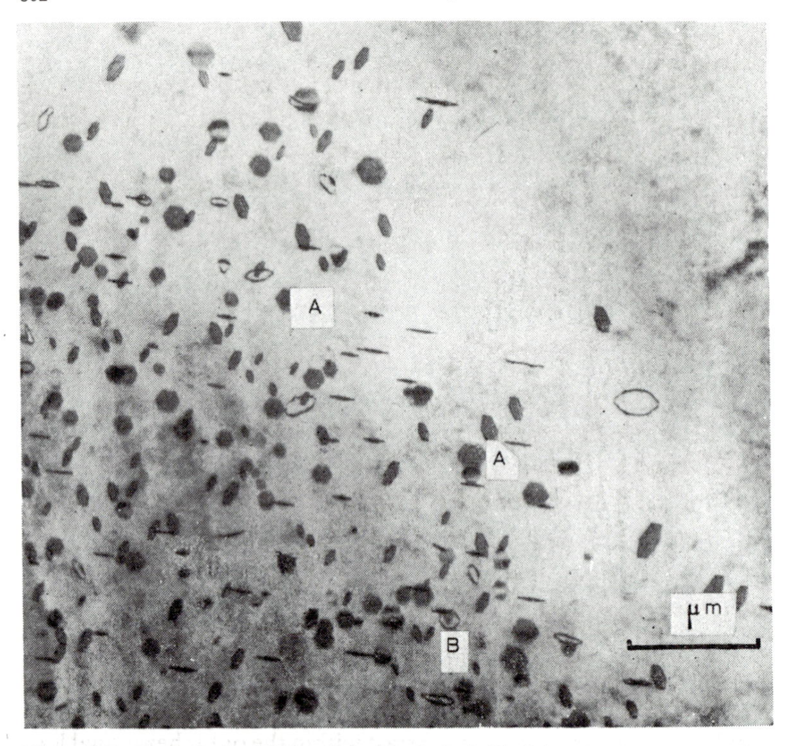

Figure 9.7 Single-faulted (*A*) and double-faulted (*B*) dislocation loops in quenched aluminium. (after Edington and Smallman, *Phil. Mag.*, 1965, **11,** 1089, courtesy of Taylor and Francis)

dislocation with larger distortion energy. An energy barrier thus exists, corresponding to a critical radius of Shockley loop for a given value of γ, which must be overcome before the Shockley loop can propagate with decrease in energy. This energy barrier is estimated to be several electron volts, and is unlikely to be overcome by thermal activation, or by the stress (γ/b) that the fault exerts on the Frank dislocation loop. The fault may be removed, however, by local shear stresses in the foil, and these are often produced during quenching. Minimizing these shear stresses by careful handling and less drastic quenching, enables the fault to be retained, as shown in *Figure 9.7*, where more than 90 per cent of the loops contain stacking faults. By stressing the foil while it is under observation in the microscope allows the unfaulting process to be observed directly (*see Figure 9.8*).

Faulted (Frank) dislocation loops have now been observed in a wide variety of f.c.c. quenched metals, e.g. nickel $\gamma \approx 200$ mJ/m^2, copper $\gamma \approx 40$ mJ/m^2, gold $\gamma \approx 30$ mJ/m^2 and silver $\gamma \approx 90$ mJ/m^2. In general, regular hexagons are not observed, instead the loops take up many polygonal forms with edges along $\langle 110 \rangle$ containing acute and obtuse corners. Faulted dislocation loops have also been produced in h.c.p. metals, such as Zn, Cd and Mg, as shown in *Figure 9.9*. The loops are 'sheared' Frank type with

$$b = A\sigma + \sigma S = \tfrac{1}{2}c + p$$

as described in Chapter 6.

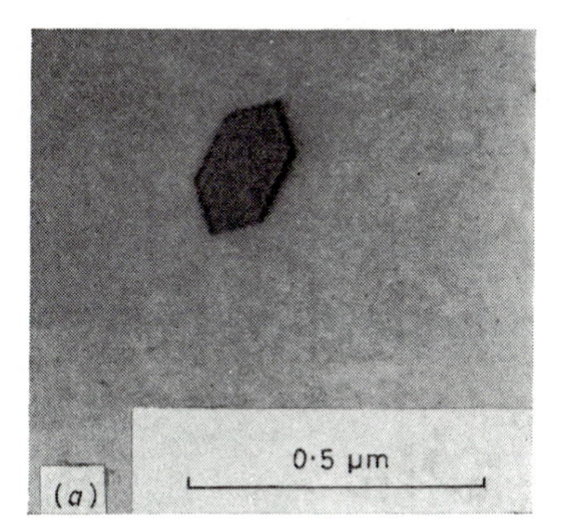

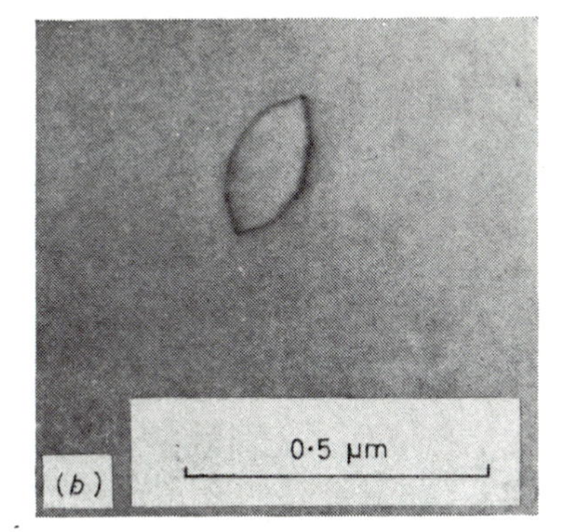

Figure 9.8 Removal of the stacking fault from a Frank sessile dislocation by stress (after Goodhew and Smallman, unpublished)

The critical vacancy supersaturation required to form vacancy loops appears to be much higher in b.c.c. metals than in f.c.c. metals and there are no unambiguous observations of vacancy loops in quenched b.c.c. metals.

Multiple loops

Many of the large Frank loops in *Figure 9.7*, for example marked **B**, contain additional trinagular-shaped loop contrast within the outer hexagonal loop. The stacking fault fringes within the triangle are usually displaced relative to those between the triangle and the hexagon by half the fringe spacing, which is the contrast expected from overlapping intrinsic stacking faults. The structural arrangement of those double-faulted loops is shown schematically in *Figure 9.10*, from which it can be seen that two intrinsic faults on next neighbouring planes are equivalent to an extrinsic fault. The observation of double-faulted loops in aluminium indicates that it is energetically more favourable to nucleate a Frank sessile loop on an existing intrinsic fault than randomly in the perfect lattice, and it therefore follows that the energy of a double or extrinsic fault is less than twice that of the intrinsic fault, i.e. $\gamma_E < 2\gamma_I$. This has been confirmed by loop annealing experiments, as discussed in section 9.6.

Slow quenching into oil favours the growth of large single Frank loops and the nucleation of double-faulted loops. The addition of a third overlapping intrinsic fault would change the stacking sequence from the perfect ABCABCABC to ABC↓B↓A↓CABC, where the arrows indicate missing planes of atoms, and produce a coherent twinned structure with two coherent twin boundaries. This structure would be energetically favourable to form, since $\gamma_{twin} < \gamma_I < \gamma_E$. It is possible, however, to reduce the energy of the crystal even further by aggregating the third layer of vacancies between the two previously formed neighbouring intrinsic faults to change the structure from an extrinsically faulted ABC↓B↓ABC to perfect ABC↓↓↓ABC structure. Such a triple layer dislocation loop is shown in *Figure 9.26*. Occasionally, a small faulted loop nucleates within the inner perfect region of the triple loop, resulting from the addition of a fourth layer of vacancies. Such observations indicate that the

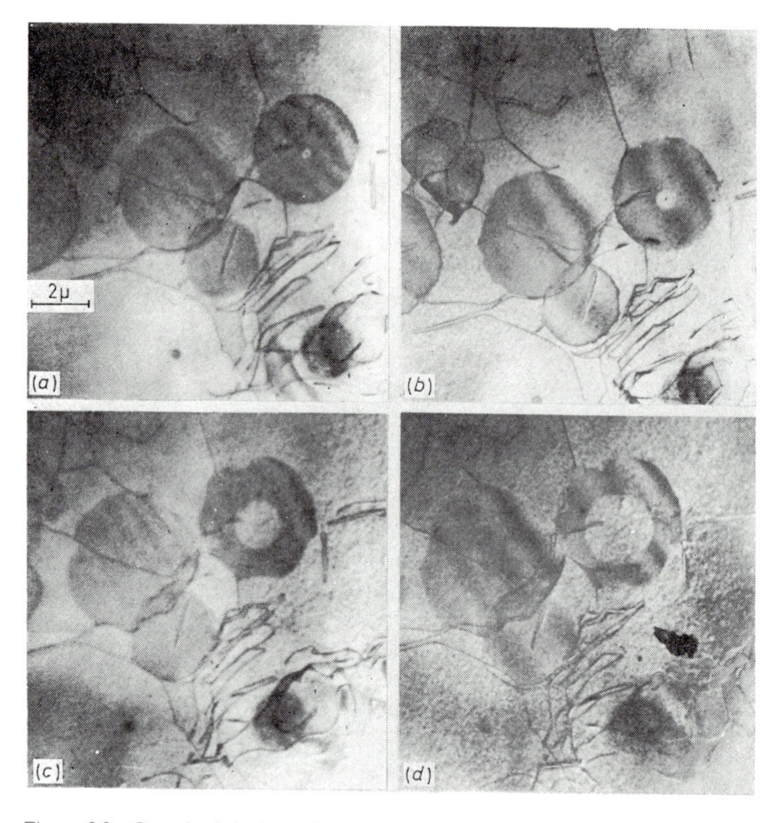

Figure 9.9 Growth of single and double faulted loops in magnesium on annealing at 175°C for (a) $t = 0$ min, (b) $t = 5$ min, (c) $t = 15$ min and (d) $t = 25$ min (after Hales, Smallman and Dobson)

nucleation of many of these vacancy defects occurs heterogeneously, probably on impurity atoms or particles, as is evident in *Figure 9.9*.

Double dislocation loops have also been found in quenched magnesium, as shown in *Figure 9.9*, in which the inner dislocation loop encloses a central region of perfect crystal and the outer loop an annulus of stacking fault. The structure of such a double loop is shown in *Figure 9.11*. The vacancy loops on adjacent atomic planes are bounded by dislocations with non-parallel Burgers vectors, i.e. $b=$

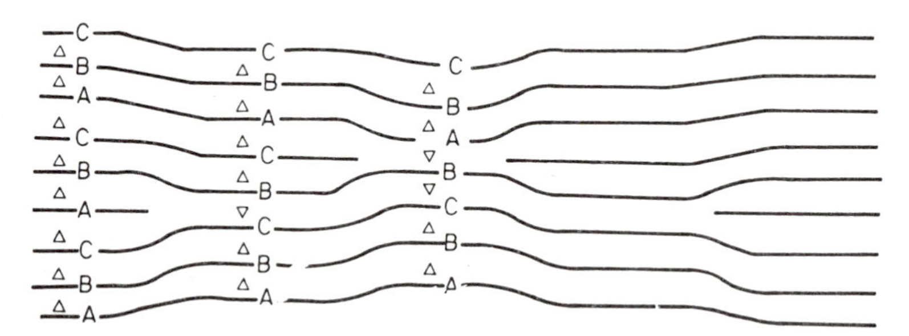

Figure 9.10 The structure of a double dislocation loop in quenched aluminium (after Edington and Smallman, *Phil. Mag.*, 1965, **11**, 1089, courtesy of Taylor and Francis)

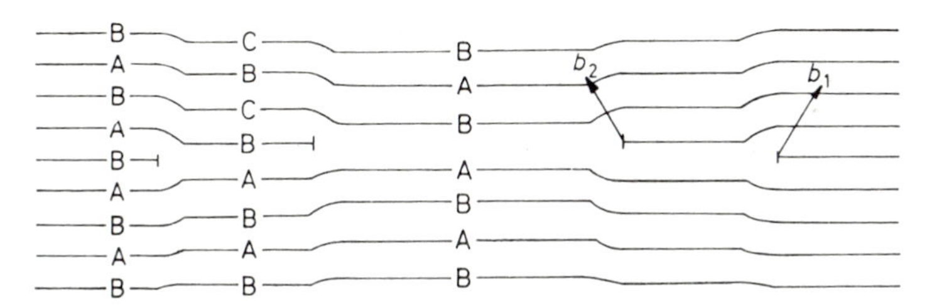

Figure 9.11 Structure of double dislocation loop in c.p.h. lattice

$(\frac{1}{2}c+p)$ and $b=(\frac{1}{2}c-p)$ respectively; the shear component of the second loop acts in such a direction as to eliminate the fault introduced by the first loop. There are six partial vectors in the basal plane p_1, p_2, p_3 and the negatives, and if one side of the loop is sheared by either p_1, p_2 or p_3 the stacking sequence is changed according to $A \rightarrow B \rightarrow C \rightarrow A$, whereas reverse shearing $A \rightarrow C \rightarrow B \rightarrow A$ results from either $-p_1$, $-p_2$ or $-p_3$. It is clear that the fault introduced by a positive partial shear can be eliminated by a subsequent shear brought about by any of the three negative partials. Three, four and more layered loops have also been observed in addition to the more common double loop. The addition of each layer of vacancies alternately introduces or removes stacking-faults, no matter whether the loops precipitate one above the other or on opposite sides of the original defect.

The formation of voids

The growth of the original vacancy cluster as a three-dimensional aggregate, i.e. void, or as a collapsed vacancy disc, i.e. dislocation loop should, in principle, depend on the relative surface to strain energy values for the respective defects. The energy of a three-dimensional void is mainly surface energy, whereas that of a Frank loop is mainly strain energy at small sizes. However, without a detailed knowledge of the surface energy of small voids and the core-energy of dislocations it is impossible to calculate, with any degree of confidence, the relative stability of these clustered vacancy defects.

The clustering of vacancies to form voids has now been observed in a number of metals with either f.c.c. or c.p.h. structure. In as-quenched specimens the voids are not spherical but bounded by crystallographic faces (*see Figure 9.12*), and usually are about 50 nm radius in size. In f.c.c. metals they are octahedral in shape with sides along $\langle 110 \rangle$, sometimes truncated by $\{100\}$ planes, and in h.c.p. metals bounded by prism and pyramidal planes. Void formation is favoured by slow quenching rates and high ageing temperatures, and the density of voids increases when gas is present in solid solution, e.g. hydrogen in copper, and either hydrogen or oxygen in silver. In aluminium and magnesium, void formation is favoured by quenching from a wet atmosphere, probably as a result of hydrogen production due to the oxidation reactions. It has been postulated that small clustered vacancy groups are stabilized by the presence of gas atoms and prevented from collapsing to a planar disc, so that some critical size for collapse can be exceeded. The voids are not conventional gas bubbles, however,

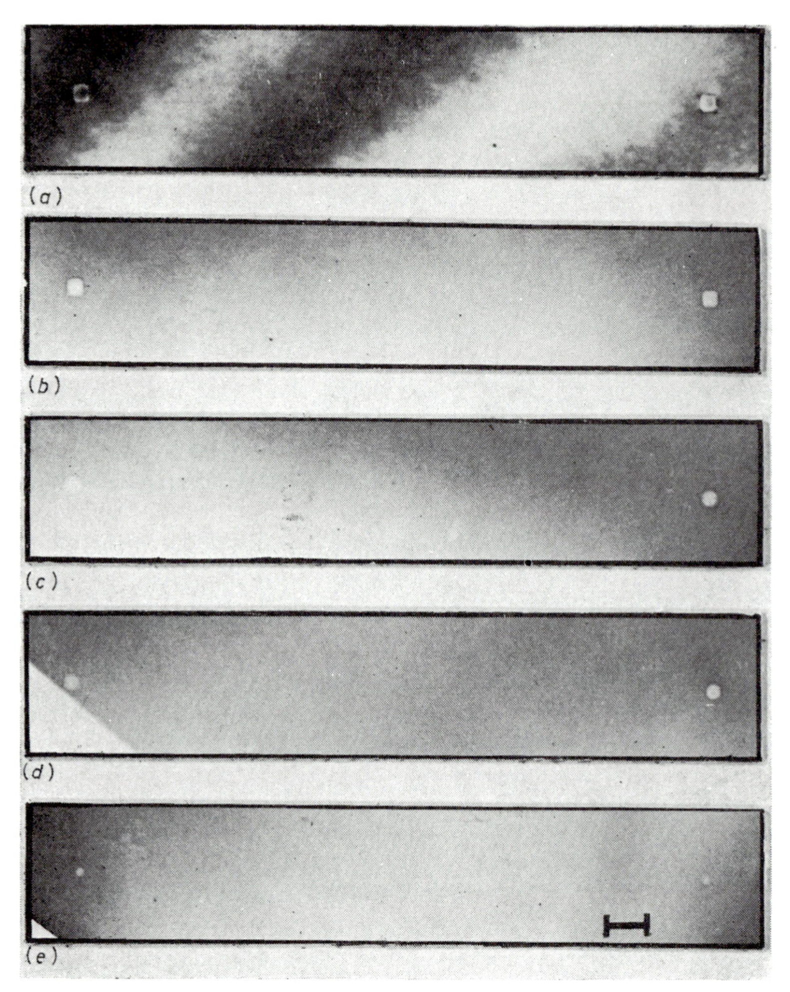

Figure 9.12 Sequence of micrographs showing the shrinkage of voids in quenched aluminium during isothermal annealing at 170°C; (a) t = 3 min, (b) t = 8 min, (c) t = 21 min, (d) t = 46 min, (e) t = 98 min. In all micrographs the scale corresponds to 0.1 μm (after Westmacott, Smallman and Dobson, *Metal Sci. J.*, 1968, **2,** 117, courtesy of the Institute of Metals)

since only a few gas atoms are required to nucleate the void after which it grows by vacancy adsorption. Nevertheless, it is probable that in the absence of gas in solution, vacancies would not cluster into three-dimensional voids but aggregate and form dislocation loops.

Stacking-fault tetrahedra

In f.c.c. metals and alloys, the vacancies may cluster into a three-dimensional defect, forming a tetrahedral arrangement of stacking faults on the four $\{111\}$ planes with the six $\langle 110 \rangle$ edges of the tetrahedron, where the stacking faults bend from one $\{111\}$ plane to another, consisting of stair-rod dislocations. The crystal

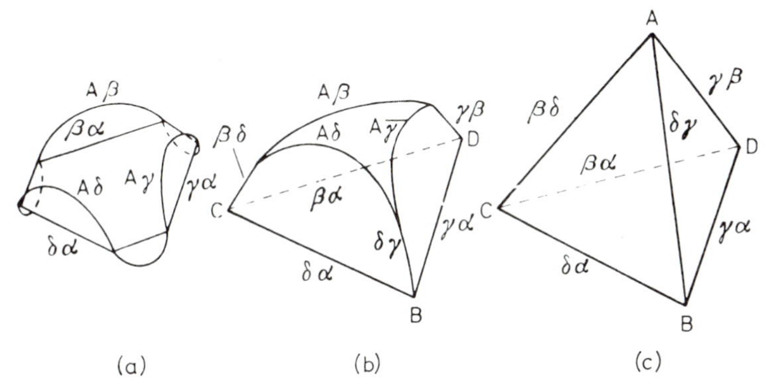

Figure 9.13 Formation of defect tetrahedron: (a) dissociation of Frank dislocations, (b) formation of new stair-rod dislocations, and (c) arrangement of the six stair-rod dislocations

structure is perfect inside and outside the tetrahedron, and the three-dimensional array of faults exhibit characteristic projected shape and contrast when seen in transmission electron micrographs as shown in *Figure 9.14*. This defect was observed originally in quenched gold and has since been observed in silver, copper, nickel–cobalt alloys and austenitic stainless steel. One mechanism for the formation of the defect tetrahedron by the dissociation of a Frank dislocation loop (*see Figure 9.13*), was first explained by Silcox and Hirsch. The Frank partial dislocation bounding a stacking fault has, because of its large Burgers vector, a high strain energy, and hence can lower its energy by dissociation according to a reaction of the type:

$$\underset{(\frac{1}{3})}{\tfrac{1}{3}a[111]} \rightarrow \underset{(\frac{1}{6})}{\tfrac{1}{6}a[121]} + \underset{(\frac{1}{18})}{\tfrac{1}{6}a[101]} \tag{9.5}$$

The figures underneath the reaction represent the energies of the dislocations, since they are proportional to the squares of the Burgers vector. This reaction is, therefore, energetically favourable. The dislocation with Burgers vector $\frac{1}{6}a[101]$ lies at the intersection of two stacking faults and is known as a stair-rod dislocation (*see Figure 6.26*). The reaction given in equation 9.5 can be seen with the aid of the Thompson tetrahedron, which shows that the Frank partial dislocation $A\alpha$ can dissociate into a Shockley partial dislocation ($A\beta$, $A\delta$ or $A\gamma$) and a low energy stair-rod dislocation ($\beta\alpha$, $\delta\alpha$ or $\gamma\alpha$) for example $A\alpha \rightarrow A\gamma + \gamma\alpha$.

The formation of the defect tetrahedron of stacking faults may now be envisaged as follows: the collapse of a vacancy disc will in the first instance lead to the formation of a Frank sessile loop bounding a stacking fault, with edges parallel to the $\langle 110 \rangle$ directions. Each side of the loop then dissociates according to equation 9.5 into the appropriate stair-rod and partial dislocations, and, as shown in *Figure 9.13(a)*, the Shockley dislocations formed by dissociation will lie on intersecting $\{111\}$ planes, above and below the plane of the hexagonal loop; the decrease in energy accompanying the dissociation will give rise to forces which tend to pull any rounded part of the loop into $\langle 110 \rangle$. Moreover, because the loop will not in general be a regular hexagon, the short sides will be eliminated

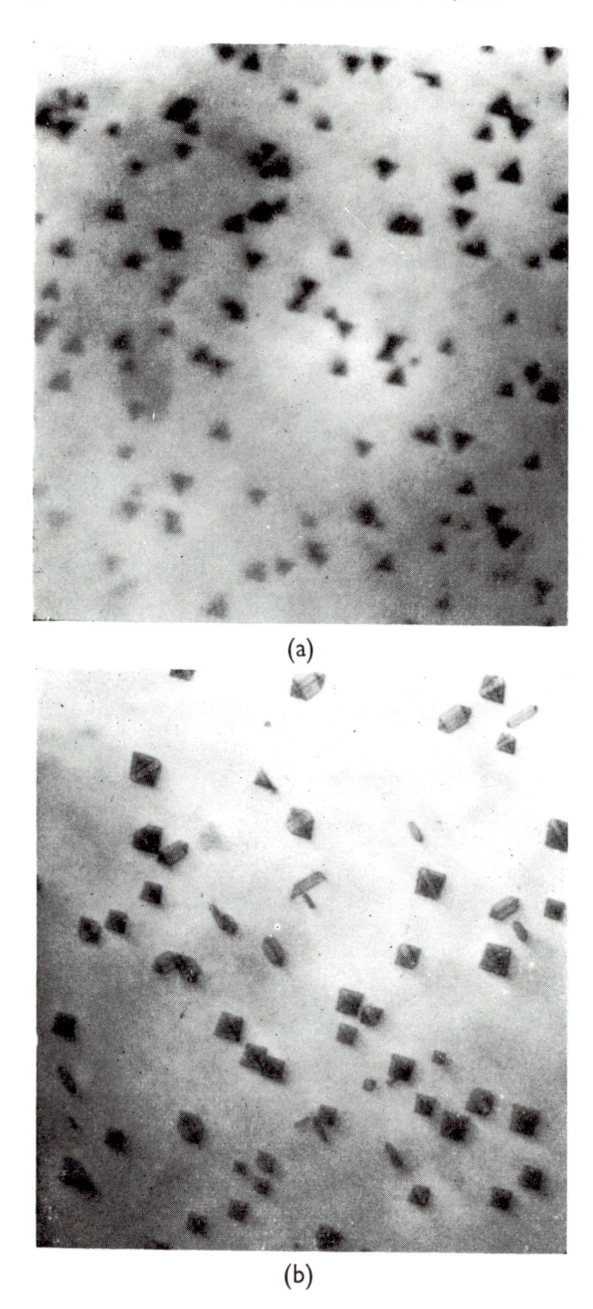

Figure 9.14 Defect tetrahedra in gold quenched from 960°C and aged for 1 hour at 100°C; (a) viewed in [110] orientation, (b) pre-annealed in oxygen and viewed in [100] orientation, (c) doped with 200 10^{-6} Mg, and (d) pre-annealed in H_2–N_2 gas (after Johnston, Dobson and Smallman)

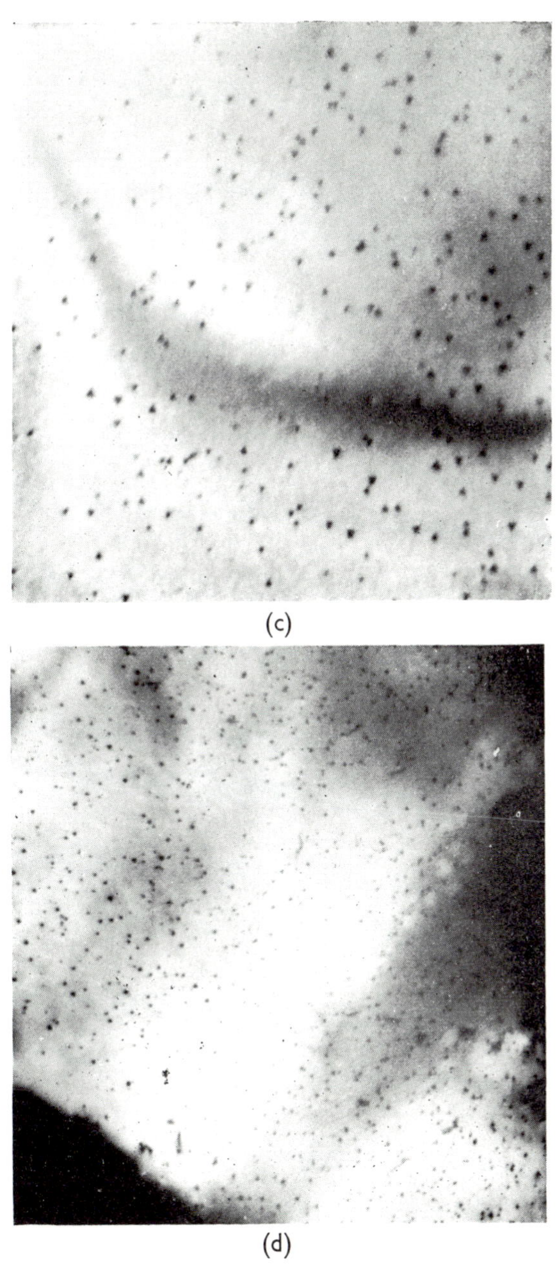

Figure 9.14 (*cont*) (c) doped with 200 10^{-6} Mg, and (d) pre-annealed in H_2–H_2 gas (after Johnston, Dobson and Smallman)

by the preferential addition of vacancies at the constricted site, and a triangular-shaped loop will form (*Figure 9.13(b)*). The partials $A\beta$, $A\gamma$ and $A\delta$ bow out on their slip plane as they are repelled by the stair-rods. Taking into account the fact that adjacent ends of the bowing loops are of opposite sign, the partials attract each other in pairs to form stair-rod dislocations along DA, BA and CA, according to the reactions

$$\gamma A + A\beta \rightarrow \gamma\beta, \quad \delta A + A\gamma \rightarrow \delta\gamma, \quad \beta A + A\delta \rightarrow \beta\delta$$

In vector notation the reactions are of the type (*see Figure 6.26(b)*)

$$\underset{(\frac{1}{6})}{\tfrac{1}{6}a[\bar{1}\bar{1}\bar{2}]} + \underset{(\frac{1}{6})}{\tfrac{1}{6}a[121]} \rightarrow \underset{(\frac{1}{18})}{\tfrac{1}{6}a[01\bar{1}]} \qquad (9.6)$$

(the reader may deduce the appropriate indices from *Figure (6.27)*, and from the addition of the squares of the Burgers vectors underneath it is clear that this reaction is also energetically favourable. The final defect will therefore be a tetrahedron made up from the intersection of stacking faults on the four $\{111\}$ planes, so that the $\langle 110 \rangle$ edges of the tetrahedron will consist of low energy stair-rod dislocations (*Figure 9.13(c)*).

The tetrahedron of stacking faults formed by the above sequence of events is essentially symmetrical, and the same configuration would have been obtained if collapse had taken place originally on any other (111) plane. The energy of the system of stair-rod dislocations in the final configuration is proportional to $6 \times \frac{1}{18} = \frac{1}{3}$, compared with $3 \times \frac{1}{3} = 1$ for the original stacking fault triangle bounded by Frank partials. Considering the dislocation energies alone, the dissociation leads to a lowering of energy to one-third of the original value. However, three additional stacking fault areas, with energies of γ per unit area, have been newly created and if there is to be no net rise in energy, these areas will impose an upper limit on the size of the tetrahedron formed. The student may wish to verify for himself that a calculation of this maximum size shows the side of the tetrahedron should be around 50 nm.

The shape of these defects, and the interference pattern within them, when observed in the electron microscope, can be explained in terms of the projection of the tetrahedral defect onto a particular plane. For example, if such a stacking fault tetrahedron is viewed along a $\langle 100 \rangle$ direction it would appear as a square in the microscope, and the contrast inside would have the form to be expected from an overlapping non-parallel stacking faults (*Figure 9.14(b)*. The other shapes observed in *Figure 9.14(b)* can be explained by truncation of the defect tetrahedra by the (100) foil surface. If, however, the tetrahedron were viewed along a $\langle 110 \rangle$ edge, it would appear as a triangle, and the contrast would take the form of fringes lying parallel to the line of intersection of the two overlapping faults. The other two planes of stacking faults have no effect on the pattern since they are parallel to the direction of observation (*Figure 9.14(a)*.

De Jong and Koehler have proposed that the tetrahedra may also form by the nucleation and growth of a three-dimensional vacancy cluster rather than by dissociation. The smallest cluster that is able to collapse to a tetrahedron and subsequently grow by the absorption of vacancies is a hexa-vacancy cluster. The growth of tetrahedra has been confirmed by examining quenched specimens of gold immediately after quenching and following (i) multiple-quenching from below the temperature at which the defects anneal-out, or (ii) various ageing treatments. Tetrahedra approaching 300 nm in size have been observed in gold, which is much larger than the 50 nm size expected from the dissociation mechanism. Growth occurs by the nucleation and propagation of jog lines across

Figure 9.15 Jog lines forming a ledge on the face of a tetrahedron

the faces of the tetrahedron, as shown in *Figure 9.15*. The jog lines are formed from a stair-rod dislocation dipole which has a vacant volume per atom site equal to one-third the volume of an atom, and hence, the jog line is often termed a one-third vacancy jog line.

The tetrahedron nucleus could form by any combination of single, di- or tri-vacancies which make a hexa-vacancy complex, but the high values for the di-vacancy binding energy indicate that it is most probably formed by di-vacancy combination. This conclusion is supported by the observation that small additions of divalent (e.g. Mg, Cd, Zn) or trivalent (e.g. Al) impurities enhance the heterogeneous nucleation of tetrahedra in quenched gold (*see Figure 9.14*(*c*)), whereas monovalent impurities (e.g. Cu) have virtually no effect at all at low concentrations. The influence of impurities on nucleation can be explained in terms of a di-vacancy/impurity interaction model. The di-vacancy in gold has a double negative charge with respect to the matrix and hence will be attracted to an impurity atom containing excess positive charge. Impurities such as Mg or Al will therefore interact strongly with di-vacancies and immobilize them, and because of the reduced negative charge the combination of an immobilized di-vacancy with a free di-vacancy is more likely to occur than the combination of two 'free' di-vacancies.

Combination of a tetra-vacancy solute atom complex with a further free-divacancy is also enhanced and, since the combining defects are not generally on the same crystallographic plane, three-dimensional tetrahedra nuclei are likely to be formed. However, if the gold contains sufficient impurity to immobilize all the di-vacancies, the nucleation of defects will be determined by the combination of migrating single vacancies with the immobilized di-vacancies which is more likely to lead to two-dimensional nuclei than di-vacancy/di-vacancy interactions. Recent doping experiments have shown that while 200 parts/10^6 of Mg increases the tetrahedra density, 500 parts/10^6 of Mg is sufficient to begin to suppress the nucleation of tetrahedra in favour of dislocation loops; Al is more effective than Mg but up to 1500 parts/10^6 of Cu has no effect.

Hydrogen in solution in the metal may also be effective in nucleating the tetrahedra, since a di-vacancy associated with a proton has a greater probability of combining with another di-vacancy, because of the reduced electrostatic repulsion, than an unassociated di-vacancy. Moreover, the di-vacancy-hydrogen complex is also mobile and will allow the tetrahedra to be nucleated irrespective of the hydrogen concentration and precludes the nucleation of loops. *Figure 9.14*(*d*) shows the increase in tetrahedra nucleation after pre-annealing in hydrogen.

9.5.2 Nuclear irradiation

Electron microscopy of irradiated metals shows that large numbers of small point defect clusters are formed on a finer scale than in quenched metals, because of the high supersaturation and low diffusion distance.

Bombardment of copper foils with 1.4×10^{21} 38 MeV α-particles m^{-2} produces about 10^{21} m^{-3} dislocation loops as shown in *Figure 9.16(a)*; a denuded region 0.8 μm wide can also be seen at the grain boundary. These loops, about 40 nm diameter, indicate that an atomic concentration of $\approx 1.5 \times 10^{-4}$ point defects have precipitated in this form. Heavier doses of α-particle bombardment produce larger diameter loops, which eventually appear as a dislocation tangle. Neutron bombardment produces similar effects to α-particle bombardment, but unless the dose is greater than 10^{21} neutrons/m^2 the loops are difficult to resolve. In copper irradiated at pile temperature the density of loops increases with dose and can be as high as 10^{14} m^{-2} in heavily bombarded metals.

The micrographs from irradiated metals reveal, in addition to the dislocation loops, numerous small centres of strain in the form of black dots somewhat less than 5 nm (50 Å) diameter, which are difficult to resolve (*see Figure 9.16(a)*). Because the two kinds of clusters differ in size and distribution, and also in their behaviour on annealing (*see* section 9.6), it is reasonable to attribute the presence of one type of defect, i.e. the large loops, to the aggregation of interstitials and the other, i.e. the small dots, to the aggregation of vacancies. This general conclusion has been confirmed by detailed contrast analysis of the defects.

The addition of an extra (111) plane in a crystal with f.c.c. structure (*see Figure 9.17*) introduces two faults in the stacking sequence and not one, as is the case when a plane of atoms is removed. In consequence, to eliminate the fault it is necessary for two partial dislocations to slip across the loop, one above the layer and one below, according to a reaction of the form

$$\tfrac{1}{3}a[\bar{1}\bar{1}\bar{1}] + \tfrac{1}{6}a[11\bar{2}] + \tfrac{1}{6}a[1\bar{2}1] \rightarrow \tfrac{1}{2}a[0\bar{1}\bar{1}] \qquad (9.7)$$

The resultant dislocation loop formed is identical to the prismatic loop produced by a vacancy cluster but has a Burgers vector of opposite sign. The size of the loops formed from interstitials increases with the irradiation dose and temperature, which suggests that small interstitial clusters initially form and subsequently grow by a diffusion process. In contrast, the vacancy clusters are much more numerous, and although their size increases slightly with dose, their number is approximately proportional to the dose and equal to the number of primary collisions which occur. This observation supports the suggestion that vacancy clusters are formed by the redistribution of vacancies created in the cascade.

Changing the type of irradiation from electron, to light charged particles such as protons, to heavy ions such as self-ions, to neutrons, results in a progressive increase in the mean recoil energy. This results in an increasingly non-uniform point defect generation due to the production of displacement cascades by primary knock-ons. During the creation of cascades, the interstitials are transported outwards (*see Figure 9.1*), most probably by focused collision sequences, i.e. along a close-packed row of atoms by a sequence of replacement

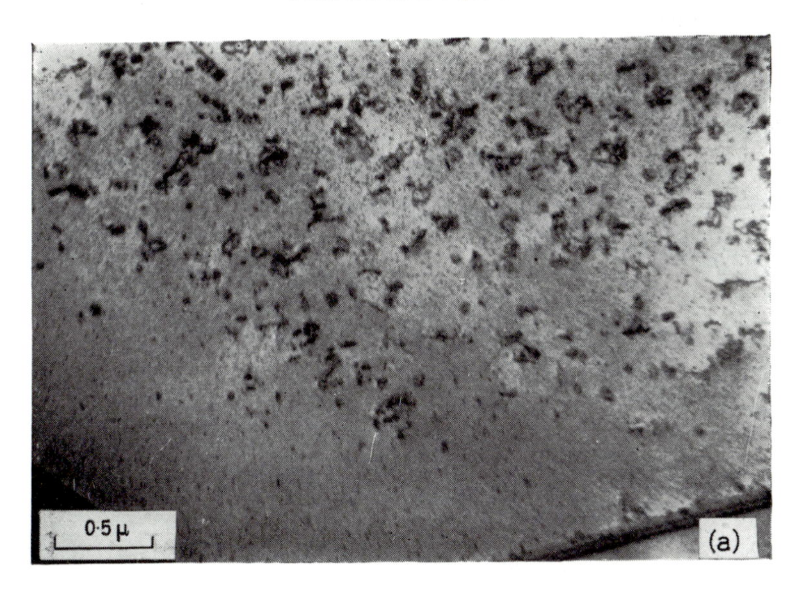

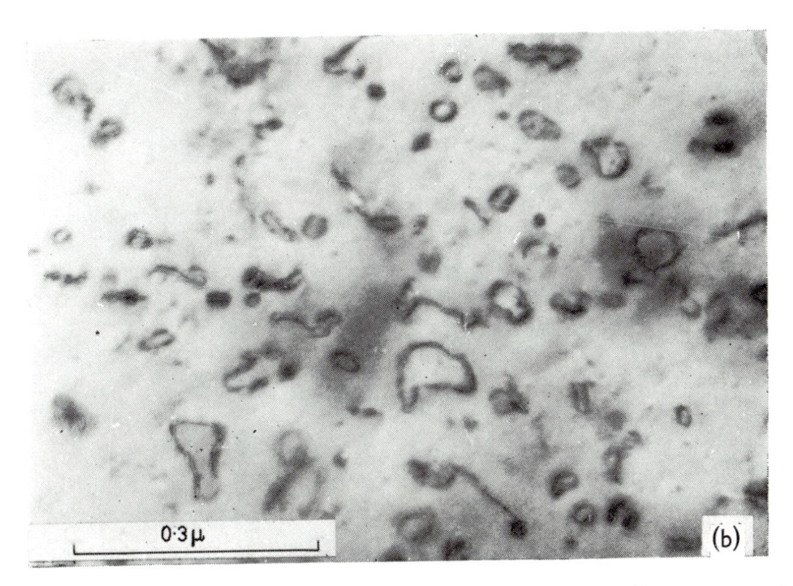

Figure 9.16 A thin film of copper after bombardment with 1.4×10^{21} α-particles m^{-2}; (a) Dislocation loops (≈40 nm dia) and small centres of strain (≈4 nm dia); (b) after a 2-hour anneal at 350°C showing large prismatic loops (after Barnes and Mazey, *Phil. Mag.* 1960, **5,** 124)

collisions, to displace the last atom in this same crystallographic direction, leaving a vacancy-rich region at the centre of the cascade which can collapse to form vacancy loops. As the irradiation temperature increases, vacancies can also aggregate to form voids.

Frank sessile dislocation loops, double-faulted loops, tetrahedra and voids

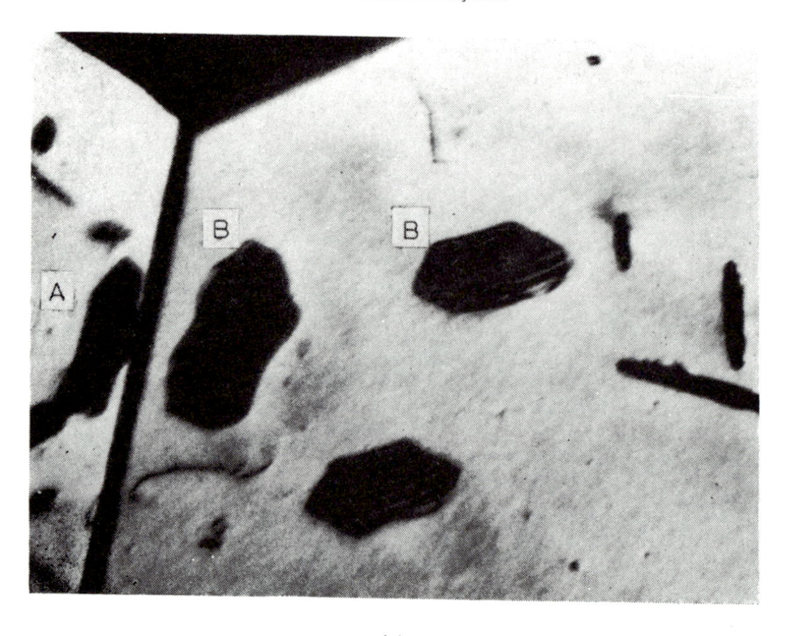

(a)

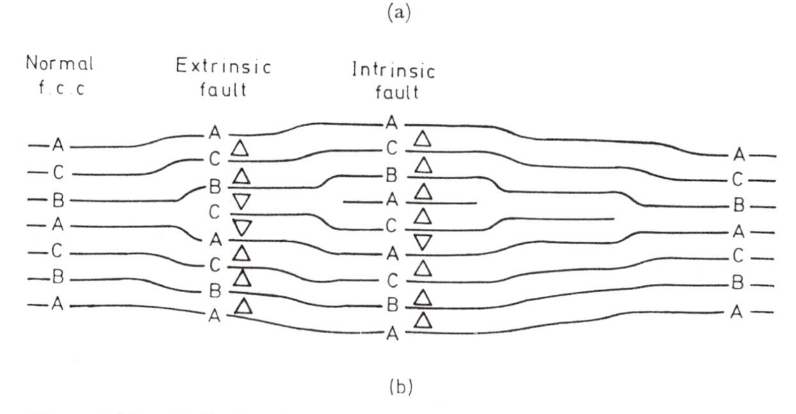

(b)

Figure 9.17 (a) Single (*A*) and double (*B*) dislocation loops in proton irradiated copper (× 43 000), (b) structure of a double dislocation loop (after Mazey and Barnes, *Phil. Mag.*, 1968, **17,** 387, courtesy of Taylor and Francis)

have all been observed in irradiated metals, but usually under different irradiation conditions.

Results from Cu, Ag and Au show that cascades collapse to form Frank loops, some of which dissociate towards stacking fault tetrahedra. The fraction of cascades collapsing to form visible loops, defined as the defect yield, is high, ≈ 0.5 in Cu to 1.0 in Au irradiated with self-ions. Moreover, the fraction of vacancies taking part in the collapse process, expressed as the cascade efficiency is also high (≈ 0.3 to 0.5). Vacancy loops have been observed on irradiation at RT in some b.c.c. metals, e.g. Mo, Nb, W, α-Fe. Generally, the loops are perfect with $b=\frac{1}{2}a\langle 111\rangle$ although they are thought to nucleate as $\frac{1}{2}a\langle 110\rangle$ faulted loops on $\{110\}$ but unfault at an early stage because of the high stacking fault energy.

Vacancy loops have also been observed in some h.c.p. metals, e.g. Zr and Ti.

Interstitial defects in the form of loops are commonly observed in all metals. In f.c.c. metals Frank loops containing extrinsic faults occur in Cu, Ag, Au, Ni, Al and austenitic steels. Clustering of interstitials on two neighbouring (111) planes to produce an intrinsically faulted defect may also occur, as shown in *Figure 9.17*. In b.c.c. metals they are predominantly perfect $\frac{1}{2}a\langle 111\rangle$.

The damage produced in h.c.p. metals by electron irradiation is very complex and for Zn and Cd ($c/a > 1.633$) several types of dislocation loops, interstitial in nature, nucleate and grow; thus $c/2$ loops i.e. with $b = [c/2]$, c-loops, $(c/2+p)$ loops, i.e. with $b = \frac{1}{6}\langle 2023$, $[c/2]+[c/2]$ loops and $\langle c/2+p\rangle + \langle c/2-p\rangle$ loops are all formed; in the very early stages of irradiation most of the loops consist of $[c/2]$ dislocations, but as they grow a second loop of $b = [c/2]$ forms in the centre, resulting in the formation of a $[c/2]+[c/2]$ loop. The $\langle c/2+p\rangle +$ $\langle c/2-p\rangle$ loops form either from the nucleation of a $\langle c/2+p\rangle$ loop inside a $\langle c/2-p$ loop or when a $[c/2]+[c/2]$ loop shears. At low dose rates and low temperatures many of the loops facet along $\langle 11\bar{2}0\rangle$ directions.

In magnesium with c/a almost ideal the nature of the loops is very sensitive to impurities, and interstitial loops with either $b = \frac{1}{3}\langle 11\bar{2}0\rangle$ on non-basal planes or basal loops with $b = (c/2+p)$ have been observed in samples with different purity. Double loops with $b = (c/2+p)+(c/2-p)$ also form but no $c/2$-loops have been observed.

In Zr and Ti ($c/a < 1.633$) irradiated with either electrons or neutrons, both vacancy and interstitials loops form on non-basal planes with $b = \frac{1}{3}\langle 11\bar{2}0\rangle$. Loops with a c-component, namely $b = \frac{1}{3}\langle 11\bar{2}3\rangle$ on $\{10\bar{1}0\}$ planes and $b = c/2$ on basal planes have also been observed; voids also form in the temperature range 0.3–0.6T_m. The fact that vacancy loops are formed on electron irradiation indicates that cascades are not essential for the formation of vacancy loops. Several factors can give rise to the increased stability of vacancy loops in these metals. One factor is the possibility of stresses arising from oxidation or anisotropic thermal expansion, i.e. interstitial loops are favoured perpendicular to a tensile axis and vacancy loops parallel. A secondary possibility is impurities segregating to dislocations and reducing the interstitial bias.

In general, interstitial loops grow during irradiation because the elastic size interaction causes dislocations to attract interstitials more strongly than vacancies. Interstitial loops are therefore intrinsically stable defects, whereas vacancy loops are basically unstable defects during irradiation. However, increasing the irradiation temperature results in vacancies aggregating to form voids. There are two important factors contributing to void formation. The first is the degree of bias the dislocation density (developed from the growth of interstitial loops) has for attracting interstitials, which suppresses the interstitial content compared to vacancies. The second factor is the important role played in void nucleation by gases, both surface-active gases such as oxygen, nitrogen and hydrogen frequently present as residual impurities, and inert gases such as helium which may be generated continuously during irradiation due to transmutation reactions. The surface-active gases such as oxygen in copper can migrate to embryo vacancy clusters and reduce the surface energy. The inert gas atoms can acquire vacancies to become gas molecules inside voids (when the gas pressure is not in equilibrium with the void surface tension) or gas bubbles when the gas pressure is considerable ($P \gtrsim 2\gamma_s/r$). Voids and bubbles can give rise to irradiation swelling and embrittlement of materials and is discussed further in section 9.8.

9.5.3 Cold work

The suggestion that large numbers of vacancies are created during fatigue has been investigated by transmission electron-microscope observations. In some areas, fatigued aluminium shows a high concentration of prismatic loops (*see Figure 9.18(a)* produced during the deformation. The role played by vacancies in the fatigue process will be discussed in Chapter 14. It is possible, however, that most of the dislocation loops observed in micrographs from cold worked metals are not formed by the collapse of a sheet of vacancies. Thus, if the jog is a multiple-jog, i.e. several Burgers vectors long, when the screw dislocation moves it may trail behind a dislocation dipole, as shown in *Figure 9.18(b)*, rather than a row of vacancies and dislocation loops may then form from the break-up of this dipole. Some of the elongated loops seen in fatigued aluminium appear to have formed this way.

Frank loops and tetrahedra have also been observed in deformed metals with low stacking-fault energy. In this case, the jogged screw dislocation is

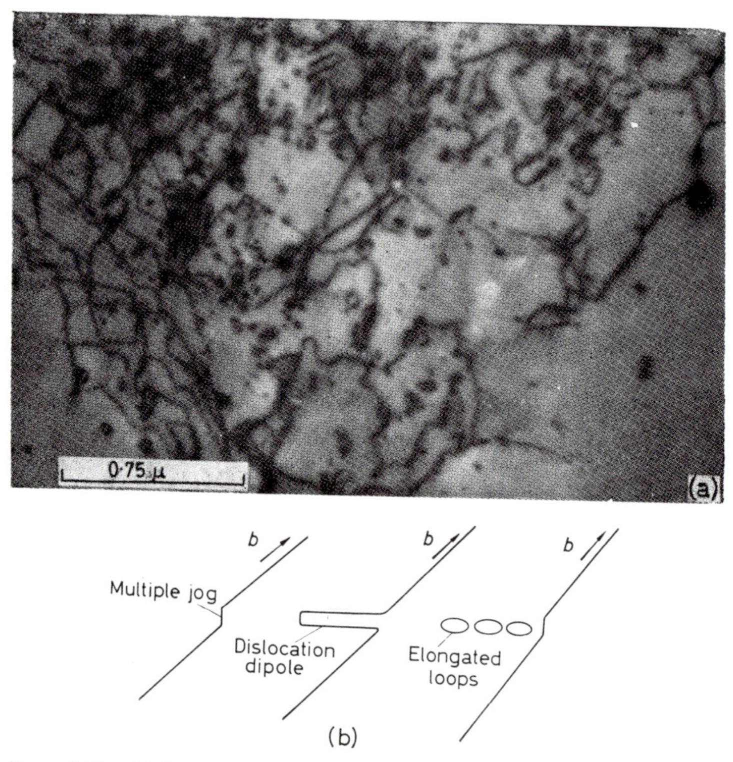

Figure 9.18 (a) Prismatic dislocation loops in fatigued aluminium (after Segall and Partridge, *Phil. Mag.* 1959, **4**, 912), (b) formation of dislocation loops from the motion of a multiple jog in a screw dislocation

believed to cross-slip to form a Frank sessile triangular-shaped loop, and some then transform to stacking fault tetrahedra by the Silcox–Hirsch dissociation mechanism. From the size of the largest tetrahedron observed and the smallest Frank loop, the stacking fault energy may be estimated.

9.5.4 Oxidation

The injection of vacancies into the underlying metal by oxidation of the surface has been discussed in section 9.2.3, and is also treated in Chapter 15. The vacancy supersaturation produced causes clustered-vacancy defects to grow by vacancy adsorption, as shown in *Figure 9.9* for magnesium, and in *Figure 15.4* for zinc which oxidizes readily at room temperature. The oxidation-rate depends on the oxygen pressure, and hence the growth-rate of loops is strongly dependent on oxygen pressure, as shown in *Figure 9.19*. Alloying elements can also markedly

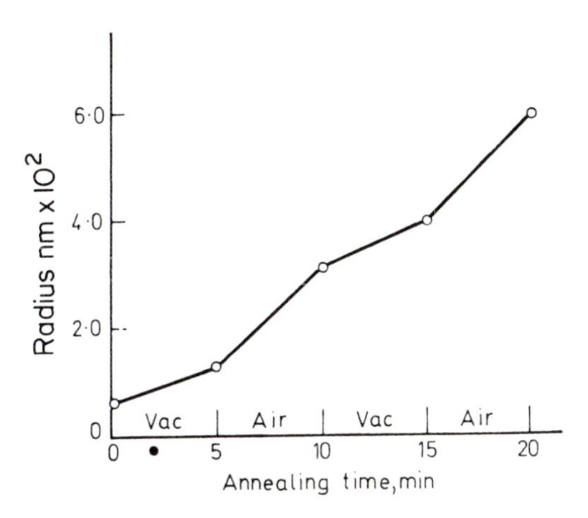

Figure 9.19 Variation of dislocation loop radius on annealing a magnesium thin film alternately in air and vacuum for 5 min at 180°C (after Hales, Dobson and Smallman, *Metal Sci. J.*, 1968, **2**, 224, courtesy of the Institute of Metals)

affect the oxidation rate (*see* Chapter 15) and the growth of loops by vacancy adsorption is correspondingly altered (*see Figure 15.4*).

9.6 The annealing of clustered defects

The electron-microscope observations confirm that the change in physical and mechanical properties on annealing is due to the clustering of point defects. For example, Stage I in the 'annealing out' of electrical resistivity of quenched pure metals is associated with the movement of vacancies to form loops, voids and other clustered defects, and the removal of the residual resistivity associated with Stage II is due to the dispersal of the loops, voids, etc. This latter process can be strikingly demonstrated by observing a heated specimen in the microscope. On heating, the dislocation loops and voids act as vacancy sources and shrink, and hence the defects annihilate themselves (*see Figures 9.20* and *9.12*); this process occurs in the temperature range where self-diffusion is rapid, and agrees

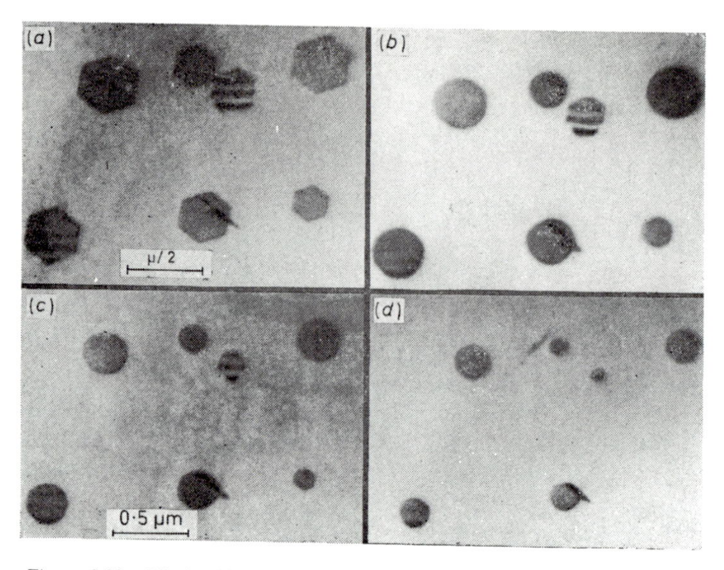

Figure 9.20 Climb of faulted loops in aluminium at 140°C; (a) $t = 0$ min, (b) $t = 12$ min, (c) $t = 24$ min, (d) $t = 30$ min, (after Dobson, Goodhew and Smallman, *Phil. Mag.*, 1967, **16,** 9, courtesy of Taylor and Francis)

closely with that where the residual resistivity anneals out. This observation indicates that the resistivity effect associated with Stage II is, in fact, due to loops and voids.

The driving force for the emission of vacancies from a vacancy defect arises in the case of (i) a prismatic loop from the line tension of the dislocation, (ii) a Frank loop from the force due to the stacking fault on the dislocation line since in intermediate and high γ-metals this force far outweighs the line tension contribution, and (iii) a void from the surface energy γ_s. The annealing of Frank loops and voids in quenched aluminium is shown in *Figures 9.20* and *9.12*, respectively. In a thin metal foil the rate of annealing is generally controlled by the rate of diffusion of vacancies away from the defect to any nearby sinks, usually the foil surfaces, rather than the emission of vacancies at the defect itself. To derive the rate equation governing the annealing, the vacancy concentration at the surface of the defect is used as one boundary condition of the diffusion equation (*see* Chapter 4) and the second boundary condition is obtained by assuming that the surfaces of a thin foil act as ideal sinks for vacancies. The rate then depends on the vacancy concentration gradient developed between the defect, where the vacancy concentration is given by

$$c = c_0 \exp\{(\mathrm{d}F/\mathrm{d}n)/\boldsymbol{k}T\} \tag{9.8}$$

with $(\mathrm{d}F/\mathrm{d}n)$ the change in free energy of the defect configuration per vacancy emitted at the temperature T, and the foil surface where the concentration is the equilibrium value c_0.

For a single, intrinsically, faulted circular dislocation loop of radius r the total energy of the defect F is given by the sum of the line energy and the fault

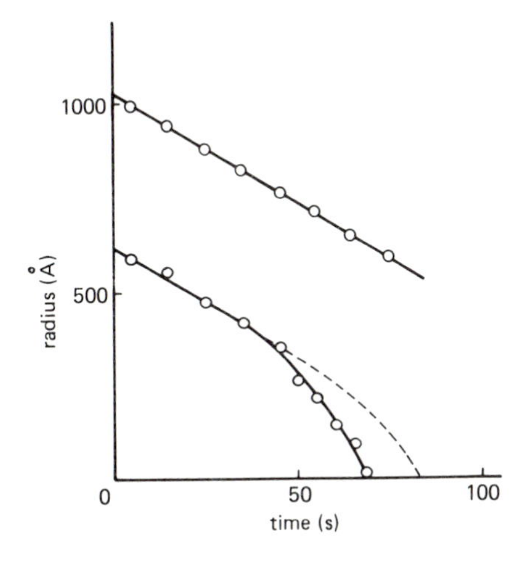

Figure 9.21 Variation of loop radius with time of annealing for Frank dislocations in Al showing the deviation from linearity at small r_0

energy, i.e.

$$F \simeq 2\pi r[\mu b^2/4\pi(1-\nu)]\ln(r/r_0)+\pi r^2\gamma_I$$

In the case of a large loop ($r > 50$ nm) in a material of intermediate or high stacking fault energy ($\gamma \gtrsim 60$ mJ/m^2) the term involving the dislocation line energy is negligible compared with the stacking fault energy term and thus, since $(dF/dn)=(dF/dr)\times(dr/dn)$, is given simply by $\gamma_1 B^2$, where B^2 is the cross-sectional area of a vacancy in the (111) plane. For large loops the diffusion geometry approximates to cylindrical diffusion* and a solution of the time-independent diffusion equation gives for the annealing rate,

$$\begin{aligned} dr/dt &= -[2\pi D/b\ln(L/b)][\exp(\gamma_I B^2/\boldsymbol{k}T)-1] \\ &= \text{const.}[\exp(\gamma_I B^2/\boldsymbol{k}T)-1] \end{aligned} \tag{9.9}$$

where $D=D_0\exp(-U_D/kT)$ is the coefficient of self-diffusion and L is half the foil thickness. The annealing rate of a prismatic dislocation loop can be similarly determined, in this case dF/dr is determined solely by the line energy, and then

$$dr/dt = -[2\pi D/b\ln(L/b)](\alpha b/r) = \text{const.}[\alpha b/r] \tag{9.10}$$

where the term containing the dislocation line energy can be approximated to $\alpha b/r$. The annealing of Frank loops is shown in *Figure 9.20* and obeys the linear relation given by equation 9.9 at large r (*Figure 9.21*); at small r the curve deviates from linearity because the line tension term can no longer be neglected and also also because the diffusion geometry changes from cylindrical to spherical symmetry. The annealing of prismatic loops is much slower, because only the line tension term is involved, and obeys an r^2 versus t relationship; comparable annealing rates to that observed for faulted loops can be obtained by raising the temperature 25°C or so. In principle equation 9.9 affords a direct determination of the stacking fault energy γ by substitution, but since U_D is usually much bigger

* For spherical diffusion geometry the pre-exponential constant is D/b.

than γB^2 this method is unduly sensitive to small errors in U_D. This difficulty may be eliminated, however, by a comparative method in which the annealing rate of a faulted loop is compared to that of a prismatic one at the same temperature. The intrinsic stacking fault energy of aluminium has been shown to be 135 mJ/m^2 by this technique.

In addition to prismatic and single-faulted (Frank) dislocation loops, double-faulted loops have also been annealed in a number of quenched f.c.c. metals. It is observed that on annealing, the intrinsic loop first shrinks until it meets the inner, extrinsically faulted region, following which the two loops shrink together as one extrinsically faulted loop. The rate of annealing of this extrinsic fault may be derived in a way similar to equation 9.9, slight numerical differences arising from the fact that two vacancies need to be emitted to reduce the fault area by B^2, giving $\mathrm{d}F/\mathrm{d}r = \gamma_E B^2/2$; twice as many vacancies have to diffuse to the surface to give the same decrease in radius as for a single fault, giving rise to the additional factor of two in the pre-exponential. Thus the annealing rate for a double, extrinsically faulted loop is given by

$$\begin{aligned} \mathrm{d}r/\mathrm{d}t &= -[\pi D/b \ln (L/b)][\exp(\gamma_E B^2/2\boldsymbol{k}T) - 1] \\ &= \text{const.} \{\exp(\gamma_E B^2/2\boldsymbol{k}T) - 1\} \end{aligned} \tag{9.11}$$

from which the extrinsic stacking fault energy may be determined. Generally γ_E is about 10–30 per cent higher in value than γ_I.

For a spherical hole in an infinite isotropic solid, the Helmholtz free energy of the solid involves the surface energy of the void γ_s and the elastic strain energy in the surrounding metal, but for voids sizes $r \geqslant 1$ nm the elastic strain energy is negligibly small. The Helmholtz free energy for a void in equilibrium with its surroundings is then $F \simeq 4\pi r^2 \gamma_s$ and since $(\mathrm{d}F/\mathrm{d}n) = (\mathrm{d}F/\mathrm{d}r)(\mathrm{d}r/\mathrm{d}n) = (8\pi r \gamma_s)(\Omega/4\pi r^2)$ where Ω is the atomic volume and n the number of vacancies in the void, equation 9.8, the concentration of vacancies in equilibrium with the void is

$$c_v = c_0 \exp(2\gamma_s \Omega / r\boldsymbol{k}T)$$

Assuming spherical diffusion geometry, the diffusion equation may be solved to give the rate of shrinkage of a void as

$$\mathrm{d}r/\mathrm{d}t = -(D/r)\{\exp(2\Omega\gamma_s/r\boldsymbol{k}T) - 1\} \tag{9.12}$$

For large r (>50 nm) the exponential term can be approximated to the first two terms of the series expansion and equation 9.12 may then be integrated to give

$$r^3 = r_i^3 - (6D\Omega\gamma_s/\boldsymbol{k}T)t \tag{9.13}$$

where r_i is the initial void radius at $t=0$. By observing the shrinkage of voids as a function of annealing time at a given temperature (*see Figure 9.12*) it is possible to obtain either the diffusivity D or the surface energy γ_s. From such observations, γ_s for aluminium is shown to be 1.14 J/m^2 in the temperature range 150–200 °C, and $D = 0.176 \exp(-1.31\ eV/\boldsymbol{k}T)$. It is difficult to determine γ_s for Al by zero creep measurements because of the oxide. This method of obtaining γ_s has been applied to other metals and is particularly useful since it gives a value of γ_s in the self-diffusion temperature range rather than near the melting point.

Loop growth can occur when the direction of the vacancy flux is towards the loop rather than away from it, as in the case of loop shrinkage. This condition can arise when the foil surface becomes a vacancy source, as for example during the growth of a surface oxide film. Loop growth is thus commonly found in Zn, Mg, Cd, although loop shrinkage is occasionally observed, presumably due to the formation of local cracks in the oxide film at which vacancies can be annihilated. *Figure 9.9* shows loops growing in Mg as a result of the vacancy supersaturation produced by oxidation. For the double loops, it is observed that stacking fault is created by vacancy absorption at the growing outer perimeter of the loop and is destroyed at the growing inner perfect loop. The perfect regions expand faster than the outer stacking fault, since the addition of a vacancy to the inner loop decreases the energy of the defect by γB^2 whereas the addition of a vacancy to the outer loop increases the energy by the same amount. This effect is further enhanced as the two loops approach each other due to vacancy transfer from the outer to inner loops. Eventually the two loops coalesce to give a perfect prismatic loop of Burgers vector $c=[0001]$ which continues to grow under the vacancy supersaturation. The outer loop growth rate is thus given by

$$\dot{r}_0=[2\pi D/B\ln(L/b)][(c_s/c_0)-\exp(\gamma B^2/\boldsymbol{k}T)] \qquad (9.14)$$

when the vacancy supersaturation term (c_s/c_0) is larger than the elastic force term tending to shrink the loop. The inner loop growth rate is

$$\dot{r}_i=[2\pi D/B\ln(L/b)][(c_s/c_0)-\exp(-\gamma B^2/\boldsymbol{k}T)] \qquad (9.15)$$

where $\exp(-\gamma B^2/\boldsymbol{k}T)\ll 1$, and the resultant prismatic loop growth rate is

$$\dot{r}_p=[\pi D/B\ln(L/b)]\{(c_s/c_0)-[(\alpha b/r)+1]\} \qquad (9.16)$$

where $(\alpha b/r)<1$ and can be neglected. By measuring these three growth rates, values for γ, (c_s/c_0) and D may be determined; Mg has been shown to have $\gamma=125$ mJ/m^2 from such measurements.

Negative sintering, i.e. growth of voids, can also occur when the driving force is reversed. One way of achieving this is by injecting vacancies into the metal as occurs during oxidation at the surface. Another way is by irradiation to produce not only vacancies but also gaseous products so that the voids contain gas. Small quantities of gas impurities are produced in uranium by fission, and it is observed that the gas atoms gather in the voids and cause swelling when the pressure $P>2\gamma_s/r$, the void then acting as a sink for vacancies. Gas producing nuclear reactions (mainly helium) can be important in other reactor materials, such as stainless steels which contain boron as an impurity. These bubbles embrittle the material along the grain boundaries.

'Annealing out' experiments have also confirmed that some of the dislocation loops produced by irradiation are formed from interstitials. Thus, in copper bombarded with α-particles, where at the end of the track of the α-particle the specimen contains large numbers of helium atoms, the vacancy clusters produced by irradiation nucleate small helium bubbles and provide a little of the space they need to form bubbles, while the interstitials cluster and form dislocation loops. When the specimen is heated at about 350 °C the gas bubbles are able to grow by acquiring vacancies from nearby vacancy sources, i.e. the dislocation loops (*Figure 9.16*). However, because these loops are formed from interstitial atoms they grow, not shrink, during the climb process and eventually become a tangled dislocation network.

Interstitial point defects have two properties important in both interstitial loop and void growth. First the elastic size interaction causes dislocations to attract interstitials more strongly than vacancies and secondly, the formation energy of an interstitial E_f^{I} is greater than that of a vacancy E_f^{v} so that the dominant process at elevated temperatures is vacancy emission. The importance of these factors to loop stability is shown by the spherical diffusion controlled rate equation

$$\frac{\mathrm{d}r}{\mathrm{d}t}=\frac{1}{b}\left\{D_{\mathrm{v}}c_{\mathrm{v}}-Z_{\mathrm{i}}D_{\mathrm{i}}c_{\mathrm{i}}-D_{\mathrm{v}}c_{0}\exp\left[\frac{(F_{\mathrm{el}}+\gamma)b^{2}}{\boldsymbol{k}T}\right]\right\} \qquad (9.17)$$

where c_{i} and c_{v} are the interstitial and vacancy concentrations respectively, D_{v} and D_{i} their diffusivities, F_{el} the elastic energy of the loop, γ the stacking fault energy and Z_{i} is a bias term defining the preferred attraction of the loops for interstitials. At low temperatures the growth is bias-driven, i.e. $Z_{\mathrm{i}}D_{\mathrm{i}}c_{\mathrm{i}}>D_{\mathrm{v}}c_{\mathrm{v}}$, whereas at higher temperatures, when $D_{\mathrm{v}}c_{0}$ becomes significant growth by vacancy emission makes an increasing contribution.

For the growth of voids during irradiation the spherical diffusion equation

$$\frac{\mathrm{d}r}{\mathrm{d}t}=\frac{1}{r}\left\{[1+(\rho r)^{1/2}]D_{\mathrm{v}}c_{\mathrm{v}}-[1+(Z_{\mathrm{i}}\rho r)^{1/2}]D_{\mathrm{i}}c_{\mathrm{i}}\right.$$

$$\left.-[1+(\rho r)^{1/2}]D_{\mathrm{v}}c_{0}\exp\left[\frac{[(2\gamma_{\mathrm{s}}/r)-P]\Omega}{\boldsymbol{k}T}\right]\right\}$$

has been developed. At low temperatures, voids undergo biased-driven growth in the presence of biased sinks, i.e. dislocation loops or network of density ρ. At higher temperatures when the thermal emission of vacancies becomes important, whether voids grow or shrink depends on the sign of $[(2\gamma_{\mathrm{s}}/r)-P]$. However, during neutron irradiation when gas is being created continuously a flux of gas atoms can arrive at the voids causing gas-driven growth.

The tetrahedral-shaped defect has been found to have an extremely stable structure. In quenched gold, for example, such defects appear to anneal out well above the self-diffusion temperature, at a temperature of about 650 °C, in silver $\approx$800 °C, and in irradiated copper $\approx$500 °C. The stability of the tetrahedra arises from the difficulty of vacancy emission, which involves the nucleation of jog lines. The tetrahedra may collapse under stress, however, particularly by interaction with a glissile dislocation.

9.7 Point defect hardening

The dependence of the yield stress, σ_y, on grain size (*Figure 9.5*) indicates that the hardening produced by point defects introduced by quenching or irradiation, is of two types; (i) an initial dislocation source hardening, and (ii) a general lattice hardening which persists after the initial yielding. The k_y term would seem to indicate that the pinning of dislocations may be attributed to point defects in the form of coarsely spaced jogs, and the electron-microscope observations of jogged dislocations would seem to confirm this. The inhomogeneity interaction (*see* Chapter 8) is considered to be the dominant effect in vacancy-dislocation interactions, with $V=-A/r^2$ where A is a constant; this compares with size effect for which $V=-A'/r$ would be appropriate for the interstitial-dislocation interaction. It is convenient, however, to write the interaction energy in the general form $V=-A/r^n$ and hence, following the treatment previously used for the kinetics of strain-ageing, the radial velocity of a point defect towards the dislocation is

$$V=(D/\boldsymbol{k}T)(nA/r^{n+1})$$

The number of a particular point defect specie that reach the dislocation in time t is

$$n(t)=\pi r^2 c_0=\pi c_0[ADn(n+2)/\boldsymbol{k}T]^{2/(n+2)}t^{2/(n+2)}$$

and when $n=2$ then $n(t)\propto t^{1/2}$, and when $n=1$, $n(t)\propto t^{2/3}$. Since the kinetics of ageing in quenched copper follow $t^{1/2}$ initially, the observations confirm the importance of the inhomogeneity interaction for vacancies.

The lattice friction term σ_i is clearly responsible for the general level of the stress–strain curve after yielding and arises from the large density of dislocation defects. However, the exact mechanisms whereby loops and tetrahedra give rise to an increased flow stress is still controversial. Vacancy clusters are believed to be formed *in situ* by the disturbance introduced by the primary collision, and hence it is not surprising that neutron irradiation at 4 K hardens the material, and that thermal activation is not essential.

Unlike dispersion-hardened alloys, the deformation of irradiated or quenched metals is characterized by a low initial rate of work hardening (*see Figure 9.4*). This has been shown to be due to the sweeping out of loops and defect clusters by the glide dislocations, leading to the formation of cleared channels. Diffusion-controlled mechanisms are not thought to be important since defect-free channels are produced by deformation at 4 K. The removal of prismatic loops both unfaulted and faulted and tetrahedra can occur as a result of the strong coalescence interactions with screws to form helical configurations (*see Figure 9.24*) and jogged dislocations when the gliding dislocations and defects make contract. Clearly, the sweeping up process occurs only if the helical and jogged configurations can glide easily. Resistance to glide will arise from jogs not lying in slip planes and also from the formation of sessile jogs, e.g. Lomer–Cottrell dislocations in f.c.c. crystals.

The temperature-dependence of the yield stress of neutron irradiated and partially annealed copper single crystals is shown in *Figure 9.22*. The defects formed are Frank loops $\lesssim 5$ nm and the gliding dislocations will form short jogs which should move readily so that the likely rate-controlling stress is the repulsive stress from the loops which has to be overcome before contact is made. After partial annealing, the smallest loops will anneal out first and sessile jogs are more likely formed from coalescence interactions with the larger remaining loops. At low temperatures the yield stress is then likely to be controlled by an

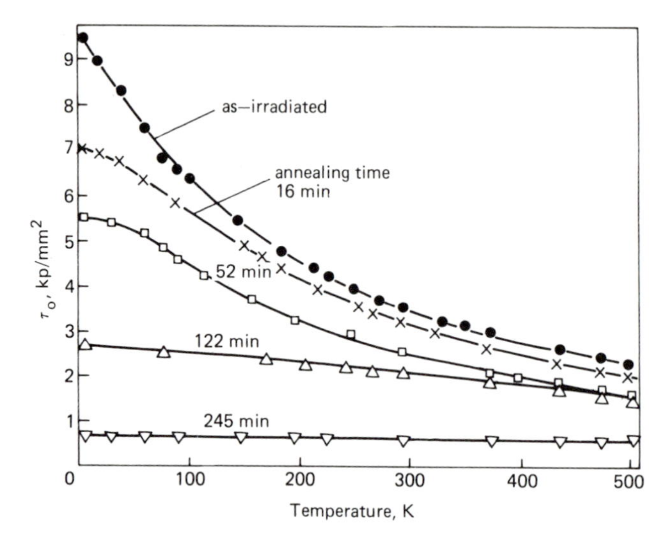

Figure 9.22 Temperature dependence of critical resolved shear stress of neutron-irradiated and subsequently partially annealed crystals of copper (nvt = 10^{19} neutrons cm^{-2}; annealing temperature 325°C; strain rate = 3×10^{-6} s^{-1}) (Rühle, *Phys Stat sal*, 1968, **26**, 661)

Orowan mechanism (with limited channelling), at high temperatures by the long-range elastic stresses, and in the intermediate temperature range by climb and glide processes.

9.8 Radiation growth and swelling

Radiation hardening arises from the distribution of vacancy and interstitial loops obstructing the motion of glissile dislocations. In non-cubic materials, partitioning of the loops on to specific habit planes can lead to an anisotropic dimensional change, known as irradiation growth. The aggregation of vacancies into a disc-shaped cavity, which collapses to form a dislocation loop will give rise to a contraction of the material in the direction of the Burgers vector. Conversely, the precipitation of a plane of interstitials will result in the growth of the material. Such behaviour could account for the growth which takes place in α-uranium single crystals during neutron irradiation, since electron micrographs from thin films of irradiated uranium show the presence of clusters of point defects.

The energy of a fission fragment is extremely high ($\approx$200 MeV) so that a high concentration of both vacancies and interstitials might be expected. A dose of 10^{24} n.m^{-2} at room temperature causes uranium to grow about 30 per cent in the [010] direction and contract in the [100] direction. However, a similar dose at the temperature of liquid nitrogen produces ten times this growth , which suggests the preservation of about 10^4 interstitials in clusters for each fission event that occurs. Growth also occurs in textured polycrystalline α-uranium and to avoid the problem a random texture has to be produced during fabrication. Similar effects can be produced in graphite.

The aggregation of vacancies to form voids has already been mentioned together with the importance of both dislocation density in biasing the removal of interstitials and gas atoms in the void nucleation process. The formation of voids leads to the phenomenon of void swelling and is of practical importance in the dimensional stability of reactor core components. The curves of *Figure 9.23* show the variation in total void volume as a function of temperature for solution-treated 316 stainless steel; the upper cut-off arises when the thermal vacancy emission from the voids exceeds the net flow into them. Comparing the ion- and electron-irradiated curves shows that increasing the recoil energy moves the lower threshold to higher temperatures and is considered to arise from the removal of vacancies by the formation of vacancy loops in cascades; cascades are not created by electron irradiation.

Voids are formed in an intermediate temperature range ≈ 0.3 to $0.6\ T_m$, above that for long-range single vacancy migration and below that for thermal vacancy emission from voids. To create the excess vacancy concentration it is also necessary to build up a critical dislocation density from loop growth to bias the interstitial flow. The sink strength of the dislocations, i.e. the effectiveness of annihilating point defects, is given by $K_i^2 = Z_i\rho$ for interstitials and $K_v^2 = Z_v\rho$ for vacancies where $(Z_i - Z_v)$ is the dislocation bias for interstitials ≈ 10 per cent and ρ is the dislocation density. As voids form they also act as sinks, and are considered neutral to vacancies and interstitials, so that $K_i^2 = K_v{}^2 = 4\pi r_v C_v$, where r_v and C_v are the void radius and concentration, respectively.

The rate theory of void swelling takes all these factors into account and (i) for moderate dislocation densities as the dislocation structure is evolving, swelling is predicted to increase linearly with irradiation dose, (ii) when ρ reaches a quasi-steady state the rate should increase as $(\text{dose})^{3/2}$, and (iii) when the void density is very high, i.e. the sink strength of the voids is greater than the sink strength of the

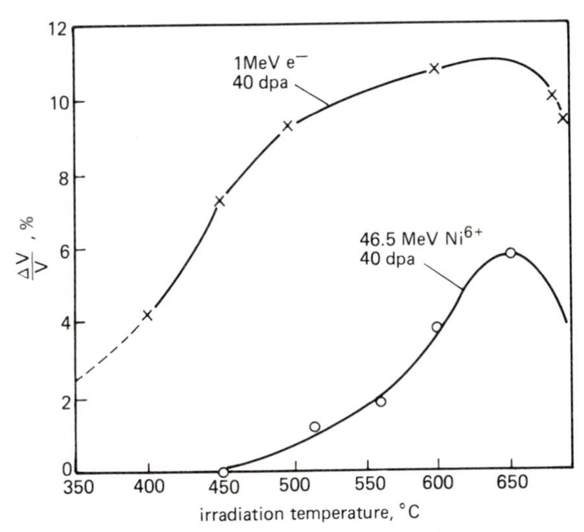

Figure 9.23 Plots of void swelling v. irradiation temperature for 1050°C solution-treated type 316 irradiated with 1 MeV electrons and 46 MeV Ni^{6+} to a dose of 40 dpa (after R. S. Nelson and J. A. Hudson)

dislocations ($K_v^2 \gg K_d^2$), the rate of swelling should again decrease. Results from electron irradiation of stainless steel show that the swelling rate is linear with dose up to 40 dpa (displacements per atom) and there is no tendency to a $(\text{dose})^{3/2}$ law, which is consistent with dislocation structure continuing to evolve over the dose and temperature range examined.

A distinctive feature of the void distribution in many b.c.c. metals is the alignment of the voids to form a 3-D lattice having the same symmetry properties as the host lattice. An important property of the void lattice is that the ratio (a_v/r_v) of the void lattice parameter to the void radius, remains approximately constant; in Mo this ratio is ≈ 8. Observations on low- and high-dose irradiations indicate that voids initially nucleate randomly and then order once C_v has saturated.

In the fuel element itself fission gas swelling can occur since uranium produces one atom of gas (Kr and Ze) for every five U atoms destroyed. This leads to $\approx 2\ m^3$ of gas (s.t.p.) per m^3 of U after a 'burn-up' of only 0.3 per cent of the U atoms.

In practice, it is necessary to keep the swelling small and also to prevent nucleation at grain boundaries when embrittlement can result. In general, variables which can affect void swelling include alloying elements together with specific impurities, and microstructural features such as precipitates, grain size and dislocation density. In ferritic steels, the interstitial solutes carbon and nitrogen are particularly effective in (i) trapping the radiation-induced vacancies and thereby enhancing recombination with interstitials, and (ii) interacting strongly with dislocations and therefore reducing the dislocation bias for preferential annihilation of interstitials, and also inhibiting the climb rate of dislocations. Substitutional alloying elements with a positive misfit such as Cr, V and Mn with an affinity for C or N, can interact with dislocations in combination with interstitials and are considered to have a greater influence than C and N alone.

These mechanisms can operate in f.c.c. alloys with specific solute atoms trapping vacancies and also elastically interacting with dislocations. Indeed the inhibition of climb has been advanced to explain the low swelling of Nimonic PE16 nickel-based alloys. In this case precipitates were considered to restrict dislocation climb. Such a mechanism of dislocation pinning is likely to be less effective than solute atoms since pinning will only occur at intervals along the dislocation line. Precipitates in the matrix which are coherent in nature can also aid swelling resistance by acting as regions of enhanced vacancy–interstitial recombination. TEM observations on θ' precipitates in Al–Cu alloys have confirmed that as these precipitates lose coherency during irradiation, the swelling resistance decreases.

9.9 Vacancy defects in alloys

On alloying aluminium, systematic changes occur in the shape and nature of the loops. A transition from faulted to rhombus-shaped perfect prismatic loops occurs at a composition which depends on the magnitude of the solute–solvent atom misfit. The internal stress which develops aids the conversion of the higher

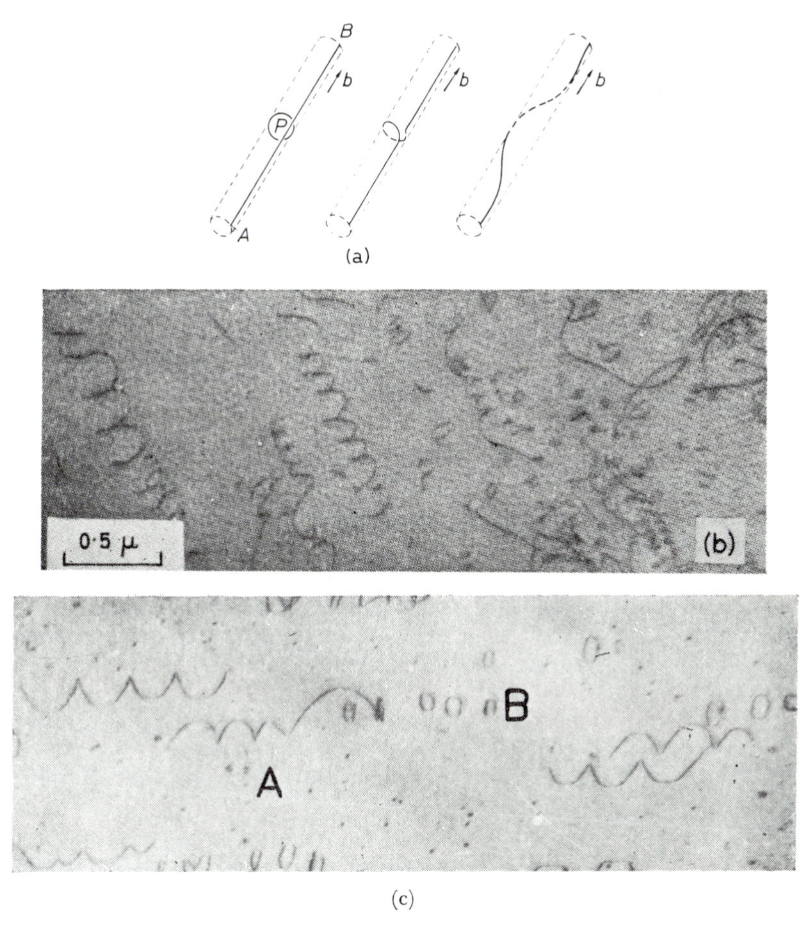

Figure 9.24 (a) Formation of helical dislocation by the condensation of vacancies *P* on the screw dislocation *AB* (after Cottrell, *Vacancies and Other Point Defects in Metals and Alloys,* courtesy of the Institute of Metals), (b) electron micrograph from aluminium–3 per cent copper, quenched from 550°C into water at room temperature, showing helical dislocations, (c) helical dislocations and prismatic loops (*A* and *B*) in aluminium–1.2 per cent silicon quenched from 550°C into water at 50°C (after Westmacott, Hull, Barnes and Smallman, *Phil. Mag.* 1961, **6,** 929)

energy Frank loop to the lower energy rhombus-shaped perfect dislocation. For alloys with higher stacking fault energy the loops may be circular, since the restriction for the dislocations to lie along $\langle 110 \rangle$ directions, to enable them to dissociate on intersecting $\{111\}$, is removed.

Observations on quenched aluminium-based alloys also show that as the solute concentration is increased not all the vacancies which precipitate out form dislocation loops; instead some of the vacancies condense on to screw dislocations, causing them to climb into helices as shown in *Figure 9.24.* It is probable that the helices also form in pure aluminium but are never observed because there are no solute atoms to lock the helical dislocations and prevent them from straightening out. A helical dislocation may dissociate into a screw dislocation to form loops; loops may also form from helices by the interaction of

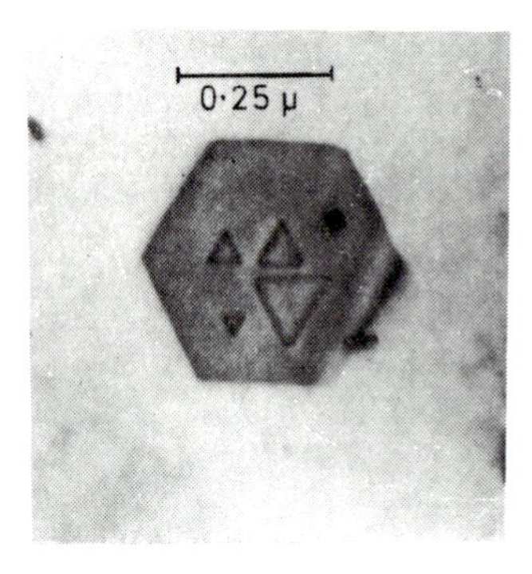

Figure 9.25 Multicomponent dislocation loop in Al–0.65% Mg quenched from 600°C into silicone oil at room temperature (after Kritzinger, Smallman and Dobson, *Acta Met.*, 1969, **17,** 49)

(i) helices of opposite hand or (ii) helices with screw dislocations of opposite Burgers vector.

When the vacancies precipitate on to an edge dislocation, a climb source may operate. The operation of such a Bardeen–Herring source is the climb equivalent of the slip source proposed by Frank and Read. The source dislocation, pinned at its ends, moves in a plane which does not contain the Burgers vector as vacancies condense onto it thereby producing climb, as shown schematically in *Figure 9.28*(*a*). Since a plane of atoms can be removed only once by vacancy precipitation, the source dislocation must be attached to dislocations having a screw component (cf. Crystal growth, Chapter 7). Such a dislocation climb source is found in quenched Al–Mg alloys (*see Figure 9.28*(*b*)) where the source dislocation is nucleated at a particle and the vacancy supersaturation produced by quenching.

In aluminium alloys, prismatic loops, and faulted dislocation loops consisting of either one, two, three or even four layers of intrinsic fault have been produced by quenching. Small additions of alloying element appear to favour the heterogeneous nucleation at impurities and *Figure 9.25* shows multi-component double loops found in Al–Mg alloys. Inner triangular-shaped loops with the same habit lies on the same side of the primary fault whilst those with an orientation differing by 60° lie on opposite sides of the fault. Since the angle between the two faults is acute, alternate segments of a hexagonal loop will be dissociated upwards and downwards respectively (*see Figure 9.13*). In the case of a second loop being nucleated on top of a primary fault, a dissociation downwards to intersect the primary fault is inhibited and thus the second loop has three sides along $\langle 110 \rangle$ favouring upward dissociation. When the second loop is nucleated on the bottom of the primary fault, the three $\langle 110 \rangle$ sides favouring downward dissociation tend to grow.

The faulted dislocation loops observed in aluminium alloys anneal in much the same way as that described for pure aluminium, and measurements show that both the intrinsic and extrinsic fault energy is lowered by alloying. In the annealing of a triple-faulted loop (*see Figure 9.26*) the perfect region expands and the faults are removed by vacancy emission and absorption, until only a loop of what resembles a perfect dislocation is left. Prismatic loops anneal out, at a somewhat higher temperature than faulted loops, in thin foils of all aluminium alloys except for Al–Mg, where loop growth is observed, due to the vacancies

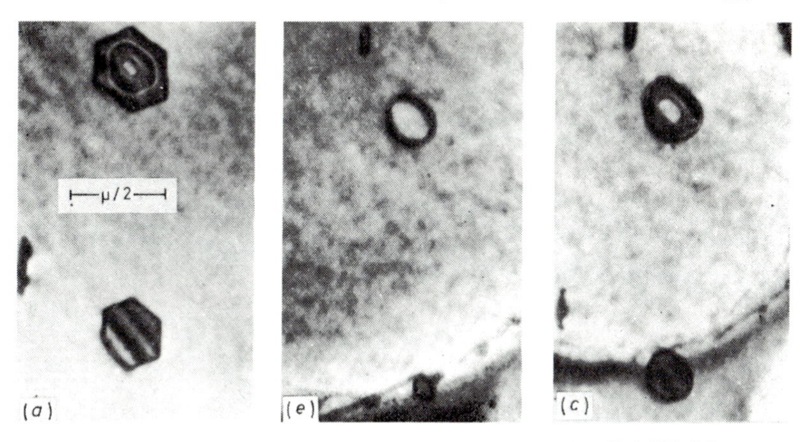

Figure 9.26 Climb of triple-loop and Frank sessile loop in Al–0.65% Mg on annealing at 150°C (after Kritzinger, Smallman and Dobson, *Acta Met.*, 1969, **17,** 49, courtesy of Pergamon Press)

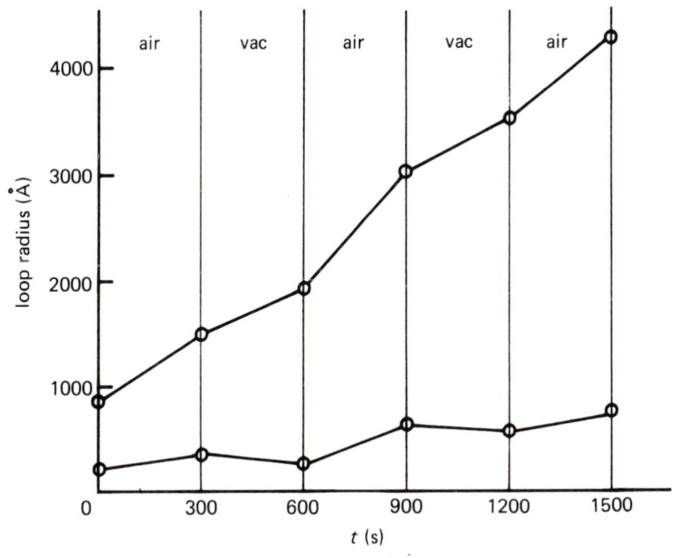

Figure 9.27 Annealing of prismatic loops in Al–Mg alloy at 200°C alternately in air and vacuum

produced by oxidation in the surface regions of the foil as the magnesium atoms become absorbed by a growing surface oxide film. In dilute Al–Mg alloys (≈ 0.5 per cent) the growth of the loops is controlled by the diffusion of Mg atoms to the surface regions, but in concentrated alloys, surface oxidation is rate controlling. *Figure 9.27* shows that the growth rate decreases when the annealing is carried out in vacuum rather than air, and for small loops where the line tension term ($\alpha b/r$) is appreciable the prismatic loops grow in air but shrink in vacuum. The production of vacancies in Al–Mg alloys by surface oxidation has been used to study the growth of successive loops of a dislocation climb source, as shown in *Figure 9.28*.

There are many other experimental observations which can only receive a

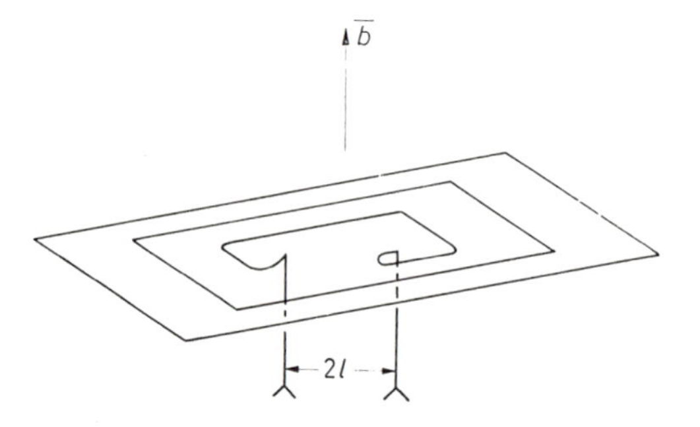

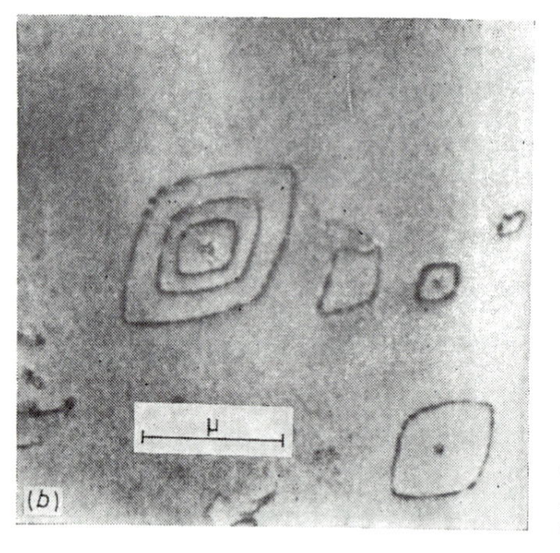

Figure 9.28 Growth of dislocation climb source, (a) schematic, (b) in Al–6.6% Mg alloy (after Kritzinger and Smallman)

satisfactory explanation in terms of the interaction of vacancies with solute atoms. This association may be important in bringing about an increase in the effective vacancy concentration near solute atoms so that the rate of diffusion is increased. The equilibrium concentration of vacancies in a dilute alloy is $c_v = c_{fv} + c_{iv}$, with c_{fv} and c_{iv} the concentrations of free and bound vacancies respectively given by

$$c_{fv} = (1 - zx)A_1 \exp(-E_f/\boldsymbol{k}T);$$

$$c_{iv} = zxA_2 \exp[-(E_f - B_{iv})/\boldsymbol{k}T]$$

where x is the solute concentration, z the co-ordination number, and A_1 and A_2 the appropriate entropy terms. Measurement of the total vacancy concentration at a given temperature by physical property and electron microscopy experiments has enabled an estimate of the solute-vacancy binding energy to be determined. As we have already seen if solute atoms are present in a metal not all the vacancies are free to migrate, and the detailed observations on structure of clustered defects in metals containing a supersaturation of vacancies indicate that specific impurities may act as preferred nucleation sites and indeed control the type of defect formed, e.g. gas atoms nucleating voids, divalent and trivalent impurities nucleating tetrahedra in gold. Perhaps the most spectacular effects of association occur in solution-treated age-hardening alloys. It is often observed that when such alloys are strained at moderate temperatures, e.g. room temperature for aluminium alloys, yielding and ageing occur successively while the specimen is under load, and the stress–strain curve breaks up into serrations, similar to those exhibited by mild steel in the blue-brittle range. This Portevin–Le Chatelier effect, as it is called, is attributed to the enhanced diffusion brought about by the vacancies created during the small amount of plastic deformation. Even in the absence of deformation, enhanced diffusion can occur in freshly quenched alloys, due to the abnormally high vacancy concentration, and both x-ray observations and electrical resistivity measurements show that the initial stages of precipitation take place at an anomalously fast rate. This behaviour, which is associated with the formation of clustered vacancy defects, will be dealt with more fully in Chapter 11.

9.10 Radiation-induced segregation, diffusion and precipitation

Radiation-induced segregation

Radiation-induced segregation is the segregation under irradiation of different chemical species in an alloy towards or away from defect sinks (free surfaces, grain boundaries, dislocations, etc.). The segregation is caused by the coupling of the different types of atom with the defect fluxes towards the sinks. There are four different possible mechanisms, which fall into two pairs, one pair connected with size effects and the other with the Kirkendall effect. With size effects the point defects drag the solute atoms to the sinks because the size of the solute atoms differs from the other types of atom present (solvent atoms). Thus interstitials drag small solute atoms to sinks and vacancies drag large solute atoms to sinks. With Kirkendall effects the faster diffusing species moves in the opposite direction to the vacancy current, but in the same direction as the interstitial

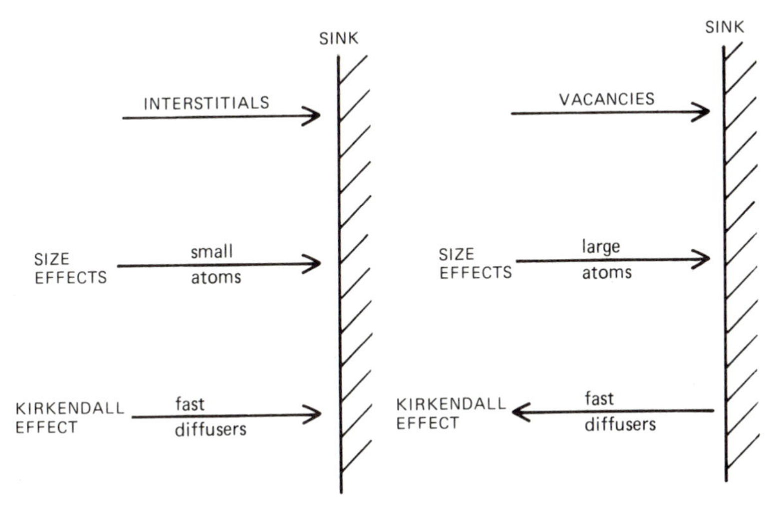

Figure 9.29 Schematic representation of radiation-induced segregation produced by interstitial and vacancy flow to defect sinks

current. The former case is usually called the 'inverse Kirkendall effect', although it is still the Kirkendall effect, but solute atoms rather than the vacancies which are of interest. The most important of these mechanisms which are summarized in *Figure 9.29* appears to be (i) the interstitial size effect mechanism – the dragging of small solute atoms to sinks by interstitials – and (ii) the vacancy Kirkendall effect – the migration away from sinks of fast diffusing atoms.

Radiation-induced segregation is technologically important in fast breeder reactors, where the high radiation levels and high temperatures cause large effects. Thus, for example, in 316 stainless steels, at temperatures in the range 350 °C–650 °C (depending on the position in the reactor) silicon and nickel segregate strongly to sinks. The small silicon atoms are dragged there by interstitials and the slow diffusing nickel stays there in increasing concentration as the other elements diffuse away by the vacancy inverse Kirkendall effect. Such diffusion (a) denudes the matrix of void-inhibiting silicon, and (b) can cause precipitation of brittle phases at grain boundaries, etc.

Radiation-enhanced diffusion

Diffusion rates may be raised by several orders of magnitude because of the increased concentration of point defects under irradiation. Thus phases expected from phase diagrams may appear at temperatures where kinetics are far too slow under normal circumstances. Many precipitates of this type have been seen in stainless steels which have been in reactors. Two totally new phases have definitely been produced and identified in alloy systems, e.g. Pd_8W and Pd_8V, and others appear likely, e.g. Cu–Ni miscibility gap.

Radiation-induced precipitation

This effect relates to the appearance of precipitates after irradiation and possibly arises from the two effects described above, i.e. segregation or enhanced diffusion. It is possible to distinguish between these two causes by post-irradiation annealing, when the segregation-induced precipitates disappear but the diffusion-induced precipitates remain, being equilibrium phases.

9.11 Radiation and ordered alloys

Ordering alloys have a particularly interesting response to the influence of point defects in excess of the equilibrium concentration. Irradiation introduces point defects and its effect on the behaviour of ordered alloys depends on two competitive processes, i.e. radiation-induced ordering and radiation-induced disordering, which can occur simultaneously. The interstitials do not contribute significantly to ordering but the radiation-induced vacancies give rise to ordering by migrating through the crystal. Disordering is assumed to take place athermally by displacements. *Figure 9.30* shows the influence of electron irradiation time and temperature on (a) initially ordered and (b) initially disordered Cu_3Au. The final state of the alloy at any irradiation temperature is

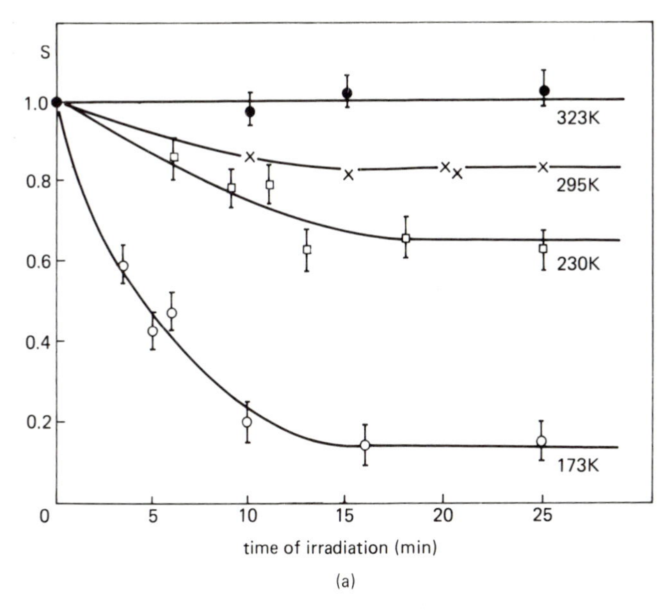

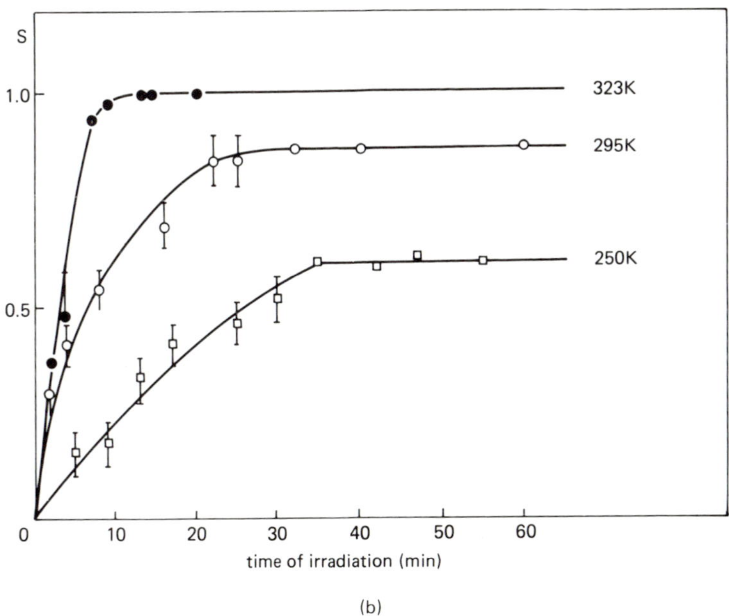

Figure 9.30 Variation in the degree of long-range order *S* for initially (a) ordered and (b) disordered Cu_3Au for various irradiation temperatures as a function of irradiation time. Accelerating voltage 600 kV (after Hameed, Loretto and Smallman, *Phil. Mag*, 1982, **46**, 707. Courtesy Taylor and Francis)

independent of the initial condition. At 323 K, Cu_3Au is fully ordered on irradiation whether it is initially ordered or not, but at low temperatures it becomes largely disordered because of the inability of the vacancies to migrate and develop order; the interstitials ($E_m^I \approx 0.1$ eV) can migrate at low temperatures.

Suggestions for further reading

Harris, J.E. and Sykes, E.C., *Physical Metallurgy of Reactor Fuel Elements*, The Metals Society, 1973

Proc. Int. Fundamental Aspects of Radiation Damage in Metals, USERDA, Gatlingburg, 1975

Smallman, R.E. and Harris, J.E. (eds), *Vacancies '76*, The Metals Society, London, 1977

Takamura, Tin-Inchi, Doyama, M. and Kiritani, M. (eds), *Point Defects and Defect Interactions in Metals*, University of Tokyo Press, 1982

Chapter 10

Work hardening and annealing

10.1 Work hardening

10.1.1 Introduction

It is well known that the properties of a metal are altered by cold working, i.e. deformation of the metal at a low temperature relative to its melting point: such a working operation may be carried out at a slightly elevated temperature for some metals, e.g. up to 500 °C in the case of molybdenum or tungsten, and at sub-zero temperatures for others, e.g. lead. However, not all the properties are improved, for although the tensile strength, yield strength and hardness are increased the plasticity and general ability of a metal to deform decreases. Moreover, the physical properties such as electrical conductivity, density and others are all lowered. Of these many changes in properties, perhaps the most outstanding are those that occur in the mechanical properties, as is illustrated by the fact that the yield stress of mild steel, for example, may be raised by cold work from 172 up to 1050 MN/m^2.

Such changes in mechanical properties are, of course, of interest theoretically, but they are also of great importance in industrial practice. This is because the rate at which a metal hardens during deformation influences both the power required and the method of working in the various shaping operations, while the magnitude of the hardness introduced governs the frequency with which metals must be annealed (always an expensive operation) to enable further working to be continued.

Since plastic flow occurs by a dislocation mechanism the fact that work hardening occurs in metals means that it becomes difficult either to generate dislocations or to move them. The ease with which the multiplication of the dislocations can occur from sources suggests that the hardening is not due to this cause but to the increased resistance these dislocations experience in moving through the lattice. All theories of work hardening depend on this assumption, and the basic idea of hardening, put forward by Taylor in 1934, is that some dislocations become 'stuck' inside the crystal and act as sources of internal stress which oppose the motion of other dislocations moving through.

One simple way in which two dislocations could become stuck is by elastic interaction. Thus, two parallel edge dislocations of opposite sign moving on

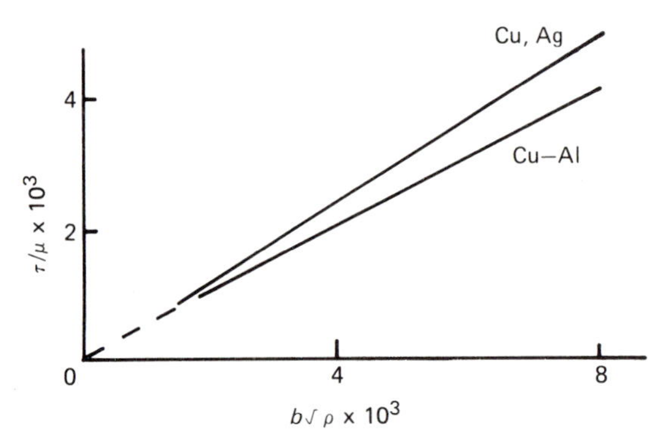

Figure 10.1 Dependence of flow stress on (dislocation density)$^{1/2}$ for Cu, Ag and Cu–Al

parallel slip planes in any sub-grain may become stuck, as a result of the interaction discussed in Chapter 6. In the first theory of work hardening due to G.I. Taylor it was assumed that dislocations are created during straining and that some of these become stuck after travelling an average distance, L, while the density of dislocations reaches ρ, i.e. work hardening is due to the dislocations getting in each other's way. The flow stress is then the stress necessary to move a dislocation in the stress field of those dislocations surrounding it. This stress τ is quite generally given by

$$\tau = \alpha\mu b/l \tag{10.1}$$

where μ is the shear modulus, b the Burgers vector, l the mean distance between dislocations which is $\approx \rho^{-1/2}$, and α a constant; in the Taylor model $\alpha = 1/8\pi(1-\nu)$. *Figure 10.1* shows such a relationship for Cu–Al single crystals and polycrystalline Ag and Cu.

Taylor in his theory only considered a two-dimensional model of a cold worked metal. However, because plastic deformation arises from the movement of dislocation loops from a source, it is more appropriate to assume that when the plastic strain is γ, N dislocation loops of side L (if we assume for convenience that square loops are emitted) have been given off per unit volume. The resultant plastic strain is then given by

$$\gamma = NL^2b \tag{10.2}$$

and l by

$$l \simeq [1/\rho^{1/2}] = [1/4LN]^{1/2} \tag{10.3}$$

Combining these equations, the stress–strain relation

$$\tau = \text{const.}\ (b/L)^{1/2}\gamma^{1/2} \tag{10.4}$$

is obtained. Taylor assumed L to be a constant, i.e. the slip lines are of constant length, which results in a parabolic relationship between τ and γ.

Taylor's assumption, that during cold work the density of dislocations increases, has been amply verified, and indeed the parabolic relationship between stress and strain is obeyed, to a first approximation, in many polycrystalline

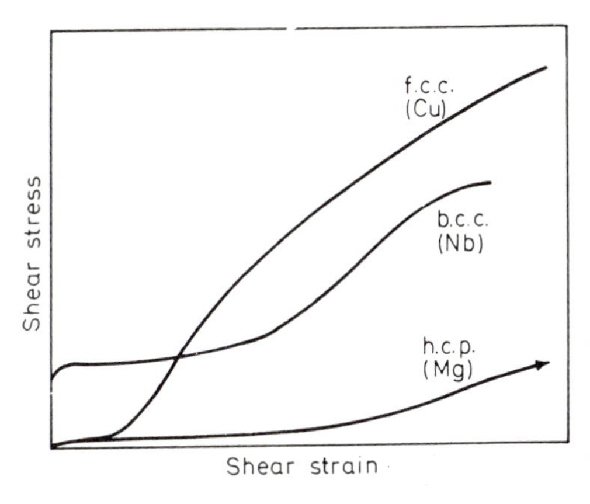

Figure 10.2 Stress–strain curves of single crystals of typical f.c.c., b.c.c. and h.c.p. metals (after Hirsch and Mitchell, *Can. J. Phys.*, 1967, **45,** 663, courtesy of the National Research Council of Canada)

aggregates where deformation in all grains takes place by multiple slip. Experimental work on single crystals shows, however, that the strain-hardening curve may deviate considerably from parabolic behaviour, and depends not only on crystal structure but also on other variables such as crystal orientation, purity and surface conditions (*see Figures 10.2* and *10.3*).

The crystal structure of the metal is important (*see Figure 10.2*) in that single crystals of some hexagonal metals slip only on one family of slip planes, those parallel to the basal plane, and these metals show a low rate of work hardening. The plastic part of the stress–strain curve is also in this case more nearly linear than parabolic with a slope which is extremely small: this slope ($d\tau/d\gamma$) becomes even smaller with increasing temperature of deformation. Cubic crystals, on the other hand, are capable of deforming in a complex manner on more than one slip system, and these metals normally show a strong work hardening behaviour. The influence of temperature depends on the stress level reached during deformation and on other factors which must be considered in greater detail. However, even in cubic crystals the rate of work hardening may be extremely small if the crystal is restricted to slip on a single slip system. Such behaviour, as mentioned in Chapter 6, points to the conclusion that strong work hardening is caused by the mutual interference of dislocations gliding on intersecting slip planes.

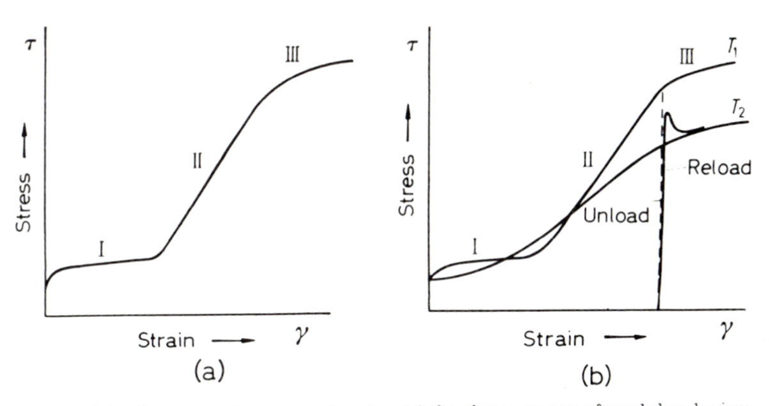

Figure 10.3 Stress–strain curves showing (a) the three stages of work hardening, and (b) work softening

Many theories of work hardening similar to that of Taylor exist but all are over-simplified, since work hardening depends not so much on individual dislocations as on the group behaviour of large numbers of them. It is clear, therefore, that a theoretical treatment which would describe the complete stress–strain relationship is impossibly difficult, and consequently the present-day approach is to examine the various stages of hardening individually, and then attempt to explain not the complete hardening curve, but instead the mechanisms which are likely to give rise to the different stages of hardening. The work hardening behaviour in metals with a cubic structure is more complex than in those with hexagonal structure because of the variety of slip systems available, and it is for this reason that much of the experimental evidence is related to these metals, particularly those with f.c.c. structures. Thus, we propose to limit our discussion to these observations in order to illustrate the various modes of hardening which can operate during cold deformation. More than one mechanism for stopping the movement of dislocations may contribute to the hardening in a given metal, and it will become evident that whether the flow stress is indeed all or only partly due to the elastic interaction between dislocations, as suggested by Taylor over 30 years ago, is still a matter of considerable discussion.

10.1.2 Three-stage hardening

The stress–strain curve of a f.c.c. single crystal is shown in *Figure 10.3* and three regions of hardening are experimentally distinguishable. The points to emphasize about these hardening stages may be summarized as follows.

(a) Stage I, or the easy glide region, immediately follows the yield point and is characterized by a low rate of work hardening θ_1 up to several per cent glide; the length of this region depends on orientation, purity and size of the crystals. The hardening rate $(\theta_1/\mu) \sim 10^{-4}$ and is of the same order as for hexagonal metals.

(b) Stage II, or the linear hardening region, shows a rapid increase in work hardening rate. The ratio $(\theta_{11}/\mu) = (\mathrm{d}\tau/\mathrm{d}\gamma)/\mu$ is of the same order of magnitude for all f.c.c. metals, i.e. 1/300 although this is $\approx 1/150$ for orientations at the corners of the stereographic triangle. In this stage short slip lines are formed during straining quite suddenly, i.e. in a short increment of stress $\Delta\tau$, and thereafter do not grow either in length or intensity. The mean length of the slip lines, $L \approx 25\ \mu$m decreases with increasing strain.

(c) Stage III, or the parabolic hardening region, the onset of which is markedly dependent on temperature, exhibits a low rate of work hardening, θ_{111}, and the appearance of coarse slip bands. This stage sets in at a strain which increases with decreasing temperature and is probably associated with the annihilation of dislocations as a consequence of cross slip.

The low stacking fault energy metals exhibit all three work hardening stages at room temperature, but metals with a high stacking fault energy often show only two stages of hardening. It is found, for example, that at 78 K aluminium behaves like copper at room temperature and exhibits all three stages, but at room temperature and above, stage II is not clearly developed and stage III starts before stage II becomes at all predominant. This difference between aluminium and the noble metals is not just due to the difference in melting point, but to the difference in stacking fault energies which affects the width of extended dislocations. The main effect of a change of temperature of deformation is, however, a change in the onset of stage III; the lower the temperature of deformation, the higher is the stress τ_{111} corresponding to the onset of stage III.

Because the flow stress of a metal may be affected by a change of temperature or strain-rate, it has been found convenient to think of the stress as made up of two parts according to the relation

$$\tau = \tau_s + \tau_g \tag{10.5}$$

where τ_s is that part of the flow stress which is dependent on temperature apart from the variation of the elastic modulus μ with temperature, and τ_g is a temperature-independent contribution. The relative importance of τ_s and τ_g can be studied conveniently by measuring the dependence of flow stress on temperature or strain rate, i.e. the change in flow stress $\Delta\tau$ on changing the temperature or strain rate, as function of deformation.

In stage I it is often observed that the ratio of τ_s/τ_g is about unity, but with increasing strain into stage II the ratio decreases and reaches a constant limit of about one-tenth. This so-called constancy of the ratio τ_s/τ_g is often referred to as the Cottrell–Stokes law. Such a law is not obeyed in b.c.c. crystals where the temperature-dependent term due to lattice friction is large. The existence of the law implies that obstacles controlling τ_s and τ_g are either proportional to each other (i.e. the scale of the dislocation arrangement changes with deformation but not its nature), or are the same, i.e. responsible for both short-range and long-range interactions.

10.1.3 Stage I

The easy glide region in cubic crystals, with its small linear hardening, corresponds closely to the hardening of c.p.h. crystals where only one glide plane operates. It occurs in crystals oriented to allow only one glide system to operate, i.e. for orientations near the [110] pole of the unit triangle (*Figure 6.8*). In this case the slip distance is large, of the order of the specimen diameter, with the probability of dislocations slipping out of the crystal. Electron microscope observations have shown that the slip lines on the surface are very long ($\approx$1 mm) and closely spaced, and that the slip steps are small corresponding to the passage of only a few dislocations. This behaviour obviously depends on such variables as sample size and oxide films, since these influence the probability of dislocations passing out of the crystal. It is also to be expected that the flow stress in easy glide will be governed by the ease with which sources begin to operate, since there is no slip on a secondary slip system to interfere with the movement of primary glide dislocations.

As soon as another glide system becomes activated there is a strong interaction between dislocations on the primary and secondary slip systems, which gives rise to a steep increase in work hardening*. It is reasonable to expect that easy glide should end, and turbulent flow begin, when the crystal reaches an orientation for which two or more slip systems are equally stressed, i.e. for orientations on the symmetry line between [100] and [111]. However, easy glide generally ends before symmetrical orientations are reached and this is principally due to the formation of deformation bands to accommodate the rotation of the glide plane in fixed grips during tensile tests. This rotation leads to a high resolved stress on the secondary slip system, and its operation gives rise to those lattice irregularities which cause some dislocations to become 'stopped' in the crystal. The transformation to Stage II then occurs.

10.1.4 Stage II

The characteristic feature of deformation in Stage II is that slip takes place on both the primary and secondary slip systems. As a result, several new lattice irregularities may be formed which will include (1) forest dislocations, (2) Lomer–Cottrell barriers, and (3) jogs produced either by moving dislocations cutting through forest dislocations, or by forest dislocations cutting through source dislocations. Consequently, the flow stress τ may be identified, in general terms, with a stress which is sufficient to operate a source and then move the dislocations against (i) the internal elastic stresses from the forest dislocations, (ii) the long-range stresses from groups of dislocations piled-up behind barriers, and (iii) the frictional resistance due to jogs. In a cold worked metal all these factors may exist to some extent, but because a linear hardening law can be derived by using any one of the various contributory factors, there have been several theories of Stage II-hardening, namely (1) the pile-up theory, (2) the forest theory and (3) the jog theory. All have been shown to have limitations in explaining various features of the deformation process, and have given way to a more phenomenological theory based on direct observations of the dislocation distribution during straining. The main features of these theories are discussed below.

The pile-up model With this model it is assumed that during the linear hardening region of deformation some of the dislocations given out by Frank–Read sources are eventually stopped at barriers, e.g. Lomer–Cottrell barriers, in the lattice and the other dislocations pile-up behind. As deformation proceeds, the number of barriers increases until each source becomes completely surrounded by barriers. This situation indicates that slip must have occurred, not only on the primary plane, but also on two different {111} planes in order to produce the barriers.

The hardening is due principally to long-range internal stresses from piled-up groups interacting with glide dislocations. The temperature-independent part of the flow stress τ_g is attributed to the stress required to overcome the long-range elastic stresses from the piled-up groups, and the temperature-dependent part of the flow stress τ_s to the process of creating jogs as the primary dislocations cut through forest dislocations threading the glide plane.

In the treatment due to Seeger, it is assumed that a piled-up group of n

* For the slip processes leading to the two different forms of hardening the rheological terms 'laminar' and 'turbulent' plastic flow have been used.

dislocations acts like a superdislocation of strength nb. The flow stress is then the stress required to push a dislocation through the softest region between neighbouring piled-up groups of n dislocations, given by

$$\tau_g = \alpha\mu(nb)\rho_p^{1/2} = \alpha\mu n^{1/2} b\rho^{1/2} \tag{10.6}$$

where ρ_p is the density of piled-up groups, ρ is the primary dislocation density and $\alpha \approx 1/2\pi$. The density of pile-ups is related to strain by

$$\frac{d\rho_p}{d\gamma} = \frac{1}{Lnb} \tag{10.7}$$

where L is the slip line length. The effective stress acting on a pile is less than the applied stress because of the screening due to other pile-ups, and is given by

$$\beta\tau = \frac{\pi\mu nb}{4L} \tag{10.8}$$

where β is somewhat less than unity. Putting $\tau = \tau_g$ the work hardening rate is found from these equations

$$\theta_{11}/\mu = \alpha^2\beta/2\pi \tag{10.9}$$

The pile-up theory is successful in explaining the observed work-hardening rate and the variation of slip line length with strain. The problem with the model arises from the reduction in the back stress from particular dislocations in the presence of many others whose stress fields have to be taken into account. There is no doubt that secondary systems play an important part in controlling the flow stress, partly by the forest intersection mechanism and partly by modifying the long-range internal stress pattern.

The forest theory The forest model arose from the early electron-microscope observation of deformed metals. In the electron microscope, piled-up groups of dislocations are not observed in the common metals of medium and low stacking fault energy, e.g. copper, silver and gold, but instead it is observed that complex networks are formed. Piled-up arrangements of dislocations are observed in metals of very low stacking fault energies, e.g. austenitic steels, α-brass, aluminium–bronze, but then only at grain boundaries and not inside the grains. This is supported by the fact that there is no evidence for Lomer–Cottrell barriers long enough to hold up piled-up groups. In all these metals, the dislocations are more characteristically arranged in thick tangled regions separating areas of the crystal relatively free from dislocations.

The dislocation tangles are thought to arise from a process of stress relaxation, as illustrated in *Figure 10.4*, in which the long-range stresses due to piled-up groups cause slip on secondary systems leading to networks; these networks are stable on removal of the stress and cause the hardening. If relaxation of the stresses from piled-up groups of dislocations occurs by network formation, and this appears to be the case for most metals, the density of forest dislocations will always be about the same as that of the primary dislocations. As a consequence—in the forest model—it is the forest dislocations not the primary dislocations which are assumed to be the most important, and the τ_g contribution

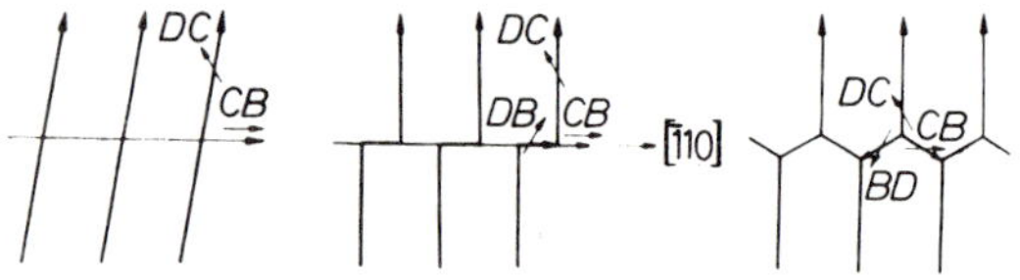

Figure 10.4 Network formation by the interaction of screw dislocation (CB) with dislocations (DC) piled up on a slip plane parallel to the plane of the paper (after Whelan, *J. Inst. Metals*, 1958–9, **87**, 409)

to the flow stress is ascribed to the stress required for the primary dislocations to overcome the elastic stress field of the forest dislocations, while the τ_s part is attributed to the work done in cutting through the tangle with consequent formation of jogs.

On such a theory the hardening in Stage II is entirely due to forest dislocations and accounts for the observation that the increase in the temperature-dependent contributions of the flow stress is proportional to the increase in the temperature-independent contribution, both effects being due to the same obstacle, the forest dislocation. The flow stress may thus be written

$$\tau = \alpha\mu b(\rho_f)^{1/2} \tag{10.10}$$

where ρ_f is the local forest density. The geometry of the dislocation arrangement is assumed to remain similar during deformation so that the forest (ρ_f) and primary (ρ_p) densities are proportional to one another, and the slip line length L is proportional to the mean forest separation, i.e. $\rho_f = k_1\rho_p$ and $L = k_2/(\rho_f)^{1/2}$, where k_1 and k_2 are constants. Using the above relationships and remembering that the change in primary dislocation density with strain γ

$$\mathrm{d}\rho_p/\mathrm{d}\gamma = 1/Lb$$

it follows that

$$\theta_{11} \simeq (k_1/k_2)\mu$$

In the process of stress relaxation, networks are formed in which the spacing of the forest dislocations is equal to that of the primary dislocations and hence $k_1 \approx 1$. Hence to obtain the correct value for θ_{11}, k_2 must be ≈ 40, which is difficult to explain.

The mesh-length theory of Kuhlmann-Wilsdorf is formally rather similar to the forest theory, the main difference being that the flow stress is controlled by the stress to operate a Frank–Read source rather than to propagate the dislocations through the forest. With increase in ρ the source length decreases and the flow stress increases. Neither the forest nor mesh-length theories are satisfactory in that no mechanism for blocking the slip line is apparent. This is a general criticism of 'homogeneous' dislocation models in which there is no clear distinction between 'hard' and 'soft' regions of the crystal.

The jog theory It was first suggested many years ago that a jog would be of importance in work hardening and that its movement would be responsible for the formation of point defects during cold work (*see* Chapter 9). Its importance was later questioned, however, when it was realized that jogs in screw dislocations are nothing but small pieces of edge dislocation which could possibly move along the dislocation conservatively. This process would then neither lead to the formation of point defects nor contribute to the resistance to dislocation movement. Such a conclusion is based on the behaviour expected from constricted jogs in relatively unextended dislocations. Hirsch has examined the structure of dissociated multiple jogs and extrapolated their properties to jogs of atomic dimension. His analysis shows that jogs may be dissociated in such a way that they become sessile (*see Figure 10.6*) and in order to move they must either first be constricted and then allowed to glide conservatively along the screw dislocation, or be dragged non-conservatively behind the gliding screw dislocation. Jogs moving in the direction that would produce interstitials can be constricted by stress, enabling them to glide conservatively. In contrast, vacancy-forming jogs cannot be constricted by stress and hence act as a frictional drag to the moving screw, except at high temperatures when the jog climbs readily.

The flow stress is determined by the number of sessile jogs on a source dislocation, the jogs being produced by secondary slip. The flow stress can be written as:

$$\tau = \alpha \mu b m \tag{10.11}$$

where m is the density of sessile jogs and $\alpha \approx 1/5$ such that $\alpha \mu b^3$ is the energy of a vacancy produced when the jog advances by $1b$. The jog density is increased by secondary slip which, being assumed proportional to the primary slip, is given by

$$\mathrm{d}m = gf\,\mathrm{d}\gamma$$

with g the ratio of the secondary to primary slip and f the fraction of the sessile jogs produced which are vacancy jogs. The hardening rate is found to be given by

$$\theta_{11} = \alpha f g \mu$$

and to obtain the correct order of magnitude $g \simeq 0.3$.

There are several unsatisfactory features of the jog theory. First, the vacancy type of jog causing the hardening is only produced by the intersection of particular slip systems, and thus the orientation dependence of θ_{11} expected is much larger than observed. Secondly, because few of the forest dislocations would produce vacancy jogs, the estimated hardening rate is less than 10 per cent of the experimentally observed rate. Finally, the jog density necessary to cause a moving dislocation to become stuck is acquired in a distance of the order of the local separation of the forest dislocations which is small compared to the slip line length.

Work hardening due to obstacles of high dislocation density None of the theories so far proposed is able to explain satisfactorily all the experimental observations. In view of these difficulties development in work hardening theory has proceeded by applying the successful parts of the individual theories to account for the detailed electron observations of dislocation structure in deformed crystals. Observations on f.c.c. and b.c.c. crystals have revealed several common features of the microstructure which include the formation of dipoles, tangles and cell structures with increasing strain. The most detailed observations have been made for copper crystals, and these are summarized below to illustrate the general pattern of behaviour. In Stage I, bands of dipoles are formed (*see Figure 10.5(a)*) elongated normal to the primary Burgers vector direction. Their formation is associated with isolated forest dislocations and individual dipoles are about 1 μm in length and $\gtrsim$10 nm wide. Different patches are arranged at spacings of about 10 μm along the line of intersection of a secondary slip plane. With increasing strain in Stage I the size of the gaps between the dipole clusters decreases and therefore the stress required to push dislocations through these gaps increases. Stage II begins (*see Figure 10.5(b)*) when the applied stress plus internal stress resolved on the secondary systems is sufficient to activate secondary sources near the dipole clusters. The resulting local secondary slip leads to local interactions between primary and secondary dislocations both in the gaps and in the clusters of dipoles, the gaps being filled with secondary dislocations and short lengths of other dislocations formed by interactions, e.g. Lomer–Cottrell dislocations in f.c.c. crystals and $a\langle 100\rangle$ type dislocations in b.c.c. crystals. Dislocation barriers are thus formed surrounding the original sources.

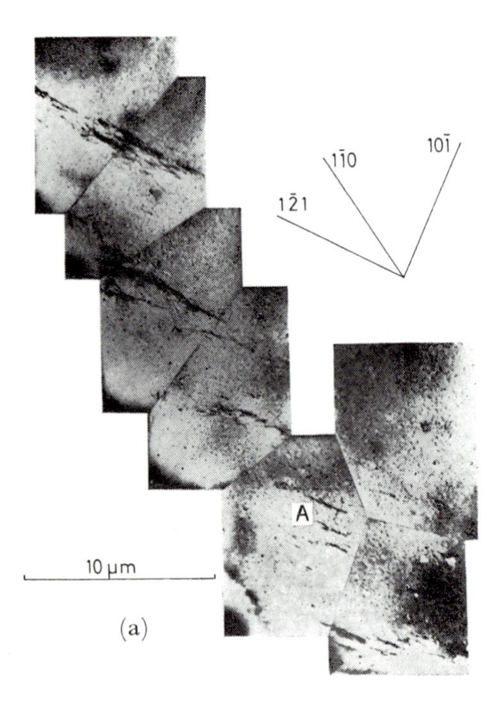

Figure 10.5 Dislocation structure observed in copper single crystals deformed in tension to (a) stage I, (b) end of easy-glide and beginning of stage II, (c) top of stage II, and (d) stage III (after Steeds, *Conference on Relation between Structure and Strength in Metals and Alloys*. Crown copyright; reproduced by permission of the Controller, H.M. Stationery Office)

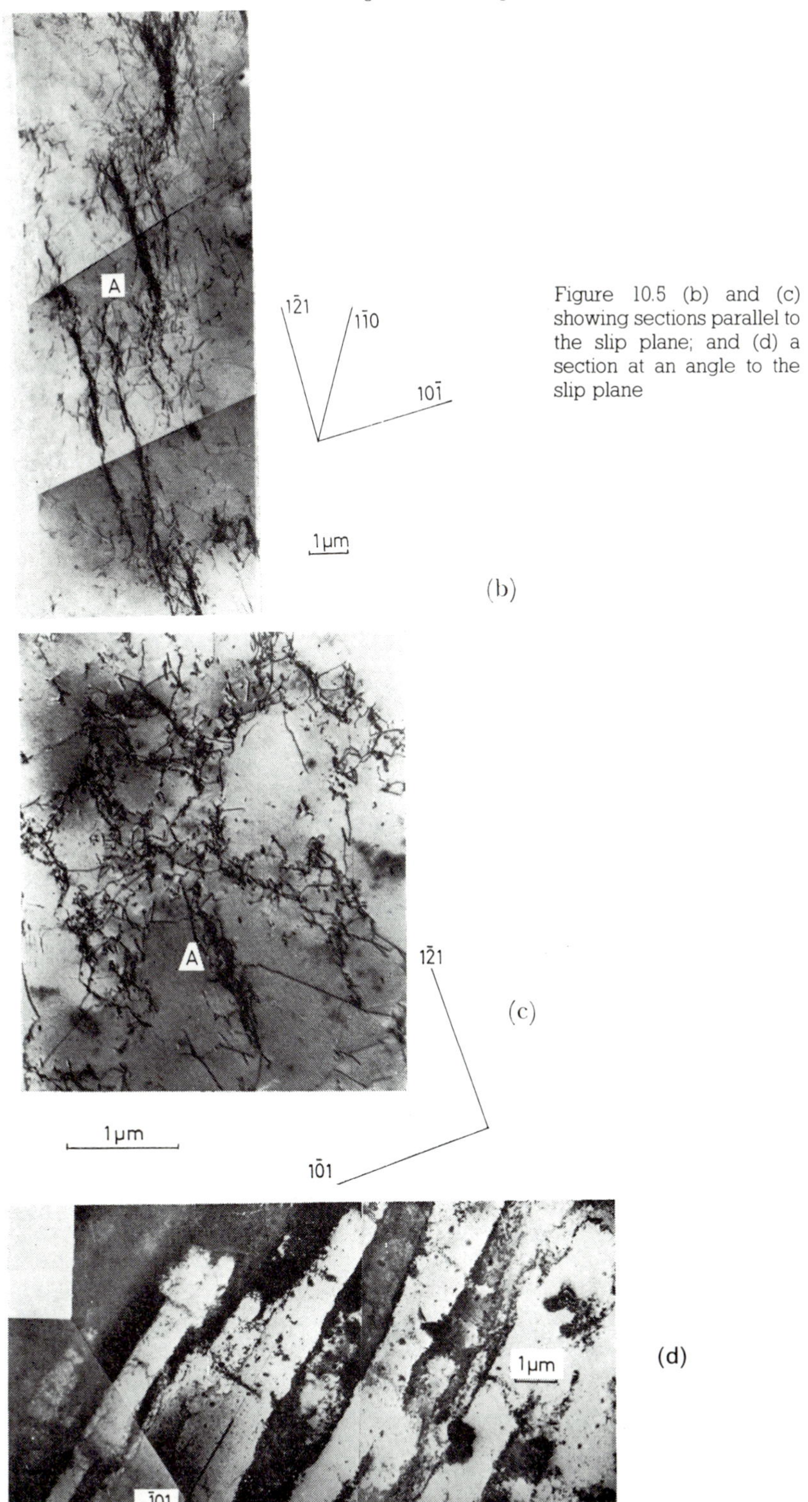

Figure 10.5 (b) and (c) showing sections parallel to the slip plane; and (d) a section at an angle to the slip plane

In Stage II (*see Figure 10.5(c)*) it is proposed that dislocations are stopped by elastic interaction when they pass too close to an existing tangled region with high dislocation density. The long-range internal stresses due to the dislocations piling up behind are partially relieved by secondary slip, which transforms the discrete pile-up into a region of high dislocation density containing secondary dislocation networks and dipoles. These regions of high dislocation density act as new obstacles to dislocation glide, and since every new obstacle is formed near one produced at a lower strain, two-dimensional dislocation structures are built up forming the walls of an irregular cell structure. With increasing strain the number of obstacles increases, the distance a dislocation glides decreases and therefore the slip line length decreases in Stage II. The structure remains similar throughout Stage II but is reduced in scale. The obstacles are in the form of ribbons of high densities of dislocations which, like pile-ups, tend to form sheets. The work hardening rate depends mainly on the effective radius of the obstacles, and this has been considered in detail by Hirsch and co-workers and shown to be a constant fraction k of the discrete pile-up length on the primary slip system. In general, the work-hardening rate is given by $\theta_{11} = k\mu/3\pi$ and for an f.c.c. crystal the small variation in k with orientation and alloying element is able to account for the variation of θ_{11} with those parameters.

The dislocation arrangement in metals with other structures is somewhat similar to that of copper with differences arising from stacking fault energy. In Cu–Al alloys the dislocations tend to be confined more to the active slip planes, the confinement increasing with decreasing γ_{SF}. In Stage I dislocation multipoles are formed as a result of dislocations of opposite sign on parallel nearby slip planes 'pairing up' with one another. Most of these dislocations are primaries. In Stage II the density of secondary dislocations is much less ($\approx 1/3$) than that of the primary dislocations. The secondary slip occurs in bands and in each band slip on one particular secondary plane predominates. In niobium, a metal with high γ_{SF}, the dislocation distribution is rather similar to copper. In Mg, typical of c.p.h. metals, Stage I is extensive and the dislocations are mainly in the form of primary edge multipoles, but forest dislocations threading the primary slip plane do not appear to be generated.

10.1.5 Stage III and the phenomenon of work softening

From the curve shown in *Figure 10.3(a)* it is evident that the rate of work hardening decreases in the later stages of the test. This observation indicates that at a sufficiently high stress or temperature the dislocations held up in Stage II are able to move by a process which at lower stresses and temperature had been suppressed. The onset of Stage III is accompanied by cross-slip, and the slip lines are broad, deep and consist of segments joined by cross-slip traces. Electron metallographic observations on sections of deformed crystal inclined to the slip plane (*see Figure 10.5(d)*) show the formation of a cell structure in the form of boundaries, approximately parallel to the primary slip plane of spacing about 1–3 μm plus other boundaries extending normal to the slip plane as a result of cross slip.

The simplest process which is in agreement with the experimental observations is that the screw dislocations held up in Stage II, cross-slip and possibly return to the primary slip plane by double cross-slip. By this mechanism, dislocations can by-pass the obstacles in their glide plane and do not have to interact strongly with them. Such behaviour leads to an increase in slip distance and a decrease in the accompanying rate of work hardening. Furthermore, it is to be expected that screw dislocations leaving the glide plane by cross-slip may also meet dislocations on parallel planes and be attracted by those of opposite sign. Annihilation then takes place and the annihilated dislocation will be replaced, partly at least, from the original source. This process if repeated can lead to slip-band formation, which is also an important experimental feature of Stage III. Hardening in Stage III is then due to the edge parts of the loops which remain in the crystal and increase in density as the source continues to operate.

The process of cross-slip is governed by the stress on the cross-slip plane and is aided by thermal activation. One mechanism of cross-slip has already been described in Chapter 6. Cross-slip may also be initiated at jogs, particularly multiple jogs. With reference to the Thompson tetrahedron, a screw dislocation BC on the δ-plane may contain a jog which can lie along one of the three directions DB, DA and DC. If it lies along DB or DC and is constricted then it can bow out on the cross-slip plane directly. The jog may, however, be extended in which case it has a rather complex structure, as indicated in *Figure 10.6*. The dissociation reactions involved are (i) dissociation of $\mathrm{BC} \rightarrow \mathrm{B}\alpha + \alpha\mathrm{C}$ on the cross-slip plane, (ii) interaction between αC and δC to produce a small length of stair-rod $\alpha\delta$ at the junction of the cross-slip and primary planes (*Figure 10.6*(*a*)), (iii) dissociation of the other Shockley partial $\mathrm{B}\alpha \rightarrow \mathrm{B}\gamma + \gamma\alpha$ and subsequent interaction of Bγ with Bδ to give stair-rod $\delta\gamma$. The resultant jog configuration is shown in *Figure 10.6*(*b*).

With sufficient stress on the cross-slip plane, the glissile Shockley partial dislocation αC bows out slightly on the cross-slip plane. Then with the aid of stress and thermal activation, the glissile partial Bδ reacts with the sessile stair-rod $\alpha\delta$ to form glissile partial Bα, constricting the jog and producing a glissile node X while the node Y remains sessile. As the node X glides, the jog lengthens until both partials αC and Bα bow out on the cross slip plane to initiate macroscopic cross-slip (*see Figure 10.6*(*d*)).

A comparison of the mechanism of cross-slip shown in *Figure 10.6* with that outlined previously in *Figure 6.25*, indicates that the jog-initiated process is considerably easier since only one constriction has to be made instead of two and part of the dislocation already lies on the cross-slip plane. Furthermore, the model based on jogs appears more satisfactory because the constriction in the extended dislocation is made by the movement of a glissile partial dislocation against a sessile stair-rod dislocation whereas in the other model both partials are free to move. If this process actually operates the small activation energy required probably makes the large stress concentrations at the head of piled-up dislocations unnecessary.

The processes such as cross-slip, annihilation etc, which lead to a lower rate of work hardening in Stage III are strongly dependent on temperature. This is strikingly demonstrated by the phenomenon of work softening illustrated in *Figure 10.3*(*b*), which shows schematically what happens when a crystal is

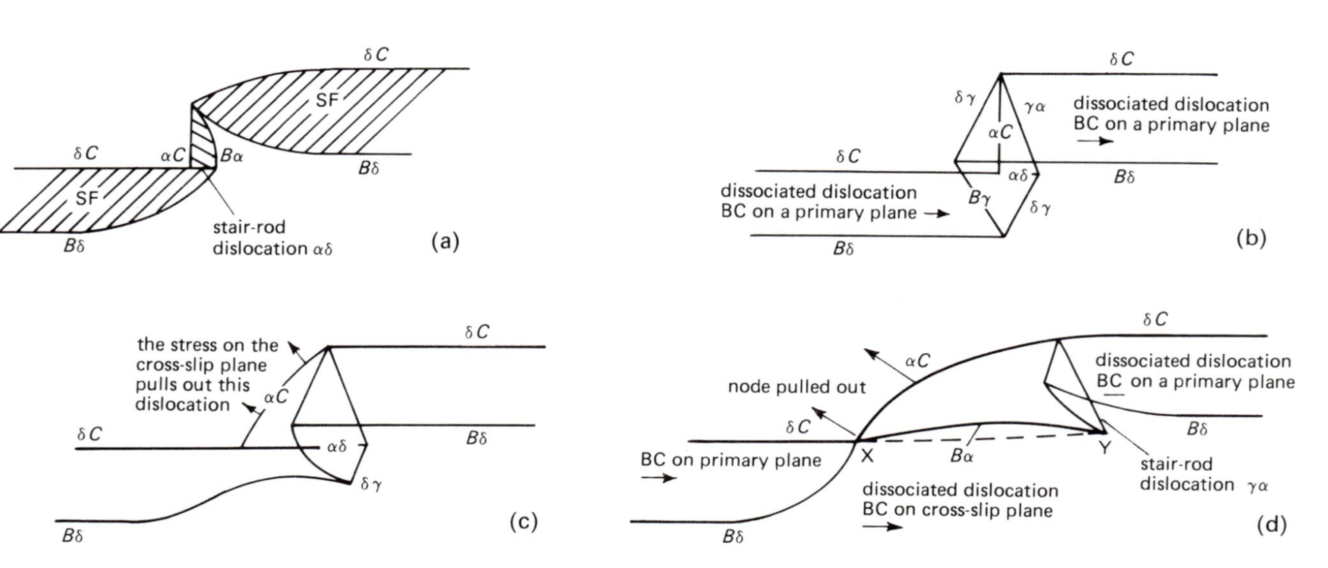

Figure 10.6 Structure of a dissociated jog (a) and (b) and the initiation of cross-slip (c) and (d)

deformed to the end of Stage II at a low temperature T_1, unloaded, and reloaded at a higher temperature T_2. The work-hardened state reached at T_1 is unstable under the combined action of applied stress and the higher temperature T_2. A catastrophic process sets in that tends to reduce the work hardening to what it would have been if the deformation had been done at the temperature T_2. The rate-determining process responsible for this work-softening behaviour is also found to be cross-slip and is similar to that which occurs at the onset of Stage III.

The importance of the value of the stacking fault energy, γ, on the stress–strain curve is evident from its importance to the process of cross-slip. Low values of γ give rise to wide stacking fault 'ribbons', and consequently cross-slip is difficult at reasonable stress levels. Thus, the screws cannot escape from their slip plane, the slip distance is small, the dislocation density is high and the transition from Stage II to Stage III is delayed. In aluminium the stacking fault ribbon width is very small because γ has a high value, and cross-slip occurs at room temperature. Stage II is, therefore, poorly developed unless testing is carried out at low temperatures. These conclusions are in agreement with the observations of dislocation density and arrangement.

10.1.6 The influence of temperature on the flow stress

An obstacle to the motion of dislocations can be overcome by a combination of applied stress and thermal activation to give a strain rate

$$\dot{\gamma} = NbA\nu_0 \exp[-U(\tau)/\boldsymbol{k}T] \tag{10.12}$$

where N is the number of dislocation slip loops with Burgers vector b each sweeping out an area A, $U(\tau)$ is the activation energy required and ν_0 is a frequency factor. In overcoming the obstacle the dislocation moves through a distance d over a length l and hence the 'activation volume' $v = bdl$. For a single activated process with energy U_0 such that $U = U_0 - v\tau_s$, the relation for the flow stress as a function of temperature and strain rate is

$$\tau = \tau_s + \tau_g = \left[\frac{U_0 - kT\ln(NbA\nu_0/\dot{\gamma})}{v}\right] + \tau_g \tag{10.13}$$

Thus at high temperatures, i.e. above $T_c = U_0/\boldsymbol{k}\ln(NbA\nu_0/\dot{\gamma})$, the stress $\tau = \tau_g$ and will be virtually independent of temperature, while below $T = T_c$ the stress will increase with decreasing temperature.

10.1.7 Work hardening in polycrystals

The dislocation structure developed during the deformation of f.c.c. and b.c.c. polycrystalline metals follows the same general pattern as that in single crystals; primary dislocations produce dipoles and loops by interaction with secondary dislocations, which give rise to local dislocation tangles gradually developing into three-dimensional networks of subboundaries. The cell size decreases with increasing strain, and the structural differences that are observed between various metals and alloys are mainly in the sharpness of the subboundaries. In b.c.c. metals, and f.c.c. metals with high stacking fault energy, the tangles rearrange into sharp boundaries but in metals of low stacking fault energy the dislocations are extended, cross slip is restricted, and sharp boundaries are not formed even at large strains. Altering the deformation temperature also has the effect of changing the dislocation distribution; lowering the deformation temperature reduces the tendency for cell formation, as shown in *Figure 10.7*. For a given dislocation distribution the dislocation density is simply related to the flow stress τ by an equation of the form

$$\tau = \tau_0 + \alpha \mu b \rho^{1/2} \tag{10.14}$$

where α is a constant at a given temperature ≈ 0.5; τ_0 for f.c.c. metals is zero (*see Figure 10.1*). The work hardening rate is determined by the ease with which tangled dislocations rearrange themselves and is high in materials with low γ, i.e. brasses, bronzes and austenitic steels compared to Al and b.c.c. metals. In some austenitic steels, work hardening may be increased and better sustained by a strain induced phase transformation (*see* Chapter 12).

Grain boundaries affect work hardening by acting as barriers to slip from one grain to the next grain. In addition, the continuity criterion of polycrystals enforces complex slip in the neighbourhood of the boundaries which spreads across the grains with increasing deformation. This introduces a dependence of work hardening rate on grain size which extends to several per cent elongation. After this stage, however, the work hardening rate is independent of grain size and for f.c.c. polycrystals is about $\mu/40$ which, allowing for the orientation factors, is roughly comparable with that found in single crystals deforming in multiple slip. Thus from the relations $\sigma = m\tau$ and $\varepsilon = \gamma/m$ the average resolved shear stress on a slip plane is rather less than half the applied tensile stress, and the average shear strain parallel to the slip plane is rather more than twice the tensile elongation. The polycrystal work hardening rate is thus related to the single crystal work hardening rate by the relation

$$\mathrm{d}\sigma/\mathrm{d}\varepsilon = m^2\, \mathrm{d}\tau/\mathrm{d}\gamma \tag{10.15}$$

For b.c.c. metals with the multiplicity of slip systems and the ease of cross slip m is more nearly 2, so that the work hardening rate is low. In polycrystalline c.p.h. metals the deformation is complicated by twinning, but in the absence of twinning $m \approx 6.5$, and hence the work hardening rate is expected to be more than an order of magnitude greater than for single crystals, and also higher than the rate observed in f.c.c. polycrystals for which $m \approx 3$.

10.1.8 Dispersion-hardened alloys

On deforming an alloy containing incoherent, non-deformable particles the rate of work-hardening is much greater than that shown by the matrix alone (*see Figure 11.10*). The dislocation density increases very rapidly with strain because the particles produce a turbulent and complex deformation pattern around them. The dislocations gliding in the matrix leave loops around particles either by bowing between the particles or by cross slipping around them; both these mechanisms are discussed in Chapter 11. The stresses in and around particles may also be relieved by activating secondary slip systems in the matrix. All these

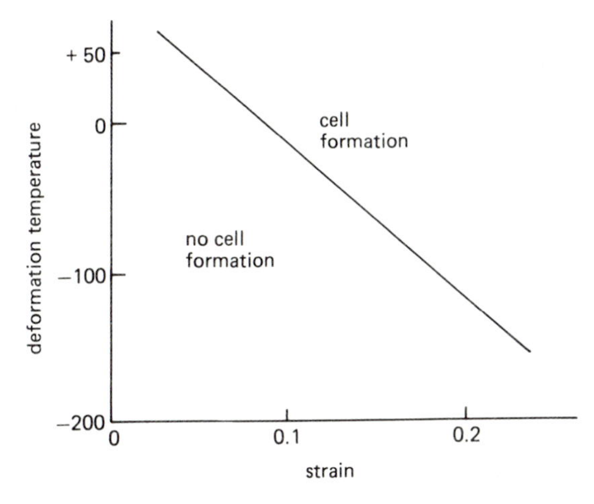

Figure 10.7 Influence of deformation strain and temperature on the formation of a cell structure in α-iron

dislocations spread out from the particle as strain proceeds and, by intersecting the primary glide plane, hinder primary dislocation motion and lead to intense work-hardening. A dense tangle of dislocations is built up at the particle and a cell structure is formed with the particles predominantly in the cell walls.

At small strains ($\lesssim 1$ per cent) work hardening probably arises from the back-stress exerted by the few Orowan loops around the particles, as described by Fisher, Hart and Pry. The stress–strain curve is reasonably linear with strain ε according to

$$\sigma = \sigma_i + \alpha\mu f^{3/2}\varepsilon$$

with the work hardening depending only on f, the volume fraction of particles. At larger strains the 'geometrically necessary' dislocations stored to accommodate the strain gradient which arises because one component deforms plastically more than the other, determine the work hardening. A determination of the average density of dislocations around the particles with which the primary dislocations interact, allows an estimate of the work hardening rate, as initially considered by Ashby. Thus, for a given strain ε and particle diameter d the number of loops per particle is

$$n \sim \varepsilon d/b$$

and the number of particles per unit volume

$$N_v = 3f/4\pi r^2, \quad \text{or} \quad 6f/\pi d^3$$

The total number of loops per unit volume is nN_v and hence the dislocation density $\rho = nN_v\pi d = 6f\varepsilon/db$. The stress–strain relationship from equation 10.14 is then

$$\sigma = \sigma_i + \alpha\mu(fb/d)^{1/2}\varepsilon^{1/2} \tag{10.16}$$

and the work hardening rate

$$d\sigma/d\varepsilon = \alpha'\mu(f/d)^{1/2}(b/\varepsilon)^{1/2} \quad (10.17)$$

Alternative models taking account of the detailed structure of the dislocation arrays, e.g. Orowan, prismatic and secondary loops, have been produced to explain some of the finer details of dispersion-hardened materials. However, this simple approach provides a useful working basis for real materials. Some additional features of dispersion-strengthened alloys are discussed in Chapter 11.

10.1.9 Work-hardening in ordered alloys

A characteristic feature of alloys with long-range order is that they work-harden more rapidly than in the disordered state. θ_{11} for Fe–Al with a B2 ordered structure is $\approx \mu/50$ at room temperature, several times greater than a typical f.c.c. or b.c.c. metal. However, the density of secondary dislocation in Stage II is relatively low and only about 1/100 of that of the primary dislocations. One mechanism for the increase in work-hardening rate is thought to arise from the generation of antiphase domain boundary (APB) tubes generated by glide of jogs on superdislocations which are not aligned along the Burgers vector direction. A possible geometry is shown in *Figure 10.8*; the superdislocation partials shown in (a) each contain a jog produced, for example, by intersection with a forest dislocation, which are non-aligned along the direction of the Burgers vector. When the dislocation glides and the jogs move non-conservatively a tube of APBs is generated, as shown in *Figure 10.8(b)*. Direct evidence for the existence of tubes from weak-beam electron microscope studies has only been reported for Fe-30 at %Al. The micrographs show faint lines along $\langle 111 \rangle$, the Burgers vector direction and are about 3 nm (30 Å) in width. The images are expected to be weak, since the contrast arises from two closely spaced overlapping faults, the second effectively cancelling the displacement caused by the first, and are visible only when superlattice reflections are excited.

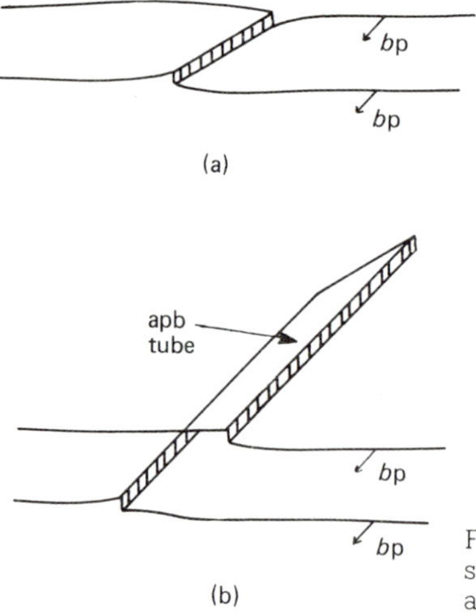

Figure 10.8 Schematic diagram of superdislocation with (a) non-aligned jogs, (b) after glide, to produce an APB-tube

Theory suggests that jogs in superdislocations in screw orientations provide a potent hardening mechanism, estimated to be about eight times as strong as that resulting from pulling out of APB tubes on non-aligned jogs on edge dislocations. The major contributions to the stress to move a dislocation are (i) τ_s, the stress to generate point defects or tubes and (ii) the interaction stress τ_i with dislocations on neighbouring slip planes, and $\tau_s+\tau_i=\frac{3}{4}\alpha_s\mu(\rho_f/\rho_p)\varepsilon$. Thus, with $\alpha_s=1.3$ and provided ρ_f/ρ_p is constant and small, linear hardening with the observed rate is obtained.

In crystals with A_3B order only one rapid stage of hardening is observed compared with the normal three-stage hardening of f.c.c. metals. Moreover, the temperature-dependence of θ_{11}/μ increases with temperature and peaks at $\sim 0.4T_m$. It has been argued that the APB tube model is unable to explain why anomalously high work-hardening rates are observed for those single crystal orientations favourable for single slip on $\{111\}$ planes alone. An alternative model to the APB tubes has generally been proposed based on cross-slip of the leading unit dislocation of the superdislocation. If the second unit dislocation cannot follow exactly in the wake of the first, both will be pinned, since to move as a single unit dislocation creates APB. Thus the cross-slip process can provide barriers to primary glide and, since cross-slip is thermally activated, the number of barriers will increase with increasing temperature. Above $0.4T_m$, the barriers recover and θ_{11}/μ decreases. In some systems, cross-slip on to cube-planes has been proposed. At room temperature the critical resolved shear stress for $\{100\}$ slip is much higher than $\{111\}$ slip, although it is argued, that the APB energy is lower on $\{100\}$. The onset of cube slip at elevated temperatures could account for the peak. In Cu_3Au single crystals with orientations favouring cube plane slip both primary cube slip and cube cross-slip have been observed in the TEM and may account for the anomalous work hardening.

10.2 Preferred orientation

When a polycrystalline metal is plastically deformed the individual grains tend to rotate into a common orientation. This preferred orientation is developed gradually with increasing deformation, and although it becomes extensive above about 90 per cent reduction in area, it is still inferior to that of a good single crystal. The degree of texture produced by a given deformation is readily shown on a monochromatic x-ray transmission photograph, since the grains no longer reflect uniformly into the diffraction rings but only into certain segments of them. The results are usually described in terms of an ideal orientation, such as $[u, v, w]$ for the fibre texture developed by drawing or swaging, and $\{hkl\}$ $\langle uvw\rangle$ for a rolling texture for which a plane of the form (hkl) lies parallel to the rolling plane and a direction of the type $\langle uvw\rangle$ is parallel to the rolling direction. However, the scatter about the ideal orientation can only be represented by means of a pole-figure which describes the spread of orientation about the ideal orientation for a particular set of (hkl) poles (*see Figure 10.9*(*b*) and (*c*)).

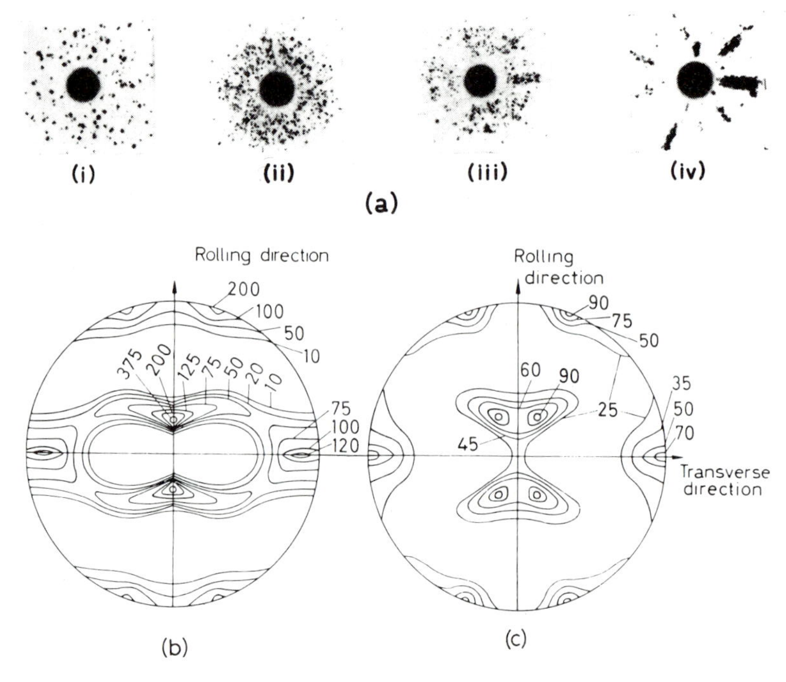

Figure 10.9 (a) Transmission photographs of copper after a deformation of (i) 2%, (ii) 50%, (iii) 80% and (iv) 95% followed by annealing for 1 h at 400°C (after Richards); (111) pole figures from (b) copper, and (c) α-brass after 95% deformation (intensities in arbitrary units)

In tension, the grains rotate in such a way that the movement of the applied stress axis is towards the operative slip direction as discussed in Chapter 6, and for compression the applied stress moves towards the slip plane normal. By considering the deformation process in terms of the particular stresses operating and applying the appropriate grain rotations it is possible to predict the stable end-grain orientation and hence the texture developed by extensive deformation. *Table 10.1* shows the predominant textures found in different metal structures for both wires and sheet.

For face-centred cubic metals a marked transition in deformation texture can be effected either by lowering the deformation temperature or by adding solid solution alloying elements which lower the stacking fault energy. The transition relates to the effect on deformation modes of reducing stacking fault energy or thermal energy, twinning becoming more prevalent and cross-slip less important at lower temperatures and stacking fault energies. This texture transition can be achieved in most f.c.c. metals by alloying additions and altering the rolling

TABLE 10.1 Deformation textures in metals with common crystal structures

Structure	*Wire (fibre texture)*	*Sheet (rolling texture)*
b.c.c.	[110]	$\{112\}\langle 1\bar{1}0\rangle$ to $\{100\}\langle 011\rangle$
f.c.c	[111], [100] double fibre	$\{110\}\langle 112\rangle$ to $\{3\bar{5}1\}\langle 112\rangle$
c.p.h.	[210]	$\{0001\}\langle 1000\rangle$

temperature. Al, however, has a high fault energy and because of the limited solid solubility it is difficult to lower by alloying. The extreme types of rolling texture, shown by copper and 70/30 brass, are given in *Figure 10.9*.

In body-centred cubic metals there are no striking examples of solid solution alloying effects on deformation texture, the preferred orientation developed being remarkably insensitive to material variables. However, material variables can affect c.p.h. textures markedly. Variations in c/a ratio alone cause alterations in the orientation developed, as may be appreciated by consideration of the twinning modes, and it is also possible that solid solution elements alter the relative values of critical resolved shear stress for different deformation modes. Processing variables are also capable of giving a degree of control in hexagonal metals. No texture, stable to further deformation, is found in hexagonal metals and the angle of inclination of the basal planes to the sheet plane varies continuously with deformation. In general, the basal plane lies at a small angle ($<45°$) to the rolling plane, tilted either towards the rolling direction (Zn, Mg) or towards the transverse direction (Ti, Zr, Be, Hf).

The deformation texture cannot, in general, be eliminated by an annealing operation even when such a treatment causes recrystallization. Instead, the formation of a new annealing texture (*see Figure 10.9(a)*) usually results, which is related to the deformation texture by standard lattice rotations.

10.3 Texture-hardening

The flow stress in single crystals varies with orientation according to Schmid's law (Chapter 6) and hence materials with a preferred orientation will also show similar plastic anisotropy, depending on the perfection of the texture. The significance of this relationship is well illustrated by a crystal of beryllium which is c.p.h. and capable of slip only on the basal plane, a compressive stress approaching ≈ 2000 MN/m^2 applied normal to the basal plane produces negligible plastic deformation. Polycrystalline beryllium sheet, with a texture such that the basal planes lie in the plane of the sheet, shows a correspondingly high strength in biaxial tension. When stretched uniaxially the flow stress is also quite high, when additional (prismatic) slip planes are forced into action even though the shear stress for their operation is five times greater than for basal slip. During deformation there is little thinning of the sheet, because the $\langle 11\bar{2}0 \rangle$ directions are aligned in the plane of the sheet. Other hexagonal metals, such as titanium and zirconium, show less marked strengthening in uniaxial tension because prismatic slip occurs more readily, but resistance to biaxial tension can still be achieved. Applications of texture-hardening lie in the use of suitably textured sheet for high biaxial strength, e.g. pressure vessels, dent resistance, etc. Because of the multiplicity of slip systems, cubic metals offer much less scope for texture-hardening. Again, a consideration of single crystal deformation gives the clue; for whereas in a hexagonal crystal m can vary from 2 (basal planes at 45° to the stress axis) to infinity (when the basal planes are normal), in an f.c.c. crystal m can vary only by a factor of 2 with orientation, and in b.c.c. crystals the variation is rather less. In extending this approach to polycrystalline material certain assumptions have to be made about the mutual constraints between grains. One approach gives $m = 3.1$ for a random aggregate of f.c.c. crystals and the calculated orientation dependence of σ/τ for fibre texture shows that a rod with $\langle 111 \rangle$ or $\langle 110 \rangle$ texture ($\sigma/\tau = 3.664$) is 20 per cent stronger than a random structure; the cube texture ($\sigma/\tau = 2.449$) is 20 per cent weaker.

If conventional mechanical properties were the sole criterion for texture-hardened materials, then it seems unlikely that they would challenge strong precipitation-hardened alloys. However, texture-hardening has more subtle benefits in sheet metal forming in optimizing fabrication performance. The variation of strength in the plane of the sheet is readily assessed by tensile tests carried out in various directions relative to the rolling direction. In many sheet applications, however, the requirement is for through thickness strength, e.g. to resist thinning during pressing operations. This is more difficult to measure and is often assessed from uniaxial tensile tests by measuring the ratio of the strain in the width direction to that in the thickness direction of a test piece. The strain ratio R is given by:

$$R = \varepsilon_w/\varepsilon_t = \ln(w_0/w)/\ln(t_0/t) = \ln(w_0/w)/\ln(wL/w_0L) \tag{10.18}$$

where w_0, L_0, t_0 are the original dimensions of width, length and thickness and w, L and t are the corresponding dimensions after straining, which is derived assuming no change in volume occurs. The average strain ratio $\bar{R}$, for tests at various angles in the plane of the sheet, is a measure of the normal anisotropy, i.e. the difference between the average properties in the plane of the sheet and that property in the direction normal to the sheet surface. A large value of R means that there is a lack of deformation modes oriented to provide strain in the through thickness direction, indicating a high through thickness strength.

In deep-drawing, schematically illustrated in *Figure 10.10*, the dominant stress system is radial tension combined with circumferential compression in the drawing zone, while that in the base and lower cup wall, i.e. central stretch forming zone, is biaxial tension. The latter stress is equivalent to a through thickness compression, plus a hydrostatic tension which does not affect the state

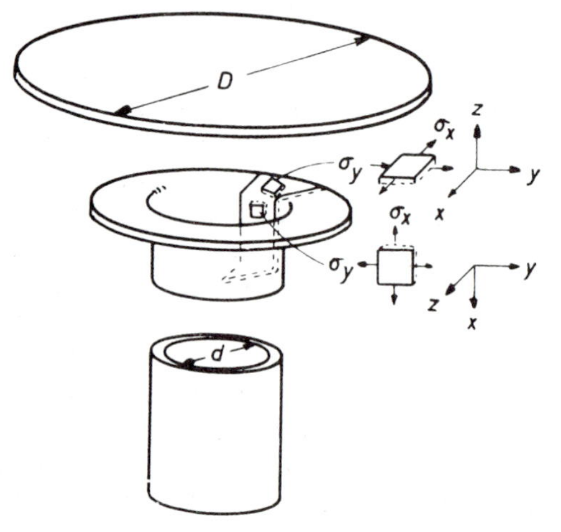

Figure 10.10 Schematic diagram of the deep-drawing operation indicating the stress systems operating in the flange and the cup wall. Limiting drawing ratio is defined as the ratio of the diameter of the largest blank which can satisfactorily complete the draw (D_{max}) to the punch diameter (d) (after Dillamore, Smallman and Wilson, *Commonwealth Mining and Metallurgy Congress*, London, 1969, courtesy of the Canadian Institute of Mining and Metallurgy)

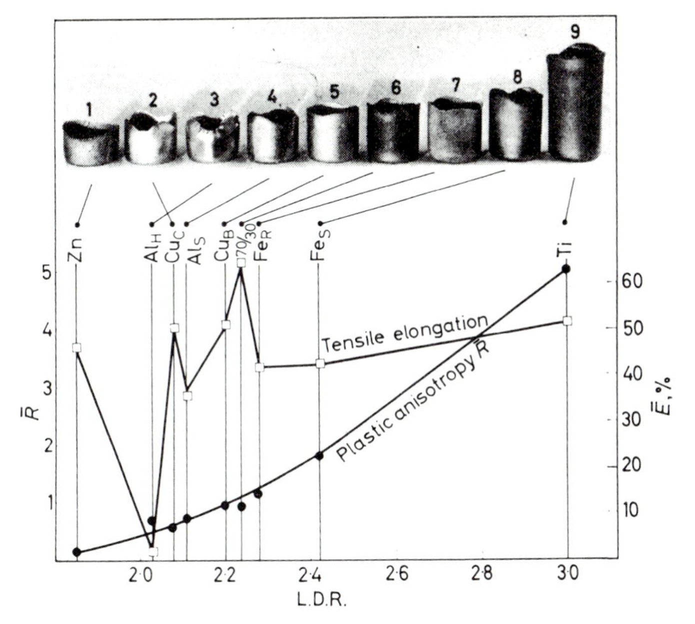

Figure 10.11 Limiting draw ratios (L.D.R.) as a function of average values of R and of elongation to fracture measured in tensile tests at 0°, 45° and 90° to the rolling direction (after Wilson, *J. Inst. Metals*, 1966, **94,** 84, courtesy of the Institute of Metals)

of yielding. Drawing failure occurs when the central stretch forming zone is insufficiently strong to support the load needed to draw the outer part of the blank through the die. Clearly differential strength levels in these two regions, leading to greater ease of deformation in the drawing zone compared with the stretching zone, would enable deeper draws to be made: this is the effect of increasing the $\bar{R}$ value, i.e. high through thickness strength relative to strength in the plane of the sheet will favour drawability. This is confirmed in *Figure 10.11* where deep drawability as determined by limiting drawing ratio, i.e. ratio of maximum drawable blank diameter to final cup diameter, is remarkably insensitive to ductility and, by inference from the wide range of materials represented in the figure, to absolute strength level. Here it is noted that for hexagonal metals slip occurs readily along $\langle 11\bar{2}0\rangle$ thus contributing no strain in the c-direction, and twinning only occurs on the $\{10\bar{1}2\}$ when the applied stress nearly parallel to the c-axis is compressive for $c/a>\sqrt{3}$ and tensile for $c/a<\sqrt{3}$. Thus titanium, $c/a<\sqrt{3}$, has a high strength in through thickness compression, whereas Zn with $c/a>\sqrt{3}$ has low through thickness strength when the basal plane is oriented parallel to the plane of the sheet. In contrast, hexagonal metals with $c/a>\sqrt{3}$ would have a high R for $\{10\bar{1}0\}$ parallel to the plane of the sheet.

Texture hardening is much less in the cubic metals, but f.c.c. materials with $\{111\}\langle 110\rangle$ slip system and b.c.c. with $\{110\}\langle 111\rangle$ are expected to increase R when the texture has component with $\{111\}$ and $\{110\}$ parallel to the plane of the sheet. The range of values of $\bar{R}$ encountered in cubic metals is much less. Face-centred cubic metals have $\bar{R}$ ranging from about 0.3 for cube-texture, $\{100\}\langle 001\rangle$, to a maximum, in textures so far attained, of just over 1.0. Higher values are sometimes obtained in body-centred cubic metals. Values of $\bar{R}$ in the range $1.4 \sim 1.8$ obtained in aluminium-killed low carbon steel are associated with significant improvements in deep drawing performance compared with rimming steel, which has $\bar{R}$-values between 1.0 and 1.4. The highest value of $\bar{R}$ in steels are associated with texture components with $\{111\}$ parallel to the surface, while crystals with $\{100\}$ parallel to the surface have a strongly depressing effect on $\bar{R}$.

In most cases it is found that the R values vary with testing direction and this has relevance in relation to the strain distribution in sheet metal forming. In particular, ear formation on pressings generally develops under a predominant uniaxial compressive stress at the edge of the pressing. The ear is a direct consequence of the variation in strain-ratio for different directions of uniaxial stressing, and a large variation in R value, where $\Delta R = (R_{max} - R_{min})$ generally correlates with a tendency to form pronounced ears. On this basis we could write a simple recipe for good deep drawing properties in terms of strain ratio measurements made in a uniaxial tensile test as high $\bar{R}$ and low ΔR. Much research is aimed at improving forming properties through texture control.

10.4 Macroscopic plasticity

In dislocation theory it is usual to consider the flow stress or yield stress of ductile metals under simple conditions of stressing. In practice, the engineer deals with metals under more complex conditions of stressing, e.g. during forming operations, and hence needs to correlate yielding under combined stresses with that in uniaxial testing. To achieve such a yield stress criterion it is usually assumed that the metal is mechanically isotropic and deforms plastically at constant volume, i.e. a hydrostatic state of stress does not affect yielding. In assuming plastic isotropy, macroscopic shear is allowed to take place along lines of maximum shear stress and crystallographic slip is ignored, and the yield stress in tension is equal to that in compression, i.e. there is no Bauschinger effect.

A given applied stress state in terms of the principal stresses $\sigma_1, \sigma_2, \sigma_3$ which act along three principal axes, X_1, X_2 and X_3, may be separated into the hydrostatic part (which produce changes in volume) and the deviatoric components (which produce changes in shape). It is assumed that the hydrostatic component has no effect on yielding and hence the more the stress state deviates from pure hydrostatic the greater the tendency to produce yield. The stresses may be represented on a stress-space plot (*see Figure 10.12(a)*) in which a line equidistant from the three stress axes represents a pure hydrostatic stress state. Deviation from this line will cause yielding if the deviation is sufficiently large, and define a yield surface which has six-fold symmetry about the hydrostatic line. This arises because the conditions of isotropy imply equal yield stresses along all three axes, and the absence of the Bauschinger effect implies equal yield stresses along σ_1 and $-\sigma_1$. Taking a section through stress space, perpendicular to the hydrostatic line gives the two simplest yield criteria satisfying the symmetry requirements corresponding to a regular hexagon and a circle.

The hexagonal form represents the Tresca criterion (*see Figure 10.12(c)*)

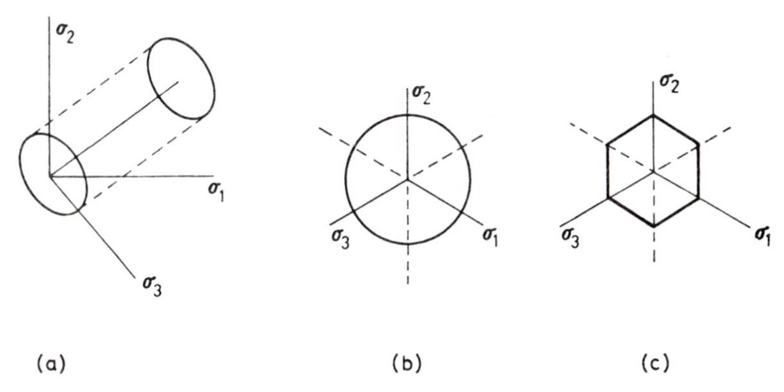

Figure 10.12 Schematic representation of the yield surface with (a) principal stresses σ_1, σ_2 and σ_3, (b) Von Mises yield criterion and (c) Tresca yield criterion

which assumes that plastic shear takes place when the maximum shear stress attains a critical value k equal to shear yield stress in uniaxial tension. This is expressed by

$$\tau_{max} = \frac{\sigma_1 - \sigma_3}{2} = k$$

where the principal stresses $\sigma_1 > \sigma_2 > \sigma_3$. This criterion is the isotropic equivalent of the law of resolved shear stress in single crystals. The tensile yield stress $Y = 2k$ is obtained by putting $\sigma_1 = Y$, $\sigma_2 = \sigma_3 = 0$.

The circular cylinder is described by the equation

$$(\sigma_1 - \sigma_2)^2 + (\sigma_2 - \sigma_3)^2 + (\sigma_3 - \sigma_1)^2 = \text{constant}$$

and is the basis of the von Mises yield criterion (*see Figure 10.12(b)*). This criterion implies that yielding will occur when the shear energy per unit volume reaches a critical value given by the constant. This constant is equal to $6k^2$ or $2Y^2$ where k is the yield stress in simple shear, as shown by putting $\sigma_2 = 0, \sigma_1 = \sigma_3$, and Y is the yield stress in uniaxial tension when $\sigma_2 = \sigma_3 = 0$. Clearly $Y = 3k$ compared to $Y = 2k$ for the Tresca criterion and, in general, this is found to agree somewhat closer with experiment.

In many practical working processes, e.g. rolling, the deformation occurs under approximately plane strain conditions with displacements confined to the X_1X_2 plane. It does not follow that the stress in this direction is zero, and, in fact, the deformation conditions are satisfied if $\sigma_3 = \frac{1}{2}(\sigma_1 + \sigma_2)$ so that the tendency for one pair of principal stresses to extend the metal along the X_3 axis is balanced by that of the other pair to contract it along this axis. Eliminating σ_3 from the von Mises criterion, the yield criterion becomes

$$(\sigma_1 - \sigma_2) = 2k$$

and the plane strain yield stress, i.e. when $\sigma_2 = 0$, given when

$$\sigma_1 = 2k = 2Y/\sqrt{3} = 1.15Y$$

For plane strain conditions, the Tresca and von Mises criteria are equivalent and

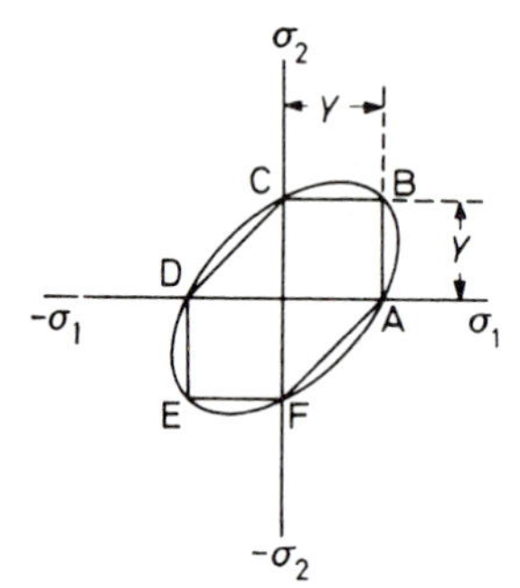

Figure 10.13 The Von Mises yield ellipse and Tresca yield hexagon

two-dimensional flow occurs when the shear stress reaches a critical value. The above condition is thus equally valid when written in terms of the deviatoric stresses σ_1', σ_2', σ_3' defined (*see* page 358) by equations of the type $\sigma_1' = \sigma_1 - \frac{1}{3}(\sigma_1 + \sigma_2 + \sigma_3)$.

Under plane stress conditions, $\sigma_3 = 0$ and the yield surface becomes two-dimensional and the von Mises criterion becomes

$$\sigma_1^2 + \sigma_1\sigma_2 + \sigma_2^2 = 3k^2 = Y^2 \tag{10.19}$$

which describes an ellipse in the stress plane. For the Tresca criterion the yield surface reduces to a hexagon inscribed in the ellipse as shown in *Figure 10.13*. Thus, when σ_1 and σ_2 have opposite signs, the Tresca criterion becomes $\sigma_1 - \sigma_2 = 2k - Y$ and is represented by the edges of the hexagon CD and FA. When they have the same sign then $\sigma_1 = 2k = Y$ or $\sigma_2 = 2k = Y$ and defines the hexagon edges AB, BC, DE and EF.

10.4.1 Effective stress and strain

For an isotropic material, a knowledge of the uniaxial tensile test behaviour together with the yield function should enable the stress–strain behaviour to be predicted for any stress-system. This is achieved by defining an effective stress-effective strain relationship such that if $\sigma = K\varepsilon^n$ is the uniaxial stress–strain relationship then we may write

$$\bar{\sigma} = K\bar{\varepsilon}^n \tag{10.20}$$

for any state of stress. The stress–strain behaviour of a thin walled tube with internal pressure is a typical example, and it is observed that the flow curves obtained in uniaxial tension and in biaxial torsion coincide when the curves are plotted in terms of effective stress and effective strain. These quantities are defined by:

$$\bar{\sigma} = \frac{\sqrt{2}}{2}[(\sigma_1 - \sigma_2)^2 + (\sigma_2 - \sigma_3)^2 + (\sigma_3 - \sigma_1)^2]^{1/2}$$

and

$$\bar{\varepsilon} = \frac{\sqrt{2}}{3}[(\varepsilon_1 - \varepsilon_2)^2 + (\varepsilon_2 - \varepsilon_3)^2 + (\varepsilon_3 - \varepsilon_1)^2]^{1/2}$$

where ε_1, ε_2 and ε_3 are the principal strains, both of which reduce to the axial normal components of stress and strain for a tensile test. It should be emphasized, however, that this generalization holds only for isotropic media and for constant loading paths, i.e. $\sigma_1 = \alpha\sigma_2 = \beta\sigma_3$ where α and β are constants independent of the value of σ_1.

To illustrate the above equivalence ideas, let us consider an example of failure by instability. The case of uniaxial tension is well known (*see* page 80) where for a stress–strain relation given by $\sigma = K\varepsilon^n$ the strain at instability is given by $\varepsilon_i = n$. Of greater interest to engineers in metal forming industries is the determination of forming limits arising from instability under more complex stressing.

Under combined stresses the instability expression can be written $\bar{\sigma}/z = \mathrm{d}\bar{\sigma}/\mathrm{d}\bar{\varepsilon}$ where $\bar{\sigma}$ and $\bar{\varepsilon}$ are the equivalent stress and strain respectively, and z is dependent on the relative stresses (*see* below). It may readily be demonstrated from considerations of the maximization of plastic work that the strain increments along principal directions are given by

$$\mathrm{d}\varepsilon_1/(\mathrm{d}\bar{\sigma}/\mathrm{d}\sigma_1) = \mathrm{d}\varepsilon_2/(\mathrm{d}\bar{\sigma}/\mathrm{d}\sigma_2) = \mathrm{d}\varepsilon_3/(\mathrm{d}\bar{\sigma}/\mathrm{d}\sigma_3) = \mathrm{d}\bar{\varepsilon}$$

or

$$\mathrm{d}\varepsilon_1/(2\sigma_1 - \sigma_2 - \sigma_3) = \mathrm{d}\varepsilon_2/(2\sigma_2 - \sigma_3 - \sigma_1)$$
$$= \mathrm{d}\varepsilon_2/(2\sigma_3 - \sigma_1 - \sigma_2) = \mathrm{d}\bar{\varepsilon}/\sigma$$

These are the Levy–Mises equations, defining the increment of equivalent strain in terms of the component increment strains for a material yielding according to the von Mises criterion. Let us now consider a plane sheet subjected to principal stresses $\sigma_2 = \alpha\sigma_1$, $\sigma_3 = 0$. Such a stress system is approximated in many sheet forming operations, and, of course, in such cases instability frequently determines the performance limits of the sheet undergoing processing. According to an analysis by Swift, instability occurs when both the loads go through zero together. Thus if $P_1 = \sigma_1 A_1$, and $P_2 = \sigma_2 A_2$, then $\mathrm{d}P_1 = \sigma_1\,\mathrm{d}A_1 + A_1\,\mathrm{d}\sigma_1 = 0$ and $\mathrm{d}\sigma_2/\sigma_2 = \mathrm{d}\varepsilon_2$, $\mathrm{d}\sigma_1/\sigma_1 = \mathrm{d}\varepsilon_1$ are now the instability conditions. These are, of course, identical to the uniaxial case, but the second principal stress has influenced the progress of work hardening and the magnitude of the principal strain ε_1 at failure $(\varepsilon_1)_i$ will be affected by the biaxiality of the stress system.

To determine this effect it is convenient to work through the equivalent stress–strain curve. On straining, a change in geometry resultant on straining causes a decrement of the individual stresses $\Delta\sigma_1$, $\Delta\sigma_2$ and a corresponding change in equivalent stress $\Delta\bar{\sigma}$. These are related through the differentiation of the von Mises criterion

$$\Delta\bar{\sigma} = \frac{(2\sigma_1 - \sigma_2)\Delta\sigma_1 + (2\sigma_2 - \sigma_1)\Delta\sigma_2}{2(\sigma_1^2 - \sigma_1\sigma_2 + \sigma_2^2)^{1/2}}$$

recalling that $\sigma_3 = 0$. Then, so long as the stress decrement resultant on straining is more than compensated for by a stress increment $\delta\bar{\sigma}$ caused by work hardening, flow will be stable. At instability $\delta\bar{\sigma} = \Delta\bar{\sigma}$ and since $\mathrm{d}\sigma_1 = \sigma_1\,\mathrm{d}\varepsilon_1$ and $\mathrm{d}\sigma_2 = \sigma_2\,\mathrm{d}\varepsilon_2$ also at instability, then

$$d\bar{\sigma}=\frac{(2\sigma_1-\sigma_2)\sigma_1\,d\varepsilon_1+(2\sigma_2-\sigma_1)\sigma_2\,d\varepsilon_2}{2(\sigma_1^2-\sigma_1\sigma_2-\sigma_2^2)^{1/2}}$$

is the instability equation. From the Levy–Mises equations $d\varepsilon_1$ and $d\varepsilon_2$ can be expressed in terms of $d\bar{\varepsilon}$, so that

$$\frac{d\bar{\sigma}}{d\bar{\varepsilon}}=\frac{(2\sigma_1-\sigma_2)\sigma_1(2\sigma_1-\sigma_2)+(2\sigma_2-\sigma_1)\sigma_2(2\sigma_2-\sigma_1)}{4(\sigma_1^2-\sigma_1\sigma_2+\sigma_2^2)}$$

Putting $\sigma_2=\alpha\sigma_1$ then

$$\frac{d\bar{\sigma}}{d\bar{\varepsilon}}=\frac{\sigma_1(4-3\alpha-3\alpha^2+4\alpha^3)}{4(1-\alpha+\alpha^2)}$$

and substituting $\bar{\sigma}=\sigma_1(1-\alpha+\alpha^3)^{1/2}$, we have

$$\frac{d\bar{\sigma}}{d\bar{\varepsilon}}=\frac{\bar{\sigma}(1+\alpha)(4-7\alpha+4\alpha^2)}{4(1-\alpha+\alpha^2)^{3/2}}$$

whence

$$z=\frac{4(1-\alpha+\alpha^2)^{3/2}}{(1+\alpha)(4-7\alpha+4\alpha^2)}$$

It may be noted that if we have an expression relating $\bar{\sigma}$ and $\bar{\varepsilon}$ of the Ludwig type, i.e. $\bar{\sigma}=K\bar{\varepsilon}^n$, the equivalent strain to instability is $(\bar{\varepsilon})_i=nz$, and for proportional loading, as assumed here, the principal strains to instability may be obtained from the initial form of the Levy–Mises equation writing ε_1, ε_2 and $\bar{\varepsilon}$ in place of $d\varepsilon_1$, $d\varepsilon_2$ and $d\bar{\varepsilon}$. This becomes, for example

$$(\varepsilon_1)_i=\frac{(2-\alpha)}{2(1-\alpha+\alpha^2)^{1/2}}\cdot n\cdot\frac{4(1-\alpha+\alpha^2)^{3/2}}{(1+\alpha)(4-7\alpha+4\alpha^2)} \qquad (10.21)$$

and the strain to instability depends on the work-hardening parameter n and the stress system applied. With $\alpha=\frac{1}{2}$, $(\varepsilon_1/n)=1$ and the deformation is equivalent to plane strain, since $(\varepsilon_2/n)=0$. When $\alpha=1$ the deformation is equivalent to a through thickness compression (*see* previous section) with $(\varepsilon_1/n)=(\varepsilon_2/n)=1$ and good stress–strain behaviour is exhibited to high strain values, the total strain being twice that for uniaxial tension. With $\alpha=-1$ we find that $(\varepsilon_1/n)=\infty$ and hence transverse compressive stress delays instability during deformation.

10.5 Annealing

10.5.1 Introduction

When a metal is cold worked, by any of the many industrial shaping operations, changes occur in both its physical and mechanical properties. While the increased hardness and strength which result from the working treatment may be of importance in certain applications, it is frequently necessary to return the metal to its original condition to allow further forming operations (e.g. deep drawing) to be carried out or for applications where optimum physical properties, such as electrical conductivity, are essential. The treatment given to the metal to bring about a decrease of the hardness and an increase in the ductility is known as annealing. This usually means keeping the deformed metal for a certain time at a temperature higher than about one-third the absolute melting point.

Cold working produces an increase in dislocation density; for most metals ρ increases from the value of 10^{10}–10^{12} lines m^{-2} typical of the annealed state, to 10^{12}–10^{13} after a few per cent deformation, and up to 10^{15}–10^{16}/lines m^{-2} in the heavily deformed state. Such an array of dislocations gives rise to a substantial strain energy stored in the lattice, so that the cold worked condition is thermodynamically unstable relative to the undeformed one. Consequently, the deformed metal will try to return to a state of lower free energy, i.e. a more perfect state. In general, this return to a more equilibrium structure cannot occur spontaneously but only at elevated temperatures where thermally activated processes such as diffusion, cross-slip and climb take place. Like all non-equilibrium processes the rate of approach to equilibrium will be governed by an Arrhenius equation of the form

$$\text{Rate} = A \exp[-Q/\boldsymbol{k}T]$$

where the activation energy Q depends on impurity content, strain etc.

The formation of atmosphere by strain ageing is one method whereby the metal reduces its excess lattice energy but this process is unique in that it usually leads to a further increase in the structure-sensitive properties rather than a reduction to the value characteristic of the annealed condition. It is necessary, therefore, to increase the temperature of the deformed metal above the strain-ageing temperature before it recovers its original softness and other properties.

The removal of the cold-worked condition occurs by a combination of three processes, namely: (i) recovery, (ii) recrystallization and (iii) grain growth. These stages have been successfully studied by using optical methods such as light microscopy, transmission electron microscopy, or x-ray diffraction; mechanical property measurements, e.g. hardness; and physical property measurements, e.g. density, electrical resistivity and stored energy. *Figure 10.14* shows the change in some of these properties on annealing. During the recovery stage the decrease in stored energy and electrical resistivity is accompanied by only a slight lowering of hardness, and the greatest simultaneous change in properties occurring during the primary recrystallization stage. However, while these measurements are no doubt striking and extremely useful, it is necessary to understand them to correlate such studies with the structural changes by which they are accompanied.

10.5.2 Recovery

This process describes the changes in the distribution and density of defects with associated changes in physical and mechanical properties which take place in worked crystals before recrystallization or alteration of orientation occurs. It will be remembered that the structure of a cold-worked metal consists of dense dislocation networks, formed by the glide and interaction of dislocations, and,

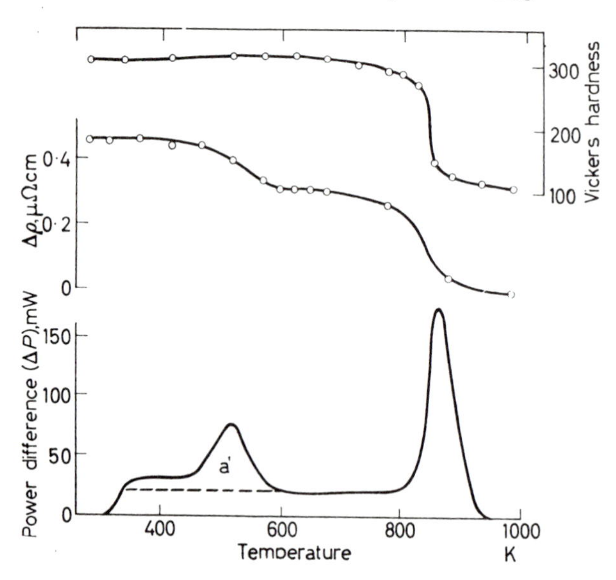

Figure 10.14 (a) Rate of release of stored energy (ΔP), increment in electrical resistivity ($\Delta\rho$) and hardness (V.P.N.) for specimens of nickel deformed in torsion and heated at 6 K/min. (Clareborough, Hargreaves and West, *Proc. Roy. Soc.* 1955, **A232,** 252)

consequently, the recovery stage of annealing is chiefly concerned with the rearrangement of these dislocations to reduce the lattice energy and does not involve the migration of large-angle boundaries. This rearrangement of the dislocations is assisted by thermal activation.

One of the most important recovery processes which leads to a resultant lowering of the lattice strain energy is rearrangement of the dislocations into cell walls. This process in its simplest form was originally termed polygonization and is illustrated schematically in *Figure 10.15*, whereby dislocations all of one sign align themselves into walls to form small-angle or sub-grain boundaries. During deformation a region of the lattice is curved, as shown in *Figure 10.15(a)*, and the observed curvature can be attributed to the formation of excess edge dislocations parallel to the axis of bending. On heating, the dislocations form a sub-boundary by a process of annihilation and rearrangement. This is shown in *Figure 10.15(b)*, from which it can be seen that it is the excess dislocations of one sign which remain after the annihilation process that align themselves into walls.

Polygonization is a simple form of sub-boundary formation and the basic movement is climb whereby the edge dislocations change their arrangement from a horizontal to a vertical grouping. This process involves the migration of vacancies to or from the edge of the half-planes of the dislocations (*see Figure 6.16*). The removal of vacancies from the lattice, together with the reduced strain energy of dislocations which results, can account for the large change in both electrical resistivity and stored energy observed during this stage, while the change in hardness can be attributed to the rearrangement of dislocations and to the reduction in the density of dislocations.

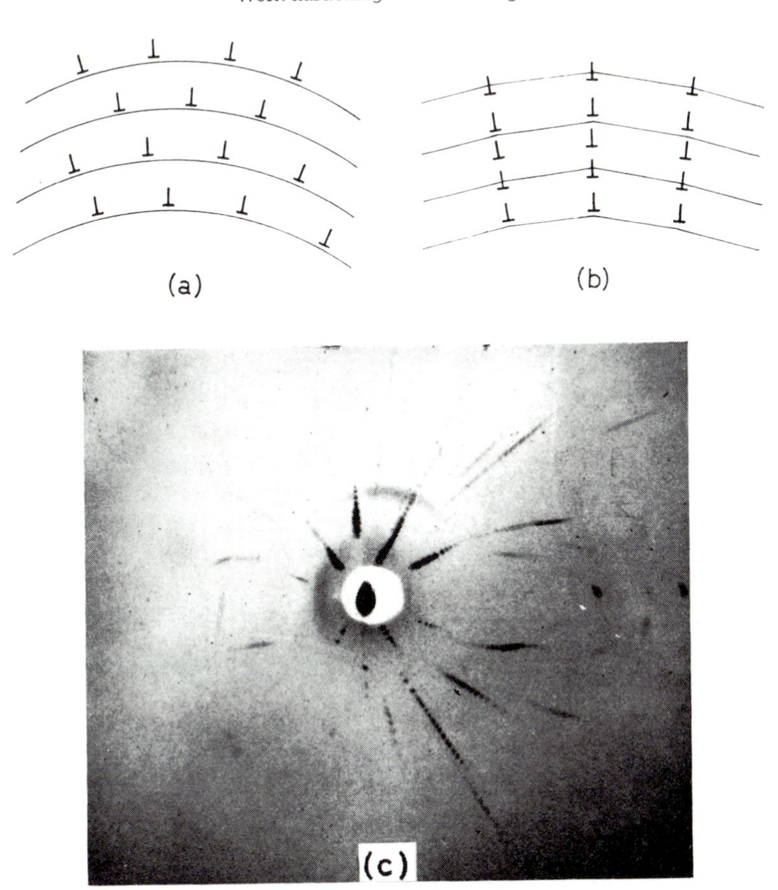

Figure 10.15 (a) Random arrangement of excess parallel edge dislocations, and (b) alignment into dislocation walls; (c) Laue photograph of polygonized zinc (after Cahn, *J. Inst. Metals*, 1949, **77,** 121)

The process of polygonization can be demonstrated using the Laue method of x-ray diffraction. Diffraction from a bent single crystal of zinc or aluminium takes the form of continuous radial streaks. On annealing, these asterisms (*see Figure 2.5*) break up into spots as shown in *Figure 10.15(c)* where each diffraction spot originates from a perfect polygonized sub-grain, and the distance between the spots represents the angular misorientation across the sub-grain boundary. The existence of these dislocation walls can, of course, be verified by delineating them with rows of etch pits, as already discussed in Chapter 7. One disadvantage of the Laue technique is its low resolving power, and it is a consequence of this that the specimen must be annealed at a high temperature, which is about 500 °C for aluminium, to allow large sub-grains to form by extensive vacancy migration. In practice, however, recovery occurs in cold-worked polycrystalline metals such as aluminium at much lower temperatures, and this observation indicates that sub-grain formation on a much finer scale than that obtained in the experiment described above must take place. In addition, deformation must occur on several independent slip systems in which case different kinds of dislocations will be produced in various parts of the grains. Thus, not only will the sub-grains be on a fine scale, but they will also be arranged in complicated patterns. This behaviour can be verified from the sharpening of x-ray diffraction lines (reduction in line broadening) as recovery proceeds, but the most direct evidence for this effect is necessarily restricted to observations in the electron microscope.

Such observations indeed show that, in heavily deformed polycrystalline aggregates at least, recovery is associated with the formation of sub-grains out of complex dislocation networks by a process of dislocation anninilation and re-arrangement. In some metals and alloys the dislocations are already partially arranged in sub-boundaries forming diffuse cell structures by dynamical recovery (*see Figure 10.5*). The conventional recovery process is then one in which these cells sharpen and grow. In other metals, dislocations are more uniformly distributed after deformation, with hardly any cell structure discernible, and the recovery process then involves formation, sharpening and growth of sub-boundaries. The sharpness of the cell structure formed by deformation depends on the stacking fault energy of the metal, the deformation temperature and the extent of deformation.

Metal crystals with hexagonal structure, such as zinc or cadmium, deformed in tension by basal slip, or crystals with cubic structure deformed in easy glide, will contain no lattice regions where the bending is sufficient to provide the excess dislocations of one sign necessary for polygonization. Nevertheless, such specimens can recover their original softness completely by the basic process of recovery, i.e. running together and mutual annihilation of dislocations.

When the two dislocations are on the same slip plane, it is possible that as they run together and annihilate they will have to cut through intersecting dislocations on other planes, i.e. 'forest' dislocations. This recovery process will, therefore, be aided by thermal fluctuations since the activation energy for such a cutting process is small. When the two dislocations of opposite sign are not on the same slip plane, climb or cross-slip must first occur, and both of these processes require thermal activation. The formation of sub-boundaries by cross-slip is schematically shown in *Figure 7.11*. In some metals, sub-grain growth has been observed to occur by coalescence of neighbouring sub-grain regions involving a process of climb and glide.

10.5.3 Recrystallization

The most significant changes in the structure-sensitive properties occur during the primary recrystallization stage. In this stage the deformed lattice is completely replaced by a new unstrained one by means of a nucleation and growth process, in which practically stress-free grains grow from nuclei formed in the deformed matrix. The orientation of the new grains differs considerably from that of the crystals they consume, so that the growth process must be regarded as incoherent, i.e. it takes place by the advance of large-angle boundaries separating the new crystals from the strained matrix.

During the growth of grains, atoms get transferred from one grain to another across the boundary. Such a process is thermally activated, as shown in *Figure*

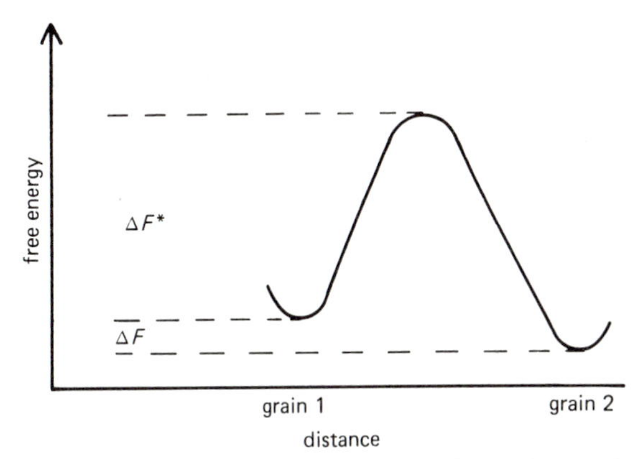

Figure 10.16 Variation in free energy during grain growth.

10.16 and by the usual reaction-rate theory the frequency of atomic transfer one way is

$$\nu \exp\left(-\frac{\Delta F}{kT}\right) s^{-1} \tag{10.22}$$

and in the reverse direction

$$\nu \exp\left(-\frac{\Delta F^* + \Delta F}{kT}\right) s^{-1} \tag{10.23}$$

where ΔF is the difference in free energy per atom between the two grains, i.e. supplying the driving force for migration, and ΔF^* is an activation energy. For each net transfer the boundary moves forward a distance b and the velocity v is given by

$$v = M\,\Delta F \tag{10.24}$$

where M is the mobility of the boundary, i.e. the velocity for unit driving force, and is thus

$$M = \frac{b\nu}{kT} \exp\left(\frac{\Delta S^*}{k}\right) \exp\left(-\frac{\Delta E^*}{kT}\right) \tag{10.25}$$

Generally, the open structure of high angle boundaries should lead to a high mobility. However they are susceptible to the segregation of impurities, low concentrations of which can reduce the boundary mobility by orders of magnitude. By contrast, special boundaries which are close to a CSL are much less affected by impurity segregation and hence can lead to higher relative mobility.

It is well known that the rate of recrystallization depends on several important factors, namely: (i) the amount of prior deformation (the greater the degree of cold work the lower the recrystallization temperature and the smaller the grain size), (ii) the temperature of the anneal (as the temperature is lowered the time to attain a constant grain size increases exponentially* and (iii) the purity of

* The velocity of linear growth of new crystals usually obeys an exponential relationship of the form $v = v_0 \exp[-Q/RT]$.

the sample (e.g. zone-refined aluminium recrystallizes below room temperature, whereas aluminium of commercial purity must be heated several hundred degrees). The role these variables play in recrystallization will be evident once the mechanism of recrystallization is known. This mechanism will now be outlined.

Measurements, using the light microscope, of the increase in diameter of a new grain as a function of time at any given temperature can be expressed as shown in *Figure 10.17*. The diameter increases linearly with time until the growing grains begin to impinge on one another, after which the rate necessarily decreases. The classical interpretation of these observations is that nuclei form spontaneously in the matrix after a so-called nucleation time, t_0, and these nuclei then proceed to grow steadily as shown by the linear relationship. The driving force for the process is provided by the stored energy of cold work contained in the strained grain on one side of the boundary relative to that on the other side. Such an interpretation would suggest that the recrystallization process occurs in two distinct stages, i.e. first nucleation and then growth.

During the linear growth period the radius of a nucleus is $R=G(t-t_0)$, where G, the growth rate is dR/dt and, assuming the nucleus is spherical, the volume of the recrystallized nucleus is $\frac{4}{3}\pi G^3(t-t_0)^3$. If the number of nuclei that form in a time increment dt is $N\,dt$ per unit volume of unrecrystallized matrix, and if the nuclei do not impinge on one another, then for unit total volume

$$f=\frac{4}{3}\pi N G^3\int_0^t (t-t_0)^3\,dt$$

or

$$f=\frac{\pi}{3}NGt^3 \tag{10.26}$$

This equation is valid in the initial stages when $f \ll 1$. When the nuclei impinge on one another the rate of recrystallization decreases and is related to the amount untransformed $(1-f)$ by

$$f=1-\exp\left(-\frac{\pi}{3}NG^3t^4\right) \tag{10.27}$$

where, for short times, equation 10.27 reduces to equation 10.26. This Johnson–Mehl equation is expected to apply to any phase transformation where there is

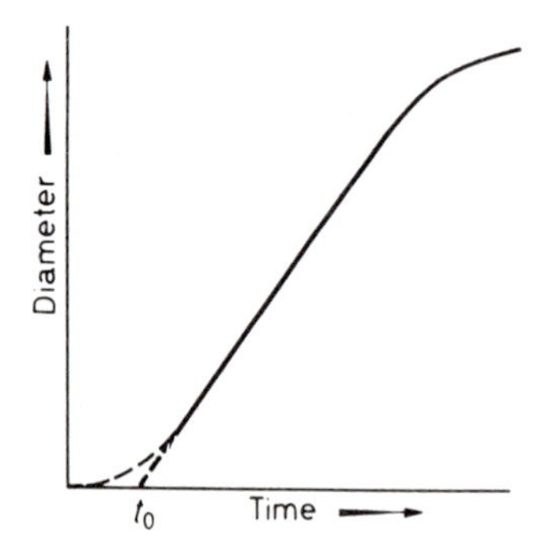

Figure 10.17 Variation of grain diameter with time at a constant temperature

random nucleation, constant N and G and small t_0. In practice, nucleation is not random and the rate not constant so that equation 10.27 will not strictly apply. For the case where the nucleation rate decreases exponentially, Avrami developed the equation

$$f = 1 - \exp(-kt^n) \tag{10.28}$$

where k and n are constants, with $n \approx 3$ for a fast, and $n \approx 4$ for a slow, decrease of nucleation rate. Provided there is no change in the nucleation mechanism, n is independent of temperature but k is very sensitive to temperature T; clearly from equation 10.27 $k = \pi N G^3/3$ and both N and G depend on T.

An alternative interpretation is that the so-called incubation time t_0, represents a period during which small nuclei, of a size too small to be observed in the light microscope, are growing very slowly. This latter interpretation follows from the recovery stage of annealing. Thus, the structure of a recovered metal consists of sub-grain regions of practically perfect crystal and, thus, one might expect the 'active' recrystallization nuclei to be formed by the growth of certain sub-grains at the expense of others.

That sub-grain formation is a necessary preliminary stage to recrystallization is amply supported by experimental observation. As mentioned earlier, it is observed that crystals of zinc or cadmium with hexagonal structure deformed in tension by basal slip, or crystals with cubic structure deformed in easy glide regain their original softness by recovery alone and, furthermore, these crystals cannot be made to recrystallize unless they are deformed inhomogeneously by, say, scratching or bending. This experiment illustrates that recrystallization nuclei are formed on heating only from strongly curved regions of the lattice where sub-grains can readily form. The importance of locally curved regions of the lattice to the nucleation process is further emphasized by the observation that when commercial metals are being recrystallized the nuclei are formed preferentially in places of high lattice distortion, such as near grain boundaries, deformation bands, or inclusions.

The process of recrystallization may be pictured as follows. After deformation, polygonization of the bent lattice regions on a fine scale occurs and this results in the formation of several regions in the lattice where the strain energy is lower than in the surrounding matrix; this is a necessary primary condition for nucleation. During this initial period when the angles between the sub-grains are small and less than one degree, the sub-grains form and grow quite rapidly. However, as the sub-grains grow to such a size that the angles between them become of the order of a few degrees, the growth of any given sub-grain at the expense of the others is very slow. Eventually one of the sub-grains will grow to such a size that the boundary it makes with the deformed matrix is fairly large angle so that the boundary mobility begins to increase with increasing angle. A large angle boundary, $\theta \approx 30$–$40°$, has a high mobility because of the large lattice irregularities or 'gaps' which exist in the boundary transition layer. The atoms on such a boundary can easily transfer their allegiance from one crystal to the other. This sub-grain is then able to grow at a much faster rate than the other sub-grains which surround it and so acts as the nucleus of a recrystallized grain. The further it grows the greater will be the difference in orientation between the nucleus and

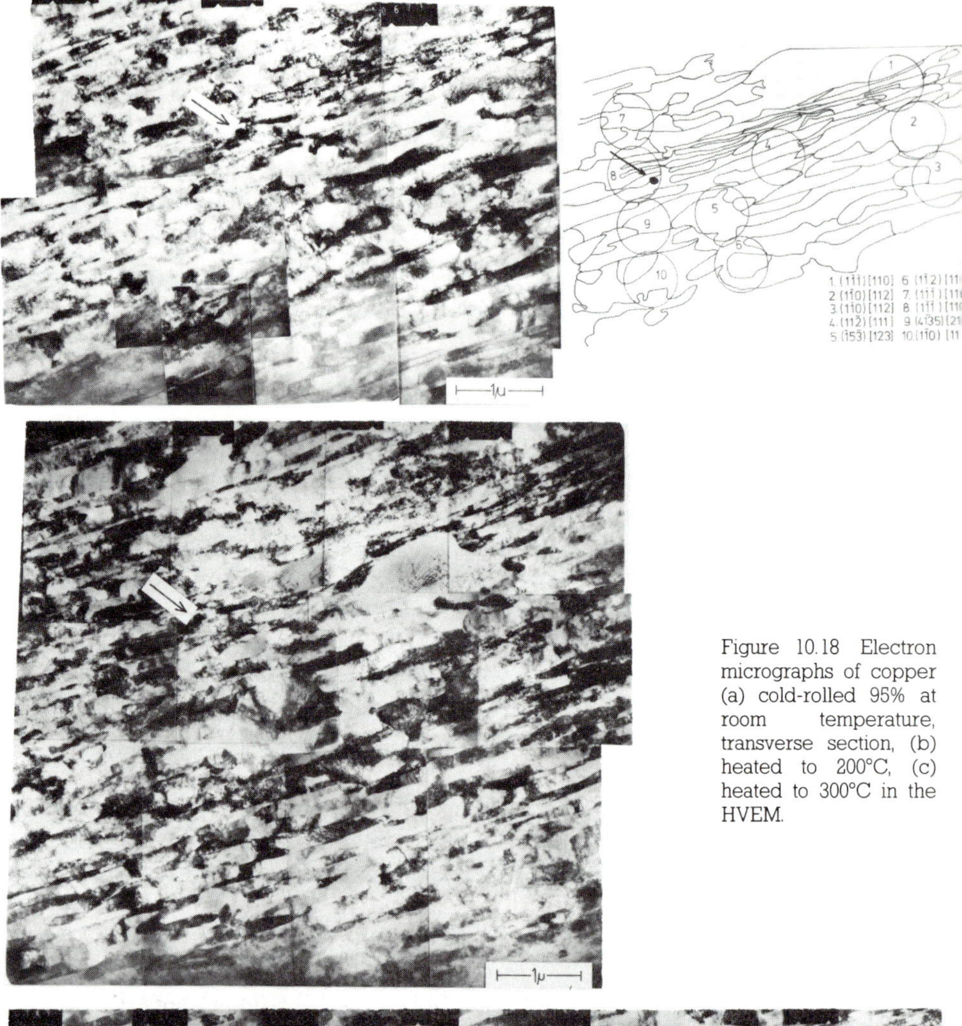

Figure 10.18 Electron micrographs of copper (a) cold-rolled 95% at room temperature, transverse section, (b) heated to 200°C, (c) heated to 300°C in the HVEM.

the matrix it meets and consumes, until it finally becomes recognizable as a new strain-free crystal separated from its surroundings by a large-angle boundary.

The recrystallization nucleus therefore has its origin as a sub-grain in the deformed microstructure. Whether it grows to become a strain-free grain depends on three factors: (i) the stored energy of cold work must be sufficiently high to provide the required driving force, (ii) the potential nucleus should have a size advantage over its neighbours, and (iii) it must be capable of continued growth by existing in a region of high lattice curvature, e.g. transition band, so that the growing nucleus can quickly achieve a high-angle boundary. *In situ* experiments in the HVEM have confirmed these factors. *Figure 10.18*(*a*) shows the as-deformed sub-structure in the transverse section of rolled copper, together with the orientations of some selected areas. The sub-grains are observed to vary in width from 50 to 500 nm, and exist between regions 1 and 8 as a transition band across which the orientation changes sharply. On heating to 200 °C, the sub-grain region 2 grows into the transition region (*Figure 10.18*(*b*)) and the orientation of the new grain well developed at 300 °C is identical to the original sub-grain (*Figure 10.18*(*c*)).

With this knowledge of recrystallization the influence of several variables known to affect the recrystallization behaviour of a metal, can now be understood. Prior deformation, for example, will control the extent to which a region of the lattice is curved. The larger the deformation the more severely will the lattice be curved and, consequently, the smaller will be the size of a growing sub-grain when it acquires a large-angle boundary. This must mean that a shorter time is necessary at any given temperature for the sub-grain to become an 'active' nucleus, or conversely, that the higher the annealing temperature the quicker will this stage be reached. In some instances, heavily cold worked metals recrystallize without any significant recovery owing to the formation of strain-free cells during deformation. The importance of impurity content on recrystallization temperature is also evident from the effect impurities have on obstructing dislocation sub-boundary and grain boundary mobility.

The intragranular nucleation of strain-free grains, as discussed above, is considered as abnormal sub-grain growth, in which it is necessary to specify that some sub-grains acquire a size advantage and are able to grow at the expense of the normal sub-grains. It has been suggested that nuclei may also be formed by a process involving the rotation of individual cells so that they coalesce with neighbouring cells to produce larger cells by volume diffusion and dislocation rearrangement.

In some circumstances, intergranular nucleation is observed in which an existing grain boundary bows out under an initial driving force equal to the difference in free energy across the grain boundary. This strain-induced boundary migration was first observed by Sperry and Beck and is shown schematically in *Figure 10.19*; the boundary movement is irregular and is from a grain with low strain, i.e. large cell size, to one of larger strain and smaller cell size. For a boundary to grow in this way the strain energy difference per unit volume across the boundary must be sufficient to supply the increase to bow out a length of boundary ≈ 1 μm.

Segregation of solute atoms to, and precipitation on, the grain boundary

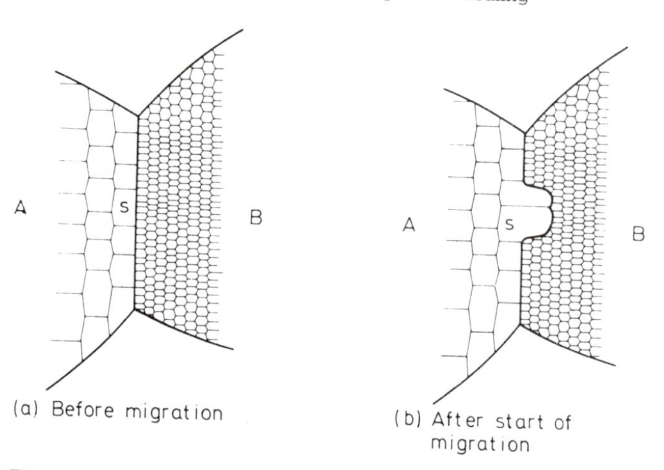

Figure 10.19 Model for strain-induced boundary migration (after Beck, *Adv. Phys.*, 1954, **3**, 245, courtesy of Taylor and Francis)

tends to inhibit intergranular nucleation and gives an advantage to intergranular nucleation, provided the dispersion is not too fine. In general, the recrystallization behaviour of two-phase alloys is extremely sensitive to the dispersion of the second phase. Small, finely dispersed particles retard recrystallization by reducing both the nucleation rate and the grain boundary mobility, whereas large coarsely dispersed particles enhance recrystallization by increasing the nucleation rate. During deformation, zones of high dislocation density and large misorientations are formed around non-deformable particles, and on annealing, recrystallization nuclei are created within these zones by a process of polygonization by sub-boundary migration. Particle-stimulated nucleation occurs above a critical particle size which decreases with increasing deformation. The finer dispersions tend to homogenize the microstructure, i.e. dislocation distribution, thereby minimizing local lattice curvature and reducing nucleation.

The formation of nuclei becomes very difficult when the spacing of second phase particles is so small that each developing sub-grain interacts with a particle before it becomes a viable nucleus. The extreme case of this is SAP (sintered aluminium powder) which contains very stable, close-spaced oxide particles. These particles prevent the rearrangement of dislocations into cell walls and their movement to form high-angle boundaries, and hence SAP must be heated to a temperature very close to the melting point before it recrystallizes.

10.5.4 Grain growth

When primary recrystallization is complete, i.e. when the growing crystals have consumed all the strained material, the metal can lower its energy further by reducing its total area of grain surface. With extensive annealing it is often found that grain boundaries straighten, small grains shrink and larger ones grow. The general phenomenon is known as grain growth, and the most important factor governing the process is the surface tension of the grain boundaries. A grain boundary has a surface tension, T (= surface free energy per unit area) because its

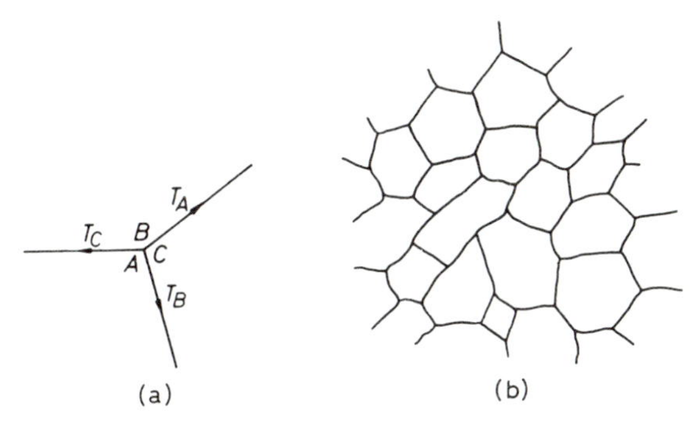

Figure 10.20 (a) Relation between angles and surface tensions at a grain boundary triple point; (b) idealized polygonal grain structure

atoms have a higher free energy than those within the grains. Consequently, to reduce this energy a polycrystal will tend to minimize the area of its grain boundaries and when this occurs the configuration taken up by any set of grain boundaries (*see Figure 10.20*) will be governed by the condition that

$$T_A/\sin A = T_B/\sin B = T_C/\sin C \tag{10.29}$$

Most grain boundaries are of the large-angle type with their energies approximately independent of orientation, so that for a random aggregate of grains $T_A = T_B = T_C$ and the equilibrium grain boundary angles are each equal to 120°. *Figure 10.20(b)* shows an idealized grain in two dimensions surrounded by others of uniform size, and it can be seen that the equilibrium grain shape takes the form of a polygon of six sides with 120° inclusive angles. All polygons with either more or less than this number of sides cannot be in equilibrium. At high temperatures where the atoms are mobile, a grain with fewer sides will tend to become smaller, under the action of the grain boundary surface tension forces, while one with more sides will tend to grow.

Second-phase particles have a major inhibiting effect on boundary migration and are particularly effective in the control of grain size. The pinning process arises from surface tension forces exerted by the particle–matrix interface on the grain boundary as it migrates past the particle. *Figure 10.21* shows that the drag exerted by the particle on the boundary, resolved in the forward direction, is

$$F = \pi r \gamma \sin 2\theta$$

where γ is the specific interfacial energy of the boundary; $F = F_{max} = \pi r \gamma$ when $\theta = 45°$. Now if there are N particles per unit volume, the volume fraction is $4\pi r^3 N/3$ and the number n intersecting unit area of boundary is given by

$$n = 3f/2\pi r^2$$

For a grain boundary migrating under the influence of its own surface tension the driving force is $2\gamma/R$, where R is the minimum radius of curvature and as the grains grow, R increases and the driving force decreases until it is balanced by the particle-drag, when growth stops. If $R \sim d$ the mean grain diameter, then the

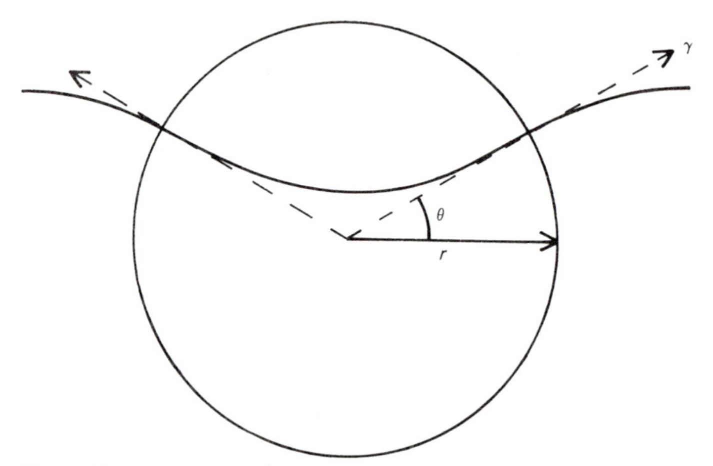

Figure 10.21 Diagram showing the drag exerted on a boundary by a particle

critical grain diameter is given by the condition

$$nF \approx 2\gamma/d_{\text{crit}}$$

or

$$d_{\text{crit}} \approx 2\gamma(2\pi r^2/3f\pi r\gamma) = 4r/3f \tag{10.30}$$

This Zener drag equation overestimates the driving force for grain growth by considering an isolated spherical grain. A heterogeneity in grain size is necessary for grain growth and taking this into account gives a revised equation

$$d_{\text{crit}} \approx \frac{\pi r}{3f}\left[\frac{3}{2} - \frac{2}{Z}\right] \tag{10.31}$$

where Z is the ratio of the diameters of growing grains to the surrounding grains. This treatment explains the successful use of small particles in refining the grain size of commercial alloys.

During the above process growth is continuous and a uniform coarsening of the polycrystalline aggregate usually occurs. Nevertheless, even after growth has finished the grain size in a specimen which was previously severely cold worked remains relatively small, because of the large number of nuclei produced by the working treatment. Exaggerated grain growth can often be induced, however, in one of two ways, namely: (i) by subjecting the specimen to a critical strain-anneal treatment, or (ii) by a process of secondary recrystallization. By applying a critical deformation (usually a few per cent strain) to the specimen the number of nuclei will be kept to a minimum, and if this strain is followed by a high temperature anneal in a thermal gradient, some of these nuclei will be made more favourable for rapid growth than others. With this technique, if the conditions are carefully controlled, the whole of the specimen may be turned into one crystal, i.e. a single crystal (*see* page 105). The term secondary recrystallization describes the process whereby a specimen which has been given a primary recrystallization treatment at a low temperature is taken to a higher temperature to enable the abnormally

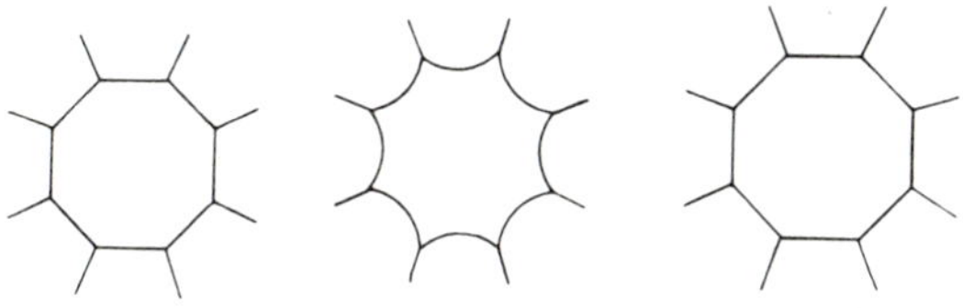

(a) (b) (c)

Figure 10.22 Grain growth during secondary recrystallization

rapid growth of a few grains to occur. The only driving force for secondary recrystallization is the reduction of grain boundary free energy, as in normal grain growth, and consequently, certain special conditions are necessary for its occurrence. One condition for this 'abnormal' growth is that normal continuous growth is impeded by the presence of inclusions, as is indicated by the exaggerated grain growth of tungsten wire containing thoria, or the sudden coarsening of deoxidized steel at about 1000 °C. A possible explanation for the phenomenon is that in some regions the grain boundaries become free (e.g. if the inclusions slowly dissolve or the boundary tears away) and as a result the grain size in such regions becomes appreciably larger than the average (*Figure 10.22(a)*). It then follows that the grain boundary junction angles between the large grain and the small ones that surround it will not satisfy the condition of equilibrium discussed above. As a consequence, further grain boundary movement to achieve 120° angles will occur, and the accompanying movement of a triple junction point will be as shown in *Figure 10.22(b)*. However, when the dihedral angles at each junction are approximately 120° a severe curvature in the grain boundary segments between the junctions will arise, and this leads to an increase in grain boundary area. Movement of these curved boundary segments towards their centres of curvature must then take place and this will give rise to the configuration shown in *Figure 10.22(c)*. Clearly, this sequence of events can be repeated and continued growth of the large grains will result.

The behaviour of the dispersed phase is extremely important in secondary recrystallization and there are many examples in metallurgical practice where the control of secondary recrystallization with dispersed particles has been used to advantage. One example is in the use of Fe–3%Si in the production of strip for transformer laminations. This material is required with (110) [001] 'Goss' texture because of the [001] easy direction of magnetization, and it is found that the presence of MnS particles favours the growth of secondary grains with the appropriate Goss texture. Another example, is in the removal of the pores during the sintering of metal and ceramic powders, such as alumina and metallic carbides. The sintering process is essentially one of vacancy creep (*see* Chapter 13) involving the diffusion of vacancies from the pore of radius r to a neighbouring grain boundary, under a driving force $2\gamma_s/r$ where γ_s is the surface energy. In practice, sintering occurs fairly rapidly up to about 95 per cent full density because there is a plentiful association of boundaries and pores. When the pores become very small, however, they are no longer able to anchor the grain boundaries against the grain growth forces, and hence the pores sinter very slowly, since they are stranded within the grains some distance from any boundary. To promote total sintering, an effective dispersion is added. The dispersion is critical, however, since it must produce sufficient drag to slow down grain growth, during which a particular pore is crossed by several migrating boundaries, but not sufficiently large to give rise to secondary recrystallization when a given pore would be stranded far from any boundary.

The relation between grain-size, temperature and strain is shown in *Figure 10.23* for both commercially pure aluminium and copper. From this diagram it is clear that either a critical strain-anneal treatment or a secondary recrystallization process may be used for the preparation of perfect strain-free single crystals.

10.5.5 Annealing twins

A prominent feature of the microstructures of most annealed f.c.c. metals and alloys is the presence of many straight-sided bands that run across grains. These bands have a twinned orientation relative to their neighbouring grain and are referred to as annealing twins. The parallel boundaries usually coincide with a (111) twinning plane with the structure coherent across it, i.e. both parts of the twin hold a single (111) plane in common.

As with formation of deformation twins, it is believed that a change in stacking sequence is all that is necessary to form an annealing twin. Such a change in stacking sequence may occur whenever a properly oriented grain boundary migrates. For example, if the boundary interface corresponds to a (111) plane, growth will proceed by the deposition of additional (111) planes in the usual stacking sequence ABCABC.... If, however, the next newly deposited layer falls into the wrong position, the sequence ABCABCB is produced which constitutes the first layer of a twin. Once a twin interface is formed, further growth may continue with the sequence in reverse order, ABCABC|BACB... until a second accident in the stacking sequence completes the twin band,

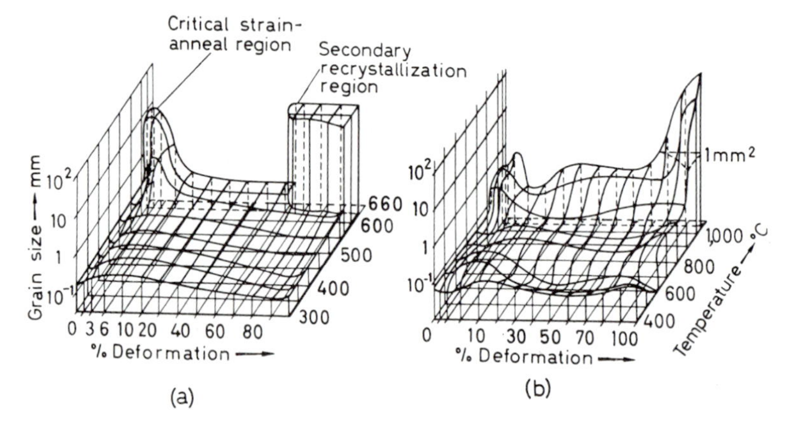

Figure 10.23 Relation between grain size, deformation and temperature for (a) aluminium and (b) copper (after Buergers, *Handbuch der Metallphysik*, by courtesy of Akadamie-Verlags-gesellschaft)

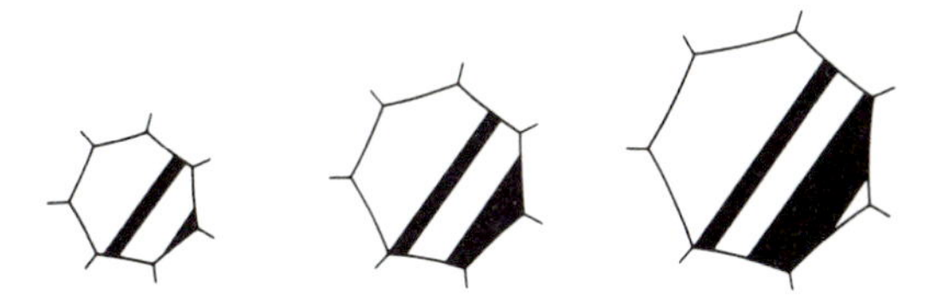

Figure 10.24 Formation and growth of annealing twins (from Burke and Turnbull, *Progress in Metal Physics 3*, 1952, courtesy of Pergamon Press)

ABCABCBACBACBABC. When a stacking error, such as that described above, occurs the number of nearest neighbours is unchanged, so that the ease of formation of a twin interface depends on the relative value of the interface energy. If this interface energy is low, as in copper where $\gamma_{gb} \gtrsim 10\gamma_{sf} \gtrsim 2\gamma_{\text{twin}}$, twinning occurs frequently while if it is high, as in aluminium, the process is rare.

Annealing twins are rarely, if ever, found in cast metals because grain boundary migration is negligible during casting. On the other hand, while worked and annealed metals show considerable twin band formation, the best examples are observed after extensive grain growth has occurred. Thus, metallography shows that a coarse grained metal often contains twins which are many times wider than any grain that was present shortly after recrystallization. This indicates that twin bands grow in width, during grain growth, by migration in a direction perpendicular to the (111) composition plane, and one mechanism whereby this can occur is illustrated schematically in *Figure 10.24*. This shows that a twin may form at the corner of a grain, since the grain boundary configuration will then have a lower interfacial energy. If this happens the twin will then be able to grow in width because one of its sides forms part of the boundary of the growing grain. Such a twin will continue to grow in width until a second mistake in the positioning of the atomic layers terminates it; a complete twin band is then formed. In copper and its alloys $\gamma_{\text{twin}}/\gamma_{\text{gb}}$ is low and hence twins occur frequently, whereas in aluminium the corresponding ratio is very much higher and so twins are rare.

Twins may develop according to the model shown in *Figure 10.25* where during grain growth a grain contact is established between grains C and D. Then if the orientation of grain D is close to the twin orientation of grain C, the nucleation of an annealing twin at the grain boundary, as shown in *Figure 10.25(d)*, will lower the total boundary energy. This follows because the twin/D interface will be reduced to about 5 per cent of the normal grain boundary energy, the energies of the C/A and twin/A interface will be approximately the same, and the extra area of interface C/twin has only a very low energy. This model indicates that the number of twins per unit grain boundary area only depends on the

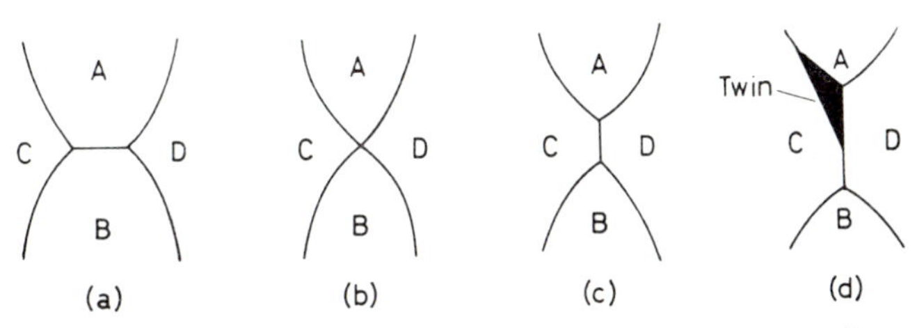

Figure 10.25 Nucleation of an annealing twin during grain growth

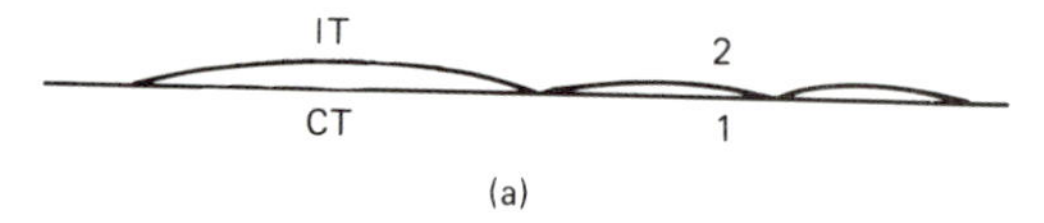

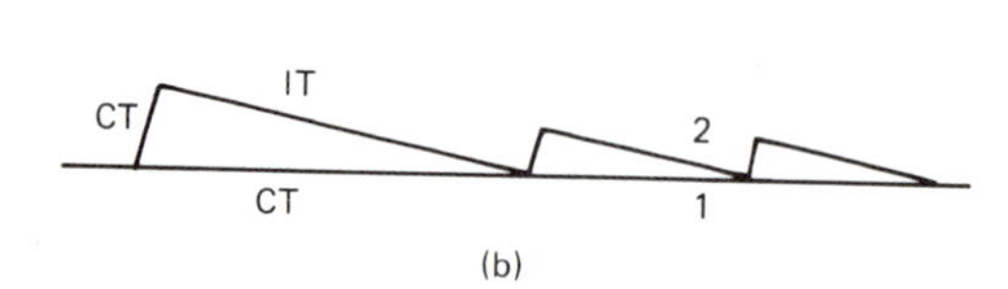

Figure 10.26 Possible mechanism for dissociation of any boundary. (a) Immediate products of dissociation are a coherent twin boundary (CT) and an almost parallel incoherent twin boundary (IT), (b) incoherent twin boundary facets to minimize total free energy (after Goodhew, *Metal Science*, March–April 1979)

number of new grain contacts made during grain growth, irrespective of grain size and annealing temperature.

Electron microscope observations on specially prepared grain boundaries, e.g. near $\Sigma 9$ and $\Sigma 11$, have shown that twins can form when boundaries dissociate to lower their energy. A possible mechanism for the dissociation of any boundary is illustrated in *Figure 10.26*. In a polycrystal most boundaries are within a degree or so of an acceptable CSL and a large fraction of them are geometrically capable of dissociating into a twin and lower Σ. The structure of boundaries, as discussed in Chapter 8, section 8.9 is an active area of experimental investigation.

10.5.6 Recrystallization textures

The preferred orientation developed by cold work often changes on recrystallization to a totally different preferred orientation. To explain this observation, Barrett and later Beck have put forward the 'oriented growth' theory of recrystallization textures in which it is proposed that nuclei of many orientations initially form but, because the rate of growth of any given nucleus depends on the orientation difference between the matrix and growing crystal, the recrystallized texture will arise from those nuclei which have the fastest growth rate in the cold worked matrix, i.e. those bounded by large-angle boundaries. It then follows that because the matrix has a texture, all the nuclei which grow will have orientations that differ by 30–40° from the cold worked texture. This explains why the new texture in f.c.c. metals is often related to the old texture, by a rotation of approximately 30–40° around $\langle 111 \rangle$ axes, in b.c.c. metals by 30° about $\langle 110 \rangle$ and in h.c.p. by 30° about $\langle 0001 \rangle$. However, while it is undoubtedly true that oriented growth provides a selection between favourable and unfavourably oriented nuclei, there are many observations to indicate that the initial nucleation is not entirely random. For instance, because of the crystallographic symmetry one would expect grains appearing in a f.c.c. texture to be related to rotations about all four $\langle 111 \rangle$ axes, i.e. eight orientations arising from two possible rotations about each of the four $\langle 111 \rangle$ axes. All these possible orientations are rarely, if ever, observed.

To account for such observations, and for those cases where the deformation texture and the annealing texture show strong similarities, oriented nucleation is considered to be important. The oriented nucleation theory assumes that the selection of orientations is determined in the nucleation stage. It is generally accepted that all recrystallization nuclei pre-exist in the deformed matrix, as sub-grains, which become more perfect through recovery processes prior to recrystallization. It is thus most probable that there is some selection of nuclei determined by the representation of the orientations in the deformation texture, and that the oriented nucleation theory should apply in some cases. In many cases the orientations which are strongly represented in the annealing texture are very weakly represented in the deformed material. The most striking example is the 'cube' texture, (100) [001], found in most f.c.c. pure metals which have been annealed following heavy rolling reductions. In this texture, the cube axes are extremely well aligned along the sheet axes, and its behaviour resembles that of a single crystal. It is thus clear that cube-oriented grains or sub-grains, must have a very high initial growth rate in order to form the remarkably strong quasi-single crystal cube texture. The percentage of cubically aligned grains increases with increased deformation, but the sharpness of the textures is profoundly affected by alloying additions. The amount of alloying addition required to suppress the texture depends on those factors which affect the stacking fault energy, such as the lattice misfit of the solute atom in the solvent lattice, valency etc, in much the same way as that described for the transition of a pure metal deformation texture.

In general, however, if the texture is to be altered a distribution of second-phase must either be present before cold rolling or be precipitated during annealing. In aluminium, for example, the amount of cube texture can be limited in favour of retained rolling texture by limiting the amount of grain growth with a precipitate dispersion of Si and Fe. By balancing the components, earing can be minimized in drawn aluminium cups. In aluminium-killed steels AlN precipitation prior to recrystallization produces a higher proportion of grains with $\{111\}$ planes parallel to the rolling plane and a high $\bar{R}$ value suitable for deep drawing (*see* page 356). The AlN dispersion affects sub-grain growth, limiting the available nuclei and increasing the orientation-selectivity, thereby favouring the high-energy $\{111\}$ grains. Improved $\bar{R}$-values in steels in general are probably due to the combined effect of particles in homogenizing the deformed micro-structure and in controlling the subsequent sub-grain growth. The overall effect is to limit the availability of nuclei with orientations other than $\{111\}$.

Suggestions for further reading

American Society for Metals, *Recrystallization, Grain Growth and Texture*, Metals Park, Ohio, 1967

Backofen, W.A., *Fundamentals of Deformation Processing*, Syracuse University Press, 1964

Dislocations and Properties of Real Materials, 50th Anniversary Conf., Metals Society, 1984

Haasen, P., Gerdd, V. and Kostorz, G. (eds), *Int. Conf. Strength of Metals and Alloys*, 1979

Honeycombe, R.W.K., *The Plastic Deformation of Metals*, Edward Arnold, 1984

National Physical Laboratory, *Symposium: Relations between the Structures and Properties of Metals*, Teddington, 1963

Martin, J.W. and Doherty, R.D., *Stability of Microstructure*, Cambridge Solid State Series, 1976

Martin, J.W., *Micromechanisms in Particle-Hardened Alloys*, C.U.P., 1980

Physics of Metals, Vol. 2, C.U.P., 1975

Chapter 11

Phase transformations I – precipitation hardening transformation

11.1 Introduction

The production of a material which possesses considerable strength at both room and elevated temperatures is of great practical importance. We have already seen how alloying and cold working can give rise to an increased yield stress, but in certain alloy systems it is possible to produce an additional increase in hardness by heat treatment alone. Such a method has many advantages, since the required strength can be induced at the most convenient production stage of the heat treatment and fabrication schedule and, moreover, the component is not sent into service in a highly stressed, plastically deformed state.

The basic requirement for such a special alloy is that it should undergo a phase transformation in the solid state. One type of alloy satisfying this requirement, already considered, is that which has an order–disorder reaction, and the hardening accompanying this process (similar in many ways to precipitation hardening) is termed order hardening. The conditions for this form of hardening are, however, quite stringent, so that the two principal hardening reactions commonly used are (a) precipitation from super-saturated solid solution, and (b) eutectoid decomposition. In this chapter we shall deal with the first of these phase transformations.

11.2 Precipitation from supersaturated solid solution

The basic requirements of a precipitation-hardening alloy system is that the solid solubility limit should decrease with decreasing temperature as shown in *Figure 11.1* for the Al–Cu system. Here an alloy exists as a homogeneous α-solid solution at high temperatures, but on cooling becomes saturated with respect to a second phase, θ; at lower temperatures the θ-phase separates out and lattice hardening often results. During the precipitation-hardening heat-treatment procedure the alloy is first solution heat-treated at the high temperature and then rapidly cooled by quenching into water or some other cooling medium. The rapid cooling suppresses the separation of the θ-phase so that the alloy exists at the low temperature in an unstable supersaturated state. If, however, after quenching, the

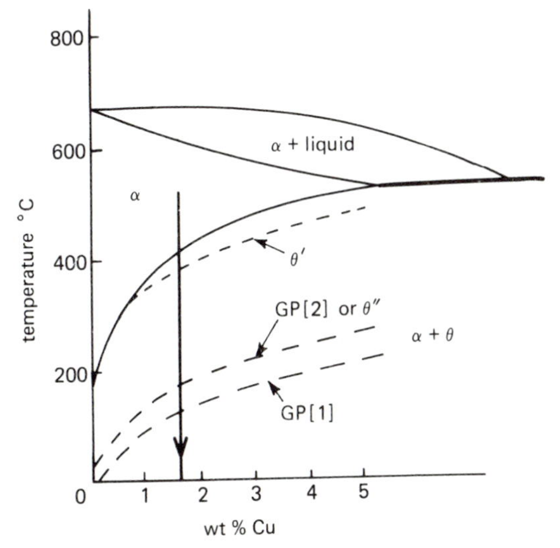

Figure 11.1 Al-rich Al–Cu binary diagram showing GP [1], θ'' and θ' solvus lines (dotted)

alloy is allowed to 'age' for a sufficient length of time, the second phase precipitates out. This precipitation occurs by a nucleation and growth process, fluctuations in solute concentration providing small clusters of atoms in the lattice which act as nuclei for the precipitate. The rate at which these nuclei grow is contolled by the rate of atomic migration, so that precipitation increases with increasing ageing temperature. However, the size of the precipitate becomes finer as the temperature at which precipitation occurs is lowered, and extensive hardening of the alloy is associated with a critical dispersion of the precipitate. If, at any given temperature, ageing is allowed to proceed too far, coarsening of the particles occurs (i.e. the small ones tend to re-dissolve, and the large ones to grow still larger as discussed in section 11.1.8), and the numerous finely dispersed, small particles are gradually replaced by a smaller number of more widely dispersed, coarser particles. In this state the alloy becomes softer, and it is then said to be in the over-aged condition (*see Figure 11.2*).

11.3 Changes in properties accompanying precipitation

The actual quenching treatment gives rise to small changes in many of the mechanical and physical properties of alloys because both solute atoms and point defects in excess of the equilibrium concentration are retained during the process, and because the quench itself often produces lattice strains. Perhaps the property most markedly affected is the electrical resistance and this is usually considerably increased. In contrast, the mechanical properties are affected relatively much less.

On ageing, the change in properties in a quenched material is more marked and, in particular, the mechanical properties often show striking modifications.

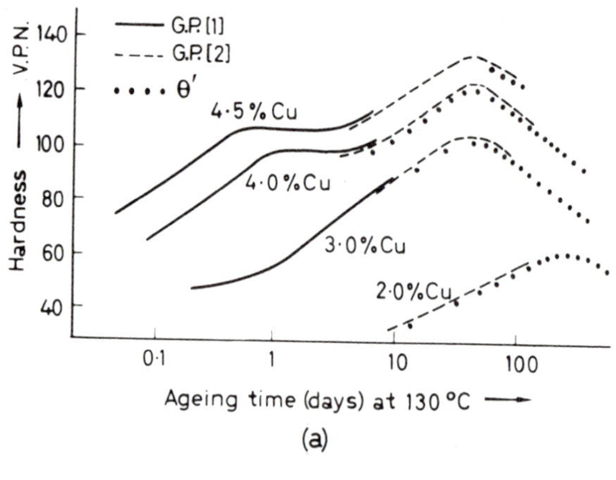

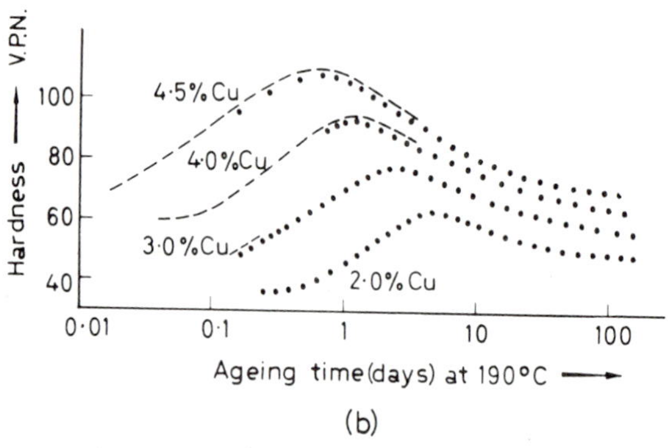

Figure 11.2 The ageing of aluminium–copper alloys at (a) 130°C and (b) at 190°C (after Silcock, Heal and Hardy, *J. Inst. Metals,* 1953–4, **82,** 239)

For example, the tensile strength of duralumin (i.e. an aluminium–4 per cent copper alloy containing magnesium, silicon and manganese) may be raised from 0.21 to 0.41 GN/m^2, while that of a copper–2 per cent beryllium alloy may be increased from 0.46 to 1.23 GN/m^2. The structure-sensitive properties such as hardness, yield stress, etc. are, of course, extremely dependent on the structural distribution of the phases and, consequently, such alloys usually exhibit softening as the finely dispersed precipitates coarsen.

A simple theory of precipitation, involving the nucleation and growth of particles of the expected new equilibrium phase, leads one to anticipate that the alloy would show a single hardening peak, the electrical resistivity a decrease, and the lattice parameter an increase (assuming the solute atom is smaller than the solvent atom) as the solute is removed from solution. Such property changes are found in practice, but only at low supersaturations and high ageing temperatures. At higher supersaturations and lower ageing temperatures the various property changes are not consistent with such a simple picture of precipitation; the alloy may show two or more age-hardening peaks, and the electrical resistivity and lattice parameter may not change in the anticipated manner. A hardening process which takes place in two stages is shown in aluminium–copper alloys (*Figure*

11.2(a)) where the initial hardening occurs without any attendant precipitation being visible in the light microscope and, more over, is accompanied by a decrease in conductivity and no change in lattice parameter. Such anomalous behaviour has led to theories which envisage precipitation as a process involving more than one stage, since the evidence suggests that some form of precipitate other than the formation of the stable new phase (e.g. $CuAl_2$ in aluminium–copper alloys) is occurring. The initial stage of precipitation, at the lower ageing temperatures, is believed to involve a clustering of solute atoms on the solvent lattice to form zones or clusters coherent with the matrix; the zones cannot be seen in the light microscope and for this reason this stage was at one time termed pre-precipitation. At a later stage of the ageing process it would be expected that these clusters would break away from the matrix lattice to form distinct particles with their own crystal structure and a definite interface. These hypotheses were confirmed originally by structural studies using x-ray diffraction techniques; the so-called pre-precipitation effects can now be observed directly in the electron microscope.

Even though clustering occurs, the general kinetic behaviour of the precipitation process is in agreement with that expected on thermodynamic grounds. From *Figure 11.2(a)* and *(b)* it is evident that the rate of ageing increases markedly with increasing temperature while the peak hardness decreases. Two-stage hardening takes place at low ageing temperatures and is associated with high maximum hardness, while single-stage hardening occurs at higher ageing temperatures, or at lower ageing temperatures for lower solute contents.

Another phenomenon commonly observed in precipitation-hardening alloys is reversion or retrogression. If an alloy hardened by ageing at low temperature is subsequently heated to a higher ageing temperature it softens temporarily, but becomes harder again on more prolonged heating. This temporary softening, or reversion of the hardening process, occurs because the very small nuclei or zones precipitated at the low temperature are unstable when raised to the higher ageing temperature, and consequently they re-dissolve and the alloy becomes softer; the temperature above which the nuclei or zones dissolve is known as the solvus temperature; *Figure 11.1* shows the solvus temperatures for GP zones, θ'', θ' and θ. On prolonged ageing at the higher temperature larger nuclei, characteristic of that temperature, are formed and the alloy again hardens. Clearly, the reversion process is reversible, provided re-hardening at the higher ageing temperature is not allowed to occur.

11.4 Structural changes

Early metallographic investigations showed that the microstructural changes which occur during the initial stages of ageing are on too fine a scale to be resolved by the light microscope, yet it is in these early stages that the most profound changes in properties are found. Accordingly, to study the process, it is necessary to employ the more sensitive and refined techniques of x-ray diffraction and electron microscopy.

The two basic x-ray techniques, important in studying the regrouping of

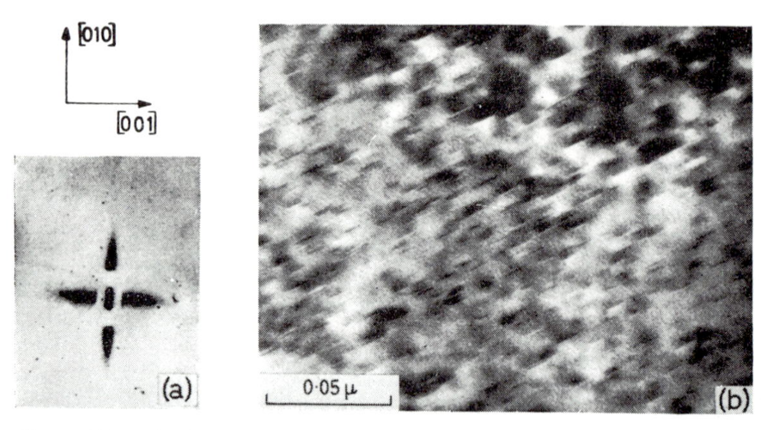

Figure 11.3 (a) Small-angle x-ray pattern from aluminium–4 per cent copper single crystal taken with molybdenum Kα radiation at a sample to film distance of 4 cm (after Guinier and Fournier, *Small-angle scattering of x-rays,* courtesy of John Wiley & Sons) (b) electron micrograph of aluminium–4 per cent copper aged 16 hours at 130°C, showing G. P. [1] zones (after Nicholson, Thomas and Nutting, *J. Inst. Metals,* 1958–59, **87,** 431)

atoms during the early stages of ageing, depend on the detection of radiation scattered away from the main diffraction lines or spots (see Chapter 2). In the first technique, developed independently by Guinier and Preston in 1938, the Laue method is used. They found that the single crystal diffraction pattern of an aluminium–copper alloy developed streaks extending from an aluminium lattice reflection along $\langle 100\rangle_{Al}$ directions. This was attributed to the formation of copper-rich regions of plate-like shape on $\{100\}$ planes of the aluminium matrix (now called Guinier–Preston zones or GP zones). The net effect of the regrouping is to modify the scattering power of, and spacing between, very small groups of $\{100\}$ planes throughout the crystal. However, being only a few atomic planes thick, the zones produce the diffraction effect typical of a two dimensional lattice, i.e. the diffraction spot gives way to a diffraction streak. In recent years the Laue method has been replaced by a single crystal oscillation technique employing monochromatic radiation, since interpretation is made easier if the wavelength of the x-rays used is known. The second technique makes use of the phenomenon of scattering of x-rays at small angles (*see* Chapter 2). Intense small-angle scattering can often be observed from age-hardening alloys (as shown in *Figures 11.3* and *11.5*) because there is usually a difference in electron density between the precipitated zone and the surrounding matrix. However, in alloys such as aluminium–magnesium or aluminium–silicon the technique is of no value because in these alloys the small difference in scattering power between the aluminium and silicon or magnesium atoms, respectively, is insufficient to give rise to appreciable scattering at small angles.

With the advent of the electron microscope the ageing of aluminium alloys was one of the first subjects to be investigated with the thin-foil transmission method.

Not only can the detailed structural changes which occur during the ageing process be followed, but electron diffraction pictures taken from selected areas of the specimen while it is still in the microscope enable further important information on the structure of the precipitated phase to be obtained. Moreover, under some conditions the interaction of moving dislocations and precipitates can be observed. This naturally leads to a more complete understanding of the hardening mechanism.

Both the x-ray and electron-microscope techniques show that in virtually all age-hardening systems the initial precipitate is not the same structure as the equilibrium phase. Instead, an ageing sequence: zones → intermediate precipitates → equilibrium precipitate is followed. This sequence occurs because the equilibrium precipitate is incoherent with the matrix, whereas the transition structures are either fully coherent, as in the case of zones, or at least partially coherent. Then, because of the importance of surface energy and strain energy factors (*see* Chapter 4) to the precipitation process, the system follows such a sequence in order to have the lowest free energy in all stages of precipitation. The surface energy term dominates the process of nucleation when the interfacial energy is large (i.e. when there is a discontinuity in atomic structure, somewhat like a grain boundary, at the interface between the nucleus and the matrix), so that for the incoherent type of precipitate the nuclei must exceed a certain minimum size before they can nucleate a new phase. To avoid such a slow mode of precipitation a coherent type of precipitate is formed instead, for which the size effect is relatively unimportant. The condition for coherence usually requires the precipitate to strain its equilibrium lattice to fit that of the matrix, or to adopt a metastable lattice. However, in spite of both a higher volume free energy and a higher strain energy, the transition structure is more stable in the early stages of precipitation because of its lower interfacial energy.

When the precipitate does become incoherent the alloy will, nevertheless, tend to reduce its surface energy as much as possible, by arranging the orientation relationship between the matrix and the precipitate so that the crystal planes which are parallel to, and separated by, the bounding surface have similar atomic spacings. Clearly, for these habit planes, as they are called, the better the crystallographic match the less will be the distortion at the interface, and the lower the surface energy. This principle governs the precipitation of many alloy phases, as shown by the frequent occurrence of the Widmanstätten structure, i.e. plate-shaped precipitates lying along prominent crystallographic planes of the matrix. It will be remembered from Chapter 4 that most precipitates are plate-shaped because the strain energy factor is least for this form.

The existence of a precipitation sequence is reflected in the ageing curves and, as we have seen in *Figure 11.2*, often leads to two stages of hardening. The zones by definition are coherent with the matrix, and as they form the alloy becomes harder. The intermediate precipitate may be coherent with the matrix, in which case a further increase of hardness occurs, or only partially coherent, when either hardening or softening may result. The equilibrium precipitate is incoherent and its formation always leads to softening. These features are best illustrated by a consideration of some actual age-hardening systems.

11.5 Some common precipitation systems

A consideration of the rules of alloying (Chapter 5) shows that conditions are particularly favourable for the occurrence of phase precipitation in aluminium-based alloys. Accordingly, both wrought and cast aluminium-based precipitation-hardening alloys are manufactured, and although the alloys aluminium–copper, aluminium–magnesium, aluminium–silicon and aluminium–zinc are the most common systems, enhanced properties are often found in ternary and quaternary systems, e.g. aluminium–magnesium–silicon, and aluminium–zinc–magnesium. Precipitation-hardening systems also occur with other base metals, notably copper, nickel and iron, some of which are listed in *Table 11.1*.

11.5.1 Aluminium–copper

The aluminium–copper alloy system exhibits the greatest number of intermediate stages in its precipitation process, and consequently is probably the most widely studied. When the copper content is high and the ageing temperature low, the sequence of stages followed, as shown in *Figure 11.2*, is GP [1], GP [2], θ' and θ ($CuAl_2$). On ageing at higher temperatures, however, one or more of these intermediate stages may be omitted and, as shown in *Figure 11.2*, corresponding differences in the hardness curves can be detected. The early stages of ageing are due to GP [1] zones, which are interpreted as plate-like clusters of copper atoms

TABLE 11.1 Some common precipitation hardening systems

Base metal	*Solute*	*Transition structure*	*Equilibrium precipitate*
Al	Cu	(i) Plate-like solute rich GP [1] zones on $\{100\}_{Al}$; (ii) ordered zones of GP [2]; (iii) θ'-phase (plates)	θ-$CuAl_2$
	Ag	(i) Spherical solute-rich zones; (ii) platelets of hexagonal γ' on $\{111\}_{Al}$.	γ-Ag_2Al
	Mg, Si	(i) GP zones rich in Mg and Si atoms on $\{100\}_{Al}$ planes; (ii) ordered zones of β'.	β-Mg_2Si (plates)
	Mg, Cu	(i) GP zones rich in Mg and Cu atoms on $\{100\}_{Al}$ planes; (ii) S′ platelets on $\{021\}_{Al}$ planes.	S-Al_2CuMg (laths)
	Mg, Zn	(i) Spherical zones rich in Mg and Zn (ii) platelets of M′ hexagonal phase on $\{111\}_{Al}$.	M-Mg_2Zn (plates or rods)
Cu	Be	(i) Be-rich regions on $\{100\}_{Cu}$ planes; (ii) γ'.	γ-CuBe
	Co	Spherical G.P. zones	β-Co plates
Fe	C	(i) Martensite (α'); (ii) martensite (α''); (iii) ε-carbide.	Fe_3C plates cementite
	N	(i) Nitrogen martensite (α'); (ii) martensite (α'') discs.	Fe_4N
Ni	Cr, M, Ti	γ' cubes.	γ-$Ni_3(AlTi)$

segregated on to {100} planes of the aluminium matrix. A typical small-angle x-ray scattering pattern and thin-foil transmission electron micrograph from GP [1] zones are shown in *Figure 11.3*. The plates are only a few atomic planes thick (giving rise to the ⟨100⟩ streaks in the x-ray pattern), but are about 10 nm long, and hence appear as bright or dark lines on the electron micrograph.

GP [2] is best described as a coherent intermediate precipitate rather than a zone, since it has a definite crystal structure; for this reason the symbol θ'' is often preferred. These precipitates, usually of maximum thickness 10 nm (100 Å) and up to 150 nm (1500 Å) diameter, have a tetragonal structure which fits perfectly with the aluminium unit cell in the *a* and *b* directions but not in the *c*. The structure postulated has a central plane which consists of 100 per cent copper atoms, the next two planes a mixture of copper and aluminium and the other two basal planes of pure aluminium, giving an overall composition of $CuAl_2$. Because of their size θ'' precipitates are easily observed in the electron microscope, and because of the ordered arrangements of copper and aluminium atoms within the structure their presence gives rise to intensity maxima on the diffraction streaks in an x-ray photograph. Since the *c* parameter 0.78 nm (7.8 Å) differs from that of aluminium 0.404 nm (4.04 Å) the aluminium planes parallel to the plate are distorted by elastic coherency strains. Moreover, the precipitate grows with the *c* direction normal to the plane of the plate, so that the strain fields become larger as it grows and at peak hardness extend from one precipitate particle to the next, as can be seen in *Figure 11.4(a)*. The direct observation of coherency strains confirms the theories of hardening based on the development of an elastically strained matrix (*see* next section).

The transition structure θ' is tetragonal; the true unit cell dimensions are $a=0.404$ and $c=0.58$ nm and the axes are parallel to $\langle 100 \rangle_{Al}$ directions. The strains around the θ' plates can be relieved, however, by the formation of a stable dislocation loop around the precipitate and such a loop has been observed around small θ' plates in the electron microscope as shown in *Figure 11.4(b)*. The long range strain fields of the precipitate and its dislocation largely cancel. Consequently, it is easier for glide dislocations to move through the lattice of the alloy containing an incoherent precipitate such as θ' than a coherent precipitate such as θ'', and the hardness falls.

The θ structure is also tetragonal, with $a=0.606$ and $c=0.487$ nm. This equilibrium precipitate is incoherent with the matrix and its formation always leads to softening, since coherency strains disappear.

11.5.2 Aluminium–silver

Investigations using x-ray diffraction and electron microscopy have shown the existence of three distinct stages in the age-hardening process, which may be summarized: silver-rich clusters → intermediate hexagonal γ' → equilibrium hexagonal γ. The hardening is associated with the first two stages in which the precipitate is coherent and partially coherent with the matrix, respectively.

During the quench and in the early stages of ageing, silver atoms cluster into small spherical aggregates and a typical small-angle x-ray picture of this stage,

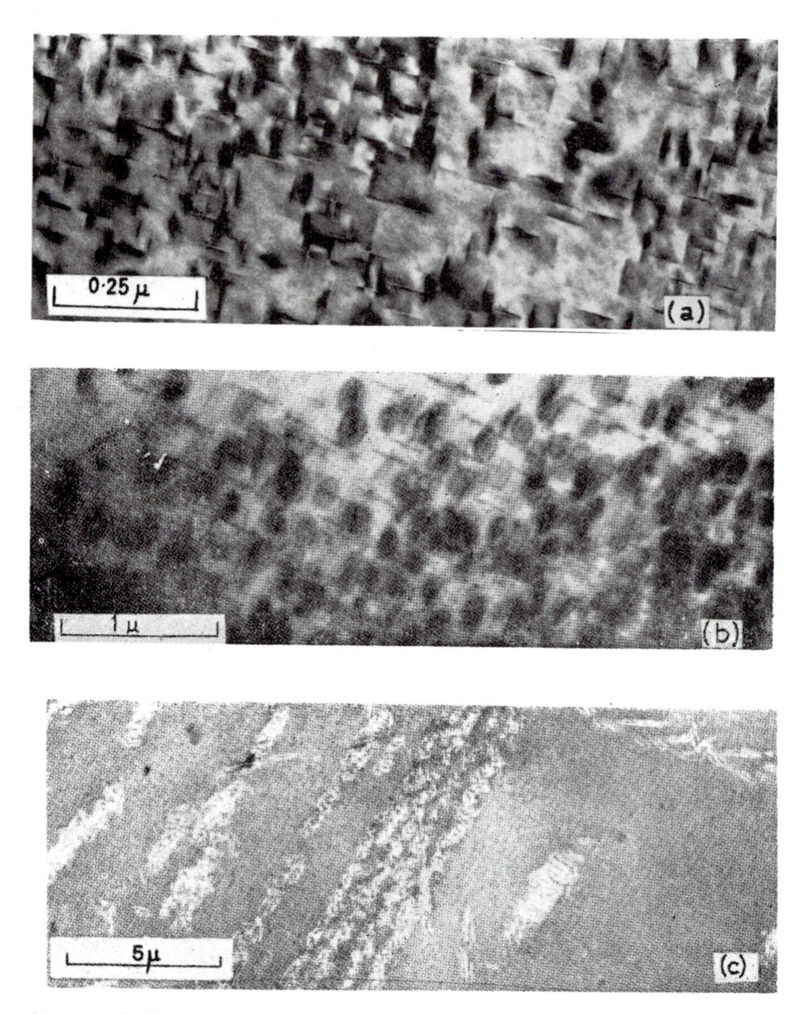

Figure 11.4 Electron micrographs from aluminium–4 per cent copper (a) aged 5 hours at 160°C showing θ″ plates, (b) aged 12 hours at 200°C, showing a dislocation ring roung θ″ plates, (c) aged 3 days at 160°C showing θ″ precipitated on helical dislocations (after Nicholson, Thomas and Nutting, *J. Inst. Metals,* 1958–59, **87,** 433)

shown in *Figure 11.5(a)*, has a diffuse ring surrounding the trace of the direct beam. The absence of intensity in the centre of the ring, i.e. at (000), is attributed to the fact that clustering takes place so rapidly that there is left a shell-like region surrounding each cluster which is low in silver content. On ageing, the clusters grow in size and decrease in number, and this is characterized by the x-ray pattern showing a gradual decrease in ring diameter. The concentration and size of clusters can be followed very accurately by measuring the intensity distribution across the ring as a function of ageing time. This intensity may be represented (*see* Chapter 2) by an equation of the form

$$I(\varepsilon)=Mn^2[\exp(-2\pi^2R^2\varepsilon^2/3\lambda^2)-\exp(-2\pi^2R_1^2\varepsilon^2/3\lambda^2)]^2 \qquad (11.1)$$

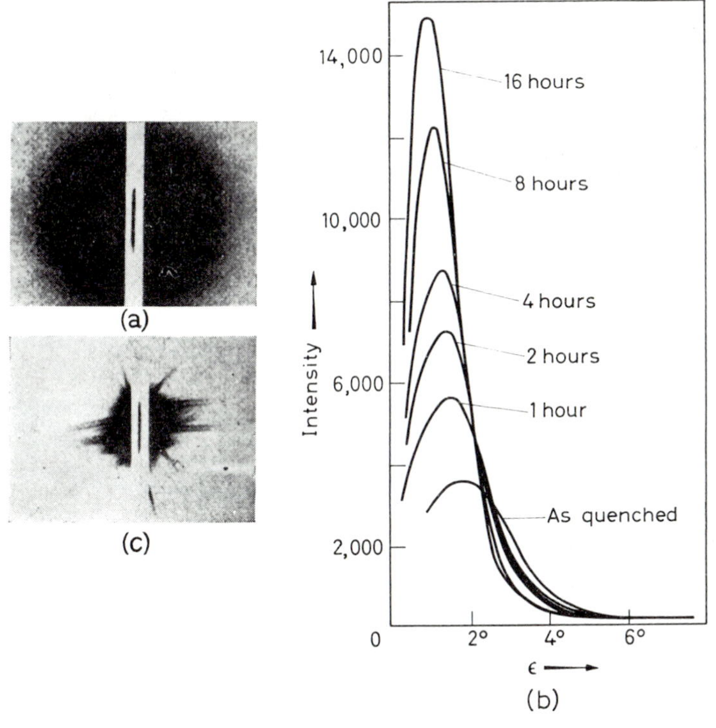

Figure 11.5 (a) Small-angle scattering of copper Kα radiation by polycrystalline aluminium–20 per cent silver in various stages of age hardening; x-ray photograph taken after quenching from 520°C (after Guinier and Walker, *Acta Metall.* 1953, **1,** 570), (b) showing the change in ring intensity and ring radius on ageing at 120°C (after Smallman and Westmacott, unpublished), (c) x-ray photograph taken after ageing at 140°C for 10 days (after Guinier and Walker)

and for values of ε greater than that corresponding to the maximum intensity, the contribution of the second term, which represents the denuded region surrounding the cluster, can be neglected. *Figure 11.5*(*b*) shows the variation in the x-ray intensity, scattered at small angles (SAS) with cluster growth, on ageing an aluminium–silver alloy at 120 °C. An analysis of this intensity distribution, using equation 11.1, indicates that the size of the zones increases from 2 to 5 nm (20 Å to 50 Å) in just a few hours at 120 °C. These zones may, of course, be seen in the electron microscope and *Figure 11.6*(*a*) is an electron micrograph showing spherical zones in an aluminium–silver alloy aged 5 hours at 160 °C; the diameter of the zones is about 10 nm (100 Å), in good agreement with that deduced by x-rays.

The shape of the zone formed in aluminium–silver alloys is different from that formed in aluminium–copper alloys, and to account for this it has been suggested that the zone shape is dependent upon the relative diameters of solute and solvent atoms. Thus, solute atoms such as silver and zinc which have atomic sizes similar to aluminium give rise to spherical zones, whereas solute atoms such as copper which have a high misfit in the solvent lattice form plate-like zones.

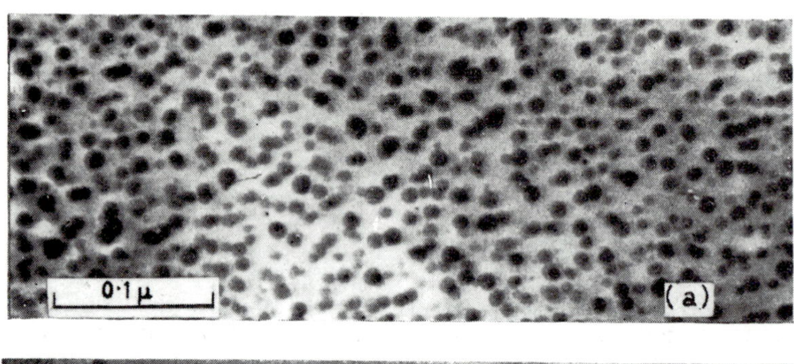

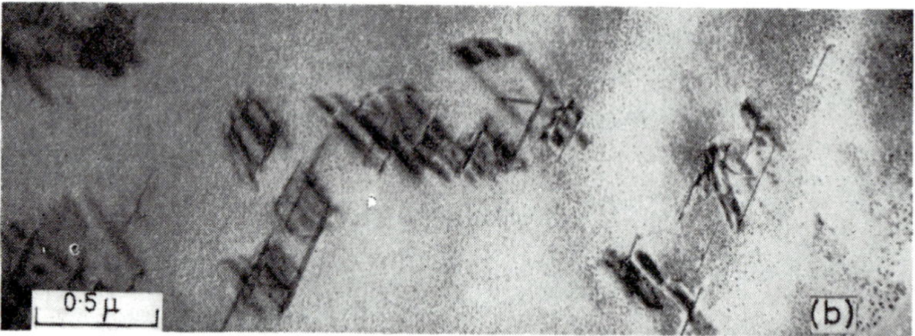

Figure 11.6 Electron micrographs from aluminium–silver alloy (a) aged 5 hours at 160°C showing spherical zones, and (b) aged 5 days at 160°C showing γ′ precipitate (after Nicholson, Thomas and Nutting, *J. Inst. Metals,* 1958–59, **87,** 431)

With prolonged annealing, the formation and growth of platelets of a new phase, γ', occur. This is characterized by the appearance in the x-ray pattern of short streaks passing through the trace of the direct beam (*Figure 11.5*(*c*)). The γ' platelet lies parallel to the $\{111\}$ planes of the matrix and its structure has lattice parameters very close to that of aluminium. However, the structure is hexagonal and, consequently, the precipitates are easily recognizable in the electron microscope by the stacking fault contrast within them, as shown in *Figure 11.6*(*b*). Clearly, these precipitates are never fully coherent with the matrix, but, nevertheless, in this alloy system, where the zones are spherical and have little or no coherency strain associated with them, and where no coherent intermediate precipitate is formed, the partially coherent γ' precipitates do provide a greater resistance to dislocation movement than zones and a second stage of hardening results.

The measurements of hardness, conductivity and other properties, indicate the presence of two stages in the precipitation process; termed cold hardening and warm hardening, since the latter stage, when γ' is formed, predominates at elevated temperatures. SAS shows that the length of anneal necessary for the appearance of the γ' streaks in the diffraction pattern varies with temperature, as follows:

T °C	*20*	*140*	*165*	*255*	*300*	*342*	*405*
Time	not observed after 3 months	100 h	22 h	18 min	3 min	1 min	15 s

Thus, it is concluded that cold hardening can be attributed to zones and that the appearance of the γ'-phase may be correlated directly to the appearance of the warm-hardening stage.

11.5.3 Complex systems

The same principles apply to the constitutionally more complex ternary and quaternary alloys as to the binary alloys. Spherical zones are found in aluminium–magnesium–zinc alloys as in aluminium–zinc, although the magnesium atom is some 12 per cent larger than the aluminium atom. The intermediate precipitate M' forms on the $\{111\}_{Al}$ planes, but it is only partially coherent with the matrix and electron micrographs show that there is no strain field associated with it. Hence, the strength of the alloy is due purely to dispersion hardening, and the alloy softens as the precipitate becomes coarser.

11.5.4 Nickel–chromium–aluminium

In nickel-based alloys the hardening phase is the ordered γ'-Ni_3Al; this γ' is an equilibrium second phase in both the binary Ni–Al and Ni–Cr–Al systems and a metastable phase in the Ni–Ti and Ni–Cr–Ti systems. These systems form the basis of the 'superalloys' which owe their properties to the close matching of the γ' and the f.c.c. matrix. The two phases have very similar lattice parameters ($\lesssim 0.25$ per cent, depending on composition) and the coherency (interfacial energy $\gamma_I \approx 10$–$20\,mJ/m^2$) confers a very low coarsening rate on the precipitate so that the alloy overages extremely slowly even at $0.7T_m$. To produce improved hot corrosion resistance the Cr content is increased and because of the reduced solubility for Al and Ti, an additional strengthener, Nb, is added. Additions of Cr, like Co, also increase the γ' solvus and lower the stacking fault energy. In alloys containing Nb, a metastable Ni_3Nb phase occurs but although ordered and coherent, it is less stable than γ' at high temperatures.

11.6 Mechanisms of hardening

The strength of an age-hardening alloy is governed by the interaction of moving dislocations and precipitates. The obstacles in precipitation-hardening alloys which hinder the motion of dislocations may be either (1) the strains around GP zones, or (2) the zones or precipitates themselves, or both. Clearly, if it is the zones themselves which are important, it will be necessary for the moving dislocations either to cut through them or go round them. Thus, merely from elementary reasoning, it would appear that there are at least three causes of hardening, namely; (1) coherency strain hardening, (2) chemical hardening, i.e. when the dislocation cuts through the precipitate, or (3) dispersion hardening, i.e. when the dislocation goes round or over the precipitate.

The relative contributions will depend on the particular alloy system but, generally, there is a critical dispersion at which the strengthening is a maximum, as shown in *Figure 11.7*. In the small particle regime the precipitates, or particles,

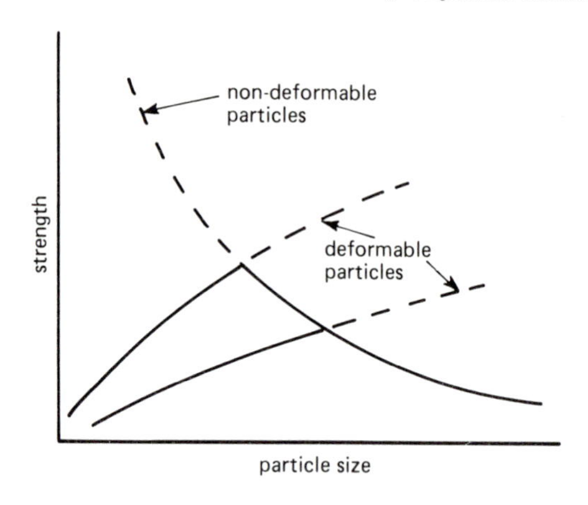

Figure 11.7 Variation of strength with particle size, defining the deformable and non-deformable particle regimes

are coherent and deformable as the dislocations cut through them, while in the larger particle regime the particles are incoherent and non-deformable as the dislocations bypass them. For deformable particles, when the dislocations pass through the particle, the intrinsic properties of the particle are of importance and alloy strength varies only weakly with particle size. For non-deformable particles, when the dislocations bypass the particles, the alloy strength is independent of the particle properties but is strongly dependent on particle size and dispersion, strength decreasing as particle size or dispersion increases. The transition from deformable to non-deformable particle controlled deformation is readily recognized by the change in microstructure, since the 'laminar' undisturbed dislocation flow for the former contrasts with the turbulent plastic flow for non-deformable particles. The latter leads to the production of a high density of dislocation loops, dipoles and other debris which results in a high rate of work hardening.

11.6.1 Coherency strain hardening

The precipitation of particles having a slight misfit in the matrix gives rise to stress fields which hinder the movement of gliding dislocations. For the dislocations to pass through the regions of internal stress the applied stress must be at least equal to the average internal stress, and for spherical particles this is given by

$$\tau = 2\mu\varepsilon f \tag{11.2}$$

where μ is the shear modulus, ε is the misfit of the particle and f is the volume fraction of precipitate. This suggestion alone, however, cannot account for the critical size of dispersion of a precipitate at which the hardening is a maximum, since equation 11.2 is independent of L, the distance between particles. To explain this, Mott and Nabarro consider the extent to which a dislocation can bow round a particle under the action of a stress τ. From page 257 it is evident that this is given by

$$r = \alpha\mu b/\tau \tag{11.3}$$

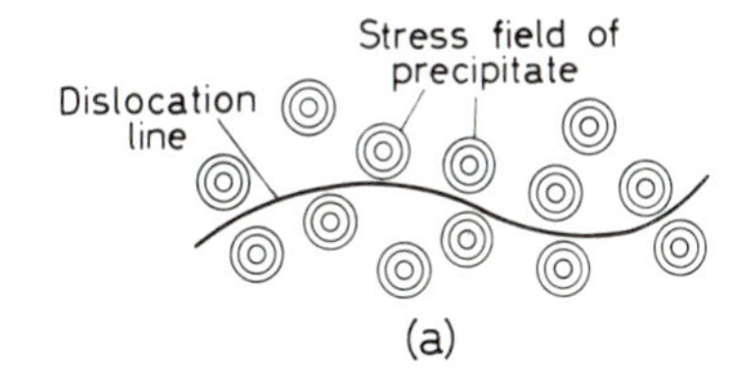

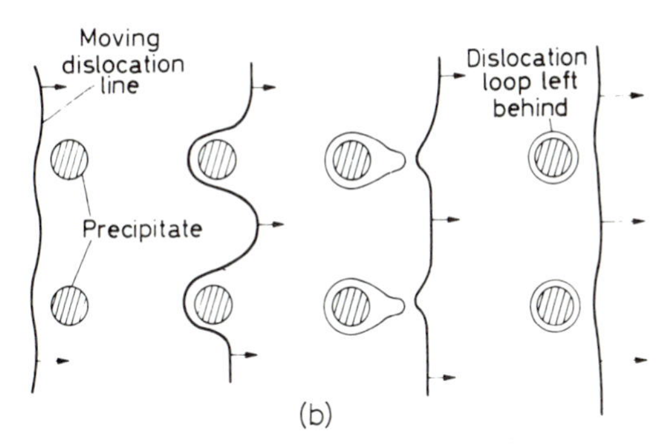

Figure 11.8 Schematic representation of a dislocation (a) curling round the stress fields from precipitates, and (b) passing between widely spaced precipitates (Orowan looping)

where r is the radius of curvature to which the dislocation is bent and $\alpha = \frac{1}{2}$. Hence, in the hardest age-hardened alloys where the yield strength is about $\mu/100$, the dislocation can bend to a radius of curvature of about 50 atomic spacings, and since the distance between particles is of the same order it would appear that the dislocation can avoid the obstacles and take a form like that shown in *Figure 11.8(a)*. With a dislocation line taking up such a configuration, in order to produce glide, each section of the dislocation line has to be taken over the adverse region of internal stress without any help from other sections of the line—the alloy is then hard. If the precipitate is dispersed on too fine a scale (e.g. when the alloy has been freshly quenched or lightly aged) the dislocation is unable to bend sufficiently to lie entirely in the regions of low internal stress. As a result, the internal stresses acting on the dislocation line largely cancel and the force resisting its movement is small—the alloy then appears soft. When the dispersion is on a coarse scale, the dislocation line is able to move between the particles, as shown in *Figure 11.8(b)*, and the hardening is again small.

For coherency strain hardening the flow stress depends on the ability of the dislocation to bend and thus experience more regions of adverse stress than of aiding stress. The flow stress therefore depends on the treatment of averaging the stress, and recent attempts separate the behaviour of small and large coherent particles. For small coherent particles the flow stress is given by

$$\tau = 4.1\mu\varepsilon^{3/2} f^{1/2} (r/b)^{1/2} \tag{11.4}$$

which predicts a greater strengthening than the simple arithmetic average of equation 11.2. For large coherent particles

$$\tau = 0.7\mu f^{1/2} (\varepsilon b^3/r^3)^{1/4} \tag{11.5}$$

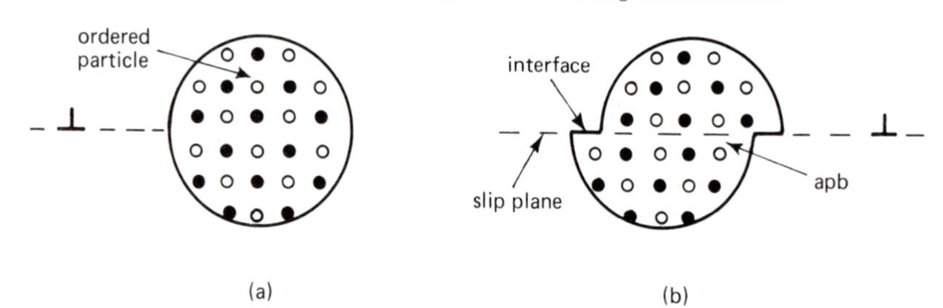

Figure 11.9 Ordered particle (a) cut by dislocations in (b) to produce new interface and apb

11.6.2 Chemical hardening

When a dislocation actually passes through a zone as shown in *Figure 11.9*, a change in the number of solvent–solute near-neighbours occurs across the slip plane. This tends to reverse the process of clustering and, hence, additional work must be done by the applied stress to bring this about. This process, known as chemical hardening, provides a short-range interaction between dislocations and precipitates and arises from three possible causes (i) the energy required to create an additional particle/matrix interface with energy γ_I per unit area which is provided by a stress

$$\tau \simeq \alpha\gamma_I^{3/2}(fr)^{1/2}/\mu b^2 \tag{11.6}$$

where α is a numerical constant, (ii) the additional work required to create an antiphase boundary inside the particle with ordered structure, given by

$$\tau \simeq \beta\gamma_{apb}^{3/2}(fr)^{1/2}/\mu b^2 \tag{11.7}$$

where β is a numerical constant, and (iii) the change in width of a dissociated dislocation as it passes through the particle where the stacking fault energy differs from the matrix (e.g. Al–Ag where $\Delta\gamma_{SF} \sim 100$ mJ/m^2 between Ag zones and Al matrix) so that

$$\tau \simeq \Delta\gamma_{SF}/b \tag{11.8}$$

Usually $\gamma_I < \gamma_b$ and so γ_I can be neglected, but the ordering within the particle requires the dislocations to glide in pairs. This leads to a strengthening given by

$$\tau = (\gamma_{apb}/2b)[(4\gamma_{apb}rf/\pi T)^{1/2} - f] \tag{11.9}$$

where T is the dislocation line tension.

11.6.3 Dispersion hardening

In dispersion hardening, it is assumed that the precipitates do not deform with the matrix and that the yield stress is the stress necessary to expand a loop of dislocation between the precipitates. This will be given by the Orowan stress

$$\tau = \alpha\mu b/L \tag{11.10}$$

where L is the separation of the precipitates. As discussed above, this process will be important in the later stages of precipitation when the precipitate becomes incoherent and the misfit strains disappear. A moving dislocation is then able to bypass the obstacles, as shown in *Figure 11.8(b)*, by moving in the clean pieces of crystal between the precipitated particles. Clearly, the flow stress will decrease as the distance between the obstacles increases, and so this effect can account for the over-aged condition which occurs in these alloys. However, even when the dispersion of the precipitate is coarse a greater applied stress is necessary to force a dislocation past the obstacles than would be the case if the obstruction were not there.

11.7 Hardening in aluminium–copper alloys

The actual hardening mechanism which operates in a given alloy will depend on several factors, such as the type of particle precipitated (e.g. whether zone, intermediate precipitate or stable phase), the magnitude of the strain and the testing temperature. In the earlier stages of ageing, i.e. before over-ageing, the coherent zones are cut by dislocations moving through the matrix and hence both coherency strain hardening and chemical hardening will be important, e.g. in such alloys as aluminium–copper, copper–beryllium and iron–vanadium–carbon. In alloys such as aluminium–silver and aluminium–zinc, however, the zones possess no strain field, so that chemical hardening will be the most important contribution. In the important high-temperature creep resistant nickel alloys the precipitate is of the Ni_3Al form which has a low particle/matrix misfit and hence chemical hardening due to dislocations cutting the particles is again predominant. To illustrate that more than one mechanism of hardening is in operation in a given alloy system, let us examine the mechanical behaviour of an aluminium–copper alloy in more detail.

Figure 11.10 shows the deformation characteristics of single crystals of an aluminium–copper (nominally 4 per cent) alloy in various structural states. The curves were obtained by testing crystals of approximately the same orientation, but the stress-strain curves from crystals containing GP [1] and GP [2] zones are quite different from those for crystals containing θ' or θ precipitates. When the crystals contain either GP [1] or GP [2] zones, the stress–strain curves are very similar to those of pure aluminium crystals, except that there is a two or three-fold increase in the flow stress. In contrast, when the crystals contain either θ' or θ precipitates the critical resolved shear stress is less than for crystals containing zones, but the initial rate of work hardening is extremely rapid. In fact, the stress–strain curves bear no similarity to those of a pure aluminium crystal. It is also observed that when θ' or θ is present as a precipitate, deformation does not take place on a single slip system but on several systems; the crystal then deforms, more nearly as a polycrystal does and the x-ray pattern develops extensive asterism. These factors are consistent with the high rate of work hardening observed in crystals containing θ' or θ precipitates.

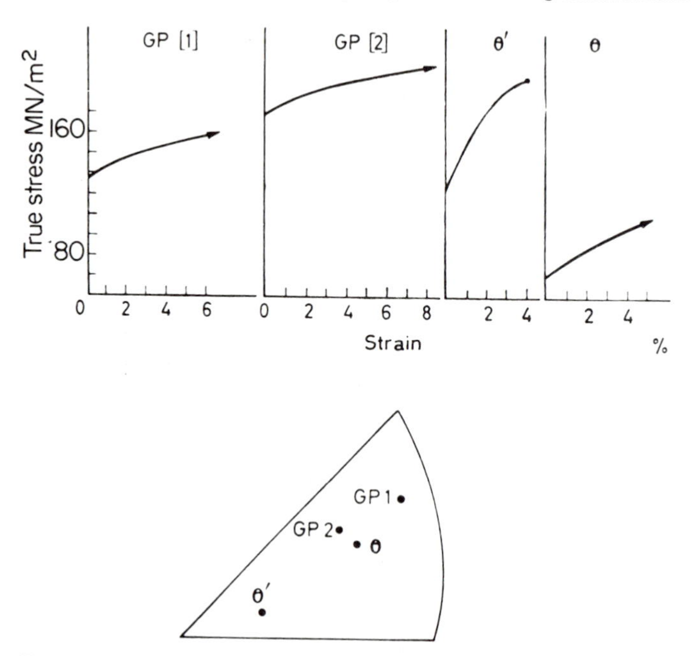

Figure 11.10 Stress–strain curves from single crystals of aluminium–4 per cent copper containing GP [1] zones, GP [2] zones, θ′-precipitates and θ′-precipitates respectively (after Fine, Bryne and Kelly)

Hardening due to GP zones

The separation of the centres of the precipitates cutting any slip plane can be deduced from both x-ray and electron microscope observations. For the crystals, relating to *Figure 11.10*, containing GP [1] zones this value is 15 nm (150 Å), and for GP [2] zones it is 25 nm (250 Å). It then follows from equation 11.3 that to avoid these precipitates the dislocations would have to bow to a radius of curvature of about 10 nm (100 Å). To do this requires a stress several times greater than the observed flow stress and, in consequence, it must be assumed that the dislocations are forced through the zones. Furthermore, if we substitute the observed values of the flow stress in the relation $\mu b/\tau = L$, it will be evident that the bowing mechanism is unlikely to operate unless the particles are about 60 nm (600 Å) apart. This is confirmed by electron microscope observations which show that dislocations pass through GP zones and coherent precipitates, but bypass non-coherent particles. Once a dislocation has cut through a zone, however, the path for subsequent dislocations on the same slip plane will be easier, so that the work-hardening rate of crystals containing zones should be low, as shown in *Figure 11.10*. The straight, well-defined slip bands observed on the surfaces of crystals containing GP [1] zones also support this interpretation.

If the zones possess no strain field, as in aluminium–silver or aluminium–zinc alloys, the flow stress would be entirely governed by the chemical hardening effect. However, the zones in aluminium copper alloys do possess strain fields, as shown in *Figure 11.3*, and, consequently, the stresses around a zone will also affect the flow stress. Each dislocation will be subjected to the stresses due to a zone at a small distance from the zone.

It will be remembered from Chapter 6 that temperature profoundly affects the flow stress if the barrier which the dislocations have to overcome is of a short-range nature. For this reason, the flow stress of crystals containing GP [1] zones will have a larger dependence on temperature than that of those containing GP [2] zones. Thus, while it is generally supposed that the strengthening effect of GP [2] zones is greater than that of GP [1], and this is true at normal temperatures (*see Figure 11.10*), at very low temperatures it is probable that GP [1] zones will have the greater strengthening effect. This has been confirmed for stress–strain curves measured at 4 K, and the large increase in flow stress at low temperatures results from the short-range interactions between zones and dislocations.

Hardening due to θ' and θ precipitates

The θ' and θ precipitates do not deform with the matrix and the critical resolved shear stress is the stress necessary to expand a loop of dislocation between them. This corresponds to the over-aged condition. The separation of the θ particles is greater than that of the θ', being somewhat greater than 1 μm and the initial flow stress is very low. In both cases, however, the subsequent rate of hardening is high because, as suggested by Fisher, Hart and Pry, the gliding dislocation interacts with the dislocation loops in the vicinity of the particles (*see Figure 11.8(b)*). The stress–strain curves show, however, that the rate of work hardening falls to a low value after a few per cent strain, and these authors attribute the maximum in the strain-hardening curve to the shearing of the particles. This process is not observed in crystals containing θ precipitates at room temperature and, consequently, it seems more likely that the particles will be avoided by cross-slip. If this is so, prismatic loops of dislocation will be formed at the particles, by the mechanism shown in *Figure 11.11*, and these will give approximately the same mean internal stress as that calculated by Fisher, Hart and Pry, but a reduced stress on the particle. The maximum in the work-hardening curve would then correspond to the stress necessary to expand these loops; this stress will be of the order of $\mu b/r$ where r is the radius of the loop which

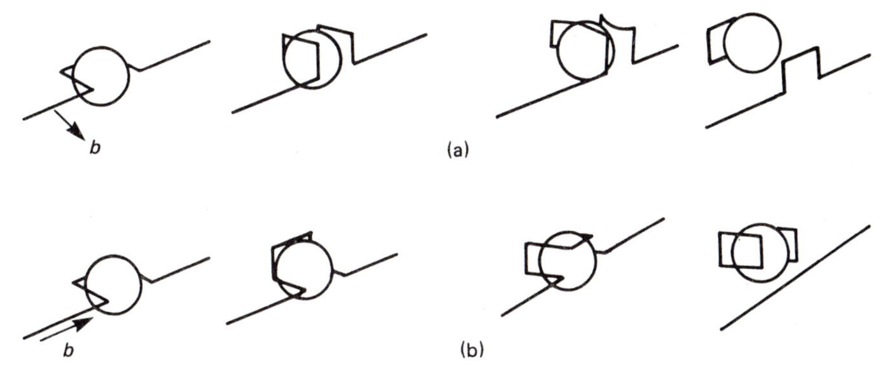

Figure 11.11 Cross-slip of (a) edge, (b) screw dislocation over a particle producing prismatic loops in the process

is somewhat greater than the particle size. At low temperatures cross-slip is difficult and the stress may be relieved either by initiating secondary slip or by fracture (*see* section 11.12).

11.8 Vacancies and precipitation

It is clear that because precipitation is controlled by the rate of atomic migration in the alloy, temperature will have a pronounced effect on the process. Moreover, since precipitation is a thermally activated process, other variables such as time of annealing, composition, grain size and prior cold work are also important. However, the basic treatment of age-hardening alloys is solution treatment followed by quenching, and the introduction of vacancies by the latter process must play an important role in the kinetic behaviour (*see* Chapter 9).

It has been recognized that near room temperature, zone formation in alloys such as aluminium–copper and aluminium–silver occurs at a rate many orders of magnitude greater than that calculated from the diffusion coefficient of the solute atoms. In aluminium–copper for example, the formation of zones is already apparent after only a few minutes at room temperature, and is complete after an hour or two, so that the copper atoms must, therefore, have moved through several atomic spacings in that time. This corresponds to an apparent diffusion coefficient of copper in aluminium of about 10^{-20}–10^{-22} $m^2 s^{-1}$, which is many orders of magnitude faster than the value of 5×10^{-29} $m^2 s^{-1}$ obtained by extrapolation of high-temperature data. Many workers have attributed this enhanced diffusion to the excess vacancies retained during the quenching treatment. Thus, since the expression for the diffusion coefficient at a given temperature contains a factor proportional to the concentration of vacancies at that temperature, if the sample contains an abnormally large vacancy concentration then the diffusion coefficient should be increased by the ratio c_Q/c_T, where c_Q is the quenched-in vacancy concentration and c_T is the equilibrium concentration. The observed clustering rate can be accounted for if the concentration of vacancies retained is about 10^{-3}–10^{-4}. The observation of loops by transmission electron microscopy allows an estimate of the number of excess vacancies to be made, and in all cases of rapid quenching the vacancy concentration in these alloys is somewhat greater than 10^{-4}, in agreement with the predictions outlined above. Clearly, as the excess vacancies are removed, the amount of enhanced diffusion diminishes, which agrees with the observations that the isothermal rate of clustering decreases continuously with increasing time. In fact, it is observed that D decreases rapidly at first and then remains at a value well above the equilibrium value for months at room temperature; the process is therefore separated into what is called the fast and slow reactions. Three mechanisms have been proposed to explain the slow reaction, namely (i) some of the vacancies quenched-in are trapped temporarily and then released slowly, (ii) the strain field around GP zones retards the diffusion of vacancies from the crystal, and (iii) the oxide film prevents the vacancies escaping from the surface or alternatively injects them as a result of oxidation-vacancy production. However, measurements show that the activation energy in the fast reaction

(≈ 0.5 eV) is smaller than in the slow reaction (≈ 1 eV) by an amount which can be attributed to the binding energy between vacancies and trapping sites. These traps are very likely small dislocation loops or voids formed by the clustering of vacancies. The equilibrium matrix vacancy concentration would then be greater than that for a well annealed crystal by a factor $\exp[\gamma\Omega/rkT]$, where γ is the surface energy, Ω the atomic volume and r the radius of the defect (*see* Chapter 9). The experimental diffusion rate can be accounted for if $r \approx 2$ nm, which is much smaller than the loops and voids usually seen, but they do exist. The activation energy for the slow reaction would then be $E_D - (\gamma\Omega/r)$ or approximately 1 eV for $r \approx 2$ nm.

Other factors known to affect the kinetics of the early stages of ageing (e.g. altering the quenching rate, interrupted quenching and cold work) may also be rationalized on the basis that these processes lead to different concentrations of excess vacancies. In general, cold working the alloy prior to ageing causes a decrease in the rate of formation of zones, which must mean that the dislocations introduced by cold work are more effective as vacancy sinks than as vacancy sources. Cold working or rapid quenching, therefore, have opposing effects on the formation of zones. Vacancies are also important in other aspects of precipitation hardening. For example, the excess vacancies, by condensing to form a high density of dislocation loops, can provide nucleation sites for intermediate precipitates. This leads to the interesting observation in aluminium–copper alloys that cold working or rapid quenching, by producing dislocations for nucleation sites, have the same effect on the formation of the θ' phase but, as we have seen above, the opposite effect on zone formation. It is also interesting to note that screw dislocations, which are not normally favourable sites for nucleation, can also become sites for preferential precipitation when they have climbed into helical dislocations by absorbing vacancies, and have thus become mainly of edge character. The long arrays of θ' phase observed in aluminium–copper alloys, shown in *Figure 11.4(c)*, have probably formed on helices in this way. In some of these alloys, defects containing stacking faults are observed, in addition to the dislocation loops and helices, and examples have been found where such defects nucleate an intermediate precipitate having a hexagonal structure. In aluminium–silver alloys it is also found that the helical dislocations introduced by quenching absorb silver and degenerate into long narrow stacking faults on $\{111\}$ planes, and these stacking fault defects then act as nuclei for the hexagonal γ' precipitate.

Many commercial alloys depend critically on the interrelation between vacancies, dislocations and solute atoms and it is found that trace impurities significantly modify the precipitation process. Thus trace elements which interact strongly with vacancies inhibit zone formation, e.g. Cd, In, Sn prevent zone formation in slowly quenched Al–Cu alloys for up to 200 days at 30 °C. This delays the age-hardening process at room temperature which gives more time for mechanically fabricating the quenched alloy before it gets too hard, thus avoiding the need for refrigeration. On the other hand, Cd increases the density of θ' precipitate by increasing the density of vacancy loops and helices which act as nuclei for precipitation and by segregating to the matrix–θ' interfaces thereby reducing the interfacial energy.

Since grain boundaries absorb vacancies in many alloys there is a grain boundary zone relatively free from precipitation. The Al/Zn/Mg alloy is one commercial alloy which suffers grain boundary weakness but it is found that trace additions of Ag have a beneficial effect in refining the precipitate structure and removing the precipitate free grain boundary zone. Here it appears that Ag atoms stabilize vacancy clusters near the grain boundary and also increase the stability of the GP zone thereby raising the GP zone solvus temperature. Similarly, in the 'Concorde' alloy, RR58, (basically Al/2.5% Cu/1.2% Mg with additions), Si addition (0.25% Si) modifies the as-quenched dislocation distribution inhibiting the nucleation and growth of dislocation loops and reducing the diameter of helices. The S-precipitate (Al_2CuMg) is homogeneously nucleated in the presence of Si rather than heterogeneously nucleated at dislocations, and the precipitate grows directly from zones, giving rise to improved and more uniform properties.

Of particular interest is the precipitation of carbon in iron, often termed quench-ageing. In this system the precipitation may be of two types, either (i) on dislocations, or (ii) in the matrix without the aid of dislocations, but in low carbon alloys (i.e. <0.05 per cent carbon) electron micrographs show that precipitation on dislocations predominates. One point of interest in these observations is the definite precipitation along $\langle 100 \rangle$ directions on individual dislocations, as shown in *Figure 11.12*. Here, the precipitate starts by saturating the dislocation and then grows in $\{100\}$ planes in the form of plates.

The precipitation of carbon on dislocations is of direct importance to the theory of strain ageing, and the subsequent development of actual precipitates from atmospheres may be responsible for the relatively small temperature-

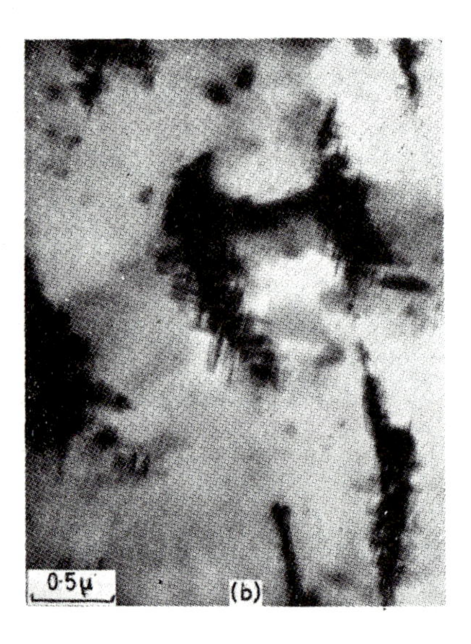

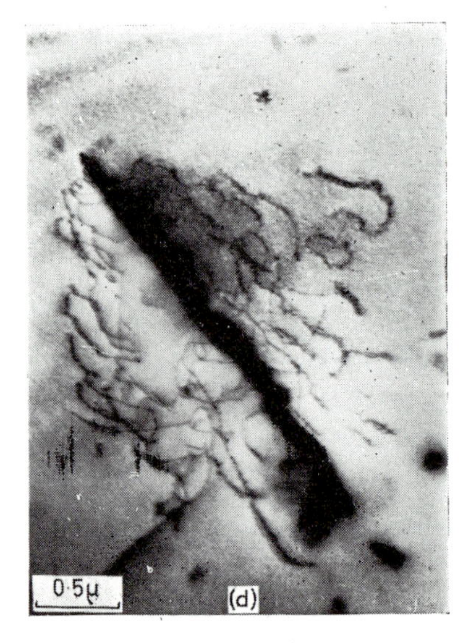

Figure 11.12 Electron micrographs from iron (<0.05%C) quenched from 690°C and (left) aged 4 days at 100°C, showing precipitates on dislocations with dendrite arms along ‹100› directions, (right) aged 3 hours at 100°C, showing precipitation on the dislocations generated around large precipitates (after Carrington, Hale and McLean, Crown copyright reserved. Reproduced by permission of the Director, National Physical Laboratory)

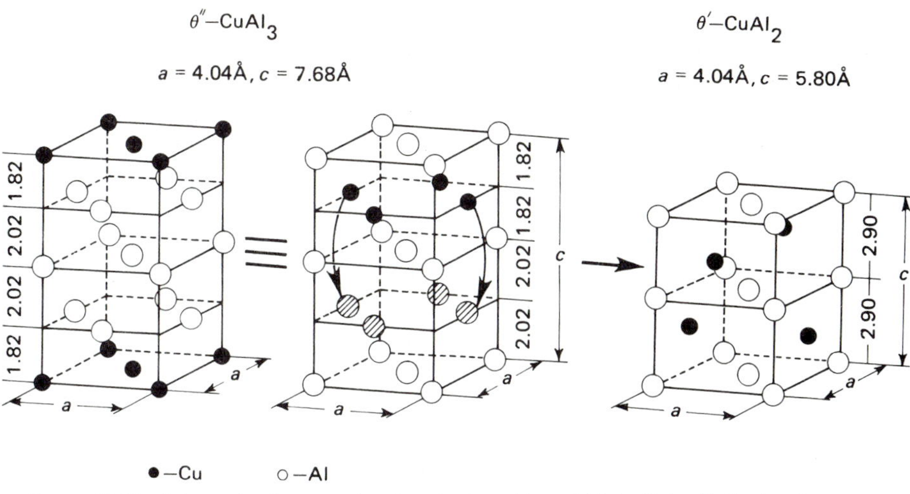

Figure 11.13 Schematic diagram showing the transition of θ″ to θ′ in Al–Cu by the vacancy mechanism. Vacancies from annealing loops are condensed on a next-nearest Al plane from the copper layer in θ″ to form the required A-A-A Al stacking. Formation of the θ′ fluorite structure then requires only slight redistribution of the copper atom layer and relaxation of the Al layer spacings (Courtesy K. H. Westmacott)

dependence of the dislocation locking observed in some steels (*see* Chapter 8). The precipitates which form without the aid of dislocations are plate-like zones also lying along {100} planes. When they become large the stresses surrounding them are often relieved by the generation of dislocations which, in turn of course, may also become anchored (*see Figure 11.12(b)*).

Apart from speeding up the kinetics of ageing, and providing dislocations nucleation sites, vacancies may play a structural role when they precipitate cooperatively with solute atoms to facilitate the basic atomic arrangements required for transforming the parent crystal structure to that of the product phase. Examples of this structural process have been documented in interstitial, substitutional, and mixed interstitial/substitutional alloy systems. In essence the process involves the systematic incorporation of excess vacancies, produced by the initial quench or during subsequent dislocation loop annealing, in a precipitate zone or plate to change the atomic stacking. A simple example of θ' formation in Al–Cu is shown schematically in *Figure 11.13*. Ideally, the structure of the θ'' phase in Al–Cu consists of layers of copper on {100} separated by three layers of aluminium atoms. If a next-nearest neighbour layer of aluminium atoms from the copper layer is removed by condensing a vacancy loop, an embryonic θ' unit cell with Al in the correct A–A–A stacking sequence is formed (*Figure 11.13(b)*). Formation of the final $CuAl_2$ θ' fluorite structure requires only shuffling half of the copper atoms into the newly-created next-nearest neighbour space and concurrent relaxation of the Al atoms to the correct θ' interplanar distances (*Figure 11.13(c)*).

The structural incorporation of vacancies in a precipitate is a non-conservative process since atomic sites are eliminated. There exist equivalent conservative processes in which the new precipitate structure is created from the old by the nucleation and expansion of partial dislocation loops with predominantly shear character. Thus, for example, the BABAB {100} plane stacking sequence of the f.c.c. structure can be changed to BAABA by the propagation of a $a/2\langle 100\rangle$ shear loop in the {100} plane, or to BAAAB by the propagation of a pair of $a/2\langle 100\rangle$ partials of opposite sign on adjacent planes. Again, the A–A–A stacking resulting from the double shear is precisely that required for the embryonic formation of the fluorite structure from the f.c.c. lattice.

In visualizing the role of lattice defects in the nucleation and growth of plate-shaped precipitates, a simple analogy with Frank and Shockley partial dislocation loops is useful. In the formation of a Frank loop, a layer of h.c.p. material is created from the f.c.c. lattice by the (non-conservative) condensation of a layer of vacancies in {111}. Exactly the same structure is formed by the (conservative) expansion of a Shockley partial loop on a {111} plane. In the former case a semi-coherent 'precipitate' is produced bounded by a $a/3\langle 111\rangle$ dislocation, and in the latter a coherent one bounded by a $a/6\langle 112\rangle$. Continued growth of precipitate plates occurs by either process or a combination of processes. Of course, formation of the final precipitate structure requires, in addition to these structural rearrangements, the long-range diffusion of the correct solute atom concentration to the growing interface.

Recent work suggests that the growth of a second-phase particle with a disparate size or crystal structure relative to the matrix is controlled by two overriding principles – the accommodation of the volume and shape change, and the optimized use of the available deformation mechanisms. Thus, in general, volumetric transformation strains are accommodated by vacancy or interstitial condensation, or prismatic dislocation loop punching, while deviatoric strains are relieved by shear loop propagation. An example is shown in *Figure. 11.14.* The formation of semicoherent Cu needles in Fe–1%Cu is accomplished by the generation of shear loops in the precipitate/matrix interface. Expansion of the loops into the matrix and incorporation into nearby precipitate interfaces leads to a complete network of dislocations interconnecting the precipitates.

11.9 Duplex ageing

In non-ferrous heat treatment there is considerable interest in double (or duplex) ageing treatments to obtain the best microstructure consistent with optimum properties. It is now realized that it is unlikely that the optimum properties will be produced in alloys of the precipitation-hardening type by a single quench and ageing treatment. For example, while the interior of grains may develop an acceptable precipitate size and density, in the neighbourhood of efficient vacancy sinks, such as grain boundaries, a precipitate-free zone (PFZ) is formed which is often associated with overageing in the boundary itself. This heterogeneous structure gives rise to poor properties, particularly under stress corrosion conditions.

Duplex ageing treatments have been used to overcome this difficulty. In Al/Zn/Mg, for example, it was found that storage at room temperature before heating to the ageing temperature leads to the formation of finer precipitate

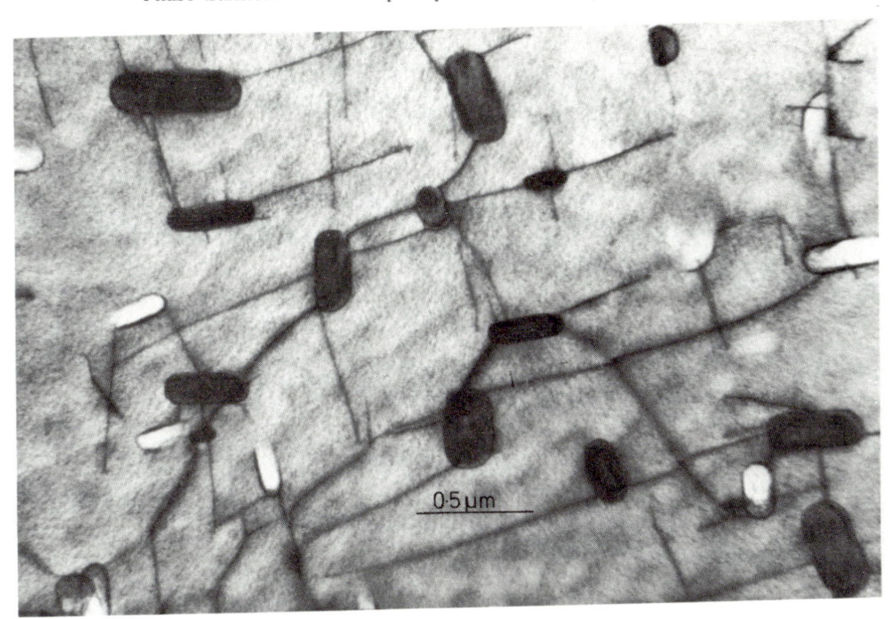

Figure 11.14 The formation of semicoherent Cu needles in Fe–1% Cu is accomplished by the generation of shear loops in the precipitate/matrix interface. Expansion of the loops into the matrix and incorporation into nearby precipitate interfaces leads to a complete network of dislocations interconnecting the precipitates (Courtesy K. H. Westmacott)

structure and better properties. This is just one special example of two-step or multiple ageing treatments which have commercial advantages and have been found to be applicable to several alloys. Duplex ageing gives better competitive mechanical properties in Al-alloys, e.g. Al/Zn/Mg alloys, with much enhanced corrosion resistance since the grain boundary zone is removed. It is possible to obtain strengths of 267–308 MN/m^2 in Mg/Zn/Mn alloys which have very good strength/weight ratio applications, and nickel alloys also develop better properties with multiple ageing treatments.

The basic idea of all heat-treatments is to 'seed' a uniform distribution of stable nuclei at the low temperature which can then be grown to optimum size at the higher temperature. In most alloys, there is a critical temperature T_c above which homogeneous nucleation of precipitate does not take place, and in some instances has been identified with the GP zone solvus. On ageing above T_c there is a certain critical zone size above which the zones are able to act as nuclei for precipitates and below which the zones dissolve.

In general, the ageing behaviour of Al/Zn/Mg alloys can be divided into three classes which can be defined by the temperature ranges involved: (a) Alloys quenched and aged above the GP zone solvus (i.e. the temperature above which the zones dissolve, which is above $\sim 155°C$ in a typical Al/Zn/Mg alloy); then, since no GP zones are ever formed during heat treatment, there are no easy nuclei for subsequent precipitation and a very coarse dispersion of precipitates results with nucleation principally on dislocations. (b) Alloys quenched and aged below the GP zone solvus, GP zones form continuously and grow to a size at which they are able to transform to precipitates. The transformation will occur rather more slowly in the grain boundary regions due to the lower vacancy concentration there but since ageing will always be below the GP zone solvus, no PFZ is formed other than a very small ($\sim$30 nm) solute-denuded zone due to precipitation in the grain boundary. (c) Alloys quenched below the GP zone solvus and aged above it, e.g. quenched to room temperature and aged at 180 °C for Al/Zn/Mg; this is the most common practical situation. The final dispersion of precipitates and the PFZ width are controlled by the nucleation treatment below 155° where GP zone size distribution is determined. A long nucleation treatment gives a fine dispersion of precipitates and a narrow PFZ.

It is possible to stabilize GP zones by addition of trace elements. These have the same effect as raising T_c, so that alloys are effectively aged below T_c. One example is Ag to Al/Zn/Mg which raises T_c from 155° to 185 °C, another is Si to Al/Cu/Mg, another Cu to Al/Mg/Si and yet another Cd or Sn to Al/Cu alloys. It is then possible to get uniform distribution and optimum properties by single ageing, and is an example of achieving by chemistry what can similarly be done with physics during multiple ageing. Whether it is best to alter the chemistry or to change the physics for a given alloy usually depends on other factors, e.g. economics.

11.10 Particle coarsening

With continued ageing at a given temperature there is a tendency for the small particles to dissolve and the resultant solute to precipitate on larger particles causing them to grow, thereby lowering the total interfacial energy. This process is termed particle coarsening, or sometimes Ostwald ripening. The driving force for particle growth is the difference between the concentration of solute (S_r) in equilibrium with small particles of radius r and that in equilibrium with larger particles. The variation of solubility with surface curvature is given (*see* Chapter 5) by the Gibbs–Thomson or Thomson–Freundlich equation

$$\ln (S_r/S) = 2\gamma\Omega/\boldsymbol{k}Tr \tag{11.11}$$

where S is the equilibrium concentration, γ the particle/matrix interfacial energy and Ω the atomic volume; since $2\gamma\Omega \ll \boldsymbol{k}Tr$ then $S_r = S[1 + 2\gamma\Omega/\boldsymbol{k}Tr]$.

To estimate the coarsening rate of a particle it is necessary to consider the rate-controlling process for material transfer. Generally the rate-limiting factor is considered to be diffusion through the matrix and the rate of change of particle radius is then derived from the equation

$$4\pi r^2(\mathrm{d}r/\mathrm{d}t) = D4\pi R^2(\mathrm{d}S/\mathrm{d}R)$$

where $\mathrm{d}S/\mathrm{d}R$ is the concentration gradient across an annulus at a distance R from the particle centre. Rewriting the equation after integration gives

$$\mathrm{d}r/\mathrm{d}t = -D(S_r - S_a)/r \tag{11.12}$$

where S_a is the average solute concentration a large distance from the particle and D is the solute diffusion coefficient. When the particle solubility is small, the total number of atoms contained in particles may be assumed constant, independent of particle size distribution, so that

$$\Sigma 4\pi r^2(\mathrm{d}r/\mathrm{d}t)=0$$

Substituting for $(\mathrm{d}r/\mathrm{d}t)$ we obtain

$$\Sigma -4\pi r_i D(S_{r_i}-S_a)=0$$

and using the Thomson–Freundlich equation to replace S_{r_i} gives

$$\Sigma 4\pi r_c D[S_a - S(1+2\gamma\Omega/\boldsymbol{k}Tr_i)]=0$$

This equation may be rearranged according to

$$(S_a - S)\Sigma r_i = \Sigma 2\gamma\Omega S/\boldsymbol{k}T = 2n\gamma\Omega S/\boldsymbol{k}T$$

where n is the total number of particles in the system at any instant and $\Sigma r_i/n = \bar{r}$, the arithmetic mean radius of particles in the system. Thus

$$(S_a - S_r)=\{2\gamma\Omega S/\boldsymbol{k}T\}[(1/\bar{r})-(1/r)]$$

and combining with equation 11.11 gives the variation of particle growth rate with radius according to

$$\mathrm{d}r/\mathrm{d}t=\{2DS\gamma\Omega/\boldsymbol{k}T\bar{r}\}[(1/\bar{r})-(1/r)] \qquad (11.13)$$

This function is plotted in *Figure 11.15* from which it is evident that particles of radius less than $\bar{r}$ are dissolving at increasing rates with decreasing values of r. All particles of radius greater than $\bar{r}$ are growing but the graph shows a maximum for particles twice the mean radius. Over a period of time the number of particles decreases discontinuously when particles dissolve, and ultimately the system would tend to form one large particle. However, before this state is reached the mean radius $\bar{r}$ increases and the growth rate of the whole system slows down.

A more detailed theory than that, due to Greenwood, outlined above has been derived by Lifshitz and Slyozov, and by Wagner taking into consideration the initial particle size distribution. They show that the mean particle radius

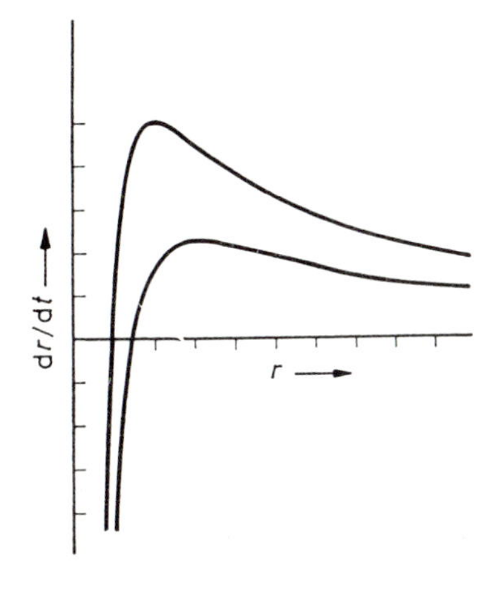

Figure 11.15 The variation of growth rate $\mathrm{d}r/\mathrm{d}t$ with particle radius r for diffusion-controlled growth, for two values of $\bar{r}$. The value of $\bar{r}$ for the lower curve is 1.5 times that for the upper curve. Particles of radius equal to the mean radius of all particles in the system at any instant are neither growing nor dissolving. Particles of twice this radius are growing at the fastest rate. The smallest particles are dissolving at a rate approximately proportional to r^{-2} (after Greenwood, *Inst. of Metals Conference on Phase Transformations,* 1968, courtesy of the Institute of Metals)

varies with time according to

$$\bar{r}^3 - \bar{r}_0^3 = Kt \tag{11.14}$$

where $\bar{r}_0$ is the mean particle radius at the onset of coarsening and K is a constant given by

$$K = 8DS\gamma\Omega/9\boldsymbol{k}T$$

This result is almost identical with the result obtained by integrating equation 11.13 in the elementary theory and assuming that the mean radius is increasing at half the rate of that of the fastest growing particle.

Coarsening rate equations have also been derived assuming that the most difficult step in the process is for the atom to enter into solution across the precipitate/matrix interface; the growth is then termed interface-controlled. The appropriate rate equation is

$$\mathrm{d}r/\mathrm{d}t = -C(S_r - S_a)$$

and leads to a coarsening equation of the form

$$\bar{r}^2 - \bar{r}_0^2 = (64CS\gamma\Omega t/81\boldsymbol{k}T) \tag{11.15}$$

where C is some interface constant.

Measurements of coarsening rates so far carried out support the analysis basis on diffusion control of the particle growth. The most detailed results have been obtained for nickel-based systems, particularly the coarsening of γ' (Ni_3Al–Ti or Si), which show a good $\bar{r}^3$ versus t relationship over a wide range of temperatures. Strains due to coherency and the fact that γ' precipitates are cube-shaped do not seriously affect the analysis in these systems. Concurrent measurements of $\bar{r}$ and the solute concentration in the matrix during coarsening have enabled values for the interfacial energy ≈ 13 mJ/m^2 to be determined. In other systems the agreement between theory and experiment is generally less precise, although generally the cube of the mean particle radius varies linearly with time, as shown in *Figure 11.16* for the growth of Mn precipitates in a Mg–Mn alloy.

Because of the ease of nucleation, particles may tend to concentrate on grain boundaries, and hence grain boundaries may play an important part in particle

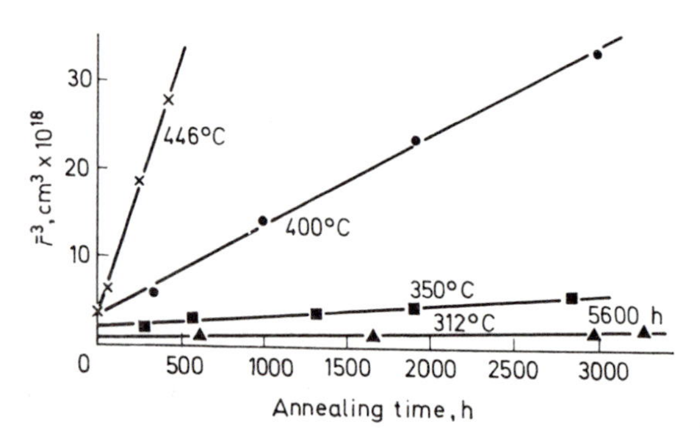

Figure 11.16 The variation of $\bar{r}^3$ with time of annealing for manganese precipitates in a magnesium matrix (after Smith, *Acta Met.*, 1967, **15,** 1867, courtesy of Pergamon Press)

growth. For such a case, the Thomson–Freundlich equation becomes

$$\ln (S_r/S) = (2\gamma - \gamma_g)\Omega/\boldsymbol{k}Tx$$

where γ_g is the grain boundary energy per unit area, and $2x$ the particle thickness, and their growth follows a law of the form

$$r_f^4 - r_0^4 = Kt \tag{11.16}$$

where the constant K includes the solute diffusion coefficient in the grain boundary and the boundary width. The activation energy for diffusion is lower in the grain boundary than in the matrix and this leads to a less-strong dependence on temperature for the growth of grain boundary precipitates and for this reason their preferential growth is likely to be predominant only at relatively low temperature.

11.11 Spinodal decomposition

In Chapter 4 it was mentioned that for any alloy composition where the free energy curve has a negative curvature, i.e. $(\mathrm{d}^2G/\mathrm{d}c^2) < 0$, small fluctuations in composition that produce A-rich and B-rich regions will bring about a lowering of the total free energy. At a given temperature the alloy must lie between two points of inflection (where $\mathrm{d}^2G/\mathrm{d}c^2 = 0$) and the locus of these points at different temperatures is depicted on the phase diagram by the chemical spinodal line (*see Figure 11.17*).

For an alloy c_0 quenched inside this spinodal, composition fluctuations increase very rapidly with time and have a time constant $\tau = -\lambda/4\pi^2 D$, where λ is the wavelength of composition modulations in one-dimension and D is the interdiffusion coefficient; for such a kinetic process, shown in *Figure 11.18* 'up-hill' diffusion takes place, i.e. regions richer in solute than the average become richer, and poorer become poorer until the equilibrium compositions c_1 and c_2 of the A-rich and B-rich regions are formed. As for normal precipitation, interfacial energy and strain energy influence the decomposition. During the early stages of decomposition the interface between A-rich and B-rich regions is diffuse and the interfacial energy becomes a gradient energy which depends on the composition gradient across the interface according to

$$\Delta G_{\mathrm{Int}} = K(\Delta c/\lambda)^2 \tag{11.17}$$

where λ is the wavelength and Δc the amplitude of the sinusoidal composition modulation, and K depends on the difference in bond energies between like and unlike atom pairs. The coherency strain energy term is related to the misfit ε between regions A and B, where $\varepsilon = (1/a)\,\mathrm{d}a/\mathrm{d}c$, the fractional change in lattice parameter a per unit composition change, and is given for an elastically isotropic solid, by

$$\Delta G_{\mathrm{strain}} = \varepsilon^2\,\Delta c^2 EV/(1-\nu) \tag{11.18}$$

with E Young's modulus, ν Poisson's ratio and V the molar volume. The total free

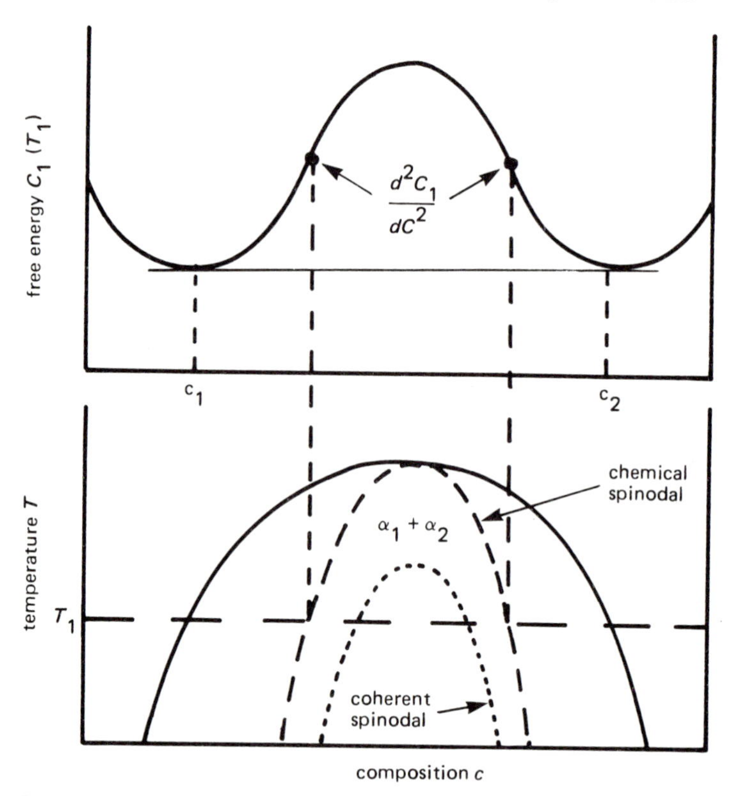

Figure 11.17 Variation of chemical and coherent spinodal with composition

energy change arising from a composition fluctuation is therefore

$$\Delta G = \left[\frac{d^2G}{dc^2} + \frac{2K}{\lambda^2} + (2\varepsilon^2 EV/(1-\nu)\right]\Delta c^2/2 \tag{11.19}$$

and a homogeneous solid solution will decompose spinodally provided

$$-(d^2G/dc^2) > (2K/\lambda^2) + (2\varepsilon^2 EV/1-\nu) \tag{11.20}$$

For $\lambda = \infty$, the condition $[(d^2G/dc^2) + (2\varepsilon^2 EV/1-\nu)] = 0$ is known as the coherent spinodal, as shown in *Figure 11.17*. The λ of the composition modulations has to satisfy the condition

$$\lambda^2 > 2K/[d^2G/dc^2 + (2\varepsilon^2 EV/1-\nu)] \tag{11.21}$$

and decreases with increasing degree of undercooling below the coherent spinodal line. A λ-value of 5–10 nm is favoured, since shorter λ's have too sharp a concentration gradient and longer λ's have too large a diffusion distance. For large misfit values, a large undercooling is required to overcome the strain energy effect. In cubic crystals, E is usually smaller along $\langle 100 \rangle$ directions and the high strain energy is accommodated more easily in the elastically soft directions, with composition modulations localized along this direction.

Spinodal decompositions have now been studied in a number of systems

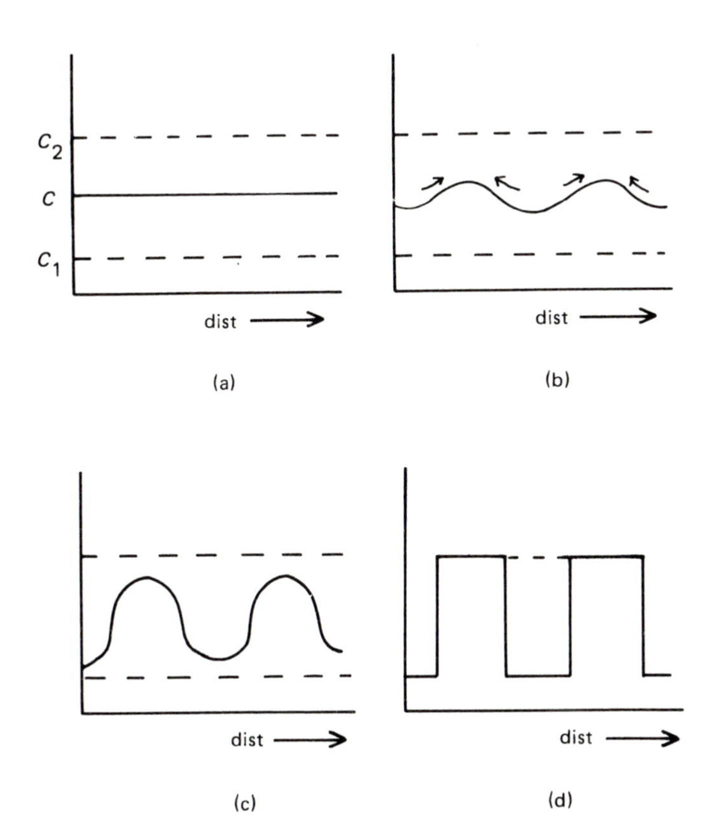

Figure 11.18 Variation of concentration with time for a spinodal system

such as Cu–Ni–Fe, Cu–Ni–Si, Ni–12%Ti, Cu–5%Ti exhibiting 'side-bands' in x-ray small-angle scattering, satellite spots in electron diffraction patterns and characteristic modulation of structure along ⟨100⟩ in electron micrographs. Many of the alloys produced by splat cooling might be expected to exhibit spinodal decomposition, and it has been suggested that in some alloy systems GP zones form in this way at high supersaturations, because the GP zone solvus (*see Figure 11.1*) gives rise to a metastable coherent miscibility gap.

The spinodally decomposed microstructure is believed to have unusually good mechanical stability under fatigure conditions.

11.12 Dispersion-hardened alloys

For an alloy containing incoherent, non-deformable, particles the rate of work-hardening is much greater than that shown by the matrix alone (*see Figure 11.10*). The dislocation density increases very rapidly with strain because the particles produce a turbulent and complex deformation pattern around them. It is possible for the dislocation gliding in the matrix to leave loops around the particle either by bowing between the particles to produce Orowan loops or cross slipping around them to produce prismatic loops, as shown in *Figure 11.11*. Generally, Orowan loops are formed only when the stacking fault energy of the matrix is low and cross-slip is difficult. The particle may also relieve the stresses in and around it by activating second slip systems by prismatic punching. All these dislocations spread out from the particle as strain proceeds and, by intersecting the primary glide plane, hinder primary dislocation motion. A dense tangle of dislocations is built up at the particle and a cell structure is formed with the particles predominantly in the cell walls. The rate of work hardening for such materials has been discussed in Chapter 10.

In precipitation hardening, the strongest alloy is produced by coupling some dispersion strengthening with a high dislocation density, as shown for Al–Cu alloys in *Figure 11.10*. This feature is also common to high strength steels produced by ausforming, where a deformation of the f.c.c. austenite prior to the transformation to martensite is used to produce a high ρ and this produces yield strengths of up to 3.1 GN/m^2. Working of maraging steels in the air-cooled condition before ageing also increases the strength considerably. Dual-phase steels also depend on the high rate of work hardening associated with martensite islands and are discussed in Chapter 12.

Above a certain temperature, dependent on strain-rate, there is a marked reduction in the work-hardening rate (*see Figure 11.19*), with a corresponding change in the microstructure resulting from a large reduction in the number of dislocation loops around particles. In this temperature regime the stress relief at particles arises by both cross-slip and climb, as shown in *Figure 11.20*. The dislocation bows around the particle and cross-slips so that the jogged loop (which may be only partially formed) anneals out by climb by vacancy transfer from jogs AA′ to jogs BB′ along the dislocation core, i.e. pipe diffusion. This depends on the spacing and height of the jogs, the exact mechanism of diffusion along the core which is related to the activation energy for pipe diffusion U_{PD}, and

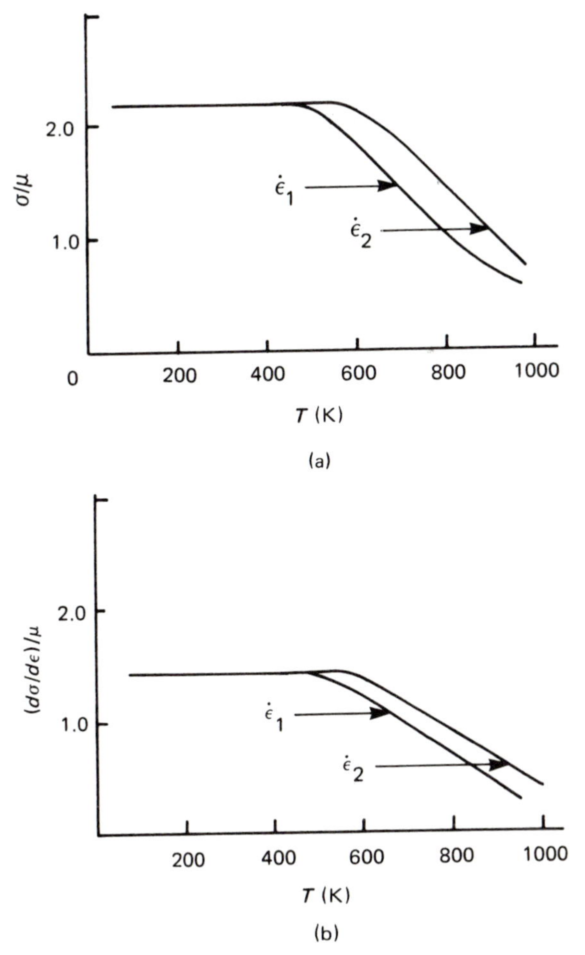

Figure 11.19 Variation of (a) flow stress, (b) work-hardening rate with temperature for Ni–Al_2O_3 alloys at strain-rate $\varepsilon_1 \simeq 3.3 \times 10^{-5}\ s^{-1}$ and $\varepsilon_2 \simeq 6.6 \times 10^{-4}\ s^{-1}$

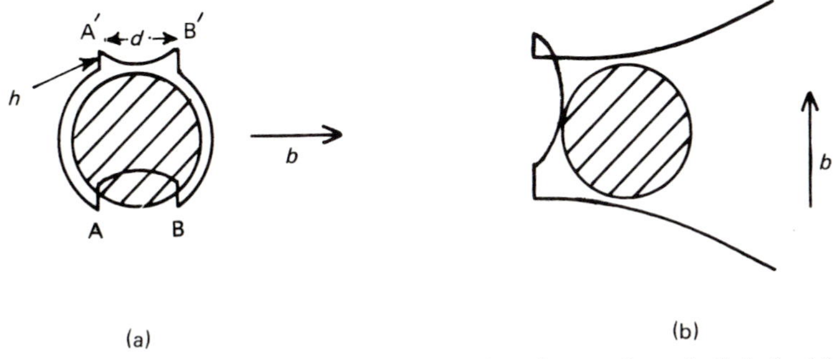

Figure 11.20 By-pass of particle by combination of cross-slip and climb for (a) Orowan loop, (b) before the formation of an Orowan loop

the driving force for the process. The loop shrinks under the climb force F_{cl} arising from the line tension of the dislocation $F_{cl} \sim \alpha \mu b^2 (\ln \ell/r_0)/2\pi\ell$ and redisappears in a time t given from climb theory (*see Chapter 9*) by

$$t = 1/[4azvb/\ell^2 h] \exp(-U_{PD}/\boldsymbol{k}T)(F_{cl}b^2/\boldsymbol{k}T) \qquad (11.22)$$

where a is the cross-section of the dislocation core, z the number of sites surrounding a vacancy into which it can jump (≈ 11) and v the frequency of atomic vibration.

During deformation of the dispersion-strengthened alloy, the number of loops produced per second per particle is $dn/dt = d\dot{\varepsilon}/b$, and the hardening effect from these will be removed if $1/t \leqslant dn/dt$, i.e. they climb out faster than they form. The climb-rate depends on temperature and T_c, the temperature where the above condition exists, is given by

$$T_c = U_{PD}/\boldsymbol{k} \ln[(4azvb^2/\ell^2 h\, d\dot{\varepsilon})(F_{cl}b^2/\boldsymbol{k}T_c)] \qquad (11.23)$$

For copper single crystals containing particles, a value for $U_{PD} \approx 0.9$ eV (which is $\approx U_{SD}/2$) has been obtained from equation 11.23, assuming $\ell = h = d/2$ and a pipe radius of $3b$. The large number of parameters in equation 11.23 leads to uncertainty in its use and U_{PD} may be more accurately determined from the measurement of the critical temperatures T_1 and T_2 at different strain-rates $\dot{\varepsilon}_1$ and $\dot{\varepsilon}_2$, thereby eliminating the parameters in equation 11.23 to give

$$U_{PD} = \boldsymbol{k}\left(\frac{1}{T_2} - \frac{1}{T_1}\right)^{-1} \ln\left(\frac{\dot{\varepsilon}_1 T_1}{\dot{\varepsilon}_2 T_2}\right) \qquad (11.24)$$

From curves such as *Figure 11.19*, U_{PD} is found to be 73 kJ mol^{-1} for pure nickel and 108 kJ mol^{-1} for Ni–67%Co; thus the activation energy of pipe diffusion appears to be dependent on stacking fault energy and is more difficult down extended dislocations than down constricted dislocations.

The observations suggest that very great strength could be produced if the load-bearing capacity of a large volume fraction of very strong particles could be used directly. With roughly equiaxed particles this is difficult to achieve since the stresses in the vicinity of the particle become large due to the large dislocation density in the matrix and failure of the particle/matrix interface occurs before the particle is loaded to its fracture stress. The particles thus do not make their maximum contribution to the strength. Substantial loading of the particles can be achieved by increasing the work of separation of the particle/matrix interface and considerable work is being carried out in this area. An alternative way is to make the dispersed phase in the form of long fibres to produce fibre-reinforced materials.

11.13 Fibre strengthening

In high-strength alloys it is not possible to attain the theoretical strength of crystals ($\sim E/10$) because of the weakening effect of dislocations and cracks.

Dislocations are readily immobilized, but this usually leads to materials with a low resistance to crack propagation. In fibre-strengthened materials cracks and dislocations still exist, but in such a way that a very high strength with useful ductility is obtained. The basic principle stems from the behaviour of bamboo wood, which even when notched and bent only develops cracks near the notch, none running into the material. A fibre composite is therefore made by using a soft ductile metal or weak resin to bind together a bundle of strong fibres. The matrix is necessary, not only to cement the fibres into a composite with transverse strength but also to produce weak, yielding longitudinal interfaces to reduce notch sensitivity.

The matrix is also important in transferring the applied load to the fibres. To understand the mechanism of load transfer, consider a strong rod of radius r embedded to a depth l in a soft matrix and suppose that the sliding friction between rod and matrix produces a shear force τ per unit area. Then, with no matrix across the bottom of the rod the total load P necessary to pull the rod out is $P=2\pi rl\tau$, which increases without limit as we increase l. If the rod has a breaking stress σ_f, the maximum load which can be applied to the rod of diameter d in terms of σ_f is given by $\sigma_f/4\tau=l/d$. The quantity (l/d) is called the aspect ratio of the fibre. Clearly if $l/d>\sigma_f/4\tau$ the rod will break before it pulls out of the matrix but if $l/d<\sigma_f/4\tau$ the rod pulls out unbroken.

The same type of reasoning can be applied to the fibre–matrix interaction in a composite, provided the yield stress of the metallic matrix is much less than the breaking stress of the fibre. Under a load applied to the composite, parallel to the fibres, the metal attempts to stretch plastically and, because of adhesion to the fibre, stretches the fibre elastically. In this case τ is taken equal to the yield stress in shear of the metallic matrix. If the aspect ratio of the fibres is larger than that given by the above condition a stress equal to the breaking stress of the fibre can be transferred to it from the matrix. In a composite, stress is transferred to the fibres from both ends so that the critical aspect ratio for a fibre to be broken is given by $(l/d)_{\text{crit}}=\sigma_f/2\tau$. Provided the fibres exceed this critical aspect ratio they need not be continuous, since large scale flow of the composite cannot occur without breaking all of the fibres simultaneously. The breaking strength of such a composite is given by $\sigma_c>\bar{\sigma}_f V_f$, with $(l/d)>(l/d)_{\text{crit}}$, where $\bar{\sigma}_f$ is the average stress in a fibre and V_f the volume fraction. V_f can easily be made equal to 0.5. With very long fibres it does not matter if τ is small, and this means that a metallic binding material can be used almost to its melting point, and hence high strength attained at very high temperatures. The highest strengths are attained when $(l/d)_{\text{crit}}$ is small and the aspect ratio of the fibres large.

The principle of fibre reinforcement has been applied in glass-reinforced plastics for about two decades. In metallurgy, covalently bound solids are being used as fibres to produce composite materials with an order of magnitude greater strength than present day constructional materials. The low density and high E of such solids is of importance in aerial transportation and rocketry, where the strength per unit mass must be maximized. The widespread use of fibre composites depends on the cheapness with which they can be produced. Long graphite fibres, for example, can be produced from commercially drawn acrylic fibres. These fibres have the strength of a modern steel, twice the modulus and one-quarter the density. Such a composite has a (modulus/density) ratio four times, and a (strength/density) ratio up to twice that of steel. Many other covalent solids can be relatively easily produced in strong whisker form by filamentary growth; these are very small (≈ 1 μm diameter but have large aspect ratios (≈ 1000). Whiskers of sapphire, silicon carbide, silicon nitride and aluminium nitride are also aivailable commercially, and means have been developed of introducing them into metallic and resin matrices. Wires of the stronger metals such as tungsten and molybdenum, which retain their strengths to very high temperatures, also provide excellent reinforcement.

11.14 Superalloys

In these alloys γ' (Ni_3Al) and γ^* (Ni_3Nb) are the principal strengtheners by chemical and coherency strain hardening. Another source of strengthening is due to solid solution hardening; Cr is a major element, Co may be added up to 20 per cent, and Mo, W and Ta up to a total of 15 per cent. These elements also dissolve in γ' so that the hardening effect may be twofold. In high-temperature service, the properties of the grain boundaries are as important as the strengthening by γ' within the grains. Grain boundary strengthening is produced mainly by precipitation of chromium and refractory metal carbides; small additions of Zr and B improve the morphology and stability of these carbides. Optimum properties are developed by multi-stage heat treatment; the intermediate stages produce the desired grain boundary microstructure of carbide particles enveloped in a film of γ' and the other stages produce two size ranges of γ' for the best combination of strength at both intermediate and high temperatures. *Table 11.2* indicates the effect of the different alloying elements.

TABLE 11.2 Influence of various alloying additions in superalloys

Influence	*Cr*	*Al*	*Ti*	*Co*	*Mo*	*W*	*B*	*Zr*	*C*	*Nb*	*Hf*
Matrix strengthening	✓			✓	✓	✓					
γ'-formers		✓	✓							✓	
Carbide-formers	✓			✓		✓				✓	✓
Grain boundary strengthening							✓	✓	✓		✓
Oxide scale formers	✓	✓									

Some of the nickel-based alloys have a tendency to form an embrittling σ-phase (based on the composition FeCr) after long term in-service applications, when composition changes occur removing σ-resisting elements such as Ni and enhancing σ-promoting elements such as Cr, Mo or W. This tendency is predicted in alloy design by a technique known as Phacomp (phase computation) based on Pauling's model of hybridization of $3d$-electrons in transition metals (*see* Chapter 5). While a fraction of the $3d$ orbitals hybridize with p and s orbitals to create the metallic bond, the remainder forms non-bonding orbitals which partly fill the electron holes in the d-shell, increasing through the transition series to give

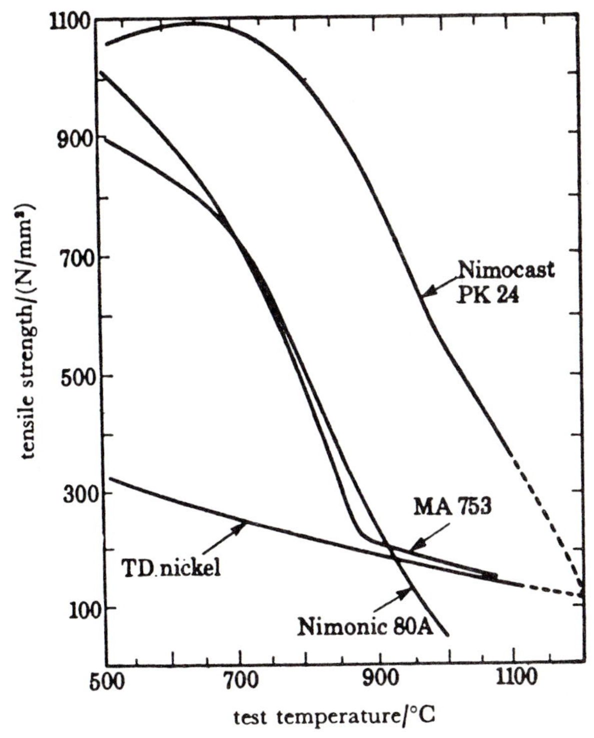

Figure 11.21 Tensile strength v. temperature for a typical wrought alloy (Nimonic 80A), a typical cast alloy (Nimocast PK 24 or IN-100), a simple dispersion hardened alloy (TD Nickel) and a combined precipitation-hardened/dispersion-hardened alloy (MA 753), (Betteridge & Heslop 1974)

electron hole numbers N_v for Cr(4.66), Mn(3.66), Fe(3.66), Co(1.71) and Ni(0.66). Computation shows that the γ/σ phase relation depends on the average hole number $\bar{N}_v$ given by

$$N_v = \sum_{i=1}^{n} m_i (N_v)_i$$

where m_i is the atomic fraction of the ith element of electron hole number N_v and n is the number of elements in the alloy. The limit of γ-phase stability is reached at $\bar{\bar{N}}_v \approx 2.5$.

All γ'-hardened alloys experience a reduction in strength at elevated temperatures because of the solution of γ' precipitate. To produce improved high-temperature strength, alloys hardened by oxides, particularly thoria have been introduced, but such thoria-dispersed (TD) nickel has poor properties at intermediate temperatures, as shown in *Figure 11.21*. Nowadays, the combined properties of γ'-hardened and dispersion hardened alloys can be obtained by a process known as mechanical alloying. In this method, the thoria dispersoid is added to the γ'-hardened alloy by a high energy input milling process, and typical properties for such an alloy MA753 are also shown in *Figure 11.21*.

Suggestions for further reading

Kelly, A. and Nicholson, R.B. (eds), *Strengthening Methods in Crystals*, Elsevier, 1971

Martin, J.W. and Doherty, R.D., *Stability of Microstructure in Metallic Systems*, Cambridge University Press, 1980

Martin, J.W., *Micromechanisms in Particle-Hardened Alloys*, Cam. Solid State Series, 1980

Porter, D.A. and Easterling, K.E., *Phase Transformations in Metals and Alloys*, Van Nostrand Reinhold, 1981

Chapter 12

Phase transformations II – the eutectoid transformation

12.1 Introduction

Eutectoid decomposition occurs in both ferrous (e.g. iron–carbon) and non-ferrous (e.g. copper–aluminium, copper–tin) alloy systems, but it is of particular importance industrially in governing the hardening of steels. In the iron–carbon system (*see Figure 3.5(b)*) the γ-phase, austenite, which is a solid solution of carbon in f.c.c. iron, decomposes on cooling to give a structure known as pearlite, composed of alternate lamellae of cementite (Fe_3C) and ferrite (α-Fe). However, when the cooling conditions are such that the alloy structure is far removed from equilibrium, an alternative transformation may occur. Thus, on very rapid cooling, a metastable phase called martensite, which is a super-saturated solid solution of carbon in ferrite, is produced. The microstructure of such a transformed steel is not homogeneous but consists of platelike needles of martensite embedded in a matrix of the parent austenite. Apart from martensite, another structure known as bainite may also be formed if the formation of pearlite is avoided by cooling the austenite rapidly through the temperature range above 550 °C, and then holding the steel at some temperature between 250°C and 550°C. A bainitic structure consists of platelike grains of ferrite, somewhat like the plates of martensite, inside which carbide particles can be seen.

The structure produced when austenite is allowed to transform isothermally at a given temperature can be conveniently represented by a diagram of the type shown in *Figure 12.1*. *Figure 12.1(a)*, for example, shows the time necessary at a given temperature to transform austenite of eutectoid composition to one of the three structures: pearlite, bainite or martensite. Such a diagram, made up from the results of a series of isothermal-decomposition experiments, is called a TTT curve, since it relates the transformation product to the time at a given temperature. It is, because of its shape, sometimes referred to as an S-curve. It will be evident from such a diagram that a wide variety of structures can be obtained from the austenite decomposition of a particular steel; the structure may range from 100 per cent coarse pearlite, when the steel will be soft and ductile, to fully martensitic, when the steel will be hard and brittle. It is because this wide range of properties can be produced by the transformation of a steel that it remains a major constructional material for engineering purposes.

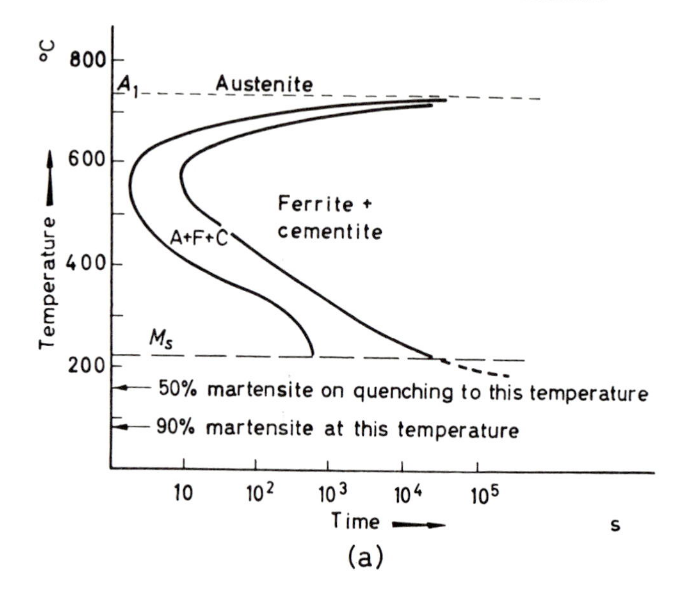

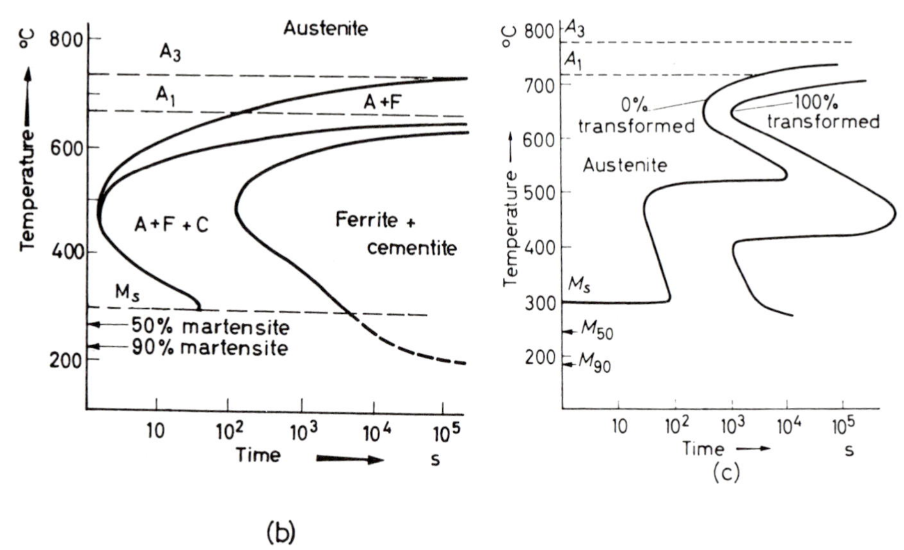

Figure 12.1 TTT curves for (a) eutectoid, (b) pro-eutectoid, (c) low alloy (e.g. Ni/Cr/Mo) steels (after *Metals Handbook*)

From the TTT curve it can be seen that just below the critical temperature, A_1, the rate of transformation is slow even though the atomic mobility must be high in this temperature range. This, as pointed out in Chapter 4, is because any phase change involving nucleation and growth (e.g. the pearlite transformation) is faced with nucleation difficulties, which arise from the surface and strain energy factors. Of course, as the transformation temperature approaches the temperature corresponding to the knee of the curve, the transformation rate increases, as discussed in Chapter 4. The slowness of the transformation below the knee of the TTT curve, when bainite is formed, is also readily understood, since atomic migration is slow at these lower temperatures and the bainite transformation depends on diffusion. The lower part of the TTT curve below about 250–300 °C indicates, however, that the transformation takes place exceedingly fast, even though atomic mobility in this temperature range must be very low. For this reason it is concluded that the martensite transformation does not depend on the speed of migration of carbon atoms and, consequently, it is often referred to as a diffusionless transformation. The austenite only starts transforming to martensite when the temperature falls below a critical temperature, usually denoted by M_S. Below M_S the percentage of austenite transformed to martensite is indicated on the S-curve by a series of horizontal lines.

The M_S temperature may be predicted for steels containing various alloying elements in weight per cent by the formula, due to Steven and Haynes, given by $M_S\,(°C) = 561 - 474\,C - 33Mn - 17Ni - 17Cr - 21Mo$.

12.2 The austenite–pearlite reaction

If a homogeneous austenitic specimen of eutectoid composition were to be transferred quickly to a bath held at some temperature between 720 °C and 550 °C, decomposition curves of the form shown in *Figure 12.2(a)* would be obtained. These curves, typical of a nucleation and growth process, indicate that the transformation undergoes an incubation period, an accelerating stage and a decelerating stage; the volume transformed into pearlite has the time-dependence described by the Avrami equation 10.28. When the transformation is in its initial stage the austenite contains a few small pearlite nodules each of which grow during the period A to B (see curve obtained at 690 °C) and at the same time further nuclei form. The percentage of austenite transformed is quite small, since

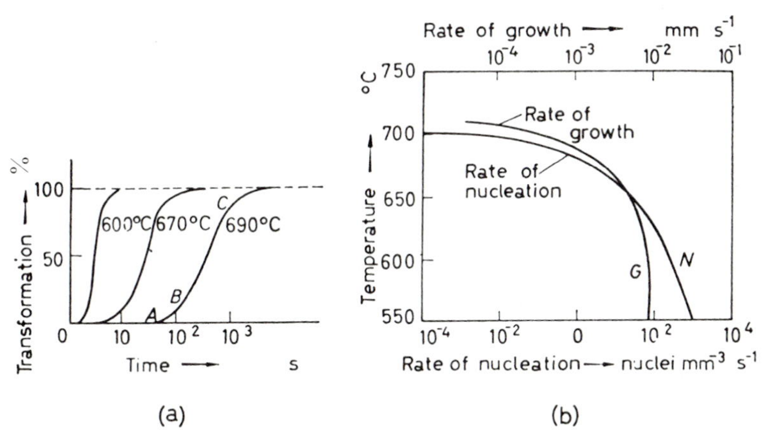

Figure 12.2 Effect of temperature on (a) amount of pearlite formed with time, (b) rate of nucleation and rate of growth of pearlite (after Mehl and Hagel, *Progress in Metal Physics, 6,* courtesy of Pergamon Press)

the nuclei are small and their total volume represents only a fraction of the original austenite. During the B to C stage the transformation rate accelerates, since as each nodule increases in size the area of contact between austenite and pearlite regions also increases: the larger the pearlite volumes the greater is the surface area upon which to deposit further transformation products. At C, the growing nodules begin to impinge on each other, so that the area of contact between pearlite and austenite decreases and from this stage onwards, the larger the nodules the lower is the rate of transformation.

12.2.1 Factors affecting nucleation and growth

Clearly, the rate of transformation depends on (a) the rate of nucleation of pearlite nodules, N (i.e. the number of nuclei formed in unit volume in unit time), and (b) the rate of growth of these nodules, $\boldsymbol{G}$ (i.e. the rate that the radius of the nodule increases with time). The variation of N and G with temperature for a eutectoid steel is shown in *Figure 12.2(b)*.

The rate of nucleation increases with decreasing temperature down to the knee of the curve and in this respect is analogous to other processes of phase precipitation where hysteresis occurs (*see* Chapter 4). In addition, the nucleation rate is very structure sensitive so that nucleation occurs readily in regions of high energy where the structure is distorted. In homogeneous austenite the nucleation of pearlite occurs almost exclusively at grain boundaries and, for this reason, the size of the austenite grains, prior to quenching, has an important effect on hardenability (a term which denotes the depth in a steel to which a fully martensitic structure can be obtained). Coarse-grained steels can be hardened more easily than fine-grained steels, because to obtain maximum hardening in a steel, the decomposition of austenite to pearlite should be avoided, and this is more easily accomplished if the grain boundary area, or the number of potential pearlite nucleation sites, is small. Thus, an increase in austenite grain size effectively pushes the upper part of the S-curve to longer times, so that, with a given cooling rate, the knee can be avoided more easily. The structure-sensitivity of the rate of nucleation is also reflected in other ways. For example, if the austenite grain is heterogeneous, pearlite nucleation is observed at inclusions as well as at grain boundaries. Moreover, plastic deformation during transformation increases the rate of transformation, since the introduction of dislocations provides extra sites for nucleation, while the vacancies produced by plastic deformation enhance the diffusion process.

The rate of growth of pearlite, like the rate of nucleation, also increases with decreasing temperature down to the knee of the curve, even though it is governed by the diffusion of carbon, which, of course, decreases with decreasing temperature. The reason for this is that the interlamellar spacing of the pearlite also decreases rapidly with decreasing temperature, and because the carbon atoms do not have to travel so far, the carbon supply is easily maintained. In contrast to the rate of nucleation, however, the rate of growth of pearlite is quite structure insensitive and, therefore, is indifferent to the presence of grain boundaries or slag inclusions. These two factors are important in governing the size of the pearlite nodules produced. If, for instance, the steel is transformed just below A_1, where the rate of nucleation is very low in comparison with the rate of growth (i.e. the ratio N/G is small), very large nodules are developed. Then, owing to the structure-insensitivity of the growth process, the few nodules formed are able to grow across grain boundaries, with the result that pearlite nodules larger than the original austenite grain size are often observed. By comparison, if the steel is transformed at a lower temperature, just above the knee of the TTT curve where N/G is large, the rate of nucleation is high and the pearlite nodule size is correspondingly small.

12.2.2 Mechanism and morphology

The growth of pearlite from austenite clearly involves two distinct processes: (a) a redistribution of carbon (since the carbon concentrates in the cementite and avoids the ferrite), and (b) a crystallographic change (since the structure of both ferrite and cementite differs from that of austenite). Of these two processes it is generally agreed that the rate of growth is governed by the diffusion of carbon atoms, and the crystallographic change occurs as readily as the redistribution of carbon will allow. The active nucleus of the pearlite nodule may be either a ferrite or cementite platelet, depending on the conditions of temperature and composition which prevail during the transformation, but usually it is assumed to be cementite. The nucleus may form at a grain boundary as shown in *Figure 12.3*(*a*), and after its formation the surrounding matrix is depleted of carbon, so that conditions favour the nucleation of ferrite plates adjacent to the cementite nucleus, *Figure 12.3*(*b*). The ferrite plates in turn reject carbon atoms into the surrounding austenite and this favours the formation of cementite nuclei, which

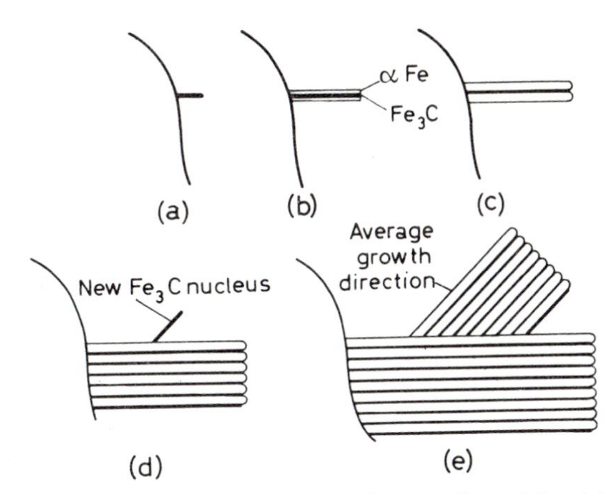

Figure 12.3 Nucleation and growth of pearlite nodules. (a) Initial Fe_3C nucleus; (b) Fe_3C plate full grown, α-Fe now nucleated; (c) α-Fe plates now full grown, new Fe_3C plates nucleated; (d) Fe_3C nucleus of different orientation forms and original nodule grows; (e) new nodule at advanced stage of growth (after Mehl and Hagel, *Progress in Metal Physics, 6*, courtesy of Pergamon Press)

then continue to grow. At the same time as the pearlite nodule grows sideways, the ferrite and cementite lamellae advance into the austenite, since the carbon atoms rejected ahead of the advancing ferrite diffuse into the path of the growing cementite, *Figure 12.3*(*c*). Eventually, a cementite plate of different orientation forms and this acts as a new nucleus as shown in *Figure 12.3*(*d*) and (*e*).

Homogeneous austenite, when held at a constant temperature, produces pearlite at a constant rate and with a constant interlamellar spacing. However, the interlamellar spacing decreases with decreasing temperature, and becomes unresolvable in the optical microscope as the temperature approaches that corresponding to the knee of the curve. An increase in hardness occurs as the spacing decreases. Zener explains the dependence of interlamellar spacing on temperature in the following way. If the interlamellar spacing is large, the diffusion distance of the carbon atoms in order to concentrate in the cementite is also large, and the rate of carbon redistribution is correspondingly slow. Conversely, if the spacing is small the area, and hence energy, of the ferrite–cementite interfaces become large. In consequence, such a high proportion of the free energy released in the austenite to pearlite transformation is needed to provide the interfacial energy, that little will remain to provide the 'driving force' for the change. Thus, a balance between these two opposing conditions is necessary to allow the formation of pearlite to proceed, and at a constant temperature the interlamellar spacing remains constant. However, because the free energy change, ΔG, accompanying the transformation increases with increasing degree of undercooling, larger interfacial areas can be tolerated as the temperature of transformation is lowered, with the result that the interlamellar spacing decreases with decreasing temperature.

12.2.3 Hypo-eutectoid steels

The majority of commercial steels are not usually of the eutectoid composition (0.8 per cent carbon), but hypo-eutectoid, i.e. <0.8 per cent carbon. In such steels, pro-eutectoid ferrite is first formed before the pearlite reaction begins and this is shown in the TTT curve by a third decomposition line. From *Figure 12.1*(*b*) it can be seen that the amount of pro-eutectoid ferrite decreases as the isothermal transformation temperature is lowered. The morphology of the precipitated ferrite depends on the usual precipitation variables, i.e. temperature, time, carbon content and grain size, and growth occurs preferentially at grain boundaries and on certain crystallographic planes. The Widmanstätten pattern with ferrite growing along $\{111\}$ planes of the parent austenite is a familiar structure of these steels.

12.2.4 The influence of alloying elements

With the exception of cobalt, all alloying elements in small amounts retard the transformation of austenite to pearlite. These elements decrease both the rate of nucleation, N, and the rate of growth, G, so that the top part of the TTT curve is displaced towards longer times. This has considerable technological importance since in the absence of such alloying elements, a steel can only transform into the harder constituents of bainite or martensite if it is in the form of very thin sections so that the cooling rate will be fast enough to avoid crossing the knee of the TTT curve during the cooling process and hence avoid pearlite transformation. For this reason most commercially heat-treatable steels contain either one or more of the elements chromium, nickel, manganese, vanadium, molybdenum or tungsten. The exact mechanism causing retardation of the transformation is still in doubt, since the most commonly accepted theory, which suggests that the slowness of diffusion of the alloying elements is the controlling factor, has not been completely substantiated. For example, it is found that cobalt increases both N and G, and that its effect on the pearlite interlamellar spacing is anomalous in that contrary to the other elements which generally increase it, additions of cobalt decrease the spacing.

With large additions of alloying elements, the simple form of TTT curve often becomes complex, as shown in *Figure 12.1*(*c*). Thus to obtain any desired structure by heat treatment, a detailed knowledge of the TTT curve is essential.

12.3 The austenite–martensite reaction

Martensite, the hardening constituent in quenched steels, is formed at temperatures below about 200 °C. The regions of the austenite which have transformed to martensite are lenticular in shape and may easily be recognized by etching, or from the distortion they produce on the polished surface of the alloy. These relief effects, shown schematically in *Figure 12.4*, indicate that the martensite needles have been formed not with the aid of atomic diffusion but by a shear process, since if atomic mobility were allowed the large strain energy associated with the transformed volume would then be largely avoided. The lenticular shape of a martensite needle is a direct consequence of the stresses produced in the surrounding matrix by the shear mechanism of the transformation and is exactly analogous to the similar effect found in mechanical twinning. The strain energy associated with martensite is tolerated because the growth of such sheared regions does not depend on diffusion, and since the regions are coherent with the matrix they are able to spread at great speed through the crystal. The large free energy change associated with the rapid formation of the new phase outweighs the strain energy, so that there is a net lowering of free energy.

Direct TEM observations of martensite plates have shown that there are two main types of martensite, one with a twinned structure (*see Figure 12.5*), known as acicular martensite, and the other with a high density of dislocations but few or no twins, called massive martensite.

12.3.1 The crystallography of the martensite transformation

In contrast to the pearlite transformation, which involves both a redistribution of carbon atoms and a structural change, the martensite transformation involves only a change in crystal structure. As discussed in

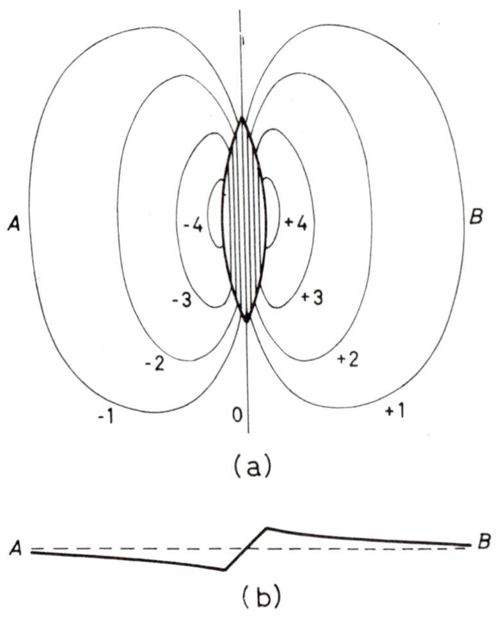

Figure 12.4 Schematic diagram of the observed shape deformation produced by a martensite plate. (a) Contour lines on an originally flat surface, (b) section of the surface through *AB*. Vertical scale much exaggerated (after Bilby and Christian, *The Mechanics of Phase Transformations in Metals,* courtesy of the Institute of Metals)

Chapter 5, the structure cell of martensite is body-centred tetragonal, which is a distorted form of a body-centred cubic structure, and hence may be regarded as a supersaturated solution of carbon in α-iron. x-ray examination shows that while the c/a ratio of the b.c.t. structure of martensite increases with increasing carbon content, the curve of c/a ratio against composition extrapolates back to $c/a=1$ for zero carbon content, and the lattice parameter is equal to that of pure α-iron (*Figure 12.6*).

From the crystallographic point of view the most important experimental data in any martensite transformation are the orientation relations of the two phases and the habit plane. In steel, there are three groups of orientations often quoted; those due to Kurdjumov and Sachs, Nishiyama, and Greninger and Troiano, respectively. According to the Kurdjumov–Sachs relation, in iron–carbon alloys with 0.5 to 1.4 per cent carbon, a $\{111\}_\gamma$ plane of the austenite lattice is parallel to the $\{110\}_\alpha$ plane of the martensite, with a $\langle 110 \rangle_\gamma$ axis of the former parallel to a $\langle 111 \rangle_\alpha$ axis of the latter; the associated habit plane is $\{225\}_\gamma$. In any one crystal there are 24 possible variants of the Kurdjumov–Sachs relationship, consisting of 12 twin pairs, both orientations of a pair having the same habit plane. However, for general discussion it is usual to choose one relation which may be written

$$(111)_\gamma \parallel (101)_\alpha \text{ with } [1\bar{1}0]_\gamma \parallel [11\bar{1}]_\alpha$$

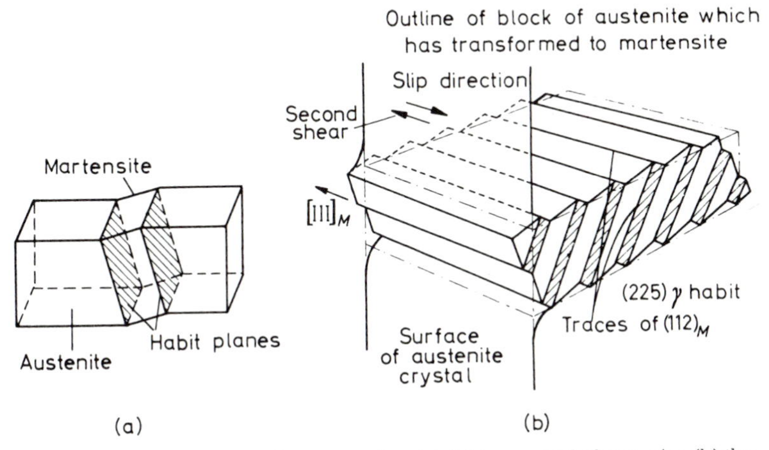

Figure 12.5 (a) Formation of a martensite platelet in a crystal of austenite, (b) the inhomogeneous twinning shear within the platelet (after Kelly and Nutting, *Proc. Roy. Soc.*, 1960, **A259,** 45, courtesy of the Royal Society)

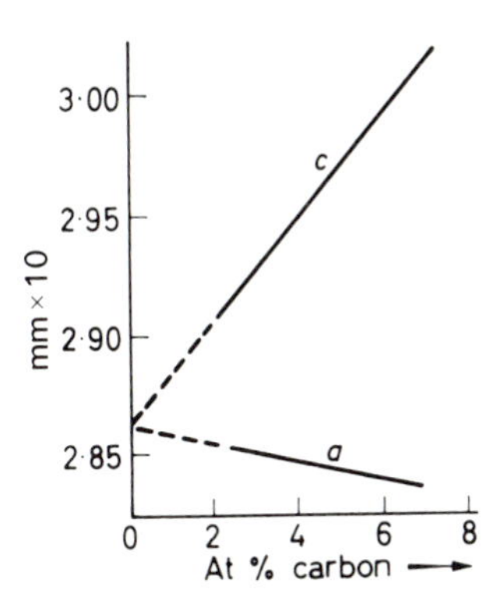

Figure 12.6 Variation of *c* and *a* parameters with carbon content in martensite (after Kurdjumov)

In the composition range 1.5 to 1.8 per cent carbon the habit plane changes to $\approx\{259\}_\gamma$ with the orientation relationship unspecified. This latter type of habit plane has also been reported by Nishiyama for iron–nickel alloys (27 to 34 per cent nickel) for which the orientation relationship is of the form

$$(111)_\gamma \parallel (101)_\alpha \text{ with } [1\bar{2}1]_\gamma \parallel [10\bar{1}]_\alpha$$

However, Greninger and Troiano have shown by precision orientation determinations that irrational relationships are very probable, and that in a ternary iron–nickel–carbon alloy (0.8 per cent carbon, 22 per cent nickel), $(111)_\gamma$ is approximately 1° from $(101)_\alpha$ with $[1\bar{2}1]_\gamma$ approximarely 2° from $[101]_\alpha$, and is associated with a habit plane about 5° from (259).

12.3.2 Mechanism of martensite formation

The martensite transformation is diffusionless, and therefore martensite forms without any interchange in the position of neighbouring atoms.

Accordingly, the observed orientation relationships are a direct consequence of the atom movements that occur during the transformation. The first suggestion of a possible transformation mechanism was made by Bain in 1934. He suggested that since austenite may be regarded as a body-centred tetragonal structure of axial ratio $\sqrt{2}$, the transformation merely involves a compression of the c-axis of the austenite unit cell and expansion of the a-axis. The interstitially dissolved carbon atoms prevent the axial ratio from going completely to unity, and, depending on composition, the c/a ratio will be between 1.08 and 1.0. Clearly, such a mechanism can only give rise to three martensite orientations whereas in practice 24 result. To account for this, Kurdjumov and Sachs proposed that the transformation takes place not by one shear process, but by a sequence of two shears (*Figure 12.7*), first along the elements $(111)_\gamma\langle 112\rangle_\gamma$, and then a minor shear along the elements $(112)_\alpha\langle 111\rangle_\alpha$; these elements are the twinning elements of the f.c.c. and b.c.c. lattice respectively. This mechanism predicts the correct orientation relations, but not the correct habit characteristics or relief effects. Accordingly, Greninger and Troiano in 1941 proposed a different two-stage transformation, consisting of an initial shear on the irrational habit plane which produces the relief effects, together with a second shear along the twinning elements of the martensite lattice. If slight adjustments in spacing are then allowed, the mechanism can account for the relief effects, habit plane, the orientation relationship and the change of structure.

Further additions to these theories have been made in an effort to produce the ideal general theory of the crystallography of martensite transformation. Bowles, for example, replaces the first shear of the Greninger–Troiano mechanism by the general type of homogeneous deformation in which the habit plane remains invariant, i.e. all directions in this plane are unrotated and unchanged in

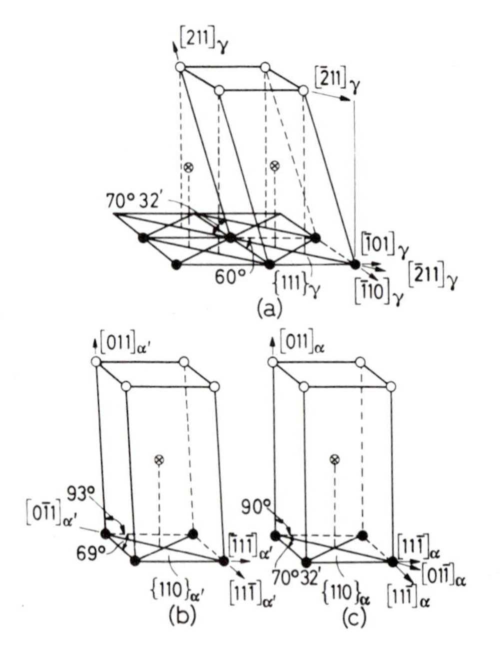

Figure 12.7 Shear mechanism of Kurdjumov and Sachs. (a) Face-centred austenite with {111}γ in horizontal plane, (b) Body-centred tetragonal martensite (α′), (c) cubic ferrite (α) (after Bowles and Barrett, *Progress in Metal Physics*, 3, courtesy of Pergamon Press)

length. However, in all such cases the problem resolves itself into one of determining whether a homogeneous strain can transform the γ-lattice into the α-lattice, while preserving coherency at the boundary between them. The homogeneous strain does not do this, so that some reasonable additional type of strain has to be added.

This shear can occur either by twinning or by slip, the mode prevailing depending on the composition and cooling rate. Between carbon contents of 0.2 and 0.5% the martensite changes from dislocated martensite arranged in thin lathes or needles to twinned acicular martensite arranged in plates. In the martensite formed at low C contents, e.g. Fe–Ni alloys, the thin laths lie parallel to each other, with a $\{111\}_\gamma$ habit, to form pockets of massive martensite with jagged boundaries due to the impingement of other nearby pockets of laths. The inhomogeneous shear produced by deformation twinning occurs on $\{112\}$ planes in the martensite, so that each martensite plate is made up of parallel twin plates of thickness 2–50 nm. By operation of such a complex transformation mode with a high index habit plane the system maintains an invariant interfacial plane.

Because of the shears involved and the speed of the transformation it is attractive to consider that dislocations play an important role in martensite formation. Some insight into the basic dislocation mechanisms has been obtained by *in situ* observations during either cooling below M_S or by straining, but unfortunately only for Ni/Cr austenitic steels with low stacking fault energy, i.e. $\gamma \approx 20$ mJ/m^2. For these alloys it has been found that stacking faults are formed either by emitting partial dislocations with $b = a/6\langle 112\rangle$ from grain boundaries or by the dissociation of unit dislocations with $b = a/2\langle 110\rangle$. In regions of the grain where on cooling or deformation a high density of stacking faults developed, the corresponding diffraction pattern revealed c.p.h. ε-martensite. On subsequent deformation or cooling, regions of ε-martensite transform rapidly into b.c.c. α-martensite, and indeed, the only way in which α-martensite was observed to form was from an ε nucleus.

Because straining or cooling can be interrupted during the *in situ* experiments it was possible to carry out a detailed analysis of the defect structure formed prior to a region becoming recognizably (from diffraction patterns) martensitic. In this way it has been shown that the interplanar spacing across the individual stacking faults in the austenite decreased to the (0001) spacing appropriate to ε-martensite. *Figure 12.8* shows micrographs which reveal this change of spacing; no contrast is expected in *Figure 12.8*(*b*) if the faulted $\{111\}$ planes remained at the f.c.c. spacing, since the condition of invisibility $g \cdot R = n$ is obeyed. The residual contrast observed arises from the supplementary displacement ΔR across the faults which, from the white outer fringe, is positive (intrinsic) in nature for both faults and $\approx 2\%$ of the $\{111\}$ spacing. The formation of regions of α from ε could also be followed although in this case the speed of the transformation precluded detailed analysis. *Figure 12.9* shows a micrograph taken after the formation of α-martensite and this, together with continuous observations, show that the martensite/matrix interface changes from $\{111\}$ to the well known $\{225\}$ as it propagates. Clearly one of the important roles that the formation of ε-martensite plays in acting as a precursor for the formation of α-martensite is in the generation of close-packed planes with ABAB stacking so that

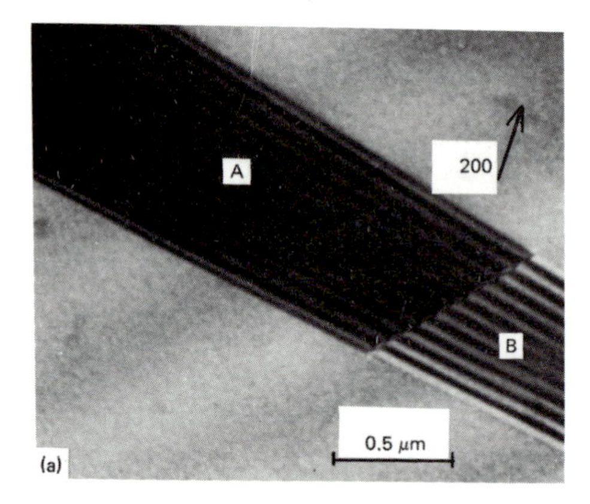

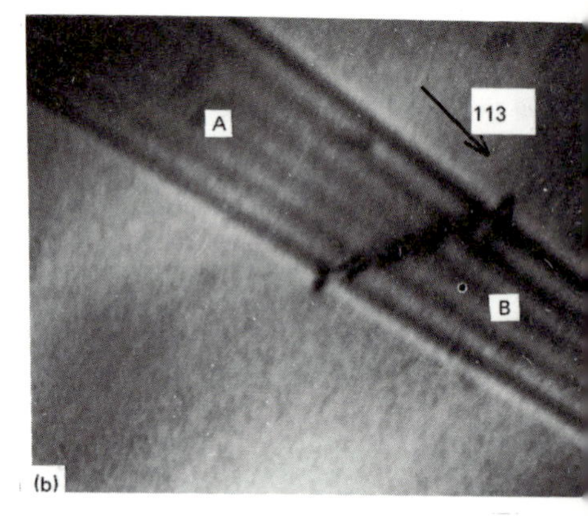

Figure 12.8 Electron micrographs showing (a) contrast from overlapping faults on $(\bar{1}11)$; A is extrinsic and B is intrinsic in nature, (b) residual contrast arising from a supplementary displacement across the faults which is intrinsic in nature for both faults A and B (after Brooks, Loretto and Smallman)

Figure 12.9 Electron micrograph showing an α-martensite plate, the austenite-martensite interface, and the faults in the austenite matrix

atomic shuffles can subsequently transform these planes to $\{110\}$ b.c.c. which are of course stacked ABAB (*see Figure 12.10*). The α-martensite forms in dislocation pile-ups where the $a/6\langle 112\rangle$ partials are forced closer together by the applied stress. The volume of effective b.c.c. material increases as more dislocations join the pile-up until the nucleus formed by this process reaches a critical size and rapid growth takes place. The martensite initially grows perpendicular to, and principally on, one side of the $\{111\}_\gamma$ slip plane associated with the nucleus, very likely corresponding to the side of the dislocations with missing half planes since α-martensite is less dense than austenite.

12.3.3 The kinetics of formation

One of the most distinctive features of the martensite transformation is that in most systems martensite is formed only when the specimen is cooling, and that the rate of martensite formation is negligible if cooling is stopped. For this reason

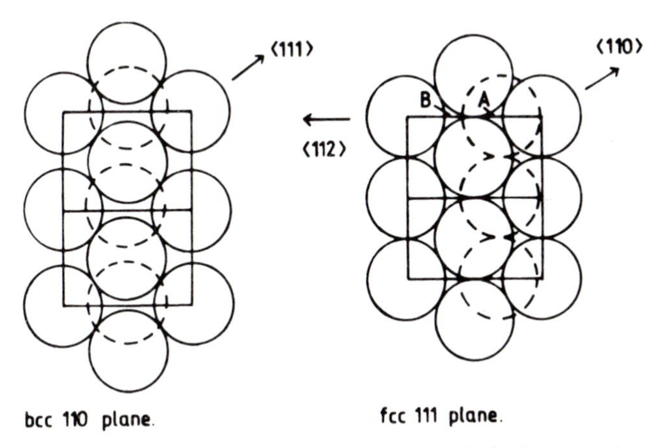

Figure 12.10 A shear of $\frac{a}{6}$<112> moves atoms in the f.c.c. structure from A sites to B sites, and after half this shear the structure has pseudo-bcc packing

the reaction is often referred to as an athermal* transformation, and the percentage of austenite transformed to martensite is indicated on the S-curve by a series of horizontal lines. The transformation begins at a temperature M_S, which is not dependent on cooling rate, but is dependent on prior thermal and mechanical history, and on composition. For example, it is well established that the M_S temperature decreases approximately linearly with increasing concentration of solutes such as carbon, nickel or manganese.

Speed of formation The observation that martensite plates form rapidly and at a rate which is temperature-independent shows that thermal activation is not required for the growth process. Electronic methods show that the martensite needles form, in iron–nickel–carbon alloys, for example, in about 10^{-7} seconds and, moreover, that the linear growth velocity is about 10^3 m s^{-1} even at very low temperatures. Such observations show that the activation energy for the growth of a martensite plate is virtually zero, and that the velocity of growth approaches the speed of sound in the matrix. Sometimes a 'burst phenomenon' is exhibited, as for example in iron–nickel alloys, when the stresses produced by one martensite plate assist in the nucleation of others. The whole process is autocatalytic and about 25 per cent of the transformation can occur in the time interval of an audible click.

The effect of applied stress Since the formation of martensite involves a homogeneous distortion of the parent structure, it is expected that externally applied stresses will be of importance. Plastic deformation is effective in forming martensite above the M_S temperature, provided the temperature does not exceed a critical value usually denoted by M_d. However, cold work above M_d may either

* In some alloys, such as iron–manganese–carbon and iron–manganese–nickel, the martensitic transformation occurs isothermally. For these systems, growth is still very rapid but the nuclei are formed by thermal activation.

accelerate or retard the transformation on subsequent cooling. Even elastic stresses, when applied above the M_S temperature and maintained during cooling, can affect the transformation; uniaxial compression or tensile stresses raise the M_S temperature while hydrostatic stresses lower the M_S temperature.

Stabilization When cooling is interrupted below M_S, stabilization of the remaining austenite often occurs. Thus, when cooling is resumed martensite forms only after an appreciable drop in temperature. Such thermal stabilization has been attributed, by some workers, to an accumulation of carbon atoms on those dislocations important to martensite formation. This may be regarded as a direct analogue of the yield phenomenon. The temperature interval before transformation is resumed increases with holding time and is analogous to the increase in yield drop accompanying carbon build-up on strain-ageing. Furthermore, when transformation in a stabilized steel does resume, it often starts with a 'burst', which in this case is analogous to the lower yield elongation.

12.4 The austenite–bainite transformation

The bainite reaction has many features common to both the pearlite and martensite reactions. The pearlite transformation involves the redistribution of carbon followed by a structure change, the martensite transformation involves the structure change alone, and by contrast the bainite transformation involves a structure change followed by the redistribution of carbon, which precipitates out as a carbide. Consequently, the austenite–bainite decomposition may be regarded as a martensite transformation involving the diffusion of carbon atoms, so that, in this case, the rate of coherent growth is necessarily slow compared with that of martensite. Lower bainite is hardly distinguishable from martensite tempered at the same temperature, while upper bainite exhibits an acicular structure. The metallographic appearance of the transformed steel is found to alter continuously between these two extremes, the actual structure exhibited being governed by the diffusion rate of the carbon, which in turn depends on the temperature of the transformation. The hardness of the reaction product also increases continuously with decreasing temperature, lower bainite being harder than upper bainite, which is harder than most fine pearlite.

The ferrite in bainite has a martensite-like appearance and is, in most cases, clearly distinguished from both ferrite and pro-eutectoid ferrite formed in the pearlite range. The bainitic ferrite exhibits the same surface relief effects as martensite, while pro-eutectoid ferrite and pearlite do not. Such surface tilting is further evidence for a shear-like transformation, but the orientation relationship between austenite and bainite is not necessarily the same as that between austenite and martensite. In fact, the bainitic ferrite has the same orientation with respect to the parent austenite as does pro-eutectoid ferrite, which suggests that ferrite may nucleate bainite.

12.5 Tempering and heat treatment

The presence of martensite in a quenched steel, while greatly increasing its hardness and UTS, causes the material to be brittle. Such behaviour is hardly surprising, since the formation of martensite is accompanied by severe matrix distortions. The hardness and strength of martensite increase sharply with increase in C content. Contributions to the strength arise from the carbon in solution, carbides precipitated during the quench, dislocations introduced during the transformation and the grain size.

Although the martensite structure is thermodynamically unstable, the steel will remain in this condition more or less indefinitely at room temperature because for a change to take place bulk diffusion of carbon, with an activation energy Q of approximately 83 kJ/mol atom is necessary. However, because there is an exponential variation of the reaction rate with temperature, the steel will be able slowly to approach the equilibrium structure at a slightly elevated temperature, i.e. rate of reaction $= A \exp[-Q/\boldsymbol{k}T]$. Thus, by a carefully controlled tempering treatment, the quenching stresses can be relieved and some of the carbon can precipitate from the supersaturated solid solution to form a finely dispersed carbide phase. In this way, the toughness of the steel can be vastly improved with very little detriment to its hardness and tensile properties.

The structural changes which occur on tempering may be considered to take place in three stages. In the primary stage, fine particles of a c.p.h. carbide phase (ε-carbide) of composition about $Fe_{2.4}C$, precipitates, with the corresponding formation of low carbon martensite. This low carbon martensite grows at the expense of the high carbon martensite until at the end of this stage the structure consists of retained austenite, ε-carbide and martensite of reduced tetragonality. During the second stage any retained austenite in the steel begins to transform isothermally to bainite, while the third stage is marked by the formation of cementite platelets. The precipitation of cementite is accompanied by a dissolution of the ε-carbide phase so that the martensite loses its remaining tetragonality and becomes b.c.c. ferrite. The degree to which these three stages overlap will depend on the temperature of the anneal and the carbon content. In consequence, the final structure produced will be governed by the initial choice of steel and the properties, and hence thermal treatment, required. Alloying elements, with the exception of Cr, affect the tempering of martensite. Plain carbon steels soften above 100 °C owing to the early formation of ε-carbide, whereas in Si-bearing steels the softening is delayed to above 250 °C, since Si stabilizes ε-carbide and delays its transformation to cementite. Alloying additions (*see Table 12.1*) thus enable the improvement in ductility to be achieved at higher tempering temperatures.

When a steel specimen is quenched prior to tempering, quenching cracks often occur. These are caused by the stresses which arise from both the transformation and the differential expansion produced when different parts of the specimen cool at different rates. To minimize such cracking, the desired properties of toughness and strength are often produced in the steel by alternative heat-treatment schedules; examples of these schedules are summarized in *Figure*

TABLE 12.1 Influence of alloying additions on tempering

Element	*Retardation in tempering per 1% addition*	*Ratio of retardation of tempering to depression of M_S*
C	−40	negative
Co	8	>8
Cr	0	0
Mn	8	0.24
Mo	17	0.8
Ni	8	0.5
Si	20	1.8
V	30	>1.0
W	10	0.9

12.11, from which it will become evident that advantage is taken of the shape of the TTT curve to economize on the time the specimen is in the furnace, and also to minimize quenching stresses. During conventional annealing, for example, the steel is heated above the upper critical temperature and allowed to cool slowly in the furnace. In isothermal annealing the steel is allowed to transform in the furnace, but when it has completely transformed, the specimen is removed from the furnace and allowed to air-cool, thereby saving furnace time. In martempering, the knee of the TTT curve is avoided by rapid cooling, but the quench is interrupted above M_S and the steel allowed to cool relatively slowly through the martensite range. With this treatment the thermal stresses set up by very rapid cooling are reduced. Such a procedure is possible because at the holding temperature there is ample time for the temperature to become equal throughout the sample before the transformation begins, and as a result the transformation occurs much more uniformly. After the transformation is complete, tempering is carried out in the usual way. In austempering, quenching is again arrested above M_S and a bainite product, having similar properties to tempered martensite, is allowed to form.

Alloying elements also lower the M_S temperatures and consequently, greater stresses and distortion are introduced during quenching. This can be minimized by austempering and martempering as discussed above, but such treatments are expensive. Alloying elements should therefore be chosen to produce the maximum retardation of tempering for minimum depression of M_S; *Table 12.1* shows that (i) C should be as low as possible, (ii) Si and Co are particularly effective, and (iii) Mo is the preferred element of the Mo, W, V group since it is easier to take into solution that V and is cheaper than W.

Some elements, particularly Mo and V, produce quite high tempering temperatures. In quantities above about 1% for Mo and ½% for V, a precipitation reaction is also introduced which has its maximum hardening effect at 550 °C. This phenomenon of increased hardness by precipitation at higher temperatures

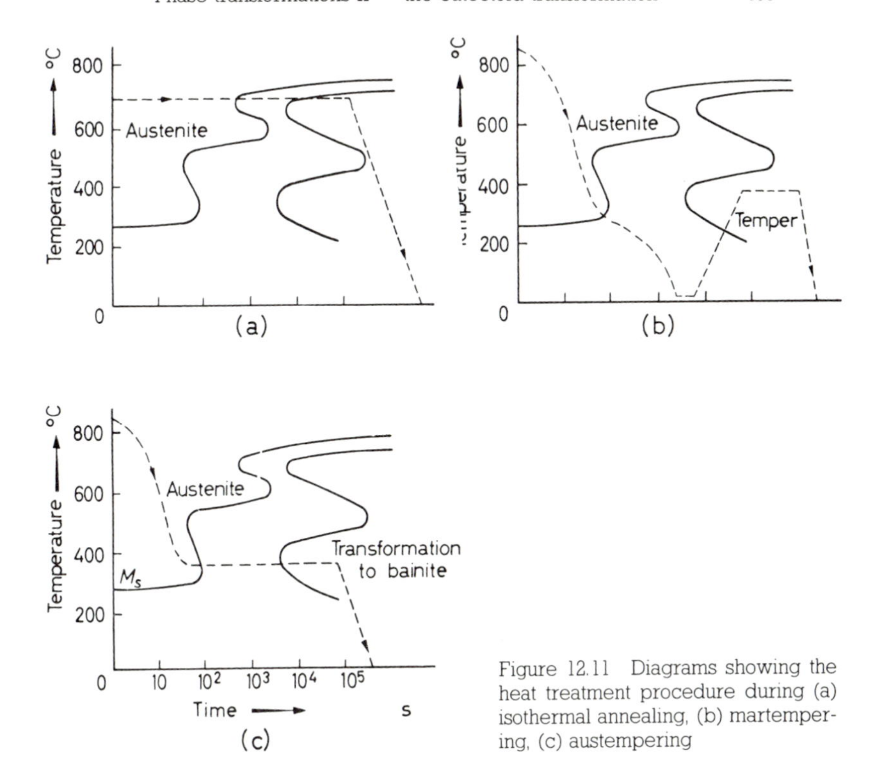

Figure 12.11 Diagrams showing the heat treatment procedure during (a) isothermal annealing, (b) martempering, (c) austempering

is known as secondary hardening and may be classified as a fourth stage of tempering. 2–2½% Mo addition produces adequate temper resistance and changes the precipitate to Mo_2C which is more resistant to overageing than Cr_7C_3 which is present in most alloy steels. High V additions lead to undissolved V_4C_3 at the quenching temperature, but 0.5 V in conjunction with 2% Mo does not form a separate carbide during tempering but dissolves in the Mo_2C. Cr also dissolved in Mo_2C but lowers the lattice parameters of the carbide and hence lowers the temper resistance by decreasing the matrix/carbide mismatch. However, 1% Cr may be tolerated without serious reduction in temper resistance and reduces the tendency to quench crack. Si decreases the lattice parameter of matrix ferrite and hence increases temper resistance. A typical secondary hardening steel usually contains 0.4% C, 2% Mo, 0.5% V, 0.5% Si and 1.5% Cr, with 1.8 GN/m^2 UTS and 15 per cent elongation.

12.6 Thermo-mechanical treatments

To produce steels with an improved strength/ductility ratio the heat-treatment may be modified to include a deformation operation in the process. The combined use of mechanical working and heat-treatment is generally called thermo-mechanical treatment (THT). Three types of treatment have proved successful with martensitic and bainitic steels. These may be classified as follows. (a) Deformation in the stable austenite range above A_3 before transformation, i.e. (HTHT). (b) Deformation below A_1 before transformation; this (LTHT) low temperature thermo-mechanical treatment is called ausforming. (c) Deformation during isothermal transformation to pearlite, i.e. below A_3, known as isoforming.

The main advantage of HTHT is in grain refinement and steels, such as silicon-steels that recrystallize slowly, are particularly suitable. It can, however, be applied to a low-alloy high carbon tool steels which are not suitable for ausforming, with significant increases in strength and toughness. The fatigue limit is also improved in many steels provided the deformation is limited to 25–30 per cent. In ausforming, the deformation is usually carried out in the range 450–550 °C and hence the steel must have a large bay in the TTT diagram to enable the deformation to be carried out. A suitable steel is En24 (0.35% C, 0.5% Mn, 1.5% Ni, 1.25% Cr, 0.25% Mo) for which the strength increases by about 4.6–7.7 MN/m^2 for each percent of deformation. The properties are improved as the deformation temperature is lowered, provided it is not below M_S, and with high deformation treatments (>70 per cent) strengths up to about E/70 with good ductility have been achieved. A very fine carbide dispersion is produced in the austenite during deformation together with a high density of dislocations. The removal of carbon from solution in the austenite means that during transformation the martensite formed is less supersaturated in C and thus has lower tetragonality and is more ductile. The carbides also pin the dislocations in the austenite, helping to retain some of them together with those formed during the transformation. The martensite formed is therefore heavily dislocated with relatively stable dislocations (compared to those which would be formed by deforming martensite at room temperature), and has superior strength and toughness. Such steels are, of course, somewhat difficult to machine.

Isoforming has potential in improving the toughness of low alloy steels. During isoforming to pearlite the normal ferrite/pearlite structure is modified, by the polygonization of sub-grains in the ferrite and the spheroidizing of cementite particles. Isoforming to bainite is also being studied.

12.7 Commercial steels and cast iron

12.7.1 Plain carbon steels

Carbon is an effective, cheap, hardening element for iron and hence a large tonnage of commercial steels contains very little alloying element. They may be divided conveniently into (i) low carbon (<0.3% C), medium carbon (0.3–0.7% C) and high carbon (0.7–1.7% C). *Figure 12.12* shows the effect of carbon on the strength and ductility. The low-carbon steels combine moderate strength with excellent ductility and are used extensively for their fabrication properties in the annealed or normalized condition for structural purposes, i.e. bridges, buildings, cars and ships. Even above about 0.2% C, however, the ductility is limiting for

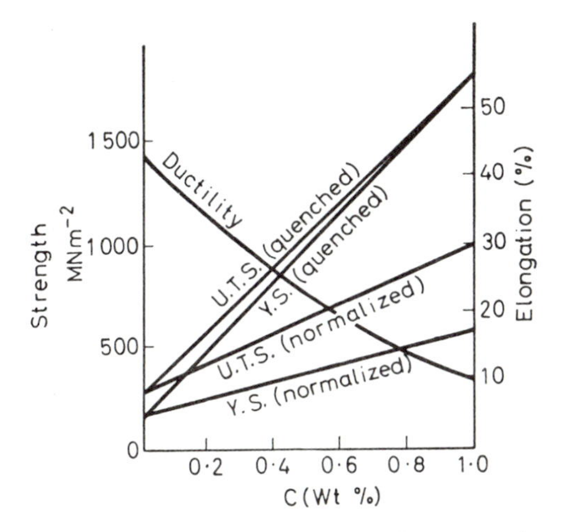

Figure 12.12 The influence of carbon content on the strength and ductility of steel

deep drawing operations, and brittle fracture becomes a problem particularly for welded thick sections. Improved low-carbon steels ($<0.2\%$ C) are produced by 'killing' the steel with Al or Si, or by adding Mn to refine the grain size. It is now more common, however, to add small amounts ($<0.1\%$) of Nb which reduces the carbon content by forming NbC particles. These particles not only restrict grain growth but also give rise to strengthening by precipitation hardening within the ferrite grains. Other carbide formers, such as Ti, may be used but because Nb does not de-oxidize it is possible to produce a semi-killed steel ingot, which because of its reduced ingot pipe gives increased tonnage yield per ingot cast.

Medium-carbon steels are capable of being quenched to form martensite and tempered to develop toughness with good strength. Tempering in higher temperature regions, i.e. 350–550 °C, produces a spheroidized carbide which toughens the steel sufficiently for use as axles, shafts, gears and rails. The process of ausforming can be applied to steels with this carbon content to produce even higher strengths without significantly reducing the ductility. The high-carbon steels are usually quench hardened and lightly tempered at 250 °C to develop considerable strength with sufficient ductility for springs, dies and cutting tools. Their limitations stem from their poor hardenability and their rapid softening properties at moderate tempering temperature.

12.7.2 Alloy steels

In low/medium alloy steels, with total alloying content up to about 5 per cent, the alloy content is governed largely by the hardenability and tempering requirements, although solid solution hardening and carbide formation may also be important. Some of these aspects have already been discussed, the main conclusions being that Mn and Cr increase hardenability and generally retard softening and tempering; Ni strengthens the ferrite and improves hardenability and toughness; copper behaves similarly but also retards tempering;

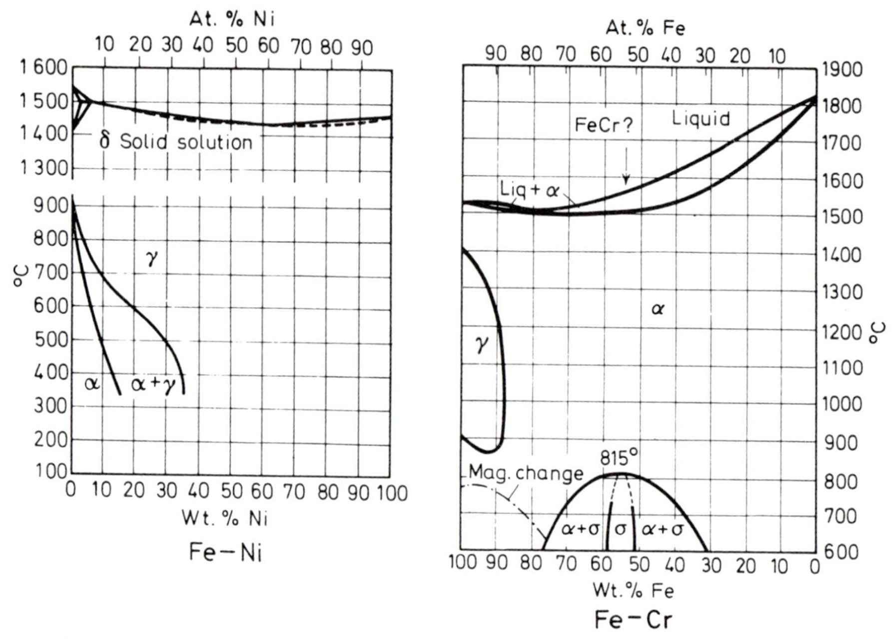

Figure 12.13 Effect of Ni and Cr on γ-field. (From Smithells, *Metals Reference Book*, Vol. 2, Butterworths, 1967)

Co strengthens ferrite and retards softening on tempering; Si retards and reduces the volume change to martensite, and both Mo and V retard tempering and provide secondary hardening.

In larger amounts, alloying elements either open up the austenite phase field, as shown in *Figure 12.13*(*a*), or close the γ-field (*Figure 12.13*(*b*)). 'Full' metals with atoms like hard spheres, e.g. Mn, Co, Ni, favour close-packed structures and open the γ-field, whereas the stable b.c.c. transition metals, e.g. Ti, V, Cr, Mo, close the field and form what is called a γ-loop. The development of austenitic steels, an important class of ferrous alloys, is dependent on the opening of the γ-phase field. The most common element added to iron to achieve this effect is Ni, as shown in *Figure 12.13*(*a*). From this diagram the equilibrium phases at lower temperatures for alloys containing 4–40% Ni are ferrite and austenite. In practice, it turns out that it is unnecessary to add the quantity of Ni to reach the γ-phase boundary at room temperature, since small additions of other elements tend to depress the γ/α transformation temperature range so making the γ metastable at room temperature. Interstitial C and N, which most ferrous alloys contain, also expand the γ-field because there are larger interstices in the f.c.c. than the b.c.c. structure. The other common element which expands the γ-field is Mn. Small amounts (<1%) are usually present in most commercial steels to reduce the harmful effect of FeS. Up to 2% Mn may be added to replace the more expensive Ni, but additions in excess of this concentration have little commercial significance until 12% Mn is reached. Hadfield's steel contains 12–14% Mn, 1% C and is noted for its toughness and is used in railway points, drilling machines and rock-crushers. The steel is water-quenched to produce austenite. The f.c.c. structure has good fracture resistance and, having a low stacking fault energy, work hardens very rapidly. During the abrasion and work hardening, the hardening is further intensified by a partial strain transformation of the austenite to martensite; this principle is used also in the sheet forming of stainless steels (*see* below).

To make the austenitic steels resistant to oxidation and corrosion (*see* Chapter 15) the element Cr is usually added in concentrations greater than 12%. Chromium closes the γ-field, however, and with very low carbon contents single-phase austenite cannot be produced with the stainless ($>12\%$) composition. These alloys form the stainless (ferritic) irons and are easily fabricated for use as furnace components. Increasing the carbon content expands the γ-loop and in the medium-carbon range Cr contents with good stainless qualities (≈ 15–18%) can be quench-hardened for cutlery purposes where martensite is required to give a hard, sharp cutting edge. The combination of both Cr and Ni, i.e. 18/8, produces the metastable austenitic stainless steel which is used in chemical plant construction, kitchenware and surgical instruments because of its ductility, toughness and cold working properties. Metastable austenitic steels have good press forming properties because the strain induced transformation to martensite provides an additional strengthening mechanism to work hardening, and moreover counteracts any drawing instability by forming martensite in the locally thinned, heavily deformed regions.

High strength transformable stainless steels with good weldability to allow fabrication of aircraft and engine components have been developed from the 0.05–0.1% C, 12% Cr, stainless by secondary hardening additions (1.5–2% Mo; 0.3–0.5% V). Small additions of Ni or Mn (2%) are also added to counteract the ferrite forming elements Mo and V to make the steel fully austenitic at the high temperatures. Air quenching to give α followed by tempering at 650 °C to precipitate Mo_2C produces a steel with high yield strength (0.75 GN/m^2) high UTS (1.03 GN/m^2) and good elongation and impact properties. Even higher strengths can be achieved with stainless (12–16% Cr; 0.05% C) steels which although austenitic at room temperature (5% Ni, 2% Mn) transform on cooling to -78 °C. The steel is easily fabricated at room temperature, cooled to control the transformation and finally tempered at 650–700 °C to precipitate Mo_2C.

12.7.3 Maraging steels

A serious limitation in producing high strength steels is the associated reduction in fracture toughness. Carbon is one of the elements which mostly affects the toughness and hence in alloy steels it is reduced to as low a level as possible, consistent with good strength. Developments in the technology of high alloy steels have produced high strengths in steels with very low carbon contents ($<0.03\%$) by a combination of martensite and age-hardening, called maraging. The maraging steels are based on an Fe–Ni containing between 18 and 25% Ni to produce massive martensite on air cooling to room temperature. Additional hardening of the martensite is achieved by precipitation of various intermetallic compounds, principally Ni_3Mo or $Ni_3(Mo, Ti)$ brought about by the addition of roughly 5% Mo, 8% Co as well as small amounts of Ti and Al; the alloys are solution heat treated at 815 °C and aged at about 485 °C. Many substitutional elements can produce age-hardening in Fe–Ni martensites, some strong (Ti, Be), some moderate (Al, Nb, Mn, Mo, Si, Ta, V) and other weak (Co, Cu, Zr) hardeners. There can, however, be rather strong interactions between elements such as Co and Mo, in that the hardening produced when these two elements are present together is much greater than if added individually. It is found that A_3B-type compounds are favoured at high Ni or (Ni + Co) contents and A_2B Laves phases at lower contents.

In the unaged condition maraging steels have a yield strength of about 0.69 GN/m^2. On ageing this increases to between 0.69 and 2.0 GN/m^2 and the precipitation strengthening is due to an Orowan mechanism according to the relation $\sigma = \sigma_0 + (\alpha\mu b/L)$ where σ_0 is the matrix strength, α a constant and L the interprecipitate spacing. The primary precipitation strengthening effect arises from the (Co + Mo) combination, but Ti plays a double role as a supplementary hardener and a refining agent to tie up residual carbon. The alloys generally have good weldability, resistance to hydrogen embrittlement (*see* page 284) and stress-corrosion but are used mainly, particularly the 18% Ni alloy, for their excellent combination of high strength and toughness.

12.7.4 High-strength, low-alloy steels (HSLA)

The requirement for structural steels to be welded satisfactorily has led to steels with lower C (<0.1%) content. Unfortunately, lowering the C content reduces the strength and this has to be compensated for by refining the grain size. This is difficult to achieve with plain C-steels rolled in the austenite range but the addition of small amounts of strong carbide-forming elements, e.g. <0.1% Nb, cause the austenite boundaries to be pinned by second-phase particles and fine-grain sizes (< 10 μm) to be produced by controlled rolling. Nitrides and carbo-nitrides as well as carbides, predominantly f.c.c. and mutually soluble in each other, may feature as suitable grain refiners in HSLA steels; examples include AlN, Nb(CN), V(CN), (NbV)CN, TiC and Ti(CN). The solubility of these particles in the austenite decreases in the order VC, TiC, NbC while the nitrides, with generally lower solubility, decrease in solubility in the order VN, AlN, TiN and NbN. Because of the low solubility of NbC, Nb is perhaps the most effective grain size controller. However, Al, V and Ti are effective in high nitrogen steels, Al because it forms only a nitride, V and Ti by forming V(CN) and Ti(CN) which are less soluble in austenite than either VC or TiC.

The major strengthening mechanism in HSLA steels is grain refinement but the required strength level is obtained usually by additional precipitation strengthening in the ferrite. VC, for example is more soluble in austenite than NbC, so if V and Nb are used in combination, then on transformation of the austenite to ferrite, NbC provides the grain refinement and VC precipitation strengthening; *Figure 12.14* shows a stress–strain curve from a typical HSLA steel.

12.7.5 Dual-phase steels

Much research into the deformation behaviour of specialty steels has been aimed at producing improved strength while maintaining good ductility. The

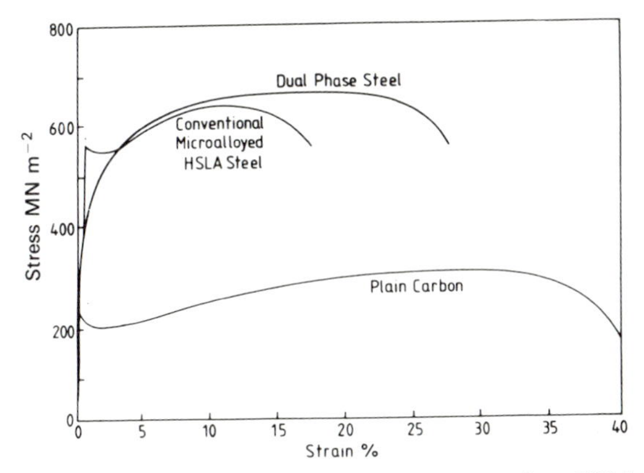

Figure 12.14 Stress–strain curves for plain carbon, HSLA and dual-phase steels

conventional means of strengthening by grain refinement, solid solution additions (Si, P, Mn) and precipitation hardening by V, Nb or Ti carbides (or carbonitrides) have been extensively explored and a conventionally treated HSLA steel would have a lower yield stress of 550 MN m^{-2}, a UTS of 620 MN m^{-2} and a total elongation of about 18 per cent. In recent years an improved strength–ductility relationship has been found for low-carbon, low-alloy steels rapidly cooled from an annealing temperature at which the steel consisted of a mixture of ferrite and austenite. Such steels have a microstructure containing principally low-carbon, fine-grained ferrite intermixed with islands of fine martensite and are known as dual-phase steels. Typical properties of this group of steels would be a UTS of 620 MN m^{-2}, a 0.2 per cent offset flow stress of 380 MN m^{-2} and a 3 per cent offset flow stress of 480 MN m^{-2} with a total elongation ≈ 28 per cent.

The implications of the improvement in mechanical properties are evident from an examination of the nominal stress–strain curves, shown in *Figure 12.14*. The dual-phase steel exhibits no yield discontinuity but work hardens rapidly so as to be just as strong as the conventional HSLA steel when both have been deformed by about 5 per cent. In contrast to ferrite–pearlite steels the work-hardening rate of dual-phase steel increases as the strength increases. The absence of discontinuous yielding in dual-phase steels is an advantage during cold pressing operations and this feature combined with the way in which they sustain work hardening to high strains makes them very attractive materials for sheet-forming operations. The flow stress and tensile strength of dual phase steels increase as the volume fraction of hard phase increases with a corresponding decrease in ductility; about 20 per cent volume fraction of martensite produces the optimum properties.

The dual phase is produced by annealing in the $(\alpha+\gamma)$ region followed by cooling at a rate which ensures that the γ-phase transforms to martensite, although some retained austenite is also usually present leading to a mixed martensite–austenite (M–A) constituent. To allow air-cooling after annealing, microalloying elements are added to low-carbon–manganese–silicon steel particularly vanadium or molybdenum and chromium. Vanadium in solid solution in the austenite increases the hardenability but the enhanced hardenability is due mainly to the presence of fine carbonitride precipitates which are unlikely to dissolve in either the austenite or the ferrite at the temperatures employed and thus inhibit the movement of the austenite/ferrite interface during the post-anneal cooling.

The martensite structure found in dual-phase steels is characteristic of plate martensite having internal microtwins. The retained austenite can transform to martensite during straining thereby contributing to the increase strength and work hardening. Interruption of the cooling, following intercritical annealing, can lead to stabilization of the austenite with an increased strength on subsequent deformation. The ferrite grains (≈ 5 μm) adjacent to the martensite islands are generally observed to have a high dislocation density resulting from the volume and shape change associated with the austenite to martensite transformation. Dislocations are also usually evident around retained austenite islands due to differential contraction of the ferrite and austenite during cooling.

Some deformation models of DP steels assume both phases are ductile and obey the Ludwig relationship, with equal strain in both phases. Measurements by several workers have, however, clearly shown a partitioning of strain between the martensite and ferrite, with the mixed (M–A) constituent exhibiting no strain until deformations well in excess of the maximum uniform strain. Models based on the partitioning of strain predict a linear relationship between yield stress, UTS and volume fraction of martensite but a linear relationship is not sensitive to the model. An alternative approach is to consider the microstructure as approximating to that of a dispersion-strengthened alloy. This would be appropriate when the martensite does not deform and still be a good approximation when the strain difference between the two phases is large. Such a model affords an immediate explanation of the high work hardening rate, as outlined in Chapter 10, arising from the interaction of the primary dislocations with the dense 'tangle' of dislocations generated in the matrix around the hard phase islands.

Several workers have examined DP steels to determine the effect of size and volume fraction of the hard phase. *Figure 12.15* shows the results at two different strain values and confirms the linear relationship between work-hardening rate ($d\sigma/d\varepsilon$) and $(f/d)^{1/2}$ predicted by the dispersion hardening theory (*see* page 352). Increasing the hard phase volume fraction while keeping the island diameter constant increases the work-hardening rate, increases the UTS but decreases the elongation. At constant volume fraction of hard phase, decreasing the mean island diameter produces no effect on the tensile strength but increases the work-hardening rate and the maximum uniform elongation (*Figure 12.16*). Thus the strength is improved by increasing the volume fraction of hard phase while the work hardening and ductility are improved by reducing the hard phase island size. Although dual-phase steels contain a complex microstructure it appears from their mechanical behaviour that they can be considered as agglomerates of non-deformable hard particles, made up of martensite and/or bainite and/or retained austenite, in a ductile matrix of ferrite. Consistent with the dispersion-strengthened model, the Bauschinger effect is rather large in dual-phase steels, as

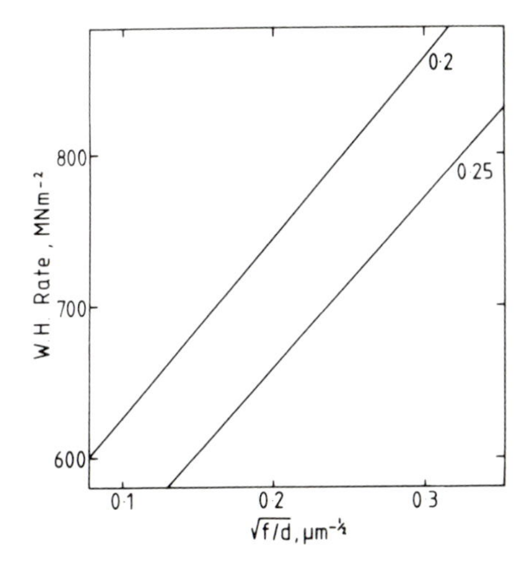

Figure 12.15 Dependence of work-hardening rate on (volume fraction f/particle size$d$)$^{1/2}$ for a dual-phase steel at strain values of 0.2 and 0.25 (after Ballinger and Gladman)

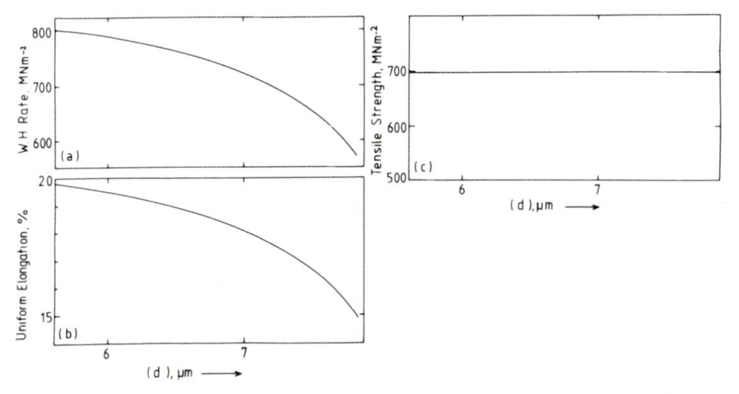

Figure 12.16 Effect of second phase particles size d at constant volume fraction f on (a) work-hardening rate, (b) elongation, (c) tensile strength (after Ballinger and Gladman)

shown in *Figure 12.17* and increases with increase in martensite content up to about 25 per cent. The Bauschinger effect arises from the long-range back stress exerted by the martensite islands, which add to the applied stress in reversed straining.

The ferrite grain size can give significant strengthening at small strains, but an increasing proportion of the strength arises from work hardening and this is independent of grain-size changes from about 3–30 μm. Solid solution strengthening of the ferrite, e.g. by silicon, enhances the work-hardening rate; P, Mn and V are also beneficial. The absence of a sharp yield point must imply that the dual-phase steel contains a high density of mobile dislocations. The microstructure exhibits such a dislocation density around the martensite islands but why these remain unpinned at ambient temperature is still in doubt, particularly as strain-ageing is significant on ageing between 423 and 573 K. Intercritical annealing allows a partitioning of the carbon to produce very low

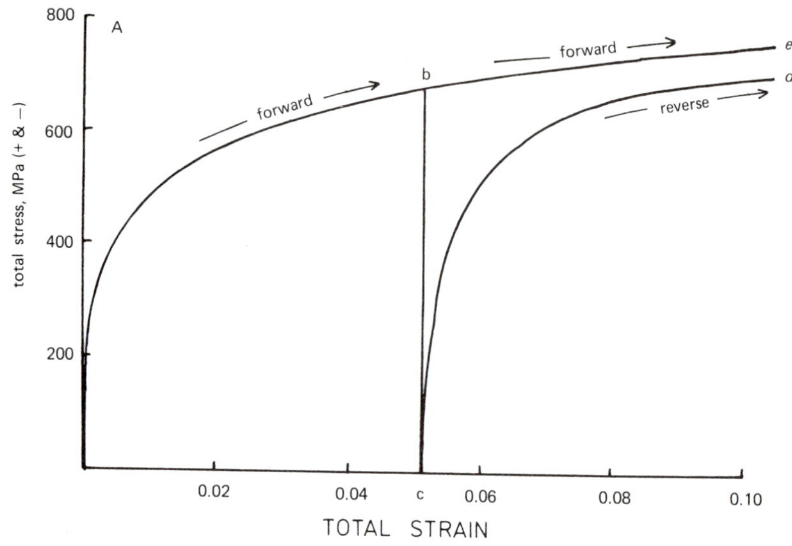

Figure 12.17 Bauschinger tests for a 0.06%C, 1.5%Mn, 0.85%Si dual-phase steel (Courtesy D. V. Wilson)

carbon ferrite, while aluminium- or silicon-killed steels have limited nitrogen remaining in solution. However, it is doubtful whether the concentration of interstitials is sufficiently low to prevent strain-ageing at low temperature; hence, it is considered more likely that continuous yielding is due to the residual stress fields surrounding second phase islands. Two possibilities then arise, (i) yielding can start in several regions at the same time rather than in one local region which initiates a general yield process catastrophically, and (ii) any local region is prevented from yielding catastrophically because the glide band has to overcome a high back stress from the M–A islands. Discontinuous yielding on ageing at higher temperatures is then interpreted in terms of the relaxation of these residual stresses, followed by classical strain ageing.

In dual-phase steels the n value ≈ 0.2 gives the high and sustained work-hardening rate required when stretch formability is the limiting factor in fabrication. However, when fracture *per se* is limiting, dual-phase steels probably perform no better than other steels with controlled inclusion content. Tensile failure of dual-phase steels is initiated either by decohesion of the martensite–ferrite interface or by cracking of the martensite islands. Improved fracture behaviour is obtained when the martensite islands are unconnected, when the martensite–ferrite interface is free from precipitates to act as stress raisers, and when the hard phase is relatively tough. The optimum martensite content is considered to be 20 per cent because above this level void formation at hard islands increases markedly.

12.7.6 Cast irons

In the iron–carbide system (page 99), the carbon is thermodynamically more stable as graphite than cementite. At the low carbon contents of typical steels, graphite is not formed, however, because of the sluggishness of the reaction to graphite. But when the carbon content is increased to that typical of cast irons (2–4% C) either graphite or cementite may separate depending on the cooling rate, chemical (alloy) composition and heat treatment. When the carbon exists as cementite the cast irons are referred to as white because of the bright fracture produced by this brittle constituent. In grey cast irons the carbon exists as flakes of graphite embedded in a ferrite–pearlite matrix and these impart a dull grey appearance to the fracture. When both cementite and graphite are present a 'mottled' iron is produced.

High cooling rates, which tend to stabilize the cementite, and the presence of carbide formers give rise to white irons. The addition of graphite forming elements (Si, Ni) produces grey irons, even when rapidly cooled if the Si is above 3%. These elements, particularly Si, alter the eutectic composition which may be taken into account by using the carbon equivalent of the cast iron, given by [total % C $\frac{1}{3}$(% Si + % P)], rather than the true carbon content. Phosphorus is present in most cast irons as a low melting point phosphide eutectic which improves the fluidity of the iron by lengthening the solidification period; this favours the decomposition of cementite. Grey cast iron is used for a wide variety of applications because of its good strength/cost ratio. It is easily cast into intricate shapes and has good machinability, since the chips break off easily at the graphite flakes. It also has a high damping capacity and hence is used for lathe and other machine frames where vibrations need to be damped out. The limited strength and ductility of grey cast iron may be improved by small additions of the carbide formers (Cr, Mo) which reduce the flake size and refine the pearlite. The main use of white irons is as a starting material for malleable cast iron, in which the cementite in the casting is decomposed by annealing. Such irons contain sufficient Si ($<1.3\%$) to promote the decomposition process during the heat treatment but not enough to produce graphite flakes during casting. Whiteheart malleable iron is made by heating the casting in an oxidizing environment, e.g. hematite iron ore at 900°C for 3–5 days. In thin sections the carbon is oxidized to ferrite, and in thick sections, ferrite at the outside gradually changes to graphite clusters in a ferrite-pearlite matrix near the inside. Blackheart malleable iron is made by annealing the white iron in a neutral packing, i.e. iron silicate slag, when the cementite is changed to rosette-shaped graphite nodules in a ferrite matrix. The deleterious cracking effect of the graphite flakes is removed by this process and a cast iron which combines the casting and machinability of grey iron with good strength and ductility, i.e. UTS 350 MN/m^2 and 5–15% elongation is produced. It is therefore used widely in engineering and agriculture where intricate shaped articles with good strength are required.

Even better mechanical properties (550 MN/m^2) can be achieved in cast irons, without destroying the excellent casting and machining properties, by the production of a spherulitic graphite. The spherulitic nodules are roughly spherical in shape and are composed of a number of graphite crystals, which grow radially from a common nucleus with their basal planes normal to the radial growth axis. This form of growth habit is promoted in an as-cast grey iron by the addition of small amounts of Mg or Ce to the molten metal in the ladle. Good strength, toughness and ductility can thus be obtained in castings that are too thick in section for malleabilizing and can replace steel castings and forgings in certain applications.

Suggestions for further reading

Chadwick, G.A., *Metallography of Phase Transformations*, Butterworths, 1972

Christian, J.W., *The Theory of Transformations in Metals*, Pergamon Press, 1965

Honeycombe, R.W.K., *Steels, Microstructure and Properties*, Edward Arnold, 1981

Petty, E.R., *Martensite – Fundamentals and Technology*, Longmans

Pickering, F.B., *Physical Metallurgy and the Design of Steels*, Applied Science Publishers, 1978

Symposium on The Mechanism of Phase Transformations in Metals, Institute of Metals, 1968

Porter, D.A. and Easterling, K.E., *Phase Transformations in Metals and Alloys*, Van Nostrand Reinhold, 1981

Chapter 13

Creep and fatigue

13.1 Creep

Creep (*see* Chapter 2) may be defined as the process by which plastic flow occurs when a constant stress is applied to a metal for a prolonged period of time. Disregarding the initial strain ε_0 which follows the application of the load, creep may be of two types: either purely transient, when only one stage of creep is exhibited, or continuous, when all three creep stages occur. The exact type of creep behaviour exhibited by the metal will depend on the stress and temperature during the test.

At low temperatures and stresses the creep curve, shown in *Figure 2.38*, exhibits only a transient flow which soon dies away almost completely. This creep behaviour is often called α or logarithmic creep, since it obeys an equation of the form

$$\varepsilon = \alpha \log t \tag{13.1}$$

Creep of this type is a low-temperature phenomenon and rarely occurs in everyday practice, where the resistance to plastic flow at high temperatures is usually being considered. Nevertheless, it is observed in polycrystalline aluminium and copper below 200 K, some hexagonal metals at low temperatures and some other materials, such as rubber and glass. Some form of transient creep will occur, of course, at all temperatures but it will not necessarily obey the logarithmic law. As a result, because it is transient creep which is mainly responsible for the initial shape of the creep curve, the creep curve starts off with the same steepness of slope at all temperatures and stresses, but bends away more gradually at high temperatures and stresses than at low ones.

At high temperatures and stresses, a specimen undergoing creep initially shows a rapid transient period of flow before it settles down to the linear steady-state stage, which eventually gives way to tertiary creep and fracture. As mentioned above, the transient creep observed under these conditions does not obey the logarithmic equation but instead one of the form

$$\varepsilon = \beta t^{1/3} \tag{13.2}$$

This type of transient creep is generally referred to as β-creep or Andrade creep,

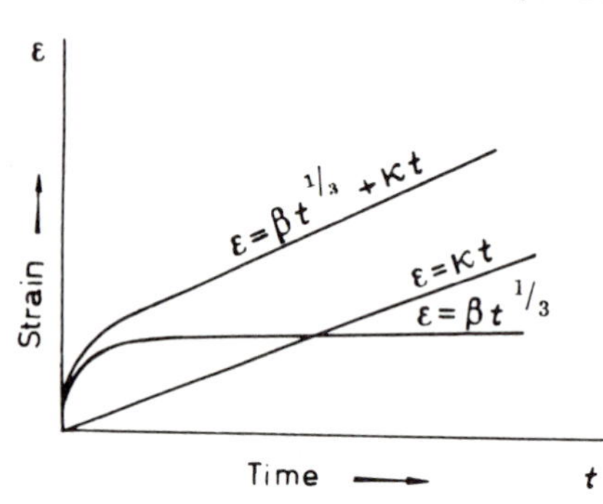

Figure 13.1 Combination of transient and steady-state creep

since Andrade demonstrated its general applicability to many polycrystalline metals when they are extended some 20 per cent in times of about 30 minutes.

The linear stage of creep is often termed steady-state or quasi-viscous creep and obeys the relation

$$\varepsilon = \kappa t \tag{13.3}$$

Consequently, because both transient and steady-state creep usually occur together during creep at high temperatures, the complete curve, *Figure 13.1*, during the primary and secondary stages of creep, fits the equation

$$\varepsilon = \beta t^{1/3} + \kappa t \tag{13.4}$$

extremely well. In contrast to transient creep, steady-state creep increases markedly with both temperature and stress. At constant stress the dependence on temperature is given by

$$\dot{\varepsilon}_{ss} = \mathrm{d}\varepsilon/\mathrm{d}t = \text{const.} \exp\left[-Q/\boldsymbol{k}T\right] \tag{13.5}$$

where Q is the activation energy for steady-state creep, while at constant temperature the dependence on stress σ (compensated for modulus E) is

$$\dot{\varepsilon}_{ss} = \text{const.}\ (\sigma/E)^n \tag{13.6}$$

Steady state creep is therefore

$$\dot{\varepsilon}_{ss} = A(\sigma/E)^n \exp\left(-Q/\boldsymbol{k}T\right) \tag{13.7}$$

The exponential dependence on temperature suggests that the predominating process during this stage of creep is one of recovery.

Accelerating creep is important in that creep fracture is most probably initiated during this stage. Both increased temperature and decreased strain-rate tend to produce intercrystalline fractures associated with creep.

13.1.1 Creep mechanisms

The general processes involved in creep are now reasonably well established. Four main deformation processes can be recognized:

1. the glide of dislocations leading to slip,
2. the climb of dislocations leading to sub-grain formation.

3. the sliding of grain boundaries.
4. the diffusion of vacancies.

Clearly, the contribution these different mechanisms make will depend on the conditions of temperature and stress prevailing during the test.

Transient creep

This form of creep, characterized by a creep rate that decreases with time, has been subdivided into α- or β-creep, depending on whether recovery does not or does occur during the test.

α-creep From the form of the decreasing creep-rate curve observed during creep it is evident that, whatever mechanism is involved, it must be one that becomes more difficult to operate as the test proceeds. One possibility is the exhaustion of dislocation sources, because at the applied stress level the most favourably oriented sources will operate first while the 'hard' sources operate later, but this concept cannot explain the structural changes which occur during creep. Most of the sources operate quite readily but some of the glissile dislocations have to surmount more difficult barriers than others. Thus, when the stress is applied to a specimen at a given temperature, those dislocations which are impeded by the smallest barriers will be able to surmount them by a process of thermal activation in a short time, and the more difficult barriers will be overcome in much longer times.

The logarithmic law then arises because augmentation of the applied stress to the value of the flow stress is necessary before plastic flow can occur during creep. This activation stress ($\Delta\sigma$ = flow stress − applied stress) occurs through thermal fluctuations, and the resultant activation energy, Q, will be an increasing function of $\Delta\sigma$. The creep rate will then be given by

$$\mathrm{d}\varepsilon/\mathrm{d}t = \text{const. } \exp\left[-Q/\boldsymbol{k}T\right]$$

and, if we suppose that Q depends upon strain,

$$\mathrm{d}\varepsilon/\mathrm{d}t = \text{const. } \exp\left[-B\varepsilon\right]$$

and integration according to

$$\int \exp(B\varepsilon)\,\mathrm{d}\varepsilon = \text{const.} \int \mathrm{d}t$$

gives the logarithmic creep formula $\varepsilon = \alpha \log \nu t$, where the constants α and ν depend upon the nature of the obstacle. By varying the details of the model various logarithmic laws can be obtained. For example, if the obstacle is a precipitate, Q is expected to vary as $(\Delta\sigma)^{3/2}$ and a time law of the form $\varepsilon = \alpha \log(\nu t)^{2/3}$ would be obeyed for dislocation loops jumping past them.

β-creep Equation 13.1 is obeyed in metals only at very low temperatures, in agreement with the belief that α-creep operates when recovery is absent. At higher temperatures, e.g. in the region of 250 K for aluminium, the observed creep rate is

represented approximately by a superposition of both α- and β-creep curves. Thus, after each thermally activated event it is found that a much larger contribution to the strain is made than can be expected from the dislocation intersecting mechanisms of α-creep alone, and this suggests that a recovery process also takes place. However, since the temperature is too low to allow normal 'static' recovery by climb, the experimental evidence favours a process of 'dynamic' recovery, involving cross-slip, for the mechanism of β-creep.

In aluminium and copper the β-term is larger the higher the temperature and the larger the applied stress. Moreover, the stress–temperature relation at that part of the test where the contribution from β-creep becomes noticeable is different for the two metals, and in both cases corresponds to the beginning of Stage III in the work hardening curve. The fact that the intersection mechanism of α-creep is the rate-controlling process up to 400 K in copper but only up to 250 K in aluminium may, from the theory of Stage III hardening (page 346), be attributed to the greater ribbon width of the stacking fault in copper compared to aluminium.

In the hexagonal metals, cadmium, zinc and magnesium, the situation is somewhat different. These metals have only one glide system, the basal plane, so that recovery is only possible by climb. The β-creep observed in zinc, cadmium and magnesium at room temperature and above is, therefore, due to the same thermally activated process that is operative in steady-state creep, i.e. climb.

Steady-state creep

The basic assumption of the mechanism of steady-state creep is that during the creep process the rate of recovery r, i.e. decrease in strength $(d\sigma/dt)$ is sufficiently fast to balance the rate of work hardening $h=(d\sigma/d\varepsilon)$. The creep rate $(d\varepsilon/dt)$ is then given by

$$d\varepsilon/dt=(d\sigma/dt)/(d\sigma/d\varepsilon)=r/h \tag{13.8}$$

However, in order to prevent work hardening, both the screw and edge parts of a glissile dislocation loop must be able to escape from tangled or piled-up regions. The edge dislocations will, of course, escape by climb, and since this process requires a higher activation energy than cross-slip, it will be the rate-controlling process in steady-state creep. The rate of recovery is governed by the rate of climb, which depends on diffusion and stress such that

$$r=A(\sigma/E)^p D=A(\sigma/E)^p D_0 \exp[-Q/\boldsymbol{k}T]$$

where D is a diffusion coefficient and the stress term arises because recovery is faster the higher the stress level and the closer dislocations are together. The work-hardening rate decreases from the initial rate h_0 with increasing stress, i.e. $h=h_0(E/\sigma)^q$, thus

$$\dot{\varepsilon}_{ss}=r/h=B(\sigma/E)^n D \tag{13.9}$$

where $B(=A/h_0)$ is a constant and n $(=p+q)$ is the stress exponent.

The structure developed in creep arises from the simultaneous work hardening and recovery. The dislocation density ρ increases with ε and the dislocation network gets finer, since dislocation spacing is proportional to $\rho^{-1/2}$.

At the same time the dislocations tend to reduce their strain energy by mutual annihilation and rearrange to form low angle boundaries and this increases the network spacing. Straining then proceeds at a rate at which the refining action just balances the growth of the network by recovery; the equilibrium network size being determined by the stress. Although dynamical recovery can occur by cross-slip, the rate controlling process in steady state creep is climb whereby edge dislocations climb out of their glide planes by absorbing or emitting vacancies the activation energy is therefore that of self-diffusion. Structural observations confirm the importance of the recovery process to steady-state creep. These show that sub-grains form within the original grains and, with increasing deformation, the sub-grain angle increases while the dislocation density within them remains constant*.

The climb process may, of course, be important in several different ways. Thus, climb may help a glissile dislocation to circumvent different barriers in the lattice, such as a sessile dislocation, or it may lead to the annihilation of dislocations of opposite sign on different glide planes. Moreover, because creep-resistant materials are rarely pure metals, the climb process may also be important in allowing a glissile dislocation to get round a precipitate or move along a grain boundary. A comprehensive analysis of steady-state creep, based on the climb of dislocations, has been given by Weertman. He assumes that dislocations are stopped by obstacles in the glide plane to form piled-up groups, and that the creep-rate is governed by the rate of climb of dislocations past the obstacle. The creep rate $\dot{\varepsilon}$ will then be $\alpha v_c/\ell$, where v_c is the speed of climb and ℓ the average climb distance. For the case where vacancy diffusion rather than emission or absorption at jogs is rate controlling, the vacancy concentration near a jog (*see* Chapter 9) is

$$c = c_0 \exp(Fb^2/\boldsymbol{k}T)$$

A steady state is reached when the vacancy flux created or destroyed at a dislocation equals the flux leaving or arriving under the influence of the concentration gradient. From Fick's Law, the flux ϕ per unit time and per unit length of dislocation is

$$\phi = D_v(c - c_0)/x = (D/c_0\Omega)[c - c_0]/x$$

where D_v is the vacancy diffusivity, Ω ($\simeq b^3$) the atomic volume, and x depends on the position r from the dislocation where $c = c_0$. The climb speed $v_c = \phi b^2$ and is given by

$$v_c = (D/\alpha b)[\exp(Fb^2/\boldsymbol{k}T) - 1]$$

with the force F for a vacancy creating dislocation and $-F$ for a vacancy annihilating dislocation. For a structure involving an equal number of each type of dislocation

$$\begin{aligned} v_c &= (D/\alpha b)[\exp(Fb^2/\boldsymbol{k}T) - \exp(-Fb^2/\boldsymbol{k}T)] \\ &= 2(D/\alpha b)\sinh(N\sigma b^3/\boldsymbol{k}T) \end{aligned} \qquad (13.10)$$

where N is a stress concentration factor. For a dislocation pile up at some barrier,

* Subgrains do not always form during creep and in some metallic solid solutions where the glide of dislocations is restrained due to the dragging of solute atoms, the steady state substructure is essentially a uniform distribution of dislocations.

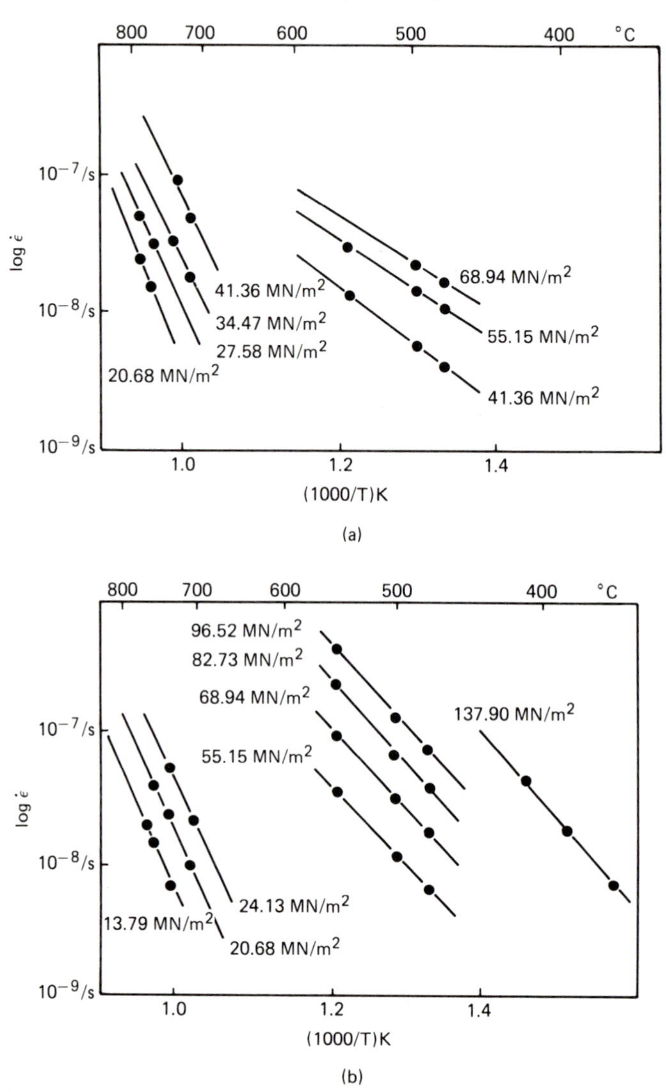

Figure 13.2 Log $\dot{\varepsilon}$ v. $^1/T$ for (a) Ni–Al_2O_3, (b) Ni-67% Co–Al_2O_3 showing the variation in activation energy above and below 0.5 T_m (after Hancock, Dillamore and Smallman)

N is proportional to σ and with $\ell\alpha(N\sigma)^{-1}$, the creep rate approximates to equation 13.9, with the stress exponent $n=4$.

The activation energy for creep Q may be obtained experimentally by plotting $\ln \dot{\varepsilon}_{ss}$ versus $1/T$, as shown in *Figure 13.2*. Usually above $0.5T_m$, Q corresponds to the activation energy for self diffusion E_{SD}, in agreement with the climb theory, but below $0.5T_m$, $Q<E_{SD}$, possibly corresponding to pipe diffusion. *Figure 13.3* shows that three creep regimes may be identified and the temperature range where $Q=E_{SD}$ can be moved to higher temperatures by increasing the strain-rate. Equation 13.9 shows that the stress exponent n can be obtained

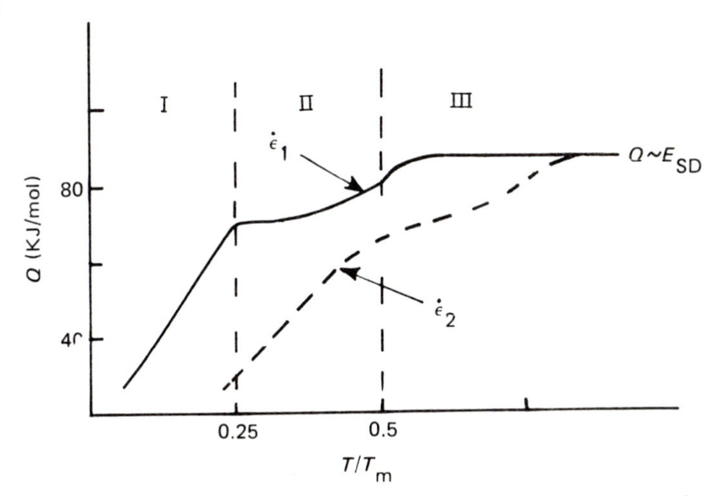

Figure 13.3 Variation in activation energy Q with temperature, for aluminium

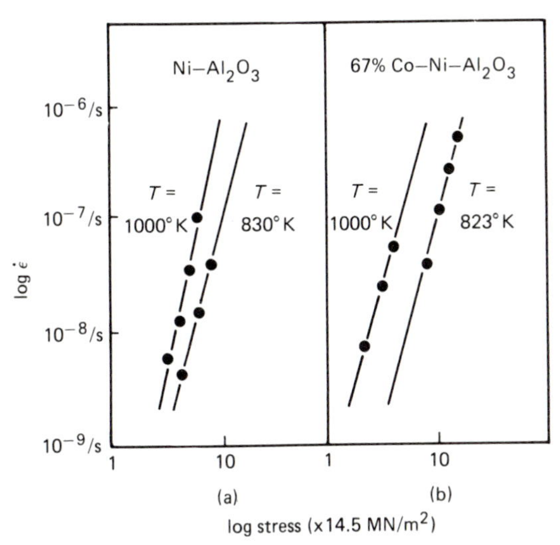

Figure 13.4 Log $\dot{\varepsilon}$ v. log σ for (a) $Ni-Al_2O_3$, (b) Ni-67% $Co-Al_2O_3$ (after Hancock, Dillamore and Smallman)

experimentally by plotting $\ln \dot{\varepsilon}_{ss}$ versus $\ln \sigma$, as shown in *Figure 13.4* where $n \approx 4$. While n is generally about 4–5 for dislocation creep, *Figure 13.5* shows that n may vary considerably from this value depending on the stress regime; at low stresses, i.e. regime I, creep occurs not by dislocation glide and climb, but by stress-directed flow of vacancies.

Creep due to grain boundaries

In the creep of polycrystals at high temperatures, the grain boundaries themselves are able to play an important part in the deformation process due to

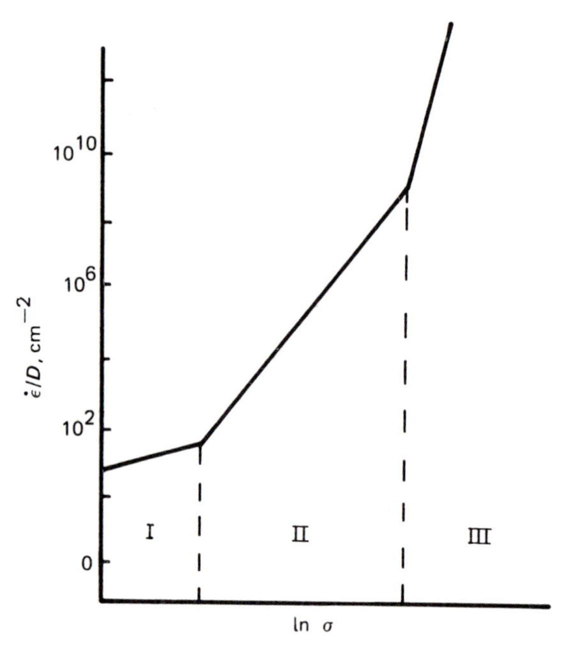

Figure 13.5 Schematic diagram showing influence of stress on diffusion-compensated steady-state creep

the fact that they may (a) slide past each other, or (b) create vacancies. Both processes involve an activation energy for diffusion and therefore may contribute to steady-state creep.

Grain boundary sliding During secondary creep, in addition to the climb of dislocations, it is possible that grains slide past each other along their boundaries. Grain boundary sliding during creep was inferred initially from the observation of steps at the boundaries, but the mechanism of sliding can be demonstrated on bi-crystals. *Figure 13.6* shows a good example of grain boundary movement in a bi-crystal of tin, where the displacement of the straight grain boundary across its middle is indicated by marker scratches.

Grain boundaries, even when specially produced for bi-crystal experiments, are not perfectly straight, and after a small amount of sliding at the boundary interface, movement will be arrested by protuberances. The grains are then locked, and the rate of slip will be determined by the rate of plastic flow in the protuberances. As a result, the rate of slip along a grain boundary is not constant with time, because the dislocations first form into piled-up groups, and later these become relaxed. Local relaxation may be envisaged as a process in which the dislocations in the pile-up climb towards the boundary, as shown in *Figure 13.7*, and in this way a vacancy flow is created along the boundary and towards the climbing dislocations. In consequence, the activation energy for grain boundary slip may be identified with that for steady-state creep. After climb, the dislocations are spread more evenly along the boundary, and are thus able to give

Figure 13.6 Grain boundary sliding on a bi-crystal tin (after Puttick and King, *J. Inst. Metals*, 1952, **81,** 537)

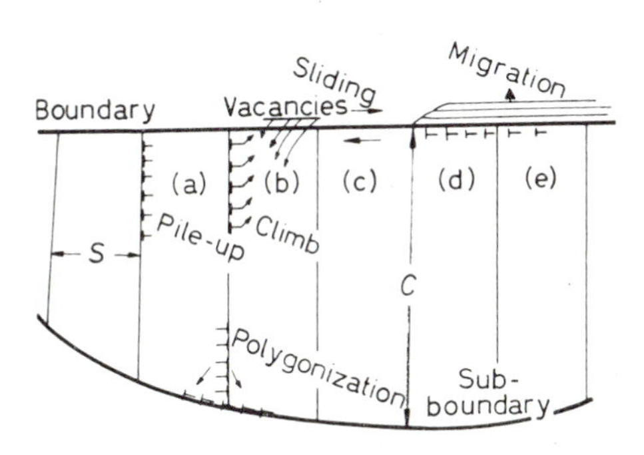

Figure 13.7 Sequence of events which may lead to boundary sliding and migration. *S*, slip spacing; *C*, sub-grain diameter (after Gifkins, *Fracture*, Technological Press, courtesy of John Wiley & Sons)

rise to grain boundary migration, when sliding has temporarily ceased, which is proportional to the overall deformation.

Diffusion creep A second creep process which also involves the grain boundaries, is one in which the boundary acts as a source and sink for vacancies. The mechanism depends on the migration of vacancies from one side of a grain to another, as shown in *Figure 13.8* and is often termed Herring–Nabarro creep, after the two workers who originally considered this process. If, in a grain of sides d under a stress σ, the atoms are transported from faces BC and AD to the faces AB and DC the grain creeps in the direction of the stress. To transport atoms in this way involves creating vacancies on the tensile faces AB and DC and destroying them on the other compressive faces by diffusion along the paths shown.

On a tensile face AB the stress exerts a force σb^2 (or $\sigma\Omega^{2/3}$) on each surface atom and so does work $\sigma b^2 \times b$ each time an atom moves forward one atomic

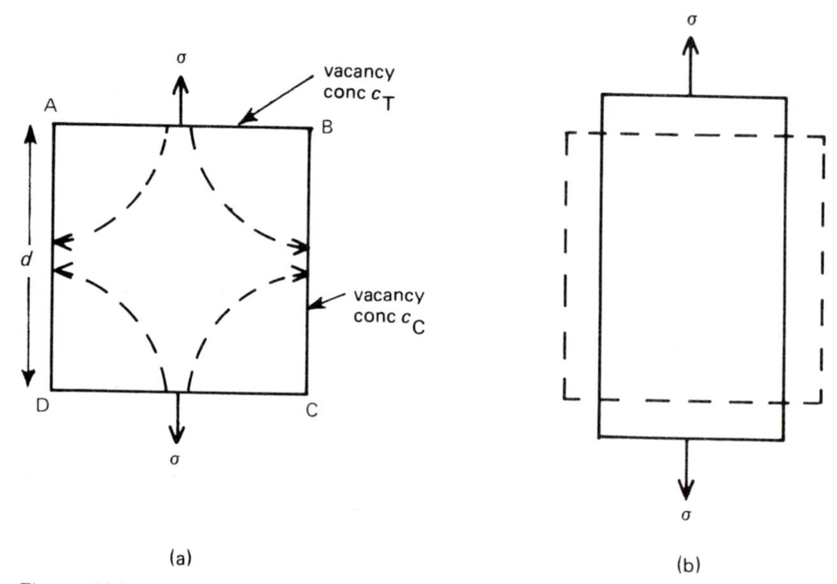

Figure 13.8 Schematic representation of Herring–Nabarro creep; with $c_c > c_T$, vacancies flow from the tensile faces to the longitudinal faces (a) to produce creep as shown in (b)

spacing b (or $\Omega^{1/3}$) to create a vacancy. The energy of formation at such a face is thus reduced to $(E_f - \sigma b^3)$ and the concentration of vacancies in equilibrium correspondingly increased to $c_T = \exp[(-E_f + \sigma b^3)/\boldsymbol{k}T] = c_0 \exp(\sigma b^3/\boldsymbol{k}T)$. The vacancy concentration on the compressive faces will be reduced to $c_c = c_0 \exp(-\sigma b^3/\boldsymbol{k}T)$. Vacancies will therefore flow down the concentration gradient, and the number crossing a face under tension to one under compression will be given by Fick's law as

$$\phi = -D_v d^2 (c_T - c_c)/\alpha d$$

where D_v is the vacancy diffusivity and α relates to the diffusion length, substituting for c_T, c_c and $D = (D_v c_0 b^3)$ leads to

$$\phi = 2dD \sinh(\sigma b^3/\boldsymbol{k}T)/\alpha b^3$$

Each vacancy created on one face and annihilated on the other produces a strain $\varepsilon = b^3/d^3$, so that the creep strain-rate $\dot{\varepsilon} = \phi(b^3/d^3)$. At high temperatures and low stresses this reduces to

$$\dot{\varepsilon}_{H\text{-}N} = 2D\sigma b^3/\alpha d^2 \boldsymbol{k}T = B_{H\text{-}N} D\sigma\Omega/d^2\boldsymbol{k}T \quad (13.11)$$

where the constant $B_{H\text{-}N} \sim 10$.

In contrast to dislocation creep, Herring–Nabarro creep varies linearly with stress and occurs at $T \approx 0.8\ T_m$ with $\sigma \approx 10^6$ N/m^2. The temperature range over which vacancy-diffusion creep is significant can be extended to much lower temperatures, i.e. $T \approx 0.5T_m$, if the vacancies flow down the grain boundaries rather than through the grains. Equation 13.11 is then modified for Coble or grain boundary diffusion creep, and is given by

$$\dot{\varepsilon}_{\text{Coble}} = B_c D_{gb} \sigma \Omega \omega / kT d^3 \tag{13.12}$$

where ω is the width of the grain boundary. Under such conditions, i.e. $T \approx 0.5$ to $0.6T_m$ and low stresses, diffusion creep becomes an important creep mechanism in a number of high-technology situations, and has been clearly identified in magnesium-based canning materials used in gas-cooled reactors.

Tertiary creep and fracture

Tertiary creep and fracture are logically considered together, since the accelerating stage represents the initiation of conditions which lead to fracture. An acceleration in the rate of creep will arise, of course, if large scale atomic movements, such as those which take place during recrystallization or when the specimen undergoes an overall change of cross-section, occur during the tests. However, these effects do not appear to be the general cause of accelerating creep because signs of recrystalliaztion are rarely observed after creep fracture, and also because accelerating creep is observed under a constant compressive load where cross-sectional changes would not lead to acceleration in the creep rate.

In many cases the onset of accelerating creep is an indication that voids or cracks are slowly but continuously forming in the material, and this has been confirmed by metallography and density measurements. The type of fracture resulting from tertiary creep is not transcrystalline but grain boundary fracture. Two types of grain boundary fracture have been observed. The first occurs principally at the triple point formed where three grain boundaries meet, and sliding along boundaries on which there is a shear stress produces stress concentrations at the point of conjunction sufficiently high to start cracks. However, under conditions of slow strain-rate for long times, which would be expected to favour recovery, small holes form on grain boundaries, especially those perpendicular to the tensile axis, and these gradually grow and coalesce.

Second-phase particles play an important part in the nucleation of cracks and cavities by concentrating stress in sliding boundaries and at the intersection of slip bands with particles but these stress concentrations are greatly reduced by plastic deformation by power-law creep and by diffusional processes. Cavity formation and early growth is therefore intimately linked to the creep process itself and the time-to-fracture correlates well with the minimum creep rate for many structural materials. Fracture occurs when the larger, more closely spaced cavities coalesce. Creep fracture is discussed further in Chapter 14.

13.1.2 Metallurgical factors affecting creep

From the theories of the various stages of creep, it is clear that there are many metallurgical factors (e.g. the grain size, the melting point of the metal, and the nature of the alloying additions) which affect the creep rate. Thus, for a particular application, the choice of metallurgical condition will depend on the service conditions of stress and temperature. Nevertheless, in creep at high temperatures it is true to say that the thermodynamically stable state is also the most stable mechanically and, consequently, any changes of phase, precipitate size or grain size must be prevented during service.

Considerations of a creep-resistant alloy

The problem of the design of engineering creep-resistant alloys is complex, and the optimum alloy for a given service usually contains several constituents in various states of solution and precipitation. Nevertheless, it is worthwhile to consider some of the physical metallurgical principles underlying creep-resistant behaviour, in the light of the preceding theories.

First, let us consider the strengthening of the solid solution by those mechanisms which cause dislocation locking, and those which contribute to lattice friction hardening. The former include solute atoms interacting with (1) the dislocation, or (2) the stacking fault. Friction hardening can arise from (1) the stress fields around individual atoms, i.e. the Mott–Nabarro effect, (2) clusters of solute atoms in solid solutions, (3) by increasing the separation of partial dislocations and so making climb, cross-slip and intersection more difficult, (4) by the solute atoms becoming attached to jogs and thereby impeding climb, and (5) by influencing the energies of formation and migration of vacancies. The alloy can also be hardened by precipitation, and it is significant that many of the successful industrial creep-resistant alloys are of this type (e.g. the Nimonic alloys, and both ferritic and austenitic steels).

The effectiveness of these various methods of conferring strength on the alloy will depend on the conditions of temperature and stress during creep. All the effects should play some part during fast primary creep, but during the slow secondary creep stage the impeding of dislocation movement by the Cottrell, Suzuki, or Mott–Nabarro effects will probably be small. This is because modern creep-resistant alloys are in service up to temperatures of about two-thirds the absolute melting point ($T/T_m \simeq 2/3$) of the parent metal, whereas above about $T/T_m \simeq 1/2$, solute atoms will migrate as fast as dislocations. Hardening which relies on clusters will be more difficult to remove than that which relies upon single atoms and should be effective up to higher temperatures. However, for any hardening mechanism to be really effective, whether it is due to solute atom clusters or actual precipitation, the rate of climb and cross-slip past the barriers these solute elements set up must be slow. Accordingly, the most probable role of solute alloying elements in modern creep-resistant alloys is in reducing the rate of climb and cross-slip processes. The three hardening mechanisms listed as 3, 4 and 5 above are all effective in this way. From this point of view, it is clear that the best parent metals on which to base creep-resistant alloys will be those in which climb and cross-slip is difficult; these include the f.c.c. and c.p.h. metals of low stacking-fault energy, for which the slip dislocations readily dissociate. That this hypothesis is, in general, true can be seen from *Table 13.1*, where the ratio of the highest service creep temperature T_s to the melting point T_m of the base metal for a number of commercial alloys is presented. Alloys based on f.c.c. iron are superior to those based on b.c.c. iron, and aluminium, which has a high stacking-fault energy, is inferior to the other close-packed metals. Generally, the creep rate is described by the empirical relation

TABLE 13.1 Ratio of T_s/T_m for various creep-resistant alloys

Base metal	Alloy	T_s/T_m
α-iron	Complex ferritic alloy	0.48
γ-iron	Austenitic alloy containing cobalt	0.59
Aluminium	Complex alloy	0.56
Nickel	Nimonic 90	0.68
Magnesium	Complex alloy	0.62

$$\dot{\varepsilon} = A(\sigma/E)^n(\gamma)^m D \qquad (13.13)$$

where A is a constant, n, m stress and fault energy exponents respectively and D the diffusivity; for f.c.c. materials $m \approx 3$ and $n \approx 5$. The reason for the good creep strength of austenitic and Ni-base materials containing Co, Cr, etc, arises from their low fault energy and also because of their relatively high melting point then D is small.

From the above discussion, it appears that a successful creep-resistant material designed in accordance with the modern physical principles would be an alloy, the composition of which gives a structure with a hardened solid-solution matrix containing a sufficient number of precipitated particles to force glissile partial dislocations either to climb or to cross-slip to circumvent them. The constitution of the Nimonic alloys, which consist of a nickel matrix containing dissolved chromium, titanium, aluminium and cobalt, is in accordance with these principles and, since no large atomic size factors are involved, it appears that one of the functions of these additions is to lower the stacking-fault energy and thus widen the separation of the partial dislocations. A second object of the titanium and aluminium alloy additions* is to produce precipitation, and in the Nimonic alloys much of the precipitate is $NiAl_3$. This precipitate is isomorphous with the matrix, and while it has a parameter difference ($\approx \frac{1}{2}$ per cent) small enough to give a low interfacial energy, it is, nevertheless, sufficiently large to give a source of hardening. Thus, since the energy of the interface provides the driving force for particle growth, this low energy interface between particle and matrix ensures a low rate of particle growth and hence, a high service temperature.

Grain boundary precipitation is advantageous in reducing grain boundary sliding (*see* p. 453). Alternatively, the weakness of the grain boundaries may be eliminated altogether by using single crystal material. Nimonic alloys used for turbine blades have been manufactured in single crystal form by directional solidification. Rolls-Royce now use blades with significantly improved creep and thermal shock resistance produced by seeding the crystal which grows in a temperature gradient achieved by computer-controlled withdrawal of the mould to maintain a macroscopically planar liquid–solid interface perpendicular to the solidification direction.

Dispersions are effective in conferring creep strength by two mechanisms. First the particle will hinder a dislocation and force it to climb and cross-slip.

* The chromium forms a spinel with NiO and hence improves the oxidation resistance.

Secondly and more important, is the retarding effect on recovery as shown by some dispersions, Cu–Al_2O_3 (extruded), SAP (sintered alumina powder), Ni–ThO_2 which retain their hardness almost to the melting point. A comparison of SAP with a 'conventional' complex aluminium alloy shows that at 250 °C there is little to choose between them but at 400 °C SAP is several times stronger. Generally, the dislocation network formed by strain-hardening interconnects the particles and is thereby anchored by them. To do this effectively, the particle must be stable at the service temperature and remain finely dispersed. This depends on the solubility C, diffusion coefficient D and interfacial energy γ_I, since the time to dissolve the particle is $t = r^4 \mathbf{k}T/DC\gamma_I R^2$. In precipitation hardening alloys, C is appreciable, and D offers little scope for adjustment; great importance is therefore placed on γ_I as for the Ni_3 (TiAl) phase in Nimonics where it is very low.

Figure 13.4 shows that $n \approx 4$ both above and below $0.5T_m$ for the Ni–Al_2O_3 and Ni/Co–Al_2O_3 alloys that were completely recrystallized, which contrasts with values very much greater than 4 for extruded TD nickel and other dispersion-strengthened alloys* containing a dislocation substructure. This demonstrates the importance of substructure and probably indicates that in completely recrystallized alloys containing a dispersoid, the particles control the creep behaviour, whereas in alloys containing a substructure the dislocation content is more important. Since $n \approx 4$ for the Ni– and Ni/Co–Al_2O_3 alloys in both temperature regimes, the operative deformation mechanism is likely to be the same, but it is clear from the activation energies, listed in *Table 13.2*, that the rate-controlling thermally activated process changes with temperature. The activation energy is greater at the higher temperature when it is also, surprisingly, composition (or stacking-fault energy) independent.

TABLE 13.2 Experimentally determined parameters from creep of Ni–Al_2O_3 and Ni–Co–Al_2O_3 alloys

Test temperature	*773 K*		*1000 K*		
Alloy	Q (kJ/mol)	A (s^{-1})	Q (kJ/mol)	A (s^{-1})	A/D_0
Ni	85	1.67×10^{16}	276	1.1×10^{28}	5.5×10^{28}
Ni–67% Co	121	9.95×10^{19}	276	2.2×10^{28}	5.8×10^{28}

Such behaviour may be explained, if it is assumed that the particles are by-passed by cross-slip (*see Figure 11.11*) and this process is easy at all temperatures, but it is the climb of the edge segments of the cross-slipped dislocations that is rate-controlling. At low temperatures, climb would proceed by pipe-diffusion so that the composition-dependence relates to the variation in the ease of pipe-diffusion along dislocations of different widths. At high temperatures, climb occurs by bulk diffusion and the absence of any composition-dependence is due to the fact that in these alloys the jog distribution is determined mainly by dislocation/particle interactions and not, as in single-phase alloys and in

* To analyse these it is generally necessary to introduce a threshold (or friction) stress σ_0, so that the effective stress is $(\sigma - \sigma_0)$.

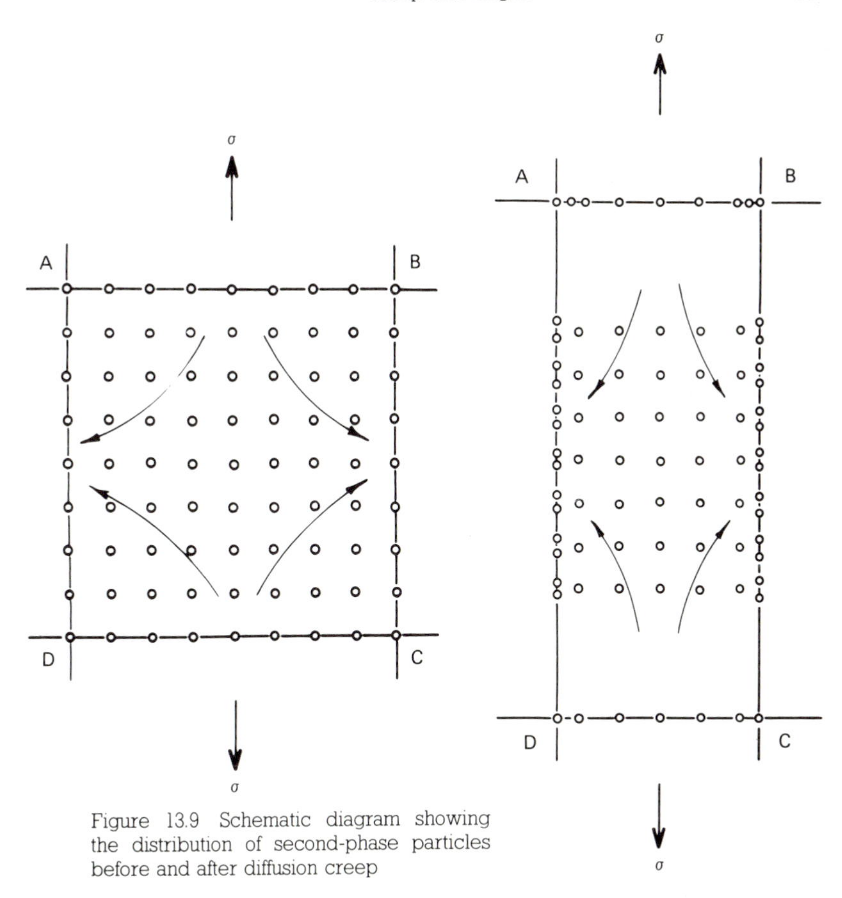

Figure 13.9 Schematic diagram showing the distribution of second-phase particles before and after diffusion creep

dispersion-strengthened alloys containing a substructure, by the matrix stacking-fault energy. The optimum creep resistance of dispersion-strengthened alloys is produced when a uniform dislocation network in a fibrous grain structure is anchored by the particles and recovery is minimized. Such a structure can reduce the creep rate by several orders of magnitude from that given in *Figure 13.4* but it depends critically upon the working and heat-treatment used in fabricating the alloy.

Second-phase particles can also inhibit diffusion creep. *Figure 13.9* shows the distribution of particles before and after diffusion creep and indicates that the longitudinal boundaries tend to collect precipitates as vacancies are absorbed and the boundaries migrate inwards, while the tensile boundaries acquire a PFZ. Such a structural change has been observed in Mg–0.5%Zr (Magnox ZR55) at 400 °C and is accompanied by a reduced creep rate. It is not anticipated that diffusion is significantly affected by the presence of particles and hence the effect is thought to be due to the particles affecting the vacancy-absorbing capabilities of the grain boundaries. Whatever mechanism is envisaged for the annihilation of vacancies at a grain boundary, the climb-glide of grain boundary dislocations is likely to be involved and such a process will be hindered by the presence of particles.

13.1.3 Deformation mechanism maps

The discussion in this chapter has emphasized that over a range of stress and temperature an alloy is capable of deforming by several alternative and independent mechanisms, e.g. dislocation creep with either pipe diffusion at low temperatures and lattice diffusion at high temperatures being the rate-controlling mechanism, and diffusional creep with either grain-boundary diffusion or lattice diffusion being important. In a particular range of temperature one of these mechanisms is dominant and it is therefore useful in engineering application to identify the operative mechanism for a given stress–temperature condition, since it is ineffective to change the metallurgical factors to influence, for example, a component deforming by power-law creep controlled by pipe diffusion if the operative mechanism is one of Herring–Nabarro creep.

The various alternative mechanisms are displayed conveniently on a deformation-mechanism map in which the appropriate stress, i.e. shear stress or equivalent stress, compensated by modulus on a log scale, is plotted against homologous temperature T/T_m as shown in *Figure 13.10* for nickel and a nickel-based superalloy with a grain size of 100 μm. By comparing the diagrams it is evident that solid solution strengthening and precipitation hardening (*see* page 414) have raised the yield stress and reduced the dislocation creep field. The shaded boxes shown in *Figure 13.10* indicate the typical stresses and temperatures to which a turbine blade would be subjected; it is evident that the mechanism of creep during operation has changed and, indeed, the creep-rate is reduced by several orders of magnitude.

13.1.4 Superplasticity

A number of materials, particularly two-phase eutectic or eutectoid alloys, have been observed to exhibit large elongations ($\approx$1000 per cent) without

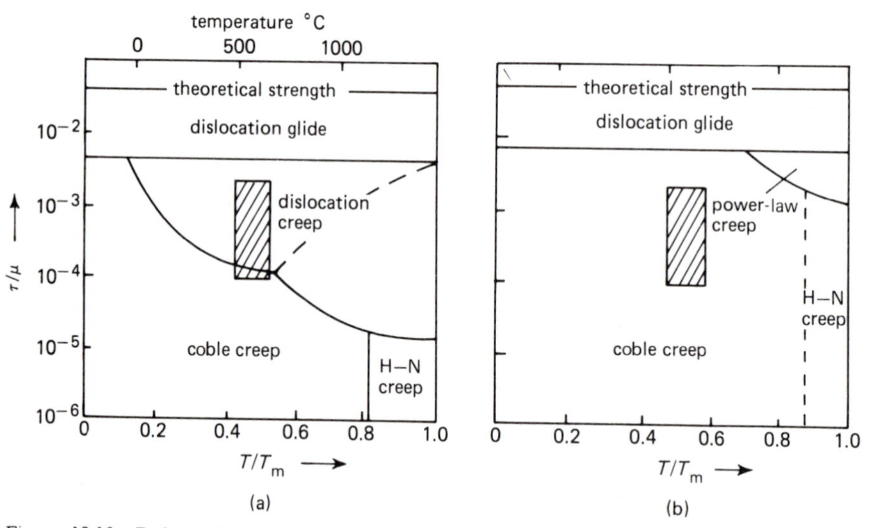

Figure 13.10 Deformation mechanism maps for (a) nickel, (b) nickel-based superalloy (after M. F. Ashby)

fracture, and such behaviour has been termed superplasticity. Several metallurgical factors have been put forward to explain superplastic behaviour and it is now generally recognized that the effect can be produced in materials either (a) with a particular structural condition, or (b) tested under special test conditions. The particular structural condition is that the material has a very fine grain size and the presence of a two-phase structure is usually of importance in maintaining this fine grain size during testing. Materials which exhibit superplastic behaviour under special test conditions are those for which a phase boundary moves through the strained material during the test, e.g. during temperature cycling.

In general, the superplastic material exhibits a high strain-rate sensitivity. Thus the plastic flow of a solid may be represented by the relation

$$\sigma = K\dot{\varepsilon}^m$$

where σ is the stress, $\dot{\varepsilon}$ the strain-rate and m an exponent generally known as the strain-rate sensitivity. When $m=1$ the flow stress is directly proportional to strain rate and the material behaves as a Newtonian viscous fluid, such as hot glass. Superplastic materials are therefore characterized by high m-values, since this leads to increased stability against necking in a tensile test. Thus, for a tensile specimen length l with cross-sectional area A under an applied load P then (see page 80) $\mathrm{d}l/l = -\mathrm{d}A/A$ and introducing the time factor we obtain

$$\dot{\varepsilon} = -(1/A)\,\mathrm{d}A/\mathrm{d}t$$

and if during deformation the equation $\sigma = K\dot{\varepsilon}^m$ is obeyed, then

$$\mathrm{d}A/\mathrm{d}t = (P/K)^{1/m} A^{\{1-(1/m)\}}$$

For most metals and alloys $m \simeq 0.1$–0.2 and the rate at which A changes is sensitively dependent on A, and hence once necking starts the process rapidly leads to failure. When $m=1$, the rate of change of area is independent of A and, as a consequence, any irregularities in specimen geometry are not accentuated during deformation. The resistance to necking therefore depends sensitively on m, and increases markedly when $m \gtrsim 0.5$. Considering, in addition, the dependence of the flow stress on strain, then

$$\sigma = K^1 \varepsilon^n \dot{\varepsilon}^m$$

and in this case, the stability against necking depends on a factor $(1-n-m)/m$, but n-values are not normally very high. Superplastic materials such as Zn–Al eutectoid, Pb–Sn eutectic, Al–Cu eutectic, etc, have m values approaching unity at elevated temperatures. Studies by Backofen and his co-workers have shown that the total elongation increases as m increases. Moreover, with increasing microstructural fineness of the material (grain-size or lamella spacing) the tendency for superplastic behaviour is increased. Two-phase structures are advantageous in maintaining a fine grain size during testing, but exceptionally high ductilities have been produced in several commercially pure metals, e.g. Ni, Zn and Mg, for which the fine grain size was maintained during testing at a particular strain-rate and temperature.

It follows that there must be several possible conditions leading to superplasticity. One situation is deformation under linear viscous creep conditions, such as Herring–Nabarro vacancy creep or grain boundary diffusion creep, when the alloy is capable of being stretched large amounts in tension at stresses too low for the usual plastic process to lead to necking. In this case, the finer the metallurgical microstructure the shorter is the migration distance for the vacancies, and the less stringent the stress conditions for superplasticity. However, the maximum possible strain rate by Herring–Nabarro creep is less than that observed experimentally. Generally, it is observed metallographically that the grain structure remains remarkably equiaxed during extensive deformation and that grain boundary sliding is a common deformation mode in several superplastic alloys. While grain boundary sliding can contribute to the overall deformation by relaxing the five independent mechanisms of slip, it cannot give rise to large elongations without bulk flow of material, e.g. grain boundary migration. In polycrystals, triple junctions obstruct the sliding process and give rise to a low m-value. Thus to increase the rate sensitivity of the boundary shear it is necessary to lower the resistance to sliding from barriers, relative to the viscous drag of the boundary; this can be achieved by grain boundary migration. Indeed, it is observed that superplasticity is controlled by grain boundary diffusion.

The complete explanation of superplasticity is still being developed, but it is already clear that during deformation individual grains or groups of grains with suitably aligned boundaries will tend to slide. Sliding continues until obstructed by a protrusion in a grain boundary, when the local stress generates dislocations which slip across the blocked grain and pile up at the opposite boundary until the back stress prevents further generation of dislocations and thus further sliding. At the temperature of the test, however, dislocations at the head of the pile-up can climb into and move along grain boundaries to annihilation sites. The continual replacement of these dislocations would permit grain boundary sliding at a rate governed by the rate of dislocation climb, which in turn is governed by grain boundary diffusion. It is important that any dislocations created by local stresses are able to traverse yielded grains and this is possible only if the 'dislocation cell size' is larger than, or at least of the order of, the grain size, i.e. a few microns. At high strain-rates and low temperatures the dislocations begin to tangle and form cell structures and superplasticity then ceases.

The above conditions imply that any metal in which the grain size remains fine during deformation could behave superplastically; this conclusion is borne out in practice. The stability of grain size can, however, be achieved more readily with a fine micro-duplex structure as observed in some Fe–20% Cr–6% Ni alloys when hot worked to produce a fine dispersion of austenite and ferrite. Such stainless steels have an attractive combination of properties (strength, toughness, fatigue strength, weldability and corrosion resistance) and unlike the usual range of two-phase stainless steels have good hot workability if $\frac{1}{2}$% Ti is added to produce a random distribution of TiC rather than $Cr_{23}C_6$ at ferrite-austenite boundaries.

Superplastic forming is now an established and growing industry largely using vacuum forming to produce intricate shapes with high draw ratios. Two alloys which have achieved engineering importance are Supral (containing Al/6% Cu/0.5%Zr) and IMI 318 (containing Ti/6%Al/4%V). Supral is deformed at

460 °C and IMI 318 at 900 °C under argon. Although the process is slow, the loads required are also low and the process can be advantageous in the press-forming field to replace some of the present expensive and complex forming technology.

13.2 Fatigue

13.2.1 Introduction

The term fatigue applies to the behaviour of a metal which, when subjected to a cyclically variable stress of sufficient magnitude (often below the yield stress) produces a detectable change in mechanical properties. In practice, a large number of service failures are due to fatigue, and so engineers are concerned mainly with fatigue failure where the specimen is actually separated into two parts. Many of these failures can be attributed to poor design of the component, and there is little or nothing that the metallurgist can do about them, but in some cases certain aspects of the failure can be ascribed to the condition of the material. Consequently, the treatment of fatigue may be conveniently divided into three aspects: (i) engineering considerations, (ii) gross metallurgical aspects, and (iii) fine-scale structural and atomic changes. It is the latter which is probably of most interest to a physical metallurgist, but although such studies have improved our knowledge of the fatigue process, they have not, so far, provided us with a complete understanding of the fatigue process.

13.2.2 Engineering considerations of fatigue

The fatigue conditions which occur in service are usually extremely complex. Common failures are found in axles where the eccentric load at a wheel or pulley produces a varying stress which is a maximum in the skin of the axle. Other examples, such as the flexure stresses produced in aircraft wings and in undercarriages during ground taxi-ing, do, however, emphasize that the stress system does not necessarily vary in a regular sinusoidal manner. The series of aircraft disasters attributed to pressurized-cabin failures is perhaps the most spectacular example of this type of fatigue failure.

In laboratory testing of materials, the stress system is usually simplified, and both the Woehler and push-pull type of test (described in Chapter 2) are in common use. The results are usually plotted on the familiar *S–N* curve (i.e. stress versus the number of cycles to failure, usually plotted on a logarithmic scale). Ferritic steels may be considered to exhibit a genuine fatigue limit with a fatigue ratio $S/\mathrm{UTS} \simeq 0.5$. However, other materials, such as aluminium or copper-based alloys, certainly those of the age-hardening variety, definitely do not show a sharp discontinuity in the *S–N* curve. For these materials no fatigue limit exists and all that can be specified is the endurance limit at *N* cycles. The importance of the effect is illustrated by the behaviour of commercial aluminium-based alloys containing zinc, magnesium and copper. Such an alloy may have a UTS of

617 MN/m^2 but the fatigue stress for a life of 10^8 cycles is only 154 MN/m^2 (i.e. a fatigue ratio at 10^8 cycles of 0.25).

Variables affecting the fatigue life

The amplitude of the stress cycle to which the specimen is subjected is the most important single variable in determining its life under fatigue conditions, but the performance of a material is also greatly affected by various other conditions, which may be summarized as follows.

Surface preparations Since fatigue cracks frequently start at or near the surface of the component, the surface condition is an important consideration in fatigue life. The removal of machining marks and other surface irregularities invariably improves the fatigue properties. Putting the surface layers under compression by shot peening or surface treatment improves the fatigue life.

Effect of temperature Temperature affects the fatigue properties in much the same way as it does the UTS; the fatigue strength is highest at low temperatures and decreases gradually with rising temperature. For mild steel the ratio of fatigue limit to UTS remains fairly constant at about 0.5, while the ratio of fatigue limit to yield stress varies over much wider limits. However, if the temperature is increased above about 100°C, both the tensile strength and the fatigue strength of mild steel show an increase, reaching a maximum value between 200 and 400 °C. This increase, which is not commonly found in other materials, has been attributed to strain-ageing.

Frequency of stress cycle In most metals the frequency of the stress cycle has little effect on the fatigue life, although lowering the frequency usually results in a slightly reduced fatigue life. The effect becomes greater if the temperature of the fatigue test is raised, when the fatigue life tends to depend on the total time of testing rather than on the number of cycles. With mild steel, however, experiments show that the normal speed effect is reversed in a certain temperature range and the number of cycles to failure increases with decrease in the frequency of the stress cycle. This effect may be correlated with the influence of temperature and strain-rate on the UTS. The temperature at which the tensile strength reaches a maximum depends on the rate of strain, and it is, therefore, not surprising that the temperature at which the fatigue strength reaches a maximum depends on the cyclic frequency.

Mean stress For conditions of fatigue where the mean stress, i.e.

$$\Delta\sigma N_f^a = (\sigma_{\max} + \sigma_{\min})/2$$

does not exceed the yield stress σ_y, then the relationship

$$\Delta\sigma N_f^a = \text{const}$$

known as Basquin's law holds over the range 10^2 to $\approx 10^5$ cycles, i.e. N less than the knee of the S–N curve, where $a \approx 1/10$ and n_f the number of cycles to failure.

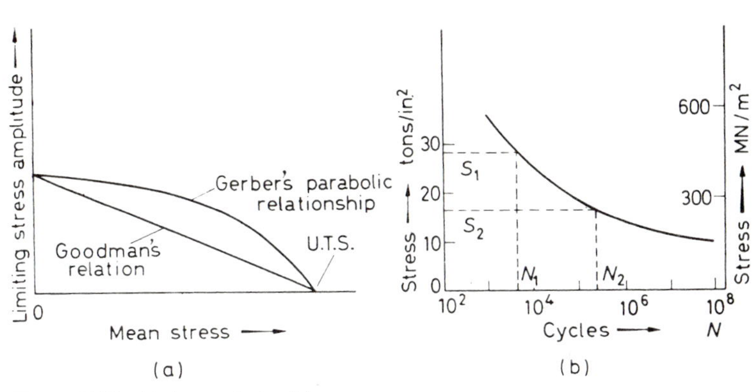

Figure 13.11 Fatigue relationships

For low cycle fatigue with $\Delta\sigma > \sigma_y$ then Basquin's law no longer holds, but a reasonable relationship

$$\Delta\varepsilon_p N_f^b = D^b = \text{constant}$$

known as the Coffin–Manson law, is found where $\Delta\varepsilon_p$ is the plastic strain range, $b \approx 0.6$, and D is the ductility of the material. If the mean stress becomes tensile a lowering of the fatigue limit results. Several relationships between fatigue limit and mean stress have been suggested, as illustrated in *Figure 13.11(a)*. However, there is no theoretical reason why a material should follow any given relationship and the only safe rule on which to base design is to carry out prior tests on the material concerned to determine its behaviour under conditions similar to those it will meet in service. Another common engineering relationship frequently used, known as Miner's concept of cumulative damage, is illustrated in *Figure 13.11(b)*. This hypothesis states that damage can be expressed in terms of the number of cycles applied divided by the number to produce failure at a given stress level. Thus, if a maximum stress of value S_1 is applied to a specimen for n_1 cycles which is less than the fatigue life N_1, and then the maximum stress is reduced to a value equal to S_2, the specimen is expected to fail after n_2 cycles, since according to Miner the following relationship will hold

$$n_1/N_1 + n_2/N_2 + \ldots = \sum n/N = 1$$

Environment Fatigue occurring in a corrosive environment is usually referred to as corrosion fatigue. It is well known that corrosive attack by a liquid medium can produce etch pits which may act as notches, but when the corrosive attack is simultaneous with fatigue stressing, the detrimental effect is far greater than just a notch effect. Moreover, from microscopic observations the environment appears to have a greater effect on crack propagation than on crack initiation. For most materials even atmospheric oxygen decreases the fatigue life by influencing the speed of crack propagation, and it is possible to obtain a relationship between fatigue life and the degree of vacuum in which the specimen has been held.

13.2.3 Metallurgical factors affecting fatigue

It is now well established that fatigue starts at the surface of the specimen. This is easy to understand in the Woehler test because, in this test, it is there that the stress is highest. However, even in push-pull fatigue, the surface is important for several reasons: (i) slip is easier at the surface than in the interior of the grains, (ii) the environment is in contact with the surface, and (iii) any specimen misalignment will always give higher stresses at the surface. Accordingly, any alteration in surface properties must bring about a change in the fatigue properties. The best fatigue resistance occurs in materials with a worked surface layer produced by emerying, shot-peening or skin-rolling the surface. This beneficial effect of a worked surface layer is principally due to the fact that the surface is put into compression, but the increased UTS as a result of work hardening also plays a part. Electropolishing the specimen by removing the surface layers usually has a detrimental effect on the fatigue properties, but other common surface preparations such as nitriding and carburizing, both of which produce a surface layer which is in compression, may be beneficial. Conversely, such surface treatments as the decarburizing of steels and the cladding of aluminium alloys with pure aluminium, increase their susceptibility to fatigue.

The alloy composition and thermal and mechanical history of the specimen are also of importance in the fatigue process. Any treatment which increases the hardness or yield strength of the material will increase the level of the stress needed to produce slip and, as we shall see later, since the fundamental processes of fatigue are largely associated with slip, this leads directly to an increase in fatigue strength. It is also clear that grain size is a relevant factor: the smaller the grain size the higher is the fatigue strength at a given temperature.

The fatigue processes in stable alloys are essentially the same as that of pure metals but there is, of course, an increase in fatigue strength. However, the processes in unstable alloys and in materials exhibiting a yield point are somewhat different. In fatigue, as in creep, metallurgical instability frequently leads to enhancement of the fundamental processes. In all cases the approach to equilibrium is more complete, so that in age-hardening materials solution-treated specimens become harder and fully aged specimens become softer. The changes which occur are local rather than general, and are associated with the enhanced diffusion brought about by the production of vacancies during the fatigue test. Clearly, since vacancy mobility is a thermally activated process such effects can be suppressed at sufficiently low temperatures.

In general, non-ferrous alloys do not exhibit the type of fatigue limit shown by mild steel. One exception to this generalization is the alloy aluminium 2–7 per cent magnesium, 0.5 per cent manganese, and it is interesting to note that this alloy also has a sharp yield point and shows Lüders markings in an ordinary tensile test. Accordingly, it has been suggested that the fatigue limit occupies a similar place in the field of alternating stresses to that filled by the yield point in unidirectional stressing. Stresses above the fatigue limit readily unlock the dislocations from their solute atom atmospheres, while below the fatigue limit most dislocations remain locked. In support of this view, it is found that when the carbon and nitrogen content of mild steel is reduced, by annealing in wet hydrogen, striking changes take place in the fatigue limit (*Figure 13.12*) as well as in the sharp yield point.

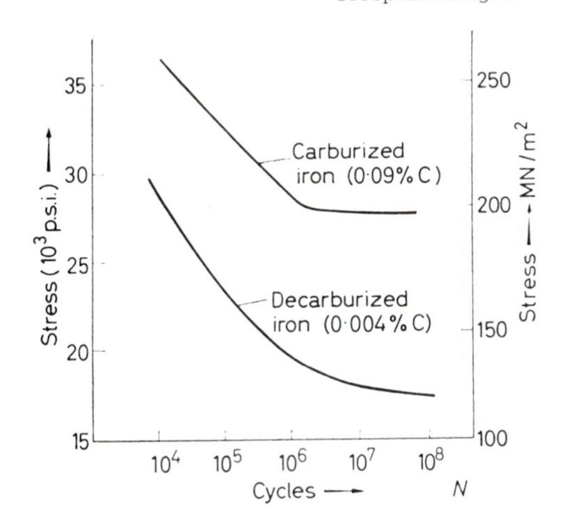

Figure 13.12 S–N curves for carburized and decarburized iron

13.2.4 The structural changes accompanying fatigue

Observations of the structural details underlying fatigue hardening show that in polycrystals large variations in slip-band distributions and the amount of lattice misorientation exist from one grain to another. Because of such variations it is difficult to typify structural changes, so that in recent years this structural work has been carried out more and more on single crystals; in particular, copper has received considerable attention as being representative of a typical metal. Such studies have now established that fatigue occurs as a result of slip, the direction of which changes with the stress cycle, and that the process continues throughout the whole of the test (shown, for example, by interrupting a test and removing the slip bands by polishing; the bands reappear on subsequent testing).

Moreover, four stages in the fatigue life of a specimen are distinguishable; these may be summarized as follows. In the early stages of the test, the whole of the specimen hardens. After about 5% of the life, slip becomes localized and persistent slip bands appear; they are termed persistent because they reappear and are not permanently removed by electropolishing. Thus, in the bulk of the metal (the matrix) reverse slip does not continue throughout the whole test. Electron microscope observations show that metal is extruded from the slip bands and that fine crevices called intrusions are formed within the band. During the third stage of the fatigue life the slip bands grow laterally and become wider, and at the same time cracks develop in them. These cracks spread initially along slip bands, but in the later stages of fracture the propagation of the crack is often not confined to certain crystallographic directions and catastrophic rupture occurs. These two important crack growth stages, i.e. stage I in the slip band and stage II roughly perpendicular to the principle stress, are shown in *Figure 13.13* and are influenced by the formation of localized (persistent) slip bands, i.e. PSBs. However PSBs are not clearly defined in low stacking fault energy, solid solution alloys.

Cyclic stressing therefore produces plastic deformation which is not fully reversible and the build-up of dislocation density within grains gives rise to fatigue hardening with an associated structure which is characteristic of the strain

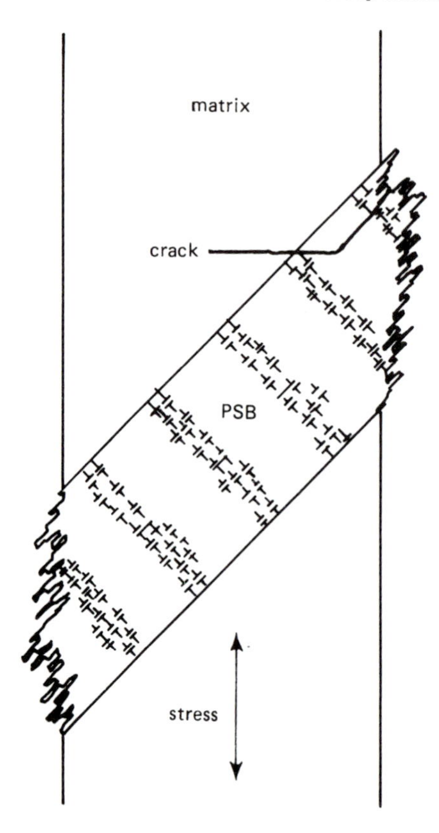

Figure 13.13 Persistent slip band (PSB) formation in fatigue, and stage I and stage II crack growth

amplitude and the ability of the dislocations to cross-slip, i.e. temperature and SFE. The non-reversible flow at the surface leads to intrusions, extrusions and crack formation in PSBs. These two aspects will now be considered separately and in greater detail.

Fatigue hardening

If a single or polycrystalline specimen is subjected to many cycles of alternating stress, it becomes harder than a similar specimen extended unidirectionally by the same stress applied only once. This may be demonstrated by stopping the fatigue test and performing a static tensile test on the specimen when, as shown in *Figure 13.14*, the yield stress is increased. During the process, persistent slip bands appear on the surface of the specimen and it is in such bands that cracks eventually form. The behaviour of a fatigue-hardened specimen has two unusual features when compared with an ordinary work-hardened material. The fatigue-hardened material, having been stressed symmetrically, has the same yield stress in compression as in tension, whereas the work-hardened specimen (e.g. pre-strained in tension) exhibits a Bauschinger effect (see p. 358). It arises from the fact that the obstacles behind the dislocation are weaker than those resisting further dislocation motion, and the pile-up stress causes it to slip back under a reduced load in the reverse direction. The other important feature is that

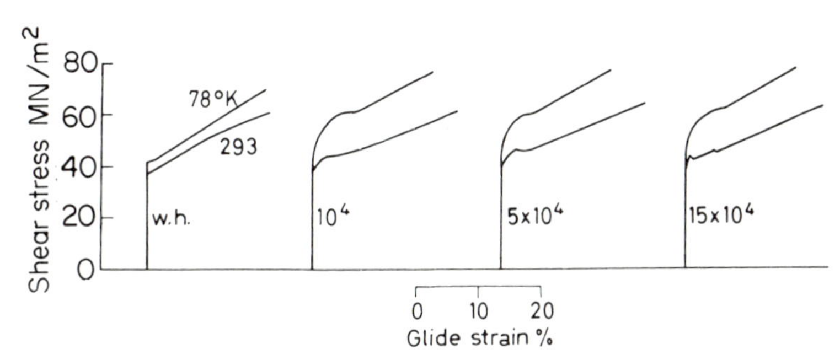

Figure 13.14 Stress–strain curves for copper after increasing amounts of fatiguing (after Broom and Ham, *Proc. Roy. Soc.* 1959, **A251,** 186)

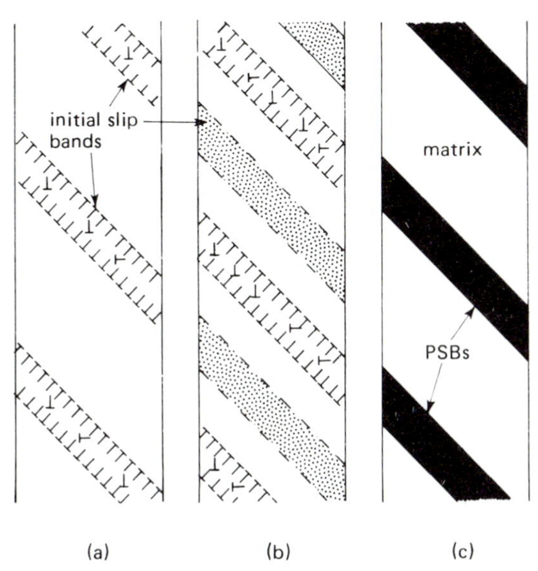

Figure 13.15 Formation of persistent slip bands (PSBs) during fatigue

the temperature-dependence of the hardening produced by fatigue is significantly greater than that of work hardening and, because of the similarity with the behaviour of metals hardened by quenching and by irradiation, it has been attributed to the effect of vacancies and dislocation loops created during fatigue.

At the start of cyclic deformation the initial slip bands, *Figure 13.15(a)*, consist largely of primary dislocations in the form of dipole and multipole arrays; the number of loops is relatively small because the frequency of cross-slip is low. As the specimen work hardens slip takes place between the initial slip bands, and the new slip bands contain successively more secondary dislocations because of the internal stress arising from nearby slip bands, *Figure 13.15(b)*. When the specimen is completely filled with slip bands, the specimen has work hardened and the softest regions are now those where slip occurred originally since these bands contain the lowest density of secondary dislocations. Further

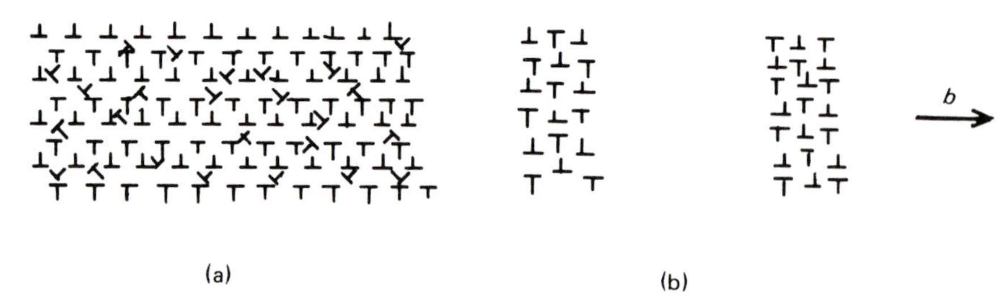

Figure 13.16 Schematic diagram showing (a) vein structure of matrix, (b) ladder structure of PSBs.

slip and the development of PSBs takes place within these original slip bands, as shown schematically in *Figure 13.15*(*c*).

As illustrated schematically in *Figure 13.16*, TEM of copper crystals shows that the main difference between the matrix and the PSBs is that in the matrix the dense arrays of edge dislocation (di- and multipoles) are in the form of large veins occupying about 50 per cent of the volume, whereas they form a 'ladder'-type structure within walls occupying about 10 per cent of the volume in PSBs. The PSBs are the active regions in the fatigue process while the matrix is associated with the inactive parts of the specimen between the PSBs. Steady-state deformation then takes place by the to-and-fro glide of the same dislocations in the matrix, whereas an equilibrium between dislocation multiplication and annihilation exists in the PSBs. Multiplication occurs by bowing-out of the walls and annihilation takes place by interaction with edge dislocations of opposite sign ($\approx$ 756b apart) on glide planes in the walls and of screw dislocations ($\approx$ 200b apart) on glide planes in the low-dislocation channels, the exact distance depending on the ease of cross-slip.

13.2.5 The formation of fatigue cracks and fatigue failure

Extrusions, intrusions and fatigue cracks can be formed at temperatures as low as 4 K where thermally activated movement of vacancies does not take place. Such observations indicate that the formation of intrusions and cracks cannot depend on either chemical or thermal action and the mechanism must be a purely geometrical process which depends on cyclic stressing.

Two general mechanisms have been suggested. The first, the Cottrell 'ratchet' mechanism, involves the use of two different slip systems with different directions and planes of slip, as is shown schematically in *Figure 13.17*. The most favoured source (e.g. S_1 in *Figure 13.17*(*a*)) produces a slip step on the surface at P during a tensile half-cycle. At a slightly greater stress in the same half-cycle, the second source S_2 produces a second step at Q (*Figure 13.17*(*b*)). During the compression half-cycle, the source S_1 produces a surface step of opposite sign at P' (*Figure 13.17*(*c*)) but, owing to the displacing action of S_2, this is not in the same plane as the first and thus an intrusion is formed. The subsequent operation of S_2 produces an extrusion at QQ' (*Figure 13.17*(*d*)) in a similar manner. Such a mechanism requires the operation of two slip systems and, in general, predicts the

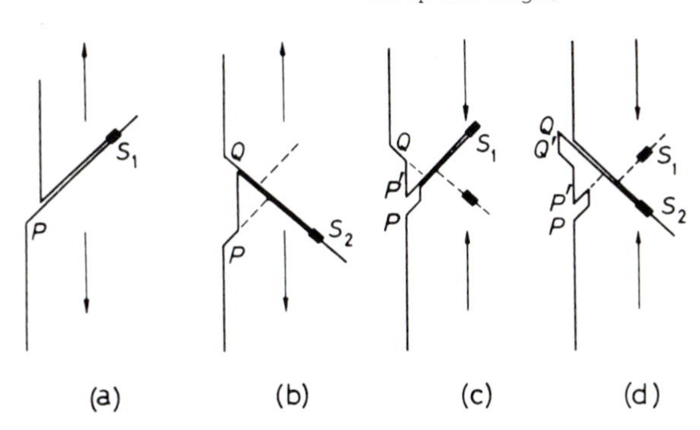

Figure 13.17 Formation of intrusions and extrusions (after Cottrell, in *Fracture*, Technological Press, courtesy of John Wiley & Sons)

occurrence of intrusions and extrusions with comparable frequency, but not in the same slip band.

The second mechanism, proposed by Mott, involves cross-slip and as a result a column of metal is extruded from the surface and a cavity is left behind in the interior of the crysral. One way in which this could happen is by the cyclic movement of a screw dislocation along a closed circuit of crystallographic planes, as shown in *Figure 13.18*. During the first half-cycle the screw dislocation glides along two faces $ABCD$ and $BB'C'C$ of the band, and during the second half-cycle returns along the faces $B'C'A'D$ and $A'D'DA$. Unlike the Cottrell mechanism this process can be operated with a single slip direction, provided cross-slip can occur.

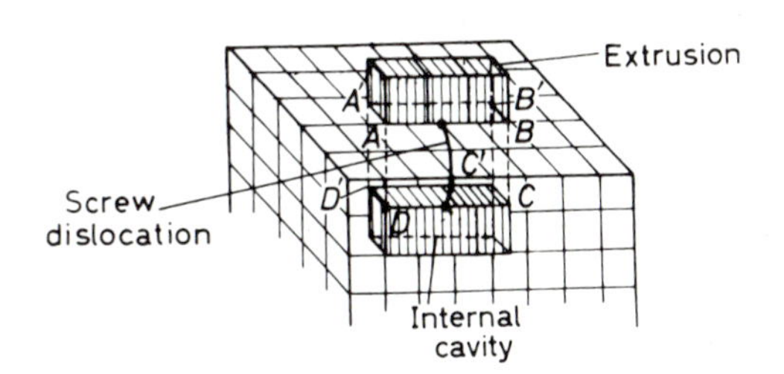

Figure 13.18 Formation of an extrusion and associated cavity by the Mott mechanism

Neither mechanism can fully explain all the experimental observations. The interacting slip mechanism predicts the occurrence of intrusions and extrusions with comparable frequency but not, as is often found, in the same slip band. With the cross-slip mechanism, there is no experimental evidence to show that cavities exist beneath the material being extruded. It may well be that different mechanisms operate under different conditions.

In a polycrystalline aggregate the operation of several slip modes is necessary and intersecting slip unavoidable. Accordingly, the widely differing fatigue behaviour of metals may be accounted for by the relative ease with which cross-slip occurs. Thus, those factors which affect the onset of stage III in the work-hardening curve will also be important in fatigue, and conditions suppressing cross-slip would, in general, increase the resistance to fatigue failure, i.e. low stacking-fault energy and low temperatures. Aluminium would be expected to have poor fatigue properties on this basis but the unfavourable fatigue characteristics of the high-strength aluminium alloys is probably also due to the unstable nature of the alloy and to the influence of vacancies.

In pure metals and alloys, transgranular cracks initiate at intrusions in PSBs or at sites of surface roughness associated with emerging planar slip bands in low SFE alloys. Often the microcrack forms at the PSB–matrix interface where the stress concentration is high. In commercial alloys containing inclusions or second-phase particles, the fatigue behaviour depends on the particle size. Small particles ≈ 0.1 μm can have beneficial effects by homogenizing the slip pattern and delaying fatigue-crack nucleation. Larger particles reduce the fatigue life by both facilitating crack nucleation by slip band/particle interaction and increasing crack growth rates by interface decohesion and voiding within the plastic zone at the crack tip. The formation of voids at particles on grain boundaries can lead to intergranular separation and crack growth. The preferential deformation of 'soft' precipitate-free zones (PFZs) associated with grain boundaries in age-hardened alloys also provides a mechanism of intergranular fatigue-crack initiation and growth. To improve the fatigue behaviour it is therefore necessary to avoid PFZs and obtain a homogeneous deformation structure and uniform precipitate distribution by heat-treatment; localized deformation in PFZs can be restricted by a reduction in grain size.

From the general appearance of a typical fatigue fracture, shown in *Figure 13.19*, one can distinguish two distinct regions. The first is a relatively smooth

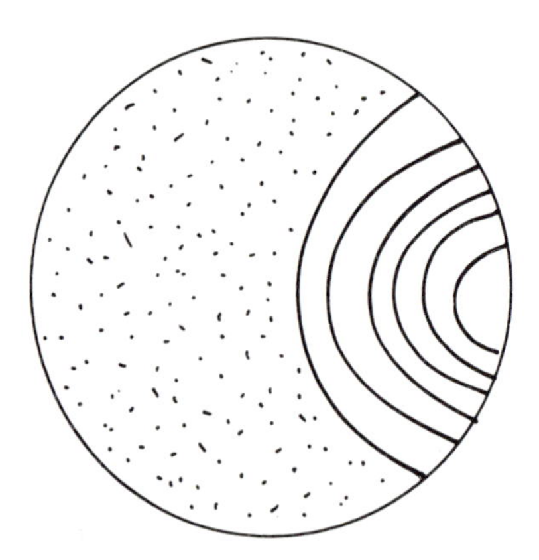

Figure 13.19 A schematic fatigue fracture

area, through which the fatigue crack has spread slowly. This area usually has concentric marks about the point of origin of the crack which correspond to the positions at which the crack was stationary for some period. The remainder of the fracture surface shows a typically rough transcrystalline fracture where the failure has been catastrophic. Electron micrographs of the relatively smooth area show that this surface is covered with more or less regular contours perpendicular to the direction of the propagation front. These fatigue striations represent the successive positions of the propagation front and are spaced further apart the higher the velocity of propagation. They are rather uninfluenced by grain boundaries and in metals where cross-slip is easy, e.g. mild steel or aluminium, may be wavy in appearance. Generally, the lower the ductility of the material the less well defined are the striations.

Stage II growth is rate-controlling in the fatigue failure of most engineering components, and is governed by the stress intensity at the tip of the advancing crack. The striations seen on the fracture surface may form by a process of plastic blunting at the tip of the crack, as shown in *Figure 13.20*. In (a) the crack under the tensile loading part of the cycle generates shear stresses at the tip. With increasing tensile load the crack opens up and new surface is created (b), separation occurs in the slip band and 'ears' are formed at the end of the crack. The plastic deformation causes the crack to be both extended and blunted (c). On the compressive part of the cycle the crack begins to close (d), the shear stresses are reversed and with increasing load the crack almost closes (e). In this part of the cycle the new surface

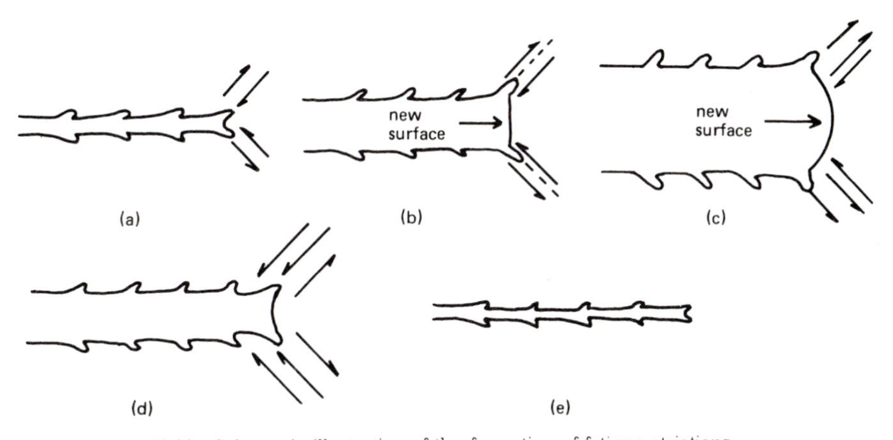

Figure 13.20 Schematic illustration of the formation of fatigue striations

folds and the ears correspond to the new striations on the final fracture surface. A one-to-one correlation therefore exists between the striations and the opening and closing with ear formation. Crack growth continues in this manner until it is long enough to cause the final instability when either brittle or ductile (due to the reduced cross-section not being able to carry the load) failure occurs. In engineering alloys, rather than pure metals, which contain inclusions or second-phase particles, cracking or voiding occurs ahead of the original crack tip rather than in the ears when the tensile stress or strain reaches a critical value (*see* Chapter 14). This macroscopic stage of fracture is clearly of importance to engineers in predicting the working life of a component and has been successfully treated by the application of fracture mechanics as discussed in Chapter 14.

13.2.6 Fatigue at elevated temperatures

At ambient temperature the fatigue process involves intracrystalline slip and surface initiation of cracks, followed by transcrystalline propagation. When fatigued at elevated tesmperatures $\gtrsim 0.5T_m$, pure metals and solid solutions show the formation of discrete cavities on grain boundaries, which grow, link up and finally produce failure. It is probable that vacancies produced by intracrystalline slip give rise to a supersaturation which causes the vacancies to condense on those grain boundaries that are under a high shear stress where the cavities can be nucleated by a sliding or ratchet mechanism. It is considered unlikely that grain boundary sliding contributes to cavity growth, increasing the grain size decreases the cavity growth because of the change in boundary area. Magnox (Mg) and alloys used in nuclear reactors up to $0.75T_m$ readily form cavities, but the high temperature nickel-base alloys to not show intergranular cavity formation during fatigue at temperatures within their normal service range, because intracrystalline slip is inhibited by γ' precipitates. Above about $0.7T_m$, however, the γ' precipitates coarsen or dissolve and fatigue then produces cavities and eventually cavity failure.

Suggestions for further reading

Frost, A. and Ashby, M.F., *Deformation Mechanism Maps*, Pergamon Press, 1982

Haasen, P., Gerdd, V., Kostorz, G. (eds), *Int. Conf. Strength of Metals and Alloys*, 1979

Honeycombe, R.W.K., *The Plastic Deformation of Metals*, Edward Arnold, 1984

Parker, E. and Colombo, U. (eds), *The Science of Material used in Advanced Technology*, Wiley Interscience, 1973

Porter, D.A. and Easterling, K.E., *Phase Transformations in Metals and Alloys*, Van Nostrand Reinhold, 1981

Chapter 14

Fracture

14.1 Brittle fracture

14.1.1 Introduction

In this section we shall deal mainly with brittle fracture of the cleavage type and with the ductile–brittle transition which occurs in metals with b.c.c. structure. The fracture observed in fatigue and creep tests is discussed in the next two sections.

The problem of brittle fracture is of great practical importance, particularly in ferrous metallurgy, because of the consequences of failure in ships' hulls, pressure vessels, bridges and pipe lines in service, especially at low temperatures. Perhaps the most spectacular examples of this type of failure occurred in welded ships during the second world war. Of 5000 US merchant ships built, more than one-fifth developed cracks in the hull within three years' service, some breaking completely in two. Welding is only significant in that this method of fabrication allows the brittle cracks to propagate uninterrupted for large distances through the structure. Other factors, more fundamental in nature, which affect the tendency to failure by brittle fracture, are equally well known. Thus, for example, the importance of impurity atoms in governing the ductility of steel was first indicated by the enhanced susceptibility to brittleness of Bessemer steels, since these contain high nitrogen contents. Refinement of the grain size of a material is also known to decrease the ductile–brittle transition temperature. However, the problem is still a contemporary one because only f.c.c. metals are commonly ductile at the lowest temperatures. The newer metals, such as the transition metals niobium, zirconium, titanium, chromium and molybdenum, are strongly embrittled by the presence of oxygen, nitrogen and carbon and stringent, and hence expensive, precautions must be taken during melting and fabrication of these metals before they become usable as structural materials. Beryllium, uranium, bismuth, antimony, germanium, silicon, intermetallic compounds (e.g. NiAl, FeAl), metallic carbides and nitrides all have little or no ductility at room temperature. Moreover, many oxides and other ceramics would be ideal creep-resistant materials, but for their extreme brittleness when cold.

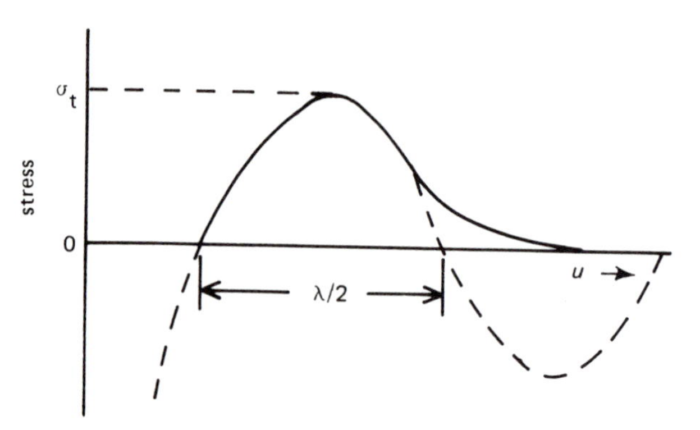

Figure 14.1 Model for estimating the theoretical fracture strength σ_t

Owing to its great practical importance, this problem has received immense theoretical and experimental attention. Here we shall outline the theoretical ideas on the mechanism of brittle fracture, and then show how these can account for the importance of such metallurgical variables as impurity content, grain size and prior cold work.

14.1.2 Griffith micro-crack criterion

Most materials break at a stress well below the theoretical fracture stress, which is that stress, σ_t, required to pull apart two adjoining layers of atoms. This stress varies with the distance between the atom planes and, as shown in *Figure 14.1*, may be approximated to a sine curve of wavelength such that

$$\sigma = \sigma_t \sin\left(\frac{2\pi u}{\lambda}\right) \simeq \sigma_t\left(\frac{2\pi u}{\lambda}\right)$$

where u is the displacement from the equilibrium spacing b. From Hooke's law $\sigma = (Eu/b)$ and hence $\sigma_t = \lambda E/2\pi b$. Now in pulling apart the two atomic planes it is necessary to supply the surface energy γ and hence

$$2\gamma = \int_0^{\lambda/2} \sigma_t \sin\left(\frac{2\pi u}{\lambda}\right) \mathrm{d}u = \frac{\lambda \sigma_t}{2\pi}$$

so that, the theoretical tensile strength is given by

$$\sigma_t = \sqrt{(E\gamma/b)} \tag{14.1}$$

Glass fibres and both metallic and non-metallic whiskers have strengths approaching this ideal value of about $E/10$, but bulk metals even when tested under favourable conditions (e.g. at 4 K) can rarely withstand stresses above $E/100$. Griffith, in 1920, was the first to suggest that this discrepancy was due to the presence of small cracks which propagate through the crystal and cause fracture. Griffith's theory deals with elastic cracks, where at the tip of the crack atomic bonds exist in every stage of elongation and fracture. As the crack grows, each of the bonds in its path take up the strain, and the work done in stretching and breaking these bonds becomes the surface energy γ of the fractured faces. If separation of the specimen between two atomic layers occurs in this way, the theoretical strength only needs to be exceeded at one point at a time, and the applied stress will be much lower than the theoretical fracture stress. The work done in breaking the bonds has to be supplied by the applied force, or by the elastic energy of the system. Assuming for a crack of length $2c$ that an approximately circular region of radius c is relieved of stress σ and hence strain energy $(\sigma^2/2E)\pi c^2$ by the presence of the crack, the condition when the elastic strain of energy balances the increase of surface energy is given by

$$\frac{\mathrm{d}}{\mathrm{d}c}\left(\frac{\pi c^2 \sigma^2}{E}\right)=\frac{\mathrm{d}}{\mathrm{d}c}(4c\gamma)$$

and leads to the well known Griffith formula

$$\sigma=\sqrt{\left(\frac{2\gamma E}{\pi c}\right)}\simeq\sqrt{\left(\frac{\gamma E}{c}\right)} \tag{14.2}$$

for the smallest tensile stress σ able to propagate a crack of length $2c$. The Griffith criterion therefore depends on the assumption that the crack will spread if the decrease in elastic strain energy resulting from an increase in $2c$ is greater than the increase in surface energy due to the increase in the surface area of the crack.

Griffith's theory has been verified by experiments on glasses and polymers at low temperatures, where a simple process of fracture by the propagation of elastic cracks occurs. In such 'weak' brittle fractures there is little or no plastic deformation and γ is mainly the surface energy (≈ 1–$10\ \mathrm{J\ m^{-2}}$) and the fracture strength $\sigma_f \approx 10^{-5}E$. In crystalline solids, however, the cracks are not of the elastic type and a plastic zone exists around the crack tip. In such specimens, fracture cannot occur unless the applied tensile stress σ is magnified to the theoretical strength σ_t. For an atomically sharp crack (where the radius of the root of the crack r is of the order of b) of length $2c$ it can be shown that the magnified stress σ_m will be given by $\sigma_m=\sigma\sqrt{(c/r)}$ which, if the crack is to propagate, must be equal to the theoretical fracture stress of the material at the end of the crack. It follows that substituting this value of σ_t in equation 14.1 leads to the Griffith formula of equation 14.2.

In such 'strong' brittle fractures γ is greatly increased by the contribution of the plastic work around the crack tip which increases the work required for crack propagation. The term γ must now be replaced by $(\gamma+\gamma_p)$ where γ_p is the plastic work term; generally $2(\gamma+\gamma_p)$ is replaced by G, the strain energy release rate, so that equation 14.2 becomes

$$\sigma=\sqrt{(EG/\pi c)} \tag{14.3}$$

Here, G might be $\sim 10^4\ \mathrm{J\ m^{-2}}$ and $\sigma_f \approx 10^{-2}$–$10^{-3}E$.

14.1.3 Micro-crack formation by plastic glide

The existence of micro-cracks in metallurgical materials, of the size required by the Griffith theory is readily acceptable, since they could easily be produced, for example, during working operations or in the neighbourhood of micro-structural constituents such as carbides in steels. However, the Griffith criterion alone cannot account completely for the behaviour of brittle metals and alloys. Perhaps the two most striking experimental observations which suggest that other factors apart from the Griffith mechanism are important are, (i) that most metals, although breaking in a brittle manner, i.e. with little plastic extension, do show, when carefully examined, some evidence of plastic deformation in the region of the fracture, and (ii) the existence of the ductile–brittle transition, in which the fracture changes from the completely ductile to the completely brittle as the temperature is lowered by a few degrees through a critical range. None of the terms in the Griffith equation varies with temperature sufficiently strongly to explain the brittle transition phenomena and, consequently, some other mechanisms must also be important. Moreover, because slip is often observed, even when the specimen fractures in a brittle manner, it is clear that slip dislocations must play some part in the fracture process.

Because of these observations it is now believed that in crystalline materials, in contrast to the glassy type of materials for which the Griffith equation is obeyed, some plastic deformation will always occur, and it is the slip dislocations generated immediately prior to fracture which give rise in certain regions of the crystal to micro-cracks. A brittle–ductile transition can then be explained on the basis of the criterion that the material is ductile at any temperature, if the yield stress at that temperature is smaller than the stress necessary for the growth of these micro-cracks, but if it is larger the material is brittle.

Several models have been suggested for the process whereby glide dislocations are converted into micro-cracks. The simplest mechanism, postulated by Stroh, is that involving a pile-up of dislocations against a barrier, such as a grain boundary, or a twin interface (the importance of twins in deformation and fracture has already been discussed in Chapter 8). The applied stress pushes the dislocations together and a crack forms beneath their coalesced half-planes, as shown in *Figure 14.2(a)*. A second mechanism of crack formation, suggested by Cottrell, is that arising from the junction of two intersecting slip planes. A dislocation gliding in the (101) plane coalesces with one gliding in the $(10\bar{1})$ plane to form a new dislocation which lies in the (001) plane according to the reaction,

$$\tfrac{1}{2}a[\bar{1}\bar{1}1]+\tfrac{1}{2}a[111]\rightarrow a[001]$$

The new dislocation, which has a Burgers vector $a[001]$, is a pure edge dislocation, and, as shown in *Figure 14.2(b)*, may be considered as a wedge, one lattice constant thick, inserted between the faces of the (001) planes. It is considered that the crack can then grow by means of other dislocations in the (101) and $(10\bar{1})$ planes running into it. Although the mechanism readily accounts for the observed (100) cleavage plane of b.c.c. metals, examples have not been directly observed in b.c.c. metals, although cracks arising from intersecting deformation twin bands often occur.

If cleavage cracks are formed by such a dislocation mechanism, the Griffith formula may be rewritten to take account of the number of dislocations, n, forming the crack. Thus, rearranging Griffith's formula we have

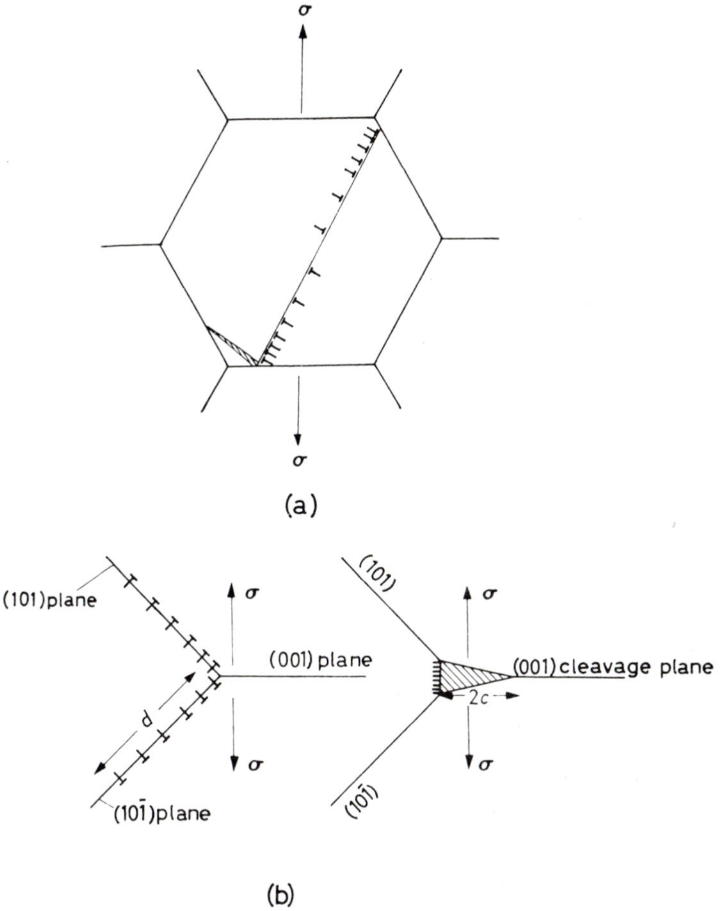

Figure 14.2 Formation of a crack (a) by the piling up of dislocations against a grain boundary after strain and (b) by the dislocations on (101) and (10$\bar{1}$) planes running together (after Cottrell, 'Brittle Fracture in Steel and Other Materials', *Trans. Amer. Inst. Mech. Engrs*, April 1958, p. 192)

$$\sigma(c\sigma/E)=\gamma$$

where the product of length $2c$ and the strain σ/E is the maximum displacement between the faces of the crack. This displacement will depend on the number of dislocations forming the cleavage wedge and may be interpreted as a pile-up of n edge dislocations, each of Burgers vector a, so that equation 14.2 becomes

$$\sigma na=2\gamma \tag{14.4}$$

and gives a general criterion for fracture. Physically, this means that a number of glide dislocations, n, run together and in doing so cause the applied stress acting on them to do some work, which for fracture to occur must be at least sufficient to supply the energy to create the new cracked faces, i.e. $2(\gamma+\gamma_p)$.

While these dislocation coalescence mechanisms may operate in single-phase materials, in two-phase alloys it is usually easier to nucleate cracks by piling up dislocations at particles, e.g. grain boundary cementite or cementite lamellae in pearlite. The pile-up stress then leads to cracking of either the particle or the particle/matrix interface.

14.1.4 The mechanism of fracture

Since it was first suggested that the susceptibility to brittle fracture of Bessemer steel might be due to its high nitrogen content, considerable evidence has accumulated to show that brittleness is associated with a yield phenomenon, which for steels can be related to the carbon or nitrogen content. From the theory outlined in Chapter 8 for the propagation of yielding in a polycrystalline metal having a b.c.c. structure, it would seem probable that the feature which is important in fracture is that yielding suddenly releases a large avalanche of dislocations, which are able to run together and form a micro-crack. If the crack spreads in a time too short to allow stress relaxation to occur by slip in neighbouring regions, brittle fracture will result.

Evidently, there are three distinct values of applied stresses at which the above processes occur. The yield stress at which slip bands trigger slip in other bands, the nucleation stress at which slip bands nucleate micro-cracks, and the growth stress at which these micro-cracks grow into a complete fracture. The nucleation process, as discussed in the previous section, depends only on the shear stress forcing the dislocations together before they coalesce and, therefore, the fracture process should also only depend on shear stresses. However, it is well known that a notched thick plate of steel remains brittle to a higher temperature than a similarly notched thin plate, which indicates that the presence of a hydrostatic tension increases the brittleness of the material. This observation implies that at the critical stage in the fracture process, a crack must already be present if the hydrostatic stress is to exert an influence. Accordingly, the nucleation of micro-cracks is considered to be an easier process than their propagation, and the growth stress the governing factor in the fracture mechanism. When the yield stress is smaller than the growth stress the material is ductile, while when it is larger the material is brittle.

Qualitatively, we would therefore expect those factors which influence the yield stress also to have an effect on the ductile–brittle fracture transition. Thus, because the yield stress increases markedly with decreasing temperature, it is not surprising that the transition from ductile to brittle behaviour takes place on lowering the temperature. The lattice 'friction' term σ_i, dislocation locking term k_y, and grain size $2d$ should also all be important because any increase in σ_i and k_y, or decrease in the grain size, will raise the yield stress with a corresponding tendency to promote brittle failure.

These conclusions have been put on a quantitative basis by Cottrell, who considered the stress needed to grow a crack at or near the tensile yield stress, σ_y, in specimens of grain diameter, $2d$. Let us consider first the formation of a micro-crack. If τ_y is the actual shear stress operating, the effective shear stress acting on a glide band is only $(\tau_y - \tau_i)$, where it will be remembered τ_i is the 'friction' stress resisting the motion of unlocked dislocations arising from the Peierls–Nabarro lattice stress, intersecting dislocations or groups of impurities. The displacement

na is then given by

$$na = [(\tau_y - \tau_i)/\mu]d \tag{14.5}$$

where μ is the shear modulus and d is the length of the slip band containing the dislocations (assumed here to be half the grain diameter). Once a micro-crack is formed, however, the whole applied tensile stress normal to the crack acts on it, so that σ can be written as ($\tau_y \times$ constant), where the constant is included to account for the ratio of normal stress to shear stress. Substituting for na and σ in the Griffith formula, equation 14.4, then fracture should be able to occur at the yield point when $\sigma = \sigma_y$ and

$$\tau_y(\tau_y - \tau_i)d = C\mu\gamma \tag{14.6}$$

where C is a constant. The importance of the avalanche of dislocations produced at the yield drop can be seen if we replace τ_y by (constant $\times \sigma_y$), τ_i by (constant $\times \sigma_i$) and use the Petch relationship $\sigma_y = \sigma_i + k_y d^{-1/2}$, when equation 14.6 becomes

$$(\sigma_i d^{1/2} + k_y)k_y = \beta\mu\gamma \tag{14.7}$$

where β is a constant which depends on the stress system; $\beta \approx 1$ for tensile deformation and $\beta \approx 1/3$ for a notched test.

This is a general equation for the propagation of a crack at the lower yield point and shows what factors are likely to influence the fracture process. Alternative models for growth-controlled cleavage fracture have been developed to incorporate the possibility of carbide particles nucleating cracks. Such models emphasize the importance of yield parameters and grain size but also carbide thickness. Coarse carbides give rise to low fracture stresses, thin carbides with high fracture stresses and ductile behaviour.

14.1.5 Factors affecting brittleness

Many of the effects of alloying, heat treatment, and condition of testing on brittle fracture can be rationalized on the basis of the above 'transition' equation.

Ductile–brittle transition Under conditions where the value of the left-hand side of equation 14.7 is less than the value of the right-hand side, ductile behaviour should be observed; when the left-hand side exceeds the right-hand side the behaviour should be brittle. Since the right-hand side of equation 14.7 varies only slowly with temperature, it is the way in which changes occur in values of the terms on the left of the equation which are important in determining the ductile–brittle transition. Thus, in a given material brittleness should be favoured by low temperatures and high strain-rate, because these give rise to large values of σ_i and k_y, and by large grain sizes.

The transition from ductile to brittle behaviour can be determined by a variety of tests, and the temperature of the transition so determined may vary in each test. This arises because even for a given structural condition the stress system applied to the specimen will vary from test to test, so that different values of β must be used in equation 14.7. Thus, for example, the typical effect of a sharp notch is to raise the transition temperature of structural steel from around 100 K for a normal tensile test into the range of 200 to 300 K, because the value of β is lowered.

Effects of composition and grain size At a constant temperature, because the values of σ_i and k_y remain fixed, the transition point will occur at a critical grain size above which the metal is brittle and below which it is ductile. This grain-size transition is well known from experimental work on steel and is shown in *Figure 14.3(a)*.

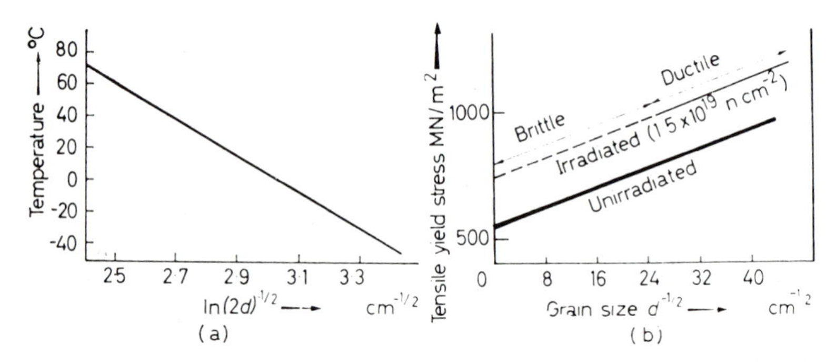

Figure 14.3 Influence of grain size on (a) transition temperature of low carbon steel (after Heslop and Petch, *Phil. Mag.* 1958, **3,** 1128) and (b) yield stress of irradiated and unirradiated low carbon steel tested at −196°C (after Hull and Mogford, *Phil. Mag.* 1958, **3,** 1213)

The appearance in equation 14.7 of the grain-size term, $d^{1/2}$, in combination with the σ_i term, enables many previous metallurgical misunderstandings to be cleared up. It shows that there is no simple connection between hardness and brittleness, since hardening produced by refining the grain size reduces the brittleness, whereas hardening due to an increase in σ_i increases the brittleness.

It is well known that 'killed' steel has very good notch toughness; this is because small aluminium additions refine the grain size and consequently reduce brittleness. Manganese reduces the grain size and by combining with carbon also reduces the k_y value so that this addition is especially beneficial in improving low temperature ductility. It is fairly evident that an improved notch toughness steel, compared with that used for welded ships in World War II, is given by increasing the manganese content and decreasing the carbon content, i.e. a high manganese-to-carbon ratio. Equation 14.7 is also useful in understanding the mechanical behaviour of the common transition metals. Thus, iron has been made ductile in tension (90 per cent reduction in area), even at 4 K, by using zone refinement (*see* Chapter 3) to lower the impurity content and hence to reduce the σ_i and k_y contributions. Of the other transition metals, niobium, vanadium and tantalum are generally ductile down to 20 K even without zone refinement, unless they are extremely coarse-grained, because of their smaller k_y terms, but chromium and molybdenum, being more susceptible to small additions of nitrogen, may be brittle at room temperature or above.

Heat-treatment and alloying Heat treatment is generally used to control the grain size of the sample and refine the structure. 'Killed' steel has very good notch toughness, because aluminium additions refine the grain size. Other additions, particularly nickel and chromium have a similar effect on low temperature ductility.

The Group 6A metals (Cr, Mo and W) are more susceptible to brittle fracture than the Group 5A metals (V, Nb and Ta). A comparison of these metals in terms of cleavage fracture is difficult, however, since Cr, Mo and W are susceptible to grain boundary fracture because segregation of impurities to such regions reduces the effective surface energy γ. However, even if this effect is eliminated by lowering the impurity level, it appears that Ta, Nb and V are more ductile than Fe, Mo, Cr and W, presumably because they have a lower k_y/μ ratio, and a higher γ value.

Work hardening and irradiation hardening Small amounts of plastic deformation at room temperature, which overcomes the yield point and unlocks some of the dislocations, improves the ductility at low temperatures. The room temperature ductility of chromium is similarly affected by small amounts of plastic deformation at 400 °C. In general, however, plastic deformation which leads to work hardening embrittles the metal because it raises the σ_i contribution, due to the formation of intersecting dislocations, vacancy aggregates and other lattice defects.

The importance of twins in fracture is not clear as there are several mechanisms other than twinning for the formation of a crack which can initiate fracture, and there is good evidence that micro-cracks form in steel in the absence of twins, and that cracks start at inclusions. Nevertheless, twinning and cleavage are generally found under similar conditions of temperature and strain-rate in b.c.c. transition metals, probably because both phenomena occur at high stress levels; the nucleation of a twin requires a higher stress than the propagation of the twin interface.

Irradiation hardening also embrittles the metal. According to the theory of this type of hardening outlined in Chapter 9, radiation damage can produce an increase in both k_y (migration to dislocations of vacancies or interstitials) and σ_i (formation of dislocation loops and other aggregated defects). However, for steel, radiation hardening is principally due to an increased σ_i contribution, presumably because the dislocations in mild steel are already too heavily locked with carbon atoms for any change in the structure of the dislocation to make any appreciable difference to k_y. Nevertheless, a neutron dose of 1.9×10^{23} n.m^{-2} will render a typical fine-grained, un-notched En2 steel, which is normally ductile at -196 °C, quite brittle (*Figure 14.3*(*b*)). Moreover, experiments on notched fine-grained steel samples (see *Figure 2.36*(*c*)) show that this dose increases the ductile–brittle transition temperature by 65 °C.

Microstructure The change in orientation at individual grain boundaries impedes the propagation of the cleavage crack by (i) creating cleavage steps, (ii) causing localized deformation, and (iii) tearing near the grain boundary (*see Figure 14.6*(*a*)). It is the extra work done (γ_p) in such processes which is thought to give rise to a value of surface energy, derived from equation 14.7, which proves to be much higher than the true one; the apparent surface energy γ is then $(\gamma_s+\gamma_p)$. It follows, therefore, that the smaller the distance a crack is able to propagate without being deviated by a change of orientation of the cleavage plane, the greater is the resistance to brittle fracture. In this respect, the coarser high temperature products of steel, such as pearlite and upper bainite, have inferior fracture characteristics compared with the finer lower bainite and martensite products. The fact that coarse carbides promote cleavage while fine carbides lead to ductile behaviour has already been discussed.

14.2 Hydrogen embrittlement

It is well known that ferritic and martensitic steels are severely embrittled by small amounts of hydrogen. The hydrogen may be introduced during melting and retained during the solidification of massive steel castings. Plating operations, e.g. Cd plating of steel for aircraft parts, may also lead to hydrogen embrittlement. Hydrogen can also be introduced during acid pickling or welding, or by exposure to H_2S atmospheres.

The chief characteristics of hydrogen embrittlement are its (a) strain-rate sensitivity, (b) temperature-dependence and (c) susceptibility to produce delayed fracture (*see Figure 14.4*(*a*)). Unlike normal brittle fracture, hydrogen embrittle-

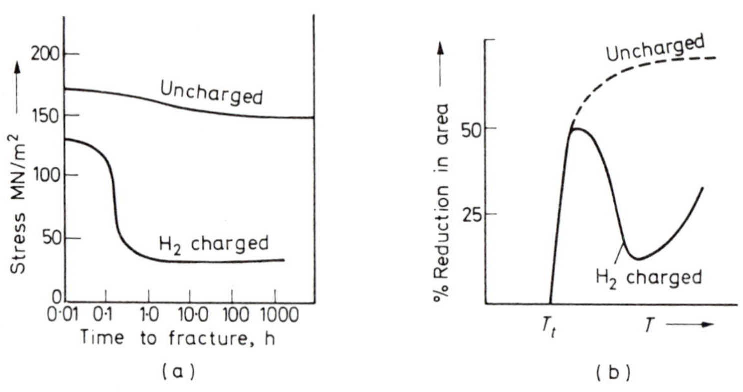

Figure 14.4 Influence of hydrogen on fracture behaviour showing (a) time-dependence and (b) temperature-dependence

ment is enhanced by slow strain-rates and consequently, notched-impact tests have little significance in detecting this type of embrittlement. Moreover, the phenomenon is not more common at low temperatures, but is most severe in some intermediate temperature range around room temperature, i.e. $-100\,°C$ to $100\,°C$. These effects have been taken to indicate that hydrogen must be present in the material and must have a high mobility in order to cause embrittlement in polycrystalline aggregates.

A commonly held concept of hydrogen embrittlement is that monatomic hydrogen precipitates at internal voids or cracks as molecular hydrogen, so that as the pressure builds up it produces fracture. Troiano and his co-workers have proposed alternatively that the critical factor is the segregation of hydrogen, under applied stress, to regions of triaxial stress just ahead of the tip of the crack, and when a critical hydrogen concentration is obtained, a small crack grows and links up with the main crack. Hydrogen may also exist in the void or crack but it is considered that this has little effect on the fracture behaviour, and it is only the hydrogen in the stressed region that causes embrittlement. Neither model considers the Griffith criterion, which must be satisfied if cracks are to continue spreading. It is also difficult to see how a dilute solution of hydrogen atoms could be responsible for the nucleation of a crack in an annealed material, or the propagation of one previously introduced by mechanical working.

An application of the fracture theory may be made to this problem. Thus, if hydrogen collects in microcracks and exerts internal pressure P, the pressure may be directly added to the external stress to produce a total stress $(P+p)$ for propagation. Thus the crack will spread when

$$(P+p)na=2(\gamma_s+\gamma_p) \tag{14.8}$$

where the surface energy is made up from a true surface energy γ_s and a plastic work term γ_p. The possibility that hydrogen causes embrittlement by becoming adsorbed on the crack surfaces thereby lowering γ is thought to be small, since the plastic work term γ_p is the major term controlling γ, whereas adsorption would mainly effect γ_s. The theory explains how cracks can spread even in the absence of applied stress, since if P is large, p can be zero and the equation for crack propagation will still be satisfied. The weaknesses of the theory are that it is not able to explain why hydrogen embrittlement does not occur at low temperatures or at high strain rates, nor the fact that hydrogen never completely embrittles polycrystalline metals, since at least 10 per cent reduction in area always occurs (*see Figure 14.4*(*b*)).

However, the following facts do emerge from considering the phenomenon of crack propagation. Supersaturated hydrogen atoms precipitate as molecular hydrogen gas at a crack nucleus, or the interface between non-metallic inclusions and the matrix. The stresses from the build-up of hydrogen pressure are then relieved by the formation of small cleavage cracks. Clearly, while the crack is propagating, an insignificant amount of hydrogen will diffuse to the crack and, as a consequence, the pressure inside the crack will drop. However, because the length of the crack has increased, if a sufficiently large and constant stress is applied, the Griffith criterion will still be satisfied and completely brittle fracture can, in theory, occur. Thus, in iron single crystals, the presence or absence of hydrogen appears to have little effect during crack propagation because the crack has little difficulty spreading through the crystal. In polycrystalline material, however, the hydrogen must be both present and mobile, since propagation occurs during tensile straining.

When a sufficiently large tensile stress is applied such that $(p+P)$ is greater than that required by the Griffith criterion, the largest and sharpest crack will start to propagate, but will eventually be stopped at a microstructural feature, such as a grain boundary, as previously discussed. The pressure in the crack will then be less than in adjacent cracks which have not been able to propagate. A concentration gradient will then exist between such cracks (since the concentration is proportional to the square root of the pressure of hydrogen) which provides a driving force for diffusion, so that the hydrogen pressure in the enlarged crack begins to increase again. The stress to propagate the crack decreases with increase in length of crack, and since p is increased by straining, a smaller increment ΔP of pressure may be sufficient to get the crack restarted. The process of crack propagation followed by a delay time for pressure build-up continues with straining until the specimen fails when the area between the cracks can no longer support the applied load. In higher strain-rate tests the hydrogen is unable to diffuse from one stopped crack to another, to help the larger crack get started before it becomes blunted by plastic deformation at the tip. The decrease in the susceptibility to hydrogen embrittlement in specimens tested at low temperatures results from the lower pressure build up at these temperatures since $PV = 3nRT$, and also because hydrogen has a lower mobility.

14.3 Fracture toughness

In engineering structures, particularly heat treated steels, cracks are likely to arise from weld defects, inclusions, surface damage, etc, and it is necessary to design structures with the knowledge that cracks are already present and capable of propagation at stresses below the macroscopic yield stress as measured in a tensile test. Since different materials show different crack propagation characteristics, e.g. hard steel and glass, it is necessary for the design engineers to find the limiting design stress in terms of some property or parameter of the material. For this reason a fracture toughness parameter is now being employed to measure the tendency of cracks of given dimensions to propagate under particular stress conditions.

In section 14.1 it was shown that $\sigma\sqrt{(\pi c)} = \sqrt{(EG)}$, which indicates that fast fracture will occur when a material is stressed to a value σ and the crack reaches some critical size, or alternatively when a material containing a crack is subjected to some critical stress σ, i.e. the critical combination of stress and crack length for fast fracture is a constant, $\sqrt{(EG)}$ for the material, where E is Young's modulus and G is the strain energy release rate. The term $\sigma\sqrt{(\pi c)}$ is given the symbol K and is called the stress intensity factor with units MN m$^{-3/2}$. Fast fracture will then occur when $K = K_c$, where $K_c\,[= \sqrt{(EG_c)}]$ is the critical stress intensity factor, or more commonly the fracture toughness parameter.

The fracture toughness of a material can alter markedly depending on whether the elastic-plastic field ahead of the crack approximates to plane strain or plane stress conditions, much larger values being obtained under plane stress conditions as operate in thin sheets. The important and critical factor is the size of the plastic zone in relation to the thickness of the section containing the crack. When this is small, as in thick plates and forgings, plane strain conditions prevail and the hydrostatic tension ahead of the crack results in a semi-brittle 'flat' fracture. When the value is large as in thin sheets of ductile metals plane stress conditions operate and the tension at the crack front is smaller, giving rise to a more ductile mode of failure. At intermediate values a mixed fracture, with a flat centre bordered by shear lips, is obtained. Thus without changing the structure or properties of the materials in any way, it is possible to produce a large difference in fracture toughness by changing the section thickness. For thick sections, when a state of complete constraint is more nearly approached, the values of K_c and G_c approach minimum limiting values. These values are denoted by K_{1c} and G_{1c} and are considered to be material constants, the subscript 1 denotes the first mode of crack extension, i.e. the opening mode (*see Figure 14.5*).

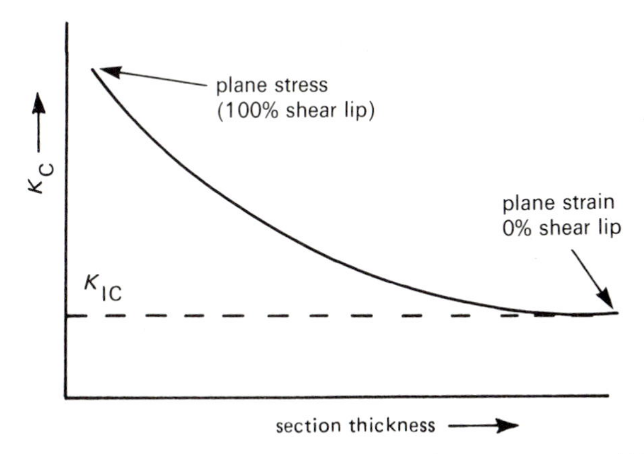

Figure 14.5 Variation in the fracture toughness parameter with section thickness

The general procedure in measuring the fracture toughness parameter is to introduce a crack of suitable size into a specimen of suitable dimension and geometry. The specimen is then loaded at a slow rate and the crack extension measured up to the critical condition. K_{1c} and G_{1c} values are now becoming increasingly specified for special applications. The first type of specimen geometry recommended by the ASTM committee is a plate containing a central through slit and the formula derived for K is

$$K = \sigma\sqrt{[W \tan(\pi c/W)]}$$

where σ is the applied tensile stress, W the width of the plate and c the half crack length. Fracture toughness requirements are likely to be written into the general specification of high technology alloys and hence it is necessary to determine the effect of heat treatment and alloying additions on fracture toughness parameters. Processes such as ausforming and controlled rolling improve the fracture toughness of certain steels. Carbon has a considerable effect and there are advantages in reducing the C-level below 0.1% where possible. HSLA steels have $C \gtrsim 0.1\%$ and the Nb, V and Ti additions form fine carbides which together with the small grain sizes enable good strength levels and acceptable fracture toughness values to be achieved. Maraging steels with high alloy and low carbon ($<0.04\%$) give very high strength combined with high toughness.

14.4 Intergranular fracture

Intergranular brittle failures are often regarded as a special class of fracture. In many alloys, however there is a delicate balance between the stress required to cause a crack to propagate by cleavage and that required to cause brittle separation along grain boundaries. Although the energy absorbed in crack propagation may be low compared to cleavage fractures, much of the analysis of cleavage is still applicable if it is considered that chemical segregation to grain boundaries or crack faces lowers the surface energy γ of the material. Fractures at low stresses are observed in austenitic chrome–nickel steels, due to the embrittling effect of intergranular carbide precipitation at grain boundaries. High transition temperatures and low fracture stresses are also common in tungsten and molybdenum as a result of the formation of thin second-phase films due to small amounts of oxygen, nitrogen or carbon. Similar behaviour is observed in the embrittlement of copper by antimony and iron by oxygen, although in some cases the second-phase films cannot be detected.

A special intergranular failure, known as temper embrittlement, occurs in some alloy steels when tempered in the range 500–600 °C. This phenomenon is associated with the segregation of certain elements or combinations of elements to the grain boundaries. The amount segregated is very small ($\sim$ a monolayer) but the species and amount has been identified by AES on specimens fractured intergranularly within the ultra-high vacuum of the Auger system. Group VIB elements are known to be the most surface active in iron but fortunately, they combine readily with Mn and Cr thereby effectively reducing their solubility. Elements in Groups IVB and VB are less surface active but often co-segregate in the boundaries with Ni and Mn. In Ni–Cr steels, the co-segregation of Ni–P and Ni–Sb occurs, but Mo additions can reduce the tendency for temper embrittlement. Since carbides are often present in the grain boundaries, these can provide the crack nucleus under the stress concentration from dislocation pile-ups either by cracking or by decohesion of the ferrite/carbide interface, particularly if the interfacial energy has been lowered by segregation.

14.5 Ductile fracture

The type of transgranular fracture which occurs in ductile metals has already been mentioned in Chapter 2. The fracture process involves three stages. First small holes or cavities nucleate usually at weak internal interfaces, e.g. particle/matrix interfaces. These cavities then expand by plastic deformation and finally coalesce by localized necking of the metal between adjacent cavities to form a fibrous fracture. A scanning electron micrograph showing the characteristics of a typical ductile failure is shown in *Figure 14.6*(*b*). This type of fracture may be regarded as taking place by the nucleation of an internal plastic cavity, rather than a crack, which grows outwards to meet the external neck which is growing inwards.

In metals such as iron, which have large yield drops, such cavities may be nucleated by piled-up glide dislocations, but in metals such as copper or

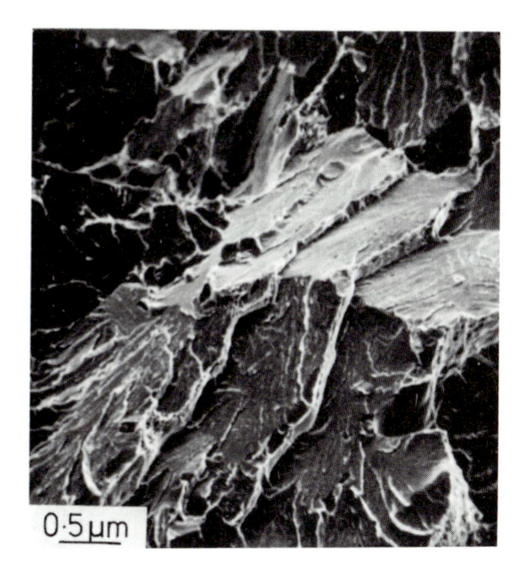

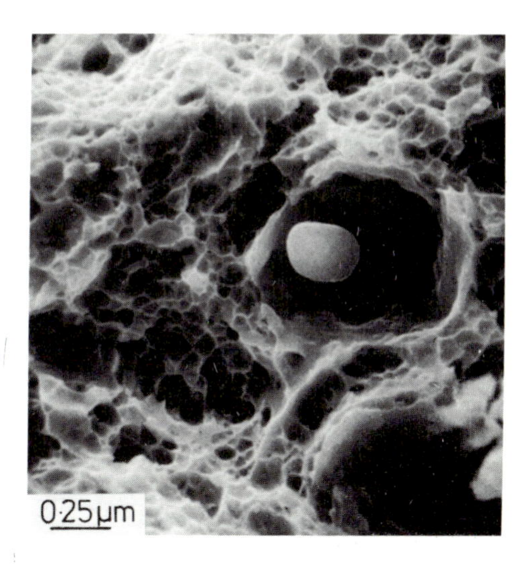

Figure 14.6 SEM micrographs of a medium carbon (0.4%) steel (a) with a pearlite–ferrite normalized structure showing the change in cleavage plane at a subgrain boundary and characteristic river pattern, (b) with a quenched and tempered martensite structure, showing large dimples associated with oxide inclusions and small dimples associated with small carbide precipitates (Courtesy Dr L. Sidanin)

aluminium this is not so. No large pile-ups are formed and experimental evidence suggests that nucleation occurs at foreign particles. For example, OFHC copper necks down to over 90 per cent reduction in area, whereas tough pitch copper shows only 70 per cent reduction in area; a similar behaviour is noted for super pure and commercial purity aluminium. Thus, if no inclusions were present, failure should occur by the specimen pulling apart entirely by the inward growth of the external neck, giving nearly 100 per cent reduction in area. Dispersion-hardened materials in general fail with a ductile fracture, the fibrous region often consisting of many dimples arising from the dispersed particle nucleating holes and causing local ductile failure.

The dislocation structure around particle inclusions leading to a high work hardening has been discussed in Chapter 10. Thus, around inclusions the local rate of work hardening can be higher than the average and the local stress on reaching some critical value σ_c will cause fracture of the inclusion or decohesion of the particle/matrix interface, thereby nucleating a void. The critical nucleation strain ε_n can be estimated and lies between 0.1 and 1.0 depending on the model. For the Ashby dispersion-hardening model outlined, the stress on the interface due to the nearest prismatic loop, at distance r, is $\mu b/r$, and this will cause separation of the interface when it reaches the theoretical strength of the interface, of order γ_w/b. The parameter r is given in terms of the applied shear strain ε, the particle diameter d and the length k equal to half the mean particle spacing as $r=4kb/\varepsilon d$. Hence, void nucleation occurs on a particle of diameter d after a strain ε, given by $\varepsilon=4k\gamma_w/\mu db$. Any stress concentration effect from other loops will increase with particle size, thus enhancing the particle size dependence of strain to voiding.

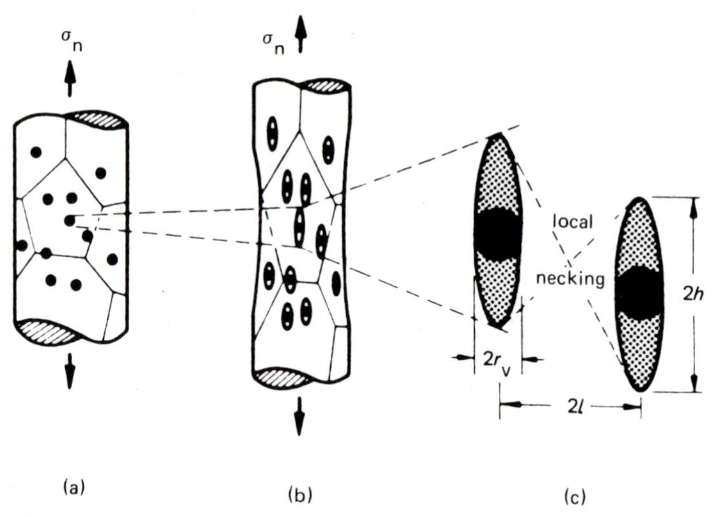

Figure 14.7 Schematic representation of ductile, and transgranular creep, fracture. (a) Voids nucleate at inclusions, (b) voids elongate as the specimen extends, (c) voids coalesce to cause fracture when their length $2h$ is about equal to their separation (after Ashby *et al.*, *Acta Met.*, **27**, 669, 1979)

Once nucleated, the voids grow until they coalesce to provide an easy fracture path. A spherical-shaped void concentrates stress under tensile conditions and, as a result, elongates initially at about $C(\approx 2)$ times the rate of the specimen, but as it becomes ellipsoidal the growth-rate slows until finally the elongated void grows at about the same rate as the specimen. At some critical strain, the plasticity becomes localized and the voids rapidly coalesce and fracture occurs. The localization of the plasticity is thought to take place when the voids reach a critical distance of approach, given when the void length $2h$ is approximately equal to the separation, as shown in *Figure 14.7*. The true strain for coalescence is then

$$\varepsilon = (1/C) \ln [\alpha(2l - 2r_v)/2r_v] \simeq (1/C) \ln [\alpha(1/f_v^{1/2}) - 1] \quad (14.9)$$

where $\alpha \approx 1$ and f_v is the volume fraction of inclusions.

Void growth leading to failure will be much more rapid in the necked portion of a tensile sample following instability than during stable deformation, since the stress system changes in the neck from uniaxial tension to approximately plane strain tension. Thus the overall ductility of a specimen will depend strongly on the macroscopic features of the stress–strain curves which (from Considère's criterion) determines the extent of stable deformation, as well as on the ductile rupture process of void nucleation and growth. Nevertheless, an equation of the form of 14.9 reasonably describes the fracture strain for cup and cone failures.

The work of decohesion influences the progress of voiding and is effective in determining the overall ductility in a simple tension test in two ways. The onset of voiding during uniform deformation depresses the rate of work-hardening which leads to a reduction in the uniform strain, and the void density and size at the onset of necking determines the amount of void growth required to cause ductile rupture. Thus for matrices having similar work-hardening properties, the one with the least tendency to wet the second phase will show both lower uniform strain and lower necking strain. For matrices with different work-hardening potential but similar work of decohesion the matrix having the lower work-hardening rate will show the lower reduction prior to necking but the greater reduction during necking, although two materials will show similar total reductions to failure.

The degree of bonding between particle and matrix may be determined from voids on particles annealed to produce an equilibrium configuration by measuring the contact angle θ of the matrix surface to the particle surface. Resolving surface forces tangential to the particle, then the specific interface energy γ_I is given approximately in terms of the matrix surface energy γ_n and the particle surface energy γ_P as $\gamma_I = \gamma_P - \gamma_n \cos\theta$. The work of separation of the interface γ_w is then given by

$$\gamma_w = \gamma_P + \gamma_m - \gamma_I = \gamma_m(1 + \cos\theta) \tag{14.10}$$

Measurements show that the interfacial energy of TD nickel is low and hence exhibits excellent ductility at room temperature. Specific additions, e.g. Zr to TD nickel, and Co to Ni–Al_2O_3 alloys are also effective in lowering the interfacial energy, thereby causing the matrix 'to wet' the particle and increase the ductility. Because of their low γ_I, dispersion-hardened materials have superior mechanical properties at high temperatures compared with conventional hardened alloys.

14.6 Fracture at elevated temperatures

Creep usually takes place above $0.3T_m$ with a rate given by $\dot{\varepsilon} = B\sigma^n$, where B and n are material parameters, as discussed in Chapter 13. Under such conditions ductile failure of a transgranular nature, similar to the ductile failure found commonly at low temperatures, may occur, when voids nucleated at inclusions within the grains, grow during creep deformation and coalesce to produce fracture. However, because these three processes are occurring at $T \approx 0.3T_m$, local recovery is taking place and this delays both the onset of void nucleation and void coalescence. More commonly at lower stresses and longer times-to-fracture, intergranular rather than transgranular fracture is observed. In this situation, grain boundary sliding leads to the formation of either wedge cracks or voids on those boundaries normal to the tensile axis, as shown schematically in *Figure 14.8(b)*. This arises because grain boundary sliding produces a higher local strain-rate on an inclusion in the boundary than in the body of the grain, i.e. $\dot{\varepsilon}_{local} \simeq \dot{\varepsilon}(fd/2r)$ where $f \approx 0.3$ is the fraction of the overall strain due to sliding. The local strain therefore reaches the critical nucleation strain ε_n much earlier than inside the grain.

The time-to-fracture t_f is observed to be $\propto (1/\dot{\varepsilon}_{ss})$ which confirms that fracture is controlled by power-law creep even though the rounded-shape of grain boundary voids indicates that local diffusion must contribute to the growth of the voids. One possibility is that the void nucleation, even in the boundary, occupies

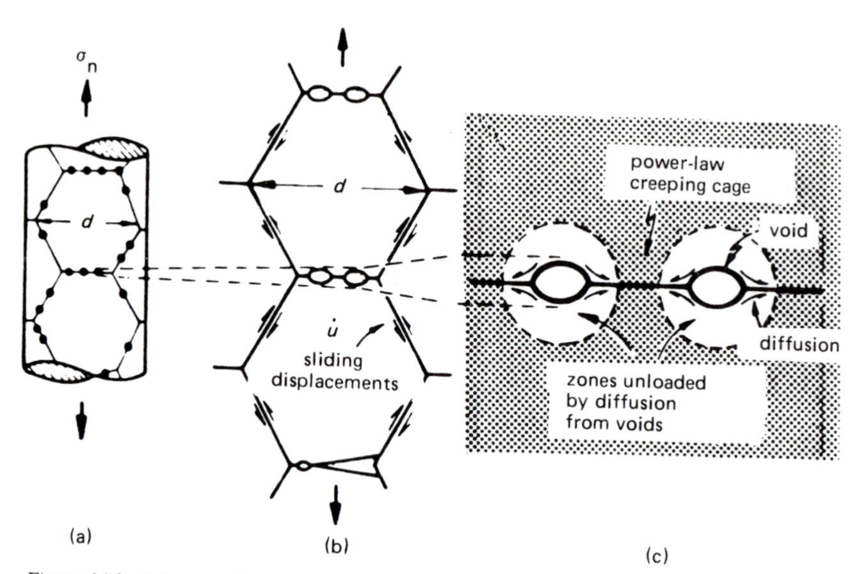

Figure 14.8 Intergranular, creep controlled, fracture. Voids nucleated by grain boundary sliding (a) and (b), grow by diffusion in (c), (after Ashby *et al.*, *Acta Met.*, **27**, 669, 1979)

a major fraction of the lifetime t_f, but a more likely general explanation is that the nucleated voids or cracks grow by local diffusion controlled by creep in the surrounding grains. *Figure 14.8*(*c*) shows the voids growing by diffusion, but between the voids the material is deforming by power-law creep, since the diffusion fields of neighbouring voids do not overlap. Void growth therefore depends on coupled diffusion and power-law creep, with the creep deformation controlling the rate of cavity growth. It is now believed that most intergranular creep fractures are governed by this type of mechanism.

At very low stresses and high temperatures where diffusion is rapid and power-law creep negligible, the diffusion fields of the growing voids overlap. Under these conditions, the grain boundary voids are able to grow entirely by boundary diffusion; void coalescence then leads to fracture by a process of creep cavitation (*Figure 14.9*). In unaxial tension the driving force arises from the process of taking atoms from the void surface and depositing them on the face of the grain that is almost perpendicular to the tensile axis, so that the specimen elongates in the direction of the stress and work is done. The vacancy concentration near the tensile boundary is $c_0 \exp(\sigma\Omega/\boldsymbol{k}T)$ and near the void of radius r is $c_0 \exp(2\gamma\Omega/r\boldsymbol{k}T)$, as discussed previously in Chapter 9, where Ω is the atomic volume and γ the surface per unit area of the void. Thus vacancies flow usually by grain boundary diffusion from the boundaries to the voids when $\sigma \geqslant 2\gamma/r$, i.e. when the chemical potential difference $(\sigma\Omega - 2\gamma\Omega/r)$ between the two sites is negative. For a void $r \simeq 10^{-6}$ m and $\gamma \approx 1$ J/m^2 the minimum stress for hole growth is ≈ 2 MN/m^2. In spite of being pure diffusional controlled growth, the voids may not always maintain their equilibrium near-spherical shape. Rapid surface diffusion is required to keep the balance between growth rate and surface redistribution and with increasing stress the voids become somewhat flattened.

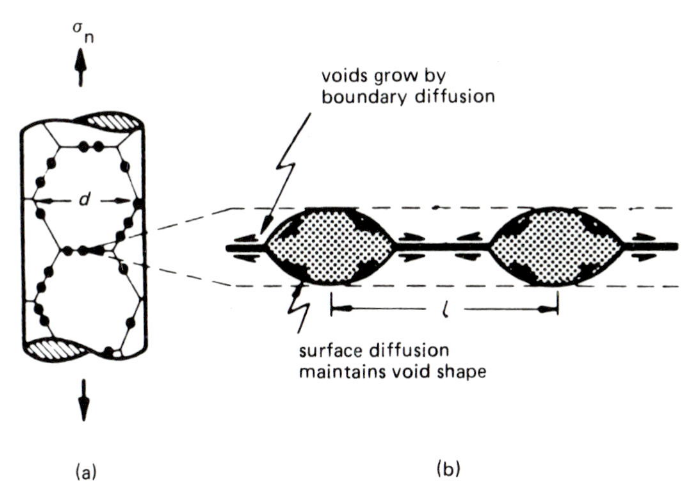

Figure 14.9 Voids lying on 'tensile' grain boundaries (a) grow by grain boundary diffusion (b), (after Ashby *et al.*, *Acta Met.*, **27**, 669, 1979)

14.7 Rupture

If the ductile failure mechanisms outlined above are inhibited then ductile rupture occurs (*see Figure 14.10*). Specimens deformed in tension ultimately reach a stage of mechanical instability when the deformation is localized either in a neck or in a shear band. With continued straining the cross-section reduces to zero and the specimen ruptures, the strain-to-rupture depending on the amount of strain before and after localization. These strains are influenced by the work-hardening behaviour and strain-rate sensitivity. Clearly, rupture is favoured when void nucleation and/or growth is inhibited. This will occur if (i) second-

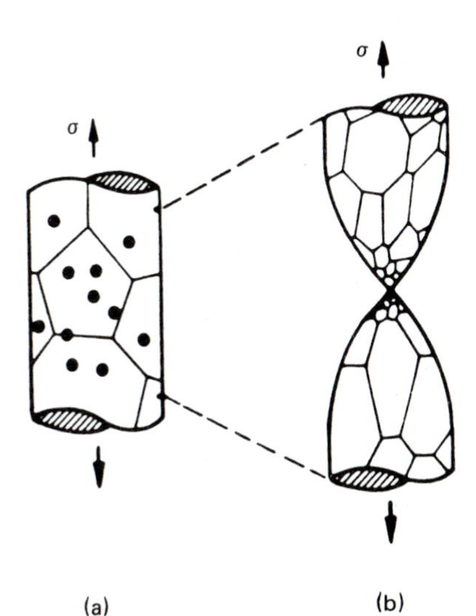

Figure 14.10 Schematic representation of rupture with dynamic recrystallization (after Ashby *et al.*, *Acta Met.*, **27**, 669, 1979)

phase particles are removed by zone-refining or dissolution at high temperatures, (ii) the matrix/particle interface is strong and ε_n is high, (iii) the stress state minimizes plastic constraint and plane strain conditions, e.g. single crystals and thin sheets, (iv) the work-hardening rate and strain-rate sensitivity is high as for superplastic materials (in some superplastic materials voids do not form but in many others they do and it is the growth and coalescence processes which are suppressed), and (v) there is stress relief at particles by recovery or dynamic recrystallization. Rupture is observed in most f.c.c. materials, usually associated with dynamic recrystallization.

14.8 Fracture mechanism maps

The fracture behaviour of a metal or alloy in different stress and temperature regimes can be summarized conveniently by displaying the dominant mechanisms on a fracture mechanism map. Seven mechanisms have been identified, three for brittle behaviour including cleavage and intergranular brittle fracture and four ductile processes. *Figure 14.11* shows schematic maps for f.c.c. and b.c.c. materials, respectively. Not all the fracture regimes are exhibited by f.c.c. materials, and even some of the ductile processes can be inhibited by altering the metallurgical variables. For example, intergranular creep fracture is absent in high-purity aluminium but occurs in commercial-purity material, and because the dispersoid suppresses dynamic recrystallization in TD nickel, rupture does not take place whereas it does in Nimonics at temperatures where the γ' and carbides dissolve.

In the b.c.c. metals, brittle behaviour is separated into three fields; a brittle failure from a pre-existing crack, well below general yield, is called either cleavage 1 or brittle intergranular fracture BIF1, depending on the fracture path. An almost totally brittle failure from a crack nucleated by slip or twinning, below general yield, is called either cleavage 2 or BIF2, and a cleavage or brittle

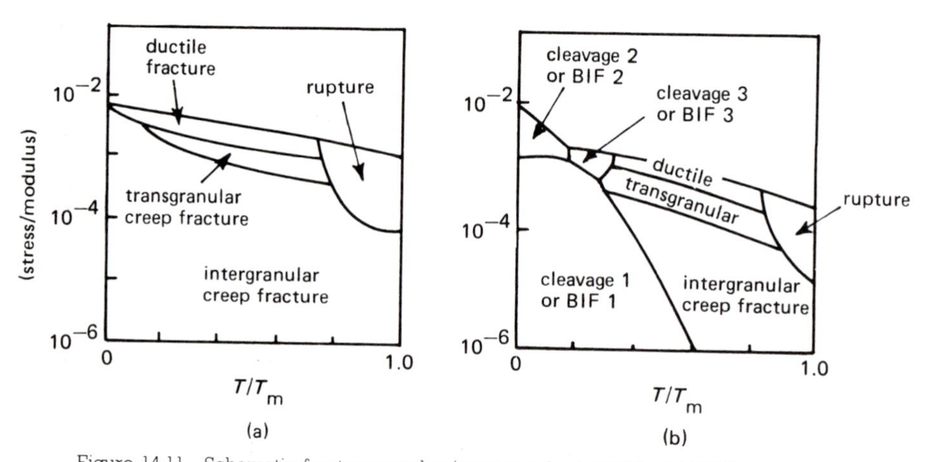

Figure 14.11 Schematic fracture mechanism maps for (a) FCC, (b) BCC materials

boundary failure after general yield and with measurable strain-to-failure is called either cleavage 3 or BIF3. In many cases, mixed transgranular and intergranular fractures are observed, as a result of small changes in impurity content, texture or temperature which cause the crack to deviate from one path to another, no distinction is then made in the regime between cleavage and BIF. While maps for only two structures are shown in *Figure 14.11*, it is evident that as the bonding changes from metallic to ionic and covalent the fracture-mechanism fields will move from left to right; refractory oxides and silicates, for example, exhibit only the three brittle regimes and intergranular creep fracture.

14.9 Fatigue crack growth

Engineering structures such as bridges, pressure vessels and oil rigs all contain cracks and it is necessary to assess the safe life of the structure, i.e. the number of stress cycles the structure will sustain before a crack grows to a critical length and propagates catastrophically. The most effective approach to this problem is by the use of fracture mechanics. Under static stress conditions the state of stress near a crack tip is described by K, the stress intensity factor measured in $\mathrm{MN\,m^{-3/2}}$, but in cyclic loading K varies over a range ΔK $(=K_{max}-K_{min})$. The cyclic stress intensity ΔK increases with time at constant load, as shown in *Figure 14.12(a)*, because the crack grows. Moreover, for a crack of length a the rate of crack growth ($\mathrm{d}a/\mathrm{d}N$) in μm per cycle varies with ΔK according to the Paris–Erdogan equation

$$\mathrm{d}a/\mathrm{d}N = C(\Delta K)^m$$

where C and m are constants, with m between 2 and 4. A typical crack growth rate curve is shown in *Figure 14.12(b)* and exhibits the expected linear relationship over part of the range. The upper limit corresponds to K_{IC}, the fracture toughness

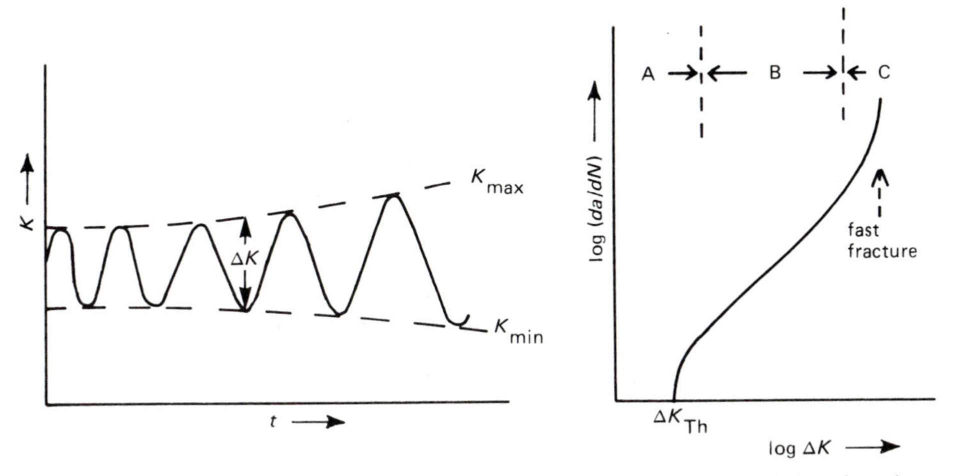

Figure 14.12 (a) Increase in stress intensity ΔK during fatigue, (b) variation of crack growth rate with increasing ΔK

of the material and the lower limit of ΔK is called the threshold for crack growth ΔK_{th}. Clearly when the stress intensity factor is less than ΔK_{th} the crack will not propagate at that particular stress and temperature, and hence ΔK_{th} is of significance in design criteria. If the initial crack length is a_0 and the critical length a_c, then the number of cycles to catastrophic failure will be given by

$$N_f = \int_{a_0}^{a_c} da/C(\Delta K)^m = \int_{a_0}^{a_c} da/C[\Delta\sigma\sqrt{(\pi a)}]^m$$

The mean stress level is known to affect the fatigue life (*see* page 464) and therefore da/dN. If the mean stress level is increased for a constant value of ΔK, K_{max} will increase and thus as K_{max} approaches K_{IC} the value of da/dN increases rapidly in practice, despite the constant value of ΔK.

A survey of fatigue fractures indicates there are four general crack growth mechanisms, (i) striation formulation, (ii) cleavage, (iii) void coalescence and (iv) intergranular separation; some of these mechanisms have been discussed in Chapter 13. The crack growth behaviour shown in *Figure 14.12(b)* can be divided into three regimes which exhibit different fracture mechanisms. In regime A, there is a considerable influence of microstructure, mean stress and environment on the crack growth rate. In regime B, failure generally occurs, particularly in steels, by a transgranular ductile striation mechanism and there is often little influence of microstructure, mean stress or mild environments on crack growth. The degree of plastic constraint which varies with specimen thickness also appears to have little effect. At higher growth rate exhibited in regime C the growth rates become extremely sensitive to both microstructure and mean stress, with a change from striation formation to fracture modes normally associated with non-cyclic deformation including cleavage and intergranular fracture.

Suggestions for further reading

Ashby, M.F., Gandhi, C. and Taplin, D.M.R., *Acta Met.*, **27**, 669, 1979

Gandhi, C. and Ashby, M.F., 'Fracture Mechanism Maps, Overview No. 5', *Acta Met.*, **27**, 1565, 1979

Knott, J., *Fundamentals of Fracture Mechanics*, Butterworth, 1973

Tetleman, A.S. and McEvily, A.J., *Fracture of Structural Materials*, John Wiley, 1967

Fracture Mechanics in Design and Service, Royal Society, 1981

Wilshire, B. and Owen, D.R.J. (eds), *Creep and Fracture of Engineering Materials and Structures*, Pineridge Press, 1981

Chapter 15

Oxidation and corrosion

15.1 Introduction

The use of nearly all metals and alloys at elevated temperatures is invariably limited by the way in which they react to their surrounding environment. The most common reaction is the oxidation of metals in air to form oxides. The process of oxidation usually involves a chemical reaction between a dry metal surface and an oxidizing gas which gives rise to a solid oxide film over the exposed surface. In practical application the solid–gaseous reaction which a metal undergoes is usually more complex than the simple metal–oxygen reaction, involving gaseous mixtures, e.g. jet engine and rocket nozzles in contact with combustion products. However, the basic principles governing these reactions are similar and hence in this chapter discussion is limited to oxidation.

At ambient temperatures, oxidation is not normally a serious problem for most commerical alloys. Instead of this dry corrosion phenomenon the major technological problem is wet, or aqueous, corrosion in which the chemical attack occurs through the medium of water. The nature and characteristics of this form of attack are also discussed.

15.2 Thermodynamics of oxidation

The tendency for a metal to oxidize, like any other spontaneous reaction, is indicated by the free energy change ΔG accompanying the formation of the oxide. Most metals readily oxidize because ΔG is negative for oxide formation. The free energy released by the combination of a fixed amount (1 mol) of the oxidizing agent with the metal is given by ΔG° and is usually termed the standard free energy of the reaction. ΔG° is, of course, related to ΔH°, the standard heat of reaction and ΔS° the standard change in entropy, by the Gibbs equation. The variation of the standard free energy change with absolute temperature for a number of metal oxides is shown in *Figure 15.1*. The noble metals which are easily reduced occur at the top of the diagram and the more reactive metals at the bottom. Some of these metals at the bottom (Al, Ti, Zr), however, resist oxidation

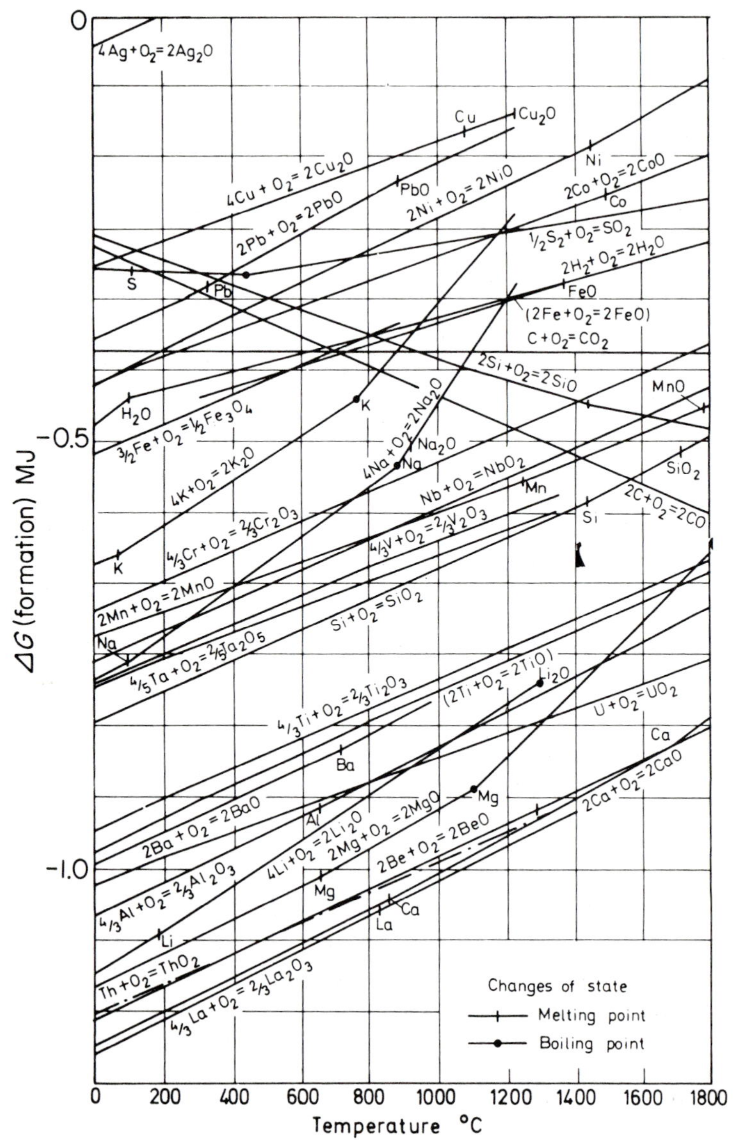

Figure 15.1 Standard free energies of formation of oxides

at room temperature owing to the impermeability of the thin coherent oxide film which first forms.

The numerical value of ΔG° for oxidation reactions decreases with increase in temperature, i.e. the oxides become less stable. This arises from the decreased entropy accompanying the reaction, solid (metal) + gas (oxygen) → solid (oxide). The metal and oxide, being solids, have roughly the same entropy values and $d(\Delta G)/dT$ is thus almost equivalent to the entropy of the oxygen, i.e. 209.3 J deg^{-1} mol^{-1}. Most of the ΔG *v.* T lines therefore slope upwards at about this value, and any change in slope is due to a change in state. As expected, melting has a small effect on ΔS and hence ΔG, but transitions through the boiling point, e.g. ZnO at 970 °C, and sublimation, e.g. Li_2O at 1330 °C, have large effects. Exceptions to the positive slope of the ΔG *v*. T line occur for carbon oxidation to CO or CO_2. In both cases the oxide product is gaseous and thus also has a high free energy. In the reaction $2C + O_2 \rightarrow 2CO$, two moles of gas are produced from one of oxygen so that

$$\Delta S = (S_{oxide} - S_{carbon} - S_{oxygen}) = (2S^{\circ}_{oxide} - S^{\circ}_{carbon} - S^{\circ}_{oxygen}) \simeq S^{\circ}_{oxide}$$

The CO free energy *v*. temperature line has a downward slope of approximately this value. For the $C + O_2 \rightarrow CO_2$ reaction, one mole of CO_2 is produced from one mole of oxygen and hence $\Delta S \simeq 0$; the CO_2 free energy line is thus almost horizontal. The carbon monoxide reaction is favoured at high temperatures and consequently carbon is a very effective reducing agent, having a greater affinity for oxygen than most oxides.

Because of the positive slope to the ΔG° *v*. T line for most oxides in *Figure 15.1*, ΔG° tends to zero at some elevated temperature. This is known as the standard dissociation temperature when the oxide is in equilibrium with the pure element and oxygen at 1 atm pressure. In the case of gold the oxide is not stable at room temperature, for silver Ag_2O dissociates when gently heated to about 200 °C, and the oxides of the Pt group of metals around 1000 °C. The other oxides dissociate at much higher temperatures. However, the temperature is affected by pressure since the free energy per mole of any gaseous phase varies with pressure P(atm) according to $G(P) = G^{\circ} + \boldsymbol{R}T \ln P$, whereas that for the solid phase is relatively unaffected. Thus, for the metal + oxygen $\rightarrow$ metal oxide reaction under standard conditions, $\Delta G^{\circ} = G^{\circ}_{oxide} - G^{\circ}_{metal} - G^{\circ}_{oxygen}$, and at P atm oxygen, $\Delta G = \Delta G^{\circ} - \boldsymbol{R}T \ln P_{O_2}$. The reaction is in equilibrium when $\Delta G = 0$ and hence

$$P_{O_2} = \exp[\Delta G^{\circ}/\boldsymbol{R}T]$$

is the equilibrium dissociation pressure of the oxide at the temperature T. If the pressure is lowered below this value the oxide will dissociate, if raised above, the oxide is stable. The common metal oxides have very low dissociation pressures $\approx 10^{-10}$ N/m^2 ($\sim 10^{-15}$ atm) at ordinary annealing temperatures and thus readily oxidize in the absence of reducing atmospheres.

The standard free energy change ΔG° is also related to the equilibrium constant K of the reaction. For the reaction discussed above, i.e.

$$Me + O_2 \rightarrow MeO_2$$

the equilibrium constant $K = [MeO_2]/[Me][O_2]$ derived from the law of mass action. The active masses of the solid metal and oxide are taken equal to unity and that of the oxygen as its partial pressure under equilibrium conditions. The equilibrium constant at constant pressure, measured in atmospheres is thus $K_P = 1/P_{O_2}$. It then follows that $\Delta G^{\circ} = -\boldsymbol{R}T \ln K_P$. To illustrate the use of these concepts, let us consider the reduction of an oxide to metal with the aid of a reducing agent, e.g. Cu_2O by steam. For the oxidation reaction

$$4Cu + O_2 \rightarrow 2Cu_2O$$

and from *Figure 15.1* at 1000 K, $\Delta G^{\circ} = -1.95 \times 10^5$ J/mol $= 1/P_{O_2}$, giving

$P_{O_2}=6.078\times10^{-6}$ N/m^2. At 1000 K the equilibrium constant for the steam reaction $2H_2O\rightarrow2H_2+O_2$ is

$$K=P_{O_2}P^2_{H_2}/P^2_{H_2O}=1.013\times10^{-15}\ \text{N/m}^2$$

Thus to reduce Cu_2O the term $P_{O_2}<6\times10^{-11}$ in the steam reaction gives $P_{H_2}/P_{H_2O}>10^{-5}$, so that an atmosphere of steam containing 1 in 10^5 parts of hydrogen is adequate to bright-anneal copper.

In any chemical reaction, the masses of the reactants and products are decreasing and increasing respectively during the reaction. As discussed on page 141, the term chemical potential μ $(=\text{d}G/\text{d}n)$ is used to denote the change of free energy of a substance in a reaction with change in the number of moles n, while the temperature, pressure and the number of moles of the other substances are kept constant. Thus,

$$\mu_i=\mu_i^0+\boldsymbol{R}T\ln P_i$$

and the free energy change of any reaction is equal to the arithmetical difference of the chemical potentials of all the phases present. So far, however, we have been dealing with ideal gaseous components and pure metals in our reaction. Generally, oxidation of alloys is of interest and we are then dealing with the solution of solute atoms in solvent metals. These are usually non-ideal solutions which behave as if they contain either more or less solute than they actually do. It is then convenient to use the activity of that component, a_i, rather than the partial pressure, P_i, or concentration, c_i. For an ideal solution $P_i=P_i^0c_i$, whereas for non-ideal solutions $P_i=P_i^0a_i$ such that a_i is an effective concentration equal to the ratio of the partial, or vapour, pressure of the ith component above the solution to its pressure in the standard state. The chemical potential may then be rewritten

$$\mu_i=\mu_i^0+\boldsymbol{R}T\ln a_i$$

where for an ideal gas mixture $a_i=P_i/P_i^0$ and by definition $P_i^0=1$. For the copper oxide reaction, the law of mass action becomes

$$K=\frac{a^2_{Cu_0O}}{a^4_{Cu}a_{O_2}}=\frac{1}{P_{O_2}}=\exp[-\Delta G^\circ/\boldsymbol{R}T]$$

where a_i^n is replaced by unity for any component present in equilibrium as a pure solid or liquid. Some solutions do behave ideally, e.g. Mn in Fe, obeying Raoult's law with $a_i=c_i$; others tend to in dilute solution, e.g. Fe in Cu, but others deviate widely with a_i approximately proportional to c_i (Henry's law).

15.3 Kinetics of oxidation

The free energy changes discussed in the previous section indicate the probable stable reaction product but make no prediction of the rate at which this product is formed. During oxidation the first oxygen molecules to be absorbed on the metal surface dissociate into their component atoms before bonding chemically to the surface atoms of the metal. This process, involving dissociation and ionization, is known as chemisorption. After the build-up of a few adsorbed layers the oxide is nucleated epitaxially on the grains of the base metal at favourable sites, such as dislocations and impurity atoms. Each nucleated region grows, impinging on one another until the oxide film forms over the whole surface. Oxides are therefore usually composed of an aggregate of individual grains or crystals, and exhibit phenomena such as recrystallization, grain growth, creep involving lattice defects, just as in a metal.

If the oxide film initially produced is porous the oxygen is able to pass through and continue to react at the oxide–metal interface. Usually, however, the film is not porous and continued oxidation involves diffusion through the oxide layer. If oxidation takes place at the oxygen–oxide surface, then metal ions and electrons have to diffuse through from the underlying metal. When the oxidation reaction occurs at the metal–oxide interface, oxygen ions have to diffuse through the oxide and electrons in the opposite direction to complete the reaction.

The growth of the oxide film may be followed by means of a thermal balance in conjunction with metallographic techniques. With the thermal balance it is possible to measure to a sensitivity of 10^{-7} g in an accurately controlled atmosphere and temperature. The most common metallographic technique is ellipsometry, which depends on the change in the plane of polarization of a beam of polarized light on reflection from an oxide surface; the angle of rotation depends on the thickness of the oxide. Interferometry is also used, but more use is being made of replicas and thin films in the transmission electron microscope and the scanning electron microscope.

The rate at which the oxide film thickens depends on the temperature and the material, as shown in *Figure 15.2*. During the initial stages of growth at low temperatures, because the oxygen atoms acquire electrons from the surface metal atoms, a strong electric field is set up across the thin oxide film pulling the metal atoms through the oxide. In this low temperature range, e.g. Fe below 200°C, the thickness increases logarithmically with time ($x \propto \ln t$), the rate of oxidation falling off as the field strength diminishes.

At intermediate temperatures, e.g. 250–1000 °C in Fe, the oxidation develops with time according to a parabolic law ($x^2 \propto t$) in nearly all metals. In this region the growth is a thermally activated process and ions pass through the oxide film by thermal movement, their speed of migration depending on the nature of the defect structure of the oxide lattice. Large stresses, either compressive or tensile,

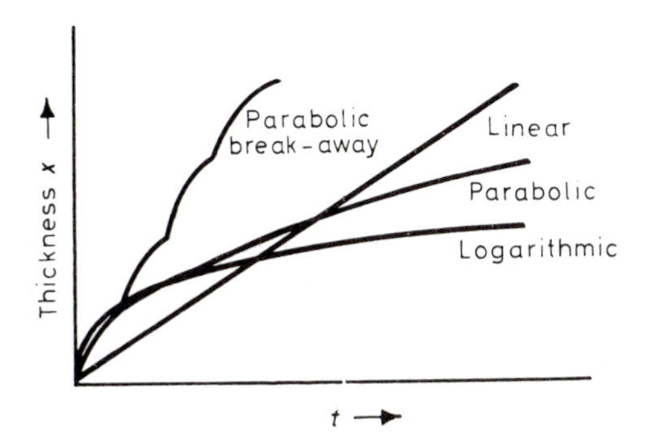

Figure 15.2 Different forms of oxidation behaviour in metals

may often build up in oxide films and lead to breakaway effects when the protective oxide film cracks and spalls. Repeated breakaway on a fine scale can prevent the development of extensive parabolic growth and the oxidation assumes an approximately linear rate or even faster. The stresses in oxide film are related to the Pilling-Bedworth (PB) ratio, defined as the ratio of the molecular volume of the oxide to the atomic volume of the metal from which the oxide is formed. If the ratio is less than unity as for Mg, Na, K, the oxide formed may be unable to give adequate protection against further oxidation right from the initial stages and under these conditions, commonly found in alkali metals, linear oxidation ($x \propto t$) is obeyed. If, however, the Pilling–Bedworth ratio is very much greater than unity, as for many of the transition metals, the oxide is too bulky and may also tend to spall.

15.4 The structure of oxides

Crystals of oxides, like metals, contains point defects at temperatures above absolute zero. However, oxides are made up of two or more different elements and the energy to form the different kinds of point defect will also be different. The defects which can be formed are anion vacancies or interstitials, and cation vacancies or interstitials. If the two defects have similar energies of formation then approximately equal concentrations of each defect will be formed, but if the energy to form one type of defect, say cation vacancies, is significantly smaller than the energies of formation of the other defects then the concentration of cation vacancies far outweighs the concentration of other defects in the crystal. When a large concentration of each type of defect is present in turn, a change in composition occurs leading to a deviation from stoichiometry. These deviations are: (1) excess metal, due to anion vacancies, (2) excess metal, due to interstitial cations, (3) excess non-metal, due to interstitial anions, (4) excess non-metal due to cation vacancies. An example of excess-metal due to anion vacancies is found in the oxidation of silicon which takes place at the metal/oxide interface. Interstitials are more likely to occur in oxides with open crystal structures and when one atom is much smaller than the other; ZnO, for example, is typical of the type 2 defect lattice having the open Wurtzite lattice, in which small zinc ions can occupy interstitial sites as shown in *Figure 15.3*(*a*). There do not appear to be any well-defined examples of oxidation involving interstitial oxygen, but the oxidation of copper to Cu_2O shown in *Figure 15.3*(*b*), is an example of (4) above, involving cation vacancies. Thus copper vacancies are created at the oxide surface and diffuse through the oxide layer and are eliminated at the oxide/metal interface.

Oxides which contain point defects of the kind described above behave as semiconductors when the electrons associated with the point defects either form positive holes or enter the conduction band of the oxide. If the electrons remain locally associated with the point defects, then charge can only be transferred by the diffusion of the charge carrying defects through the oxide. Both *p*- and *n*-type semiconductors are formed when oxides deviate from stoichiometry, the former arises from a deficiency of cations and the latter from an excess of cations.

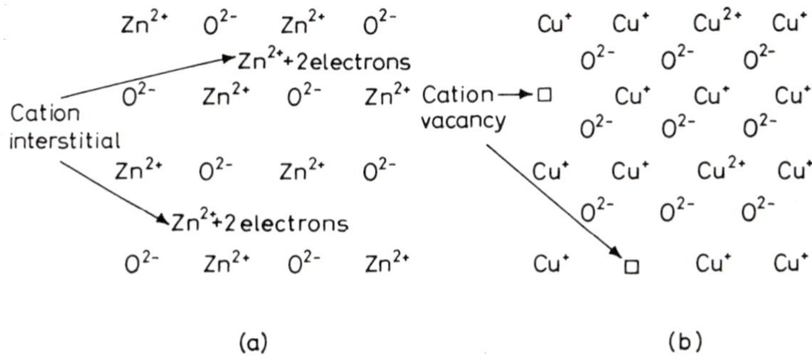

Figure 15.3 Schematic arrangement of ions in two typical oxides (a) $Zn_{>1}O$, with excess metal due to cation interstitials and (b) $Cu_{<2}O$, with excess non-metal due to cation vacancies

Examples of *p*-type semiconducting oxides are NiO, PbO and Cu_2O whilst the oxides of Zn, Cd and Be are *n*-type semiconductors.

15.5 Wagner's theory of oxidation

At high temperatures oxide films thicken according to the parabolic rate law, $x^2 \propto t$ and the mechanism by which thickening proceeds has been explained by Wagner. Point defects, of the kind discussed in the previous section, diffuse through the oxide under the influence of a constant concentration gradient. The defects are annihilated at one of the interfaces causing a new lattice site to be formed. Specifically, zinc oxide thickens by the diffusion of zinc interstitials created at the metal/oxide interface through the oxide to the oxide/oxygen interface where they are removed by the reaction

$$2Zn_i^{++} + 4e + O_2 \rightarrow 2ZnO$$

The concentration of zinc interstitials at the metal/oxide interface is maintained by the reaction

$$Zn_{(metal)} \rightarrow Zn_i^{++} + 2e$$

with the creation of vacancies in the zinc lattice. The migration of charged interstitial defects is accompanied by the migration of electrons and for thick oxide films it is reasonable to assume that the concentrations of the two migrating species are constant at the two surfaces of the oxide, i.e. oxide/gas and oxide/metal, governed by local thermodynamic equilibria. There is thus a constant concentration difference Δc across the oxide and the rate of transport through unit area will be $D\,\Delta c/x$, where D is a diffusion coefficient and x the film thickness. The rate of growth is then

$$dx/dt \propto D(\Delta c/x)$$

and the film thickens parabolically according to the relation

$$x^2 = kt$$

where $\boldsymbol{k}$ is a constant involving several structural parameters. Wagner has shown that the oxidation process can be equated to an ionic plus an electronic current, and obtained a rate equation for oxidation in equivalents/cm^2 s involving the transport numbers of anions and electrons respectively, the conductivity of the oxide, the chemical potentials of the diffusing ions at the interfaces and the thickness of the oxide film. Many oxides thicken according to a parabolic law over some particular temperature range. It is a thermally activated process and the rate constant $\boldsymbol{k}=\boldsymbol{k}_0 \exp[-Q/\boldsymbol{R}T]$ with Q equal to the activation energy for the rate controlling diffusion process.

At low temperatures and for thin oxide films a logarithmic rate law is observed and in order to account for this the Wagner mechanism has been modified by Cabrera and Mott. The Wagner mechanism is only applicable when the concentrations of point defects and electrons are equal throughout the film; for thin oxide films this is not so, a charged layer is established at the oxide/oxygen interface. Here the oxygen atoms on the outer surface become negative ions by extracting electrons from the metal underneath the film and so exert an electrostatic attraction on the positive ions in the metal. When the oxide thickness is $\gtrsim 10$ nm this layer results in an extremely large electric field being set up which pulls the diffusing ions through the film and accelerates the oxidation process. As the film thickens the field strength decreases as the distance between positive and negative ions increases and the oxidation rate approximates to that predicted by the Wagner theory.

As the scale thickens, according to a parabolic law, the resultant stress at the interface increases and eventually the oxide layer can fail either by fracture parallel to the interface or by a shear or tensile fracture through the layer. In these regions the oxidation rate is then increased until the build-up in stress is again relieved by local fracture of the oxide scale. Unless the scale fracture process occurs at the same time over the whole surface of the specimen then the repeated parabolic nature of the oxidation rate will be smoothed out and an approximately linear law observed. This breakaway parabolic law is sometimes called paralinear, and is common in oxidation of titanium when the oxide reaches a critical thickness. In some metals, however, such as U, W and Ce the linear rate process is associated with an interface reaction converting a thin protective inner oxide layer to a non-protective porous oxide.

15.6 Parameters affecting oxidation rates

The Wagner theory of oxidation and its dependence on the nature of the defect structure has been successful in explaining the behaviour of oxides under various conditions, notably the effects of small alloying additions and oxygen pressure variations. The observed effects can be explained by reference to typical *n*- and *p*-type semiconducting oxides. For oxidation of Zn to ZnO the zinc atoms enter the oxide interstitially at the oxide/metal interface ($Zn \rightarrow Zn_i^{++} + 2e$), and diffuse to the oxide/oxygen interface. The oxide/oxygen interface reaction ($2Zn_i^{++} + 4e + O_2 \rightarrow 2ZnO$) is assumed to be a rapid (equilibrium) process and consequently, the concentration of defects at this interface is very small, the dissociation pressure for the reaction at room temperature is very small (1.3×10^{-118} N/m^2) compared with the reaction pressure. The defect gradient across the oxide layer is approximately independent of oxygen pressure and, because this gradient effectively determines the oxidation rate, the rate is independent of oxygen pressure. This is found to be the case experimentally for oxide thicknesses in the Wagner region. By considering the oxide as a semiconductor with a relatively small defect concentration, the law of mass action can be applied to the defect species. For the oxide/oxygen interface reaction this means that

$$[Zn^{++}]^2[e]^4 = \text{constant}$$

The effect of small alloying additions can be explained (Wagner–Hauffe) by considering this equation. Suppose an alloying element is added to the metal that enters the oxide on the cation lattice. Since there are associated with each cation site only two electron sites available in the valence band of the oxide, if the element is trivalent the excess electrons enter the conduction band, increasing the concentration of electrons. For a dilute solution the equilibrium constant remains unaffected and hence, from the above equation, the net effect of adding the element will be to decrease the concentration of zinc interstitials and thus the rate of oxidation. Conversely, addition of a monovalent element will increase the oxidation rate. Experimentally it is found that Al decreases and Li increases the oxidation rate. During the growth of the oxide, zinc interstitials are formed at the oxide/metal interface and migrate into the oxide. Correspondingly, vacancies are formed which migrate into the zinc lattice of the underlying metal. This mechanism of vacancy production has been discussed in Chapter 9. Vacancy-type dislocation loops in the metal therefore grow by climb as a result of the production of vacancies by oxidation. For pure zinc at room temperature and atmospheric pressure, the mean growth rate of dislocation loops has been found to be (≈ 0.02 nm/s) and on adding 0.1 atomic per cent Al the rate decreased to (≈ 0.0025 nm/s), whereas the same amount of Li increased the rate to (≈ 0.11 nm/s) (*see Figure 15.4*).

For Cu_2O, a *p*-type semiconductor, the oxide formation and cation vacancy ($Cu^+\square$) creation takes place at the oxide/oxygen interface, according to

$$O_2 + 4Cu = 2Cu_2O + 4(Cu^+\square) + 4(e\square)$$

The defect diffuses across the oxide and is eliminated at the oxide/metal interface; the equilibrium concentration of defects is at the metal/oxide interface and the excess at the oxide/oxygen interface. It follows therefore that the reaction rate is pressure dependent. Applying the law of mass action to the oxidation reaction gives

$$[Cu^+\square]^4[e\square]^4 = \text{const } P_{O_2}$$

and, since electrical neutrality requires $[Cu^+\square] = [e\square]$, then

$$[Cu^+\square] = \text{Const. } P_{O_2}^{1/8}$$

and the reaction rate should be proportional to the 1/8th power of the oxygen pressure. In practice, it varies as $P_{O_2}^{1/7}$, and the discrepancy is thought to be due to the defect concentration not being sufficiently low to neglect any interaction

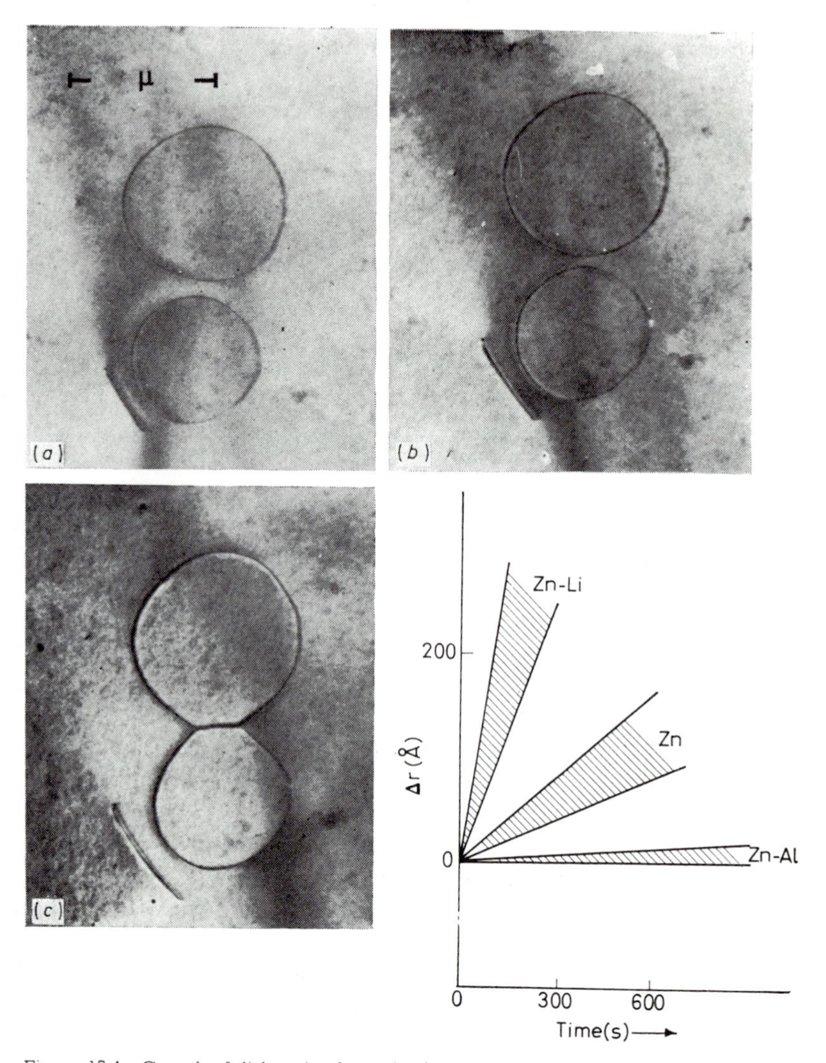

Figure 15.4 Growth of dislocation loops in zinc at room temperature (a) t = 0, (b) t = 32 min, (c) t = 39 min, (d) shows the effect of dilute alloying addition on loop growth (after Hales, Dobson and Smallman, *Metal Sci. J.*, 1968, **2,** 224, courtesy of the Institute of Metals)

effects. The addition of lower valency cations, e.g. transition metals, would contribute fewer electrons and thereby increase the concentration of holes, decreasing the vacancy concentration and hence the rate. Conversely, higher valency cations increase the rate of oxidation.

15.7 Oxidation resistance

The addition of alloying elements according to the Wagner–Hauffe rule just considered, is one way in which the oxidation rate may be changed to give increased oxidation resistance. The alloying element may be added, however, because it is a strong oxide former and forms its own oxide on the metal surface in preference to that of the solvent metal. Chromium, for example, is an excellent additive, forming a protective Cr_2O_3 layer on a number of metals, e.g. Fe, Ni, but is detrimental to Ti which forms an *n*-type anion defective oxide. Aluminium additions to copper similarly improve the oxidation behaviour by preferential oxidation to Al_2O_3. In some cases, the oxide formed is a compound oxide of both the solute and solvent metals. The best known examples are the spinels which have the general formula $M'O\ M''_2O_3$ and a cubic structure, e.g. $NiO.Cr_2O_3$ and $FeO.Cr_2O_3$. It is probable that the spinel formation is temperature-dependent, with Cr_2O_3 forming at low temperatures and the spinel at higher temperatures.

Stainless steels (ferritic, austenitic or martensitic) are among the best oxidation resistant alloys and are based on Fe–Cr. When iron is heated above about 570 °C the oxide scale which forms consists of Wustite, FeO (a *p*-type semiconductor) next to the metal, Magnetite Fe_3O_4 (a *p*-type semiconductor) next and Haematite Fe_2O_3 (an *n*-type semiconductor) on the outside. When Cr is added at low concentrations the Cr forms a spinel $FeO.CrO_3$ with the Wustite and later with the other two oxides. However, a minimum Cr addition of 12 per cent is required before the inner layer is replaced by Cr_2O_3 below a thin outer layer of Fe_2O_3. Heat-resistant steels for service at temperatures above 1000 °C usually contain 18% Cr or more, and austenitic stainless steels 18% Cr, 8% Ni. The growth of Cr_2O_3 on austenitic stainless steels containing up to 20% Cr appears to be rate-controlled by chromium diffusion. Kinetic factors determine whether Cr_2O_3 or a duplex spinel oxide form, the nucleation of Cr_2O_3 is favoured by higher Cr levels, higher temperatures and by surface treatments, e.g. deformation, which increase the diffusivity. Surface treatments which deplete the surface of Cr promote the formation of spinel oxide. Once Cr_2O_3 is formed, if this film is removed or disrupted, then spinel oxidation is favoured because of the local lowering of Cr.

Essentially, some of the coatings, e.g. silicides used for the protection of refractory metals such as niobium are examples of compound oxides but these are not produced by alloying. Such coatings are produced by spraying and other means, and involve severe practical difficulties, particularly with respect to cracking.

15.8 Intergranular voiding – stress v vacancy injection

With nickel, oxidation at ≈ 1000°C gives rise to prolific void formation which has been attributed by various workers to (i) vacancy injection, (ii) oxide induced stress or (iii) CO_2 gas produced by oxidation of carbon. With the oxide-induced stress theory, it is assumed that the tensile stress due to the PB ratio >1 causes the growth of small voids on grain boundaries normal to the stress axis by creep. The mechanism has gained some favour since stresses $\approx 10^6\ N/m^2$ can be produced but recent control experiments on nickel single crystals still exhibit voiding and indicate that void growth by 'Hull–Rimmer' creep is not a likely mechanism. It is also observed that the amount of voiding produced is sensitive to

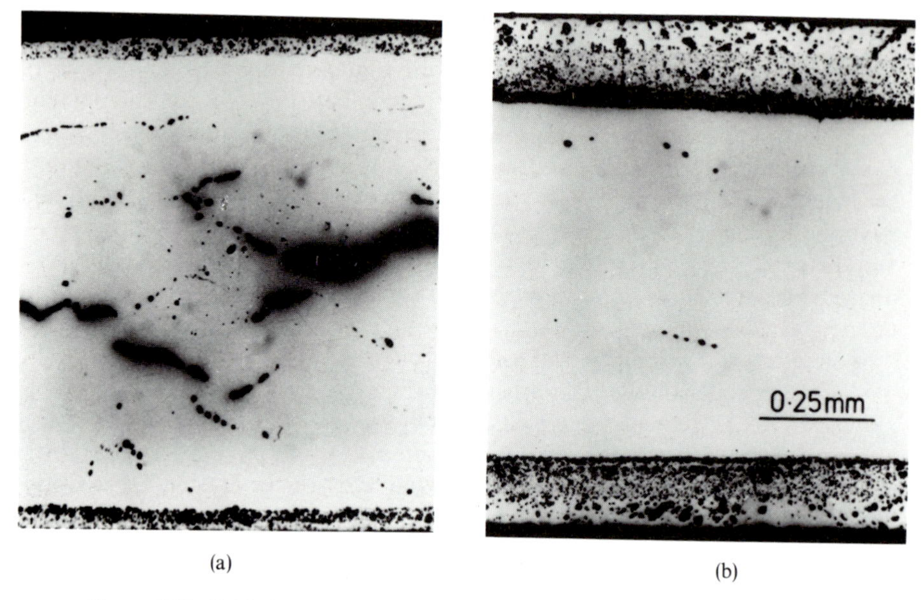

Figure 15.5 Oxidation at 1000°C of nickel (a) containing carbon and (b) decarburised (after Deacon, Loretto and Smallman)

carbon content and for de-carburized samples the void density is very much reduced, but this is accompanied by a much increased oxidation rate (*see Figure 15.5*). It is therefore concluded that oxidation gives rise to vacancy injection and in the case of carbon-containing nickel the vacancies diffuse to carbon, mainly on boundaries, to relieve the stress due to gas formation thereby producing a large void/bubble density. For de-carburized nickel the vacancies are not required by the $C \rightarrow CO_2$ gas reaction and are annihilated at the oxide/metal interface, giving rise to a non-protective oxide and increased oxidation, as observed in *Figure 15.5(b)*.

While vacancies are required for void growth, it is not essential for these to be supplied by surface oxidation. Indeed, it has been demonstrated quite clearly that voids can be produced in the absence of oxide film formation by oxidizing in a very low partial pressure of oxygen. In this case the vacancies for the voids do not come from vacancy injection, since there is no surface oxidation, but must be produced by the climb of the dislocation sub-structure the only remaining source of vacancies. Experiments with the HVEM on nickel, support this contention. Interstitial dislocation loops produced by irradiation in the HVEM are observed to decrease in size on oxidizing at 500 °C as a result of the vacancy injection by oxide film formation, but are observed to grow on annealing in a very low partial pressure of oxygen. The interstitital loops grow by emitting vacancies to accommodate the requirement for vacancies by the carbon reaction and void formation. It is concluded therefore that vacancy injection is an important process during oxidation but there are associated stress effects due either to gas reactions or to the oxide that can determine the fate of the point defects injected.

15.9 Breakaway oxidation

The oxidation behaviour of mild and low alloy steels in high pressure (≈ 1.4 MN m^{-2}) CO_2 (containing small amounts of CO and water vapour) as used in gas-cooled reactors at ≈ 750 K, indicates the importance of vacancy injection and oxide–metal adhesion. The initial oxidation with the formation of Fe_3O_4 follows the normal 'protective', parabolic relationship. However, after an incubation period, which depends on temperature, gas pressure, water vapour content and alloy composition a transition to a faster 'non-protective' oxidation occurs; this has been termed breakaway oxidation and exhibits approximately linear kinetics. The incubation time is increased at lower temperatures and lower water vapour contents and in Fe–Cr alloys. Small additions of silicon have a profound effect on decreasing breakaway and reducing the gas pressure to 0.1 MN m^{-2} (1 atm) seems to prevent the phenomenon entirely.

The microstructural changes which result from high-pressure CO_2 oxidation are complex and naturally depend on alloying, etc, with quite significant variations occurring in Cr content and carbide distribution within the oxide layers formed and in the underlying metal substrate. In mild steel, however, the structure is less complicated and a working model, developed by Gibbs shown in *Figure 15.6*, serves to illustrate the important features of the

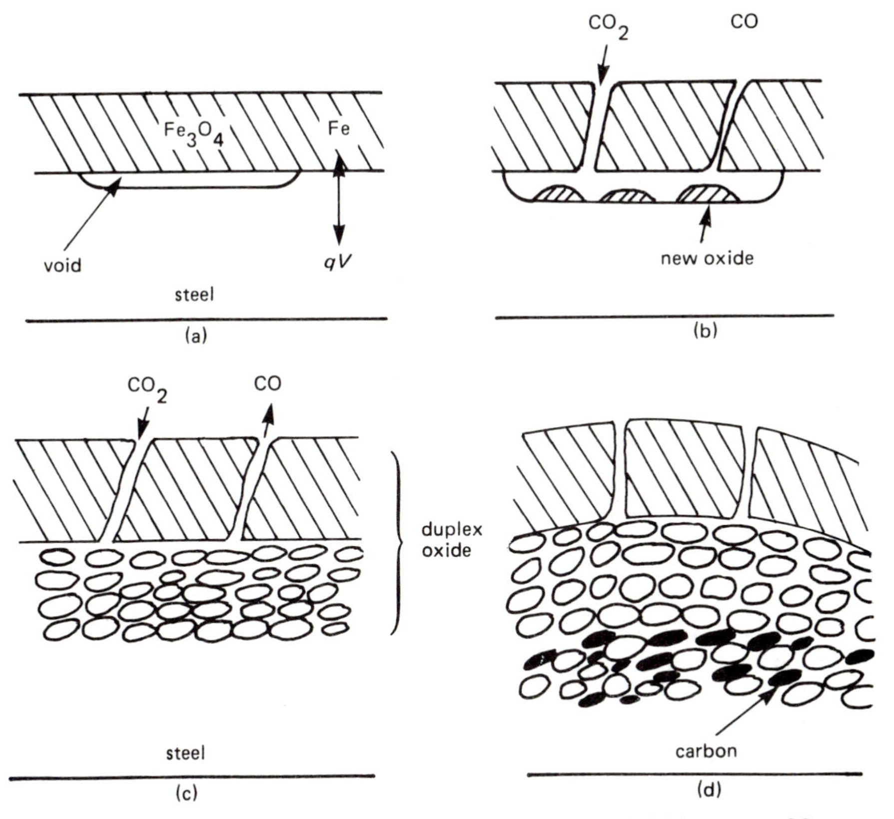

Figure 15.6 Model for the process of breakaway oxidation in high pressure CO_2

phenomenon. Initially the Fe_3O_4 oxide scale grows by cation diffusion outwards, possibly by short-circuit paths such as grain boundaries and dislocations. This process gives rise to the injection of vacancies into the metal, some of which condense at the interface to give loss of scale–metal adhesion, as shown in *Figure 15.6(a)*. Gas access to this 'fresh' metal is then maintained through microchannels developed in the oxide above the voidage by cracking or along grain boundaries. New oxide then forms in the space provided by the voidage by the reaction

$$3Fe + 4CO_2 \rightarrow Fe_3O_4 + 4CO$$

and continues to grow in the volume continually made available by further vacancy condensation.

Some of the CO liberated escapes through the microchannels, but eventually these are restricted by growing oxide crystals and the CO/CO_2 ratio rises, reducing the local oxygen potential and decreasing the oxide crystal growth rate, so that a microporous structure is perpetuated (*Figure 15.6(c)*). The build up of CO therefore regulates the growth of the inner layer of the duplex oxide structure. Eventually, breakaway occurs with the inner layer rapidly growing when the CO level decreases rapidly as a result of a catalysed Boudouard reaction $2CO \rightarrow CO_2 + C$. The carbon is deposited between the crystals (*Figure 15.6(d)*) maintaining a very porous structure.

At low CO_2 gas pressures the carbon activity at the scale–metal interface is too low to support the Boudouard catalyst, while water vapour promotes and silicon delays the formation of a Boudouard catalyst.

15.10 Aqueous corrosion

Metals corrode in aqueous environments by an electrochemical mechanism involving the dissolution of the metal as ions, e.g. $Fe \rightarrow Fe^{2+} + 2e$. The excess electrons generated in the electrolyte either reduce hydrogen ions (particularly in acid solutions) according to

$$2H^+ + 2e \rightarrow H_2$$

so that gas is evolved from the metal, or create hydroxyl ions by the reduction of dissolved oxygen according to

$$O_2 + 4e + 2H_2O \rightarrow 4OH^-$$

The corrosion rate is therefore associated with the flow of electrons or an electrical current. The two reactions involving oxidation (in which the metal ionizes) and reduction occur at anodic and cathodic sites respectively on the metal surface. Generally, the metal surface consists of both anodic and cathodic sites, depending on segregation, microstructure, stress, etc, but if the metal is partially immersed there is often a distinct separation of the anodic and cathodic areas with the latter near the waterline where oxygen is readily dissolved (differential aeration). *Figure 15.7* illustrates the formation of such a differential aeration cell; Fe^{2+} ions pass into solution from the anode and OH^- ions from the cathode, and where they meet they form ferrous hydroxide $Fe(OH)_2$. However,

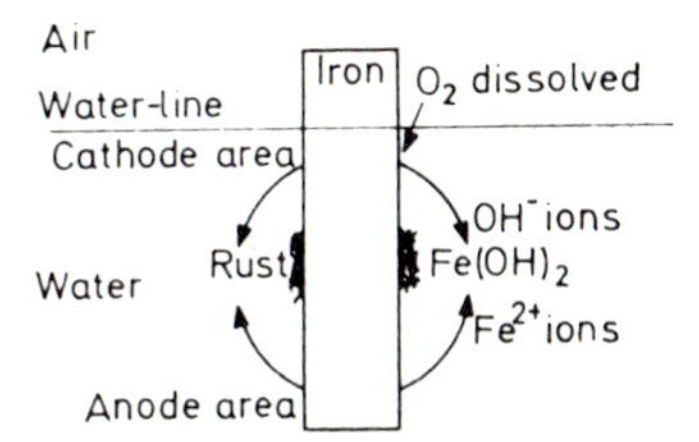

Figure 15.7 Corrosion of iron by differential aeration

depending on the aeration, this may oxidize to $Fe(OH)_3$, red-rust $Fe_2O_3 . H_2O$, or black magnetite Fe_3O_4. Such a process is important when water, particularly sea-water, collects in crevices formed by service, manufacture or design. In this form of corrosion the rate-controlling process is usually the supply of oxygen to the cathodic areas and, if the cathodic area is large, can often lead to intense local attack of small anode areas, such as pits, scratches, crevices, etc.

15.11 The electrochemical series

In the absence of differential aeration, the formation of anodic and cathodic areas depends on the ability to ionize. Some metals ionize easily, others with difficulty and consequently anodic and cathodic areas may be produced for example by segregation, or the joining of dissimilar metals. When any metal is immersed in an aqueous solution containing its own ions, positive ions go into solution until the resulting electromotive force (e.m.f.) is sufficient to prevent any further solution; this e.m.f. is the electrode potential or half-cell potential. To measure this e.m.f. it is necessary to use a second reference electrode in the solution, usually a standard hydrogen electrode. With no current flowing the applied potential cancels out the extra potential developed by the spontaneous ionization at the metal electrode over and above that at the standard hydrogen electrode. With different metal electrodes a table of potentials (E_0) can be produced for the half-cell reactions

$$M \rightarrow M^{n+} + ne$$

where E_0 is positive. The usual convention is to write the half-cell reaction in the reverse direction so that the sign of E_0 is also reversed, i.e. E_0 is negative; E_0 is referred to as the standard electrode potential.

It is common practice to express the tendency of a metal to ionize in terms of this voltage, or potential, E_0, rather than free energy, where $\Delta G = -nFE_0$ for the half-cell reaction with nF coulombs of electrical charge transported per mole. The half-cell potentials are given in *Table 15.1* for various metals, and refer to the potential developed in a standard ion concentration of one mole of ions per litre, i.e. unit activity, relative to a standard hydrogen electrode at 25 °C which is assigned a zero voltage. The voltage developed in any galvanic couple, i.e. two half cells, is given by the difference of the electrode potentials. If the activity of the solution is increased then the potential increases according to the Nernst equation

TABLE 15.1 Electrochemical series

Electrode reaction		*Standard electrode potential E_0 (V)*
$Cs = Cs^+ + e$	↑ Reactive metals	−3.02
$Li = Li^+ + e$		−3.02
$K = K^+ + e$		−2.92
$Na = Na^+ + e$		−2.71
$Ca = Ca^{2+} + 2e$		−2.50
$Mg = Mg^{2+} + 2e$		−2.34
$Al = Al^{3+} + 3e$		−1.07
$Ti = Ti^{2+} + 2e$		−1.67
$Zn = Zn^{2+} + 2e$		−0.76
$Cr = Cr^{+3} + 3e$		−0.50
$Fe = Fe^{2+} + 2e$		−0.44
$Cd = Cd^{2+} + 2e$		−0.40
$Ni = Ni^{2+} + 2e$		−0.25
$Sn = Sn^{2+} + 2e$		−0.136
$Pb = Pb^{2+} + 2e$		−0.126
$H = 2H^+ + 2e$		0.00
$Cu = Cu^{+2} + 2e$	Noble metals ↓	+0.34
$Hg = Hg^{2+} + 2e$		+0.80
$Ag = Ag^+ + e$		+0.80
$Pt = Pt^{2+} + 2e$		+1.20
$Au = Au^+ + e$		+1.68

$$E = E_0 + \frac{RT}{nF} \ln a$$

The easily ionizable 'reactive' metals have large negative potentials and dissolve even in concentrated solutions of their own ions, whereas the noble metals have positive potentials and are deposited from solution. These differences show that the valency electrons are strongly bound to the positive core in the noble metals because of the short distance of interaction, i.e. $d_{atomic} \simeq d_{ionic}$. A metal will therefore displace from solution the ions of a metal more noble than itself in the series. When two dissimilar metals are connected in neutral solution to form a cell the more metallic metal becomes the anode and the metal with the lower tendency to ionize becomes the cathode. The electrochemical series indicates which metal will corrode in the cell but gives no information on the rate of reactions. When an anode M corrodes, its ions enter into the solution initially low in M^+ ions, but as current flows the concentration of ions increases. This leads to a change in electrode potential known as polarization, as shown in *Figure 15.8(a)*, and corresponds to a reduced tendency to ionize. The current density in the cell is a maximum when the anode and cathode potential curves intersect. Such a condition would exist if the two metals were joined together or anode and cathode regions existed on the same metal, i.e. differential aeration; this potential is referred to as the corrosion potential and the current, the corrosion current.

In many reactions particularly in acid solutions, hydrogen gas is given off at the cathode rather than the anode metal deposited. In practice, the evolution of hydrogen gas at the cathode requires a smaller additional overvoltage, the

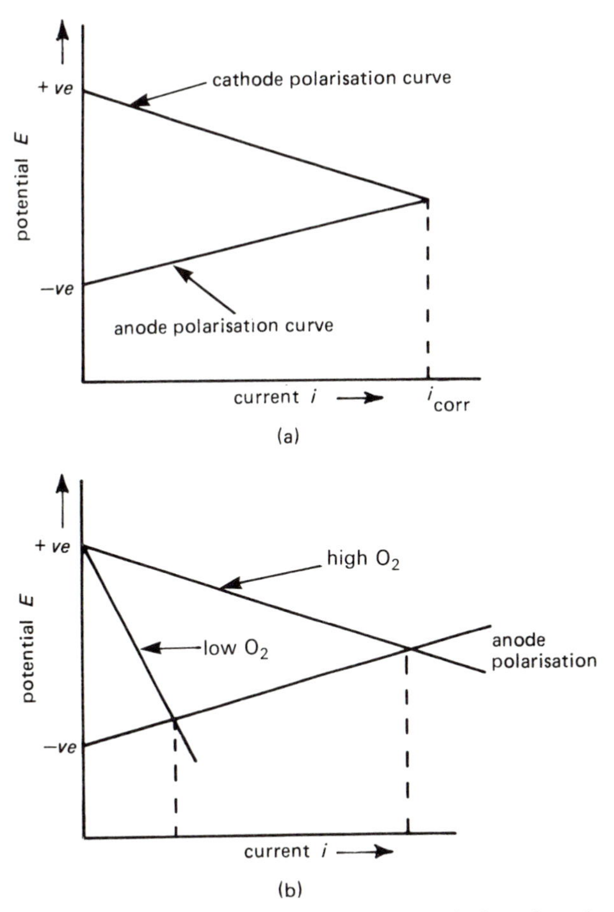

Figure 15.8 Schematic representation of (a) cathode and anode polarization curves, (b) influence of oxygen concentration on cathode polarization

magnitude of which varies considerably from one cathode metal to another, and is high for Pb, Sn and Zn and low for Ag, Cu, Fe and Ni; this overvoltage is clearly of importance in electrodeposition of metals. In corrosion, the overvoltage arising from the activation energy opposing the electrode reaction decreases the potential of the cell, i.e. hydrogen atoms effectively shield or polarize the cathode. The degree of polarization is a function of current density and the potential E to drive the reaction decreases because of the increased rate of H_2 evolution, as shown in *Figure 15.9* for the corrosion of zinc and iron in acid solutions. Corrosion can develop up to a rate given by the current when the potential difference required to drive the reaction is zero; for zinc this is i_{Zn} and for iron i_{Fe}. Because of its large overvoltage zinc is corroded more slowly than iron, even though there is a bigger difference between zinc and hydrogen than iron and hydrogen in the electrochemical series. The presence of Pt in the acid solution, because of its low overvoltage, increases the corrosion rate as it plates out on the cathode metal surface. In neutral or alkaline solutions, depolarization is brought

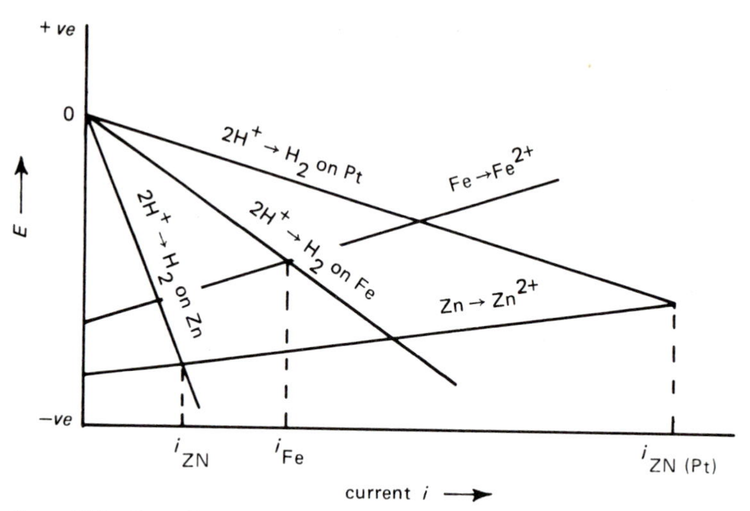

Figure 15.9 Corrosion of zinc and iron and the effect of polarization

about by supplying oxygen to the cathode area which reacts with the hydrogen ions as shown in *Figure 15.8(b)*. In the absence of oxygen both anodic and cathodic reactions experience polarization and corrosion finally stops; it is well known that iron doesn't rust in oxygen-free water.

It is apparent that the cell potential depends on the electrode material, the ion concentration of the electrolyte passivity and polarization effects. Thus it is not always possible to predict the precise electrochemical behaviour merely from the electrochemical series, i.e. which metals will be anode or cathode, and the magnitude of the cell voltage. Therefore it is necessary to determine the specific behaviour of different metals in solutions of different acidity. The results are displayed usually in Pourbaix diagrams as shown in *Figure 15.10*. With stainless steel, for example, the anodic polarization curve is not straightforward as discussed previously, but takes the form shown in *Figure 15.11*, where the low-current region corresponds to the condition of passivity. The corrosion rate depends on the position at which the cathode polarization curve for hydrogen evolution crosses this anode curve, and can be quite high if it crosses outside the passive region. Pourbaix diagrams map out the regions of passivity for solutions of different acidity. *Figure 15.10* shows that the passive region is restricted to certain conditions of pH; for Ti this is quite extensive but Ni is passive only in very acid solutions and Al in neutral solutions. Interestingly, these diagrams indicate that for Ti and Ni in contact with each other in corrosive conditions then Ni would corrode, and that passivity has changed their order in the electrochemical series. In general, passivity is maintained by conditions of high oxygen concentration but is destroyed by the presence of certain ions such as chlorides.

The corrosion behaviour of metals and alloys can therefore be predicted with certainty only by obtaining experimental data under simulated service conditions. For practical purposes, the cell potentials of many materials have been obtained in a single environment, the most common being sea water. Such data in tabular form are called a galvanic series, as illustrated in *Table 15.2*. If a

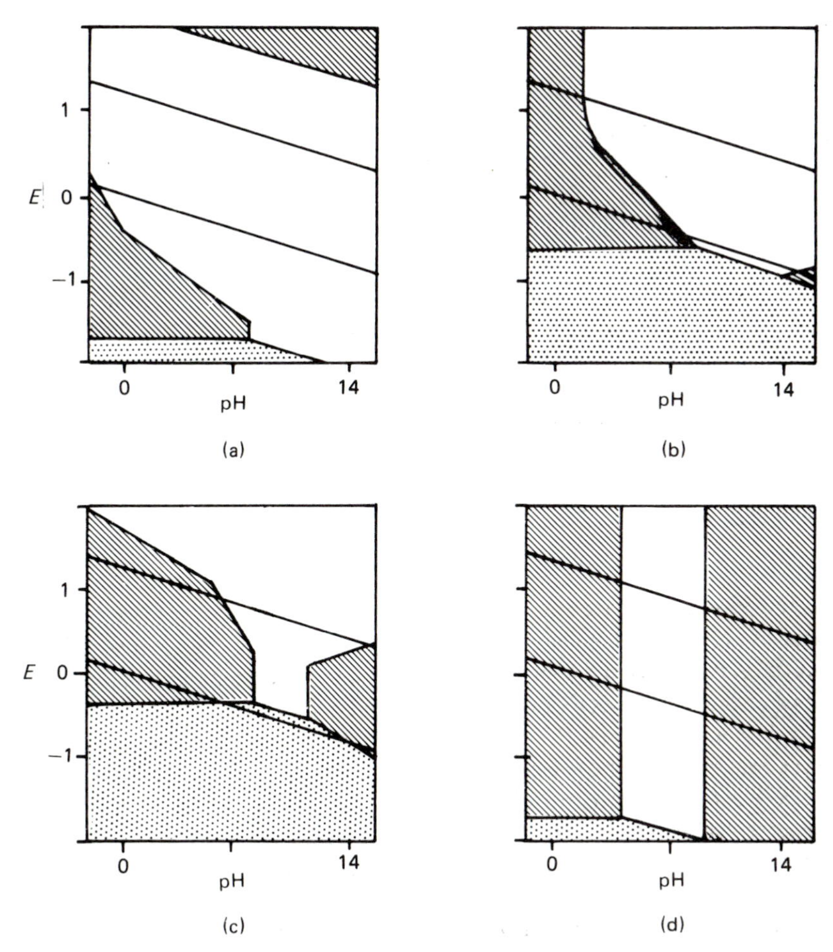

Figure 15.10 Pourbaix diagrams for (a) Ti, (b) Fe, (c) Ni, (d) Al. The clear regions are passive, the heavily-shaded regions corroding and the lightly-shaded regions immune. The sloping lines represent the upper and lower boundary conditions in service

pair of metals from this series were connected together in sea water, the metal which is higher in the series would be the anode and corrode, and the farther they are apart the greater the corrosion tendency. Similar data exist for other environments.

15.12 Corrosion protection

The principles of corrosion outlined indicate several possible methods of controlling corrosion. Since current must pass for corrosion to proceed, any factor, such as cathodic polarization which reduces the current will reduce the corrosion rate. Metals having a high overvoltage should be utilized where possible. In neutral and alkaline solution de-aeration of the electrolyte to remove oxygen is beneficial in reducing corrosion, e.g. heating the solution or holding

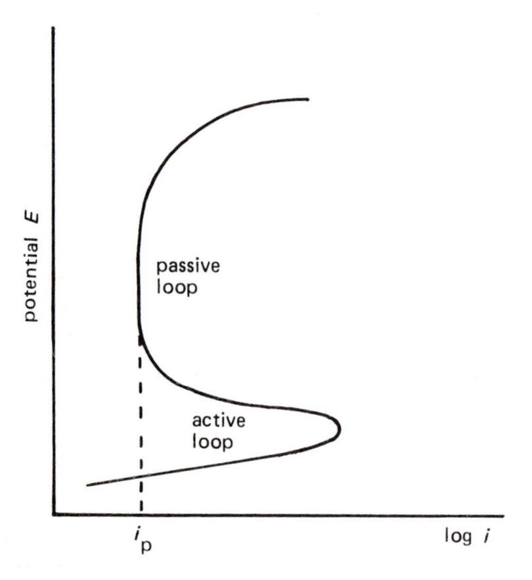

Figure 15.11 Anode polarization curve for stainless steel

TABLE 15.2 Galvanic series in sea water

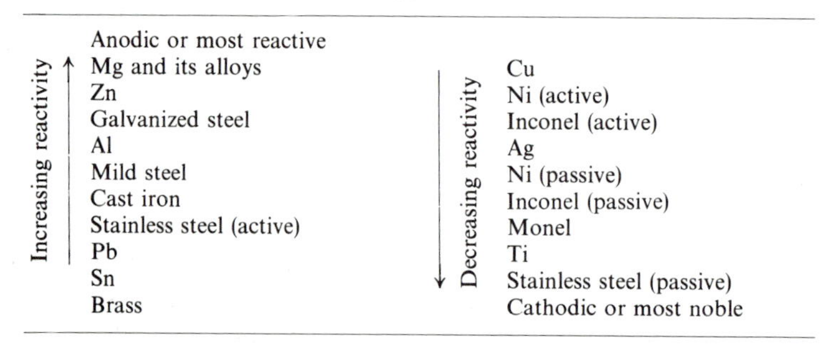

Increasing reactivity ↑	Decreasing reactivity ↓
Anodic or most reactive	
Mg and its alloys	Cu
Zn	Ni (active)
Galvanized steel	Inconel (active)
Al	Ag
Mild steel	Ni (passive)
Cast iron	Inconel (passive)
Stainless steel (active)	Monel
Pb	Ti
Sn	Stainless steel (passive)
Brass	Cathodic or most noble

under a reduced pressure preferably of an inert gas. It is sometimes possible to reduce both cathode and anode reactions by 'artificial' polarization, for example by adding inhibitors which stifle the electrode reaction. Calcium bicarbonate, naturally present in hard water, deposits calcium carbonate on metal cathodes and stifles the reaction. Soluble salts of magnesium and zinc act similarly by precipitating hydroxide in neutral solutions.

Anodic inhibitors for ferrous materials include potassium chromate and sodium phosphate, which convert the Fe^{2+} ions to insoluble precipitates stifling the anodic reaction. This form of protection has no effect on the cathodic reaction and hence if the inhibitor fails to seal off the anode completely intensive local attack occurs leading to pitting. Moreover, the small current density at the cathode leads to a low rate of polarization and the attack is maintained. Sodium benzoate is often used as an anodic inhibitor in radiators because of its good sealing qualities, with little tendency for pitting.

Some metals are naturally protected by their adherent oxide films; metal oxides are poor electrical conductors and so insulate the metal from solution. For the reaction to proceed, metal atoms have to diffuse through the oxide to the metal/liquid interface and electrons back through the high resistance oxide. The corrosion current is very much reduced by the formulation of such protective or passive oxide films. Al is cathodic to zinc in sea water even though the electrochemical series shows it to be more active. Materials which are passivated in this way are chromium, stainless steels, inconel and nickel in oxidizing conditions. Reducing environments, e.g. stainless steels in HCl, destroy the passive film and render the materials active to corrosion attack. Certain materials may be artificially passivated by painting. The main pigments used are red lead, zinc oxide and chromate, usually suspended in linseed oil and thinned with white spirit. Slightly soluble chromates in the paint passivate the underlying metal when water is present. Red lead reacts with the linseed oil to form lead salts of various fatty acids which are good anodic inhibitors.

In practice, sacrificial or cathodic protection is widely used. A typical example is galvanized steel sheet when the steel is protected by sacrificial corrosion of the zinc coating. Any regions of steel exposed by small flaws in the coating, polarize rapidly since they are cathodic and small in area. Such cathodic protection is also used for ships and steel pipe-lines buried underground. Auxiliary sacrificial anodes are placed at frequent intervals in the corrosive medium in contact with the pipe. Protection may also be achieved by impressing a d.c. voltage to make it a cathode, with the negative terminal of the d.c. source connected to a sacrificial anode.

15.13 Corrosion failures

In service, there are many types of corrosion attack which lead to rapid failure of components. A familiar example is intergranular corrosion and is associated with the tendency for grain boundaries to undergo localized anodic attack. Some materials are, however, particularly sensitive. The common example of this sensitization occurs in 18/8 stainless steel, which is normally protected by a passivating Cr_2O_3 film, after heating to 500–800 °C and slowly cooling. During cooling, chromium near the grain boundaries precipitates as chromium carbide. As a consequence, these regions are depleted in Cr to levels below 12 per cent and are no longer protected by the passive oxide film. They become anodic relative to the interior of the grain and, being narrow, are strongly attacked by the corrosion current generated by the cathode reactions elsewhere. Sensitization may be avoided by rapid cooling, but in large structures that is not possible, particularly after welding when the phenomenon (called weld decay) is common. The effect is then overcome by stabilizing the stainless steel by the addition of a small amount (0.5%) of a strong carbide former such as Nb or Ti which associates with the carbon in preference to the Cr.

Other forms of corrosion failure require the component to be stressed, either directly or by internal stress. Common examples include stress-corrosion cracking and corrosion fatigue. Hydrogen embrittlement is sometimes included in this category but this type of failure has somewhat different characteristics and has been considered previously. These failures have certain features in common.

In chemically active environments susceptible alloys develop deep fissures along active slip planes, particularly alloys with low stacking fault energy with wide dislocations and planar stacking faults, or along grain boundaries. For such selective chemical action, the free energy of reaction can provide almost all the surface energy for fracture, which may then spread under extremely low stresses.

Stress corrosion cracking was first observed in α-brass cartridge cases stored in ammoniacal environments. The phenomenon, called season cracking since it occurred more frequently during the monsoon season in the tropics, was prevented by giving the cold-worked brass cases a mild annealing treatment to relieve the internal stress. Occurrence of the phenomenon has since been extended to many alloys in different environments, e.g. Al–Cu, Al–Mg, Ti–Al, magnesium alloys, stainless steels in the presence of chloride ions; mild steels with hydroxyl ions (caustic embrittlement) and copper alloys with ammonia ions.

Stress corrosion cracking can be either transgranular or intergranular. There appears to be no unique mechanism of transgranular stress corrosion cracking, since no single factor is common to all susceptible alloys. In general, however, all susceptible alloys are unstable in the environment concerned but are largely protected by a surface film that is locally destroyed in some way. The variations on the basic mechanism arise from the different ways in which local activity is generated. Breakdown in passivity may occur as a result of the emergence of dislocation pile-ups, stacking faults, micro-cracks, precipitates (such as hydrides in Ti alloys) at the surface of the specimen, so that a highly localized anodic attack then takes place. The gradual opening of the resultant crack occurs by plastic yielding at the tip and as the liquid is sucked in also prevents any tendency to polarize.

Many alloys exhibit coarse slip and have similar dislocation substructures, e.g. co-planar arrays of dislocations or wide planar stacking faults, but are not equally susceptible to stress-corrosion. The observation has been attributed to the time necessary to repassivate an active area. Additions of Cr and Si to susceptible austenitic steels for example, do not significantly alter the dislocation distribution but are found to decrease the susceptibility to cracking, probably by lowering the repassivation time.

The susceptibility to transgranular stress corrosion of austenitic steels, α-brasses, titanium alloys, etc, which exhibit co-planar arrays of dislocations and stacking faults may be reduced by raising the stacking fault energy by altering the alloy compositions. Cross-slip is then made easier and deformation gives rise to fine slip, so that the narrower, fresh surfaces created have a less severe effect. The addition of elements to promote passivation or, more important, the speed of repassivation should also prove beneficial.

Intergranular cracking appears to be associated with a narrow soft zone near the grain boundaries. In α-brass this zone may be produced by local dezincification. In high strength Al-alloys there is no doubt that it is associated with the grain boundary precipitate-free zones, i.e. PFZs. In such areas the strain-rate may be so rapid, because the strain is localized, that repassivation cannot occur. Cracking then proceeds even though the slip steps developed are narrow, the crack dissolving anodically as discussed for sensitized stainless steel. In practice there are many examples of intergranular cracking, including cases (i) that depend strongly on stress, e.g. Al-alloys, (ii) where stress has a comparatively minor role, e.g. steel cracking, in nitrate solutions and (iii) which occur in the absence of stress, e.g. sensitized 18Cr–8Ni steels; the latter case is the extreme example of failure to repassivate for purely electrochemical reasons. In some materials the crack propagates, as in ductile failure, by internal necking between inclusions which occurs by a combination of stress and dissolution processes. The stress sensitivity depends on the particle distribution and is quite high for fine-scale and low for coarse-scale distributions. The change in precipitate distribution in grain boundaries produced, for example, by duplex ageing can thus change the stress-dependence of intergranular failure.

In conditions where the environment plays a role, the crack growth rate V varies with stress intensity K in the manner shown in *Figure 15.12*. In region I the crack velocity shows a marked dependence on stress, in region II the velocity is independent of the stress intensity and in region III the rate becomes very fast as K_{IC} is approached. K_{ISC} is extensively quoted as the threshold stress intensity below which the crack growth rate is negligible, e.g. $\gtrsim 10^{-10}$ m s^{-1} but, like the endurance limit in fatigue, does not exist for all materials. In region I the rate of crack growth is controlled by the rate at which the metal dissolves and the time for which the metal surface is exposed. While anodic dissolution takes place on the exposed metal at the crack tip, cathodic reactions occur at the oxide film on the crack sides leading to the evolution of hydrogen which diffuses to the region of triaxial tensile stress and hydrogen-induced cracking. At higher stress intensities (region II) the strain-rate is higher and faster processes become rate-controlling,

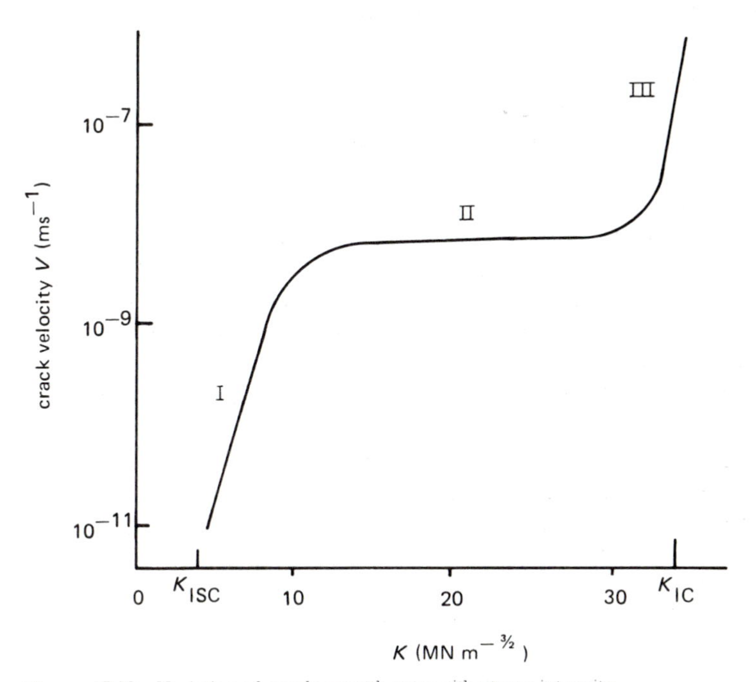

Figure 15.12 Variation of crack growth rate with stress intensity

generally the rate of diffusion of new reactants into the crack tip region. In hydrogen embrittlement this is probably the rate of hydrogen diffusion.

Corrosion fatigue

The influence of a corrosive environment, even mildly oxidizing, in reducing the fatigue life has been briefly mentioned in Chapter 13. The S–N curve shows no tendency to level out, but falls to low *S*-values. The damage ratio, i.e. corrosion fatigue strength divided by the normal fatigue strength, in salt water environments, is only about 0.5 for stainless steels and 0.2 for mild steel. The formation of intrusions and extrusions give rise to fresh surface steps which form very active anodic sites in aqueous environments, analogous to the situation at the tip of a stress corrosion crack. This form of fatigue is influenced by those factors affecting normal fatigue but in addition, involves electrochemical factors. It is normally reduced by plating, cladding and painting but difficulties may arise in localizing the attack to a small number of sites, since the surface is continually being deformed. Anodic inhibitors may also reduce the corrosion fatigue but their use is more limited than in the absence of fatigue because of the probability of incomplete inhibition leading to increased corrosion.

Very often fretting corrosion, caused by two surfaces rubbing together, is associated with fatigue failure. The oxidation and corrosion product is continually removed, so that the problem must be tackled by improving the mechanical linkage of moving parts and by the effective use of lubricants.

With corrosion fatigue and fracture mechanics the threshold ΔK_{th} is reduced and the rate of crack propagation is usually increased by a factor of two or so. Much larger increases in crack growth rate are produced, however, in low-frequency cycling when stress-corrosion fatigue effects become important.

Suggestions for further reading

Chilton, J.P., *Principles of Metallic Corrosion*, Chemical Society, London, 1973

Gibbs, G.B., Influence of chemical environment on high temperature mechanical properties, *Conf. Physical Metallurgy of Reactor Fuel Elements*, Metals Society, London, 1975

Kubaschewski, O. and Hopkins, B.E., *Oxidation of Metals and Alloys*, Butterworths, 1962

Swann, P.R., Ford, F.P., Westwood, A.R.C. (eds), *Mechanisms of Environment Sensitive Cracking of Materials*, Metals Society, London, 1977

West, J.M., *Basic Corrosion and Oxidation*, Ellis Horwood, 1980

Appendix: Units and useful factors

The traditional scientific units used throughout this book have been replaced in the UK by SI units (Système International d'Unités). These units are derived from the following base-units:

Unit	*Name*	*Symbol*
Electric current	ampere	A
Length	metre	m
Luminous intensity	candela	cd
Mass	kilogram	kg
Thermodynamic temperature	kelvin	K
Time	second	s
Amount of substance	mole	mol

From these base-units other units are derived as follows:

Force	newton	$N = kg\ m\ s^{-2}$
Work, energy, quantity of heat	joule	$J = N\ m$
Power	watt	$W = J\ s^{-1}$
Electric charge	coulomb	$C = A\ s$
Electric potential	volt	$V = W\ A^{-1}$
Electric capacitance	farad	$F = A\ s\ V^{-1}$
Electric resistance	ohm	$\Omega = V\ A^{-1}$
Frequency	hertz	$Hz = s^{-1}$
Magnetic flux	weber	$Wb = V\ s$
Magnetic flux density	tesla	$T = Wb\ m^{-2}$
Inductance	henry	$H = V\ s\ A^{-1}$
Luminous flux	lumen	$lm = cd\ sr$
Illumination	lux	$lx = lm\ m^{-2}$

Multiples and sub-multiplies of these units may be used where found to be convenient:

Multiplication Factor	*Prefix*	*Symbol*
10^{12}	tera	T
10^{9}	giga	G
10^{6}	mega	M
10^{3}	kilo	k
10^{2}	hecto	h
10^{1}	deca	da
10^{-1}	deci	d
10^{-2}	centi	c
10^{-3}	milli	m
10^{-6}	micro	μ
10^{-9}	nano	n
10^{-12}	pico	p
10^{-15}	femto	f
10^{-18}	atto	a

It must be noted, however, that when the multiple or sub-multiple prefix immediately precedes the symbol for a unit raised to some power other than unity, then it is convention that the prefix is also incorporated in the power, e.g. $mm^2 = 10^{-6}\ m^2$ and not $10^{-3}\ m^2$.

To aid the conversion of traditional units to SI units, some useful constants and factors in common use are given below.

Quantity	*Symbol*	*Traditional Unit*	*SI Unit*
1 Atmosphere (pressure)	atm		$101\,325\ N\ m^{-2}$
Avogadro constant	N_A	6.0225×10^{23}	$6.0225 \times 10^{23}\ mol^{-1}$
1 Angstrom	Å	10^{-8} cm	10^{-10} m
1 barn	b	$10^{-24}\ cm^2$	$10^{-28}\ m^2$
1 bar	bar		$10^5\ N\ m^{-2}$
Boltzmann constant	k	1.380×10^{-16} erg deg^{-1}	$1.380 \times 10^{-23}\ J\ K^{-1}$
1 calorie	cal	2.61×10^{19} eV	4.184 J
		4.186 joules	
1 dyne	dyn		10^{-5} N
1 dyne cm^{-2}		$1.45 \times 10^{-5}\ lbf/in^2$	$10^{-1}\ N\ m^{-2}$
1 day		86 400 s	86.4 ks
1 degree (angle)		0.017 rad	17 m rad
1 erg		6.24×10^{11} eV	10^{-7} J
		2.39×10^{-8} cal	
1 erg cm^{-2}		$6.24 \times 10^{11}\ eV\ cm^{-2}$	$10^{-3}\ J\ m^{-2}$
gas constant	R	8.3143×10^7 erg g-atom^{-1}	$8.3143\ J\ mol^{-1}\ K^{-1}$
		1.987 cal deg^{-1} g-atom^{-1}	
density of Al	ρ	$2.7\ g\ cm^{-3}$	$2.7 \times 10^3\ kg\ m^{-3}$
density of Fe		$7.9\ g\ cm^{-3}$	$7.9 \times 10^3\ kg\ m^{-3}$
density of Cu		$8.9\ g\ cm^{-3}$	$8.9 \times 10^3\ kg\ m^{-3}$
density of Ni		$8.9\ g\ cm^{-3}$	$8.9 \times 10^3\ kg\ m^{-3}$
electronic charge	e	4.8×10^{-10} e.s.u.	1.6021×10^{-19} C
1 electron volt	eV	3.83×10^{-20} cal	
		1.6021×10^{-12} erg	1.6021×10^{-19} J

Quantity		*Symbol*	*Traditional Unit*	*SI Unit*
Faraday constant		$F = N_A e$		9.6487×10^4 C mol^{-1}
1 inch		in	2.54 cm	25.4 mm
1 kilocalorie		kcal	4.186×10^{10} erg	
1 kilogramme		kg	2.21 lb	1 kg
1 kilogramme cm^{-2}		kg cm^{-2}	14.22 lbf/in^2	98.06 kN m^{-2}
1 litre		l		1 dm^3
mass of electron		m_e	9.1091×10^{-28} g	9.1091×10^{-31} kg
1 micron		μm	10^4 Ångstroms 10^{-4} cm	10^{-6} m
1 minute (angle)			2.91×10^{-4} radians	min $= 2.91 \times 10^{-4}$ rad
Planck's constant		h	6.6256×10^{-27} erg s	6.6256×10^{-34} J s
1 pound		lb	453.59 g	0.453 59 kg
1 pound (force)		lbf		4.448 22 N
1 p.s.i.		lbf/in^2		6894.76 N m^{-2}
1 radian		rad	57.296 degrees	1 rad
shear modulus	Al	μ	2.7×10^{11} dyn cm^{-2}	2.7×10^{10} N m^{-2}
(average) for	Fe		8.3×10^{11} dyn cm^{-2}	8.3×10^{10} N m^{-2}
	Ni		7.4×10^{11} dyn cm^{-2}	7.4×10^{10} N m^{-2}
	Cu		4.5×10^{11} dyn cm^{-2}	4.5×10^{10} N m^{-2}
	Au		2.99×10^{11} dyn cm^{-2}	2.999×10^{10} N m^{-2}
1 ton (force)		1 tonf		9.964 02 kN
1 t.s.i.		1 tonf/in^2	1.5749 kgf/mm^2	15.4443 MN m^{-2} 15.4443 MPa (Pa = Pascal = N m^{-2})
1 tonne		t	1000 kg	10^3 kg
1 torr		torr	1 mm Hg	133.322 N m^{-2}
velocity of light		c	$2.997\,925 \times 10^{10}$ cm/s	$2.997\,925 \times 10^8$ m s^{-1}

Index